# Computational Intelligence in Remote Sensing—Volume II

# Computational Intelligence in Remote Sensing—Volume II

Editors

**Yue Wu**
**Kai Qin**
**Maoguo Gong**
**Qiguang Miao**

Basel • Beijing • Wuhan • Barcelona • Belgrade • Novi Sad • Cluj • Manchester

*Editors*
Yue Wu
Xidian University
Xi'an
China

Kai Qin
Swinburne University of
Technology
Hawthorn, VIC
Australia

Maoguo Gong
Xidian University
Xi'an
China

Qiguang Miao
Xidian University
Xi'an
China

*Editorial Office*
MDPI
St. Alban-Anlage 66
4052 Basel, Switzerland

This is a reprint of articles from the Topic published online in the open access journals *Applied Sciences* (ISSN 2076-3417), *Electronics* (ISSN 2079-9292), *Remote Sensing* (ISSN 2072-4292), and *AI* (ISSN 2673-2688) (available at: https://www.mdpi.com/topics/Remote_Sensing).

For citation purposes, cite each article independently as indicated on the article page online and as indicated below:

Lastname, A.A.; Lastname, B.B. Article Title. *Journal Name* **Year**, *Volume Number*, Page Range.

**Volume II**
**ISBN 978-3-7258-0413-9 (Hbk)**
**ISBN 978-3-7258-0414-6 (PDF)**
**doi.org/10.3390/books978-3-7258-0414-6**

**Set**
**ISBN 978-3-7258-0377-4 (Hbk)**
**ISBN 978-3-7258-0378-1 (PDF)**

# Contents

 *remote sensing*

*Editorial*

# Computational Intelligence in Remote Sensing

Yue Wu [1,*], Maoguo Gong [2], Qiguang Miao [1] and Kai Qin [3]

1   Department of Computer Science and Technology, Xidian University, Xi'an 710071, China; qgmiao@xidian.edu.cn
2   Key Laboratory of Intelligent Perception and Image Understanding, Xidian University, Xi'an 710071, China; gong@ieee.org
3   Department of Computer Science and Software Engineering, Swinburne University of Technology, Victoria 3122, Australia; kqin@swin.edu.au
*   Correspondence: ywu@xidian.edu.cn

## 1. Introduction

With the development of Earth observation techniques, vast amounts of remote sensing data with a high spectral–spatial–temporal resolution are captured all the time, and remote sensing data processing and analysis have been successfully used in numerous fields, including geography, environmental monitoring, land survey, disaster management, mineral exploration and more. For the processing, analysis and application of remote sensing data, there are many challenges, such as the vast amount of data, complex data structures, small labeled samples and nonconvex optimization. In recent years, the convergence of computational intelligence (CI) and remote sensing has ushered in a new era of possibilities for understanding and harnessing the wealth of information that Earth observation satellites provide. Computational intelligence methods, such as deep neural networks, evolutionary optimization and swarm intelligence, have demonstrated remarkable capabilities in unveiling intricate patterns within satellite images, time series data and multispectral/hyperspectral information. In the future, CI will produce effective solutions to the challenges in remote sensing.

## 2. Recent Research and Progress

This Topic series aims to highlight the latest research and advances in the application of computational intelligence in the field of remote sensing. In total, this Topic series contains 12 papers written by research experts on topics of interest. Based on the synthesis of these latest achievements, they can be categorized into four sections: computational intelligence methods in hyperspectral remote sensing images; object detection techniques in remote sensing images; deep learning approaches in remote sensing image classification and intelligent optimization and control in satellite image applications.

### 2.1. Computational Intelligence Methods in Hyperspectral Remote Sensing Images

This section consists of three papers. The first paper is written by A.C.P. Silva, K.T.Z. Coimbra, L.W.R. Filho, G. Pessin and R.E. Correa-Pabón. They mainly explore the possibility of applying machine learning models to monitor the quality of iron ore [1]. The second paper, written by W. Shuai, F. Jiang, H. Zheng and J. Li, mainly proposes a new method with high processing efficiency for change detection in remote sensing images, called MSGATN [2]. The last work studies SAR image segmentation based on fuzzy c-means and is by J. Zhu, F. Wang and H. You. Experiments show that the framework can achieve more than 97% segmentation accuracy [3].

### 2.2. Object Detection Techniques in Remote Sensing Images

The following three papers mainly utilize deep learning techniques to solve practical problems in the field of remote sensing image object detection. The first paper,

**Citation:** Wu, Y.; Gong, M.; Miao, Q.; Qin, K. Computational Intelligence in Remote Sensing. *Remote Sens.* **2023**, *15*, 5325. https://doi.org/10.3390/rs15225325

Received: 23 October 2023
Accepted: 24 October 2023
Published: 12 November 2023

by R. Chen and S. Liu et al., proposes an effective infrared object detection method based on source model guidance [4]. They show two explicit examples based on CenterNet and YOLOv3, respectively, and experimentally demonstrate that the method can achieve powerful performance with limited samples. The second paper, by L. Yu and X. Zhou et al., proposes a method for boundary-aware salient object detection in optical remote sensing images [5]. The method uses a graph convolutional network-based feature extraction module and a boundary-aware attention-based module to improve the accuracy and robustness of boundary-aware salient object detection. The third paper, by F. Zhou and H. Deng et al., studies deep learning-based aircraft detection [6]. The paper proposes an enhanced YOLOv5 model in which a ConvNext-based feature extraction module and a Transformer-based feature fusion module are used to improve the detection performance.

### 2.3. Deep Learning Approaches in Remote Sensing Image Classification

This section includes three papers. The first paper is authored by H. Toriya and A. Dewan et al., who primarily explore the key point matching problem in image features. They propose using a deep neural network (DNN) to construct an image translator and introduce a new edge enhancement filter methodology within the conditional generative adversarial network (cGAN) structure to tackle this issue [7]. The second paper, written by Z. Wei and Z. Zhang, describes a network built on multi-level strip pooling and a feature enhancement module (MSPFE-Net). Here, deep learning is effectively applied to address the challenge of road extraction [8]. In the third paper, L. Zeng and Y. Huo et al. develop the high-quality seed instance mining (HSIM) module, alongside the dynamic pseudo-instance label assignment (DPILA), to address the issue of weakly supervised detection in remote sensing images [9].

### 2.4. Intelligent Optimization and Control in Satellite Image Applications

This section includes three state-of-the-art papers for reference focusing on different research directions in satellite images. The first paper is authored by T. Zheng, Y. Dai, C. Xue and L. Zhou. They propose a method for solving near-lossless hyperspectral data compression using recursive least squares. They use the linear combination of previous pixels to predict the target pixel values while using a recursive least squares filter to iteratively update the weight matrix for prediction, which effectively removes spatial and spectral redundancy information [10]. The second paper is written by N. Andrijević, V. Urošević, B. Arsić, D. Herceg and B. Savić. This paper designs a time prediction model for bee influx and outflow in a bee colony ecosystem with a large number of sensors by simulating the correlation between the environment and bee colony activity to simulate the bee colony ecosystem [11]. L. Li, D. Yin, Q. Li, Q. Zhang and Z. Mao propose a verification method for ultraviolet imagers using the seeker optimization algorithm. This method can effectively use ultraviolet imagers to conduct authenticity check studies on ocean surface radiation data [12].

## 3. Discussion

The papers provide an exchange platform for researchers in the field of remote sensing images, covering topics such as hyperspectral remote sensing image processing, remote sensing image classification, segmentation, object detection and intelligent optimization and control in satellite image applications. These themes represent a series of key issues in the field of remote sensing images. The research papers in this journal not only delve into these issues, but also propose new methods and ideas, providing strong support for future research directions.

In this issue of the journal, we have seen a series of important developments in the field of hyperspectral remote sensing image processing. Researchers have utilized the rich information of hyperspectral data to not only improve the performance of segmentation, but also provide new tools for application fields such as resource management and environmental monitoring. In addition, remote sensing image classification, segmentation and object detection have always been research hotspots. Research in this journal shows that deep learning technology has made significant progress in the application of these tasks.

The papers in this research Topic showcase the innovative and influential contributions of researchers in this field. Researchers have not only delved into various issues, but also proposed many new methods and technologies, demonstrating the potential of computational intelligence in advancing our understanding of remote sensing images and providing strong support for future research directions. In the future, we can look forward to more interdisciplinary cooperation, combining remote sensing image research with application fields such as environmental science, agriculture and urban planning to solve complex real-world problems. We encourage readers to further explore the cutting-edge research and novel applications presented in these papers to provide new impetus for scientific and technological innovation.

**Author Contributions:** Conceptualization, Y.W. and M.G.; writing—original draft preparation, Y.W. and M.G.; writing—review and editing, Q.M. and K.Q. All authors have read and agreed to the published version of the manuscript.

**Data Availability Statement:** No new data were created or analyzed in this study. Data sharing is not applicable to this article.

**Acknowledgments:** Thanks to all authors, peer reviewers and editorial team members for their valuable contributions. Their dedication and hard work have been instrumental in the outcome of this Topic series. Herewith, congratulations to all the authors for their outstanding achievements on relevant topics.

**Conflicts of Interest:** The authors declare no conflict of interest.

## References

1. Silva, A.C.P.; Coimbra, K.T.Z.; Filho, L.W.R.; Pessin, G.; Correa-Pabón, R.E. Monitoring of Iron Ore Quality through Ultra-Spectral Data and Machine Learning Methods. *AI* **2022**, *3*, 554–570. [CrossRef]
2. Shuai, W.; Jiang, F.; Zheng, H.; Li, J. MSGATN: A Superpixel-Based Multi-Scale Siamese Graph Attention Network for Change Detection in Remote Sensing Images. *Appl. Sci.* **2022**, *12*, 5158. [CrossRef]
3. Zhu, J.; Wang, F.; You, H. SAR Image Segmentation by Efficient Fuzzy C-Means Framework with Adaptive Generalized Likelihood Ratio Nonlocal Spatial Information Embedded. *Remote Sens.* **2022**, *14*, 1621. [CrossRef]
4. Chen, R.; Liu, S.; Mu, J.; Miao, Z.; Li, F. Borrow from Source Models: Efficient Infrared Object Detection with Limited Examples. *Appl. Sci.* **2022**, *12*, 1896. [CrossRef]
5. Yu, L.; Zhou, X.; Wang, L.; Zhang, J. Boundary-Aware Salient Object Detection in Optical Remote-Sensing Images. *Electronics* **2022**, *11*, 4200. [CrossRef]
6. Zhou, F.; Deng, H.; Xu, Q.; Lan, X. CNTR-YOLO: Improved YOLOv5 Based on ConvNext and Transformer for Aircraft Detection in Remote Sensing Images. *Electronics* **2023**, *12*, 2671. [CrossRef]
7. Toriya, H.; Dewan, A.; Ikeda, H.; Owada, N.; Saadat, M.; Inagaki, F.; Kawamura, Y.; Kitahara, I. Use of a DNN-Based Image Translator with Edge Enhancement Technique to Estimate Correspondence between SAR and Optical Images. *Appl. Sci.* **2022**, *12*, 4159. [CrossRef]
8. Wei, Z.; Zhang, Z. Remote Sensing Image Road Extraction Network Based on MSPFE-Net. *Electronics* **2023**, *12*, 1713. [CrossRef]
9. Zeng, L.; Huo, Y.; Qian, X.; Chen, Z. High-Quality Instance Mining and Dynamic Label Assignment for Weakly Supervised Object Detection in Remote Sensing Images. *Electronics* **2023**, *12*, 2758. [CrossRef]
10. Zheng, T.; Dai, Y.; Xue, C.; Zhou, L. Recursive Least Squares for Near-Lossless Hyperspectral Data Compression. *Appl. Sci.* **2022**, *12*, 7172. [CrossRef]
11. Andrijević, N.; Urošević, V.; Arsić, B.; Herceg, D.; Savić, B. IoT Monitoring and Prediction Modeling of Honeybee Activity with Alarm. *Electronics* **2022**, *11*, 783. [CrossRef]
12. Li, L.; Yin, D.; Li, Q.; Zhang, Q.; Mao, Z. An Exploratory Verification Method for Validation of Sea Surface Radiance of HY-1C Satellite UVI Payload Based on SOA Algorithm. *Electronics* **2023**, *12*, 2766. [CrossRef]

*Article*

# High-Quality Instance Mining and Dynamic Label Assignment for Weakly Supervised Object Detection in Remote Sensing Images

**Li Zeng , Yu Huo *, Xiaoliang Qian and Zhiwu Chen ***

College of Electrical and Information Engineering, Zhengzhou University of Light Industry, Zhengzhou 450002, China; zengli_qxl@163.com (L.Z.); qxl_sunshine@163.com (X.Q.)
* Correspondence: yuhuo_henry2022@163.com (Y.H.); chenyyj@163.com (Z.C.)

**Abstract:** Weakly supervised object detection (WSOD) in remote sensing images (RSIs) has attracted more and more attention because its training merely relies on image-level category labels, which significantly reduces the cost of manual annotation. With the exploration of WSOD, it has obtained many promising results. However, most of the WSOD methods still have two challenges. The first challenge is that the detection results of WSOD tend to locate the significant regions of the object but not the overall object. The second challenge is that the traditional pseudo-instance label assignment strategy cannot adapt to the quality distribution change of proposals during training, which is not conducive to training a high-performance detector. To tackle the first challenge, a novel high-quality seed instance mining (HSIM) module is designed to mine high-quality seed instances. Specifically, the proposal comprehensive score (PCS) that consists of the traditional proposal score (PS) and the proposal space contribution score (PSCS) is designed as a novel metric to mine seed instances, where the PS indicates the probability that a proposal pertains to a certain category and the PSCS is calculated by the spatial correlation between top-scoring proposals, which is utilized to evaluate the wholeness with which a proposal locates an object. Consequently, the high PCS will encourage the WSOD model to mine the high-quality seed instances. To tackle the second challenge, a dynamic pseudo-instance label assignment (DPILA) strategy is developed by dynamically setting the label assignment threshold to train high-quality instances. Consequently, the DPILA can better adapt the distribution change of proposals according to the dynamic threshold during training and further promote model performance. The ablation studies verify the validity of the proposed PCS and DPILA. The comparison experiments verify that our method obtains better performance than other advanced WSOD methods on two popular RSIs datasets.

**Keywords:** weakly supervised object detection; remote sensing images; proposal comprehensive score; dynamic label assignment

**Citation:** Zeng, L. ; Huo, Y.; Qian, X.; Chen, Z. High-Quality Instance Mining and Dynamic Label Assignment for Weakly Supervised Object Detection in Remote Sensing Images. *Electronics* **2023**, *12*, 2758. https://doi.org/10.3390/electronics12132758

Academic Editor: George A. Tsihrintzis

Received: 23 May 2023
Revised: 17 June 2023
Accepted: 19 June 2023
Published: 21 June 2023

## 1. Introduction

Object detection in RSIs is a pivotal task of imagery interpretation, its purpose is to identify and locate high-value geographical objects in RSIs. Object detection in RSIs has wide applications in various fields, such as environmental monitoring [1,2], urban planning [3], agriculture [4,5], anomaly detection [6,7], and so on. With the progression of machine learning [8–14], object detection acquires satisfactory performance. The advanced performance is obtained by the fully supervised object detection (FSOD) [15–19] methods. However, the FSOD method needs category and location labels for instances to drive model training. Obviously, manually annotating the location labels for each instance of each RSI is laborious. In order to alleviate the burdensome annotated costs, weakly supervised object detection (WSOD) methods [20,21] have gradually entered the view of researchers because WSOD methods only require image-level category labels to drive model training.

At present, most of the WSOD models are trained based on the paradigm of multiple instance learning (MIL) [22–25]. Specifically, the training image is treated as a bag of latent instances, and then the latent instances are utilized to train the instance detector under the MIL constraints. Among these, a pioneering weakly supervised deep detection network (WSDDN) [26] has been developed, which first introduces MIL into the WSOD model. On the basis of WSDDN, an online instance classifier refinement (OICR) model [27] is developed by adding $K$ instance classifier refinement (ICR) branches, which further improves the performance of the WSOD model. Subsequently, some works have been developed to further enhance the performance of WSOD through employing spatial correlation [28], initialization models [29], collaborative learning [30], etc.

Although the performance of classical WSOD has made significant progress, there are still two main challenges to be solved. The first challenge is that most of the WSOD methods [27,31] merely employ the proposal score (PS) to mine seed instances, however, high PS usually locates the remarkable region of an object but not the overall object. Unfortunately, these methods will obtain worse performance in RSIs with noisy background. The second challenge is that the traditional pseudo-instance labels assignment (PILA) strategy [27,31] cannot adapt to the quality distribution change of proposals during training. Specifically, the traditional PILA strategy sets a fixed label assignment threshold to determine the attribute (i.e., belonging to a positive or negative instance) of each instance. However, along with the training, the fixed threshold setting and dynamic model training are not matched, which is not conducive to training high-quality instances.

In order to tackle the first challenge, a novel high-quality seed instances mining (HSIM) module is designed to mine high-quality seed instances, as shown in Figure 1. Specifically, the proposal comprehensive score (PCS) is first designed and is composed of traditional proposal score (PS) and proposal space contribution score (PSCS). The PS indicates the probability that a proposal pertains to one category; the PSCS is calculated by considering the spatial relationships between top-scoring proposals and is utilized to measure the extent to which the proposal locates an object. Consequently, seed instances mined by PCS can better locate an object than traditional mined strategy, which merely utilize the PS.

In order to tackle the second challenge, an innovative dynamic pseudo instance label assignment (DPILA) strategy is developed to better adapt to the quality distribution change of proposals during training and, meanwhile, raise the number of positive instances in the initial training stage. Specifically, a label assignment threshold is dynamically calculated via elaborately designing a function that increases with the number of iterations. Consequently, the DPILA strategy can dynamically assign pseudo instance label for each instance, and further improves the performance of WSOD.

Our contributions can be summed up as follows. The first contribution is that a novel HSIM module is designed to mine high-quality seed instances. Specifically, a PCS is first designed, which is composed of traditional PS and proposed PSCS, where the PSCS is calculated by considering the spatial relationships between top-scoring proposals to estimate the wholeness with which the proposal locates an object. The seed instances mined by PCS can more completely locate an object than traditional mined strategies, which merely utilizes the PS; The second contribution is that a DPILA strategy is proposed to better adapt to the quality distribution change of proposals during training. Specifically, a dynamic label assignment threshold is defined by elaborately designing a function that increases with the number of iterations. The proposed DPILA strategy can dynamically assign a pseudo-instance label for each instance, which is conducive to model training; The third contribution is that the ablation studies verify the validity of PCS and DPILA. The comparison experiments display that our method obtains higher performance than other advanced WSOD methods on two popular RSIs datasets. Specifically, our method surpasses separately the state-of-the-art WSDDN, OICR, PCL, and MELM methods by 12.2% (8.3%), 12.8% (5.1%), 7.9% (3.4%) and 5.0% (2.9%) in terms of mAP on the NWPU VHR-10.v2 (DIOR) dataset, and surpasses them by 23.2% (11.9%), 18.4% (9.5%), 13.3% (2.8%) and 8.5% (1.0%) in terms of CorLoc on the NWPU VHR-10.v2 (DIOR) dataset.

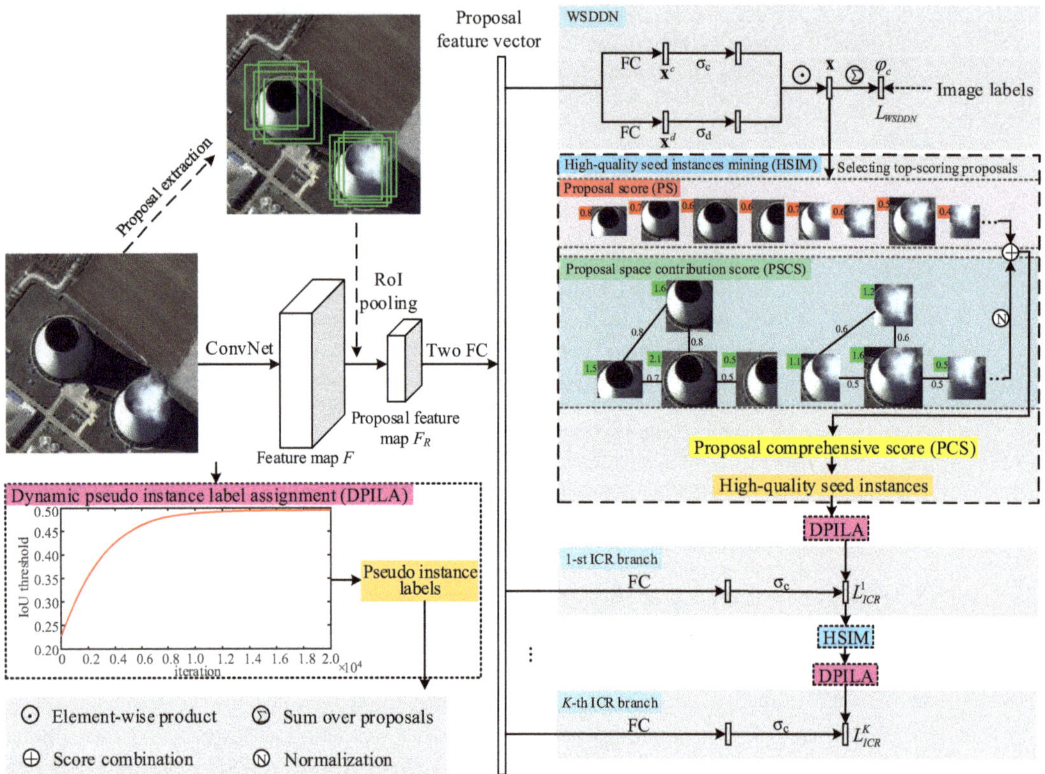


**Figure 1.** The overall framework of our method, which is established on the OICR network [27] by introducing two proposed modules including high-quality seed instance mining (HSIM) module and dynamic pseudo instance labels assignment (DPILA) strategy. Here, the HSIM is designed to mine high-quality seed instances. The DPILA strategy is proposed to better adapt to the quality distribution change of proposals during training.

## 2. Related Work

### 2.1. State-of-the-Art Weakly Supervised Object Detection Methods

Fully supervised object detection (FSOD) methods have achieved satisfactory performance. However, it needs category and location labels to drive model training, which is time-consuming to annotate with these precise labels. WSOD methods, which only require image-level labels to drive model training, have gradually entered the view of researchers. For example, Feng et al. [32] proposed a progressive contextual instance refinement strategy that can highlight more object parts and relieve the part domination problem. Yao et al. [33] proposed a dynamic curriculum learning strategy to robustly improve the performance. Feng et al. [34] proposed a triple context-aware network that can learn complementary and discriminative features and improve the performance of WSOD. Chen et al. [30] introduced the collaborative learning strategy into the WSOD model to improve its performance of WSOD. Feng et al. [35] proposed a self-supervised adversarial and equivariant network, that could learn complementary and consistent instance features, and promote the performance of WSOD. Chen et al. [36] proposed a full-coverage collaborative network, which could enhance the ability of multiscale feature extraction for WSOD detector.

### 2.2. Pseudo Instance Labels Mining

There are no instance-level labels to drive the model training in the WSOD. Therefore, it is a challenge to mine pseudo-instance labels for each instance. The current mainstream pseudo-instance labels mining strategy can be divided into two steps, namely, seed instances mining and pseudo-instance label assignment. The details of the two steps are as follows.

### 2.2.1. Seed Instances Mining

Most of the seed instance mining strategies [27,37,38] select the proposal with the highest score in category $c$ as seed instance. However, the strategy ignores the plain fact that RSIs usually contain multiple instances in the same category, and it is unreasonable to only select the proposal with the highest score as the seed instance in category $c$. Therefore, some improvements have been proposed. For instance, Tang et al. [39] use the $k$-means method to split the proposals into several clusters according to proposal score, select the proposal with the highest score in each cluster, and then utilize graph-based method to choose multiple seed instances with same category. Lin et al. [40] consider that the same category instance should have a similar appearance feature. Specifically, by selecting the highest-score proposal as a seed instance in category $c$, then calculating the similarity between the seed instance and other instances, if the similarity of a certain proposal is greater than the pre-set threshold, the proposal is selected as another seed instance. Cheng et al. [41] proposed a self-guided proposal generation strategy to generate directly high-quality seed instances. Qian et al. [42] proposed a novel seed instance mining strategy by employing the supplemental segmentation information. Ren et al. [31] sort all of the proposals from high to low according to the PS of existing categories in an image and then select proposals with the top $p\%$ score as the candidate seed instances. Finally, a similar non-maximum suppression (NMS) [43] operation is utilized to choose ultimate seed instances.

### 2.2.2. Pseudo-Instance Labels Assignment

Most of the WSOD methods [27,31,39,44] assign a pseudo-instance labels according to the fixed labels assignment threshold. Concretely, suppose an image contains category label $c$, the seed instance $R_{si}$ belonging to category $c$ can be mined according to the abovementioned methods. Furthermore, the $R_{si}$ is labeled category $c$, i.e., $y^k_{cR_{si}} = 1$ and $y^k_{c'R_{si}} = 0, c \neq c'$, where $k$ indicates the $k$-th ICR branch. Inspired by the reality that the proposals that have high spatial coverage with the seed instance should be assigned the same label. Specifically, if the maximum intersection over union (IoU) between a certain proposal and seed instances is greater than the fixed label assignment threshold of 0.5, then the proposals as neighbor positive instances are also labelled to category $c$ and denote it to $R_{npi}$, namely, $y^k_{cR_{npi}} = 1$ and $y^k_{c'R_{npi}} = 0, c \neq c'$, otherwise the proposals are labelled to background instance and denote it to $R_{bi}$, namely, $y^k_{(C+1)R_{bi}} = 1$ and $y^k_{cR_{bi}} = 0, c \neq C + 1$.

However, aforementioned methods merely employ the PS to mine seed instances, which leads to the mined instances inclining to locate discriminative regions of objects rather than overall objects. In addition, the fixed label assignment strategy cannot adapt to the quality distribution change of proposals, which is not conducive to training high-quality instances. These are also the problems to be solved in this paper.

## 3. Materials and Methods

As shown in Figure 1, the OICR framework [27] is employed as the baseline framework of the proposed method. On the basis of OICR, a novel high-quality seed instance mining (HSIM) module is designed to mine high-quality seed instances. Specifically, the PCS is first designed, which is composed of traditional PS and PSCS. The PS indicates the probability that a proposal pertains to a certain category; the PSCS is calculated by considering the spatial relationships between top-scoring proposals, which is utilized to measure the extent to which the proposal locates an object. In addition, a novel dynamic pseudo instance labels assignment (DPILA) strategy is proposed to better adapt to the quality distribution change of proposals during training and, meanwhile, raise the number of positive instances in the initial training stage. Specifically, a label assignment threshold is dynamically calculated by elaborately designing a function that increases with the number of iterations.

### 3.1. Basic Weakly Supervised Object Detection Network

Bilen et al. [26] put forward a path-breaking weakly supervised deep detection network (WSDDN), which is the footstone of WSOD. The details of the WSDDN are as follows. Firstly, preset an image $I$ and image-level category labels $Y = [y_1, \ldots y_c, \ldots, y_C]$, where $y_c \in \{1, 0\}$ denotes present or absent object category $c$ in an image, and $C$ expresses the quantity of object category. For each image, a range of proposals $R = \{r_1, r_2, \ldots, r_{|R|}\}$ are produced via employing edge boxes (EB) [45] or selective search (SS) [46] algorithms, where $|R|$ expresses the quantity of proposals. Secondly, the feature maps $F \in \mathbb{R}^{W \times H \times C}$ are obtained by sending the image $I$ into the convolutional network (ConvNet), where $C$, $H$, and $W$ indicate the channels, height, and width of the feature maps $F$. Thirdly, the feature maps $F$ and the proposals $R$ are sent into the region of interest (RoI) pooling layer to obtain the proposal feature maps $F_R$ with a fixed size. Fourthly, the proposal feature vectors are obtained via two fully connected (FC) layers. These proposal feature vectors are then sent into two side-by-side branches, i.e., classification branch and detection branch, to produce two matrices $\mathbf{x}^c, \mathbf{x}^d \in \mathbb{R}^{C \times |R|}$ through respective FC layers. The classification score and detection score of each proposal are obtained by performing a softmax operation on the two matrices $\mathbf{x}^c$ , $\mathbf{x}^d$ along different directions; the details are as follows:

$$[\sigma(\mathbf{x}^c)]_{cr} = \frac{e^{x^c_{cr}}}{\sum_{c'=1}^{C} e^{x^c_{c'r}}}, [\sigma(\mathbf{x}^d)]_{cr} = \frac{e^{x^d_{cr}}}{\sum_{r'=1}^{|R|} e^{x^d_{cr'}}} \tag{1}$$

where $[\sigma(\mathbf{x}^c)]_{cr}$ indicates the probability that the proposal $r$ pertains to category $c$, $[\sigma(\mathbf{x}^d)]_{cr}$ represents the dedication of the proposal $r$ to category $c$. The 'dedication' indicates the contribution of a proposal $r$ to the image being classified in category $c$. Therefore, the $[\sigma(\mathbf{x}^d)]_{cr}$ also belongs to the probability to a certain extent; namely, the higher the $[\sigma(\mathbf{x}^d)]_{cr}$ value, the greater the probability of belonging to a positive instance. The proposal score is calculated via element-wise product between $\sigma(\mathbf{x}^c)$ and $\sigma(\mathbf{x}^d)$, which is denoted as follows:

$$\mathbf{x} = \sigma(\mathbf{x}^c) \odot \sigma(\mathbf{x}^d) \tag{2}$$

where $\mathbf{x} \in \mathbb{R}^{C \times |R|}$ represents the proposal score. Furthermore, image-level prediction score $\varphi_c$ of category $c$ can be acquired by the sum of all proposals as follows:

$$\varphi_c = \sum_{r=1}^{|R|} x_{cr} \tag{3}$$

Finally, the loss function $L_{WSDDN}$ of WSDDN is defined as follows:

$$L_{WSDDN} = - \sum_{c=1}^{C} (y_c \log \varphi_c + (1 - y_c) \log(1 - \varphi_c)) \tag{4}$$

where $y_c \in \{1, 0\}$ expresses the image-level category label, which indicates present or absent object category $c$ in an image.

To further promote the performance of the WSOD model, Tang et al. [27] introduced multi-stage instance classifier refinement (ICR) branches to improve the WSOD network. Specifically, we added $K$ parallel ICR branches on the WSDDN, and each ICR branch consists of a FC layer and a softmax layer, and the output $(C + 1)$ dimension score matrix $\mathbf{x}^k \in \mathbb{R}^{(C+1) \times |R|}$, where $k \in 1, 2, \ldots, K$, and the $(C + 1)$-th dimension denotes background. The $k$-th ICR branch is supervised through the previous $(k - 1)$-th branch, excluding the

1-st ICR branch from WSDDN (i.e., **x**). Finally, $K$ ICR branches are trained by utilizing the cross-entropy loss, which is formulated as follows:

$$L_{ICR}^k = -\frac{1}{|R|} \sum_{r=1}^{|R|} \sum_{c=1}^{C+1} w_r^k y_{cr}^k \log x_{cr}^k \tag{5}$$

where the $w_r^k$ denotes the loss weight, the $y_{cr}^k \in \{1, 0\}$ indicates the pseudo instance label. For more details, please refer to [27].

However, most of the existing methods [27,31,39] merely employ the proposal score (PS) of proposal to mine seed instances, where the PS indicates the probability that a proposal pertains to one category. Specifically, the proposal with the highest PS in a certain category is selected as the seed instance. However, the proposal (seed instance) with the highest PS usually locates the remarkable region of object but not the overall object. Therefore, existing methods are not able to mine high-quality seed instances.

### 3.2. High-Quality Seed Instance Mining Guided by Proposal Comprehensive Score

To overcome the above challenge, the proposal comprehensive score (PCS) is designed, which comprehensively considers the traditional proposal score (PS) and the proposed proposal space contribution score (PSCS). The PSCS is calculated by considering the spatial relationships between top-scoring proposals and is utilized to measure the extent to which the proposal locates an object. Consequently, seed instances mined by PCS can more completely locate an object than the traditional mined strategies, which merely utilize the PS. The details of PCS are as follows.

Firstly, the proposals are sorted from high to low based on their corresponding PS in the existing category. Secondly, the proposals with the top $p\%$ PS in category $c$ are selected as top-scoring proposals and defined them as an assembly $R_c' = \{r_1', \ldots, r_n', \ldots, r_N'\}$, where the $N$ expresses the quantity of top-scoring proposals in class $c$. Thirdly, the PSCS of each top-scoring proposal is calculated pursuant to the spatial relationship between the top-scoring proposals. Fourthly, the PCS is calculated by combining the PS and PSCS, which are defined as follows:

$$PCS_{cn} = \alpha PS_{cn} + (1 - \alpha) PSCS_{cn} \tag{6}$$

where $PS_{cn}$ indicates proposal score of the $n$-th proposal $r_n'$ in category $c$, $PSCS_{cn}$ denotes the proposal space contribution score of $r_n'$ in category $c$, $\alpha$ is the hyper-parameter to balance the contribution of PS and PSCS. The details of PSCS are as follows.

The undirected weighted graph $G_c^s = (V_c^s, E_c^s)$ is first constructed according to the spatial correlation of $R_c'$, where the vertexes $V_c^s$ denotes top-scoring proposals, each edge $E_c^s = \{\sigma_c^{nn'}\}$ denotes the spatial correlation between vertexes. As shown in Figure 2, the weight of each edge is obtained via calculating the IoU between vertexes, which is defined as follows:

$$\sigma_c^{r_n' r_{n'}'} = \begin{cases} \text{IoU}(r_n', r_{n'}'), & \text{if IoU}(r_n', r_{n'}') \geq T \\ 0, & \text{otherwise} \end{cases} \tag{7}$$

where the $T$ indicates hyper-parameter, the $\text{IoU}(r_n', r_{n'}')$ indicates the IoU value between $r_n'$ and $r_{n'}'$, $n \neq n'$. Based on this, the $PSCS_{cn}$ can be calculated as follows:

$$PSCS_{cn} = N(\sum_{r_{n'}' \in R_c'} \sigma_c^{r_n' r_{n'}'}), n \neq n' \tag{8}$$

where $N(\cdot)$ indicates the normalization operator. Finally, following the mining strategy [31], the PCS is utilized to mine high-quality seed instances, and denotes them as a assemble $R_c^s = \{r_1^s, \ldots, r_m^s, \ldots, r_M^s\}$, where the $M$ denotes the number of $R_c^s$ in category $c$.

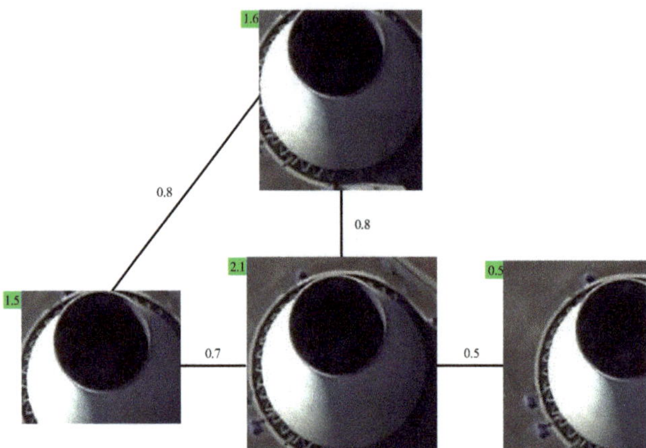

**Figure 2.** The details of weighted graph. Here, the graph is not undirected but has weighted. Specifically, the vertexes of graph denote top-scoring proposals, each edge denotes the spatial correlation (i.e., IoU) between vertexes.

### 3.3. Dynamic Pseudo Instance Label Assignment for Each Instance

Most of the WSOD methods usually set a fixed instance label assignment threshold (i.e., IoU value) to determine whether a certain proposal belongs to the positive or negative instance. If the IoU value between the proposal $r$ and its nearest seed instance $r_m^s$ greater than or equal to the default threshold $T_{IoU}$, the proposal is labeled as a positive instance; otherwise, the proposal is assigned a negative instance. Specifically, the label is defined as follows:

$$\text{label} = \begin{cases} 1, & \text{if IoU}(r, r_m^s) \geq T_{IoU} \\ 0, & \text{otherwise} \end{cases} \tag{9}$$

where $r \notin R_c^s$ indicates a certain proposal, $T_{IoU}$ is a fixed value and usually set to 0.5, which cannot adapt to the quality distribution change of proposals. In addition, setting a high $T_{IoU}$ may lead to the loss of some potential positive instances at the early stage of model training.

To overcome this issue, a dynamic pseudo instance label assignment (DPILA) strategy is proposed. The dynamic means that the label assignment threshold changes as the training progresses. Specifically, a growth function is designed to gradually adjust the IoU threshold as training goes on. The dynamic IoU threshold $T_{IoU}^d$ is defined as follows, and its variation curve is also demonstrated in Figure 3.

$$T_{IoU}^d = \frac{1}{1 + e^{-l \times t - m}} - 0.5 \tag{10}$$

where $l$ and $m$ denote hyper-parameters, $t$ indicates the number of current iterations. Therefore, the label is redefined as follows:

$$\text{label} = \begin{cases} 1, & \text{if IoU}(r, r_m^s) \geq T_{IoU}^d \\ 0, & \text{otherwise} \end{cases} \tag{11}$$

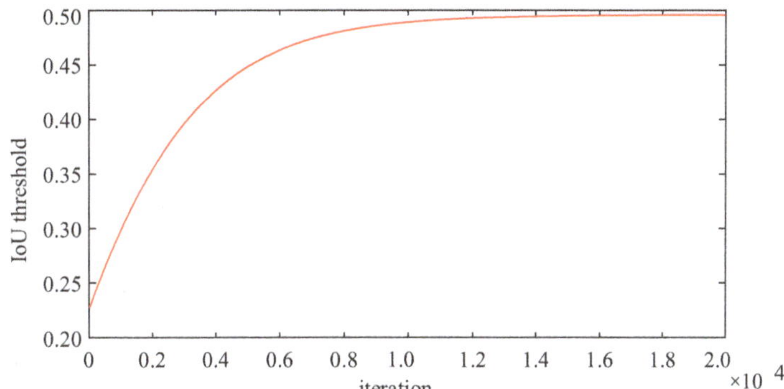

**Figure 3.** The variation curve of dynamic IoU threshold. The horizontal axis represents the number of iterations, the vertical axis represents the IoU threshold.

During testing, the DPILA strategy is discarded (i.e., all experiment results are from the mean output of 3 ICR branches), and the threshold is a fixed value (i.e., 0.5) following the WSOD criterion [27,31,39].

## 4. Experiment

### 4.1. Experiment Setup

#### 4.1.1. Datasets

Extensive experiments are implemented to measure the validity of the proposed methods on the NWPU VHR-10.v2 dataset [47,48] and DIOR dataset [49]. The NWPU VHR-10.v2 dataset comprises 1172 images, each with dimensions of $400 \times 400$ pixels, which has 879 trainval images and 293 test images and includes 10 object categories and 2775 instances. The DIOR dataset has a greater level of difficulty and includes 23,463 images, each with dimensions of $800 \times 800$ pixels. The DIOR dataset is partitioned into a trainval set, consisting of 11,725 images, and a testing set, comprising 11,738 images, which includes 20 object categories and 192,472 instances.

#### 4.1.2. Evaluation Metric

We employed two standard metrics to evaluate the performance of our method, which are widely used and accepted evaluation metrics in WSOD, namely, mean average precision (mAP) and correct localization (CorLoc) [50], where mAP evaluates the accuracy of detection on the testing set and CorLoc assesses the accuracy of localization on the trainval set. The two evaluation metrics comply with the PASCAL protocol.

#### 4.1.3. Implementation Details

The OICR network serves as the baseline framework for the proposed method. Similar to refs. [27,39,51], the VGG-16 [52] is utilized as the backbone network, which has undergone pre-training on the large-scale ImageNet dataset [8], in accordance with standard practice. The quantity of ICR branches is configured as 3. Following the standard of WSOD, merely image-level category labels of the trainval set are employed to train our model. We utilized the stochastic gradient descent (SGD) strategy to optimize our WSOD model, configuring values of 0.9 and 0.0001 for the momentum and weight decay hyperparameters, respectively. The initial learning rate and batch size is separately set at 0.01 and 8. We conducted a total of 20K and 60K training iterations on the NWPU VHR-10.v2 and DIOR datasets, respectively. The decay weight of the learning rate is set to 0.1, and the step size are separately set at 18K and 50K iterations on the NWPU VHR-10.v2 and DIOR datasets. The hyper-parameters $l$, $m$ and $p$ are separately set to 0.0002, 1 and 15. For data augmentation, all training images are augmented via rotating $90°$, $180°$ and horizontal flipping [32,33]. In addition, following the mainstream methods [27,39], the images are resized into five distinct scales {480, 576, 688, 864, and 1200} for training and testing. Inferential results are post-processed via implementing NMS operation, whose threshold is set at 0.3 [32,39,53,54]. The training details can also be seen in Table 1. The region proposals are generated via using the image segmentation algorithm (i.e., the selective search algorithm [46]). Specifically, the algorithm consists of the following three steps: (1) Initial segmentation: the image is segmented into small regions based on pixel intensity and texture similarity. (2) Similarity measure: all adjacent region pairs are combined and assigned a similarity score based on color, texture, size, and shape differences. (3) Proposals generation: the most similar regions are merged repeatedly until the desired number of proposals is obtained. Following the paradigm of WSOD, about 2000 region proposals are generated via a selective search algorithm. The scale of image segmentation is not fixed, which is determined according to the merger of similar regions in step (3).

All experiments are implemented on 8 TITAN RTX GPUs with the PyTorch framework.

**Table 1.** The training details of our method, which includes training setting and parameter setting.

| | Learning Rate | Batch Size | Momentum | Weight Decay | Iteration Numbers |
|---|---|---|---|---|---|
| Training Setting | 0.01 | 8 | 0.9 | 0.0001 | 20 K/60 K |
| Parameter setting | $K$<br>3 | $l$<br>0.0002 | $m$<br>1 | $p$ (%)<br>15 | NMS threshold<br>0.3 |

*4.2. Parameter Analyses*

4.2.1. Parameter Analysis of $\alpha$

As previously discussed, the parameter $\alpha$ plays a critical role in determining the relative contributions of PS and PSCS. To objectively assess this relationship, we conducted a quantitative analysis of the DIOR dataset. As demonstrated in Figure 4, our approach achieved the highest mAP when $\alpha$ is 0.5. Based on these results, we adopted $\alpha = 0.5$ as the optimal value for this paper.

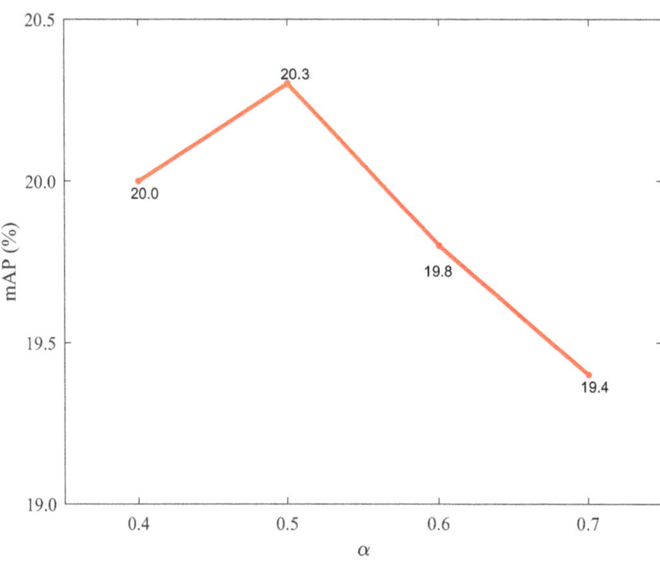


**Figure 4.** Parameter analysis of $\alpha$ on the DIOR dataset. The horizontal axis represents different $\alpha$ values, the vertical axis represents the mAP values.

### 4.2.2. Parameter Analysis of $T$

As mentioned before, $T$ is the threshold to determine the value of $\sigma_c^{r'_n r'_{n'}}$, which is analyzed quantitatively on the DIOR dataset. As demonstrated in Figure 5, our approach achieved the highest mAP when $T$ is 0.7. Based on these results, we adopted $T = 0.7$ as the optimal value for this paper.

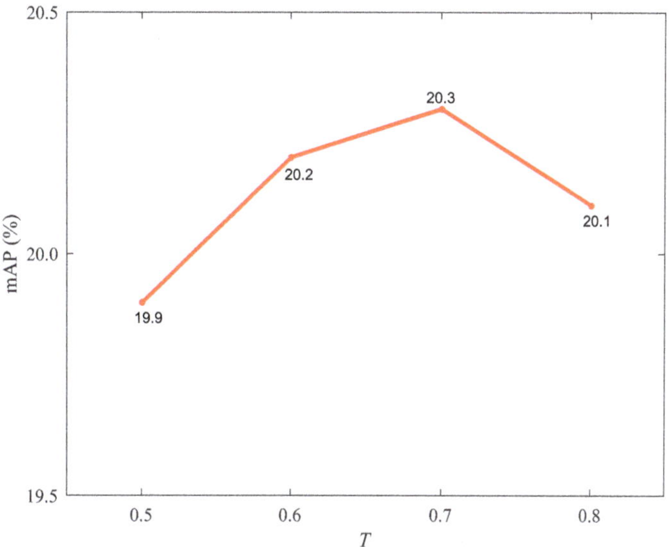

**Figure 5.** Parameter analysis of $T$ on the DIOR dataset. The horizontal axis represents different $T$ values, the vertical axis represents the mAP values.

### 4.3. Ablation Studies

Ablation studies are constructed to verify the validity of the proposed PCS and DPILA. Specifically, as shown in Table 2, the baseline, baseline+PCS, baseline+DPILA, and baseline+PCS+DPILA experiments are implemented on the DIOR dataset.

**Table 2.** Ablation studies of our method on the DIOR dataset.

| Baseline (OICR) | PCS | DPILA | DIOR | |
| --- | --- | --- | --- | --- |
| | | | **mAP** | **CorLoc** |
| | | | 16.5 | 34.8 |
| | ✓ | | 20.3 | 42.2 |
| ✓ | | ✓ | 18.9 | 41.0 |
| | ✓ | ✓ | **21.6** | **44.3** |

Bold entities denote best results.

### 4.3.1. Influence of PCS

The baseline+PCS experiment is constructed to validate the influence of the proposed PCS. As shown in Table 2, the baseline+PCS method obtains 20.3% mAP and 42.2% CorLoc on the DIOR dataset, which surpasses the baseline method 3.8% mAP and 7.4% CorLoc. Therefore, the validity of PCS is verified obviously. The major reason for performance enhancement is that the proposed PCS can effectively guide the WSOD model to mine high-quality seed instances, which further encourage model to locate more complete object.

### 4.3.2. Influence of DPILA

The baseline+DPILA experiment is constructed to validate the influence of the proposed DPILA. As shown in Table 2, the baseline+DPILA method obtains 18.9% mAP and

41.0% CorLoc, which outperforms the baseline method 2.4% mAP and 6.2% CorLoc on the DIOR dataset. Therefore, the validity of DPILA is verified obviously. The major reason for performance enhancement is that the proposed DPILA strategy can adapt to the quality distribution change of proposals during training and mine some potential positive instances at the early stage of model training. Consequently, the DPILA strategy can dynamically assign a pseudo-instance label for each instance, which further improves the performance of WSOD.

The baseline+PCS+DPILA experiment is constructed to verify the influence of the combination of PCS and DPILA. As shown in Table 2, the baseline+PCS+DPILA method obtains 21.6% mAP and 44.3% CorLoc on the DIOR dataset, which outperforms the other three methods. Therefore, the validity of the combination of PCS and DPILA is verified effectively.

### 4.4. Comparison with Other Advanced WSOD Methods

To further validate the integrated performance of our method, we reported the comprehensive results and provided comparisons with seven WSOD methods and four fully supervised object detection (FSOD) methods on two popular RSIs datasets. Specifically, the 4 WSOD methods, including WSDDN [26], OICR [27], min-entropy latent model (MELM) [53], and proposal cluster learning (PCL) [39], were compared with our method on two RSIs datasets. The other 3 WSOD methods, including dynamic curriculum learning (DCL) [33], full-coverage collaborative Network (FCC-Net) [36], and collaborative learning-based network (CLN) [30], were compared with our method on the DIOR dataset. The 4 FSOD methods include region-based convolutional neural networks (R-CNN) [55], Fast R-CNN [56], Faster R-CNN [57], and rotation-invariant convolutional neural networks (RICNN) [47].

### 4.4.1. Comparison in Terms of mAP

Tables 3 and 4 demonstrate the comparison in terms of mAP between our approach and other advanced WSOD methods. Specifically, as shown in Table 3, our approach obtains 47.3% mAP on the NWPU VHR-10.v2 dataset. Compared with other advanced WSOD methods, our method significantly exceeds the WSDDN, OICR, PCL, and MELM by 12.2%, 12.8%, 7.9%, and 5.0% in terms of mAP, respectively, on the NWPU VHR-10.v2 dataset. As shown in Table 4, our method obtains 21.6% mAP on the DIOR dataset. Compared with the other advanced WSOD methods, our method significantly exceeds the WSDDN, OICR, PCL, MELM, DCL, FCC-Net and CLN-RSOD methods on the DIOR dataset, with an increase in mAP of 8.3%, 5.1%, 3.4%, 2.9%, 1.4%, 3.3% and 3.3%, respectively. Compared with the FSOD methods, our approach further decreases the performance gap between FSOD method and WSOD method.

**Table 3.** Comparisons with other advanced methods in terms of AP (%) and mAP (%) on the NWPU VHR-10.v2 dataset.

| Method | Airplane | Ship | Storage Tank | Baseball Diamond | Tennis Court | Basketball Court | Ground Track Field | Harbor | Bridge | Vehicle | mAP |
|---|---|---|---|---|---|---|---|---|---|---|---|
| R-CNN [55] | 85.4 | 88.9 | 62.8 | 19.7 | 90.7 | 58.2 | 68.0 | 79.9 | 54.2 | 49.9 | 65.8 |
| RICNN [47] | 88.7 | 78.3 | 86.3 | 89.1 | 42.3 | 56.9 | 87.7 | 67.5 | 62.3 | 72.0 | 73.1 |
| Fast R-CNN [56] | 90.9 | 90.6 | 89.3 | 47.3 | 100.0 | 85.9 | 84.9 | 88.2 | 80.3 | 69.8 | 82.7 |
| Faster R-CNN [57] | 90.9 | 86.3 | 90.5 | 98.2 | 89.7 | 69.6 | 100.0 | 80.1 | 61.5 | 78.1 | 84.5 |
| WSDDN [26] | 30.1 | 41.7 | 35.0 | 88.9 | 12.9 | 23.9 | 99.4 | 13.9 | 1.9 | 3.6 | 35.1 |
| OICR [27] | 13.7 | 67.4 | 57.2 | 55.2 | 13.6 | 39.7 | 92.8 | 0.2 | 1.8 | 3.7 | 34.5 |
| PCL [39] | 26.0 | 63.8 | 2.5 | 89.8 | 64.5 | 76.1 | 77.9 | 0.0 | 1.3 | 15.7 | 39.4 |
| MELM [53] | 80.9 | 69.3 | 10.5 | 90.2 | 12.8 | 20.1 | 99.2 | 17.1 | 14.2 | 8.7 | 42.3 |
| Ours | 77.9 | 32.0 | 48.1 | 90.9 | 28.5 | 62.4 | 88.6 | 40.2 | 1.2 | 3.6 | **47.3** |

Bold entities denote best results.

**Table 4.** Comparisons with other advanced methods in terms of AP (%) and mAP (%) on the DIOR dataset.

| Method | Airplane | Airport | Baseball Field | Basketball Court | Bridge | Chimney | Dam | Expressway Service Area | Expressway Toll Station | Golf Field |
|---|---|---|---|---|---|---|---|---|---|---|
| R-CNN [55] | 35.6 | 43.0 | 53.8 | 62.3 | 15.6 | 53.7 | 33.7 | 50.2 | 33.5 | 50.1 |
| RICNN [47] | 39.1 | 61.0 | 60.1 | 66.3 | 25.3 | 63.3 | 41.1 | 51.7 | 36.6 | 55.9 |
| Fast R-CNN [56] | 44.2 | 66.8 | 67.0 | 60.5 | 15.6 | 72.3 | 52.0 | 65.9 | 44.8 | 72.1 |
| Faster R-CNN [57] | 50.3 | 62.6 | 66.0 | 80.9 | 28.8 | 68.2 | 47.3 | 58.5 | 48.1 | 60.4 |
| WSDDN [26] | 9.1 | 39.7 | 37.8 | 20.2 | 0.3 | 12.2 | 0.6 | 0.7 | 11.9 | 4.9 |
| OICR [27] | 8.7 | 28.3 | 44.1 | 18.2 | 1.3 | 20.2 | 0.1 | 0.7 | 29.9 | 13.8 |
| PCL [39] | 21.5 | 35.2 | 59.8 | 23.5 | 3.0 | 43.7 | 0.1 | 0.9 | 1.5 | 2.9 |
| MELM [53] | 28.1 | 3.2 | 62.5 | 28.7 | 0.1 | 62.5 | 0.2 | 28.4 | 13.1 | 15.2 |
| DCL [33] | 20.9 | 22.7 | 54.2 | 11.5 | 6.0 | 61.0 | 0.1 | 1.1 | 31.0 | 30.9 |
| FCC-Net [36] | 20.1 | 38.8 | 52.0 | 23.4 | 1.8 | 22.3 | 0.2 | 0.6 | 28.7 | 14.1 |
| CLN [30] | 10.1 | 33.2 | 43.9 | 23.4 | 0.8 | 38.8 | 0.7 | 1.1 | 19.3 | 11.6 |
| Ours | 10.5 | 32.4 | 64.2 | 28.0 | 1.1 | 13.3 | 0.3 | 0.3 | 29.9 | 50.9 |

| Method | Ground Track Field | Harbor | Overpass | Ship | Stadium | Storage Tank | Tennis Court | Train Station | Vehicle | Windmill | mAP |
|---|---|---|---|---|---|---|---|---|---|---|---|
| R-CNN [55] | 49.3 | 39.5 | 30.9 | 9.1 | 60.8 | 18.0 | 54.0 | 36.1 | 9.1 | 16.4 | 37.7 |
| RICNN [47] | 58.9 | 43.5 | 39.0 | 9.1 | 61.1 | 19.1 | 63.5 | 46.1 | 11.4 | 31.5 | 44.2 |
| Fast R-CNN [56] | 62.9 | 46.2 | 38.0 | 32.1 | 71.0 | 35.0 | 58.3 | 37.9 | 19.2 | 38.1 | 50.0 |
| Faster R-CNN [57] | 67.0 | 43.9 | 46.9 | 58.5 | 52.4 | 42.4 | 79.5 | 48.0 | 34.8 | 65.4 | 55.5 |
| WSDDN [26] | 42.4 | 4.7 | 1.1 | 0.7 | 63.0 | 4.0 | 6.1 | 0.5 | 4.6 | 1.1 | 13.3 |
| OICR [27] | 57.4 | 10.7 | 11.1 | 9.1 | 59.3 | 7.1 | 0.7 | 0.1 | 9.1 | 0.4 | 16.5 |
| PCL [39] | 56.4 | 16.8 | 11.1 | 9.1 | 57.6 | 9.1 | 2.5 | 0.1 | 4.6 | 4.6 | 18.2 |
| MELM [53] | 41.1 | 26.1 | 0.4 | 9.1 | 8.6 | 15.0 | 20.6 | 9.8 | 0.0 | 0.5 | 18.7 |
| DCL [33] | 56.5 | 5.1 | 2.7 | 9.1 | 63.7 | 9.1 | 10.4 | 0.0 | 7.3 | 0.8 | 20.2 |
| FCC-Net [36] | 56.0 | 11.1 | 10.9 | 10.0 | 57.5 | 9.1 | 3.6 | 0.1 | 5.9 | 0.7 | 18.3 |
| CLN [30] | 48.9 | 19.6 | 9.5 | 13.0 | 54.5 | 10.8 | 10.3 | 0.5 | 9.2 | 6.7 | 18.3 |
| Ours | 55.4 | 12.4 | 15.0 | 34.0 | 33.9 | 30.0 | 1.3 | 4.1 | 14.8 | 0.8 | **21.6** |

Bold entities denote best results.

### 4.4.2. Comparison in Terms of CorLoc

Tables 5 and 6 demonstrate the comparison in terms of CorLoc between our approach and other advanced WSOD methods. Specifically, as shown in Table 5, our approach acquires 58.4% CorLoc on the NWPU VHR-10.v2 dataset. Compared with the other advanced WSOD methods, our method significantly exceeds the WSDDN, OICR, PCL, and MELM methods on the NWPU VHR-10.v2 dataset, with an increase in CorLoc of 23.2%, 18.4%, 13.3%, and 8.5%, respectively. As shown in Table 6, our method obtains 44.3% CorLoc on the DIOR dataset. In comparison to other advanced WSOD methods, our approach significantly exceeds the WSDDN, OICR, PCL, MELM, DCL and FCC-Net methods by 11.9%, 9.5%, 2.8%, 1.0%, 2.1%, and 2.6% CorLoc, respectively, on the DIOR dataset.

**Table 5.** Comparisons with other advanced methods in terms of CorLoc (%) on the NWPU VHR-10.v2 dataset.

| Method | WSDDN [26] | OICR [27] | PCL [39] | MELM [53] | Ours |
|---|---|---|---|---|---|
| NWPU VHR-10.v2 | 35.2 | 40.0 | 45.1 | 49.9 | **58.4** |

Bold entities denote best results.

**Table 6.** Comparisons with other advanced methods in terms of CorLoc (%) on the DIOR dataset. '-' denotes the CorLoc value has not been reported in their study.

| Method | WSDDN [26] | OICR [27] | PCL [39] | MELM [53] | DCL [33] | FCC-Net [36] | CLN [30] | Ours |
|---|---|---|---|---|---|---|---|---|
| DIOR | 32.4 | 34.8 | 41.5 | 43.3 | 42.2 | 41.7 | - | **44.3** |

Bold entities denote best results.

### 4.4.3. Subjective Comparison

In addition, to further evaluate our method, Four advanced WSOD methods that provide source codes are subjectively compared with our method on two RSI datasets in

Figures 6 and 7, respectively. Figure 6 shows the visual comparison results on the NWPU VHR-10.v2 dataset, and the objects with different categories are enclosed by utilizing the bounding boxes with different colors. Figure 7 displays the visual comparison results on the DIOR dataset, and the objects are enclosed by utilizing green bounding boxes. What is more, the category of object is attached to the bounding box. As shown in Figures 6 and 7, the detection results of our approach can completely locate and correctly identify objects.

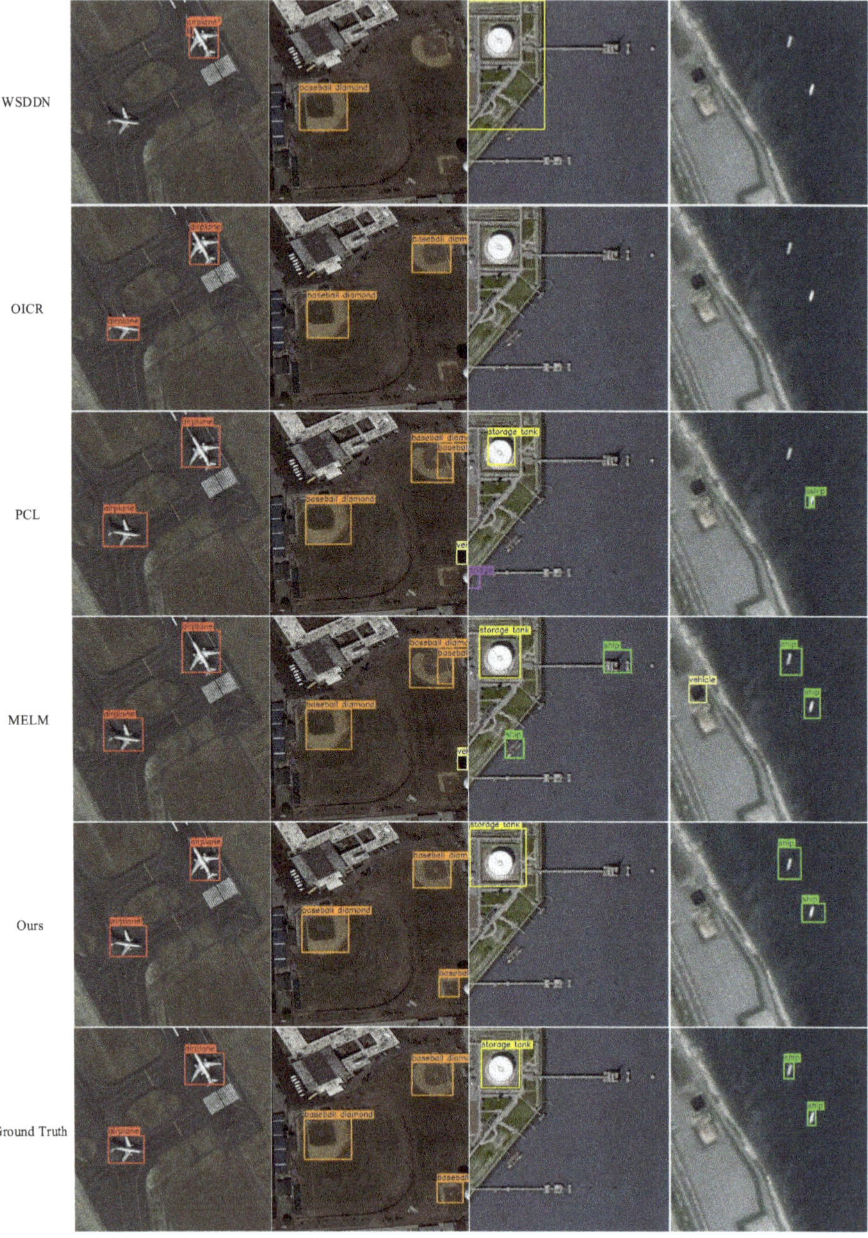

**Figure 6.** Four advanced WSOD methods that provide source codes are subjectively compared with our method on the NWPU VHR-10.v2 dataset.

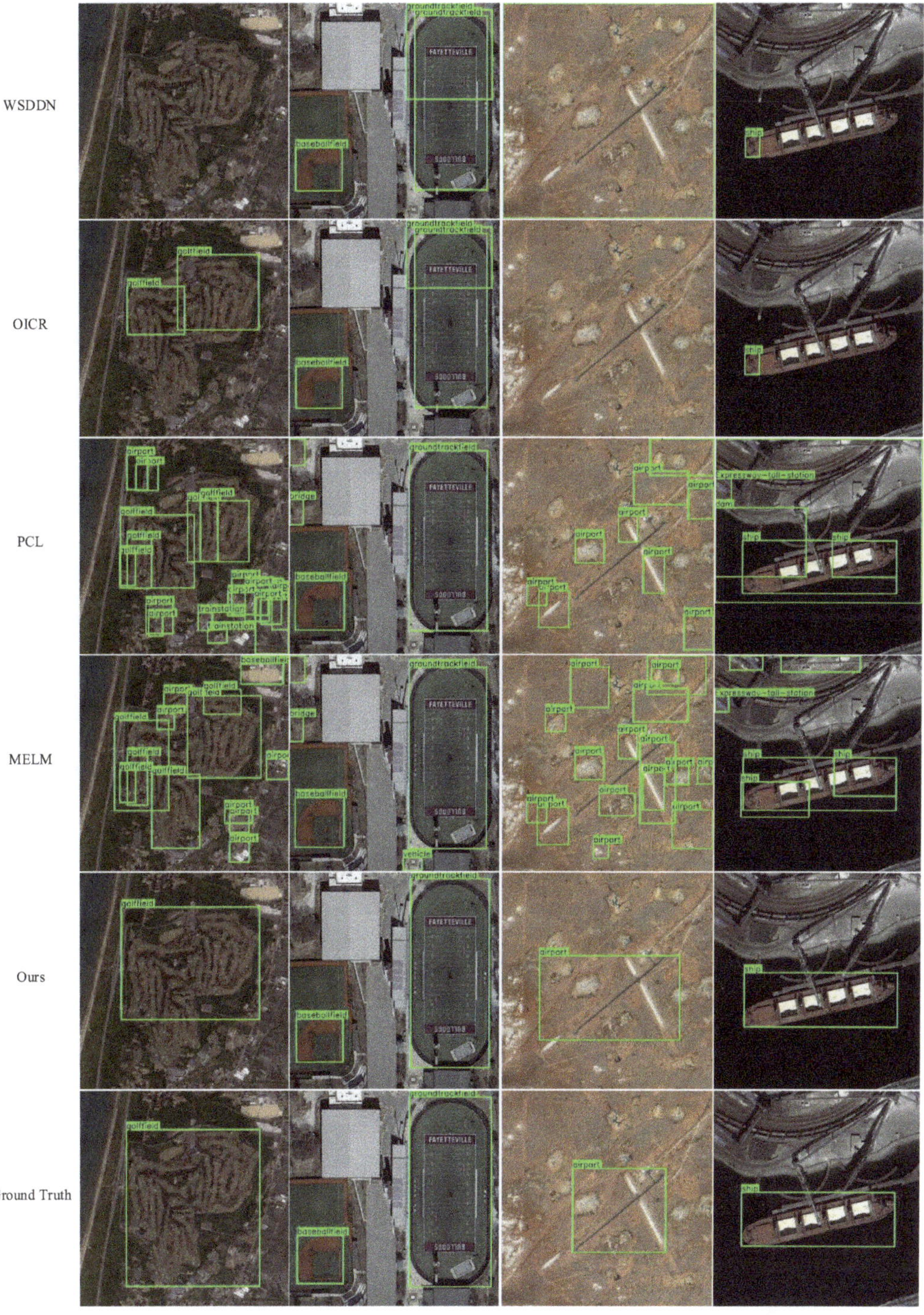


**Figure 7.** Four advanced WSOD methods that provide source codes are subjectively compared with our method on the DIOR dataset.

*4.5. Runtime Analysis*

In order to assess the practicality of the proposed approach in real-world scenarios, we further reported the runtime of the proposed method in terms of training and inference. As shown in Table 7, during training, compared with the baseline method, the computational time increases from 24.8 to 30.4 h by incorporating the HSIM into the baseline method. The additional complexity is mainly introduced because HSIM is added. Furthermore, when we incorporate the DPILA into the baseline method, the computational time increased from 24.8 to 25.0 h, which is caused by the calculation of DPILA. During inference, the HSIM module and calculation of DPILA are discarded; namely, all experiment results are from the mean output of 3 ICR branches (as shown in the lower right of Figure 1). Therefore, all methods have the same complexity, which costs the same inference time (i.e., 2.2 h) during inference. Although the training time of the baseline method is less than ours (24.8 versus 30.7 h), its performance is reduced by 5.1% compared with ours.

**Table 7.** The Complexity analysis of our method on the DIOR Dataset. All experiments are implemented on ubuntu16.04 and NVIDIA TITAN RTX GPU.

| Method | Training Time (Hours) | Inference Time (Hours) | mAP (%) |
| --- | --- | --- | --- |
| Baseline (OICR) | 24.8 | 2.2 | 16.5 |
| +HSIM (PCS) | 30.4 | 2.2 | 20.3 |
| +DPILA | 25.0 | 2.2 | 18.9 |
| +HSIM+DPILA | 30.7 | 2.2 | 21.6 |

## 5. Discussion

To tackle the first challenge, the detection results of WSOD tend to locate the significant regions of the object but not the overall object. The PCS, which consists of traditional PS and PSCS, is designed as a novel metric to mine high-quality seed instances. To tackle the second challenge, traditional pseudo-instance label assignment strategies cannot adapt to the quality distribution changes of proposals during training, which is not conducive to training a high-performance detector. A DPILA strategy is developed via dynamically setting the label assignment threshold to train high-quality instances. Consequently, collaborating on the proposed PCS with DPILA achieves better performance than other advanced WSOD methods on two popular RSIs datasets. Specifically, our method surpasses separately WSDDN, OICR, PCL, and MELM methods by 12.2% (8.3%), 12.8% (5.1%), 7.9% (3.4%), and 5.0% (2.9%) in terms of mAP on the NWPU VHR-10.v2 (DIOR) dataset, and surpasses separately WSDDN, OICR, PCL, and MELM methods by 23.2% (11.9%), 18.4% (9.5%), 13.3% (2.8%) , and 8.5% (1.0%) in terms of CorLoc on the NWPU VHR-10.v2 (DIOR) dataset.

## 6. Conclusions

In this paper, a novel HSIM module is designed to tackle the challenge that the detection results of WSOD detector tend to locate the significant regions of an object but not the overall object. Specifically, the PCS is first designed and is composed of traditional PS and proposed PSCS. The PSCS is utilized to evaluate the wholeness with which a proposal locates an object. Consequently, high PCS will encourage the WSOD model to mine high-quality seed instances. A DPILA strategy is developed to tackle the challenge that traditional pseudo-instance label assignment strategies cannot adapt to the quality distribution change of proposals during training. Specifically, a dynamic label assignment threshold is defined by elaborately designing a function that increases with the number of iterations. Consequently, the DPILA strategy can dynamically assign a pseudo instance label for each instance, which further improves the performance of WSOD. The ablation studies verify the validity of the proposed PCS and DPILA. The comparison experiments verify that our approach obtains better performance than other advanced WSOD detectors on two popular RSIs datasets. The subjective comparison straightforwardly demonstrates that our method can completely locate and correctly identify objects.

The shortcomings of the proposed model are that it achieves poor performance in individual classes such as Dam, Windmill, etc. The possible reason is that our model is susceptible to interference from complex backgrounds. For instance, the Dam is disturbed by the large reservoir, so the reservoir is often mistakenly identified as Dam. The Windmill is disturbed by the shadow of Windmill, so the shadow of Windmill is often mistakenly identified as Windmill. To improve the anti-interference ability of our model, we plan to design a novel feature enhancement module to enhance the feature extraction ability of WSOD. The high-quality feature is conducive to correctly identifying the object and enhances the robustness of the WSOD model.

**Author Contributions:** Conceptualization, L.Z., Y.H. and X.Q.; methodology, L.Z. and Y.H.; software, Y.H.; validation, X.Q. and Z.C.; formal analysis, L.Z., X.Q. and Z.C.; resources, Z.C.; writing—original draft, Y.H.; writing—review and editing, L.Z.; supervision, Z.C.; project administration, Z.C.; funding acquisition, X.Q. All authors have read and agreed to the published version of the manuscript.

**Funding:** This work was supported in part by the National Natural Science Foundation of China under Grant 62076223, in part by the Key Science and Technology Program of Henan Province under Grant 232102211018.

**Institutional Review Board Statement:** Not applicable.

**Informed Consent Statement:** Not applicable.

**Data Availability Statement:** The NWPU VHR-10.v2 and DIOR datasets are available at following URLs: https://drive.google.com/file/d/15xd4TASVAC2irRf02GA4LqYFbH7QITR-/view (accessed on 15 October 2022) and https://drive.google.com/drive/folders/1UdlgHk49iu6WpcJ5467iT-UqNPpx__CC (accessed on 15 October 2022), respectively.

**Conflicts of Interest:** The authors declare no conflict of interest.

# References

1. Li, Z.; Ma, Z.; van der Kuijp, T.J.; Yuan, Z.; Huang, L. A review of soil heavy metal pollution from mines in China: Pollution and health risk assessment. *Sci. Total Environ.* **2014**, *468*, 843–853. [CrossRef]
2. Sanaei, F.; Amin, M.M.; Alavijeh, Z.P.; Esfahani, R.A.; Sadeghi, M.; Bandarrig, N.S.; Fatehizadeh, A.; Taheri, E.; Rezakazemi, M. Health risk assessment of potentially toxic elements intake via food crops consumption: Monte Carlo simulation-based probabilistic and heavy metal pollution index. *Environ. Sci. Pollut. Res.* **2021**, *28*, 1479–1490. [CrossRef]
3. Oliveira, V.; Pinho, P. Evaluation in urban planning: Advances and prospects. *J. Plan. Lit.* **2010**, *24*, 343–361. [CrossRef]
4. Wosner, O.; Farjon, G.; Bar-Hillel, A. Object detection in agricultural contexts: A multiple resolution benchmark and comparison to human. *Comput. Electron. Agric.* **2021**, *189*, 106404. [CrossRef]
5. Zhao, W.; Yamada, W.; Li, T.; Digman, M.; Runge, T. Augmenting crop detection for precision agriculture with deep visual transfer learning—A case study of bale detection. *Remote Sens.* **2020**, *13*, 23. [CrossRef]
6. Lin, S.; Zhang, M.; Cheng, X.; Wang, L.; Xu, M.; Wang, H. Hyperspectral anomaly detection via dual dictionaries construction guided by two-stage complementary decision. *Remote Sens.* **2022**, *14*, 1784. [CrossRef]
7. Cheng, X.; Zhang, M.; Lin, S.; Zhou, K.; Wang, L.; Wang, H. Multiscale superpixel guided discriminative forest for hyperspectral anomaly detection. *Remote Sens.* **2022**, *14*, 4828. [CrossRef]
8. Krizhevsky, A.; Sutskever, I.; Hinton, G.E. ImageNet Classification with Deep Convolutional Neural Networks. In Proceedings of the Conference and Workshop on Neural Information Processing Systems, Lake Tahoe, NV, USA, 3–6 December 2012; pp. 1097–1105.
9. Qian, X.; Zeng, Y.; Wang, W.; Zhang, Q. Co-Saliency Detection Guided by Group Weakly Supervised Learning. *IEEE Trans. Multimed.* **2023**, *25*, 1810–1818. [CrossRef]
10. Lin, S.; Zhang, M.; Cheng, X.; Zhou, K.; Zhao, S.; Wang, H. Dual Collaborative Constraints Regularized Low-Rank and Sparse Representation via Robust Dictionaries Construction for Hyperspectral Anomaly Detection. *IEEE J. Sel. Top. Appl. Earth Obs. Remote Sens.* **2023**, *16*, 2009–2024. [CrossRef]
11. Cheng, X.; Zhang, M.; Lin, S.; Zhou, K.; Zhao, S.; Wang, H. Two-Stream Isolation Forest Based on Deep Features for Hyperspectral Anomaly Detection. *IEEE Geosci. Remote Sens. Lett.* **2023**, *20*, 1–5. [CrossRef]
12. Kuo, W.; Hariharan, B.; Malik, J. DeepBox: Learning Objectness with Convolutional Networks. In Proceedings of the IEEE International Conference on Computer Vision (ICCV), Santiago, Chile, 11–18 December 2015; pp. 2479–2487.

13. Qian, X.; Cheng, X.; Cheng, G.; Yao, X.; Jiang, L. Two-stream encoder GAN with progressive training for co-saliency detection. *IEEE Signal Process. Lett.* **2021**, *28*, 180–184. [CrossRef]
14. Lin, S.; Zhang, M.; Cheng, X.; Zhou, K.; Zhao, S.; Wang, H. Hyperspectral Anomaly Detection via Sparse Representation and Collaborative Representation. *IEEE J. Sel. Top. Appl. Earth Obs. Remote Sens.* **2022**, *16*, 946–961. [CrossRef]
15. Han, X.; Zhong, Y.; Zhang, L. An efficient and robust integrated geospatial object detection framework for high spatial resolution remote sensing imagery. *Remote Sens.* **2017**, *9*, 666. [CrossRef]
16. Qian, X.; Wu, B.; Cheng, G.; Yao, X.; Wang, W.; Han, J. Building a bridge of bounding box regression between oriented and horizontal object detection in remote sensing images. *IEEE Trans. Geosci. Remote Sens.* **2023**, *61*, 1–9. [CrossRef]
17. Deng, Z.; Sun, H.; Zhou, S.; Zhao, J.; Lei, L.; Zou, H. Multi-scale object detection in remote sensing imagery with convolutional neural networks. *ISPRS J. Photogramm. Remote Sens.* **2018**, *145*, 3–22. [CrossRef]
18. Zhang, Y.; Ma, C.; Zhuo, L.; Li, J. Arbitrary-Oriented Object Detection in Aerial Images with Dynamic Deformable Convolution and Self-Normalizing Channel Attention. *Electronics* **2023**, *12*, 2132. [CrossRef]
19. Qian, X.; Lin, S.; Cheng, G.; Yao, X.; Ren, H.; Wang, W. Object detection in remote sensing images based on improved bounding box regression and multi-level features fusion. *Remote Sens.* **2020**, *12*, 143. [CrossRef]
20. Fasana, C.; Pasini, S.; Milani, F.; Fraternali, P. Weakly Supervised Object Detection for Remote Sensing Images: A Survey. *Remote Sens.* **2022**, *14* 5362. [CrossRef]
21. Zhang, X.; Yu, W.; Ma, X.; Kang, X. Weakly Supervised Local-Global Anchor Guidance Network for Landslide Extraction With Image-Level Annotations. *IEEE Geosci. Remote Sens. Lett.* **2023**, *20*, 6005505. [CrossRef]
22. Ren, W.; Huang, K.; Tao, D.; Tan, T. Weakly supervised large scale object localization with multiple instance learning and bag splitting. *IEEE Trans. Pattern Anal. Mach. Intell.* **2015**, *38*, 405–416. [CrossRef]
23. Wang, X.; Zhu, Z.; Yao, C.; Bai, X. Relaxed multiple-instance SVM with application to object discovery. In Proceedings of the IEEE International Conference on Computer Vision, Santiago, Chile, 7–13 December 2015; pp. 1224–1232.
24. Cinbis, R.G.; Verbeek, J.; Schmid, C. Weakly supervised object localization with multi-fold multiple instance learning. *IEEE Trans. Pattern Anal. Mach. Intell.* **2016**, *39*, 189–203. [CrossRef] [PubMed]
25. Hong, D.; Yokoya, N.; Chanussot, J.; Zhu, X.X. An augmented linear mixing model to address spectral variability for hyperspectral unmixing. *IEEE Trans. Image Process.* **2018**, *28*, 1923–1938. [CrossRef]
26. Bilen, H.; Vedaldi, A. Weakly supervised deep detection networks. In Proceedings of the IEEE Conference on Computer Vision and Pattern Recognition, Las Vegas, NV, USA, 27–30 June 2016; pp. 2846–2854.
27. Tang, P.; Wang, X.; Bai, X.; Liu, W. Multiple instance detection network with online instance classifier refinement. In Proceedings of the IEEE Conference on Computer Vision and Pattern Recognition, Honolulu, HI, USA, 21–26 July 2017; pp. 2843–2851.
28. Kantorov, V.; Oquab, M.; Cho, M.; Laptev, I. Contextlocnet: Context-aware deep network models for weakly supervised localization. In Proceedings of the European Conference on Computer Vision, Amsterdam, The Netherlands, 11–14 October 2016; Springer: Berlin/Heidelberg, Germany, 2016; pp. 350–365.
29. Li, D.; Huang, J.B.; Li, Y.; Wang, S.; Yang, M.H. Weakly supervised object localization with progressive domain adaptation. In Proceedings of the IEEE Conference on Computer Vision and Pattern Recognition, Las Vegas, NV, USA, 27–30 June 2016; pp. 3512–3520.
30. Chen, S.; Wang, H.; Mukherjee, M.; Xu, X. Collaborative Learning-based Network for Weakly Supervised Remote Sensing Object Detection. *IEEE J. Sel. Top. Appl. Earth Obs. Remote Sens.* 2022, *Early access.* [CrossRef]
31. Ren, Z.; Yu, Z.; Yang, X.; Liu, M.Y.; Lee, Y.J.; Schwing, A.G.; Kautz, J. Instance-aware, context-focused, and memory-efficient weakly supervised object detection. In Proceedings of the IEEE/CVF Conference on Computer Vision and Pattern Recognition, Seattle, WA, USA, 13–19 June 2020; pp. 10598–10607.
32. Feng, X.; Han, J.; Yao, X.; Cheng, G. Progressive contextual instance refinement for weakly supervised object detection in remote sensing images. *IEEE Trans. Geosci. Remote Sens.* **2020**, *58*, 8002–8012. [CrossRef]
33. Yao, X.; Feng, X.; Han, J.; Cheng, G.; Guo, L. Automatic weakly supervised object detection from high spatial resolution remote sensing images via dynamic curriculum learning. *IEEE Trans. Geosci. Remote Sens.* **2020**, *59*, 675–685. [CrossRef]
34. Feng, X.; Han, J.; Yao, X.; Cheng, G. TCANet: Triple Context-Aware Network for Weakly Supervised Object Detection in Remote Sensing Images. *IEEE Trans. Geosci. Remote Sens.* **2021**, *59*, 6946–6955. [CrossRef]
35. Feng, X.; Yao, X.; Cheng, G.; Han, J.; Han, J. SAENet: Self-Supervised Adversarial and Equivariant Network for Weakly Supervised Object Detection in Remote Sensing Images. *IEEE Trans. Geosci. Remote Sens.* **2022**, *60*, 5610411. [CrossRef]
36. Chen, S.; Shao, D.; Shu, X.; Zhang, C.; Wang, J. FCC-Net: A Full-Coverage Collaborative Network for Weakly Supervised Remote Sensing Object Detection. *Electronics* **2020**, *9*, 1356. [CrossRef]
37. Kosugi, S.; Yamasaki, T.; Aizawa, K. Object-aware instance labeling for weakly supervised object detection. In Proceedings of the IEEE International Conference on Computer Vision, Seoul, Republic of Korea, 27 October–2 November 2019; pp. 6064–6072.
38. Zeng, Z.; Liu, B.; Fu, J.; Chao, H.; Zhang, L. Wsod2: Learning bottom-up and top-down objectness distillation for weakly-supervised object detection. In Proceedings of the IEEE International Conference on Computer Vision, Seoul, Republic of Korea, 27 October–2 November 2019; pp. 8292–8300.
39. Tang, P.; Wang, X.; Bai, S.; Shen, W.; Bai, X.; Liu, W.; Yuille, A. Pcl: Proposal cluster learning for weakly supervised object detection. *IEEE Trans. Pattern Anal. Mach. Intell.* **2018**, *42*, 176–191. [CrossRef]

40. Lin, C.; Wang, S.; Xu, D.; Lu, Y.; Zhang, W. Object instance mining for weakly supervised object detection. *Proc. AAAI Conf. Artif. Intell.* **2020**, *34*, 11482–11489. [CrossRef]
41. Cheng, G.; Xie, X.; Chen, W.; Feng, X.; Yao, X.; Han, J. Self-Guided Proposal Generation for Weakly Supervised Object Detection. *IEEE Trans. Geosci. Remote Sens.* **2022**, *60*, 1–11. [CrossRef]
42. Qian, X.; Huo, Y.; Cheng, G.; Yao, X.; Li, K.; Ren, H.; Wang, W. Incorporating the Completeness and Difficulty of Proposals Into Weakly Supervised Object Detection in Remote Sensing Images. *IEEE J. Sel. Top. Appl. Earth Obs. Remote Sens.* **2022**, *15*, 1902–1911. [CrossRef]
43. Hosang, J.; Benenson, R.; Schiele, B. Learning non-maximum suppression. In Proceedings of the IEEE Conference on Computer Vision and Pattern Recognition, Honolulu, HI, USA, 21–26 July 2017; pp. 4507–4515.
44. Huo, Y.; Qian, X.; Li, C.; Wang, W. Multiple Instances Complementary Detection and Difficulty Evaluation for Weakly Supervised Object Detection in Remote Sensing Images. *IEEE Geosci. Remote Sens. Lett.* 2023, *Early access.* [CrossRef]
45. Zitnick, C.L.; Dollár, P. Edge boxes: Locating object proposals from edges. In Proceedings of the European Conference on Computer Vision, Zurich, Switzerland, 6–12 Sepetmber 2014; Springer: Berlin/Heidelberg, Germany, 2014; pp. 391–405.
46. Uijlings, J.R.; Van De Sande, K.E.; Gevers, T.; Smeulders, A.W. Selective search for object recognition. *Int. J. Comput. Vis.* **2013**, *104*, 154–171. [CrossRef]
47. Cheng, G.; Zhou, P.; Han, J. Learning rotation-invariant convolutional neural networks for object detection in VHR optical remote sensing images. *IEEE Trans. Geosci. Remote Sens.* **2016**, *54*, 7405–7415. [CrossRef]
48. Li, K.; Cheng, G.; Bu, S.; You, X. Rotation-insensitive and context-augmented object detection in remote sensing images. *IEEE Trans. Geosci. Remote Sens.* **2017**, *56*, 2337–2348. [CrossRef]
49. Li, K.; Wan, G.; Cheng, G.; Meng, L.; Han, J. Object detection in optical remote sensing images: A survey and a new benchmark. *ISPRS J. Photogramm. Remote Sens.* **2020**, *159*, 296–307. [CrossRef]
50. Deselaers, T.; Alexe, B.; Ferrari, V. Weakly supervised localization and learning with generic knowledge. *Int. J. Comput. Vis.* **2012**, *100*, 275–293. [CrossRef]
51. Qian, X.; Li, C.; Wang, W.; Yao, X.; Cheng, G. Semantic segmentation guided pseudo label mining and instance re-detection for weakly supervised object detection in remote sensing images. *Int. J. Appl. Earth Obs. Geoinf.* **2023**, *119*, 103301. [CrossRef]
52. Simonyan, K.; Zisserman, A. Very deep convolutional networks for large-scale image recognition. In Proceedings of the International Conference on Learning Representations, San Diego, CA, USA, 7–9 May 2015; pp. 1–13.
53. Wan, F.; Wei, P.; Jiao, J.; Han, Z.; Ye, Q. Min-entropy latent model for weakly supervised object detection. In Proceedings of the IEEE Conference on Computer Vision and Pattern Recognition, Salt Lake City, UT, USA, 18–23 June 2018; pp. 1297–1306.
54. Wang, B.; Zhao, Y.; Li, X. Multiple instance graph learning for weakly supervised remote sensing object detection. *IEEE Trans. Geosci. Remote Sens.* **2021**, *60*, 5613112. [CrossRef]
55. Girshick, R.; Donahue, J.; Darrell, T.; Malik, J. Rich feature hierarchies for accurate object detection and semantic segmentation. In Proceedings of the IEEE Conference on Computer Vision and Pattern Recognition, Columbus, OH, USA, 23–28 June 2014; pp. 580–587.
56. Girshick, R. Fast r-cnn. In Proceedings of the IEEE International Conference on Computer Vision, Santiago, Chile, 7–13 December 2015; pp. 1440–1448.
57. Ren, S.; He, K.; Girshick, R.; Sun, J. Faster R-CNN: Towards real-time object detection with region proposal networks. *IEEE Trans. Pattern Anal. Mach. Intell.* **2016**, *39*, 1137–1149. [CrossRef] [PubMed]

*remote sensing*

*Article*

# TPENAS: A Two-Phase Evolutionary Neural Architecture Search for Remote Sensing Image Classification

Lei Ao [1,2], Kaiyuan Feng [2], Kai Sheng [1,2,3,*], Hongyu Zhao [2], Xin He [1] and Zigang Chen [4]

[1] Guangzhou Institute of Technology, Xidian University, Guangzhou 510555, China; leiao@stu.xidian.edu.cn (L.A.); hexin@xidian.edu.cn (X.H.)

[2] Key Laboratory of Collaborative Intelligence Systems, Ministry of Education, School of Electronic Engineering, Xidian University, Xi'an 710071, China; 18021110260@stu.xidian.edu.cn (K.F.); hongyuz@stu.xidian.edu.cn (H.Z.)

[3] Academy of Advanced Interdisciplinary Research, Xidian University, Xi'an 710071, China

[4] School of Cyber Security and Information Law, Chongqing University of Posts and Telecommunications, Chongqing 400065, China; chenzg@cqupt.edu.cn

* Correspondence: kaisheng@xidian.edu.cn

**Abstract:** The application of deep learning in remote sensing image classification has been paid more and more attention by industry and academia. However, manually designed remote sensing image classification models based on convolutional neural networks usually require sophisticated expert knowledge. Moreover, it is notoriously difficult to design a model with both high classification accuracy and few parameters. Recently, neural architecture search (NAS) has emerged as an effective method that can greatly reduce the heavy burden of manually designing models. However, it remains a challenge to search for a classification model with high classification accuracy and few parameters in the huge search space. To tackle this challenge, we propose TPENAS, a two-phase evolutionary neural architecture search framework, which optimizes the model using computational intelligence techniques in two search phases. In the first search phase, TPENAS searches for the optimal depth of the model. In the second search phase, TPENAS searches for the structure of the model from the perspective of the whole model. Experiments on three open benchmark datasets demonstrate that our proposed TPENAS outperforms the state-of-the-art baselines in both classification accuracy and reducing parameters.

**Keywords:** computational intelligence; neural architecture search (NAS); remote sensing image classification; multi-objective optimization; convolutional neural network (CNN)

**Citation:** Ao, L.; Feng, K.; Sheng, K.; Zhao, H.; He, X.; Chen, Z. TPENAS: A Two-Phase Evolutionary Neural Architecture Search for Remote Sensing Image Classification. *Remote Sens.* **2023**, *15*, 2212. https://doi.org/10.3390/rs15082212

Academic Editor: Andrzej Stateczny

Received: 25 March 2023
Revised: 15 April 2023
Accepted: 19 April 2023
Published: 21 April 2023

## 1. Introduction

With the advancement of remote sensing technology, more and more abundant ground information can be obtained from remote sensing images, which facilitates many research directions and applications, such as change detection [1–6], land use classification [7,8], remote sensing image classification [9,10], etc. As a basic task of remote sensing image processing [11], remote sensing image classification is the classification of remote sensing scene images into a group of semantic categories, which has been widely used in environmental monitoring [12], geospatial object detection [13], and urban planning [14].

In recent decades, with the advancement of deep learning [15–18], many algorithms [9,10,19] have been proposed to solve the remote sensing image classification problem. These algorithms can roughly be categorized into traditional and deep learning-based algorithms, which mainly differ in the way of feature extraction. The former extracts the features of remote sensing images by manually designing feature extraction operators, such as improved fisher kernel (IFK) [20], spatial pyramid matching (SPM) [21], and bag-of-visual-words (BoVW) [22] algorithms. The latter automatically extracts remote sensing image features through deep learning methods such as the autoencoder

(AE) [23–25], CNN [26], and generative adversarial network (GAN) [27–29]. The traditional methods need to specially design a feature extraction operator for the remote sensing image. The extracted features are low-level features, such as texture, color, shape, and gradient, resulting in low classification accuracy on remote sensing image classification tasks. In contrast, deep learning-based methods can automatically learn high-level semantic features of remote sensing images without the need for special feature extractors and achieve high overall accuracy on the remote sensing image classification task. Otávio et al. [30] compared the overall accuracy of deep learning methods with traditional methods on the UC Merced Land-use (UCM21) dataset [22] and demonstrated that deep learning methods outperform traditional methods.

In the past ten years, convolutional neural networks (CNNs) have made a significant breakthrough in image classification. A large number of excellent CNN models have emerged, such as AlexNet [31], VGGNet [32], ResNet [33], GoogleNet [34], and DensNet [35]. However, when applied to remote sensing image classification, these classical CNN models do not perform as well due to the unique characteristics of remote sensing images, such as big intra-class diversity, high inter-class similarity, and coexistence of multiple ground objects. Therefore, many deep learning models [36–39] are tailored for remote sensing image classification. Yu et al. [37] proposed the HABFNet framework to alleviate the problems of high intra-class diversity and high inter-class similarity in remote sensing images. HABFNet uses ResNet50 to extract image features, then enhances features at different levels through a channel attention scheme, and fuses features through bilinear pooling. The fused features have a stronger discriminative ability, which improves the classification accuracy of the algorithm in remote sensing image classification. Wei et al. [38] proposed a novel CAD network that uses an attention mechanism to extract more discriminative features, which alleviates the difficulty of classification caused by large changes in object scale. Gong et al. [39] proposed D-CNN to alleviate the problems of high intra-class diversity and high inter-class similarity in remote sensing images, thereby further improving remote sensing image classification accuracy. Wang et al. [40] proposed a semi-supervised classification framework by designing the inner-class dense neighbors (IDN) algorithm to reduce the reliance on the labels of the samples and simultaneously improve the classification accuracy of the model. CNN-based algorithms perform very well on remote sensing image classification tasks.

Although deep learning methods have achieved high classification accuracy in remote sensing image classification, it is extremely difficult for those without professional knowledge about deep learning to design a model with high classification accuracy. In recent years, NAS has emerged as a promising alternative method, which can automatically design a CNN model with high classification accuracy without prior knowledge. The existing NAS methods can be divided into three categories: NAS based on reinforcement learning (NAS-RL) [41–43], evolutionary neural architecture search (ENAS) [44–46], and NAS based on gradient (NAS-G) [47–49]. The evolutionary algorithm (EA) [50,51] is a heuristic global optimization algorithm. Due to its powerful optimization ability and easy parallel computation, the evolutionary algorithm has attracted more and more scholars' attention in the automatic design of deep neural network structure. Real et al. [44] first proposed using evolutionary computation to optimize the structure of a CNN. This method does not require any human operations after the algorithm is executed and finally outputs a fully trained CNN model. The algorithm achieves competitive classification accuracy on the CIFAR-10 and CIFAR-100 datasets, but at a prohibitively high computational cost. Since then, many researchers have proposed many schemes to reduce computational costs. Elsken et al. [52] proposed a simple and efficient NASH algorithm, which uses network morphisms [53] to generate weight-inherited sub-networks and efficiently optimizes an excellent CNN architecture through a simple hill-climbing algorithm. Hui et al. [54] proposed the EENA algorithm, which uses prior knowledge to guide the evolutionary process, thereby accelerating the search process. Wang et al. [55] evaluated individuals with some batch data randomly selected on the validation set, and the evaluation results of each batch data were averaged as the fitness value of the individual, which significantly improves the evaluation speed of the individual. A population-based optimization algorithm, such as a genetic algorithm, is one of the most commonly used evolutionary algorithms. The evaluation of each individual in the algorithm is independent, so the population-based optimization algorithm easily performs parallel computing. Based on this feature, Xie et al. [56] built the BenchENAS platform. When evaluating individual fitness, individuals in the population can be evaluated parallelly in a common lab environment, which significantly speeds up population evaluation and promotes the development of ENAS. These methods have achieved excellent performance on natural image classification tasks.

Many methods [57–59] have been proposed to utilize NAS to solve object recognition in satellite imagery tasks. In remote sensing image classification, gradient-based NAS methods are the most commonly used methods. The general idea is to first search for an optimal cell and then form a CNN model by stacking multiple cells. The main difference between these algorithms is the optimal cell search scheme. Zhang et al. [57] proposed a more efficient search algorithm for remote sensing image classification, named RS-DARTS, which improves the model classification accuracy and speeds up the search for optimal cells by adding noise and sampling the neural network. Peng et al. [58] proposed the GPAS algorithm, which uses greedy and aggressive strategies to search for the optimal cell. Chen et al. [59] proposed the CIPAL framework for remote sensing image classification, which utilizes channel compression to reduce the time of structure search. Ma et al. [60] proposed the SceneNet algorithm, which yields a competitive set of remote sensing image classification models by optimizing the architecture of the model. Wan et al. [61] proposed an efficient neural network architecture search method for remote sensing image classification. By designing a two-step evolutionary search method, cells were constructed from the eight kinds of lightweight operators, and the remote sensing image classification model was constructed by stacking cells. Povilas et al. [62] proposed the NAS-MACU algorithm for object recognition in satellite imagery. NAS-MACU automatically searches for high-performance cell topologies using the NAS algorithm and then constructs an object recognition model by stacking multiple candidate cells. These methods, with the exception of SceneNet, first search for an optimal cell and then construct the final model by stacking multiple identical cells, which will bring two problems. The classification accuracy of a model would deteriorate if there were too few stacked cells, but if there were too many, the model would become redundant and have more parameters and floating point operations (FLOPs). On the other hand, all stacked cells are the same, and the network structure is not considered globally. The impact of the number of blocks on the performance of a model is not taken into account by SceneNet, despite the fact that it globally searches the structure of a model. In addition, in the practical application of remote sensing image classification, the CNN model is also limited by classification accuracy, computing power, memory capacity, and so on. Therefore, designing a CNN model must strike a balance between these limiting conditions.

To this end, we propose TPENAS, which can automatically build a model with optimal depth and output multiple alternative models for remote sensing image classification. Specifically, users with limited deep learning knowledge can obtain a model with excellent performance for remote sensing image classification. The algorithm is run once to generate a set of models from which the most suitable one can be selected based on the limiting conditions. The difficulty of remote sensing image classification tasks varies with different scenarios. As a result, the depth of the CNN model should also be different. Therefore, we design the first search phase to solve this problem. The depth and classification accuracy of the CNN model are used to formulate a multi-objective optimization problem, and then a population-based multi-objective optimization algorithm is used to solve this problem, in which individuals representing CNN models with different depths are initialized in the population and the diversity of the depth of the model is maintained during the population update process. the depth of the CNN model is then determined according to the optimal solution in the optimized population. In order to let the model output a set of models and search for the structure of the model globally, we design the second search phase. A multi-objective optimization problem is formulated according to the complexity and overall accuracy of the model, and we design a population-based multi-objective optimization algorithm to solve the problem, in which individuals in the population are encoded into the entire CNN model. By solving this multi-objective optimization problem, a set of models with superior performance can be obtained.

The experimental results on three open benchmark datasets show the superiority of our algorithm over other classic deep learning classification models and NAS algorithms. The main contributions of this paper are as follows:

(1) We propose a two-phase evolutionary multi-objective neural architecture search (TPE-NAS) framework for remote sensing image classification. The first search phase explores the optimal the depth of the model, and the second search phase finds the most suitable structure for the model. Our algorithm can automatically design a CNN model suitable for remote sensing image classification, which eases the heavy burden posed by manually designing a CNN model.
(2) We propose the first search phase that determines the depth of the CNN model. A multi-objective optimization problem is established with the depth and classification accuracy of the model as optimization goals. This problem is solved by a heuristic multi-objective optimization algorithm to find the optimal the depth of model.
(3) We propose the second search phase that globally searches the structure of the CNN model. We encode the entire CNN model as a binary string, allowing population evolution to optimize the CNN structure globally. Furthermore, we simultaneously optimize the classification error and complexity of the model so that the final result can provide a set of Pareto solutions, giving users more options in practical applications.
(4) The effectiveness of the proposed TPENAS is verified on three public benchmark datasets. Extensive experiments show that the model searched by the TPENAS outperforms the classic classification CNN model. Compared with other NAS methods, TPENAS not only has higher classification accuracy but also has advantages in the GFLOPs and parameters of the model.

The remainder of this paper is organized as follows. Section 2 describes the proposed TPENAS algorithm in detail. Section 3 describes the experimental settings and experimental results. Section 4 analyses the number of models that the TPENAS should evaluate as well as the implication of model depth on test performance. The conclusion of this paper is given in Section 5.

## 2. Materials and Methods

In Section 2.1, we establish the optimization model of two search phases and give the optimization algorithm framework. In Section 2.2, we introduce the algorithm of the first search phase in detail, including encoding scheme, initialization, population evolution, and solution selection. In Section 2.3, we discuss how to use the first search phase algorithm to optimize the optimization problem in the second search phase and give a summary of the overall algorithm.

### 2.1. The Overall Framework

The existing remote sensing image classification models can be summarized in two parts. The first part is the image feature extractor, and the second part is the feature classifier. The result of image feature extraction seriously affects the classification accuracy of the model. As we all know, CNNs are one of the most commonly used image feature extractors, and feature extractors with different structures will have a significant impact on classification accuracy. Therefore, TPENAS focuses on developing efficient feature extractors.

The purpose of our algorithm is to solve two problems, the first is to reduce the difficulty of manually designing a classification model, and the second is to automatically design an appropriate classification model in different scenarios. Our algorithm is divided into two phases, the purpose of the first search phase is to find the appropriate depth of the model, and the second search phase is to find the appropriate structure of the model. Therefore, we formulate the multi-objective optimization problem in two phases, respectively.

$$min\{F_1, F_2\}, \begin{cases} F_1 = \frac{n\_incorrected\_sample}{n\_all\_sample} \\ F_2 = n\_block \end{cases} \tag{1}$$

$$min\{F_1, F_3\}, \begin{cases} F_1 = \frac{n\_incorrected\_sample}{n\_all\_sample} \\ F_3 = GFLPOs \end{cases} \tag{2}$$

In the first search phase, we regard $F_1$ and $F_2$ of the model as two optimization objectives, as shown in Equation (1). $F_1$ represents the overall accuracy of the model, which is the misclassified samples divided by all samples in the test dataset. $F_2$ represents the depth of the model, which is the number of blocks of the model. By optimizing Equation (1), we are able to select the appropriate number of blocks and consequently find the appropriate depth of the model. Similarly, in the second search phase, we regard the $F_1$ and $F_3$ of the model as two optimization objectives. $F_3$ represents the GFLOPs of the model, as shown in Equation (2). By optimizing Equation (2), we are able to obtain a set of optimal solutions, that is, there does not exist a solution that is better than the optimal solution on both OA and GFLOPs.

We cannot confirm whether this is a convex optimization problem or a non-convex optimization problem. Therefore, we use a genetic algorithm to design optimization algorithms to solve these two optimization problems. A genetic algorithm is a heuristic optimization algorithm that can solve both convex and non-convex optimization problems. Therefore, we design the TPENAS algorithm, as shown in Algorithm 1, to optimize these two optimization problems.

Algorithm 1 shows the pseudocode of TPENAS, which consists of two parts: the first search phase and the second search phase. On the remote sensing classification problem $D$, the first search phase (see lines 1–10) explores the depth of the model and the second search phase (see lines 11–22) explores the structure of the model. In the first search phase, $N$ individuals are randomly initialized as the initial population, and each individual in the population is evaluated on the problem $D$ to obtain the encoding length and classification error rate (see line 2) for each individual. Population $P_0$ is optimized for $T_1$ iterations through population evolution (see lines 3–7). The optimal solution front $\vec{\alpha}$ is chosen from population $P_{T_1}$, and the optimal solution is then chosen based on $\vec{\alpha}$ (see lines 8–9). By calculating the length of the optimal solution, we determine that the individual code length of the second search phase is $l$ (see line 10). In the second search phase, similar to the first search phase, $R$ individuals are first randomly initialized as the initial population, where each individual has an encoding length of $l$ (see line 11). $E$ is an external population, and its role is to collect the population's individuals in each generation. The population evolution updates the population $T_2$ times, and the external population obtains $R \times T_2$ individuals (see lines 12–19). The optimal Pareto front is computed from $E$, and the most suitable individual is selected to decode it to the corresponding CNN model for the remote sensing classification (see lines 20–22).

---

**Algorithm 1** The Pseudocode of TPENAS

**Input:**

      $T_1$: the maximum population iterations during the first search phase;

      $T_2$: the maximum population iterations during the second search phase;

      $N$: the population size in the first search phase.

      $R$: the population size in the second search phase.

      $D$: remote sensing image classification problem.

**Output:**

      The best model.

1: **First Search Phase:**

2: $P_0 \leftarrow$ Initialize and evaluate the population with the size of $N$;

3: $i \leftarrow 1$;

4: **while** $i \leq T_1$ **do**

5:    $P_i \leftarrow$ population evolution $(P_{i-1}, D)$;

6:    $i \leftarrow i + 1$;

7: **end while**

8: $\vec{\alpha} \leftarrow$ Calculate the best solution front from $P_{T_1}$;

9: $\mathrm{p}^* \leftarrow$ Select the best individual from $\vec{\alpha}$;

10: $l \leftarrow$ Calculate the length of the code in individual $\mathrm{p}^*$;

11: **Second Search Phase:**

12: $Q_0 \leftarrow$ Initialize and evaluate the population with the size of $R$, where the length of the code in each population member is $l$;

13: $E \leftarrow \varnothing$;

14: $t \leftarrow 1$;

15: **while** $t \leq T_2$ **do**

16:    $Q_t \leftarrow$ population evolution $(Q_{t-1}, D)$;

17:    $E \leftarrow E \bigcup Q_t$;

18:    $t \leftarrow t + 1$;

19: **end while**

20: $\vec{\beta} \leftarrow$ Calculate the Pareto front from $E$;

21: $\mathrm{q}^* \leftarrow$ Choose the best individual from $\vec{\beta}$;

22: Decoding individual $\mathrm{q}^*$ to the corresponding remote sensing image classification model.

---

### 2.2. The First Search Phase

The number of layers and structure of CNN greatly affect the ability to extract features. Therefore, we designed the first search phase with the aim of exploring the effect of the depth of the CNN model on classification accuracy in remote sensing image classification. Below, we detail the design of the first search phase.

#### 2.2.1. Encoding Schedule

In order to optimize the depth and structure of the model using the genetic algorithms, we need to represent the remote sensing image classification model as a binary string in order to optimize Equation (1). The topology of a block can be regarded as a directed acyclic graph, and its encoding rules corresponding to binary strings must meet the following three rules.

(1)   A block with $n$ nodes is represented by $n$ groups of binary strings.

(2)   The $i$-th group of codes is represented by $i + 1$ bit binary. The $j$-th bit of the $i$-th group indicates whether the $(i + 1)$-th node is connected to the $j$-th node ($i > j$ and $i = n - 1$), 1 means connection, 0 means disconnection.

(3)   The last group has only one bit, which indicates whether there is a direct connection from the input to the output.

For the convenience of identification, each group of binary strings is connected with the symbol "-". The formula for calculating the coding length of the feature extraction block is $L = \frac{n(n-1)}{2} + 1$ , where $L$ represents the coding length of the block and $n$ represents the number of nodes contained in the block.

Figure 1 shows the coding diagram of a block with 5 nodes. IFM and OFM represent the input feature map and the output feature map, respectively. Each node represents a $3 \times 3$ convolution operation followed by batch normalization (BN) and a rectified linear unit (ReLU). The dashed arrows point out the correspondence between the binary bit "1" in the binary string and the edge of the directed acyclic graph. Population evolution in Section 2.2.3 can be used to conveniently optimize the network structure using binary strings.

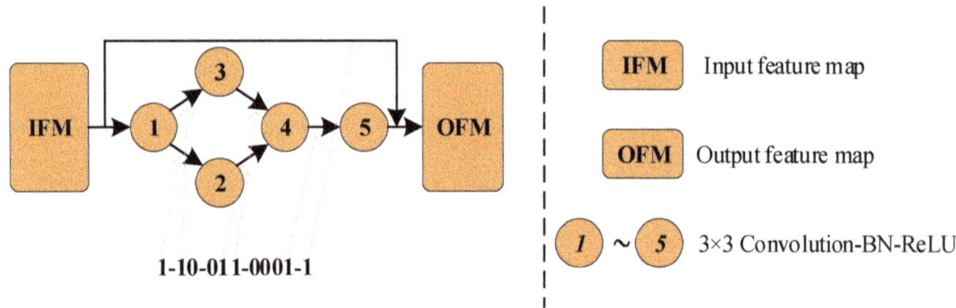


**Figure 1.** The encoding diagram of a feature extraction block.

### 2.2.2. Initialization

It is clear from the coding scheme described in Section 2.2.1 that a block with $n$ nodes needs to be represented by $\frac{n(n-1)}{2} + 1$ binary bits. Therefore, an individual with $m$ blocks is represented by $m \cdot \left(\frac{n(n-1)}{2} + 1\right)$ binary bits.

The search space of individuals in the first search phase can be obtained as shown in Equation (3).

$$\Omega = \sum_{i=1}^{m} 2^{\frac{in(n-1)}{2}+i} \tag{3}$$

where $\Omega$ represents the search space and $m$ represents the number of individuals with different numbers of blocks. An individual with $n$ nodes and $m$ blocks in each block is represented by $l_m = \frac{m(n^2-n+2)}{2}$ binary bits. The starting point of the optimization algorithm is the initialization population. The population represents a collection of individuals, each of which represents a remote sensing image classification model. We represent an individual using a vector. Therefore, we randomly initialize $K$ vectors that have length $l_i$ ($i$ = 1, 2, …, $m$), each of which has a value of 0 or 1 as the initial population. Each vector represents an individual in the population, thus the population size is $mK$. The initial population serves as the starting point for population evolution in Section 2.2.3. We decode each individual in the population and test the individual's classification error on the testing dataset after training on the training dataset. At the same time, we also calculate the number of blocks in the individual.

### 2.2.3. Population Evolution

Figure 2 depicts a schematic diagram of population evolution. Consistent with the paradigm of a genetic algorithm, the population evolution is primarily made up of crossover and mutation, evaluation, as well as selection. First, individuals in the initial population are randomly selected for crossover. Then, crossover and mutation operations are performed on the selected individuals to obtain offspring individuals. Finally, all new individuals are evaluated and a new generation population is selected. This process is looped until the stop condition is met. We describe crossover, mutation, evaluation, and environmental selection in detail below.

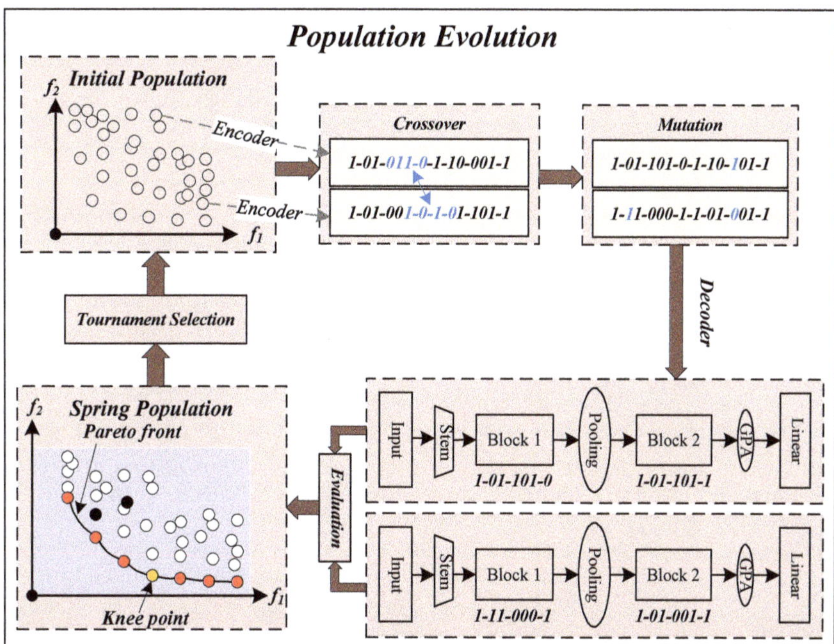



**Figure 2.** The diagram of population evolution in TPENAS. Stem represents a convolution operation; Block represents a feature extraction block; Pooling represents a pooling operation; GPA represents a global average pooling operation; Linear represents a fully connected layer.

(1) Crossover and Mutation

Crossover and mutation operations are used to generate better-quality individuals, which are common operations in genetic algorithms. We randomly pick two individuals from the population and perform a crossover on them with probability $p_c$. The crossover operation involves selecting a continuous binary string of the same length from two individuals and generating two new individuals by exchanging the binary string segments. The two new individuals perform mutation operations respectively to generate new individuals. The mutation operation is practiced by inverting each binary bit with probability $p_m$ in turn. In the experiment, the crossover probability and mutation probability are set to $p_c = 0.5$ and $p_m = \frac{1}{l}$, respectively, where $l$ represents the code length of the individual.

(2) Evaluation

In the first search phase, we need to evaluate the overall classification accuracy and the length of the individual. To meet the minimum optimization problem, we use the overall classification error rate of the model on the testing dataset to evaluate the individual's overall classification accuracy. We use the number of blocks to evaluate an individual's length. Before evaluating an individual, we need to decode the binary string representing the individual into the corresponding CNN model. The model is trained on the training dataset and then tested on the testing dataset to obtain the overall classification error rate of the model. It is worth noting that in the whole optimization process, we save the binary code of the individual, the overall classification error rate, and the number of blocks of the model into the external population $E$, and, before evaluating each individual, we first query the individual in the set $E$. If it exists, the overall classification error rate of the individual and the number of blocks of the model are directly copied without retraining the model, which saves time in the first search phase.

(3) Environmental Selection

We select the offspring population by binary tournament selection. Specifically, two individuals are selected firstly from the parent population, and then the most suitable one from the two individuals is chosen and added to the offspring population. Repeat $N$ times to select $N$ individuals as the offspring population.

### 2.2.4. Solution Selection

To determine the optimal depth in the remote sensing image classification model, we first select the highest classification accuracy from individuals outputted by population evolution in Section 2.2.3. This will form the optimal solution front. Then, the knee point method [63] is used to select the optimal solution from the optimal solution front. Finally, we determine the optimal depth of the model by calculating the number of blocks in the optimal solution.

### 2.3. The Second Search Phase

During the second search phase, we explore the impact of the network's structure on the classification accuracy of remote sensing images. We consider both classification accuracy and the complexity of the model. In the second search phase, we use the GFLOPs of the model to represent the complexity of the model. Similar to the first search phase, we build a multi-objective optimization problem using the classification error rate and GFLOPs of the model. Because the number of blocks and evaluation metrics of the individual in the second search phase differ from those in the first search phase, we can use the heuristic-based multi-objective optimization algorithm designed in the first search phase to solve the multi-objective optimization problem in the second search phase. Therefore, we can modify some parts of the first search phase to implement the second search phase process. There are three differences from the first search phase, as follows:

(1) In the first search phase, we determine the optimal number of blocks of individuals. In the second search phase, we optimize the classification error rate and GFLOPs of the model and no longer optimize the block number of the model. Therefore, when initializing the population as in Section 2.2.2, $M$ individuals with the same number of blocks are randomly initialized.

(2) In the second search phase, the two optimization objectives are the classification error rate and GFLOPs of the model. Therefore, when evaluating individuals as in Section 2.2.3, we evaluate the individual's classification error rate and calculate the individual's GFLOPs.

(3) We do not select the optimal individual from the final population as in Section 2.2.4. This is because we use the binary tournament selection method when choosing the offspring population, which may overlook some Pareto solutions. As a result, we aggregate all of the individuals from each generation into an external population $\Omega$ and then select the Pareto front from $\Omega$.

As mentioned above in Section 2.1, we specifically set these two problems as multi-objective optimization problems and designed the TPENAS algorithm to solve these problems employing a genetic algorithm paradigm. TPENAS solves for the depth of the model in the first search phase and produces a set of solutions that balance overall accuracy and GFLOPs in the second search phase. The result of TPENAS in two phases is a set of solutions, and we can choose the appropriate one according to our practical needs.

## 3. Results

In this section, we discuss experimental details to validate TPENAS. Section 3.1 introduces the datasets used in the experiments. Section 3.2 describes the experimental settings. Section 3.3 shows the experimental results.

### 3.1. Datasets

The proposed method is verified on three datasets, namely UCM21 [22], PatternNet [64], and NWPU45 [65] datasets. The characteristics of the three datasets are summarized in Table 1. Table 1 shows the three obvious characteristics of the three datasets. First, the large variation in the scene classes between the UCM21 dataset and NWPU45 dataset. The number of scenes in the NWPU45 dataset is more than double that of the UCM21 dataset. Second, the large variation in the size of the three datasets. NWPU45 datasets are 15 times larger than UCM21 datasets. PatternNet dataset has eight times the number of images per class as the UCM21 dataset. Third, the large variation in the spatial resolution of the three datasets.

**Table 1.** Characteristics of the three datasets in our experiments.

| Dataset | Scene Classes | Total Image | Image per Class | Spatial Resolution (m) | Image Size |
|---|---|---|---|---|---|
| UCM21 | 21 | 2100 | 100 | 0.3 | 256 × 256 |
| PatternNet | 38 | 30,400 | 800 | 0.06~4.69 | 256 × 256 |
| NWPU45 | 45 | 31,500 | 700 | 0.2~30 | 256 × 256 |

The spatial resolution of the UCM21 dataset is fixed at 0.3 m. The spatial resolution of the PatternNet dataset has a small range of 4.63 m, while the NWPU45 dataset has a large range of 28.8 m. Some samples are shown from the three datasets in Figures 3–5, respectively.

**Figure 3.** Some samples from the UCM21 dataset.

**Figure 4.** Some samples from the PatternNet dataset.

**Figure 5.** Some samples from the NWPU45 dataset.

*3.2. Experimental Settings*

3.2.1. Parameter Setting

The experiment is divided into two parts. The first part is the search phase, including the first search phase and the second search phase. The second part is the retraining phase. The hyperparameters of the two parts are shown in Table 2. In total, 80%, 40%, and 20% of samples of the UCM21, PatternNet, and NWPU45 datasets are split into training datasets, and the rest are used as testing datasets. The hardware configuration and software version of the experimental environment are shown in Table 3.

**Table 2.** Hardware configuration and software version of the experimental environment.

|  | Versions |
| --- | --- |
| CPU | Inter(R) Core(TM) i7-10700 |
| GPU | NVIDIA GeForce 3090 |
| Pytorch | 1.11.0 |
| Python | 3.10.4 |

**Table 3.** Hyperparameters of the proposed algorithm.

| Phase | Hyperparameter Name | Hyperparameter Value |
| --- | --- | --- |
| First search phase | Population size | 64 |
|  | Number of blocks | from 1 to 8 |
|  | Number of nodes | 6 |
|  | Crossover probability | 0.5 |
|  | Mutation probability | $\frac{1}{Encoding\_length}$ |
|  | Batch size | 16 |
|  | Optimizer | SGD |
|  | Momentum | 0.9 |
|  | Weight decay | $5e^{-4}$ |
|  | Learning strategy | Cosine |
|  | Learning rate | 0.03 |
|  | Epoch_UCM21 | 50 |
|  | Epoch_PatternNet | 20 |
|  | Epoch_NWPU45 | 50 |
| Second search phase | Population size | 40 |
|  | Number of nodes | 6 |
|  | Crossover probability | 0.5 |
|  | Mutation probability | $\frac{1}{Encoding\_length}$ |
|  | Batch size | 16 |
|  | Optimizer | SGD |
|  | Momentum | 0.9 |
|  | Weight decay | $5e^{-4}$ |
|  | Learning strategy | Cosine |
|  | Learning rate | 0.03 |
|  | Epoch_UCM21 | 50 |
|  | Epoch_PatternNet | 15 |
|  | Epoch_NWPU45 | 20 |
| Retraining phase | Eopch | 1000 |
|  | Batch size | 16 |
|  | Optimizer | SGD |
|  | Momentum | 0.9 |
|  | Weight decay | $5e^{-4}$ |
|  | Learning strategy | Cosine |
|  | Learning rate | 0.03 |
|  | Loss function | Cross Entropy Loss |

### 3.2.2. Evaluation Metrics

In order to evaluate the effectiveness of the proposed algorithm, we use the overall accuracy (OA) and confusion matrix (CM) as the evaluation metrics for the classification accuracy of the model and use the FLOPs and parameters (Params) as evaluation metrics to assess the computational cost and parameters of the remote sensing image classification model.

OA indicates the overall classification accuracy of the model, which represents the ratio of correctly classified samples to all samples in the testing dataset. OA and CM are calculated using Equations (4) and (5), respectively, where $S$ and $S_c$ denote all samples and samples of category $c$ in the testing dataset, respectively; $K$ indicates the number of categories in the testing dataset; $I()$ is the indicator function; $f()$ denotes the remote sensing image classification model; $x$ denotes the input sample; $y_c$ denotes the label of the category $c$ sample. CM is a matrix with $K$ rows and $K$ columns. $CM_{i,j}$ denotes the proportion of samples of category $i$ misclassified as samples of category $j$ among all samples of category $i$ in the testing dataset.

$$OA = \tfrac{1}{S} \sum_{c=1}^{K} \sum_{t=1}^{S_c} I(f(x_{c,t}) = y_c) \tag{4}$$

$$CM_{i,j} = \sum_{c=1}^{K} \sum_{t=1}^{N_c} \tfrac{1}{S_c} I(f(x_{c,t}) = y_j) \tag{5}$$

We calculate FLOPs and Params by using Equations (6) and (7), where $H_{in}$ and $W_{in}$ denote the height and width of the input feature map, respectively; $C_{in}$ and $C_{out}$ denote the number of input channels and output channels of the convolution kernel, respectively; $I$ and $O$ denote the number of input and output nodes in the fully connected layer, respectively; $k$ denotes the size of the convolution kernel.

$$FLOPs = \begin{cases} FLOPs_{conv.} = 2H_{in}W_{in}(C_{in}k^2 + 1)C_{out} \\ FLOPs_{FC.} = (2I - 1)O \end{cases} \tag{6}$$

$$Params = \begin{cases} Params_{conv.} = C_{out}(k^2C_{in} + 1) \\ Params_{FC.} = (I + 1)O \end{cases} \tag{7}$$

*3.3. Comparison of the Proposed TPENAS with Other Methods*

3.3.1. Results on UCM21 Dataset

We conducted experiments on the UCM21 dataset, the first publicly available remote sensing image classification dataset.

In the first search phase, 80% of data in the UCM21 dataset are randomly divided as the training dataset, and the rest as the testing dataset by a stratified sampling algorithm. The search results of the first search phase are shown in Figure 6. The red dotted line shows a downward trend as the number of blocks increases and stabilizes when the number of blocks equals five. We can see that as the number of blocks increases, the classificatin error rate of the model decreases. However, after the number of blocks is equal to five, the classificatin error rate of the model does not decrease significantly, but increases slightly. As a result, in the first search phase, the optimal solution is chosen when the number of blocks equals five.

According to the optimal solution obtained in the first search phase, using the same training dataset and testing dataset as the first search phase, we further search for the structure of the network in the second search phase. Through the second search phase, we obtain the Pareto front, as shown in Figure 7. It can be seen from the Pareto front that the GFLOPs of the solution vary from 0.5~4.5, which provides a variety of solutions. In order to select the model with the lowest classification error rate, we train each network in the Pareto solution set from scratch for 1000 epochs using the same dataset as the second search phase and select the individual with the lowest classification error as our chosen solution. Because the solutions in the Pareto solution set are only trained for 50 epochs, they are not fully trained. After full training, in the Pareto solution set, individuals with a high classification error rate may obtain a lower classification error rate than those with a low classification error rate. We will discuss this in Section 4.

In the retraining phase, we use five-fold cross-validation, with four folds as the training dataset and one fold as the testing dataset, to evaluate selected individuals. A total of 5 independent models are trained for 1000 epochs and tested, respectively. The average of the test accuracy of the five models is compared to other algorithms, including classic classification models such as AlexNet, VGG16, ResNet50, and others, as well as NAS-based methods such as NASNet, SGAS, DARTS, and so on. Additionally, we also compared their GFLOPs and Params, as shown in Table 4. Compared with classic classification models, TPENAS achieves the highest OA, and the GFLOPs are only higher than AlexNet, while the parameters are significantly lower than other models. Compared with NAS-based methods, the OA of TPENAS is 9.91% higher than NASNet and 2.03% higher than RSNet, and the parameters of TPENAS are at least half lower than those of NASNet, SGAS, and DARTS.

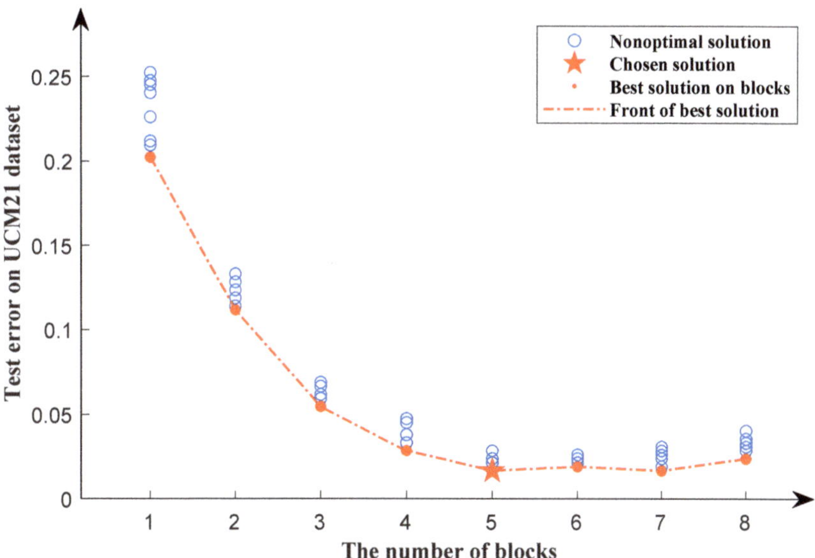

**Figure 6.** Search results of the first search phase for the UCM21 dataset. The red dots represent the optimal solution in a particular block, while the blue circles represent nonoptimal solutions. The pentagram represents the chosen solution. The red dotted line represents the front of the best solution.

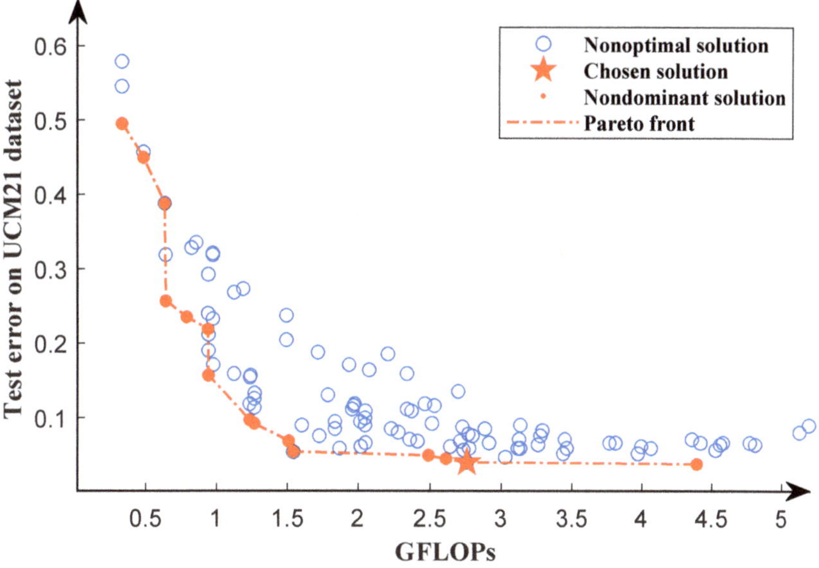

**Figure 7.** Pareto front of the second search phase for the UCM21 dataset. The red dots represent the optimal solution, while the blue circles indicate the nonoptimal solutions. The Pareto front is represented by the red dotted line. The pentagram indicates the chosen solution.

**Table 4.** The OA, GFLOPs, and Params of TPENAS are compared with the other methods on the UCM21 dataset (the ratio of training samples to test samples is 8:2). The upward arrow (↑) indicates that the larger the number, the better the result. The downward arrow (↓) indicates that the smaller the number, the better the result.

| Method | OA (%) ↑ | GFLOPs ↓ | Params (M) ↓ | Search Strategy |
|---|---|---|---|---|
| AlexNet [66] | 81.19 | 0.92 | 57.09 | manual |
| VGG16 [32] | 78.57 | 20.18 | 134.35 | manual |
| ResNet50 [33] | 85.24 | 5.37 | 23.56 | manual |
| ConvNeXt [67] | 84.29 | 20.07 | 88.57 | manual |
| DenseNet161 [35] | 86.19 | 10.17 | 26.52 | manual |
| Fine-tuned AlexNet [66] | 92.14 | 0.92 | 57.09 | manual |
| Fine-tuned VGG16 [32] | 95.48 | 20.18 | 134.35 | manual |
| Fine-tuned ResNet50 [33] | 98.57 | 5.37 | 23.56 | manual |
| Fine-tuned ConvNeXt [67] | 97.86 | 20.07 | 88.57 | manual |
| Fine-tuned DenseNet161 [35] | 98.33 | 10.17 | 26.52 | manual |
| NASNet [43] | 89.62 | 0.77 | 4.26 | NAS |
| SGAS [68] | 92.05 | 0.81 | 4.69 | NAS |
| MNASNet [69] | 94.52 | 0.43 | 3.13 | NAS |
| RTRMM [70] | 96.76 | 0.38 | 0.82 | NAS |
| DARTS [47] | 95.19 | 0.71 | 3.97 | NAS |
| PDARTS [71] | 91.52 | 0.73 | 4.19 | NAS |
| RSNet [72] | 96.78 | 1.19 | 1.22 | NAS |
| CIPAL [59] | 96.58 | - | 1.58 | NAS |
| ALP [73] | 93.43 | - | 2.63 | NAS |
| TPENAS (ours) | 98.81 | 2.76 | 1.80 | NAS |

The classification confusion matrix are shown in Figure 8. C01~C21 represent the categories agricultural, airplane, baseball diamond, beach, buildings, chaparral, dense residential, forest, freeway, golf course, harbor, intersection, medium residential, mobile home park, overpass, parking lot, river, runway, sparse residential, storage tanks and tennis court, respectively. The classification accuracy of TPENAS in the scenes "buildings" and "storage tanks" is 90%, the classification accuracy of "tennis court" is 95%, and the other scenes are 100%, which proves the excellent performance of our proposed algorithm.

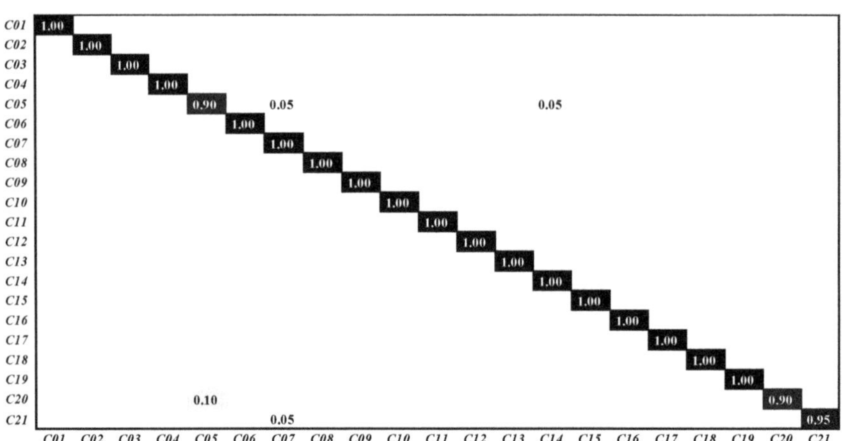


**Figure 8.** The classification confusion matrix on UCM21 dataset.

### 3.3.2. Result on PatternNet Dataset

To further verify the performance of TPENAS, we validate experiments on the PatternNet dataset, which contains more scenes and more samples than the UCM21 dataset. During the first search phase, 40% of the data is selected at random as the training dataset and the remainder as the testing dataset. Figure 9 shows the search results in the first search phase. We can see that as the number of blocks increases, the red dotted line gradually decreases and flattens out after block equals 3. This demonstrates that as the number of blocks increases, the classification error rate of the model gradually decreases. However, when the number of blocks exceeds 3, the classification error rate of the model does not decrease significantly. As a result, the individual's block in the second search phase is set to 3. On the UCM21 dataset, the optimal number of blocks in the first search phase is 5. It demonstrates that the optimal number of blocks for the model varies depending on the dataset.

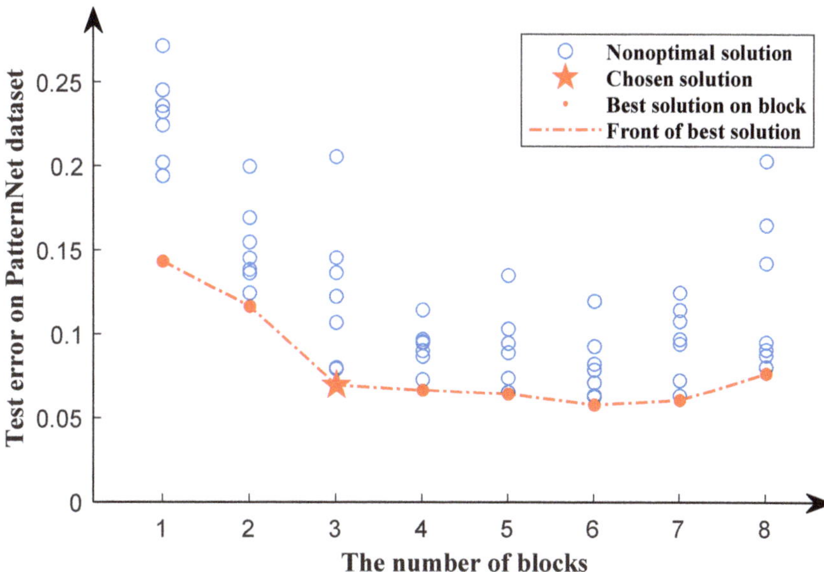


**Figure 9.** Search results of the first search phase for the PatternNet dataset. The red dots indicate the optimal solution in a particular block, while the blue circles indicate nonoptimal solutions. The pentagram represents the chosen solution. The red dotted line represents the front of the best solution.

We searched the architecture of the model in the second search phase using the same dataset as the first search phase, and the result is shown in Figure 10. Each solution in the Pareto solution set has a unique network structure, and each solution in the Pareto solution set dominates at least one nonoptimal solution. Therefore, the second search phase is able to provide multiple models for remote sensing images classification. Each solution in the Pareto solution set is trained from scratch for 1000 epochs. The solution with the highest classification accuracy in the testing dataset is chosen for comparison with other algorithms.

During the retraining phase, 40% of the data in the PatternNet dataset is randomly selected to train the selected solutions from scratch and tested on the remaining 60% of the data. The experiment was repeated 5 times and the average overall accuracy was calculated, and the result is shown in Table 5. The OA of TPENAS is higher than classic classification models. Compared with NAS-based methods, the OA of TPENAS is higher than other algorithms, except that it is slightly lower than PDARTS. TPENAS has lower GFLOPs than classic classification models and roughly twice the GFLOPs of NAS-based methods. It is worth noting that parameters of TPENAS is at least one-twentieth of the classic classification models and NAS-based methods.

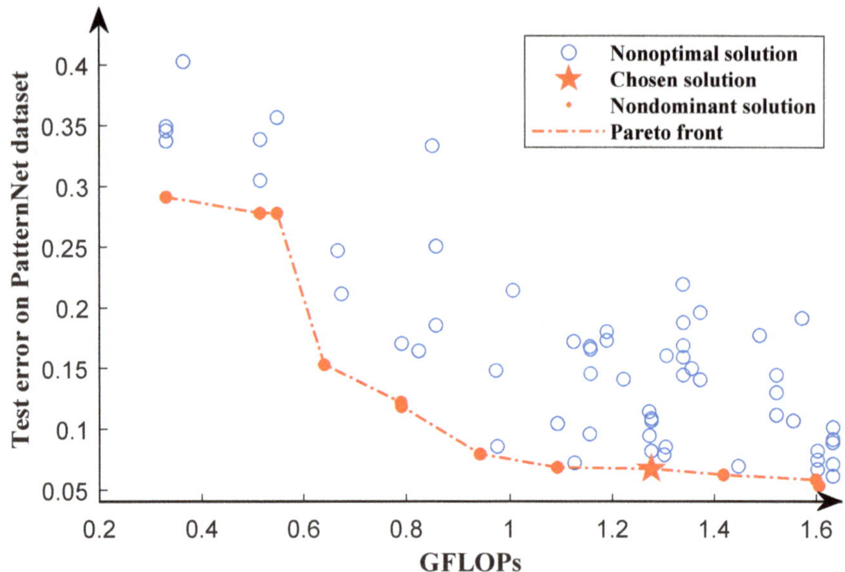

**Figure 10.** Pareto front of the second search phase for the PatternNet dataset. The red dots represent the optimal solution, while the blue circles represent the nonoptimal solution. The Pareto front is represented by the red dotted line. The pentagram indicates the chosen solution.

**Table 5.** The OA, GFLOPs, and Params of TPENAS are compared with the other methods on the PatternNet dataset (the ratio of training samples to test samples is 4:6). The upward arrow (↑) indicates that the larger the number, the better the result. The downward arrow (↓) indicates that the smaller the number, the better the result.

| Method | OA (%) ↑ | GFLOPs ↓ | Params (M) ↓ | Search Strategy |
|---|---|---|---|---|
| VGG16 [32] | 97.31 | 20.18 | 134.42 | manual |
| GoogLeNet [34] | 96.12 | 1.96 | 56.64 | manual |
| ResNet50 [33] | 96.71 | 5.37 | 235.96 | manual |
| Fine-tuned VGG16 [32] | 98.31 | 20.18 | 134.42 | manual |
| Fine-tuned GoogLeNet [34] | 97.56 | 1.96 | 56.64 | manual |
| Fine-tuned ResNet50 [33] | 98.23 | 5.37 | 23.59 | manual |
| DARTS [47] | 95.58 | 0.71 | 3.98 | NAS |
| PDARTS [71] | **99.10** | 0.73 | 4.21 | NAS |
| Fair DARTS [74] | 98.88 | **0.53** | 3.32 | NAS |
| GPAS [58] | 99.01 | - | 3.72 | NAS |
| TPENAS (ours) | 99.05 | 1.30 | **0.15** | NAS |

The confusion matrix is shown in Figure 11. C01~C38 represent airplane, baseball field, basketball court, beach, bridge, cemetery, chaparral, christmas tree farm, closed road, coastal mansion, crosswalk, dense residential, ferry terminal, football field, forest, freeway, golf course, harbor, intersection, mobile home park, nursing home, oil gas field, oil well, overpass, parking lot, parking space, railway, river, runway, runway marking, shipping yard, solar panel, sparse residential, storage tank, swimming pool, tennis court, transformer station and wastewater treatment plant, respectively. Each scene has a classification accuracy greater than 97%, and more than half of the scenes have a classification accuracy of 100%. The experimental results on the PatternNet dataset show that the proposed TPENAS method can find acceptable depth and structure of the CNN model.

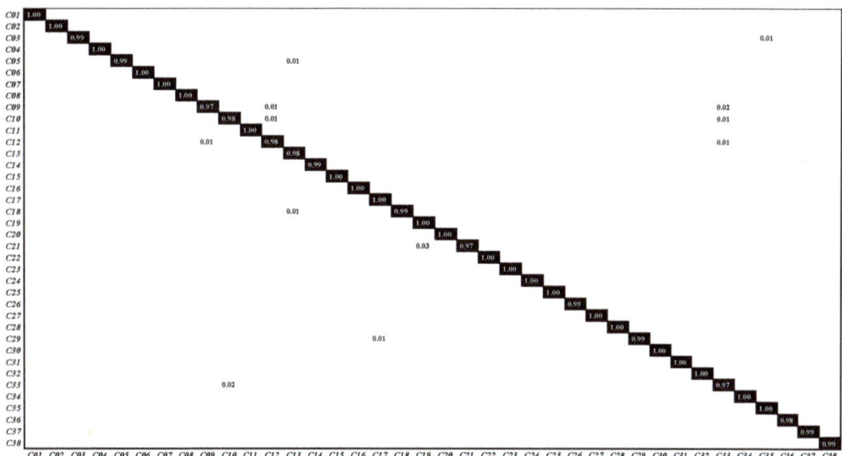

**Figure 11.** The classification confusion matrix on PatternNet dataset.

### 3.3.3. Result on NWPU45 Dataset

Following experiments on the UCM21 and PatternNet datasets, we tested the TP-MEANS method on the NWPU45 dataset, which is currently the largest remote sensing image classification dataset. In the first search phase, 20% of the data are randomly selected as a training dataset, and the remaining 80% are used as a testing dataset. The results of the first search phase are shown in Figure 12.

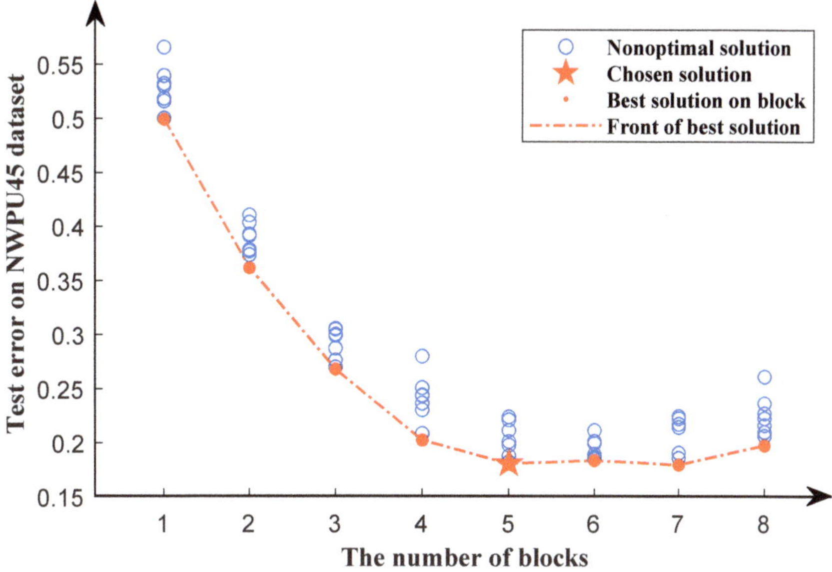


**Figure 12.** Search results of the first search phase for the NWPU45 dataset. The red dots indicate the optimal solution in a particular block, while the blue circles indicate nonoptimal solutions. The pentagram represents the chosen solution. The red dotted line represents the front of the best solution.

Consistent with the experiments on the UCM21 dataset, the overall accuracy of the model decreases as the number of blocks increases until it reaches five. When the number of blocks exceeds five, the overall accuracy of the model becomes stable. As a result, we chose five blocks as the optimal individual length for the second search phase.

The second search phase is performed using the same dataset as the first search phase, and the obtained Pareto front is shown in Figure 13. The solutions in the Pareto solution set have different structures and are not superior to other solutions in the Pareto solution set in terms of GFLOPs and test error. This demonstrates that by running the algorithm once, we can generate multiple competing models for remote sensing image classification. Similar to the UCM21 dataset, we train all solutions from scratch for 1000 epochs and select the solution with the lowest test error to compare with other algorithms.

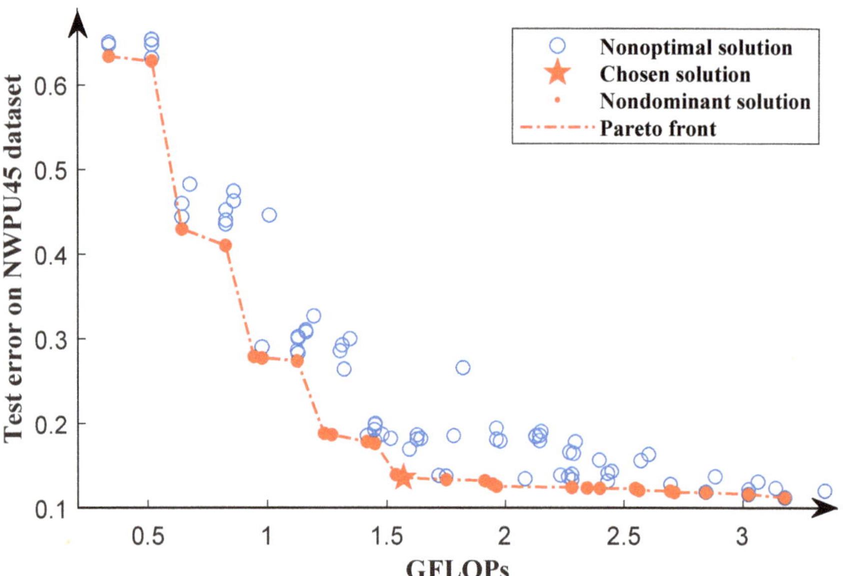


**Figure 13.** Pareto front of the second search phase for the NWPU45 dataset. The red dots represent the optimal solution, while the blue circles represent the nonoptimal solutions. The Pareto front is represented by the red dotted line. The pentagram indicates the chosen solution.

In the retraining phase, we use five-fold cross-validation to evaluate the selected solutions. Different from the UCM21 dataset, we chose one fold as a training dataset and the remaining four folds as a testing dataset. A total of 5 independent models were trained from scratch for 1000 epochs, and their test results were averaged. We also compare classic classification models and NAS-based methods, as shown in Table 6. TPENAS_large denotes that the solution with the highest classification accuracy is selected from the Pareto solution set, and TPENAS_small denotes that the solution selected from the Pareto solution has higher classification accuracy and fewer parameters than other NAS-based models. The OA of TPENAS_large is 90.38%, which is better than both classic classification models and NAS-based methods. The GFLOPs of TPENAS_large are lower than VGG16, GoogleNet, and ResNet50, and slightly higher than AlexNet. With the exception of the RTRMM method, the parameters of TPENAS_large are significantly lower than those of classic classification models such as AlexNet and VGG16 and are at least half that of other NAS-based methods. Table 6 shows that the parameters of TPENAS_small are half that of RTRMM and the OA is higher than that of NAS-based methods.

**Table 6.** The OA, GFLOPs, and Params of TPENAS are compared with the other methods on the NWPU45 dataset (the ratio of training samples to test samples is 2:8). The upward arrow (↑) indicates that the larger the number, the better the result. The downward arrow (↓) indicates that the smaller the number, the better the result.

| Method | OA (%) ↑ | GFLOPs ↓ | Params (M) ↓ | Search Strategy |
| --- | --- | --- | --- | --- |
| AlexNet [66] | 79.85 | 0.92 | 57.19 | manual |
| VGGNet16 [32] | 79.79 | 20.18 | 134.44 | manual |
| GoogleNet [34] | 78.48 | 1.97 | 5.65 | manual |
| ResNet50 [33] | 83.00 | 5.37 | 23.60 | manual |
| Fine-tuned AlexNet [66] | 85.16 | 0.92 | 57.19 | manual |
| Fine-tuned VGG16 [32] | 90.36 | 20.18 | 134.44 | manual |
| Fine-tuned GoogLeNet [34] | 86.02 | 1.96 | 5.65 | manual |
| NASNet [43] | 67.48 | 0.77 | 4.28 | NAS |
| SGAS [68] | 75.87 | 0.81 | 4.70 | NAS |
| DARTS [47] | 67.48 | 0.77 | 3.41 | NAS |
| MNASNet [69] | 81.92 | 0.43 | 3.16 | NAS |
| PDARTS [71] | 82.14 | 0.73 | 4.21 | NAS |
| RTRMM [70] | 86.32 | **0.39** | 0.83 | NAS |
| TPENAS_*large* (ours) | **90.38** | 1.65 | 1.67 | NAS |
| TPENAS_*small* (ours) | 87.79 | 1.27 | **0.41** | NAS |

Figure 14 depicts the classification confusion matrix. C01~C45 represent airplane, airport, baseball diamond, basketball court, beach, bridge, chaparral, church, circular farmland, cloud, commercial area, dense residential, desert, forest, freeway, golf course, ground track field, harbor, industrial area, intersection, island, lake, meadow, medium residential, mobile home park, mountain, overpass, place, parking lot, railway, railway station, rectangular farmland, river, roundabout, runway, sea ice, ship, snow berg, sparse residential, stadium, storage tank, tennis court, terrace, thermal power station and wetland, respectively. As shown in Figure 14, 20% of the place images are misclassified as church images and 9% of the church images are misclassified as place images. As shown in Figure 15, the category church is very similar to the category place, making it extremely difficult for the model to extract discriminative features from these images. TPENAS achieves high classification accuracy in other categories.

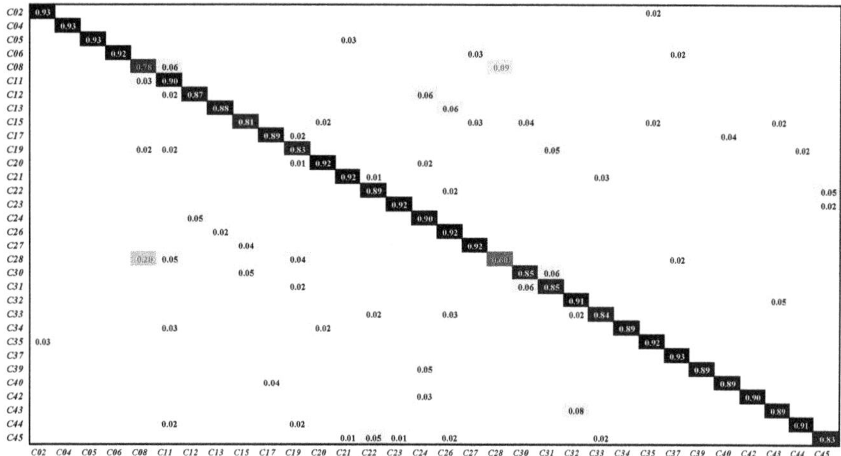

**Figure 14.** The classification confusion matrix on the NWPU45 dataset. We removed categories with classification accuracy higher than 95% in the confusion matrix and did not display numbers less than 0.01.

**Figure 15.** Some samples in the categories *church* and *place*. The four images in the first row are in the category *church*, and the four images in the second row are in the category *place*.

### 3.3.4. Compared to Other CNN-Based Methods

Our proposed method is compared with other CNN-based remote sensing image classification methods, as shown in Table 7. The OA of TPENAS is higher than other CNN-based methods, which shows that the OA of our algorithm on remote sensing image classification tasks is satisfactory.

**Table 7.** The OA of TPENAS is compared with the other CNN-based methods on the UCM21 dataset and NWPU45 dataset (the ratio of training samples to test samples is 8:2).

| Method | UCM21 | NWPU45 |
| --- | --- | --- |
| MARTA GANs [75] | 94.86 | 75.03 |
| Attention GANs [76] | 97.69 | 77.99 |
| VGG-16-CapsNet [77] | 98.81 | 89.18 |
| GBN [78] | 98.57 | - |
| MSCP [79] | 98.36 | 88.93 |
| TPENAS (ours) | **98.81** | **90.38** |

### 3.3.5. Compared to Other ENAS Methods

TPENAS was compared with other ENAS methods, as shown in Table 8. On the NWPU45 dataset, SceneNet, E2SCNet, and TPENAS were all trained using 80% of the dataset and then tested on the remaining 20% of the dataset. Experimental results show that our algorithm has a higher OA than SceneNet and E2SCNet, and that Params and GFLOPs are not the worst among the three algorithms.

**Table 8.** The OA, Params, and GFLOPs of TPENAS are compared with the ENAS methods on the NWPU45 dataset (the ratio of training samples to test samples is 8:2). The upward arrow (↑) indicates that the larger the number, the better the result. The downward arrow (↓) indicates that the smaller the number, the better the result.

| Method | OA ↑ | Params (M) ↓ | GFLOPs ↓ |
| --- | --- | --- | --- |
| SceneNet [60] | 95.22 | **1.02** | 9.47 |
| E2SCNet [61] | 95.23 | 3.88 | **0.60** |
| TPENAS_large (ours) | **95.70** | 1.65 | 1.67 |

## 4. Discussion

### 4.1. Analysis of the Number of Evaluated Models

We compare the number of evaluated models for the proposed TPENAS and traversal search. In this paper, traversal search means that models with different numbers of blocks perform the second search phase respectively, and then the best model is selected from all the search results. In our experimental setup, the models have 8 different block numbers, so the traversal search needs to evaluate 32,000 models. TPENAS only needs to evaluate 4640 models. According to the current experimental settings, the number of models to be evaluated by TPENAS is $\vartheta_1 = 8 \times 10 \times \vartheta + 40 \times 100$, and the number of models to be evaluated by traversal search is $\vartheta_2 = 40 \times 100 \times \vartheta$, therefore, TPENAS evaluates $\vartheta_d$ fewer models than traversal search, where $\vartheta_d = \vartheta_2 - \vartheta_1 = 3920 \times \vartheta - 4000$, and $\vartheta$ is the number of blocks that can be selected. It can be seen that as $\vartheta$ increases, $\vartheta_d$ increases linearly. This demonstrates that our algorithm is more time-efficient.

*4.2. Analysis of the Depth of the Model in the Second Search Phase*

The aim of the first search phase is to find the optimal depth of the model, which is the number of blocks of the model. We use the overall accuracy of the model and the number of blocks of the model as two optimization objectives to build a multi-objective optimization problem, as shown in Equation (1). The optimal solution to this optimization problem is the appropriate number of blocks. We designed the first search phase algorithm to optimize this optimization problem. On the UCM21 dataset, the number of blocks output by the first search phase algorithm is five, but this does not indicate an optimal solution to this optimization problem.

We selected five different block numbers to conduct ablation experiments on the UCM21 dataset to validate the effectiveness of the first search phase. Figure 16 shows the Pareto front of the experimental results. As we can see, it is not true that the greater the number of blocks, the better the Pareto front of the individual. For example, the Pareto front with seven blocks is worse than the Pareto front with six blocks, indicating that the former is superior. We discover that the Pareto front is optimal when the block equals five, which matches the results in our first search phase.

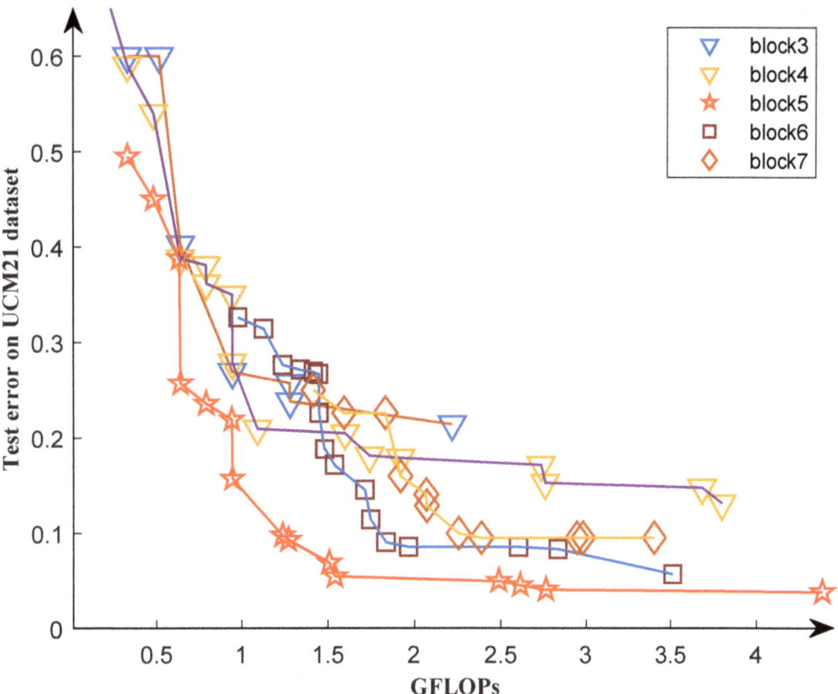


**Figure 16.** The Pareto fronts obtained by experiments with five different block numbers during the second search phase.

### 4.3. Analysis of Fully Trained Models and Non-Fully Trained Models

In the Pareto front obtained from the second search phase, we fully train each solution from scratch and then select the solution with the lowest classification error to compare with other methods. The reason is that after fully training from scratch, the solution with the lowest classification error may not be the solution with the lowest classification error in the Pareto solution set. Therefore, we conducted four sets of comparative experiments, as shown in Figure 17. In the upper left figure, after the solutions in the Pareto solution set with blocks equal to three are fully trained, the solution with the lowest classification error is the optimal solution in the Pareto set. However, the other three subplots show that after training completely from scratch, the solution with the lowest classification error is not the solution with the lowest classification error in the Pareto solution set. This further illustrates the rationality of our choice of the final solution.

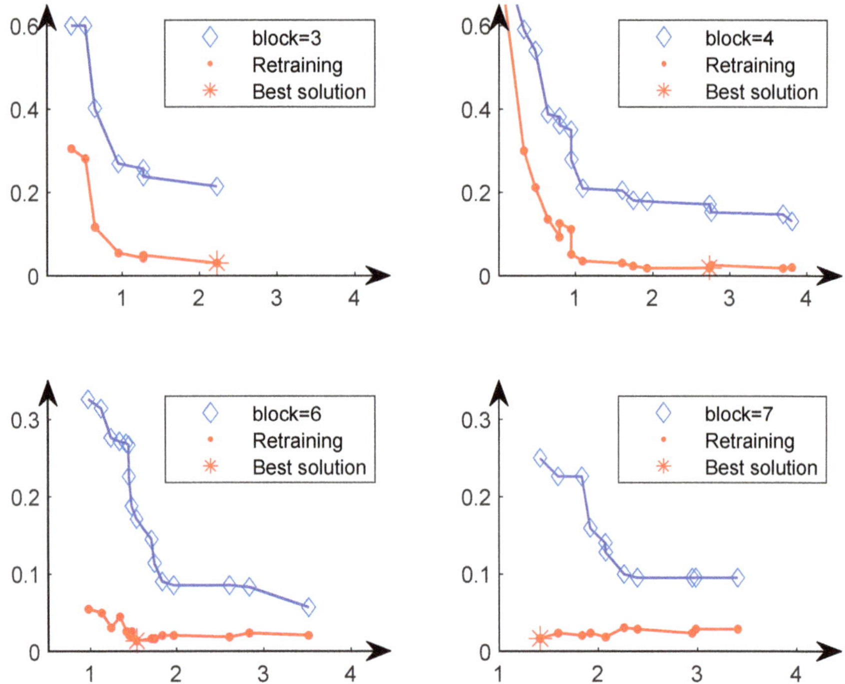

**Figure 17.** The horizontal axis represents the GFLOPs of the individual, and the vertical axis represents the test error of the individual on the UCM21 dataset. The blue curve represents the Pareto front on the second phase. The red curve represents the result obtained by fully training the solution on the Pareto front from scratch.

### 4.4. Analysis of TPENAS Algorithm with Fewer Training Samples

Fewer training samples have two effects on the TPENAS algorithm. Firstly, it decreases the runtime of the TPENAS algorithm. Since the runtime of TPENAS is spent primarily on evaluating the model, reducing the training samples will linearly reduce the evaluation time of the model. Specifically, a $\Gamma$-fold reduction in training samples will reduce the time to evaluate a single model by approximately a factor of $\Gamma$. Furthermore, it also enables the algorithm to reduce its reliance on sample labels. Secondly, it is not conducive to TPENAS outputting well-structured models. This is because too few training samples will make the model easily overfitted during training. In the second search phase, the overfitted model will not evaluate the model accurately, resulting in low OA for well-structured individuals in the population. Therefore, a reasonable selection of the number of training samples is able to both reduce the running time of the TPENAS algorithm and search for structurally appropriate models. Such individuals can easily be eliminated in selecting the next-generation population, which leads to difficulties for the TPENAS algorithm to output a well-structured model.

## 5. Conclusions

In this paper, we propose TPENAS, a two-phase evolutionary neural architecture search for remote sensing image classification, which overcomes the shortcomings of manually designed CNN and NAS algorithms. In the first search phase, we optimize the classification accuracy and depth of the model to determine the maximum depth of the model on the benchmark dataset. In the second search phase, we optimize the classification accuracy and GFLOPs of the model to obtain a set of models for remote sensing image classification. The experimental results on the NWPU45 dataset show that TPENAS improves overall classification accuracy by 4.02% when compared to other NAS algorithms. Furthermore, it reduces the parameters by at least 13 times when compared to classic classification methods. In future work, we will explore how to design a more discriminative deep learning method to greatly promote the classification of similar images. In addition, in practical application scenarios, enough training samples can sometimes be difficult to obtain, and how to design high-accuracy remote sensing image classification models with small samples remains an open research question.

**Author Contributions:** Conceptualization, L.A. and K.F.; methodology, L.A. and K.F.; validation, L.A. and K.F.; investigation, L.A.; writing—original draft preparation, L.A. and K.F.; writing—review and editing, K.S., K.F., H.Z., X.H. and Z.C. All authors have read and agreed to the published version of the manuscript.

**Funding:** This work was supported by the National Key R&D program of China (No. 2022YFB4300700), Qin Chuangyuan cited the high-level innovative and entrepreneurial talent project (QCYRCXM-2022-21) and Guangdong High Level Innovation Research Institution Project (2021B0909050008).

**Institutional Review Board Statement:** Not applicable.

**Informed Consent Statement:** Not applicable.

**Data Availability Statement:** The NWPU-RESISC45 dataset used in this research is openly and freely available at https://1drv.ms/u/s!AmgKYzARBl5ca3HNaHIlzp$_$IXjs, accessed on 3 April 2017. The UC Merced Land-use dataset used in this research is openly and freely available at http://weegee.vision.ucmerced.edu/datasets/landuse.html, accessed on 28 October 2010. The PatternNet dataset used in this research is openly and freely available at https://sites.google.com/view/zhouwx/dataset, accessed on 25 May 2018.

**Conflicts of Interest:** The authors declare no conflict of interest.

## References

1. Gong, M.; Jiang, F.; Qin, A.K.; Liu, T.; Zhan, T.; Lu, D.; Zheng, H.; Zhang, M. A spectral and spatial attention network for change detection in hyperspectral images. *IEEE Trans. Geosci. Remote Sens.* **2022**, *60*, 5521614. [CrossRef]
2. Gong, M.; Zhou, Z.; Ma, J. Change detection in synthetic aperture radar images based on image fusion and fuzzy clustering. *IEEE Trans. Image Process.* **2012**, *21*, 2141–2151. [CrossRef] [PubMed]
3. Gong, M.; Zhao, J.; Liu, J.; Miao, Q.; Jiao, L. Change detection in synthetic aperture radar images based on deep neural networks. *IEEE Trans. Neural Netw. Learn. Syst.* **2016**, *27*, 125–138. [CrossRef]
4. Liu, T.; Gong, M.; Lu, D.; Zhang, Q.; Zheng, H.; Jiang, F.; Zhang, M. Building change detection for VHR remote sensing images via local–global pyramid network and cross-task transfer learning strategy. *IEEE Trans. Geosci. Remote Sens.* **2022**, *60*, 4704817. [CrossRef]
5. Gong, M.; Su, L.; Jia, M.; Chen, W. Fuzzy clustering with a modified MRF energy function for change detection in synthetic aperture radar images. *IEEE Trans. Fuzzy Syst.* **2014**, *22*, 98–109. [CrossRef]
6. Wu, Y.; Li, J.; Yuan, Y.; Qin, A.; Miao, Q.G.; Gong, M.G. Commonality autoencoder: Learning common features for change detection from heterogeneous images. *IEEE Trans. Neural Netw. Learn. Syst.* **2021**, *33*, 4257–4270. [CrossRef] [PubMed]
7. Castelluccio, M.; Poggi, G.; Sansone, C.; Verdoliva, L. Land use classification in remote sensing images by convolutional neural networks. *arXiv* **2015**, arXiv:1508.00092.

8.  Zhu, Q.; Sun, Y.; Guan, Q.; Wang, L.; Lin, W. A weakly pseudo-supervised decorrelated subdomain adaptation framework for cross-domain land-use classification. *IEEE Trans. Geosci. Remote Sens.* **2022**, *60*, 5623913. [CrossRef]
9.  Pires de Lima, R.; Marfurt, K. Convolutional neural network for remote-sensing scene classification: Transfer learning analysis. *Remote Sens.* **2019**, *12*, 86. [CrossRef]
10. Gong, M.; Li, J.; Zhang, Y.; Wu, Y.; Zhang, M. Two-path aggregation attention network with quad-patch data augmentation for few-shot scene classification. *IEEE Trans. Geosci. Remote Sens.* **2022**, *60*, 4511616. [CrossRef]
11. Wang, Z.; Li, J.; Liu, Y.; Xie, F.; Li, P. An adaptive surrogate-assisted endmember extraction framework based on intelligent optimization algorithms for hyperspectral remote sensing images. *Remote Sens.* **2022**, *14*, 892. [CrossRef]
12. Huang, X.; Wen, D.; Li, J.; Qin, R. Multi-level monitoring of subtle urban changes for the megacities of China using high-resolution multi-view satellite imagery. *Remote Sens. Environ.* **2017**, *196*, 56–75. [CrossRef]
13. Li, Y.; Zhang, Y.; Huang, X.; Yuille, A.L. Deep networks under scene-level supervision for multi-class geospatial object detection from remote sensing images. *ISPRS J. Photogramm. Remote Sens.* **2018**, *146*, 182–196. [CrossRef]
14. Longbotham, N.; Chaapel, C.; Bleiler, L.; Padwick, C.; Emery, W.J.; Pacifici, F. Very high resolution multiangle urban classification analysis. *IEEE Trans. Geosci. Remote Sens.* **2012**, *50*, 1155–1170. [CrossRef]
15. Gong, M.; Liang, Y.; Shi, J.; Ma, W.; Ma, J. Fuzzy c-means clustering with local information and kernel metric for image segmentation. *IEEE Trans. Image Process.* **2012**, *22*, 573–584. [CrossRef]
16. Wu, Y.; Ma, W.; Gong, M.; Su, L.; Jiao, L. A novel point-matching algorithm based on fast sample consensus for image registration. *IEEE Geosci. Remote Sens. Lett.* **2014**, *12*, 43–47. [CrossRef]
17. Zhang, Y.; Gong, M.; Li, J.; Zhang, M.; Jiang, F.; Zhao, H. Self-supervised monocular depth estimation with multiscale perception. *IEEE Trans. Image Process.* **2022**, *31*, 3251–3266. [CrossRef]
18. Li, H.; Li, J.; Zhao, Y.; Gong, M.; Zhang, Y.; Liu, T. Cost-sensitive self-paced learning with adaptive regularization for classification of image time series. *IEEE J. Sel. Top. Appl. Earth Obs. Remote Sens.* **2021**, *14*, 11713–11727. [CrossRef]
19. Li, J.; Gong, M.; Liu, H.; Zhang, Y.; Zhang, M.; Wu, Y. Multiform ensemble self-supervised learning for few-shot remote sensing scene classification. *IEEE Trans. Geosci. Remote Sens.* **2023**, *61*, 4500416. [CrossRef]
20. Perronnin, F.; Sánchez, J.; Mensink, T. Improving the fisher kernel for large-scale image classification. In Proceedings of the European Conference on Computer Vision, Crete, Greece, 5–11 September 2010; pp. 143–156.
21. Lazebnik, S.; Schmid, C.; Ponce, J. Beyond bags of features: Spatial pyramid matching for recognizing natural scene categories. In Proceedings of the IEEE Computer Society Conference on Computer Vision and Pattern Recognition (CVPR'06), New York, NY, USA, 17–22 June 2006; pp. 2169–2178.
22. Yang, Y.; Newsam, S. Bag-of-visual-words and spatial extensions for land-use classification. In Proceedings of the 18th SIGSPATIAL International Conference on Advances in Geographic Information Systems, San Jose, CA, USA, 2–5 November 2010; pp. 270–279.
23. Zhang, F.; Du, B.; Zhang, L. Saliency-guided unsupervised feature learning for scene classification. *IEEE Trans. Geosci. Remote Sens.* **2015**, *53*, 2175–2184. [CrossRef]
24. Cheng, G.; Han, J.; Guo, L.; Liu, T. Learning coarse-to-fine sparselets for efficient object detection and scene classification. In Proceedings of the IEEE Conference on Computer Vision and Pattern Recognition, Boston, MA, USA, 7–12 June 2015; pp. 1173–1181.
25. Han, X.; Zhong, Y.; Zhao, B.; Zhang, L. Scene classification based on a hierarchical convolutional sparse auto-encoder for high spatial resolution imagery. *Int. J. Remote Sens.* **2017**, *38*, 514–536. [CrossRef]
26. Shi, C.; Zhang, X.; Sun, J.; Wang, L. Remote sensing scene image classification based on self-compensating convolution neural network. *Remote Sens.* **2022**, *14*, 545. [CrossRef]
27. Zhu, L.; Chen, Y.; Ghamisi, P.; Benediktsson, J.A. Generative adversarial networks for hyperspectral image classification. *IEEE Trans. Geosci. Remote Sens.* **2018**, *56*, 5046–5063. [CrossRef]
28. Gu, S.; Zhang, R.; Luo, H.; Li, M.; Feng, H.; Tang, X. Improved singan integrated with an attentional mechanism for remote sensing image classification. *Remote Sens.* **2021**, *13*, 1713. [CrossRef]
29. Miao, W.; Geng, J.; Jiang, W. Semi-supervised remote-sensing image scene classification using representation consistency siamese network. *IEEE Trans. Geosci. Remote Sens.* **2022**, *60*, 5616614. [CrossRef]
30. Penatti, O.A.; Nogueira, K.; Dos Santos, J.A. Do deep features generalize from everyday objects to remote sensing and aerial scenes domains? In Proceedings of the IEEE Conference on Computer Vision and Pattern Recognition Workshops, Boston, MA, USA, 7–12 June 2015; pp. 44–51.
31. Krizhevsky, A.; Sutskever, I.; Hinton, G.E. Imagenet classification with deep convolutional neural networks. In *Advances in Neural Information Processing Systems 25*; Curran Associates, Inc.: Red Hook, NY, USA, 2012.
32. Simonyan, K.; Zisserman, A. Very deep convolutional networks for large-scale image recognition. *arXiv* **2014**, arXiv:1409.1556.
33. He, K.; Zhang, X.; Ren, S.; Sun, J. Deep residual learning for image recognition. In Proceedings of the IEEE Conference on Computer Vision and Pattern Recognition, Amsterdam, The Netherlands, 8–16 October 2016; pp. 770–778.
34. Szegedy, C.; Liu, W.; Jia, Y.; Sermanet, P.; Reed, S.; Anguelov, D.; Erhan, D.; Vanhoucke, V.; Rabinovich, A. Going deeper with convolutions. In Proceedings of the IEEE Conference on Computer Vision and Pattern Recognition, Boston, MA, USA, 7–12 June 2015; pp. 1–9.
35. Huang, G.; Liu, Z.; Van Der Maaten, L.; Weinberger, K.Q. Densely connected convolutional networks. In Proceedings of the IEEE Conference on Computer Vision and Pattern Recognition, Honolulu, HI, USA, 21–26 July 2017; pp. 4700–4708.

36. Zhang, B.; Zhang, Y.; Wang, S. A lightweight and discriminative model for remote sensing scene classification with multidilation pooling module. *IEEE J. Sel. Top. Appl. Earth Obs. Remote Sens.* **2019**, *12*, 2636–2653. [CrossRef]
37. Yu, D.; Guo, H.; Xu, Q.; Lu, J.; Zhao, C.; Lin, Y. Hierarchical attention and bilinear fusion for remote sensing image scene classification. *IEEE J. Sel. Top. Appl. Earth Obs. Remote Sens.* **2020**, *13*, 6372–6383. [CrossRef]
38. Tong, W.; Chen, W.; Han, W.; Li, X.; Wang, L. Channel-attention-based densenet network for remote sensing image scene classification. *IEEE J. Sel. Top. Appl. Earth Obs. Remote Sens.* **2020**, *13*, 4121–4132. [CrossRef]
39. Cheng, G.; Yang, C.; Yao, X.; Guo, L.; Han, J. When deep learning meets metric learning: Remote sensing image scene classification via learning discriminative CNNs. *IEEE Trans. Geosci. Remote Sens.* **2018**, *56*, 2811–2821. [CrossRef]
40. Wang, C.; Tang, X.; Li, L.; Tian, B.; Zhou, Y.; Shi, J. IDN: Inner-class dense neighbours for semi-supervised learning-based remote sensing scene classification. *Remote Sens. Lett.* **2023**, *14*, 80–90. [CrossRef]
41. Zoph, B.; Le, Q.V. Neural architecture search with reinforcement learning. *arXiv* **2016**, arXiv:1611.01578.
42. Baker, B.; Gupta, O.; Naik, N.; Raskar, R. Designing neural network architectures using reinforcement learning. *arXiv* **2016**, arXiv:1611.02167.
43. Zoph, B.; Vasudevan, V.; Shlens, J.; Le, Q.V. Learning transferable architectures for scalable image recognition. In Proceedings of the IEEE Conference on Computer Vision and Pattern Recognition, Salt Lake City, UT, USA, 18–22 June 2018; pp. 8697–8710.
44. Real, E.; Moore, S.; Selle, A.; Saxena, S.; Suematsu, Y.L.; Tan, J.; Le, Q.V.; Kurakin, A. Large-scale evolution of image classifiers. In Proceedings of the International Conference on Machine Learning, Sydney, NSW, Australia, 6–11 August 2017; pp. 2902–2911.
45. Xie, L.; Yuille, A. Genetic cnn. In Proceedings of the the IEEE International Conference on Computer Vision, Venice, Italy, 22–29 October 2017; pp. 1379–1388.
46. Sun, Y.; Xue, B.; Zhang, M.; Yen, G.G. Evolving deep convolutional neural networks for image classification. *IEEE Trans. Evol. Comput.* **2020**, *24*, 394–407. [CrossRef]
47. Liu, H.; Simonyan, K.; Yang, Y. Darts: Differentiable architecture search. *arXiv* **2018**, arXiv:1806.09055.
48. Xie, S.; Zheng, H.; Liu, C.; Lin, L. SNAS: Stochastic neural architecture search. *arXiv* **2018**, arXiv:1812.09926.
49. Tanveer, M.S.; Khan, M.U.K.; Kyung, C.M. Fine-tuning darts for image classification. In Proceedings of the IEEE International Conference on Pattern Recognition, Los Alamitos, CA, USA, 11–17 October 2021; pp. 4789–4796.
50. Wu, Y.; Ding, H.; Gong, M.; Qin, A.; Ma, W.; Miao, Q.; Tan, K.C. Evolutionary multiform optimization with two-stage bidirectional knowledge transfer strategy for point cloud registration. *IEEE Trans. Evol. Comput.* 2022. [CrossRef]
51. Li, J.; Li, H.; Liu, Y.; Gong, M. Multi-fidelity evolutionary multitasking optimization for hyperspectral endmember extraction. *Appl. Soft Comput.* **2021**, *111*, 107713. [CrossRef]
52. Elsken, T.; Metzen, J.H.; Hutter, F. Simple and efficient architecture search for convolutional neural networks. *arXiv* **2017**, arXiv:1711.04528.
53. Chen, T.; Goodfellow, I.; Shlens, J. Net2net: Accelerating learning via knowledge transfer. *arXiv* **2015**, arXiv:1511.05641.
54. Zhu, H.; An, Z.; Yang, C.; Xu, K.; Zhao, E.; Xu, Y. EENA: Efficient evolution of neural architecture. In Proceedings of the IEEE/CVF International Conference on Computer Vision Workshops, Los Alamitos, CA, USA, 27–28 October 2019; pp. 1891–1899.
55. Wang, B.; Sun, Y.; Xue, B.; Zhang, M. Evolving deep convolutional neural networks by variable-length particle swarm optimization for image classification. In Proceedings of the IEEE Congress on Evolutionary Computation, Rio de Janeiro, Brazil, 8–13 July 2018; pp. 1–8.
56. Xie, X.; Liu, Y.; Sun, Y.; Yen, G.G.; Xue, B.; Zhang, M. BenchENAS: A benchmarking platform for evolutionary neural architecture search. *IEEE Trans. Evol. Comput.* **2022**, *26*, 1473–1485. [CrossRef]
57. Zhang, Z.; Liu, S.; Zhang, Y.; Chen, W. RS-DARTS: A convolutional neural architecture search for remote sensing image scene classification. *Remote Sens.* **2021**, *14*, 141. [CrossRef]
58. Peng, C.; Li, Y.; Jiao, L.; Shang, R. Efficient convolutional neural architecture search for remote sensing image scene classification. *IEEE Trans. Geosci. Remote Sens.* **2020**, *59*, 6092–6105. [CrossRef]
59. Chen, J.; Huang, H.; Peng, J.; Zhu, J.; Chen, L.; Tao, C.; Li, H. Contextual information-preserved architecture learning for remote-sensing scene classification. *IEEE Trans. Geosci. Remote Sens.* **2022**, *60*, 5602614. [CrossRef]
60. Ma, A.; Wan, Y.; Zhong, Y.; Wang, J.; Zhang, L. SceneNet: Remote sensing scene classification deep learning network using multi-objective neural evolution architecture search. *ISPRS J. Photogramm. Remote Sens.* **2021**, *172*, 171–188. [CrossRef]
61. Wan, Y.; Zhong, Y.; Ma, A.; Wang, J.; Zhang, L. E2SCNet: Efficient multiobjective evolutionary automatic search for remote sensing image scene classification network architecture. *IEEE Trans. Neural Netw. Learn. Syst.* 2022, *Early Access*.
62. Gudzius, P.; Kurasova, O.; Darulis, V.; Filatovas, E. AutoML-based neural architecture search for object recognition in satellite imagery. *Remote Sens.* **2022**, *15*, 91. [CrossRef]
63. Zhang, X.; Tian, Y.; Jin, Y. A knee point-driven evolutionary algorithm for many-objective optimization. *IEEE Trans. Evol. Comput.* **2015**, *19*, 761–776. [CrossRef]
64. Zhou, W.; Newsam, S.; Li, C.; Shao, Z. PatternNet: A benchmark dataset for performance evaluation of remote sensing image retrieval. *ISPRS J. Photogramm. Remote Sens.* **2018**, *145*, 197–209. [CrossRef]
65. Cheng, G.; Han, J.; Lu, X. Remote sensing image scene classification: Benchmark and state of the art. *Proc. IEEE* **2017**, *105*, 1865–1883. [CrossRef]
66. Krizhevsky, A.; Sutskever, I.; Hinton, G.E. ImageNet classification with deep convolutional neural networks. *Commun. ACM* **2017**, *60*, 84–90. [CrossRef]

67. Liu, Z.; Mao, H.; Wu, C.Y.; Feichtenhofer, C.; Darrell, T.; Xie, S. A convnet for the 2020s. In Proceedings of the IEEE/CVF Conference on Computer Vision and Pattern Recognition, New Orleans, LA, USA, 21–24 June 2022; pp. 11976–11986.
68. Li, G.; Qian, G.; Delgadillo, I.C.; Muller, M.; Thabet, A.; Ghanem, B. Sgas: Sequential greedy architecture search. In Proceedings of the IEEE/CVF Conference on Computer Vision and Pattern Recognition, Los Alamitos, CA, USA, 25 June 2020; pp. 1620–1630.
69. Tan, M.; Chen, B.; Pang, R.; Vasudevan, V.; Sandler, M.; Howard, A.; Le, Q.V. Mnasnet: Platform-aware neural architecture search for mobile. In Proceedings of the IEEE/CVF Conference on Computer Vision and Pattern Recognition, Long Beach, CA, USA, 15–20 June 2019; pp. 2820–2828.
70. Li, J.; Weinmann, M.; Sun, X.; Diao, W.; Feng, Y.; Fu, K. Random topology and random multiscale mapping: An automated design of multiscale and lightweight neural network for remote-sensing image recognition. *IEEE Trans. Geosci. Remote Sens.* **2022**, *60*, 5610917. [CrossRef]
71. Chen, X.; Xie, L.; Wu, J.; Tian, Q. Progressive differentiable architecture search: Bridging the depth gap between search and evaluation. In Proceedings of the IEEE/CVF International Conference on Computer Vision, Seoul, Republic of Korea, 27–28 October 2019; pp. 1294–1303.
72. Wang, J.; Zhong, Y.; Zheng, Z.; Ma, A.; Zhang, L. RSNet: The search for remote sensing deep neural networks in recognition tasks. *IEEE Trans. Geosci. Remote Sens.* **2020**, *59*, 2520–2534. [CrossRef]
73. Chen, J.; Huang, H.; Peng, J.; Zhu, J.; Chen, L.; Li, W.; Sun, B.; Li, H. Convolution neural network architecture learning for remote sensing scene classification. *arXiv* **2020**, arXiv:2001.09614.
74. Chu, X.; Zhou, T.; Zhang, B.; Li, J. Fair darts: Eliminating unfair advantages in differentiable architecture search. In Proceedings of the European Conference on Computer Vision, Glasgow, UK, 23–28 August 2020; pp. 465–480.
75. Lin, D.; Fu, K.; Wang, Y.; Xu, G.; Sun, X. MARTA GANs: Unsupervised representation learning for remote sensing image classification. *IEEE Geosci. Remote Sens. Lett.* **2017**, *14*, 2092–2096. [CrossRef]
76. Yu, Y.; Li, X.; Liu, F. Attention GANs: Unsupervised deep feature learning for aerial scene classification. *IEEE Trans. Geosci. Remote Sens.* **2019**, *58*, 519–531. [CrossRef]
77. Zhang, W.; Tang, P.; Zhao, L. Remote sensing image scene classification using CNN-CapsNet. *Remote Sens.* **2019**, *11*, 494. [CrossRef]
78. Sun, H.; Li, S.; Zheng, X.; Lu, X. Remote sensing scene classification by gated bidirectional network. *IEEE Trans. Geosci. Remote Sens.* **2020**, *58*, 82–96. [CrossRef]
79. He, N.; Fang, L.; Li, S.; Plaza, A.; Plaza, J. Remote sensing scene cassification using multilayer stacked covariance pooling. *IEEE Trans. Geosci. Remote Sens.* **2018**, *56*, 6899–6910. [CrossRef]

*Article*

# Remote Sensing Image Road Extraction Network Based on MSPFE-Net

Zhiheng Wei [1,2] and Zhenyu Zhang [1,2,*]

1 School of Information Science and Engineering, Xinjiang University, Urumqi 830017, China; wei577245796@163.com
2 Xinjiang Key Laboratory of Multilingual Information Technology, Xinjiang University, Urumqi 830017, China
* Correspondence: zhangzhenyu@xju.edu.cn

**Abstract:** Road extraction is a hot task in the field of remote sensing, and it has been widely concerned and applied by researchers, especially using deep learning methods. However, many models using convolutional neural networks ignore the attributes of roads, and the shape of the road is banded and discrete. In addition, the continuity and accuracy of road extraction are also affected by narrow roads and roads blocked by trees. This paper designs a network (MSPFE-Net) based on multi-level strip pooling and feature enhancement. The overall architecture of MSPFE-Net is encoder-decoder, and this network has two main modules. One is a multi-level strip pooling module, which aggregates long-range dependencies of different levels to ensure the connectivity of the road. The other module is the feature enhancement module, which is used to enhance the clarity and local details of the road. We perform a series of experiments on the dataset, Massachusetts Roads Dataset, a public dataset. The experimental data showed that the model in this paper was better than the comparison models.

**Keywords:** road extraction; convolutional neural network; remote sensing images; strip pooling

**Citation:** Wei, Z.; Zhang, Z. Remote Sensing Image Road Extraction Network Based on MSPFE-Net. *Electronics* **2023**, *12*, 1713. https://doi.org/10.3390/electronics12071713

Academic Editors: Yue Wu, Kai Qin, Maoguo Gong and Qiguang Miao

Received: 11 March 2023
Revised: 27 March 2023
Accepted: 31 March 2023
Published: 4 April 2023

## 1. Introduction

The continuous progress of remote sensing and artificial intelligence has laid a solid theoretical and technical foundation for road extraction. As an important ground object, roads play a crucial role in intelligent transportation [1,2], urban planning [3,4], and emergency tasks [5,6]. Traditional remote sensing road extraction uses manual interpretation, which consumes a lot of workforce and time. The automatic extraction of roads is helpful to reduce the manual workload, and it also accelerates the speed of road extraction, so road extraction has excellent research value. Due to the different shapes of roads, the influence of background factors, and being blocked by trees or shadows, the road extraction task is difficult and challenging [7–9].

Deep learning promotes the progress of computer vision, especially in object detection, semantic segmentation, image classification, and other aspects, and it has a good effect. Scholars also began to use deep learning technology to complete remote sensing image road extraction [10–13]. Although the model based on deep learning has achieved good results in extracting road tasks, and many road extraction algorithms have problems, such as road breaking caused by occlusion, difficult extraction of the narrow road, and incorrect identification of roads and background. In order to solve these problems, road extraction models need strong long-distance dependencies or global context information, and the road extraction algorithm usually uses attention mechanism or atrous convolution technology to obtain long-distance dependencies or global context information. The attention mechanism is a method to improve the ability of global context modeling. However, it consumes a lot of memory. Other methods include atrous convolution and spatial pyramid pooling, which can expand the receptive field of convolutional neural network, but the strip target features extracted by the square window may be mixed with irrelevant target information.

Inspired by the idea of strip pooling proposed by Hou, Feng et al. [14], this paper introduces and improves upon it. The strip pooling has several distinct characteristics. Firstly, the strip pooling has a long and banded shape at a dimension, so it can capture long-range relationships of isolated regions. Then, strip pooling maintains a narrow shape along a spatial dimension, which helps to capture the local feature of targets and can reduce the interference of irrelevant target information. The network combined with strip pooling has the ability to obtain multiple types of contexts. It is fundamentally different from traditional spatial pooling. The idea of strip pooling is well applied in remote sensing image road extraction scenes. Therefore, a multi-level strip pooling and feature enhancement network (MSPFE-Net) called MSPFE-Net is designed in this paper. In MSPFE-Net, the multi-level strip pooling module is responsible for fully extracting long-range context information. The feature enhancement module is used to enhance the clarity and local details of the road.

The main content of this article consists of the following:

1.  A multi-level strip pooling module (MSPM) was designed to extract global context information to ensure the connectivity and integrity of road extraction.
2.  A feature enhancement module (FEM) was proposed, which mainly enhanced the clarity and local details of the road
3.  MSPFE-Net is designed and implemented for road extraction tasks. The effectiveness of MSPFE-Net was verified on the Massachusetts Roads Dataset.

The chapter structure of this article is arranged as follows: Section 2 introduces the related work. Section 3 shows the structure of MSPFE-Net and explains the rationale for each module. Section 4 shows experimental contents, containing experimental datasets, evaluation methods, experimental settings, and experimental results. Sections 5 and 6 introduce this paper's discussion and conclusion, respectively.

## 2. Related Works

In recent years, a series of algorithms for road extraction have been proposed. According to the characteristics of various algorithms, there are mainly two types: traditional type and type of deep learning.

### 2.1. Traditional Type

Traditional types have the following methods: template matching method, knowledge-driven method, and object-oriented method [15]. Template matching is a method that applies geometric, topological, and radial features of road images. According to the template type, it can be divided into rule templates and variable templates. The advantages of the rule template are less computation, good stability, and simplicity, while the disadvantages are affected by the transformation of radiation characteristics. The advantage of a variable template is that it can be applied to images with irregular road edges and irregular road radiation information. The disadvantage is that it requires a large amount of computation. The primary process is to design the template according to the rules, obtain the regional extreme value through the template using the measure function and update the road location. Haverkamp [16] used the comparative analysis of multiple rectangular templates to rotate at certain angular intervals to form a group of discrete rectangular templates In the knowledge-driven method, knowledge can be divided into geometric knowledge, contextual knowledge, and auxiliary knowledge. Wenfeng Wang et al. [17] put forward a straight-line detection algorithm using the property of parallel edges of roads, recognized parallel features using principal component analysis, and direction consistency criteria The object-oriented method is to obtain the output results by segmentation, classification, and post-processing of the input image. The segmentation methods include threshold segmentation, graph segmentation, and edge segmentation. Classification methods include geometric feature classification and SVM; post-processing includes tensor voting and mathematical morphology. Maboudi et al. [18] used guided filtering to eliminate the inconsistency of pavement image texture and then used a method including color and shape data for road extraction.

### 2.2. Type of Deep Learning

Using a convolutional neural network (CNN) to obtain road features from many image sample data. Mnih and Hinton [19] combined deep learning with road extraction for the first time, using a restricted Boltzmann machine to detect road areas from images. P. Li et al. [20] designed a network combining CNN and linear integral convolution to extract roads. Wei et al. [21] applied an improved cross entropy loss function to the CNN, which can help improve the topological information of the road. The fully convolutional network (FCN) was proposed in 2014 [22]. The deconvolution of FCN can make the final feature map have the same size as the input image after up-sampling and predict the category of each pixel. Although FCN plays a pioneering role in semantic segmentation, its accuracy is low. Since the U-Net model was applied in the task of medical image segmentation, it achieved good results. The U-Net network is an improved fully convolutional network [23]. It includes the skip connection. In the process of up-sampling, the feature map in the process of down-sampling is fused in concatenate. Many subsequent image segmentation models have adopted this idea and improved on it. In the field of road extraction, there is no exception. Singh et al. [24] proposed their improved U-Net model to realize the function of road extraction. He et al. [25] designed the deep residual network to make the number of network layers deeper. Zhang et al. [26] realized road extraction by introducing ResNet into U-Net and combining the advantages of both. In order to accomplish road extraction at different scales, Gao et al. [27] introduced a feature pyramid and proposed a multi-feature pyramid network (MFPN). Cheng et al. [28] proposed a cascading end-to-end network (CasNet), which completes the road detection task and the road centerline extraction task simultaneously through two cascading networks. However, these methods have an insufficient receptive field to capture effective and rich long-distance context information, which is crucial for road extraction. The lack of long-distance context information will directly lead to the discontinuity of road extraction results or even the phenomenon that roads cannot be extracted completely. In order to connect discontinuous broken roads, many researchers have considered various schemes to capture long-distance context information to model the topological relationship between broken roads [29]. The main method is to use atrous convolution [30]. Atrous convolution can effectively expand the receptive field without increasing the amount of computation. Taking into account the natural connectivity and large span of the road, Zhou et al. [31] added the concatenation mode and parallel mode of different atrous convolution to form D-LinkNet for road extraction. In order to extract road features at different scales, He et al. [32] introduced the atrous spatial pyramid pooling (ASPP) module. According to the above analysis, compared with the traditional method, the deep learning method can greatly improve the accuracy and automation of road extraction, but there are still problems of road breaking caused by occlusion in the road extraction results [33]. Although many researchers have offered some solutions, there is still a lack of high-performance, end-to-end road extraction networks that can solve this problem. Tao et al. [34] proposed to integrate a well designed spatial information inference structure into the deeplabv3+ network to maintain the continuity of road extraction by realizing multi-direction transmission of information between pixels. Zhou et al. [35] proposed a boundary and topologically-aware road extraction network (BT-RoadNet) in order to improve the quality of road boundaries and solve the problem of road discontinuity. The network is divided into thick and thin prediction modules to obtain detailed boundary information, and the spatial context module is designed to solve the problem of discontinuous road results. Lu et al. [36] proposed GAMSNet, which uses multi-scale residual learning to extract multi-scale features, and then it uses global perception operations to capture long-distance relationships. Tan et al. [37] proposed a new end-to-end encoder-decoder architecture network to solve the problem of road location information loss due to reduced spatial resolution. This network obtains different levels of features by encoders, and the decoder is composed of a scale fusion module and a scale sensitive module, respectively, achieving the task of fusing features and assigning weights. Zhu et al. [38] designed a global context-aware batch processing independent network

(GCB-Net), which effectively integrates global context features by using the improved non-local module as a global context-aware module. Wang et al. [39] designed a model combining global attention, and it can enhance the performance of road segmentation.

In conclusion, Although the method based on deep learning effectively extracts roads, it is still difficult to extract roads especially in complex scenes. Therefore, it is necessary to fully consider the structural characteristics of roads and improve the accuracy of road extraction in complex scenes.

## 3. Methods

The details of the proposed network model are introduced in this section. Section 3.1 shows the architecture of the MSPFE-Net; Section 3.2 shows the multi-level strip pooling module. Section 3.3 introduces the feature enhancement module in detail. Section 3.4 introduces the loss function.

### 3.1. MSPFE-Net Model

The MSPFE-Net is shown in Figure 1, it is mainly composed of the encoder, multi-level strip pooling module (MSPM), feature enhancement module (FEM), and decoder. The encoder uses the Resnet50 network. The output results of the encoder in the first four layers are used as inputs to MSPM, which is used to strengthen the long-range dependencies of the model. The output results of the encoder in the fifth layer are used as inputs to the feature enhancement module, and it focuses on collecting various types of contexts through different pooling operations to make the road feature representation more discriminating. The output feature of MSPM will be added with the up-sampled feature map in the decoder.

### 3.2. Multi-Level Strip Pooling Module

Figure 2 explains the theory of the strip pooling. The strip pooling window performs pooling at horizontal or vertical dimensions, and the input feature is a two-dimensional tensor $x \in R^{H \times W}$. Different from two-dimensional average pooling, the method of strip pooling is to sum the value of a row or column and divide it by the number of rows or columns, respectively. Therefore, the horizontal strip pooling output $y^h \in R^H$ can be expressed as:

$$y_i^h = \frac{1}{W} \sum_{j=0}^{W-1} x_{i,j} \tag{1}$$

Similarly, the vertical strip pooling outputs $y^v \in R^W$ can be expressed as:

$$y_i^v = \frac{1}{H} \sum_{i=0}^{H-1} x_{i,j} \tag{2}$$

The strip pooling in Figure 2 is similar to the traditional pooling method, and the pixel values are averaged over the locations on the feature maps corresponding to the pooling kernels. A feature map is an input, here actually $C \times H \times W$. For ease of description, only one channel is drawn. The feature map input processing principle of C channels is the same as that of one channel operation shown here. After horizontal and vertical strip pooling, the input feature map becomes $H \times 1$ and $1 \times W$. The element values within the pooling window are averaged, and the value is used as the pooling output value. Subsequently, one-dimensional convolution is conducted on the output values, and the two output feature maps are expanded along the horizontal and vertical directions. Then, the two feature maps had the same size, and the expanded feature maps corresponding to the same position were summed to obtain the $H \times W$ feature map.

One of the difficulties of road extraction is maintaining the road's continuity. In order to reduce the impact of this problem, MSPM is proposed in this paper to fully obtain long-range dependencies to keep the connectivity and integrity of road topology. The core idea of MSPM is to extract different features by strip pooling at different levels and fuse these features. MSPM is added to the skip connection section of MSPFE-Net to extract long-range context information at different levels.

The framework of MSPM is shown in Figure 3, which contains three strip pooling sub-blocks of different sizes, L1, L2, and L3, respectively. Input feature $x$ is processed by the L1, L2, and L3 sub-blocks, respectively, then the model obtains three output features $y^{L1}$, $y^{L2}$, and $y^{L3}$ and add the corresponding position of $y^{L1}$, $y^{L2}$, and $y^{L3}$. The last operation is to multiply the summed result with the input $x$. Finally, there is the output result of MSPM, $y^{out}$ can be expressed as:

$$y^{out} = x \otimes \left( y^{L1} + y^{L2} + y^{L3} \right) \tag{3}$$

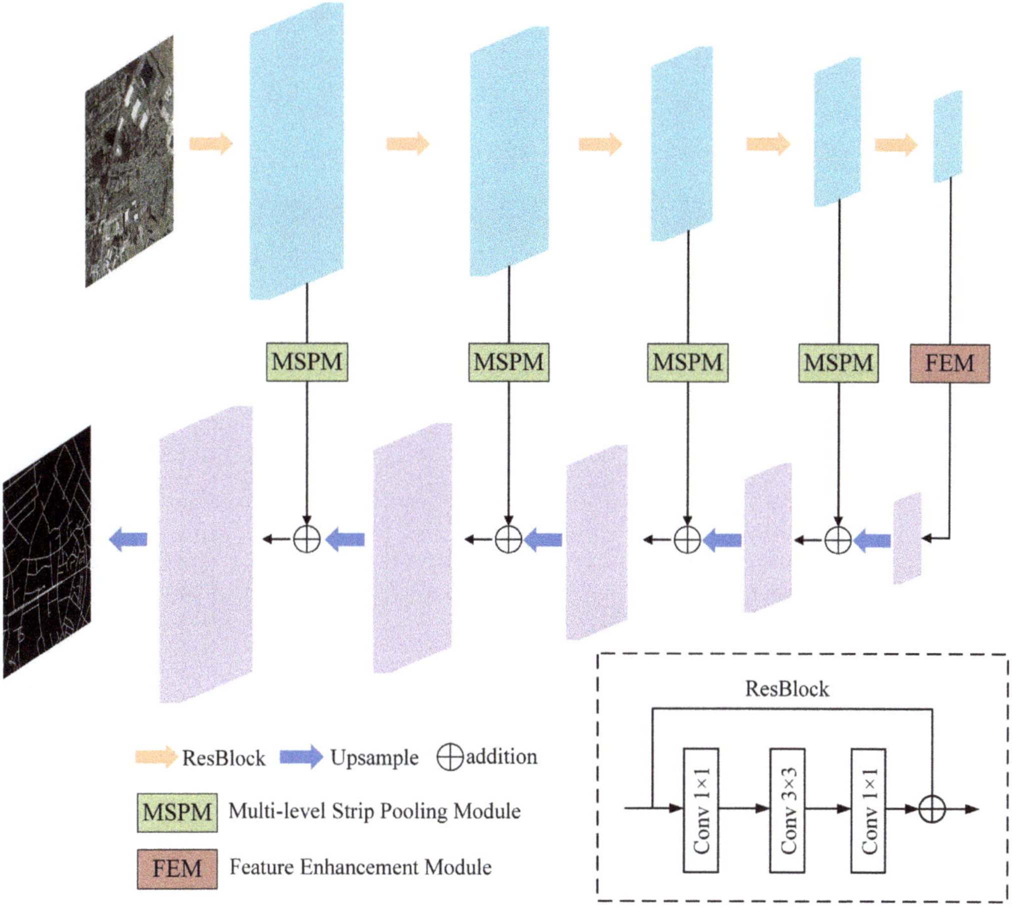

**Figure 1.** The architecture of MSPFE-Net.

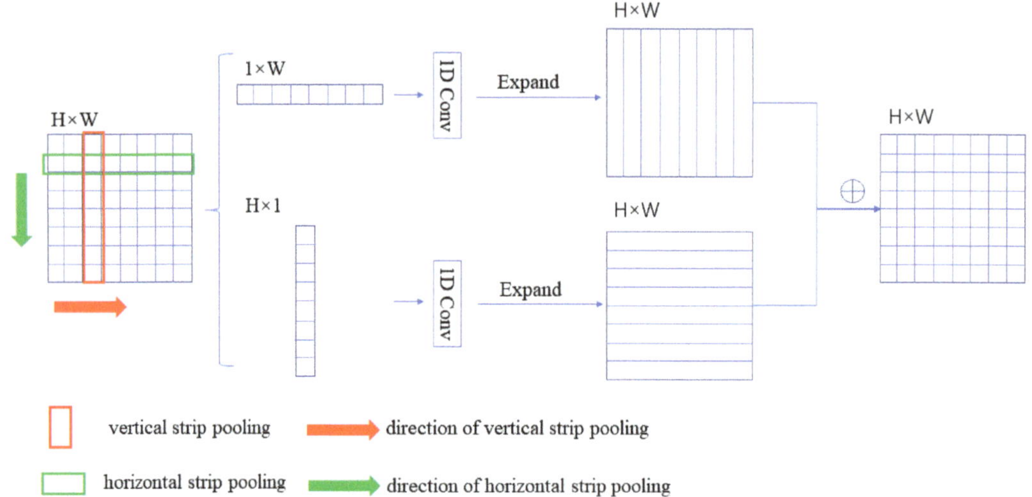


**Figure 2.** Illustration of strip pooling.

### 3.3. Feature Enhancement Module

The FEM, shown in Figure 4, focuses on collecting various types of contexts through different pooling operations to make feature representation more discriminating. The advantage is that it can be used continuously to extend long-range dependencies and reinforce local details.

The feature enhancement module is composed of four sub-modules, namely, $3 \times 3$ convolution, $3 \times 3$ maximum pooling, horizontal strip pooling, and vertical strip pooling. These four sub-modules are represented as $f_1$, $f_2$, $f_3$, and $f_4$. The input of FEM is represented as x, and the output of FEM is represented as y. Firstly, $3 \times 3$ convolution and $3 \times 3$ maximum pooling were carried out to add the above result features to obtain the feature map, namely, $y_1$. Similarly, horizontal strip pooling and vertical strip pooling were carried out, and the results were added to obtain the feature map, namely, $y_2$, which was splicing $y_1$ and $y_2$. After $1 \times 1$ conv, the final result feature $y$ was obtained. They capture both long-range and local dependencies information, and it is essential for remote sensing image road extraction scene resolution networks. FEM can be expressed as:

$$y_1 = f_1(x) + f_2(x) \tag{4}$$

$$y_2 = f_3(x) + f_4(x) \tag{5}$$

$$y = C_{1 \times 1}(CONCAT(y_1, y_2)) \tag{6}$$

where $C_{1 \times 1}$ is $1 \times 1$ convolution, and CONCAT is a concatenate operation.

For long-range dependencies, unlike previous work using a global averaging pooling layer, we capture context information by using both horizontal and vertical strip pooling operations. At the same time, strip pooling makes it possible to connect discrete areas and code areas with strip structures throughout the road scene. However, in the case of tight distribution of semantic regions, capturing road local context information also requires spatial pooling. Considering this, convolution operation and pooling layer are used to obtain short-range dependencies.

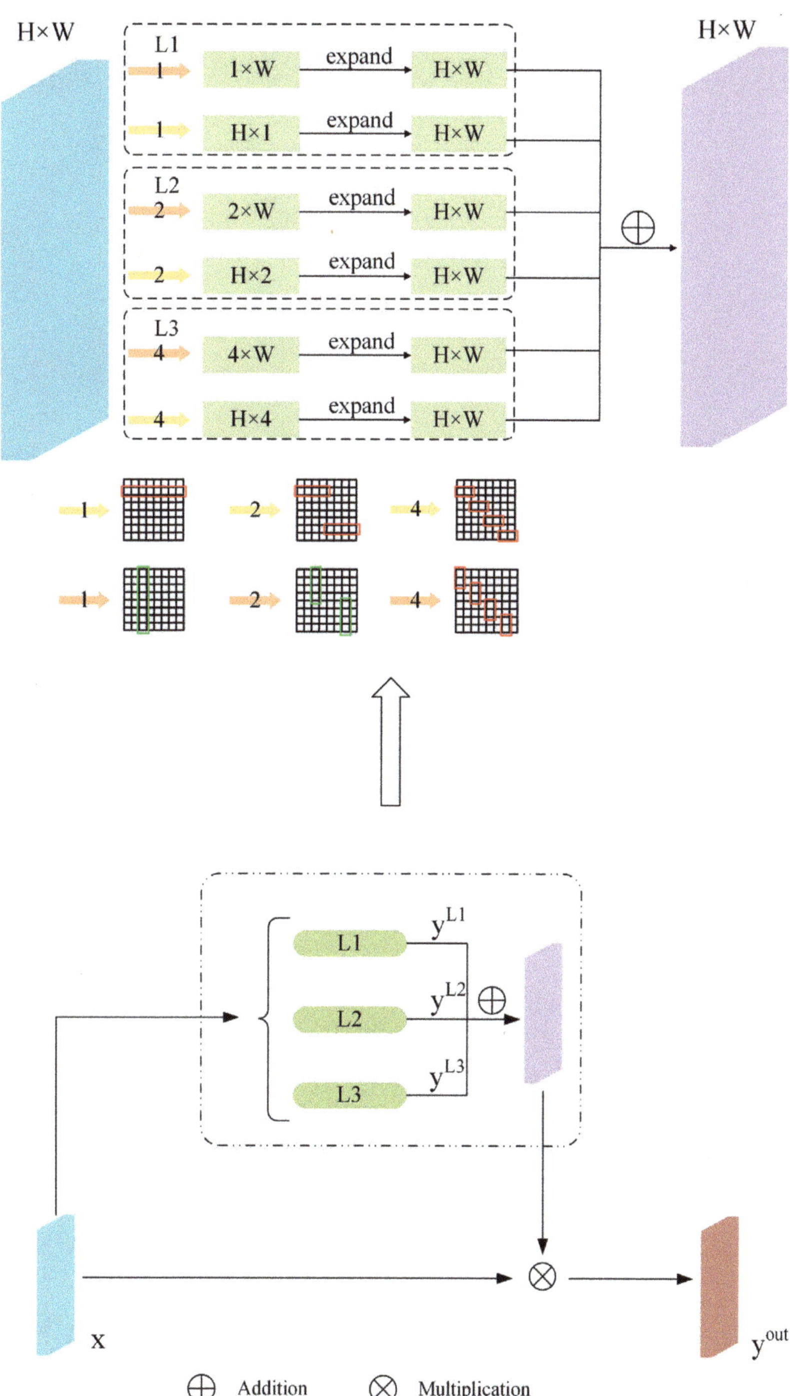


**Figure 3.** The structure of MSPM.

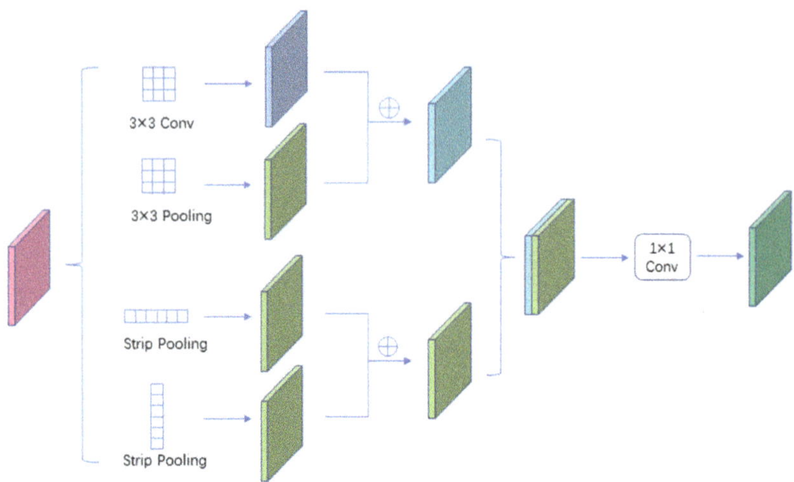


**Figure 4.** Illustration of feature enhancement module.

*3.4. Loss Function*

The binary cross entropy loss function is applied to most pixel-level segmentation tasks. However, when the number of pixels on the target is much smaller than the number of pixels in the background, that is, the samples are highly unbalanced, and the loss function has the disadvantage of misleading the model to seriously bias the background. In this paper, it is necessary to judge whether the pixels predicted by the model are roads or backgrounds. The road area is small, and the background area is too large. If the binary cross entropy loss function is used, this will make the model deviate from the optimal direction during the training process. To reduce the impact of this problem, the dice coefficient loss function and the focal loss function are used together as the loss function.

The dice coefficient loss function is calculated as follows:

$$L_d = 1 - \frac{2|X \cap Y|}{|X| + |Y|} \tag{7}$$

In the formula, $X$ is the generated prediction map, $Y$ is the label, $|X \cap Y|$ is the intersection of label and prediction, $|X|$ is the number of elements of the label, and $|Y|$ represents the number of predicted elements.

The focal loss function is based on the binary cross entropy loss. It is a dynamically scaled cross entropy loss. Through a dynamic scaling factor, the weight of easily distinguishable samples can be dynamically reduced in the training process to focus on those indistinguishable samples quickly. The focal loss function is as follows:

$$L_f = FL(p_t) = -\alpha_t(1 - p_t)^\gamma \log(p_t) \tag{8}$$

Among them, $-\log(p_t)$ is the initial cross entropy loss function, $\alpha$ is the weight parameter between categories, $(1 - p_t)^\gamma$ is the easy/hard sample adjustment factor, and $\gamma$ is the focusing parameter.

The final loss function is the sum of the dice coefficient loss function and the focal loss function, namely:

$$L_{loss} = L_f + L_d \tag{9}$$

## 4. Results

### 4.1. Dataset

Massachusetts Roads Dataset [40] is used in the experimental dataset as shown in Figure 5. The pixel size of the Massachusetts Roads Dataset is $1500 \times 1500$, and there are 1171 pairs of images and labels. In this experiment, the number of training images, test images, and validation images is 1108, 49, and 14, respectively.

(a)

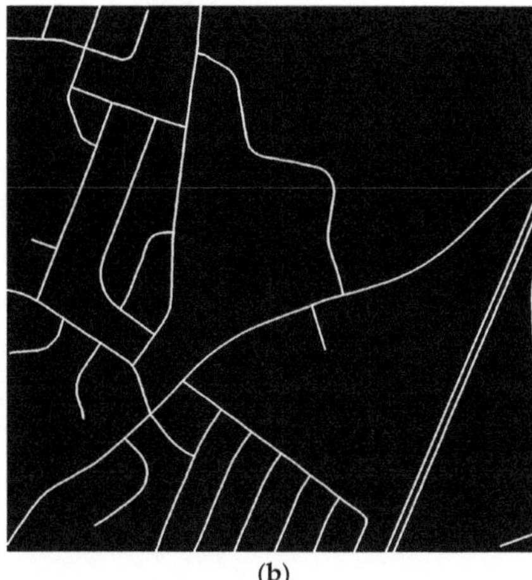

(b)

**Figure 5.** Massachusetts Road Dataset. (**a**) Image; (**b**) Label.

### 4.2. Evaluation Metrics

Selecting appropriate evaluation metrics is of great reference significance for evaluating the model's performance. This paper adopts Recall, Precision, *F1-Score* (F1), and intersection over union (*IoU*), and these evaluation metrics commonly used in semantic segmentation. *F1-Score* is calculated by two indicators. Intersection over union (*IoU*) refers to the ratio between the intersection of predicted road pixels and real labeled road pixels and their union. The specific calculation method is:

$$Precision = \frac{TP}{TP + FP} \tag{10}$$

$$Recall = \frac{TP}{TP + FN} \tag{11}$$

$$F1 - Score = 2 \times \frac{Precision \times Recall}{Precision + Recall} \tag{12}$$

$$IoU = \frac{TP}{TP + FP + FN} \tag{13}$$

In the formula, $TP$ is the number of pixels correctly classified as roads, $TN$ is the number of pixels correctly classified as non-roads, $FP$ represents the number of pixels wrongly classified as roads, and $FN$ represents the number of pixels wrongly classified as non-roads. The values of these evaluation metrics are in the range of [0,1]. The best effect is that the evaluation metrics value is equal to 1

### 4.3. Experimental Settings

In the process of model training, the batch size of each training input network sample is 4. The initial learning rate is 0.001, and the decay of the learning rate is adjusted by the cosine annealing algorithm. The maximum training epoch is 100. The optimizer uses the

Adam algorithm with momentum set to 0.9. The loss function combines the dice coefficient loss function and focal loss function.

### 4.4. Experimental Results and Analysis

In order to test and prove the feasibility and rationality of MSPFE-Net, the mainstream semantic segmentation network is applied to the task of road segmentation, and the MSPFE-Net is compared. Table 1 shows the comparison of each model in the road segmentation task. Analysis showed that: (1) the F1-Score of the proposed model was 12.03%, 6.58%, 3.35%, 2.17%, and 1.40% higher than that of Deeplabv3+, U-Net, HRNetV2, D-LinkNet, and RefineNet, respectively. (2) The IoU of the proposed model was 13.89%, 7.91%, 4.13%, 2.70%, and 1.74% higher than that of Deeplabv3+, U-Net, HRNetV2, D-LinkNet, and RefineNet, respectively. (3) From F1-Score and IoU, the MSPFE-Net is better than Deeplabv3+, U-Net, HRNetV2, D-LinkNet, and RefineNet.

**Table 1.** Numerical results of different networks.

| Networks | Precision (%) | Recall (%) | F1 (%) | IoU (%) |
| --- | --- | --- | --- | --- |
| Deeplabv3+ | 54.13% | 73.25% | 62.26% | 45.20% |
| U-Net | 77.41% | 60.17% | 67.71% | 51.18% |
| HRNetV2 | 65.48% | 77.38% | 70.94% | 54.96% |
| D-LinkNet | 71.72% | 72.51% | 72.12% | 56.39% |
| RefineNet | 69.57% | 76.54% | 72.89% | 57.35% |
| MSPFE-Net(ours) | 73.11% | 75.50% | 74.29% | 59.09% |

Figure 6 shows the effect of each model. According to the road extraction result map based on Deeplabv3+, the connectivity and geometric topological relationship of the road remain relatively complete, but the details of the road edge are rough, and more areas are misjudged as roads in the background map. This phenomenon is mainly due to the fact that the Deeplabv3+ model focuses on extracting deep semantic information, while the overall shape of the road is thin, and road details will be lost through the Deeplabv3+ backbone network. The road extraction effect based on the U-Net model is relatively clear in the details of road edges, and adjacent roads can be accurately displayed and separated. However, because of the limitation of the traditional convolution receptive field, the long-range features cannot be captured effectively. Therefore, the overall connectivity of roads extracted by U-Net is poor, and roads have more fracture zones, especially thin and narrow roads that have long fracture zones. The road extraction effect based on the D-LinkNet model is generally good, but D-LinkNet does not obtain enough long-range context information, so some roads appear discontinuous, and edge details are not clear. The overall structure of the road extracted by HRNetV2 is relatively complete, but the edge of the road is rough, and adjacent roads cannot be distinguished. The road extracted by RefineNet has good continuity, but there are some misjudgments. By means of incorporating the MSPM and FEM, the presented model in this paper has proficiently conserved intricate details whilst capturing the long-range feature relationships pertaining to the road network. Evidently, based on the visual analysis of Figure 6, the extracted road has exhibited substantial preservation of overall framework and connectivity whilst manifesting enhanced clarity of edge details. Consequently, this breakthrough endeavor promotes the depiction of intricate details and effective feature recognition, thus paving the way for an improved comprehension of road and their networks.

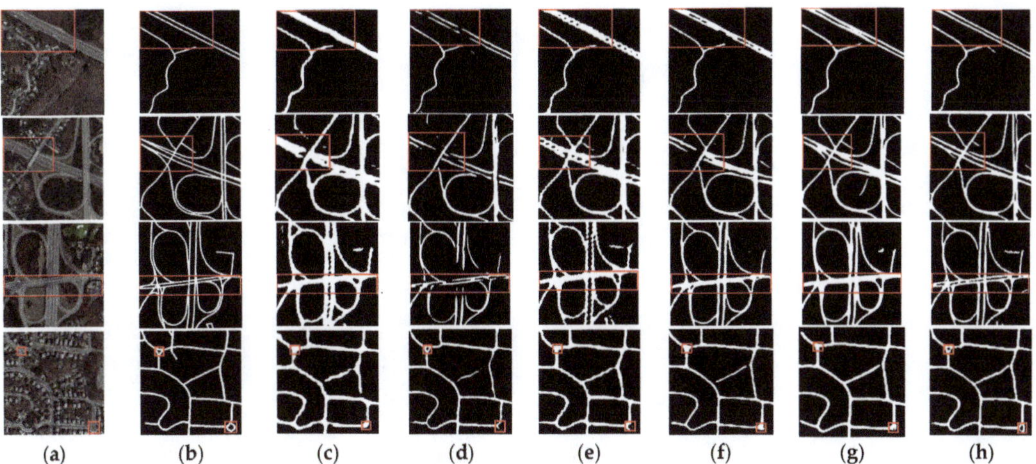

**Figure 6.** Results of MSPFE-Net and other methods. (**a**) Image; (**b**) Label; (**c**) DeepLabv3+; (**d**) U-Net; (**e**) HRNetV2; (**f**) D-LinkNet; (**g**) RefineNet (**h**) MSPFE-Net.( Red squares represent key areas.).

In Figure 7, it can be found that the road is blocked by trees, the MSFPE-Net accurately restored the road, while the road extracted by other comparison models is discontinuous or even not extracted.

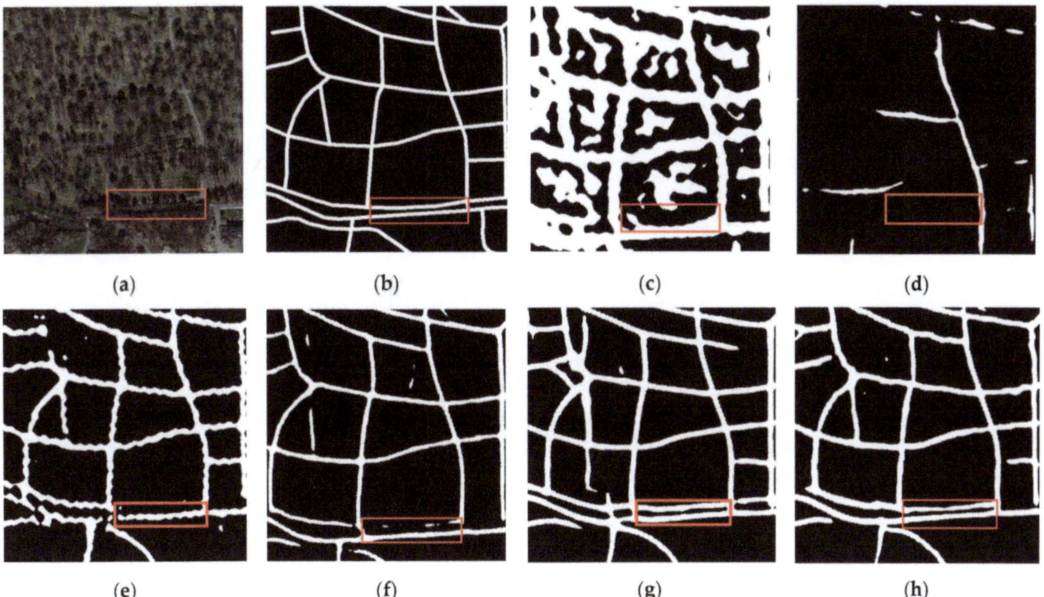

**Figure 7.** Results of road extraction in occlusion scene. (**a**) Image; (**b**) Label; (**c**) DeepLabv3+; (**d**) U-Net; (**e**) HRNetV2; (**f**) D-LinkNet; (**g**) RefineNet; and (**h**) MSPFE-Net.( Red squares represent key areas.).

In Figure 8, it can be found that the background of some areas is similar to the road, and the MSFPE-Net basically accurately extracts the road, while other comparison models mistakenly regard the road as the background.

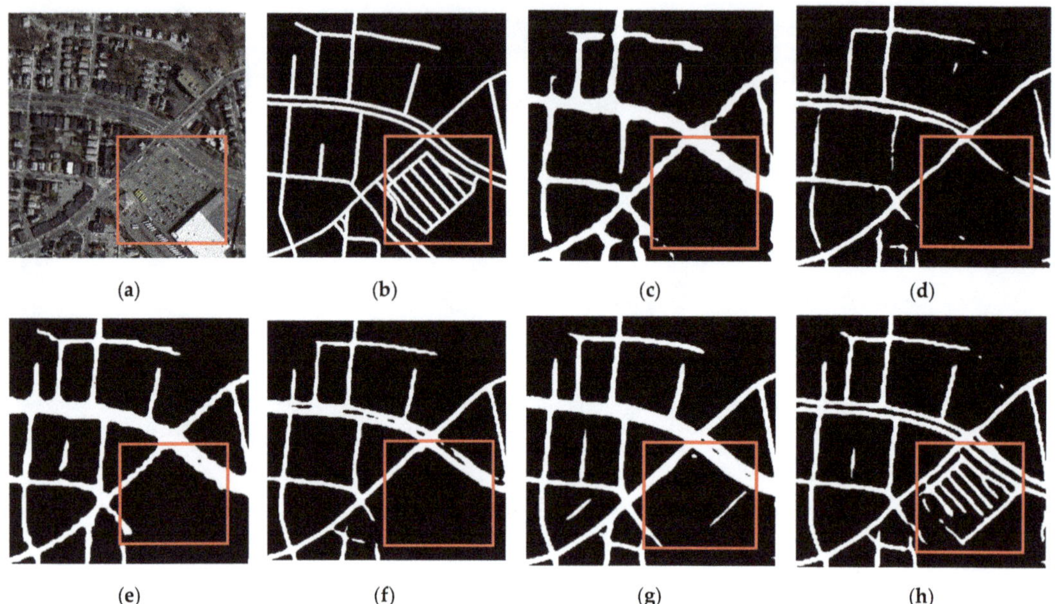

**Figure 8.** Results of road extraction (the road is similar to the background). (**a**) Image; (**b**) Label; (**c**) DeepLabv3+; (**d**) U-Net; (**e**) HRNetV2; (**f**) D-LinkNet; (**g**) RefineNet; and (**h**) MSPFE-Net.( Red squares represent key areas.).

This method uses multi-level strip pooling combined with a feature enhancement module to ensure road connectivity and road edge details. The goal of the multi-level strip pooling module is to obtain the global context information and long-range dependencies and connect the discretely distributed paths in the image. The feature enhancement module is used to obtain the road's local context information and improve the road edge's segmentation effect.

## 5. Discussion

### 5.1. Ablation Experiments

5.1.1. Influence of MSPM and FEM

We conducted a series of ablation experiments on MSPFE-NET using the Massachusetts Road Dataset. In order to prove the effectiveness of each module, the baseline is U-Net with ResNet50 as the backbone, and then each module is added separately. Table 2 shows the experimental data of all main modules.

**Table 2.** Ablation experiments of MSPM and FEM.

| Networks | Precision (%) | Recall (%) | F1 (%) | IoU (%) |
| --- | --- | --- | --- | --- |
| Baseline | 75.92% | 63.28% | 69.02% | 52.70% |
| Baseline + MSPM (L1) | 72.54% | 70.23% | 71.37% | 55.48% |
| Baseline + MSPM (L1 + L2) | 68.87% | 77.95% | 73.13% | 57.64% |
| Baseline + MSPM (L1 + L2 + L3) | 69.87% | 77.65% | 73.55% | 58.17% |
| Baseline + FEM | 68.71% | 71.76% | 70.20% | 54.08% |

After adding MSPM (L1) to the baseline, Recall, F1, and IoU were enhanced by 6.95%, 2.35%, and 2.78%, respectively. After adding L2 to Baseline + MSPM (L1), the Recall of F1 and IoU of Baseline + MSPM (L1 + L2) increased by 7.72%, 1.76%, and 2.16% compared with Baseline + MSPM (L1), respectively. After adding L3 to Baseline + MSPM (L1 + L2), the Precision, F1, and IoU of Baseline + MSPM (L1 + L2) increased by 1.00%, 0.42%, and 0.53% over Baseline + MSPM (L1 + L2), respectively. According to all the results, as shown in Figure 9, with the addition of strip pooling of different levels (L1, L2, L3) into the model, the overall connection of the road is better, and the form of the slender road is higher, which fully verifies the effectiveness of MSPM. After adding FEM to the baseline, Recall, F1, and IoU enhanced by 8.48%, 1.18%, and 1.38%, respectively. This proves that FEM plays a certain role in improving road extraction capability.

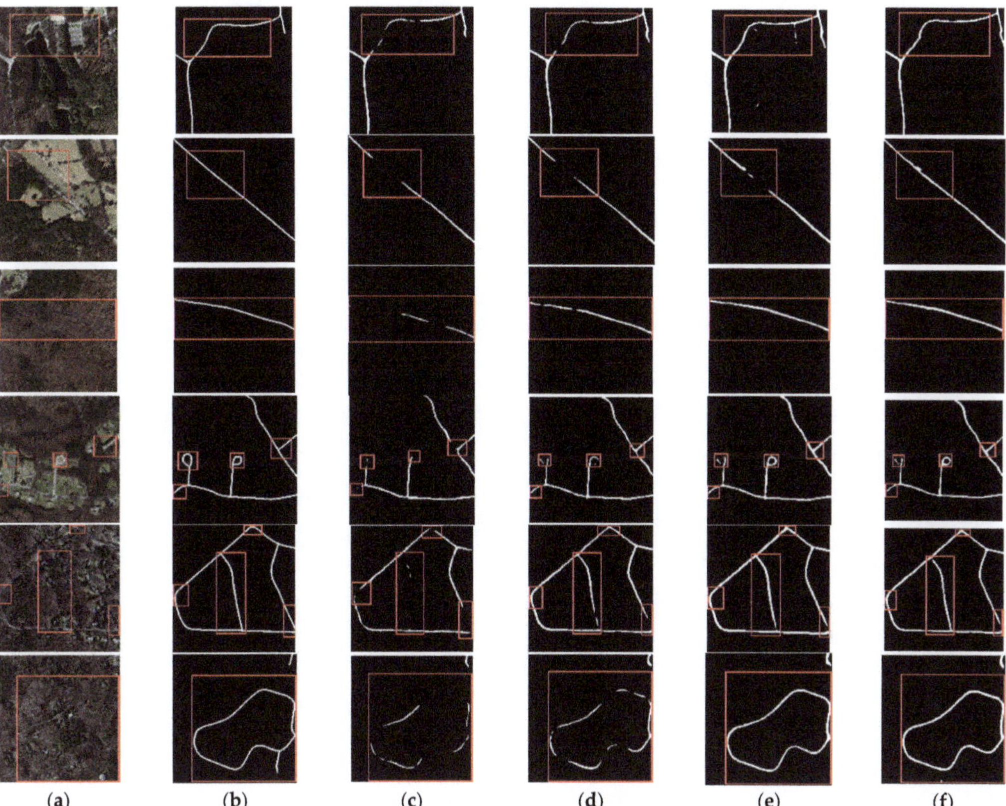

| (a) | (b) | (c) | (d) | (e) | (f) |

**Figure 9.** Comparison of adding MSPM and FEM. (**a**) Image; (**b**) Label; (**c**) Baseline; (**d**) Baseline + MSPM (L1); (**e**) Baseline + MSPM (L1 + L2); and (**f**) Baseline + MSPM (L1 + L2 + L3).( Red squares represent key areas.).

## 5.1.2. Comparison of Loss Function

In order to prove the effectiveness of the dice coefficient loss function and focal loss function, the baseline is MSPFE-Net. Table 3 shows the experimental results.

**Table 3.** Ablation experiments of MSPM and FEM.

| Networks | Precision (%) | Recall (%) | F1 (%) | IoU (%) |
| --- | --- | --- | --- | --- |
| Baseline + cross entropy loss function | 79.62% | 65.84% | 72.08% | 56.34% |
| Baseline + focal loss function | 77.12% | 68.04% | 72.30% | 56.61% |
| Baseline + dice coefficient loss function | 73.33% | 74.37% | 73.84% | 58.53% |

According to Table 3, after the addition of the focal loss function, the evaluation metrics have not changed much. When the dice coefficient loss function was used as a loss function alone, the evaluation metrics increased significantly, with IoU increasing by 2.19%.

## 6. Conclusions

MSPFE-Net is designed and implemented to extract roads, which can extract narrow roads and also restore roads that are covered by trees or shadows. When the road is similar to the background, the MSPFE-Net basically accurately extracts the road. MSPFE-Net ensures the connectedness and accuracy of the road. MSPFE-Net utilizes a multi-level strip pooling module to collect context information for road extraction. This module incorporates both horizontal and vertical strip pooling operations to gather context information of different levels and long-range dependencies. Due to the full acquisition of context information, the continuity of the road is improved. In areas with dense roads, the MSPFE-Net uses a feature enhancement module to collect local context information and enhance the segmentation effect of the road edge. Experimental results show that MSPFE-Net is better than other comparative models in experiments on evaluation metrics and results from images. Although MSPFE-Net has basically completed the task of road segmentation, roads are similar to the background in some areas, and there are also a few discontinuous roads

**Author Contributions:** Conceptualization, Z.W. and Z.Z.; methodology, Z.W.; software, Z.W.; validation, Z.W.; formal analysis, Z.W. and Z.Z.; investigation, Z.W. and Z.Z.; writing—original draft preparation, Z.W.; writing—review and editing, Z.W.; supervision, Z.Z. All authors have read and agreed to the published version of the manuscript.

**Funding:** This research received no external funding.

**Institutional Review Board Statement:** Not applicable.

**Informed Consent Statement:** Not applicable.

**Data Availability Statement:** Not applicable.

**Acknowledgments:** The authors appreciate the constructive comments and suggestions from the reviewers that helped improve the quality of this manuscript.

**Conflicts of Interest:** The authors declare no conflict of interest.

## References

1. Xu, Y.; Xie, Z.; Feng, Y.; Chen, Z. Road Extraction from High-Resolution Remote Sensing Imagery Using Deep Learning. *Remote Sens.* **2018**, *10*, 1461. [CrossRef]
2. Li, Y.; Guo, L.; Rao, J.; Xu, L.; Jin, S. Road Segmentation Based on Hybrid Convolutional Network for High-Resolution Visible Remote Sensing Image. *IEEE Geosci. Remote Sens. Lett.* **2019**, *16*, 613–617. [CrossRef]
3. Bong, D.; Lai, K.C.; Joseph, A. Automatic Road Network Recognition and Extraction for Urban Planning. *Int. J. Appl. Sci. Eng Technol.* **2009**, *5*, 209–215.
4. Hinz, S.; Baumgartner, A.; Ebner, H. Modeling Contextual Knowledge for Controlling Road Extraction in Urban Areas. In Proceedings of the IEEE/ISPRS Joint Workshop on Remote Sensing and Data Fusion over Urban Areas (Cat. No.01EX482), Rome, Italy, 8–9 November 2001; pp. 40–44.
5. Ma, H.; Lu, N.; Ge, L.; Li, Q.; You, X.; Li, X. Automatic Road Damage Detection Using High-Resolution Satellite Images and Road Maps. In Proceedings of the 2013 IEEE International Geoscience and Remote Sensing Symposium—IGARSS, Melbourne, VIC, Australia, 21–26 July 2013; pp. 3718–3721.
6. Li, Q.; Zhang, J.; Wang, N. Damaged Road Extraction from Post-Seismic Remote Sensing Images Based on Gis and Object-Oriented Method. In Proceedings of the 2016 IEEE International Geoscience and Remote Sensing Symposium (IGARSS), Beijing, China 10–15 July 2016; pp. 4247–4250.
7. Das, S.; Mirnalinee, T.T.; Varghese, K. Use of Salient Features for the Design of a Multistage Framework to Extract Roads From High-Resolution Multispectral Satellite Images. *IEEE Trans. Geosci. Remote Sens.* **2011**, *49*, 3906–3931. [CrossRef]
8. Lv, X.; Ming, D.; Chen, Y.; Wang, M. Very High Resolution Remote Sensing Image Classification with SEEDS-CNN and Scale Effect Analysis for Superpixel CNN Classification. *Int. J. Remote Sens.* **2018**, *40*, 506–531. [CrossRef]
9. Cheng, G.; Zhu, F.; Xiang, S.; Pan, C. Accurate Urban Road Centerline Extraction from VHR Imagery via Multiscale Segmentation and Tensor Voting. *Neurocomputing* **2016**, *205*, 407–420. [CrossRef]

10. Yang, M.; Yuan, Y.; Liu, G. SDU-Net: Road Extraction via Spatial Enhanced and Densely Connected U-Net. *Pattern Recognit.* **2022**, *126*, 108549. [CrossRef]

11. Xie, Y.; Miao, F.; Zhou, K.; Peng, J. HsgNet: A Road Extraction Network Based on Global Perception of High-Order Spatial Information. *ISPRS Int. J. Geo-Inf.* **2019**, *8*, 571. [CrossRef]

12. Zhang, X.; Ma, W.; Li, C.; Wu, J.; Tang, X.; Jiao, L. Fully Convolutional Network-Based Ensemble Method for Road Extraction From Aerial Images. *IEEE Geosci. Remote Sens. Lett.* **2020**, *17*, 1777–1781. [CrossRef]

13. Chen, Z.; Deng, L.; Luo, Y.; Li, D.; Marcato Junior, J.; Nunes Gonçalves, W.; Awal Md Nurunnabi, A.; Li, J.; Wang, C.; Li, D. Road Extraction in Remote Sensing Data: A Survey. *Int. J. Appl. Earth Obs. Geoinf.* **2022**, *112*, 102833. [CrossRef]

14. Hou, Q.; Zhang, L.; Cheng, M.-M.; Feng, J. Strip Pooling: Rethinking Spatial Pooling for Scene Parsing. In Proceedings of the 2020 IEEE/CVF Conference on Computer Vision and Pattern Recognition (CVPR), Seattle, WA, USA, 13–19 June 2020.

15. Raziq, A.; Xu, A.; Li, Y. Automatic Extraction of Urban Road Centerlines from High-Resolution Satellite Imagery Using Automa-tic Thresholding and Morphological Operation Method. *J. Geogr. Inf. Syst.* **2016**, *8*, 517–525. [CrossRef]

16. Haverkamp, D.S. Extracting Straight Road Structure in Urban Environments Using IKONOS Satellite Imagery. *Opt. Eng.* **2002**, *41*, 2107–2110. [CrossRef]

17. Wenfeng, W.; Shuhua, Z.; Yihao, F.; Weili, D. Parallel Edges Detection from Remote Sensing Image Using Local Orientation Co-ding. *Acta Opt. Sin.* **2012**, *32*, 0315001. [CrossRef]

18. Maboudi, M.; Amini, J.; Hahn, M.; Saati, M. Road Network Extraction from VHR Satellite Images Using Context Aware Object Feature Integration and Tensor Voting. *Remote Sens.* **2016**, *8*, 637. [CrossRef]

19. Mnih, V.; Hinton, G.E. Learning to Detect Roads in High-Resolution Aerial Images. In Proceedings of the Computer Vision—ECCV 2010, Heraklion, Greece, 5–11 September 2010; Daniilidis, K., Maragos, P., Paragios, N., Eds.; Springer: Berlin, Heidelberg, 2010; pp. 210–223.

20. Li, P.; Zang, Y.; Wang, C.; Li, J.; Cheng, M.; Luo, L.; Yu, Y. Road Network Extraction via Deep Learning and Line Integral Convolution. In Proceedings of the 2016 IEEE International Geoscience and Remote Sensing Symposium (IGARSS), Beijing, China, 10–15 July 2016; pp. 1599–1602.

21. Wei, Y.; Wang, Z.; Xu, M. Road Structure Refined CNN for Road Extraction in Aerial Image. *IEEE Geosci. Remote Sens. Lett.* **2017**, *14*, 709–713. [CrossRef]

22. Long, J.; Shelhamer, E.; Darrell, T. Fully Convolutional Networks for Semantic Segmentation. *IEEE Trans. Pattern Anal. Mach. Intell.* **2017**, *39*, 640–651.

23. Ronneberger, O.; Fischer, P.; Brox, T. U-Net: Convolutional Networks for Biomedical Image Segmentation. In Proceedings of the International Conference on Medical Image Computing and Computer-Assisted Intervention, Munich, Germany, 5–9 October 2015.

24. Singh, P.; Dash, R. A Two-Step Deep Convolution Neural Network for Road Extraction from Aerial Images. In Proceedings of the 2019 6th International Conference on Signal Processing and Integrated Networks (SPIN), Noida, India, 7–8 March 2019; pp. 660–664.

25. He, K.; Zhang, X.; Ren, S.; Sun, J. Deep Residual Learning for Image Recognition. In Proceedings of the 2016 IEEE Conference on Computer Vision and Pattern Recognition (CVPR), Las Vegas, NV, USA, 27–30 June 2016; pp. 770–778.

26. Zhang, Z.; Liu, Q.; Wang, Y. Road Extraction by Deep Residual U-Net. *IEEE Geosci. Remote Sens. Lett.* **2018**, *15*, 749–753. [CrossRef]

27. Gao, X.; Sun, X.; Zhang, Y.; Yan, M.; Xu, G.; Sun, H.; Jiao, J.; Fu, K. An End-to-End Neural Network for Road Extraction From Remote Sensing Imagery by Multiple Feature Pyramid Network. *IEEE Access* **2018**, *6*, 39401–39414. [CrossRef]

28. Cheng, G.; Wang, Y.; Xu, S.; Wang, H.; Xiang, S.; Pan, C. Automatic Road Detection and Centerline Extraction via Cascaded End-to-End Convolutional Neural Network. *IEEE Trans. Geosci. Remote Sens.* **2017**, *55*, 3322–3337. [CrossRef]

29. Yingxiao, X.; Chen, H.; Du, C.; Li, J. MSACon: Mining Spatial Attention-Based Contextual Information for Road Extraction. *IEEE Trans. Geosci. Remote Sens.* **2021**, *60*, 1–17. [CrossRef]

30. Yu, F.; Koltun, V. Multi-Scale Context Aggregation by Dilated Convolutions. *arXiv* **2015**, arXiv:1511.07122.

31. Zhou, L.; Zhang, C.; Wu, M. D-LinkNet: LinkNet with Pretrained Encoder and Dilated Convolution for High Resolution Satellite Imagery Road Extraction. In Proceedings of the 2018 IEEE/CVF Conference on Computer Vision and Pattern Recognition Workshops (CVPRW), IEEE, Salt Lake City, UT, USA, 18–22 June 2018; pp. 192–1924.

32. He, H.; Yang, D.; Wang, S.; Wang, S.; Li, Y. Road Extraction by Using Atrous Spatial Pyramid Pooling Integrated Encoder-Decoder Network and Structural Similarity Loss. *Remote Sens.* **2019**, *11*, 1015. [CrossRef]

33. Chen, R.; Hu, Y.; Wu, T.; Peng, L. Spatial Attention Network for Road Extraction. In Proceedings of the IGARSS 2020—2020 IEEE International Geoscience and Remote Sensing Symposium, Waikoloa, HI, USA, 26 September–2 October 2020; pp. 1841–1844.

34. Tao, C.; Qi, J.; Li, Y.; Wang, H.; Li, H. Spatial Information Inference Net: Road Extraction Using Road-Specific Contextual Information. *ISPRS J. Photogramm. Remote Sens.* **2019**, *158*, 155–166. [CrossRef]

35. Zhou, M.; Sui, H.; Chen, S.; Wang, J.; Chen, X. BT-RoadNet: A Boundary and Topologically-Aware Neural Network for Road Extraction from High-Resolution Remote Sensing Imagery. *ISPRS J. Photogramm. Remote Sens.* **2020**, *168*, 288–306. [CrossRef]

36. Lu, X.; Zhong, Y.; Zheng, Z.; Zhang, L. GAMSNet: Globally Aware Road Detection Network with Multi-Scale Residual Learning. *ISPRS J. Photogramm. Remote Sens.* **2021**, *175*, 340–352. [CrossRef]

37. Tan, X.; Xiao, Z.; Wan, Q.; Shao, W. Scale Sensitive Neural Network for Road Segmentation in High-Resolution Remote Sensing Images. *IEEE Geosci. Remote Sens. Lett.* **2021**, *18*, 533–537. [CrossRef]

38. Zhu, Q.; Zhang, Y.; Wang, L.; Zhong, Y.; Guan, Q.; Lu, X.; Zhang, L.; Li, D. A Global Context-Aware and Batch-Independent Network for Road Extraction from VHR Satellite Imagery. *ISPRS J. Photogramm. Remote Sens.* **2021**, *175*, 353–365. [CrossRef]
39. Wang, S.; Yang, H.; Wu, Q.; Zheng, Z.; Wu, Y.; Li, J. An Improved Method for Road Extraction from High-Resolution Remote-Sensing Images That Enhances Boundary Information. *Sensors* **2020**, *20*, 2064. [CrossRef]
40. Mnih, V. Machine Learning for Aerial Image Labeling. Ph.D. Thesis, University of Toronto, Toronto, ON, Canada, 2013.

*electronics*

*Article*

# Boundary-Aware Salient Object Detection in Optical Remote-Sensing Images

Longxuan Yu [1], Xiaofei Zhou [2,*], Lingbo Wang [2] and Jiyong Zhang [2,*]

[1]  School of Computer Science and Technology, Hangzhou Dianzi University, Hangzhou 310018, China
[2]  School of Automation, Hangzhou Dianzi University, Hangzhou 310018, China
*  Correspondence: zxforchid@outlook.com (X.Z.); jzhang@hdu.edu.cn (J.Z.)

**Abstract:** Different from the traditional natural scene images, optical remote-sensing images (RSIs) suffer from diverse imaging orientations, cluttered backgrounds, and various scene types. Therefore, the object-detection methods salient to optical RSIs require effective localization and segmentation to deal with complex scenarios, especially small targets, serious occlusion, and multiple targets. However, the existing models' experimental results are incapable of distinguishing salient objects and backgrounds using clear boundaries. To tackle this problem, we introduce boundary information to perform salient object detection in optical RSIs. Specifically, we first combine the encoder's low-level and high-level features (i.e., abundant local spatial and semantic information) via a feature-interaction operation, yielding boundary information. Then, the boundary cues are introduced into each decoder block, where the decoder features are directed to focus more on the boundary details and objects simultaneously. In this way, we can generate high-quality saliency maps which can highlight salient objects from optical RSIs completely and accurately. Extensive experiments are performed on a public dataset (i.e., ORSSD dataset), and the experimental results demonstrate the effectiveness of our model when compared with the cutting-edge saliency models.

**Keywords:** remote-sensing images; salient objects; boundary details; edge

**Citation:** Yu, L.; Zhou, X.; Wang, L.; Zhang, J. Boundary-Aware Salient Object Detection in Optical Remote-Sensing Images. *Electronics* **2022**, *11*, 4200. https://doi.org/10.3390/electronics11244200

Academic Editors: Yue Wu, Kai Qin, Maoguo Gong and Qiguang Miao

Received: 11 November 2022
Accepted: 10 December 2022
Published: 15 December 2022

**Publisher's Note:** MDPI stays neutral with regard to jurisdictional claims in published maps and institutional affiliations.

## 1. Introduction

Human vision systems tend to focus more on prominent areas in images, which is the visual attention mechanism, with many efforts made to design various methods to highlight salient objects in images or videos [1,2].

Formally, salient object detection is employed to automatically highlight the most visually distinctive regions in a scene [2], and has been applied to many research fields, such as object detection [3], image segmentation [4], image quality assessment [5], visual categorization [6], and medical image processing [7], to name a few.

Over the last twenty years, many saliency models have been designed [1,8]. The early efforts mainly focus on hand-crafted features, including the prior heuristic-based efforts [9–11] and the traditional machine-learning-based methods [12–14]. With the rapid development of deep-learning technologies, many deep-learning-based saliency models [15–21] have significantly improved the performance of saliency detection. Among the existing deep-learning-based models, some models [16–19] attempt to introduce edge information into their networks to provide precise boundary details for inference results. The current saliency models are obviously applicable to traditional RGB images (natural scenes) [22], RGB-T images [23], RGB-D images [24], light-field images [25], and optical remote-sensing images [26,27]. Among these, the saliency detection of optical remote-sensing images has gained increasing attention, because it has been widely employed in a variety of domains, including military, agriculture, and disaster relief.

However, there is a significant difference between the traditional natural-scene images and optical remote-sensing images. Salient objects in traditional natural-scene images are usually with high-contrast, a single object, and a prior center. In contrast, the objects in optical RSIs are usually of diverse types, various scales, and different orientations. Meanwhile, the optical RSIs have low contrast between salient objects and background. In addition, optical RSIs are usually photographed by high-altitude aircrafts or satellites. Recently, there have been many efforts devoted to this research field [26,28–31]. However, their performance degrades to some degree when dealing with some complex scenes.

Motivated by the aforementioned descriptions, we propose an innovative boundary-aware saliency model to detect salient objects in optical RSIs, as shown in Figure 1. Our model is built using an encoder–decoder architecture, and we focus on the extraction and usage of salient boundaries. Concretely, our model contains three modules, including the feature-extraction module (i.e., encoder), the edge module, and the feature-integration module (i.e., decoder). Particularly, our model first extracts salient boundaries via an edge module. Similarly to [16], the edge module simultaneously incorporates low-level and high-level deep features to generate boundary cues, where the low-level features convey spatial details and high-level deep features provide rich semantic information. In contrast to [16], our edge module also investigates the interaction effect [19] of salient boundaries and salient objects, which further elevates the boundary features and object features. Here, different from [19], we do not adopt stack modules to iteratively refine deep features. Instead, the salient boundaries are endowed with position information about salient objects, which promotes the completeness of saliency inference. Then, we pass on the boundary information to each decoding process, in which the salient boundary progressively refines the salient reasoning. In this way, we can obtain high-quality saliency maps to highlight salient regions in optical RSIs.

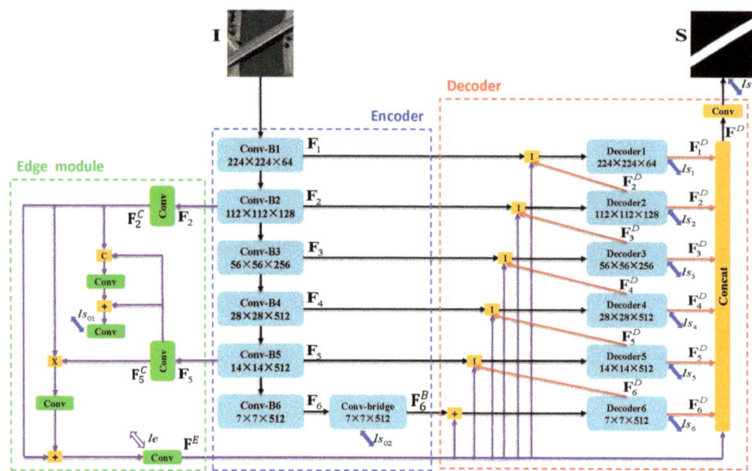


**Figure 1.** Illustration of the proposed saliency model.

Overall, the main contributions of the proposed network can be presented as follows:

1. We propose a novel boundary-aware saliency model for salient object detection, in which our model tries to introduce the salient boundaries to precisely segment the salient objects from optical RSIs.
2. We propose an effective edge module to provide boundary information for saliency detection, where boundary cues and object features are enhanced by the interaction between low-level spatial features and high-level semantic features.
3. Extensive experiments are performed on the public dataset ORSSD, and the results show that our model performs better than the state-of-the-art saliency models, which demonstrates the effectiveness of the proposed model.

The remaining of this paper is organized as follows. The related works are reviewed in Section 2. Section 3 gives a detailed introduction to the proposed saliency model. The experimental results and the analyses are detailed in Section 4. Finally, we draw a conclusion for this paper in Section 5.

## 2. Related Works

In recent years, we have seen significant advances in the research on salient object detection. Particularly, from the heuristic prior-based models [2,9,13] to the deep-learning-based models [17–19], significant efforts for salient object detection have been made. Meanwhile, due to the imaging method and image complexity, there is considerable difference between salient object detection in optical RSIs and salient object detection targeting natural-scene images. Therefore, this section mainly gives a brief review of the two different tasks.

### 2.1. Saliency Detection of Nature-Scene Images

Early efforts were usually constructed based on heuristic priors such as center prior, contrast prior, morphology prior, background prior, and so on. The pioneering work was proposed by Itti et al. [2], where the center-surround difference is designed to compute saliency scores using color, orientation, and intensity features. Subsequently, Cheng et al. [9] proposed a global contrast-based saliency model which gives a good segmentation for the prominent objects in natural images. In [32], Wei et al. paid more attention to the background rather than salient objects, and proposed a geodesic saliency model by exploiting boundary and connectivity priors of natural images. After that, machine-learning algorithms have been devoted to saliency models. For example, Liu et al. [33] attempted to aggregate multiple feature-based saliency maps using the conditional random forest. In [34], Huang et al. exploited multiple-instances learning algorithms to compute the saliency values of different regions. In [3], the Adaboost algorithm was adopted to fuse various hand-crafted feature-based saliency maps. Abdusalomov et al. [35] presented a unique saliency cutting method, which exploits local adaptive thresholding to generate four regions from a given saliency map.

Recently, deep-learning technologies have achieved remarkable progress, and they have also been applied to the saliency detection task. For example, Hou et al. [15] inserted short connections into the skip-layer structures of holistically nested edge detectors which effectively improve the quality of saliency inference results. In [22], Deng et al. proposed a recurrent residual refinement network to perform saliency detection, in which the residual refinement blocks learn the complementary saliency information by using low-level and high-level features. Qin et al. [36] designed a two-level nested U structure to capture more context information from different sizes of receptive fields, for which the rise in the network depth does not increase the memory and computation cost. In [37], Wang et al. exploited the recurrent fully convolutional networks to integrate saliency prior cues for more precise saliency prediction. In [38], Liu et al. designed a pooling-based structure to reduce the aliasing effect, where the global guidance module and the feature-aggregation module are deployed to progressively refine the high-level semantic information. Meanwhile, the authors also employed edge cues to sharpen the boundary details of salient objects. Similarly, in [16,17], the edge information was implicity and explicitly used to give a good depiction of the salient edges, respectively.

### 2.2. Saliency Detection of Optical Remote-Scene Images

Remarkable progress has been made in the saliency models aiming at natural-scene images, and lots of methods are proposed to enhance the effect of optical RSIs. In [28], Zhao et al. proposed a sparsity-guided saliency model which used bayesian theory to combine global and background features. In [29], the structure tensor and background contrast are employed to generate superpixel feature maps which are fused into the final pixel-level saliency map. In [39], Li et al. proposed a two-step building extraction method from remote-sensing images by fusing saliency information, of which the rooftops are more likely to attract visual attention than surrounding objects. In [40], the fully convolutional network is utilized to address the issue of inshore ship detection, where the deep layer of the network conducts detection and the shallow layer supplements with precise localization. In [30], Zhang et al. constructed a saliency-oriented active contour model, where the contour information is adopted to assist object detection. In [41], a self-adaptive multiple-feature fusion method is employed to explore the internal connection in optical RSIs, where the dual-tree complex wavelet transform is used to obtain the texture features. In [42], Liu et al. tried to adopt an unsupervised method to solve the oil-tank detection problem using the color Markov chain. In [43], a multi-level ship-detection method is proposed to detect various types of offshore ships using all possible imaging variations. Recently, in [26,27], the authors articulated further concerns regarding the integration of multi-level deep features, which can effectively highlight salient objects in optical RSIs.

Though the existing saliency models targeting natural-scene images have achieved satisfactory performance, it is inappropriate to extend the existing models to optical RSIs directly. The reason behind this lies in the differences between the two kinds of images. In addition, the existing saliency models targeting optical RSIs suffer from low-quality boundary details due to the neglecting of edge information. Therefore, in this paper, we attempt to integrate the edge cues into the entire network.

## 3. The Proposed Method

This part first gives an overview of the proposed saliency model in Section 3.1. Then, the detailed description of the proposed saliency model is presented in the following sections. Section 3.2 details the feature extraction (i.e., encoder). Section 3.3 presents the edge module. Section 3.4 provides the details of feature integration (i.e., decoder). Lastly, the training and implementation details will be outlined in Section 3.5.

### 3.1. Overall Architecture

The overall architecture of our model is shown in Figure 1. The main part of our model is constructed based on a U-shape structure. Given an optical remote-sensing image $\mathbf{I}$, we first feed it into the encoder part to extract multi-level features $\{\mathbf{F}_i\}_{i=1}^6$. Then, the feature $\mathbf{F}_6$ will feed into a bridge module "Conv-bridge" to further capture effective global semantic information $\mathbf{F}_6^B$. After that, we use the decoder to aggregate the multi-level deep features $\{\{\mathbf{F}_i\}_{i=1}^5, \mathbf{F}_6^B\}$, yielding the final saliency map $\mathbf{S}$. During the decoding process, to give a precise saliency inference, we introduce the edge information into the decoding process, where the edge feature $\mathbf{F}^E$ generated by the edge module combines with the multi-level deep features $\{\mathbf{F}_2, \mathbf{F}_5\}$.

### 3.2. Feature Extraction

Salient objects in optical RSIs vary different sizes. This phenomenon will degrade the performance of saliency models. Meanwhile, we find that many efforts [17,36] have sufficiently extracted and fused multi-level deep features including low-level information and high-level information. Inspired by SegNet [36] and BASNet [17], our saliency model, shown in Figure 1, is designed as an encoder–decoder structure with multi-level feature extraction. Meanwhile, following the deeply supervised efforts [16,44], the side outputs of our decoder are also supervised by the ground truth. The encoder of our model consists of six convolution blocks, where the first four blocks are the same as ResNet-34 [45]. The fifth block and the sixth block are all composed of three basic residual blocks [45] with 512 filters after a non-overlapping max pooling layer (size = 2). Based on the encoder, we can obtain six deep features $\{\mathbf{F}_i\}_{i=1}^6$. In addition, referring to BASNet [17], we also add a bridge module after the sixth convolutional block, which endows our model with more representative semantic features. The bridge module consists of three convolution layers with 512 dilated (dilation=2) $3\times3$ convolutional layers, and each convolution layer is followed by a batch-normalization (BN) layer and a ReLU activation function. Based on the bridge module, we can generate a more effective semantic feature $\mathbf{F}_6^B$. Following this encoder architecture, we can obtain the six levels of deep features $\{\{\mathbf{F}_i\}_{i=1}^{5}, \mathbf{F}_6^B\}$.

### 3.3. Edge Module

Edge information is useful for optimizing segmentation [16]. Many saliency models [16,17,38] have introduced edge information to conduct salient object detection, which gives accurate boundary details for saliency maps. Although the existing methods use edge information to guide the inference network, the extraction of edge features is not efficient. Furthermore, the existing models do not give sufficient usage of edge information. Therefore, we propose a novel edge module, which has two streams, as shown in Figure 1.

Specifically, the feature $\mathbf{F}_2$ from the second convolutional block and the feature $\mathbf{F}_5$ from the fifth convolutional block are first processed by convolutional layers, which can be written as

$$\begin{cases} \mathbf{F}_2^C = Conv(\mathbf{F}_2) \\ \mathbf{F}_5^C = Conv(\mathbf{F}_5) \end{cases}' \tag{1}$$

where $Conv$ means convolutional block.

Then, the convolutional block results $\mathbf{F}_2^C$ and $\mathbf{F}_5^C$ are used to generate the saliency and edge maps. Firstly, the feature $\mathbf{F}_2^C$ first concatenates with the feature $\mathbf{F}_5^C$, and then the concatenated feature is processed by a convolutional layer. In addition, the output of the convolutional layer and $\mathbf{F}_5^C$ are further combined via an element-wise summation. Lastly, we predict a saliency map $\mathbf{S}^*$ on the combined feature via convolutional layers. This process can be defined as follows

$$\mathbf{S}^* = \delta(Conv(Conv([\mathbf{F}_2^C, \mathbf{F}_5^C]) + \mathbf{F}_5^C)), \tag{2}$$

where $\delta$ denotes the sigmoid activation function.

In addition, the features $\mathbf{F}_2^C$ and $\mathbf{F}_5^C$ are first combined in an element-wise multiplication fashion, and then the multiplication result is processed by a convolutional layer. The fused feature further combines the feature $\mathbf{F}_2^C$. Finally, we predict the edge $\mathbf{E}$ from the fused feature via convolutional layers, which are further supervised by the salient edge maps. The entire process can be defined as

$$\mathbf{E} = \delta(Conv(Conv([\mathbf{F}_2^C \odot \mathbf{F}_5^C]) + \mathbf{F}_5^C)). \tag{3}$$

Therefore, in our edge module, the edge information not only contains spatial details but is also endowed with semantic information about salient objects. This is beneficial for the following decoding process (i.e., feature integration).

### 3.4. Feature Integration

Many saliency inference networks [17,46] adopt the encoder–decoder architecture to conduct salient object detection, which progressively recovers the spatial details of saliency maps and achieves promising results. Inspired by this, we also adopt the encoder–decoder architecture while introducing the edge information. Here, our decoder contains six decoder blocks, namely, Decoder$i$ ($i = 1, \cdots, 6$). Each decoder block consists of three convolutional blocks, where each convolutional block contains a convolutional layer, a batch-normalization (BN) layer, and a ReLU layer. During the decoding process, the input of each decoder block is the decoder feature $\{\mathbf{F}_i^D\}_{i=2}^{6}$ from the previous decoder block and the encoder feature $\{\mathbf{F}_i\}_{i=1}^{5}$ from the current-level encoder block. Meanwhile, to endow our model with accurate spatial details, we combine the edge information $\mathbf{F}^E$ with each encoder feature, and the enhanced encoder feature will take part in the decoding process. Therefore, we define each decoding process as follows

$$\mathbf{F}_i^D = Conv([\mathbf{F}_{i+1}^D, \mathbf{F}_i, \mathbf{F}^E]). \tag{4}$$

In this way, we can obtain six levels of decoder features, namely, $\{\mathbf{F}_i^D\}_{i=1}^6$. After that, we combine all decoder features and the edge cues, which can be written as

$$\mathbf{F}_D = [\mathbf{F}_1^D, \mathbf{F}_2^D, \mathbf{F}_3^D, \mathbf{F}_4^D, \mathbf{F}_5^D, \mathbf{F}_6^D, \mathbf{F}^E], \tag{5}$$

where $\mathbf{F}_D$ is the fused feature. Based on $\mathbf{F}_D$, we deploy a 3×3 convolutional layer and sigmoid activation function to generate the final saliency map **S**. The entire process can be defined as

$$\mathbf{S} = \delta(Conv(\mathbf{F}_D)). \tag{6}$$

*3.5. Model Learning and Implementation*

Deep supervision has been successfully adopted by many vision tasks [36,47], where the deep supervision can promote the training process and improve the performance of saliency models [44,48]. Inspired by the existing saliency object-detection models [17,38,44], we give deep supervision for six decoder blocks using the hybrid loss [17]. The total loss $l$ of our model can be denoted as:

$$l = \sum_i (l_{bce}^i + l_{IoU}^i + l_{ssim}^i + l_{bce}^{e,i}). \tag{7}$$

where $l_{bce}^i$, $l_{IoU}^i$ and $l_{ssim}^i$ denote the BCE loss, IoU loss and SSIM loss of the $i$th sample. $l_{bce}^{e,i}$ is used to compute the edge loss.

To train our model, we adopt the same training set as LV-Net [26], where 600 images selected from the ORSSD dataset [26] are used for the training set and the remaining 200 images are treated as the testing set. Furthermore, to train the proposed model, the training set is augmented by performing rotation with angles 90, 180, and 270 and conducting flipping on the rotated images. Following this, the training set contains 4800 samples.

We implemented our model with Pytorch on a PC with an Intel i7-6700 CPU, 32GB RAM, and a NVIDIA GeForce RTX2080Ti (with 11GB memory). We set our epoch number and batch size to 200 and 4, respectively. The input images were resized to $256 \times 256$. Our optimizer is Adam, where the initial learning rate lr = $10^{-3}$, betas = (0.9, 0.999), eps = $10^{-8}$, and weight decay = 0.

## 4. Experimental Results

This section first presents the ORSSD datasets and evaluation metrics in Section 4.1 Then, in Section 4.2, we compare our model with the state-of-the-art optical RSIs saliency models from the perspective of quantitative and qualitative views. Lastly, the detailed ablation studies are shown in Section 4.3.

*4.1. Datasets and Evaluation Metrics*

To comprehensively validate our model, we adopt the public challenging optical RSIs dataset, namely, ORSSD [26]. Concretely, the ORSSD dataset contains 600 images with pixel-wise annotations. The images have diverse resolutions such as $256 \times 256$, $300 \times 300$ and $800 \times 600$. They contain lots of scenes, such as house, airplane, car, ship, bridge, sea river, and bay, etc.

To quantitatively compare all the models, we employed four evaluation metrics, S-measure (S) [49], max F-measure (maxF), max E-measure (maxE) [50] and mean absolute error (MAE), to evaluate the performance of all models.

To perform a subjective comparison of all models, we employed a method of subjective comparison. Concretely, we first randomly selected some images and their corresponding ground truths. Then, we visually presented the saliency maps of our model and other state-of-the-art models.

*4.2. Comparison with the State of the Art*

To validate the performance of our model, we drew a comparison between our model and 19 state-of-the-art saliency models containing five optical RSIs saliency models (PDFNet [27], LVNet [26], SSD [28], SPS [29], ASD [30]); four unsupervised saliency models majoring in natural-scene images (RBD [11], RCRR [14], DSG [51], MILPS [34]); and 10 deep-learning-based saliency models targeting natural-scene images (R3Net [22], DSS [15], RADF [52], RFCN [37], PoolNet [38], BASNet [17], EGNet [16], CPD [18], SCRN [19], U2Net [36]) on the ORSSD dataset. Meanwhile, for a fair comparison, we retrained the existing deep-learning-based models by running the source codes or obtaining the results provided by the authors. Next, we show the quantitative and qualitative comparisons, successively.

Table 1 reports the quantitative results of our model and the 19 latest methods on the benchmark dataset. According to the evaluation results, it can be seen that our model performs best. Specifically, the performance of our model is better than the top-two models including SCRN, PDFNet in terms of S-measure and MAE, and our model performs slightly lower than SCRN in terms of F-measure and E-measure. In addition, we also present the PR curves and F-measure curves of different models in Figure 2. It can be clearly seen that our model outperforms other models.

**Table 1.** Quantitative comparison results of S-measure, max F-measure, max E-measure, and MAE on the ORSSD dataset. Here, "↑" ("↓") means that the larger (smaller) the better. The best three results in each row are marked in red, green, and blue, respectively.

| | ORSSD Dataset | | | |
|---|---|---|---|---|
| | $S \uparrow$ | $F_\beta \uparrow$ | $E_\beta \uparrow$ | $MAE \downarrow$ |
| PDFNet [27] | 0.9112 | 0.8726 | 0.9608 | 0.0149 |
| LVNet [26] | 0.8815 | 0.8263 | 0.9456 | 0.0207 |
| SSD [28] | 0.5838 | 0.4460 | 0.7052 | 0.1126 |
| SPS [29] | 0.5758 | 0.3820 | 0.6472 | 0.1233 |
| ASD [30] | 0.5477 | 0.4701 | 0.7448 | 0.2119 |
| RBD [11] | 0.7662 | 0.6579 | 0.8501 | 0.0626 |
| RCRR [14] | 0.6849 | 0.5591 | 0.7651 | 0.1277 |
| DSG [51] | 0.7195 | 0.6238 | 0.7912 | 0.1041 |
| MILPS [34] | 0.7361 | 0.6519 | 0.8265 | 0.0913 |
| R3Net [22] | 0.8141 | 0.7456 | 0.8913 | 0.0399 |
| DSS [15] | 0.8262 | 0.7467 | 0.8860 | 0.0363 |
| RADF [52] | 0.8259 | 0.7619 | 0.9130 | 0.0382 |
| RFCN [37] | 0.8437 | 0.7742 | 0.9157 | 0.0293 |
| PoolNet [38] | 0.8551 | 0.8229 | 0.9368 | 0.0293 |
| BASNet [17] | 0.8963 | 0.8282 | 0.9346 | 0.0204 |
| EGNet [16] | 0.8774 | 0.8187 | 0.9165 | 0.0308 |
| CPD [18] | 0.8627 | 0.8033 | 0.9115 | 0.0297 |
| SCRN [19] | 0.9061 | 0.8846 | 0.9647 | 0.0157 |
| U2Net [36] | 0.9162 | 0.8738 | 0.9539 | 0.0166 |
| **Ours** | 0.9233 | 0.8786 | 0.9581 | 0.0120 |

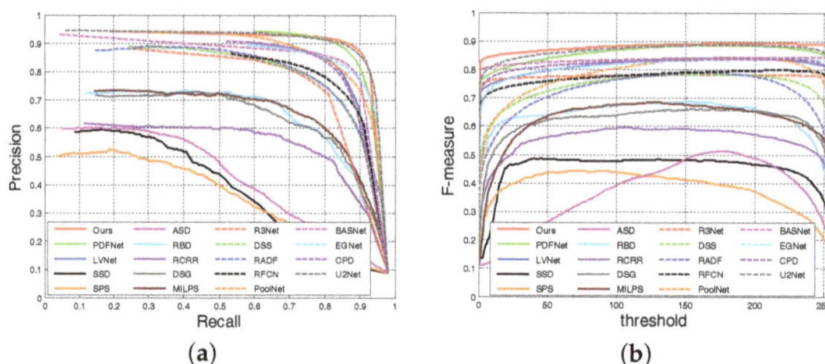

**Figure 2.** (Better viewed in color) quantitative evaluation of different saliency models: (**a**) P-R curves of different methods of the ORSSD dataset, and (**b**) F-measure curves of different methods of the ORSSD dataset.

In addition, Figure 3 shows the qualitative analysis of different models. It can be seen that our model can more completely highlight irregular objects and multiple objects. For example, in Figure 3f,g, the large object is not completely detected, whereas the saliency map is incomplete. In stark contrast, our model can give an accurate saliency prediction, and the object is completely highlighted. In Figure 3m,q,r, the saliency maps cannot completely highlight salient objects. Similarly, in Figure 3, the existing models cannot highlight the fifth row of objects, whereas the saliency maps only detect parts of salient objects. Meanwhile, we can find that our model, Figure 3c, can completely and accurately highlight salient objects. This is mainly a benefit of the edge module of our model, which provides accurate edge information for salient objects.

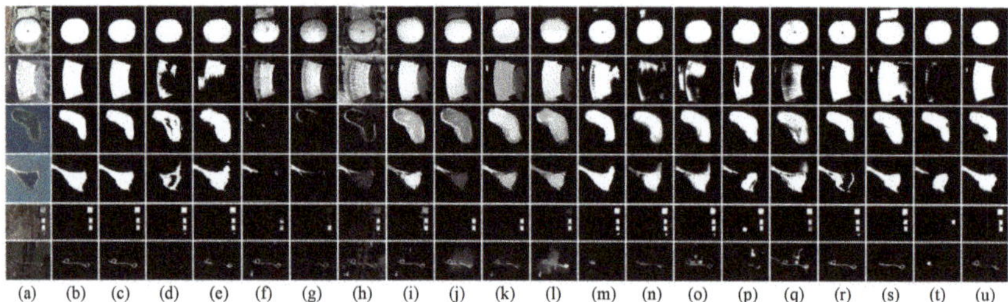

**Figure 3.** Visualization comparison of different optical RSI saliency models on several challenging scenes. (**a**): RGB, (**b**): GT, (**c**): Ours, (**d**): PDFNet, (**e**): LVNet, (**f**): RBD, (**g**): RCRR, (**h**): DSG, (**i**): MILPS, (**j**): SSD, (**k**): SPS, (**l**): ASD, (**m**): R3Net, (**n**): DSS, (**o**): RADF, (**p**): RFCN, (**q**): PoolNet (**r**): BASNet (**s**): EGNet (**t**): CPD, (**u**): U2Net.

### 4.3. Ablation Studies

This section profoundly analyzes some important components of our model through quantitative and qualitative comparisons. Specifically, the crucial components of our model include the edge module and the fusion module. Our model without edge information is denoted as "w/o Edge". Our model does not fuse the side outputs of the decoder, and performs saliency inference on the first decoder block, which is denoted as "w/o Fusion". In addition, we also explore the BCE loss in the supervision of saliency maps, and, thus, we remove the BCE loss in the hybrid loss, which is marked as "w/o bceloss".

According to the quantitative comparison results shown in Table 2 and Figure 4, we can find that our model outperforms the three variations including w/o Edge, w/o Fusion, and w/o bceloss in terms of S-measure, max F-measure, max E-measure, and MAE. From the qualitative comparison results shown in Figure 5, we can find that the Edge module, Fusion module, and BCE loss can effectively improve the performance of our model to a certain extent. This clearly demonstrates the effectiveness of the three components in our model.

**Table 2.** Ablation analysis on ORSSD dataset. The best result is marked in **boldface**. ↑ and ↓ represent smaller and larger is better.

| Model | $S-measure$ ↑ | $MAE$ ↓ | $maxE$ ↑ | $maxF$ ↑ |
|---|---|---|---|---|
| *w/o bceloss* | 0.8980 | 0.0182 | 0.9482 | 0.8612 |
| *w/o Edge* | 0.9054 | 0.0160 | 0.9501 | 0.8523 |
| *w/o Fusion* | 0.9142 | 0.0149 | 0.9514 | 0.8690 |
| ***Ours*** | **0.9233** | **0.0120** | **0.9581** | **0.8786** |

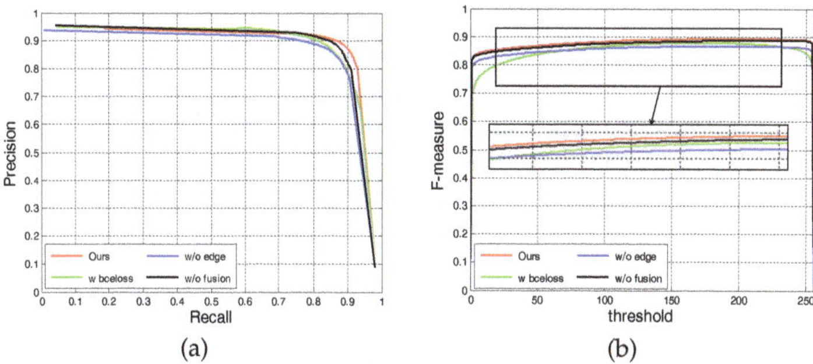

**Figure 4.** (better viewed in color) Quantitative evaluation of our network and ablation network: (**a**) P-R curves, and (**b**) F-measure curves.

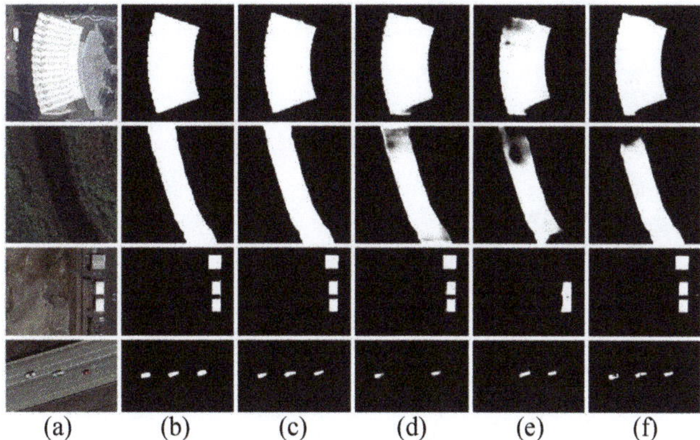

(a)　(b)　(c)　(d)　(e)　(f)

**Figure 5.** Qualitative Visual results of ablation analysis. (**a**): RGB, (**b**): GT, (**c**): Ours, (**d**): w/o Edge, (**e**): w/o Fusion, (**f**): w/o bceloss.

### 4.4. Failure Cases and Analysis

According to the aforementioned descriptions, the proposed model can accurately highlight salient objects in optical RSIs. However, our model is still incapable of generating satisfactory results when dealing with the different scales of salient objects shown in Figure 6. For instance, the two examples in the first and second rows of Figure 6 present two salient objects, i.e., bridges and tiny vehicles. As presented in Figure 6c, our model falsely highlights the background regions around the salient objects. It can be seen that the predicted saliency maps cannot completely highlight salient objects when dealing with the tiny white vehicles. For the bottom two examples in Figure 6a, a roof with patterned lines and a stadium with shadows are where the contrast between salient objects and background is low. As presented in Figure 6c, our model is incapable of highlighting salient objects. Therefore, we can conclude that the scene with different-sized salient objects are still challenging for our model. To address this issue, we should pay more attention to the design of the effective integration methods for multi-level deep features, providing more discriminative representations for salient objects.

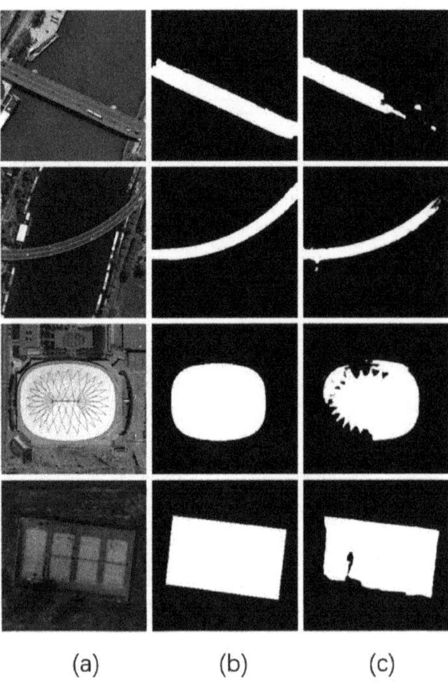

**Figure 6.** Some failure examples. (**a**) Optical RSIs. (**b**) Ground truth. (**c**) Saliency maps generated by our model.

## 5. Conclusions

This paper introduces the boundary information into our model for salient object detection in optical RSIs. Concretely, the edge module is first designed to acquire the edge cues. Here, we combine the low-level feature and the high-level feature to interactively obtain the edge features. Then, we endow the generated multi-level deep features with the edge cues, using the edge information to enhance the decoding process. This can direct the features and give more options for salient regions in optical RSIs. Following this, we can obtain high-quality saliency maps which can highlight salient objects from optical RSIs entirely and accurately. Experiments are conducted on the public dataset and the comprehensive comparison results show that our model performs better than the state-of-the-art models. In our future work, we will address more concerns on designing more effective saliency models targeting optical RSIs, which will be endowed with powerful characterization ability for salient objects and equipped with an effective feature fusion module.

**Author Contributions:** Methodology, X.Z.; Software, L.Y. and L.W.; Supervision, J.Z.; Writing—original draft, L.Y. and L.W.; Writing—review and editing, L.Y. and X.Z. All authors have read and agreed to the published version of the manuscript.

**Funding:** This research was supported in part by the National Natural Science Foundation of China under Grants 62271180 and 61901145, in part by the Fundamental Research Funds for the

Provincial Universities of Zhejiang under Grants GK229909299001-009, in part by the Zhejiang Province Nature Science Foundation of China under Grant LZ22F020003, and in part by the Hangzhou Dianzi University (HDU) and the China Electronics Corporation DATA (CECDATA) Joint Research Center of Big Data Technologies under Grant KYH063120009.

**Conflicts of Interest:** The authors declare no conflict of interest.

# References


1. Borji, A.; Cheng, M.M.; Hou, Q.; Jiang, H.; Li, J. Salient Object Detection: A Survey. *Comput. Vis. Pattern Recognit.* **2014**, *5*, 117–150. [CrossRef]
2. Itti, L.; Koch, C.; Niebur, E. A model of saliency-based visual attention for rapid scene analysis. *IEEE Trans. Pattern Anal. Mach. Intell.* **1998**, *20*, 1254–1259. [CrossRef]
3. Zhou, X.; Liu, Z.; Sun, G.; Ye, L.; Wang, X. Improving saliency detection via multiple kernel boosting and adaptive fusion. *IEEE Signal Process. Lett.* **2016**, *23*, 517–521. [CrossRef]
4. Chen, L.C.; Papandreou, G.; Kokkinos, I.; Murphy, K.; Yuille, A.L. Deeplab: Semantic image segmentation with deep convolutional nets, atrous convolution, and fully connected crfs. *IEEE Trans. Pattern Anal. Mach. Intell.* **2017**, *40*, 834–848. [CrossRef] [PubMed]
5. Wang, X.; Ma, L.; Kwong, S.; Zhou, Y. Quaternion representation based visual saliency for stereoscopic image quality assessment. *Signal Process.* **2018**, *145*, 202–213. [CrossRef]
6. Han, J.; Yao, X.; Cheng, G.; Feng, X.; Xu, D. P-CNN: Part-Based Convolutional Neural Networks for Fine-Grained Visual Categorization. *IEEE Trans. Pattern Anal. Mach. Intell.* **2020**, *44*, 579–590. [CrossRef]
7. Khosravan, N.; Celik, H.; Turkbey, B.; Cheng, R.; McCreedy, E.; McAuliffe, M.; Bednarova, S.; Jones, E.; Chen, X.; Choyke, P.; et al. Gaze2Segment: A pilot study for integrating eye-tracking technology into medical image segmentation. In *Medical Computer Vision and Bayesian and Graphical Models for Biomedical Imaging*; Springer: Berlin/Heidelberg, Germany, 2016; pp. 94–104.
8. Wang, W.; Lai, Q.; Fu, H.; Shen, J.; Ling, H.; Yang, R. Salient object detection in the deep learning era: An in-depth survey. *IEEE Trans. Pattern Anal. Mach. Intell.* **2021**, *44*, 3239–3259. [CrossRef]
9. Cheng, M.M.; Zhang, G.X.; Mitra, N.J.; Huang, X.; Hu, S.M. Global contrast based salient region detection. In Proceedings of the IEEE Conference on Computer Vision and Pattern Recognition (CVPR), Colorado Springs, CO, USA, 20–25 June 2011 ; IEEE: New York, NY, USA, 2011; pp. 409–416.
10. Perazzi, F.; Krähenbühl, P.; Pritch, Y.; Hornung, A. Saliency filters: Contrast based filtering for salient region detection. In Proceedings of the IEEE Conference on Computer Vision and Pattern Recognition (CVPR), Providence, RI, USA, 16–21 June 2012; IEEE: New York, NY, USA, 2012; pp. 733–740.
11. Zhu, W.; Liang, S.; Wei, Y.; Sun, J. Saliency optimization from robust background detection. In Proceedings of the IEEE Conference on Computer Vision and Pattern Recognition (CVPR), Columbus, OH, USA, 23–28 June 2014; IEEE: New York, NY, USA, 2014; pp. 2814–2821.
12. Tong, N.; Lu, H.; Ruan, X.; Yang, M.H. Salient object detection via bootstrap learning. In Proceedings of the IEEE Conference on Computer Vision and Pattern Recognition (CVPR), Columbus, OH, USA, 23–28 June 2014; IEEE: New York, NY, USA, 2015; pp. 1884–1892.
13. Jiang, H.; Wang, J.; Yuan, Z.; Wu, Y.; Zheng, N.; Li, S. Salient object detection: A discriminative regional feature integration approach. In Proceedings of the IEEE Conference on Computer Vision and Pattern Recognition (CVPR), Portland, OR, USA, 23–28 June 2013; IEEE: New York, NY, USA, 2013; pp. 2083–2090.
14. Yuan, Y.; Li, C.; Kim, J.; Cai, W.; Feng, D.D. Reversion correction and regularized random walk ranking for saliency detection. *IEEE Trans. Image Process.* **2017**, *27*, 1311–1322. [CrossRef]
15. Hou, Q.; Cheng, M.M.; Hu, X.; Borji, A.; Tu, Z.; Torr, P.H. Deeply supervised salient object detection with short connections. In Proceedings of the IEEE Conference on Computer Vision and Pattern Recognition (CVPR), Honolulu, HI, USA, 21–26 July 2017; IEEE: New York, NY, USA, 2017; pp. 3203–3212.
16. Zhao, J.X.; Liu, J.J.; Fan, D.P.; Cao, Y.; Yang, J.; Cheng, M.M. EGNet: Edge guidance network for salient object detection. In Proceedings of the International Conference on Computer Vision (ICCV), Seoul, Republic of Korea, 27 October–2 November 2019; IEEE: New York, NY, USA, 2019; pp. 8779–8788.
17. Qin, X.; Zhang, Z.; Huang, C.; Gao, C.; Dehghan, M.; Jagersand, M. Basnet: Boundary-aware salient object detection. In Proceedings of the IEEE Conference on Computer Vision and Pattern Recognition (CVPR), Long Beach, CA, USA, 15–20 June 2019; IEEE: New York, NY, USA, 2019; pp. 7479–7489.
18. Wu, Z.; Su, L.; Huang, Q. Cascaded partial decoder for fast and accurate salient object detection. In Proceedings of the IEEE Conference on Computer Vision and Pattern Recognition (CVPR), Long Beach, CA, USA, 15–20 June 2019; IEEE: New York, NY, USA, 2019; pp. 3907–3916.
19. Wu, Z.; Su, L.; Huang, Q. Stacked cross refinement network for edge-aware salient object detection. In Proceedings of the International Conference on Computer Vision (ICCV), Seoul, Republic of Korea, 27 October–2 November 2019; IEEE: New York, NY, USA, 2019; pp. 7264–7273.
20. Chen, C.; Wei, J.; Peng, C.; Zhang, W.; Qin, H. Improved Saliency Detection in RGB-D Images Using Two-Phase Depth Estimation and Selective Deep Fusion. *IEEE Trans. Image Process.* **2020**, *29*, 4296–4307. [CrossRef]

21. Chen, C.; Wang, G.; Peng, C.; Fang, Y.; Zhang, D.; Qin, H. Exploring Rich and Efficient Spatial Temporal Interactions for Real-Time Video Salient Object Detection. *IEEE Trans. Image Process.* **2021**, *30*, 3995–4007. [CrossRef]

22. Deng, Z.; Hu, X.; Zhu, L.; Xu, X.; Qin, J.; Han, G.; Heng, P.A. R3net: Recurrent residual refinement network for saliency detection. In Proceedings of the 27th International Joint Conference on Artificial Intelligence (IJCAI), Stockholm, Sweden, 13–19 July 2018; AAAI Press: Menlo Park, CA, USA, 2018; pp. 684–690.

23. Tu, Z.; Xia, T.; Li, C.; Wang, X.; Ma, Y.; Tang, J. RGB-T Image Saliency Detection via Collaborative Graph Learning. *IEEE Trans. Multimed.* **2020** , *22*, 160–173. [CrossRef]

24. Cong, R.; Lei, J.; Fu, H.; Huang, Q.; Cao, X.; Hou, C. Co-saliency Detection for RGBD Images Based on Multi-constraint Feature Matching and Cross Label Propagation. *IEEE Trans. Image Process.* **2017**, *27*, 568–579. [CrossRef] [PubMed]

25. Piao, Y.; Li, X.; Zhang, M.; Yu, J.; Lu, H. Saliency Detection via Depth-Induced Cellular Automata on Light Field. *IEEE Trans. Image Process.* **2020**, *29*, 1879–1889. [CrossRef] [PubMed]

26. Li, C.; Cong, R.; Hou, J.; Zhang, S.; Qian, Y.; Kwong, S. Nested network with two-stream pyramid for salient object detection in optical remote sensing images. *IEEE Trans. Geosci. Remote Sens.* **2019**, *57*, 9156–9166. [CrossRef]

27. Li, C.; Cong, R.; Guo, C.; Li, H.; Zhang, C.; Zheng, F.; Zhao, Y. A parallel down-up fusion network for salient object detection in optical remote sensing images. *Neurocomputing* **2020**, *415*, 411–420. [CrossRef]

28. Zhao, D.; Wang, J.; Shi, J.; Jiang, Z. Sparsity-guided saliency detection for remote sensing images. *J. Appl. Remote Sens.* **2015**, *9*, 095055. [CrossRef]

29. Ma, L.; Du, B.; Chen, H.; Soomro, N.Q. Region-of-interest detection via superpixel-to-pixel saliency analysis for remote sensing image. *IEEE Geosci. Remote Sens. Lett.* **2016**, *13*, 1752–1756. [CrossRef]

30. Zhang, Q.; Zhang, L.; Shi, W.; Liu, Y. Airport extraction via complementary saliency analysis and saliency-oriented active contour model. *IEEE Geosci. Remote Sens. Lett.* **2018**, *15*, 1085–1089. [CrossRef]

31. Zhang, Q.; Cong, R.; Li, C.; Cheng, M.M.; Fang, Y.; Cao, X.; Zhao, Y.; Kwong, S. Dense Attention Fluid Network for Salient Object Detection in Optical Remote Sensing Images. *IEEE Trans. Image Process.* **2020**, *30*, 1305–1317. [CrossRef]

32. Wei, Y.; Wen, F.; Zhu, W.; Sun, J. Geodesic saliency using background priors. In Proceedings of the European Conference on Computer Vision, Florence, Italy, 7–13 October 2012; Springer: Berlin/Heidelberg, Germany, 2012; pp. 29–42.

33. Liu, T.; Yuan, Z.; Sun, J.; Wang, J.; Zheng, N.; Tang, X.; Shum, H.Y. Learning to detect a salient object. *IEEE Trans. Pattern Anal. Mach. Intell.* **2010**, *33*, 353–367.

34. Huang, F.; Qi, J.; Lu, H.; Zhang, L.; Ruan, X. Salient object detection via multiple instance learning. *IEEE Trans. Image Process.* **2017**, *26*, 1911–1922. [CrossRef] [PubMed]

35. Abdusalomov, A.; Mukhiddinov, M.; Djuraev, O.; Khamdamov, U.; Whangbo, T.K. Automatic salient object extraction based on locally adaptive thresholding to generate tactile graphics. *Appl. Sci.* **2020**, *10*, 3350. [CrossRef]

36. Qin, X.; Zhang, Z.; Huang, C.; Dehghan, M.; Zaiane, O.R.; Jagersand, M. U2-Net: Going deeper with nested U-structure for salient object detection. *Pattern Recognit.* **2020**, *106*, 107404. [CrossRef]

37. Wang, L.; Wang, L.; Lu, H.; Zhang, P.; Ruan, X. Salient object detection with recurrent fully convolutional networks. *IEEE Trans. Pattern Anal. Mach. Intell.* **2018**, *41*, 1734–1746. [CrossRef]

38. Liu, J.J.; Hou, Q.; Cheng, M.M.; Feng, J.; Jiang, J. A simple pooling-based design for real-time salient object detection. In Proceedings of the IEEE Conference on Computer Vision and Pattern Recognition (CVPR), Long Beach, CA, USA, 15–20 June 2019; IEEE: New York, NY, USA, 2019; pp. 3917–3926.

39. Li, E.; Xu, S.; Meng, W.; Zhang, X. Building extraction from remotely sensed images by integrating saliency cue. *IEEE J. Sel. Top. Appl. Earth Obs. Remote Sens.* **2016**, *10*, 906–919. [CrossRef]

40. Lin, H.; Shi, Z.; Zou, Z. Fully convolutional network with task partitioning for inshore ship detection in optical remote sensing images. *IEEE Geosci. Remote Sens. Lett.* **2017**, *14*, 1665–1669. [CrossRef]

41. Zhang, L.; Liu, Y.; Zhang, J. Saliency detection based on self-adaptive multiple feature fusion for remote sensing images. *Int. J. Remote Sens.* **2019**, *40*, 8270–8297. [CrossRef]

42. Liu, Z.; Zhao, D.; Shi, Z.; Jiang, Z. Unsupervised Saliency Model with Color Markov Chain for Oil Tank Detection. *Remote Sens.* **2019**, *11*, 1089. [CrossRef]

43. Dong, C.; Liu, J.; Xu, F.; Liu, C. Ship Detection from Optical Remote Sensing Images Using Multi-Scale Analysis and Fourier HOG Descriptor. *Remote Sens.* **2019**, *11*, 1529. [CrossRef]

44. Liu, Y.; Cheng, M.M.; Zhang, X.Y.; Nie, G.Y.; Wang, M. DNA: Deeply supervised nonlinear aggregation for salient object detection. *IEEE Trans. Cybern.* **2021**. 10.1109/TCYB.2021.3051350. [CrossRef]

45. He, K.; Zhang, X.; Ren, S.; Sun, J. Deep residual learning for image recognition. In Proceedings of the IEEE Conference on Computer Vision and Pattern Recognition (CVPR), Las Vegas, NA, USA, 27–30 June 2016; IEEE: New York, NY, USA, 2016; pp. 770–778.

46. Ronneberger, O.; Fischer, P.; Brox, T. U-Net: Convolutional Networks for Biomedical Image Segmentation. *Med. Image Comput. Comput. Assist. Interv.* **2015**, *9351*, 234–241.

47. Liu, N.; Han, J.; Yang, M.H. PiCANet: Learning Pixel-wise Contextual Attention for Saliency Detection. *Comput. Vis. Pattern Recognit.* **2017**, 3089–3098.

48. Xie, S.; Tu, Z. Holistically-Nested Edge Detection. *Int. J. Comput. Vis.* **2015**, 1395–1403.

.   Fan, D.P.; Cheng, M.M.; Liu, Y.; Li, T.; Borji, A. Structure-measure: A new way to evaluate foreground maps. In Proceedings of the International Conference on Computer Vision (ICCV), Venice, Italy, 22–29 October 2017; pp. 4548–4557.

.   Fan, D.P.; Gong, C.; Cao, Y.; Ren, B.; Cheng, M.M.; Borji, A. Enhanced-alignment Measure for Binary Foreground Map Evaluation. In Proceedings of the 27th International Joint Conference on Artificial Intelligence (IJCAI), Stockholm, Sweden, 13–19 July 2018; AAAI Press: Menlo Park, CA, USA, 2018; pp. 698–704.

.   Zhou, L.; Yang, Z.; Zhou, Z.; Hu, D. Salient region detection using diffusion process on a two-layer sparse graph. *IEEE Trans. Image Process.* **2017**, *26*, 5882–5894. [CrossRef]

.   Hu, X.; Zhu, L.; Qin, J.; Fu, C.W.; Heng, P.A. Recurrently aggregating deep features for salient object detection. In Proceedings of the Thirty-second AAAI conference on Artificial Intelligence (AAAI), New Orleans, LA, USA, 2–7 February 2018.

 *remote sensing*

*Article*

# Improved One-Stage Detectors with Neck Attention Block for Object Detection in Remote Sensing

Kaiqi Lang [1,2], Mingyu Yang [1], Hao Wang [1], Hanyu Wang [1,2], Zilong Wang [1,2], Jingzhong Zhang [3] and Honghai Shen [1,*]

1   Key Laboratory of Airborne Optical Imaging and Measurement, Changchun Institute of Optics, Fine Mechanics and Physics, Chinese Academy of Sciences, Changchun 130033, China
2   University of Chinese Academy of Sciences, Beijing 100049, China
3   Forest Protection Research Institute of Heilongjiang Province, Harbin 150040, China
*   Correspondence: shenhh@ciomp.ac.cn

**Abstract:** Object detection in remote sensing is becoming a conspicuous challenge with the rapidly increasing quantity and quality of remote sensing images. Although the application of Deep Learning has obtained remarkable performance in Computer Vision, detecting multi-scale targets in remote sensing images is still an unsolved problem, especially for small instances which possess limited features and intricate backgrounds. In this work, we managed to cope with this problem by designing a neck attention block (NAB), a simple and flexible module which combines the convolutional bottleneck structure and the attention mechanism, different from traditional attention mechanisms that focus on designing complicated attention branches. In addition, Vehicle in High-Resolution Aerial Imagery (VHRAI), a diverse, dense, and challenging dataset, was proposed for studying small object detection. To validate the effectiveness and generalization of NAB, we conducted experiments on a variety of datasets with the improved YOLOv3, YOLOv4-Tiny, and SSD. On VHRAI, the improved YOLOv3 and YOLOv4-Tiny surpassed the original models by 1.98% and 1.89% mAP, respectively. Similarly, they exceeded the original models by 1.12% and 3.72% mAP on TGRS-HRRSD, a large multi-scale dataset. Including SSD, these three models also showed excellent generalizability on PASCAL VOC.

**Keywords:** remote sensing; multi-scale object detection; small object detection; attention mechanism; YOLOv3; YOLOv4-Tiny; SSD

**Citation:** Lang, K.; Yang, M.; Wang, H.; Wang, H.; Wang, Z.; Zhang, J.; Shen, H. Improved One-Stage Detectors with Neck Attention Block for Object Detection in Remote Sensing. *Remote Sens.* **2022**, *14*, 5805. https://doi.org/10.3390/rs14225805

Academic Editors: Yue Wu, Kai Qin, Qiguang Miao and Maoguo Gong

Received: 17 October 2022
Accepted: 14 November 2022
Published: 17 November 2022

**Publisher's Note:** MDPI stays neutral with regard to jurisdictional claims in published maps and institutional affiliations.

## 1. Introduction

In remote sensing, multiple satellites and aircraft are used to capture images that contain significant information, such as the characteristics and changes of landscape, man-made targets, and traces. Object detection is a critical approach to extracting useful information from remote sensing images. It plays a vital role in environmental monitoring, geological hazard detection, land-use/land-cover mapping, geographic information system update, military reconnaissance and location, and land planning [1].

Traditional object detectors, which are usually composed of region proposal, feature extraction, feature fusion, and classifier training, require elaborately hand-made features and must be trained step by step. Therefore, these methods have inferior efficiency, accuracy, and generalizability. Especially with the rapid advancement of the quantity and quality of optical remote sensing images, these methods can not meet the requirement of practical applications by degrees.

In the last decades, convolutional neural networks (CNNs) have made tremendous breakthroughs in various computer vision tasks, including image classification, object detection, and semantic segmentation. The application of CNNs in object detection for remote sensing images achieves better accuracy, higher efficiency, and more powerful generalizability than traditional methods. A common CNN detector is composed of a backbone, which is pretrained with large datasets and used to extract feature maps; a neck, which can enhance feature representation and make feature transition smooth from feature maps to output; and a head, which is used to generate regression and classification predictions.

The backbone, the most significant part of a CNN model, determines the fundamental performance of a CNN model. Since the advent of AlexNet [2], a variety of backbones have been designed for improving the capability of feature extraction, such as VGG16 [3], Inception [4], ResNet [5], ResNeXt [6], and Darknet53 [7]. In these backbones, an important research direction is to increase the depth and width of the network. AlexNet only has five convolutional layers, and VGG16 has sixteen convolutional layers. After the creation of a residual block, ResNet-152 contains 152 convolutional layers. Meanwhile, the set of Inception structures. which concentrates on increasing the width of a model, also obtains excellent performance.

The purpose of the neck is to refine feature maps from the backbone and transmit them to the head. In order to aggregate bottom and top features, a Feature Pyramid Network (FPN) [8] is designed to combine low-resolution features and high-resolution features by adding a top-down path. To address the shortcoming that top feature maps lack location information in FPN, a Path Aggregation Network (PAN) [9] further adds a down-top path on the basis of the FPN. Although the neck has a significant function in enhancing feature representation and making feature transition from feature maps to output smooth, the research for the neck is still inadequate. Most CNN models neglect its essentiality; for example, SSD [10] directly transmits feature maps from the backbone to the head, while YOLOv3 and RetinaNet [11] simply append several convolutional layers after FPN.

The head is a simple structure that only contains several convolutional layers. It can generate regression and classification predictions, including the coordinates of bounding boxes and the class probabilities.

Most well-known object detectors, which are composed of the above modules, could be split into two-stage detectors and one-stage detectors. The central idea of two-stage detectors is to generate region proposals by the region proposal network, then predict sparse output by detecting each proposal, such as R-CNN [12], Fast R-CNN [13], Faster R-CNN [14], and Mask R-CNN [15]. Compared with two-stage detectors, one-stage ones predict dense output straight from CNNs with the goal of improving detection speed while maintaining comparative performance. YOLOv1 [16], YOLOv2 [17], YOLOv3, YOLOv4 [18], SSD, and RetinaNet are examples of one-stage detectors.

Although these detectors are designed for nature images, their applications in remote sensing have made unprecedented progress. For instance, Yuanxin Ye et al. developed a model with the adaptive feature fusion mechanism based on EfficientDet [19]; the authors of [20] improved YOLOv3 by combining DenseNet with YOLOv3 for multi-scale detection; Ke Li et al. proposed DetectIon in Optical Remote sensing images (DIOR), a large-scale dataset, and compared various detectors in DIOR [21]; Zhenfang Qu et al. designed an auxiliary network with CBAM to improve YOLOv3 [22]; the authors of [23] modified YOLOv4 with MobileNet v2 and depth-wise separable convolution to achieve the tradeoff between detection accuracy and speed; and Yafei Jing et al. introduced the vision transformer and Bi-Directional FPN into YOLOv5s [24]. In remote sensing, multi-scale object detection has made obvious advances by transferring and improving existing detectors. However, it still cannot meet the requirements of practical applications, especially in small object detection.

To address the aforementioned problem, we concentrate on the neck of detectors and carefully design neck attention block (NAB), a simple and flexible module which combines the attention mechanism and the convolutional bottleneck structure to enhance the feature representation capability and promote feature transition from feature maps to dense output. It can extract global information and calibrate the channels of feature maps. It can be inserted straightforwardly after the feature maps generated by the backbone or the path aggregation structure. In addition, we propose a publicly dataset, Vehicle in High

Resolution Aerial Imagery (VHRAI) for small object detection. YOLOv3, YOLOv4-Tiny, and SSD were modified simply with NAB, and the improved models were validated on various datasets. By conducting experiments compared with the original models, we demonstrate that NAB is beneficial to small object detection and multi-scale object detection in remote sensing. In addition, it had excellent generalizability on various datasets and models.

The rest of this paper is organized as follows. In Section 2, we introduce some papers about one-stage detectors, attention mechanisms, and small object detection. Section 3 describes NAB, the improved one-stage detectors with NAB, and VHRAI created for small object detection in detail. Section 4 shows the experiments of the improved models on various datasets. Section 5 discusses NAB and the improved models. Lastly, the conclusion is shown in Section 6.

## 2. Related Work

### 2.1. One-Stage Detector

In remote sensing, most applications, such as target tracking, military reconnaissance, and disaster relief, have an increased demand for real-time detection. To balance the accuracy and speed of object detection, we concentrate on the research of one-stage detectors.

One-stage detectors can be divided into anchor-based ones and anchor-free ones. For improving recall rate, anchor-based detectors set pre-defined boxes with different scales and ratios for predictions, such as YOLOv2-v4, SSD, and RetinaNet. SSD appends several layers after VGG16 to produce muti-scale output. Based on YOLOv2, YOLOv3 selects the more powerful Darknet-53 as the backbone and uses the FPN to generate multi-scale predictions. YOLOv4 chooses many measures, including CSPNet [25], CIOU [26], and Mosaic, to modify YOLOv3. For real-time detection, YOLOv4-Tiny obtains an extremely higher speed by decreasing the parameters of YOLOv4. Anchor-free detectors directly predict the boxes without the limitation of anchor boxes, such as CornerNet [27], FCOS [28], and YOLOX [29]. FCOS, based on RetinaNet, takes the location, which falls into any ground-truth box, as a positive sample and adds the center-ness branch to depress low-quality predictions. YOLOX proposes more powerful SimOTA as label assignment. Although anchor-free detectors do not need to search for the hyperparameters of anchor boxes and have less complexity, they have lower precision in detecting remote sensing images whose scale of instances changes enormously. By comparing many detectors, we decide to select YOLOv3 and SSD as our baselines to analyze NAB. In addition, we improved YOLOv4-Tiny with NAB for real-time detection.

### 2.2. Attention Mechanism

Inspired by human vision, attention mechanisms, which enhance meaningful features and depress noise, have shown remarkable improvement in deep learning. In this paper, we focus on the attention mechanisms about CNNs rather than Scaled Dot-Product Attention in Transformer [30]. This method can be well combined with the convolution operation and has lower computational complexity and a faster convergence rate. It can be divided into channel attention and spatial attention. Channel attention focuses on the importance of different channels, and spatial attention focuses on the importance of different locations. In the past few years, many representative blocks have emerged in attention mechanisms, such as SE [31], ECA [32], CA [33], and CBAM [34]. SE is the paradigm of channel attention which adaptively rescales the channels by utilizing the information of feature maps. Figure 1 shows the structure of the SE block in detail. It obtains global information by GAP (global average pooling) and then utilizes two fully connected layers to produce the response of each channel. Finally, channel-wise multiplication is implemented between the response and the original feature map. ECA thinks the two fully connected layers of SE are unnecessary and adopts one-dimensional convolution to achieve local cross-channel interaction. CBAM combines channel attention and spatial attention to acquire the importance of every channel and location. It concatenates the results of GAP and

GMP (global maximum pooling) to extract more robust information, as shown in Figure 2. CA proposes coordinate attention to calculate the width attention and height attention, respectively. Then, CA implements channel-wise multiplication between them. In remote sensing, AAFM based on CBAM is proposed to create the basic block of EfficientNet. MCA-YOLOv5-Light adopts the MCA attention mechanism to extract more productive information [35].

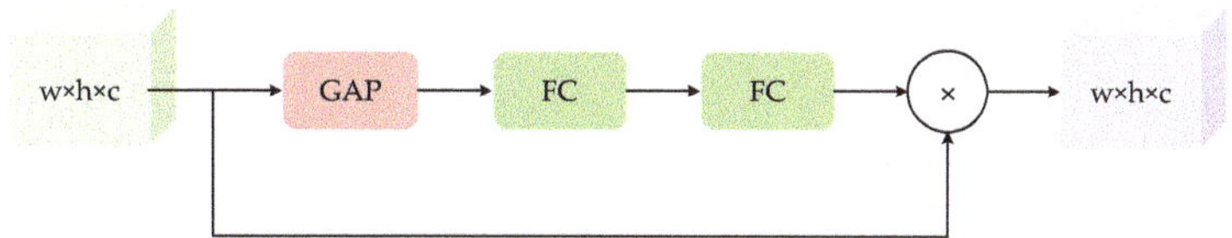


**Figure 1.** The structure of the SE block. w, h, and c denote the width, height, and channel of a feature map, respectively. 'GAP' is the average-pooling operation along the weight and height axes. 'FC' represents a fully connected layer with an activation function.

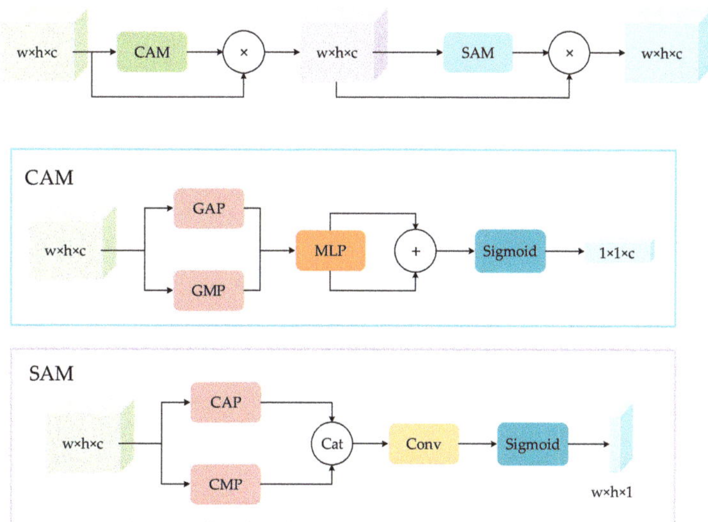


**Figure 2.** The structure of CBAM. 'CAM' and 'SAM' denote the channel and spatial attention modules, respectively. Similar to 'GAP', 'GMP' is the max-pooling operation along the spatial dimension. Similarly, 'CAP' and 'CMP' are the operations along the channel axis, respectively. The results of 'GAP' and 'GMP' use the identical 'MLP', which is composed of sequential fully connected layers.

Although current studies about attention mechanisms design various architectures in the attention branch, they have a common characteristic that they obtain attention by performing some operations on the feature map and then utilizing the attention to rescale the original feature map. However, NAB, proposed by us, introduces an extra branch to adaptively enhance the information of feature maps, as illustrated in Section 3.1.

### 2.3. Small Object Detection

In multi-scale detection, detecting small targets which have limited features, diverse distributions, and arbitrary orientations is a big problem. First, there is no uniform definition of small objects. The most universal definition is from MS COCO, which regards objects less than $32 \times 32$ pixels as small objects [36]. In DOTA [37], an object whose height of the horizontal bounding box ranges from 10 to 50 pixels is defined as a small object. TinyPerson takes an object that ranges from 20 to 32 pixels as a small object [38]. Chen et al.

established a small object dataset whose ratio of the bounding box over the image of all instances was between 0.08% and 0.58% [39]. With the consideration of limited receptive field and down-sampling rate, we selected the above definition of MS COCO for small objects in this paper.

For small object detection in remote sensing, the datasets and related research are inadequate. The significant targets of remote sensing images usually contain almost 20 categories, such as soccer ball fields, vehicles, planes. In these categories, vehicles, ships and planes, which generally have a large number of small instances. TAS [40], VEDAI [41] and COWC [42] only focus on vehicles; HRSC2016 only contains ships [43]; and UCAS AOD is concerned with vehicles and planes [44]. These datasets have many instances that do not meet our definition of small objects. DOTA and DIOR contain enormous multi-scale instances, but they do not specialize in annotating small objects.

With respect to object detectors, YOLO-fine, which is based on YOLOv3, increases the resolution of feature maps for detecting small targets [45]. Deconv R-CNN introduces a deconvolution layer to recover more details [46], and SOON constructs a receptive field enhancement module to extract spatial information [47]. Most research ignores the importance of the neck. Our proposed NAB is a flexible module which is used to extract global information and propel the transition of features in the neck.

## 3. Materials and Methods

### 3.1. NAB

In the early stages of CNNs, the neck of an object detector, which is usually used to transmit feature maps generated by the backbone to the head, is generally composed of several convolutional layers. With the advent of FPN and PAN, the neck plays another important role in producing multi-scale feature maps that possess strong sematic information by appending top-down and down-top paths. Then, these feature maps are sent into the head via the identical layers. The way of stacking layers in the neck has a large burden for the models with multi-scale output. For example, every output of FPN is connected with 5 convolutional layers in the neck of YOLOv3. This way increases the parameters of YOLOv3 and causes overfitting in the training process, especially for remote sensing datasets that contain inadequate images.

In order to enhance representation capability in the neck, we carefully designed NAB which combines the channel attention and the convolutional bottleneck structure. It consists of an attention branch, which adopts attention mechanisms to learn where and what to focus on, and a bottleneck branch, which utilizes the convolutional bottleneck structure to refine features and obtain robust feature representation adaptively.

Whereas attention mechanisms contain channel attention and spatial attention, we only utilized channel attention in the attention branch. There is an empirical explanation why we excluded spatial attention: for the dense output of one-stage detectors: Each grid cell predicts the result of the corresponding region in an input image. Every region should be weighted equally. Because the neck is close to the final output, spatial attention would breach this equality and result in bad performance. However, the channels of a grid cell denote different properties, such as the coordinates of a bounding box and the categories. Using channel attention can propel feature representation and convergence.

Inspired by SE, the attention branch adopts GAP to aggregate global information as illustrated in Equation (1). The input is assumed as X, and $X_{i,j}$ denotes the value of a specific spatial location. Then, the information is forwarded to successive multi-layer perceptrons (MLPs) composed of two fully connected layers. It is notable that the last layer in the branch follows a Sigmoid function to generate factors which are restricted to the range of 0–1.

$$F_{GAP}(X) = \frac{1}{h \times w} \sum_{i=1}^{w} \sum_{j=1}^{h} X_{i,j} \tag{1}$$

The purpose of appending a bottleneck branch is to enhance the adaptive ability of the attention mechanism and to produce a more robust feature map via convolutional layers. We opted to utilize the factors generated by the attention branch to recalibrate the output of the bottleneck branch, which is the most distinctive point compared with traditional attention mechanisms that remodify the original feature maps with the factors. The reasons why this novel method feasible are as follows: The outputs of both branches originate from the identical feature map. This can increase the flexibility of the attention mechanism and make the block refine features adaptively. Our proposed NAB can acquire more robust features and decrease the extra complexity introduced by traditional attention mechanisms.

The structure of NAB is shown in Figure 3. In NAB, the first and second lines denote the attention branch and the bottleneck branch, respectively. The bottleneck branch is composed of 'BB', which contains $3 \times 3$ and $1 \times 1$ convolution layers. It is notable that each convolutional layer is connected with BN (Batch Normalization) [48] and ReLU. The attention branch is composed of 'GAP' and several 'MLP'. We set 'BB' and 'MLP' to have an identical number, denoted by m. The output feature map has the same size and channel as the input one. Our proposed NAB is an innovation for traditional attention mechanisms which rescale the original feature map. It can decrease the parameters of neck and enhance feature representation ability. If m is 1, then the attention branch and the bottleneck branch can be represented as Equations (2) and (3), respectively. '$F_{fc}$' denotes one FC layer with an activation function. '$F_{1c}$' and '$F_{3c}$' represent $1 \times 1$ and $3 \times 3$ convolutional layers with Batch Normalization and a ReLU function, respectively. By implementing channel-wise multiplication, the output of NAB can be obtained, as shown in Equation (4). In Section 4, we show the excellent performance of NAB for small object detection and multi-scale object detection on various datasets.

$$F_{\text{attention}}(X) = F_{fc}(F_{fc}(F_{\text{GAP}}(X))) \tag{2}$$

$$F_{\text{bottleneck}}(X) = F_{1c}(F_{3c}(X)) \tag{3}$$

$$F_{\text{NAB}}(X) = F_{\text{attention}}(X) \otimes F_{\text{bottleneck}}(X) \tag{4}$$

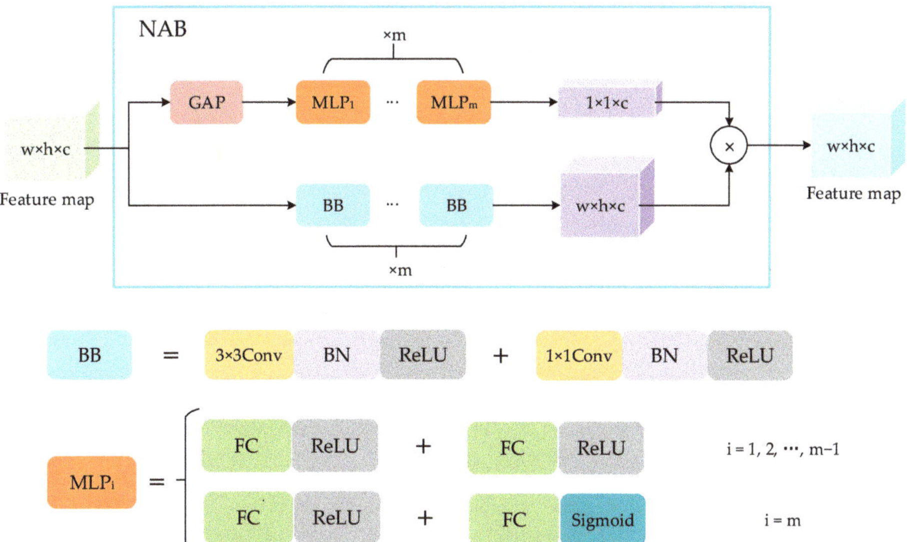


**Figure 3.** The structure of NAB. 'BB' is the abbreviation of bottleneck block. The number of 'MLP' and 'BB' denoted by m is equal. The second 'FC' in the last MLP uses a Sigmoid activation function to scale the factors into the range of 0–1.

### 3.2. Improved Models

In remote sensing, the instances, which usually have complicated backgrounds, uneven distributions, and diverse scales, bring enormous computational complexity for object detectors. To balance the accuracy and speed of object detection, we concentrated on the research of one-stage detectors. In addition, NAB, which assigns different attributes to the channels of a feature map, is consistent with the output of one-stage detectors whose every channel denotes a kind of attribute, such as the coordinates of bounding boxes and the probabilities of classes. We selected YOLOv3, YOLOv4-Tiny, and SSD as the improved models from various one-stage detectors.

YOLOv3 is the baseline of many detectors, including YOLOv4, YOLOv5, and YOLOX. It has important values for researching one-stage detectors; therefore, we selected YOLOv3 to verify the effectiveness of NAB. On the basis of YOLOv2, YOLOv3 adopts more powerful Darknet53 as the backbone to enhance the capability of feature extraction and FPN to generate multi-scale output. Due to the way that YOLOv3 transmits semantic information to finer-grained feature maps by the top-down path, it obtains salient performance on small object detection. In the neck, it is notable that YOLOv3 has five sequential 'CBL' blocks before each head. It has an inferior ability in fusing feature maps generated by FPN and Darknet53 and introduces redundant parameters to increase the risk of overfitting. Aiming at achieving higher performance while decreasing computational complexity, the original five 'CBL' were replaced with our proposed NAB and $1 \times 1$ 'CBL' which was used to reduce the channels of the feature map. Figure 4 depicts the modification in the neck of YOLOv3. In Section 4, we contrast different models and show the highlighted performance of NAB on a variety of datasets.

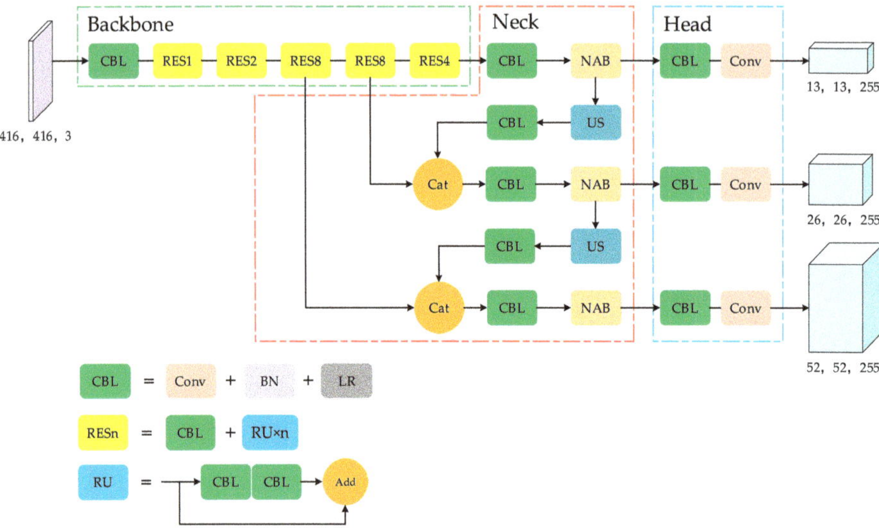


**Figure 4.** The network of the improved YOLOv3. It can be divided into the backbone, called Darknet53; the neck, which contains FPN and NAB; and the head, which is composed of two convolutional layers. 'Cat' and 'US' denote the operations of concat and up-sampling, respectively. 'LR' is the abbreviation of Leaky ReLU.

SSD is another paradigm of one-stage detectors. The backbone is composed of the truncated VGG16 and several auxiliary convolutional layers. It selects six multi-scale feature maps that are generated by different convolutional blocks of the backbone. Then, these feature maps are transmitted to the corresponding detection layers in the head. Each layer has two convolutional layers, one for predicting the probabilities of classes and the other for predicting the information of bounding boxes. If the input size is $300 \times 300$, then SSD will generate 8732 outputs. Because of the large size of optical remote sensing images, the application of SSD in remote sensing has extremely tremendous computation complexity and low detection efficiency. As a result, we improved SSD for validating the generality of NAB in nature images rather than remote sensing images. In the original SSD, the author introduced 'L2_norm' to scale the feature map of 'Conv_4', which is different from others. Because NAB also has the same function, we concisely removed the 'L2_norm'. We inserted NAB and 'CBL' between the backbone and the head of SSD to enhance the capability of feature representation and facilitate the feature transition, as depicted in Figure 5. Section 4.3 shows the excellent generalizability of NAB in detecting nature images.

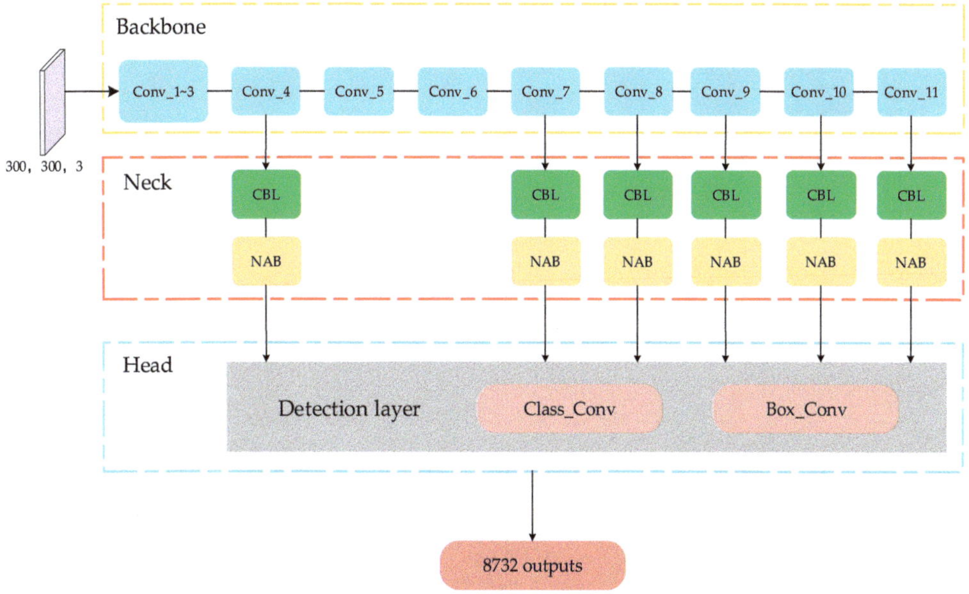



**Figure 5.** The network of the improved SSD. 'Conv_1~3' is the first three blocks of VGG16.

For real-time detection, YOLOv4-Tiny, which has an excellent balance between accuracy and speed, was modified with NAB. It is a simple version of YOLOv4 and has only about one-tenth of YOLOv4's parameters. YOLOv4-Tiny introduces the idea of CSPnet, which is the largest difference between it and YOLOv3. It only has two outputs for reducing the parameters. We improved it by inserting NAB into the neck, as shown in Figure 6. Because YOLOv4-Tiny has fewer channels than YOLOv3 and SSD, we did not append 'CBL' before NAB. By conducting experiments on various datasets, we found that the improved YOLOv4-Tiny has a more powerful capability in multi-scale object detection than the original one, though it increases computational complexity slightly.

### 3.3. Datasets

Deep learning is a science driven by data. Thanks to substantial datasets that are available in remote sensing, multi-scale object detection with CNNs has made remarkable progress. However, small object detection remains a challenge. Besides the characteristics of small targets, another reason is the lack of appropriate datasets that specialize in detecting small instances. In order to boost the performance of small object detection, we created Vehicle in High-Resolution Aerial Imagery (VHRAI), a dataset that contains 900 aerial images with $960 \times 540$ pixels captured at a height of 1000 m for vehicle detection. We utilized LabelImg, an open-source image annotation tool, to annotate instances [49]. Each object instance was manually labeled by a horizontal bounding box which was composed of the coordinates of the central point, the size of the box, and the category.

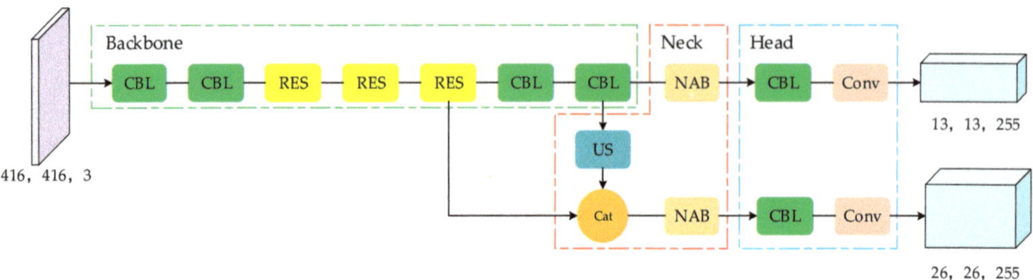


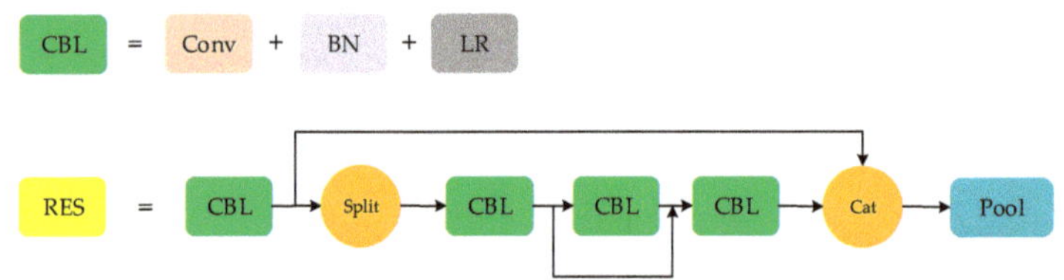


**Figure 6.** The network of the improved YOLOv4-Tiny. 'Split' is the operation that divides the feature map into two portions along the channel axis. 'Pool' denotes $2 \times 2$ max-pooling.

Because VHRAI is created for small object detection, we compared it with some well-known datasets which mainly concentrate on researching vehicles and ships, including TAS, UCAS-AOD, HRSC2016, DLR-MVDA [50], COWC, and VEDAI, as listed in Table 1. The average area per instance of DLR-MVDA and VHRAI is far smaller than other datasets. DLR-MVDA and VHRAI are annotated with oriented bounding boxes (OBB) and horizontal bounding boxes (HBB), respectively. Both have important value in object detection. Compared with VEDAI (512), VHRAI has more instances and smaller bounding boxes. However, VHRAI has fewer instances than some large datasets, including UCAS-AOD and COWC. In the future, we will further capture more images to enlarge VHRAI.

**Table 1.** Comparisons between the proposed VHRAI and several publicly available datasets in remote sensing. VEDAI (512) denotes the version of VEDAI, whose image width is 512. Because the annotations of the testing set in DLR-MVDA are unavailable, we only display the properties of the training set. '#' represents the meaning of 'the number of'.

| Datasets | # Categories | # Images | # Instances | Image Width | Average Area per Instance |
|---|---|---|---|---|---|
| TAS | 1 | 30 | 1319 | 792 | 805 |
| UCAS-AOD | 2 | 1510 | 14,597 | 1280 | 4888 |
| HRSC2016 | 1 | 1070 | 2976 | ~1000 | 56,575 |
| DLR-MVDA | 2 | 10 | 3505 | 5616 | 239 |
| COWC | 1 | 53 | 32,716 | 2000~19,000 | 1024 |
| VEDAI (512) | 9 | 1250 | 3757 | 512 | 3108 |
| VHRAI (ours) | 1 | 900 | 5589 | 960 | 369 |

Figure 7 shows the characteristics of VHRAI. It has diverse backgrounds, uneven distributions, and tiny scales. In the Earth observation community, VHRAI is a challenging dataset for small object detection.

**Figure 7.** Examples in VHRAI.

To validate the effect of NAB for multi-scale object detection in remote sensing, we selected TGRS-HRRSD, a public dataset which has 21,761 images and 13 categories [51]. This elaborate dataset achieves an excellent balance between all categories. The average scale per category of TGRS-HRRSD ranges from 41.96 to 276.50 pixels. In addition, aiming at indicating the generalizability of NAB, we conducted experiments on PASCAL VOC [52], a commonly used natural scene dataset. The entire results are displayed in the next section.

## 4. Results

### 4.1. Evaluation Criteria

The output of an object detector can be divided into four categories: True Positive (TP), False Positive (FP), True Negative (TN), and False Negative (FN). TP and FP denote a positive sample that is classified correctly and incorrectly, respectively. TN and FN represent a negative sample that is classified correctly and incorrectly, respectively. Through analyzing these categories, we can obtain Precision, which illustrates the proportion of TP in all positive samples, and Recall, which indicates the proportion of TP in all positive ground-truth samples, depicted in Equations (5) and (6). These two indicators have some limitations as evaluation criteria. Confidence is the threshold that estimates a sample is positive or negative. Different Confidence can generate different Precision and Recall. Precision increases and Recall decreases in general as Confidence gradually increases.

$$\text{Precision} = \frac{\text{TP}}{\text{TP} + \text{FP}} \tag{5}$$

$$\text{Recall} = \frac{\text{TP}}{\text{TP} + \text{FN}} \tag{6}$$

We can acquire the Precision/Recall Curve by setting a different Confidence. Average Precision (AP) denotes the area under Precision/Recall Curve for a class. The mean

Average Precision (mAP) is the mean value of every class's AP. The mAP is a significant evaluation criterion in object detection. In this paper, we considered Precision and Recall for a comprehensive comparison. AP_s proposed in MS COCO was also adopted to show the performance of detecting small objects. In addition, the number of parameters was used to evaluate computational complexity and detection speed.

### 4.2. VHRAI

VHRAI, whose average size of instances is $19.22 \times 19.19$ pixels, is a valuable dataset for small object detection. On this dataset, we validated the effectiveness of NAB by comparing the improved YOLOv3 and YOLOv4-Tiny with the original ones. We also compared traditional methods that contained SE and CBAM with NAB to reveal the importance of the bottleneck branch in NAB, as shown in Figure 8. Furthermore, we analyzed the influence of 'm', a hyperparameter in NAB. It is notable that these improvements were adopted in all multi-scale paths.

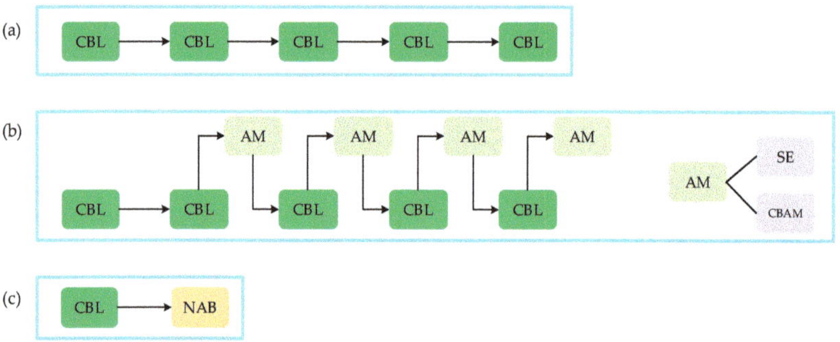


**Figure 8.** (**a**) The original 'CBL5' in the neck. (**b**) The improved 'CBL5' with the attention mechanism in the neck. (**c**) NAB in the neck. The structure of 'CBL' is shown in Figure 4. 'AM', which is the abbreviation of attention mechanism, could be SE or CBAM.

All experiment results on VHRAI are listed in Table 2. YOLOv3-NAB (m = 1) obtained the best AP 90.29% among all models, surpassing YOLOv3 by 1.94%. The accuracy and recall of YOLOv3-NAB were also better than the original model. At the same time, YOLOv3-NAB (m = 1) reduced parameters by almost 11% compared with YOLOv3. We also compared the loss and mAP curves of YOLOv3-NAB (m = 1) and YOLOv3 to acquire a more robust conclusion, as shown in Figure 9. With respect to traditional attention mechanisms, YOLOv3-SE had a slightly poorer performance than YOLOv3, and YOLOv3-CBAM exceeded YOLOv3 by 0.74%. However, YOLOv3-NAB (m = 1), which improved the attention mechanism by appending a bottleneck branch, achieved more salient performance and had fewer parameters compared with YOLOv3-CBAM and YOLOv3-SE. The reason why YOLOv3-NAB (m = 1) had fewer parameters is that traditional attention mechanisms only can rescale the feature map generated by the layer and cannot serve as an independent module. With this limitation, YOLOv3-SE had more convolutional layers than YOLOv3-NAB (m = 1). These results clearly show the effectiveness of NAB, which is a better way to utilize the attention mechanism in the neck. In addition, when we set 'm' = 2, YOLOv3-NAB (m = 2) obtained worse AP and had more parameters than YOLOv3-NAB (m = 1). This may be attributed to the fact that more parameters increase the risk of overfitting with the limitation of inadequate images. In the next experiments, the default value of 'm' in NAB was 1.

**Table 2.** Detection results on VHRAI. For a fair comparison, the models based on YOLOv3 had the same configuration. The models based on YOLOv4-Tiny also had the same configuration. YOLOv3-SE and YOLOv3-CBAM adopted the improved 'CBL5' with attention mechanisms in Figure 8b.

| Model | # Parameters | Precision (%) | Recall (%) | AP (%) | AP_s (%) |
|---|---|---|---|---|---|
| YOLOv3 | 61.52 M | 83.73 | 83.87 | 88.35 | 41.9 |
| YOLOv3-SE | 63.24 M | 84.44 | 83.5 | 88.15 | 39.5 |
| YOLOv3-CBAM | 63.24 M | 85.88 | 83.76 | 89.09 | 41.4 |
| YOLOv3-NAB (m = 1) | 54.81 M | 86.62 | 84.36 | 90.29 | 42.9 |
| YOLOv3-NAB (m = 2) | 61.87 M | 81.01 | 87.45 | 89.16 | 41.9 |
| YOLOv4-Tiny | 5.87 M | 71.35 | 58.77 | 63.99 | 20.6 |
| YOLOv4-Tiny-NAB | 7.05 M | 72.05 | 60.39 | 65.82 | 21.6 |

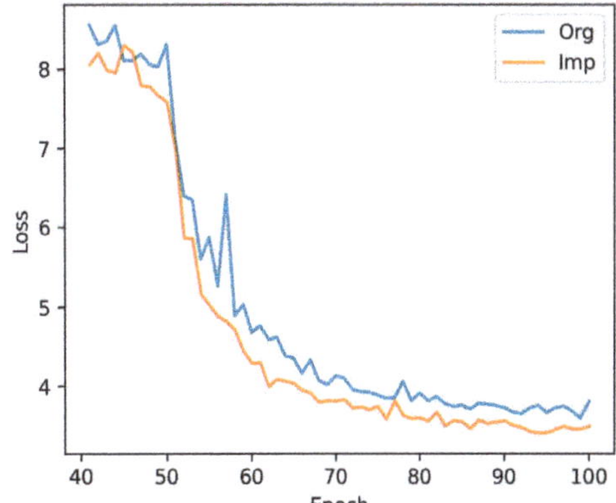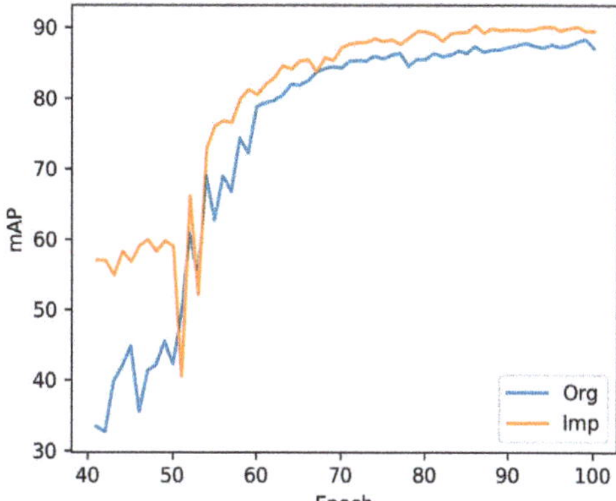

**Figure 9.** The loss and mAP curves of YOLOv3 and YOLOv3-NAB (m = 1). 'Org' and 'Imp' denote YOLOv3 and YOLOv3-NAB (m = 1), respectively.

For real-time detection, we conducted experiments on YOLOv4-Tiny. YOLOv4-Tiny-NAB achieved better precision, recall, and AP, exceeding the original model by 0.7%, 1.62% and 1.83%, respectively, despite having slightly more parameters. Furthermore, due to YOLOv4-Tiny, which cut an important path for small object detection, we found that YOLOv4-Tiny-NAB had a large gap in performance compared with YOLOv3-NAB (m = 1). It is notable that YOLOv4-Tiny-NAB had an extremely fast speed in detection.

In addition, we compared the AP_s of the above models, which is used to evaluate the performance for small object detection precisely in MS COCO. Undoubtedly, YOLOv3-NAB (m = 1) obtained the best AP_s, outperforming YOLOv3-SE and YOLOv3-CBAM by 3.4% and 1.5%, respectively. In addition, YOLOv4-Tiny-NAB was better than YOLOv4-Tiny. These experiments on VHRAI apparently demonstrate the effectiveness of NAB in small object detection. Different from traditional attention mechanisms, we introduced an extra branch to enhance the ability of adaptively extracting features rather than focusing on designing a more complicated attention branch. Furthermore, NAB can be inserted into a model flexibly as an independent structure, similar to the above models.

### 4.3. TGRS-HRRSD

Although we proved the effectiveness of NAB in small object detection, which is a crucial part of multi-scale detection, it is necessary to conduct experiments on a multi-scale dataset to acquire a reliable conclusion. TGRS-HRRSD is a large dataset for multi-scale object detection. It has 13 categories, and the average scale per category ranges from 41.96 pixels to 276.50 pixels. We selected TGRS-HRRSD as the dataset and compared YOLOv3-NAB, YOLOv3-SE, YOLOv3-CBAM, and YOLOv4-Tiny-NAB with the original models.

Table 3 shows the detection results. YOLOv3-NAB, which had fewer parameters than YOLOv3, scored 92.16% mAP, surpassing YOLOv3 by 1.06%. With respect to traditional attention mechanisms, YOLOv3-SE had an inferior performance than YOLOv3-NAB, and YOLOv3-CBAM was comparable with YOLOv3-NAB, but its parameters increased by 13%. Compared with YOLOv4-Tiny, YOLOv4-Tiny-NAB, which was improved with NAB in the neck, obtained a remarkable performance that exceeded the original model by 3.72% mAP. It also outperformed in all categories. Its mAP was even close to YOLOv3, though it only had about one-tenth of YOLOv3's parameters. These experiments clearly illustrate that NAB can obtain robust feature representation and is helpful for multi-scale object detection as a flexible module.

**Table 3.** Detection results on TGRS-HRRSD.

| Model | Ship | Bridge | Ground Track Field | Storage Tank | Basketball Court | Tennis Court | Airplane | Baseball Diamond | Harbor | Vehicle | Crossroad | T Junction | Parking Lot | mAP (%) |
|---|---|---|---|---|---|---|---|---|---|---|---|---|---|---|
| YOLOv3 | 92.65 | 92.04 | 98.40 | 93.99 | 83.17 | 96.06 | 99.57 | 93.05 | 95.02 | 92.69 | 93.92 | 83.62 | 70.09 | 91.10 |
| YOLOv3-SE | 94.37 | 92.72 | 98.31 | 94.47 | 84.05 | 95.81 | 98.73 | 93.62 | 92.79 | 96.91 | 92.31 | 82.86 | 70.67 | 91.35 |
| YOLOv3-CBAM | 94.63 | 92.78 | 98.71 | 96.89 | 82.35 | 95.12 | 99.54 | 93.63 | 97.27 | 97.02 | 92.51 | 84.72 | 72.03 | 92.09 |
| YOLOv3-NAB | 94.59 | 93.33 | 98.33 | 96.12 | 82.84 | 95.83 | 99.02 | 93.34 | 96.79 | 97.05 | 94.08 | 85.03 | 71.79 | 92.16 |
| YOLOv4-Tiny | 86.34 | 73.18 | 92.31 | 97.20 | 69.60 | 93.53 | 98.88 | 89.90 | 84.36 | 90.27 | 87.13 | 68.85 | 53.36 | 83.44 |
| YOLOv4-Tiny-NAB | 89.72 | 85.31 | 95.89 | 97.28 | 71.30 | 93.61 | 98.94 | 91.61 | 92.11 | 93.43 | 89.98 | 73.15 | 60.79 | 87.16 |

### 4.4. PASCAL VOC

NAB had excellent performance in multi-scale remote sensing images, and we speculate that it is not limited in remote sensing. The experiments on PASCAL VOC, a well-known dataset that contains 21504 nature images, were conducted to validate the generalizability of NAB. All models were trained on the union of VOC2007 and VOC2012 trainval, and they were evaluated with the VOC2007 test. The detection results are shown in Table 4. The improved models, including YOLOv3-NAB, YOLOv4-Tiny-NAB, and SSD-NAB, acquired a better performance than the original models, surpassing them by 0.88%, 1.98% and 0.82% mAP, respectively. Compared with traditional mechanisms, YOLOv3-NAB outperformed YOLOv3-SE and YOLOv3-CBAM by 0.86% and 0.57%, respectively, while decreasing 13% parameters. Consequently, we confirm that NAB can be generalized to natural scenes and applied to various one-stage detectors.

**Table 4.** Detection results on PASCAL VOC. We chose 10 categories at random to show the comparisons of their AP (%). We retrained YOLOv3, YOLOv4-Tiny, and SSD for a fair comparison using the same configuration of the improved models.

| Model | Aero | Bike | Bird | Bottle | Car | Cow | Dog | Horse | Sofa | Train | mAP (%) |
|---|---|---|---|---|---|---|---|---|---|---|---|
| YOLOv3 | 88.84 | 85.88 | 80.11 | 63.78 | 90.90 | 84.40 | 86.62 | 86.99 | 73.91 | 88.86 | 80.47 |
| YOLOv3-SE | 89.28 | 86.81 | 81.81 | 63.21 | 90.89 | 83.38 | 86.82 | 86.46 | 78.58 | 89.80 | 80.49 |
| YOLOv3-CBAM | 89.18 | 87.26 | 82.68 | 62.37 | 91.32 | 84.31 | 85.54 | 89.97 | 80.94 | 89.86 | 80.78 |
| YOLOv3-NAB | 89.65 | 87.78 | 81.74 | 65.97 | 91.02 | 86.89 | 86.38 | 87.18 | 71.55 | 88.70 | 81.35 |
| YOLOv4-Tiny | 84.06 | 85.29 | 74.35 | 62.55 | 90.54 | 81.41 | 77.86 | 86.55 | 73.10 | 84.39 | 77.08 |
| YOLOv4-Tiny-NAB | 87.21 | 87.09 | 76.97 | 66.72 | 91.77 | 80.75 | 79.52 | 87.67 | 73.18 | 86.75 | 79.06 |
| SSD | 77.81 | 84.93 | 75.35 | 42.32 | 86.50 | 77.35 | 86.99 | 88.60 | 73.52 | 84.93 | 75.8 |
| SSD-NAB | 79.75 | 85.06 | 77.55 | 45.79 | 85.61 | 77.15 | 85.67 | 87.80 | 74.46 | 86.66 | 76.62 |

## 5. Discussion

In this paper, we presented NAB, an architectural module designed to enhance the capability of feature representation and promote feature transition in the neck by combining the attention mechanism and the convolutional bottleneck structure. Since the output of NAB has the same dimensions as the input, it is simple and flexible to utilize in the neck of one-stage detectors. In addition, VHRAI, whose instances have an extremely small size, was created to address the challenge of small object detection, which is a crucial part of multi-scale object detection in remote sensing.

We improved several classic one-stage detectors with NAB and conducted many experiments on various datasets. YOLOv3-NAB, YOLOv4-Tiny-NAB, and SSD-NAB achieved a better performance than the original models. Figure 10 shows some detection examples of YOLOv3-NAB on different datasets. VHRAI is a diverse and challenging dataset whose average area per instance is 369 pixels. It perfectly meets our definition of small objects and is smaller than other famous datasets, such as VEDAI (512), HRSC2016, UCAS-AOD, and TAS. The results of YOLOv3-NAB and YOLOv4-Tiny-NAB on VHRAI clearly illustrate the effectiveness of NAB for small object detection. Furthermore, we compared NAB with SE and CBAM in the neck. NAB, which appends a convolutional bottleneck branch, showed better flexibility and performance while decreasing the model's parameters. The experiments on TGRS-HRRSD demonstrate the capability of NAB for multi-scale detection in remote sensing. YOLOv4-Tiny had poor feature representation with the limitation of parameters. However, the result that YOLOv4-Tiny-NAB exceeded the original model by 3.72% mAP clearly verifies NAB can enhance the capability of feature representation and propel feature transition. These improved models also obtained better detection precision on PASCAL VOC. This fact indicates that NAB is not limited in remote sensing and can be generalized to detect nature images.

Whereas NAB brings promising results in remote sensing and natural scenes, some issues still remain and call for further research. NAB is restricted in the neck of a model. Whether it can replace traditional attention mechanisms in the backbone or not remains unknown. In addition, there are some similarities in the neck of one-stage and two-stage detectors. In the future, we will further focus on the backbone and improve two-stage detectors with NAB for object detection.

**Figure 10.** *Cont.*

**Figure 10.** Detection examples of YOLOv3-NAB on various datasets.

## 6. Conclusions

In this paper, we designed a simple and flexible module for the neck of a model, called NAB. Unlike traditional attention mechanisms which focus on designing a more complicated attention branch, NAB appends a convolutional bottleneck branch with the attention branch for enhancing feature representation capability and promoting feature transition. In addition, VHRAI, a challenging dataset whose instances have an extremely small size, was proposed for small object detection. The improved models, including YOLOv3-NAB, YOLOv4-Tiny-NAB, and SSD-NAB, achieved excellent performance on various datasets, which clearly proves the effectiveness and generalizability of NAB in small object detection and multi-scale object detection. In the future, we will focus on improving the backbone of a model and two-stage detectors on the basis of NAB.

**Author Contributions:** Conceptualization, K.L. and M.Y.; methodology, K.L. and M.Y.; software, K.L. and M.Y.; validation K.L. and Z.W.; formal analysis, K.L. and H.S.; investigation, H.W. (Hao Wang) and K.L.; resources, K.L. and H.W. (Hao Wang); data curation, K.L. and H.W. (Hao Wang); writing—original draft preparation, K.L.; writing—review and editing, M.Y.; visualization, K.L. and H.W. (Hanyu Wang); supervision, H.S.; project administration, H.S.; funding acquisition, J.Z. All authors have read and agreed to the published version of the manuscript.

**Funding:** This research was funded by the Scientific Research Business Fee Fund of Heilongjiang Provincial Scientific Research Institutes, Research on Key Technologies of Wide Area Forest and Grass Fire Aerial Monitoring and Early Warning (No. CZKYF2020B009).

**Data Availability Statement:** All data used during the study have been uploaded at https://github.com/youjiaowuya/NAB (accessed on 1 June 2020).

**Acknowledgments:** This research was supported by Foundation items: The National Key Research and Development Program of China.

**Conflicts of Interest:** The authors declare no conflict of interest.

## References

1. Cheng, G.; Han, J. A survey on object detection in optical remote sensing images. *ISPRS J. Photogramm. Remote Sens.* **2016**, *117*, 11–28. [CrossRef]
2. Krizhevsky, A.; Sutskever, I.; Hinton, G. ImageNet Classification with Deep Convolutional Neural Networks. *Commun. ACM* **2017**, *60*, 84–90. [CrossRef]
3. Simonyan, K.; Zisserman, A. Very deep convolutional networks for large-scale image recognition. *arXiv* **2014**, arXiv:1409.1556.
4. Szegedy, C.; Wei, L.; Yangqing, J.; Sermanet, P.; Reed, S.; Anguelov, D.; Erhan, D.; Vanhoucke, V.; Rabinovich, A. Going deeper with convolutions. In Proceedings of the 2015 IEEE Conference on Computer Vision and Pattern Recognition (CVPR), Boston, MA, USA, 7–12 June 2015; pp. 1–9.
5. He, K.; Zhang, X.; Ren, S.; Sun, J. Deep residual learning for image recognition. In Proceedings of the 2016 IEEE Conference on Computer Vision and Pattern Recognition (CVPR), Las Vegas, NV, USA, 27–30 June 2016; pp. 770–778.
6. Xie, S.; Girshick, R.; Dollár, P.; Tu, Z.; He, K. Aggregated residual transformations for deep neural networks. In Proceedings of the 2017 IEEE Conference on Computer Vision and Pattern Recognition (CVPR), Honolulu, HI, USA, 21–26 July 2017; pp. 5987–5995.

7. Redmon, J.; Farhadi, A. Yolov3: An incremental improvement. *arXiv* **2018**, arXiv:1804.02767.
8. Lin, T.Y.; Dollár, P.; Girshick, R.; He, K.; Hariharan, B.; Belongie, S. Feature pyramid networks for object detection. In Proceedings of the 2017 IEEE Conference on Computer Vision and Pattern Recognition (CVPR), Honolulu, HI, USA, 21–26 July 2017; pp. 936–944.
9. Liu, S.; Qi, L.; Qin, H.; Shi, J.; Jia, J. Path aggregation network for instance segmentation. In Proceedings of the 2018 IEEE/CVF Conference on Computer Vision and Pattern Recognition, Salt Lake City, UT, USA, 18–23 June 2018; pp. 8759–8768.
10. Liu, W.; Anguelov, D.; Erhan, D.; Szegedy, C.; Reed, S.; Fu, C.-Y.; Berg, A.C. SSD: Single shot MultiBox detector. In Proceedings of the Computer Vision—ECCV 2016, Amsterdam, The Netherlands, 11–14 October 2016; Springer: Cham, Switzerland; pp. 21–37.
11. Lin, T.Y.; Goyal, P.; Girshick, R.; He, K.; Dollár, P. Focal Loss for Dense Object Detection. *IEEE Trans. Pattern Anal. Mach. Intell.* **2020**, *42*, 318–327. [CrossRef]
12. Girshick, R.; Donahue, J.; Darrell, T.; Malik, J. Rich feature hierarchies for accurate object detection and semantic segmentation. In Proceedings of the 2014 IEEE Conference on Computer Vision and Pattern Recognition, Columbus, OH, USA, 23–28 June 2014; pp. 580–587.
13. Girshick, R. Fast R-CNN. In Proceedings of the 2015 IEEE International Conference on Computer Vision (ICCV), Santiago, Chile, 7–13 December 2015; pp. 1440–1448.
14. Ren, S.; He, K.; Girshick, R.; Sun, J. Faster R-CNN: Towards Real-Time Object Detection with Region Proposal Networks. *IEEE Trans. Pattern Anal. Mach. Intell.* **2017**, *39*, 1137–1149. [CrossRef]
15. He, K.; Gkioxari, G.; Dollár, P.; Girshick, R. Mask R-CNN. *IEEE Trans. Pattern Anal. Mach. Intell.* **2020**, *42*, 386–397. [CrossRef]
16. Redmon, J.; Divvala, S.; Girshick, R.; Farhadi, A. You only look once: Unified, real-time object detection. In Proceedings of the 2016 IEEE Conference on Computer Vision and Pattern Recognition (CVPR), Las Vegas, NV, USA, 27–30 June 2016; pp. 779–788.
17. Redmon, J.; Farhadi, A. YOLO9000: Better, faster, stronger. In Proceedings of the 2017 IEEE Conference on Computer Vision and Pattern Recognition (CVPR), Honolulu, HI, USA, 21–26 July 2017; pp. 6517–6525.
18. Bochkovskiy, A.; Wang, C.-Y.; Liao, H.-Y.M. Yolov4: Optimal speed and accuracy of object detection. *arXiv* **2020**, arXiv:2004.10934.
19. Ye, Y.; Ren, X.; Zhu, B.; Tang, T.; Tan, X.; Gui, Y.; Yao, Q. An adaptive attention fusion mechanism convolutional network for object detection in remote sensing images. *Remote Sens.* **2022**, *14*, 516. [CrossRef]
20. Xu, D.; Wu, Y. Improved YOLO-V3 with DenseNet for multi-scale remote sensing target detection. *Sensors* **2020**, *20*, 4276. [CrossRef]
21. Li, K.; Wan, G.; Cheng, G.; Meng, L.; Han, J. Object detection in optical remote sensing images: A survey and a new benchmark. *ISPRS J. Photogramm. Remote Sens.* **2020**, *159*, 296–307. [CrossRef]
22. Qu, Z.; Zhu, F.; Qi, C. Remote Sensing Image Target Detection: Improvement of the YOLOv3 Model with Auxiliary Networks. *Remote Sens.* **2021**, *13*, 3908. [CrossRef]
23. Zhang, M.; Xu, S.; Song, W.; He, Q.; Wei, Q. Lightweight Underwater Object Detection Based on YOLO v4 and Multi-Scale Attentional Feature Fusion. *Remote Sens.* **2021**, *13*, 4706. [CrossRef]
24. Jing, Y.; Ren, Y.; Liu, Y.; Wang, D.; Yu, L. Automatic Extraction of Damaged Houses by Earthquake Based on Improved YOLOv5: A Case Study in Yangbi. *Remote Sens.* **2022**, *14*, 382. [CrossRef]
25. Wang, C.Y.; Liao, H.Y.M.; Wu, Y.H.; Chen, P.Y.; Hsieh, J.W.; Yeh, I.H. CSPNet: A new backbone that can enhance learning capability of CNN. In Proceedings of the 2020 IEEE/CVF Conference on Computer Vision and Pattern Recognition Workshops (CVPRW), Seattle, WA, USA, 14–19 June 2020; pp. 1571–1580.
26. Zheng, Z.; Wang, P.; Liu, W.; Li, J.; Ye, R.; Ren, D. Distance-IoU loss: Faster and better learning for bounding box regression. In Proceedings of the AAAI 2020 Conference, New York, NY, USA, 7–12 February 2020.
27. Law, H.; Deng, J. CornerNet: Detecting Objects as Paired Keypoints. *Int. J. Comput. Vis.* **2020**, *128*, 642–656. [CrossRef]
28. Tian, Z.; Shen, C.; Chen, H.; He, T. FCOS: Fully convolutional one-stage object detection. In Proceedings of the 2019 IEEE/CVF International Conference on Computer Vision (ICCV), Seoul, Republic of Korea, 27 October–2 November 2019; pp. 9626–9635.
29. Ge, Z.; Liu, S.; Wang, F.; Li, Z.; Sun, J. YOLOX: Exceeding YOLO Series in 2021. *arXiv* **2021**, arXiv:2107.08430.
30. Vaswani, A.; Shazeer, N.M.; Parmar, N.; Uszkoreit, J.; Jones, L.; Gomez, A.N.; Kaiser, L.; Polosukhin, I. Attention is All you Need. *arXiv* **2017**, arXiv:1312.4400.
31. Hu, J.; Shen, L.; Sun, G. Squeeze-and-Excitation networks. In Proceedings of the 2018 IEEE/CVF Conference on Computer Vision and Pattern Recognition, Salt Lake City, UT, USA, 18–23 June 2018; pp. 7132–7141.
32. Wang, Q.; Wu, B.; Zhu, P.; Li, P.; Zuo, W.; Hu, Q. ECA-Net: Efficient channel attention for deep convolutional neural networks. In Proceedings of the 2020 IEEE/CVF Conference on Computer Vision and Pattern Recognition (CVPR), Seattle, WA, USA, 13–19 June 2020; pp. 11531–11539.
33. Hou, Q.; Zhou, D.; Feng, J. Coordinate attention for efficient mobile network design. In Proceedings of the 2021 IEEE/CVF Conference on Computer Vision and Pattern Recognition (CVPR), Nashville, TN, USA, 20–25 June 2021; pp. 13708–13717.
34. Woo, S.; Park, J.; Lee, J.Y.; Kweon, I.S. CBAM: Convolutional block attention module. In Proceedings of the European Conference on Computer Vision, Munich, Germany, 8–14 September 2018.
35. Sun, C.; Zhang, S.; Qu, P.; Wu, X.; Feng, P.; Tao, Z.; Zhang, J.; Wang, Y. MCA-YOLOV5-Light: A Faster, Stronger and Lighter Algorithm for Helmet-Wearing Detection. *Appl. Sci.* **2022**, *12*, 9697. [CrossRef]

36. Lin, T.-Y.; Maire, M.; Belongie, S.; Hays, J.; Perona, P.; Ramanan, D.; Dollár, P.; Zitnick, C.L. Microsoft COCO: Common objects in context. In Proceedings of the Computer Vision—ECCV 2014, Zurich, Switzerland, 6–12 September 2014; Springer: Cham, Switzerland, 2014; pp. 740–755.

37. Xia, G.S.; Bai, X.; Ding, J.; Zhu, Z.; Belongie, S.; Luo, J.; Datcu, M.; Pelillo, M.; Zhang, L. DOTA: A large-scale dataset for object detection in aerial images. In Proceedings of the 2018 IEEE/CVF Conference on Computer Vision and Pattern Recognition, Salt Lake City, UT, USA, 18–23 June 2018; pp. 3974–3983.

38. Yu, X.; Gong, Y.; Jiang, N.; Ye, Q.; Han, Z. Scale match for tiny person detection. In Proceedings of the IEEE Winter Conference on Applications of Computer Vision, Snowmass Village, CO, USA, 1–5 March 2020; pp. 1246–1254.

39. Chen, C.; Liu, M.-Y.; Tuzel, O.; Xiao, J. R-CNN for small object detection. In Proceedings of the ACCV 2016—Asian Conference on Computer Vision, Taipei, Taiwan, 21–23 November 2016.

40. Heitz, G.; Koller, D. Learning spatial context: Using stuff to find things. In Proceedings of the ECCV 2008—10th European Conference on Computer Vision, Marseille, France, 12–18 October 2008.

41. Razakarivony, S.; Jurie, F. Vehicle detection in aerial imagery: A small target detection benchmark. *J. Vis. Commun. Image Represent.* **2016**, *34*, 187–203. [CrossRef]

42. Mundhenk, T.N.; Konjevod, G.; Sakla, W.A.; Boakye, K. A Large contextual dataset for classification, detection and counting of cars with deep learning. In Proceedings of the ECCV—Computer Vision, Amsterdam, The Netherlands, 11–14 October 2016.

43. Liu, Z.; Wang, H.; Weng, L.; Yang, Y. Ship Rotated Bounding Box Space for Ship Extraction From High-Resolution Optical Satellite Images With Complex Backgrounds. *IEEE Geosci. Remote Sens. Lett.* **2016**, *13*, 1074–1078. [CrossRef]

44. Zhu, H.; Chen, X.; Dai, W.; Fu, K.; Ye, Q.; Jiao, J. Orientation robust object detection in aerial images using deep convolutional neural network. In Proceedings of the 2015 IEEE International Conference on Image Processing (ICIP), Quebec City, QC, Canada, 27–30 September 2015; pp. 3735–3739.

45. Pham, M.-T.; Courtrai, L.; Friguet, C.; Lefèvre, S.; Baussard, A. YOLO-Fine: One-Stage Detector of Small Objects Under Various Backgrounds in Remote Sensing Images. *Remote Sens.* **2020**, *12*, 2501. [CrossRef]

46. Zhang, W.; Wang, S.-P.; Thachan, S.; Chen, J.; Qian, Y.-t. Deconv R-CNN for small object detection on remote sensing images. In Proceedings of the IGARSS—IEEE International Geoscience Remote Sensing Symposium, Valencia, Spain, 22–27 July 2018; pp. 2483–2486.

47. Qin, H.; Li, Y.; Lei, J.; Xie, W.; Wang, Z. A Specially Optimized One-Stage Network for Object Detection in Remote Sensing Images. *IEEE Geosci. Remote Sens. Lett.* **2021**, *18*, 401–405. [CrossRef]

48. Ioffe, S.; Szegedy, C. Batch normalization: Accelerating deep network training by reducing internal covariate shift. In Proceedings of the 32nd International Conference on International Conference on Machine Learning—Volume 37, Lille, France, 6–11 July 2015; pp. 448–456.

49. Tzutalin. LabelImg. Git Code. 2015. Available online: https://github.com/tzutalin/labelImg (accessed on 10 September 2020).

50. Liu, K.; Mattyus, G. Fast Multiclass Vehicle Detection on Aerial Images. *IEEE Geosci. Remote Sens. Lett.* **2015**, *12*, 1938–1942. [CrossRef]

51. Zhang, Y.; Yuan, Y.; Feng, Y.; Lu, X. Hierarchical and Robust Convolutional Neural Network for Very High-Resolution Remote Sensing Object Detection. *IEEE Trans. Geosci. Remote Sens.* **2019**, *57*, 5535–5548. [CrossRef]

52. Everingham, M.; Eslami, S.M.; Van Gool, L.; Williams, C.K.I.; Winn, J.; Zisserman, A. The pascal visual object classes challenge: A retrospective. *Int. J. Comput. Vis.* **2015**, *111*, 98–136. [CrossRef]

 *remote sensing*

*Article*

# MANet: A Network Architecture for Remote Sensing Spatiotemporal Fusion Based on Multiscale and Attention Mechanisms

**Huimin Cao, Xiaobo Luo *, Yidong Peng and Tianshou Xie**

College of Computer Science and Technology, Chongqing University of Posts and Telecommunications, Chongqing 400065, China
* Correspondence: luoxb@cqupt.edu.cn

**Abstract:** Obtaining high-spatial–high-temporal (HTHS) resolution remote sensing images from a single sensor remains a great challenge due to the cost and technical limitations. Spatiotemporal fusion (STF) technology breaks through the technical limitations of existing sensors and provides a convenient and economical solution for obtaining HTHS resolution images. At present, most STF methods use stacked convolutional layers to extract image features and then obtain fusion images by using a summation strategy. However, these convolution operations may lead to the loss of feature information, and the summation strategy results in poorly fused images due to a lack of consideration of global spatial feature information. To address these issues, this article proposes a STF network architecture based on multiscale and attention mechanisms (MANet). The multiscale mechanism module composed of dilated convolutions is used to extract the detailed features of low-spatial resolution remote sensing images at multiple scales. The channel attention mechanism adaptively adjusts the weights of the feature map channels to retain more temporal and spatial information in the upsampling process, while the non-local attention mechanism adjusts the initial fusion images to obtain more accurate predicted images by calculating the correlation between pixels. We use two datasets with different characteristics to conduct the experiments, and the results prove that the proposed MANet method with fewer parameters obtains better fusion results than the existing machine learning-based and deep learning-based fusion methods.

**Keywords:** multiscale mechanism; STF; non-local attention; dilated convolution

**Citation:** Cao, H.; Luo, X.; Peng, Y.; Xie, T. MANet: A Network Architecture for Remote Sensing Spatiotemporal Fusion Based on Multiscale and Attention Mechanisms. *Remote Sens.* **2022**, *14*, 4600. https://doi.org/10.3390/rs14184600

Academic Editor: Qiangqiang Yuan

Received: 22 July 2022
Accepted: 7 September 2022
Published: 15 September 2022

**Publisher's Note:** MDPI stays neutral with regard to jurisdictional claims in published maps and institutional affiliations.

## 1. Introduction

HTHS resolution remote sensing images are significant for remote sensing application fields such as urban land cover mapping [1], disaster warning [2], surface change detection [3], assessment of the area affected by an earthquake [4], and urban heat island monitoring [5]. The temporal and spatial resolutions of remote sensing images acquired by different sensors are mutually limited, and these sensors are broadly divided into two main types. One type is equipped on the Landsat series, Gaofen series, Sentinel, and other satellites, and the other is the Moderate-resolution Imaging Spectroradiometer (MODIS). The Landsat series contains a diverse range of advanced thermal infrared sensors and mappers for mapping, which have different sensitivities to different bands. Remote sensing images required by the American Landsat series have a high-spatial resolution of 15–30 m and a revisit cycle of approximately 16 days. In contrast, remote sensing images obtained by MODIS on Terra/Aqua have a low-spatial resolution of 250 m–1 km and a revisit cycle of one day. However, it is difficult to obtain cloud-free image data for some months in many areas because of the interference of cloudy weather, which reduces the temporal resolution of the images to some extent. As sensor technology and deep learning improve by leaps and bounds, the research that uses STF methods to obtain HTHS resolution images has also attracted increasing attention [6]. STF is an effective method that can combine high-temporal–low-spatial (HTLS) remote sensing images with low-temporal–high-spatial (LTHS) remote sensing images to generate HTHS remote sensing images [7]. Although unmanned aerial vehicles (UAVs) can easily obtain HTHS resolution images, they do not apply to practical remote sensing applications for monitoring large surface areas because the image size they obtain is relatively small. In addition, it is difficult for UAVs to obtain images of depopulated zones, and most of the images obtained by UAVs are not publicly available, while remote sensing images obtained by satellites not only cover a wide area, but also most of them are free. Therefore, the major way to obtain HTHS images is through STF methods.

## 2. Related Work

In recent years, a large number of studies have been performed on STF methods for remote sensing images. According to different optimization strategies, STF methods can be roughly classified into four categories: transform-based STF methods, image reconstruction-based STF methods, hybrid pixel decomposition-based STF methods, and learning-based STF methods [8].

Transform-based STF methods involve wavelet transform and principal component analysis methods. STF methods based on wavelet transforms use wavelet transform technology to perform wavelet decomposition on remote sensing images and then fuse each decomposed layer, and the fusion results are ultimately acquired by the inverse wavelet transform [9–11]. In addition, methods based on principal component analysis first use a principal component method to separate the first principal component of high-spatial resolution remote sensing images and then extract the brightness component, and finally merge the extracted brightness image with the resampled low-spatial resolution remote sensing images to obtain fusion images [12].

The principle of the STF method based on image reconstruction is to calculate the weights of the similar adjacent pixels in input images and then obtain the target fusion images through interpolation according to the synthesis weights, including time and space. For example, Gao et al. [13] proposed a STF method STARFM, which is a new model that estimates adjacent pixels' contribution to the reflectance of central pixels by calculating the weights of spectral difference, temporal difference, and pixel location distance. It is a relatively effective method for a study area where the reflectance of adjacent pixels varies little. To boost the pixel reconstruction of STARFM for nonuniform areas, Zhu et al. [14] proposed a STF method ESTARFM, which is an enhanced version of STARFM, that also searches for similar pixels first and calculates the weights of candidate pixels. The difference is that the ESTARFM calculates the weights of similar image pixels and transformation coefficients fully considering the internal relationship of the hybrid image pixels, which makes the experimental results of the algorithm in the region with high heterogeneity perform well compared with the STARFM method. A new STF model based on image reflectance changes (STAARCH) [15] proposed by Hilker et al., which is also inspired by STARFM, detects reflectance changes and denotes disturbances using Tasseled Cap transformations [16,17] of both Landsat images and MODIS image reflectance data.

The essence of the STF method based on unmixing is to unmix the spectral details of high-spatial resolution images at the prior time, and then predict the corresponding HTHS resolution remote sensing images [18]. For example, Zhukov et al. [19] proposed UMMF in 1999, which is a new STF model that first decomposes the spectrum of low-spatial resolution images and then fuses them with high-spatial resolution images to generate HTHS resolution remote sensing images. Based on UMMF, Wu et al. [20] proposed a new STF method STDFA, which considers the spatial and temporal variations in the calculation of the model and finally achieves good fusion results. These two methods require multiple high-spatial resolution images to guarantee fusion accuracy. However, the number of high-spatial resolution images obtained by sensors is limited due to cloud pollution in practical remote sensing applications. To solve this problem, Zhu et al. [21] proposed a flexible STF method FSDAF in 2016, which performs well in heterogeneous regions with a high speed by inputting a cloud-free and high-spatial resolution image.

Learning-based fusion methods can be roughly divided into dictionary-pair learning-based methods [8,22–24] and deep learning-based fusion methods. The algorithms based on dictionary-pair learning predict images by establishing the correspondence mainly according to the structural similarity between low- and high-spatial resolution images. For example, Huang et al. [22] proposed a STF network, SPSTFM, in 2012, a new model based on sparse representation, which is the first time to train dictionary pairs between high-spatial resolution residual images and low-spatial resolution residual images. However, this method is not practical in remote sensing applications because this STF method predicts HTHS images by using multiple high-spatial resolution images. Therefore, Wei et al. [23] proposed an optimization STF model in 2016, which predicts images based on semi-coupled dictionary-pair learning and structural sparsity. In 2021, Peng et al. [25] proposed a STF method, SCDNTSR, based on dictionary learning, which first considers the spectral correlation of image bands and further improves the accuracy of fusion results.

In recent years, deep learning has demonstrated its particular strengths in various fields. Inspired by the super-resolution structure of SRCNN [26] proposed by Dong et al., Tan et al. [27] proposed a STF model, DCSTFN, to predict images by using two branches dealing with spatial and temporal variation information separately. As the convolutional operation in feature extraction leads to the loss of details, EDCSTFN [28] was proposed based on DCSTFN, which added residual coding blocks and designed a compound loss function to improve the ability of extracted features. Considering the nonlinear mapping and super-resolution mapping between the input images, Song et al. [29] proposed the STFDCNN network, which designs two convolutional network branches to learn these two mappings separately and finally obtains the fused images through a weighting strategy. In addition, Liu et al. [30] proposed a fusion approach, StfNet, in 2019, which establishes the temporal dependence between low-spatial resolution images and predicts high-spatial resolution images according to the temporal consistency and the super-resolution technology. Tan et al. [31] proposed a STF model, GAN-STFM, based on unsupervised learning and obtained HTHS resolution images through only two images.

At present, there are still some problems with deep learning-based STF methods. First, the temporal change information and spatial features extracted from low-spatial resolution images by stacked convolutional layers are insufficient [18,27], and some details are lost during the upsampling process [32,33]. Second, a summation fusion strategy may result in poorly fused images due to a lack of consideration of global spatial feature information. To address the above issues, we propose a STF network architecture MANet based on multiscale and attention mechanisms. In MANet, we adopt three images for fusion. First, we obtained a residual image by performing a subtraction operation on two low-spatial resolution images, and then we input it into the whole network with a high-spatial resolution image. Our main contributions in this article are summarized as follows:

1.  A multiscale mechanism is used to extract temporal and spatial change information from low-spatial resolution images at multiple scales, which is to provide more detailed information for the subsequent fusion process.
2.  A residual channel attention upsampling (RCAU) module is designed to upsample the low-spatial resolution image. Inspired by DenseNet [34] and FPN [35] structures, the rich spatial details of high-spatial resolution images are used to complement the spatial loss of low-spatial resolution images during the upsampling process. This collaborative network structure makes the spatial and spectral information of the reconstructed images more accurate.
3.  A non-local attention mechanism is proposed to reconstruct the fused image by learning the global contextual information, which can improve the accuracy of the temporal and spatial information of the fused image.

The rest of the manuscript is organized as follows. Section 3 introduces the overall structure and internal modules of the MANet method. Section 4 describes the experimental part of the model, including the introduction of the datasets, the display of experimental results, and comparisons with other classical STF methods. Section 5 is our discussion, and Section 6 is the conclusion.

## 3. Materials and Methods

### 3.1. MANet Architecture

Figure 1 shows the overall architecture of MANet, in which cubes with different colors represent different convolution operations, ReLU activation functions and other specific operations. The MODIS image at time $t_i$ ($i = 1, 2$) is represented by $M_i$, and the Landsat image at time $t_i$ is represented by $L_i$. The MANet architecture contains three main parts:

- A sub-network is used to process residual low-spatial resolution images, extracting the temporal and spatial variation information.
- A sub-network is used to process high-spatial resolution images, extracting spatial and spectral information.
- To obtain more accurate fused images, a new fusion strategy is introduced to further learn the global temporal and spatial change information of the fused image.

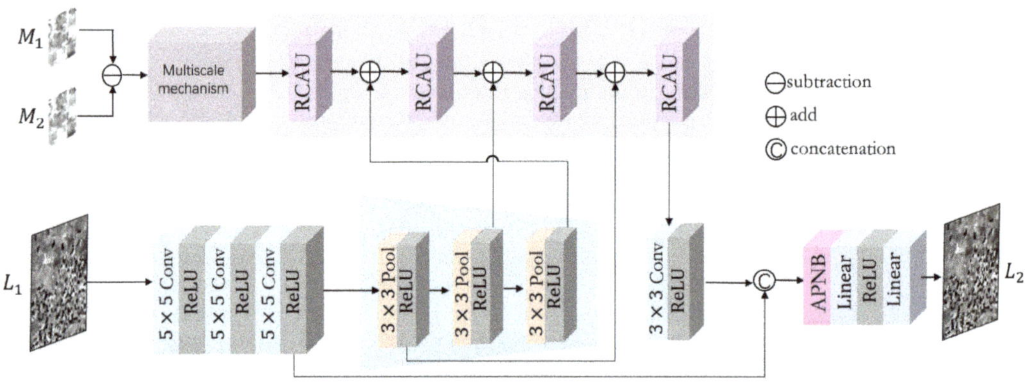


**Figure 1.** The overall architecture of MANet.

We first obtain residual image $M_{12}$ by subtracting $M_1$ from $M_2$, which contains temporal and spatial variation information from time $t_1$ to time $t_2$. We then send this residual image to a multiscale mechanism to extract temporal and spatial change information at multiple scales. Since the size of the MODIS images we input is one-sixteenth that of Landsat images, we need to upsample them to the same size as the Landsat images for subsequent feature fusion. In addition, MODIS images contain fewer spatial details and may lose temporal and spatial information during upsampling. We design a new upsampling module named RCAU, which maintains more temporal and spatial detail information in useful channels during upsampling. Since MODIS images contain less spatial information than Landsat images, we use the rich spatial information of Landsat images to compensate for the loss of spatial feature information during the upsampling operation of MODIS images. We downsample Landsat images and then add the feature maps after downsampling with those of the upsampled MODIS images, and this operation helps to extract the spatial information of MODIS images in the upsampling process. We obtain high-spatial resolution images upsampled by 16 times with four RCAU modules. Meanwhile, we input the Landsat image at a prior time into three 5 × 5 convolution kernel sequences to extract spatial details. Then, we fuse the upsampled feature maps containing temporal and spatial variation information with the feature map extracted from the Landsat image and obtain preliminary feature maps containing temporal and spatial information. In the process of feature fusion, the local temporal and spatial information of the feature map may be wrong. Therefore, we use an asymmetrical pyramid non-local block (APNB) [36] module to learn the global temporal and spatial information from the preliminary feature map and obtain the enhanced feature map. Finally, the feature maps obtained by the APNB module are sent to the two fully connected layers to obtain the final fusion image $L_2$, which integrates all the temporal and spatial information.

### 3.2. Multiscale Mechanism

The spatial structures of remote sensing images are very complex. In addition, convolution layers with a single receptive field are directly used to extract information, which may result in the loss of detailed information due to the limitation of the receptive field of convolutional layers. To address this issue, we use a multiscale mechanism [37] composed of convolutional kernels with different receptive fields to simultaneously extract temporal and spatial change features, which can improve the fusion accuracy. We input the residual feature maps obtained by subtracting MODIS images into this multiscale mechanism and then concatenate the obtained feature maps at different scales to acquire a feature map containing temporal and spatial variation information, as shown in Figure 2. This multiscale mechanism is composed of three $3 \times 3$ convolution kernels, and their dilation rates are 1, 2, and 3, respectively. The larger the dilation rate is, the larger the receptive field of convolution layers, and the spatial and temporal change information may be more comprehensive. We extract features using three convolution layers with different dilation rates in parallel and then obtain the detailed feature information at different scales. In this article, the feature maps obtained by these three convolutional layers contain 12 channels, respectively, and then these feature maps are concatenated to acquire a feature map with 36 channels.

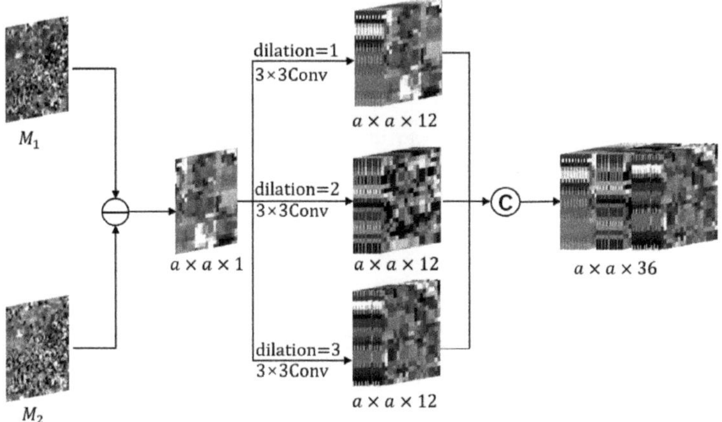


**Figure 2.** Network architecture of the multiscale mechanism.

### 3.3. Attentional Mechanism

#### 3.3.1. RCAU Module

Some spatial and temporal variation information in remote sensing images may be lost during the upsampling process. To retain more feature information in the upsampling process, we design an RCAU module to upsample remote sensing images and input the upsampled feature maps to a channel attention mechanism to adaptively assign weight to each channel according to the importance of channel details. This RCAU module is similar to the RCAB module [38], except that RCAU changes this first convolution layer to a deconvolution layer, which is to achieve the upsampling operation of MODIS images. The RCAB module integrates the channel attention mechanism and a residual block. The channel attention mechanism (CA) can adaptively assign weight to each channel according to the importance of channel details [39]. The residual block is used to combine deep features with shallow features in the network structure to reduce feature loss. The RCAB module has been proven to have a good effect on the application of single RGB image super-resolution [38]. Since the upsampling operation of remote sensing images is different from the super-resolution of natural images, the resolution difference of remote sensing images is approximately 16 times. In the RCAU module, a deconvolution layer is first used to double the spatial resolution, and the resulting feature maps are then sent to the ReLU activation and a convolution layer to further extract features. To reduce the spatial feature loss during upsampling, we use the channel attention mechanism to extract details more efficiently by acquiring dependencies between channels and restraining unnecessary information [40], as shown in Figure 3. Finally, we use a residual structure to add the feature map after the deconvolution operation to the feature map after the channel attention mechanism to fuse the information from shallow and deep network layers. To achieve the 16-times resolution scale fusion of remote sensing images, we need to use four RCAU modules in the MANet structure. For the layer $n$th ($n$ = 1, 2, 3, 4) RCAU module, we have:

$$F_{n,b} = D_{n,b}(F_{n,b-1}) + C_{n,b}(X_{n,b}) \times X_{n,b} \tag{1}$$

where $C_{n,b}$ denotes the channel attention function, $F_{n,b}$ and $F_{n,b-1}$ are the input and output of the RCAU module, respectively, $D_{n,b}$ is the function that acts on the input feature map in the RCAU module, which contains the deconvolution and ReLU operations, and the RCAU learns the residual component $X_{n,b}$ from the input feature map. $X_{n,b}$ is composed of a Conv layer, which can be defined as:

$$X_{n,b} = W_{n,b}^1 \times D_{n,b} \tag{2}$$

where $W_{n,b}^1$ represents the weight of the the Conv layer. $D_{n,b}$ is multiplied by the weight to obtain the residual component $X_{n,b}$. Therefore, the RCAU module not only increases the size of low-spatial resolution images by two times, but also the detailed texture information of low-spatial resolution images can be restored by using the rich spatial details of Landsat images, which is achieved by adding the feature maps of Landsat images that have been downsampled to the low-spatial resolution images. We then input the feature maps obtained after the four RCAU modules into the next convolutional layer to further extract detailed features.

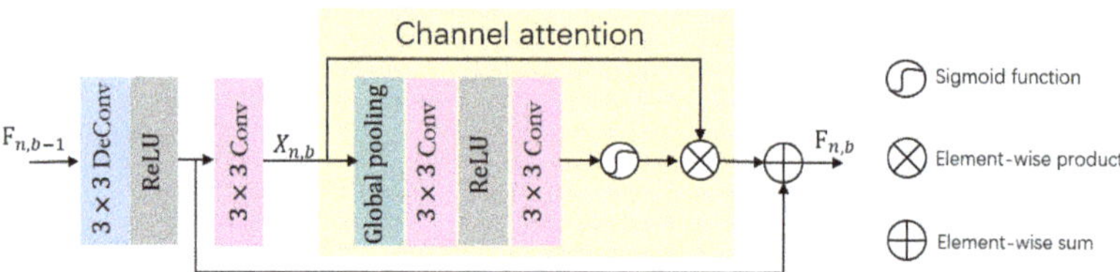


**Figure 3.** The architecture of RCAU module.

### 3.3.2. APNB Architecture

The initial fused image was obtained by a simple addition operation, containing unitary information of spatial details and temporal changes, and the pixels of the image are independent of each other, which may produce jagged edges and noise. If we directly send it to a fully connected layer, the fusion image will contain more noise, and the fusion effect will be worse. Therefore, we used the non-local autocorrelation of the image to restore the non-local information of fusion image and improve the fusion result. This refers to an asymmetrical pyramid non-local block (APNB) module used in the MANet structure, which is an improved non-local model [41]. It realizes remote dependence by calculating the relationship between each query pixel and all the other pixels and aggregating the features of all pixels in the image. Thus, the relationship between pixels in the initial fusion image can be considered from the perspective of global details, making the fusion result close to the real image. It has been proven that APNB can be used to improve segmentation performance in semantic segmentation [42]. Figure 4 shows the network architecture of APNB module, where X is the input initial fusion image. The channels of this image are halved by three $1 \times 1$ convolution layers, and the feature vectors Key, Value and Query are separately generated by flattening. Key and Query are used to calculate the similarity of pixels. Value represents the feature vector directly input to the network. To exploit multiscale correlations, the pyramid pooling layer structure was used for Key and Value to handle correlations at different scales. The adaptive average pooling layer was used to generate $1 \times 1$, $3 \times 3$, $6 \times 6$, and $8 \times 8$ matrices, which were flattened and connected into a vector. This vector was multiplied by the transposed Query to obtain a matrix containing correlations between different pixels. Afterward, the similarity weight was obtained by the softmax operation of this matrix, and then we multiplied the similarity weight by Value to obtain the feature map with global attention. Finally, to add the relationship between global pixels to the fused image, we sent the feature map to a reshape layer and a convolution layer, and then, the feature map was added to the initial fused image X. The latest fused image Y with a global relationship was obtained through two fully connected layers.

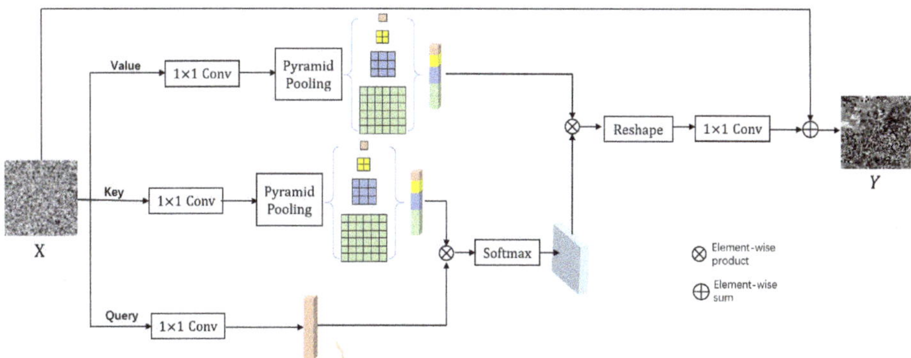


**Figure 4.** The architecture of APNB module.

### 3.4. Loss Function

The MSE loss function is often used to evaluate the error between a predicted image and a real image in a STF model, which ignores the global quality of an image during the training process, we designed a compound loss function, which includes content loss and vision loss. The formula is:

$$L_{MANet} = L_{content} + \alpha \times L_{vision} \tag{3}$$

where $\alpha$ represents the weighting coefficient of vision loss. After many experiments, setting $\alpha$ to 0.8 worked best. The content loss is often used to ensure the pixel-level supervision of an image in a STF model, we use a Charbonnier loss [43] to calculate the content loss by calculating the pixel error between two images in this experiment. In content loss, the similarity between the real image and the predicted image is enforced by enhancing pixel-wise reconstruction, which can better process the outliers in the predicted image that are very different from the pixels in the real image. It can be defined as:

$$L_{content} = \sqrt{(y - x)^2 + \epsilon^2} \tag{4}$$

where $x$ and $y$ are the predicted value and the real observed value, respectively. The $\epsilon$ is used to prevent the error backpropagation, which is empirically set to $1 \times 10^{-3}$. In the compound loss function, the content loss is to improve the similarity of the texture details between the predicted image and the real image, while the vision loss is to measure the visual similarity between images [44]. In the STF model, the multiscale structural similarity (MS-SSIM) [28,45] is used to calculate the vision loss, which is the multiscale version of SSIM. The SSIM index is used to comprehensively evaluate the similarity of images based on three parts: structure, contrast, and luminance, and it can also evaluate the structural similarity of images by calculating the mean, variance, and covariance between the real image and the predicted image. MS-SSIM is used to calculate the structural similarity of multiple levels after reducing the image to different scales, which reduces noise and blur around edges to obtain more accurate predicted images. Vision loss can be obtained by MS-SSIM, which can be defined as:

$$
L_{vision} = 1 - \prod_{m=1}^{M} \left( \frac{2\mu_x\mu_y + c_1}{\mu_x^2 + \mu_y^2 + c_1} \right)^{\beta_m} \left( \frac{2\sigma_{xy} + c_2}{\sigma_x^2 + \sigma_y^2 + c_2} \right)^{\gamma_m}
\tag{5}
$$

where $M$ represents the highest scale, $\beta_m$ and $\gamma_m$ represent the proportion of the two fractions, $\mu_x$ and $\mu_y$ represent the mean of the predicted image $x$ and the real image $y$, respectively. $\sigma_x^2$ and $\sigma_y^2$ represent the variance of the predicted image $x$ and the real image $y$, respectively. $\sigma_{xy}$ represents the covariance of the predicted image $x$ and the real image $y$. $c_1$ and $c_2$ are two constants to ensure the stability of the formula.

The experimental results show that MS-SSIM can effectively restore the high-frequency characteristics of the predicted image [32]. Therefore, adding a vision loss function to the compound loss function can obtain more accurate prediction results.

## 4. Experiments

### 4.1. Datasets

To verify the effect of the proposed fusion model, we use two datasets to conduct the experiment. Figure 5a shows the Lower Gwydir Catchment (LGC) [46], which is located in northern New South Wales, Australia (NSW, 149.2815°E, 29.0855°S). This dataset contains 14 pairs of cloud-free MODIS-Landsat images from 16 April 2004 to 3 April 2005. MODIS images were obtained from MODIS Terra MOD09GA Collection 5 Data, and Landsat images were obtained from Landsat-5 TM and were atmospherically corrected using the algorithm [47] proposed by Li et al. The LGC dataset mainly takes the land cover area as the experimental area, including arid farmland, irrigated paddy fields, and forest land and the spectral information in the area is more variable; thus, we mainly observe spectral changes [18]. The original LGC dataset image size is 3200 × 2720 and consists of six bands

Figure 5b shows the Coleambally Irrigation Area (CIA) study cite [46], which is located in southern New South Wales, Australia (NSW, 34.0034°E, 145.0675°S). This dataset contains 17 pairs of cloud-free MODIS-Landsat images from 7 October 2001 to 17 May 2002. The MODIS images were obtained by MODIS Terra MOD09GA Collection 5 data, and the Landsat images were obtained by Landsat-7 ETM+ and were atmospherically corrected using MODTRAN4 [48] as outlined inVan Niel and McVicar [49]. On the CIA dataset farmlands are mainly selected as the experimental area, and the phenological changes on different dates were obvious; thus, we take it as the dataset with high-spatial heterogeneity [18]. The original CIA dataset size is 1720 × 2040 and contains a total of six bands.

Before training the network, we first cropped all these images from the center to a size of 1200 × 1200. The resolution difference between the original Landsat and MODIS images is 16 times, and we scale all the MODIS images to a size of 75 × 75 for reducing training parameters. From Figures 6 and 7, we can see the changes in the MODIS-Landsat image pairs of the CIA and LGC datasets on different dates, and the two datasets were input into the MANet structure for training. In these two datasets, we arranged the MODIS-Landsat image pairs in chronological order, and the temporally closest two image pairs were grouped in a data group according to the temporal distance. The time of the reference image is always before, and the time of the predicted image is always after. Finally, there are 16 data groups available in the CIA dataset and 13 data groups available in the LGC dataset

The grouped data is then randomly assigned to 60% of the dataset as the training dataset, 20% as the validation dataset, and the remaining 20% as the test dataset. In the whole experiment, the three parts of the datasets were selected assuming there was no intersection.

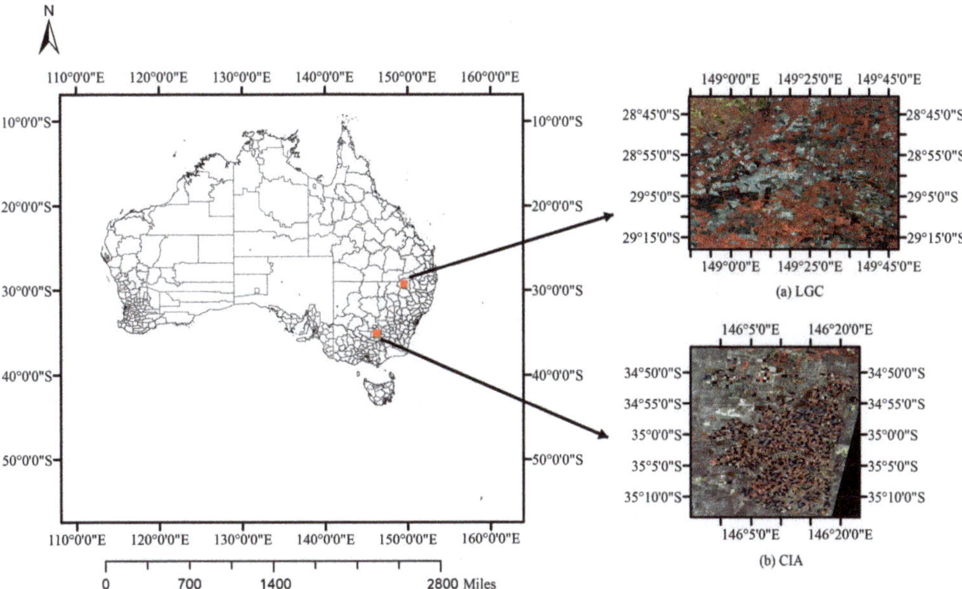

**Figure 5.** Location of the Coleambally Irrigation Area (CIA) and the Lower Gwydir Catchment (LGC).

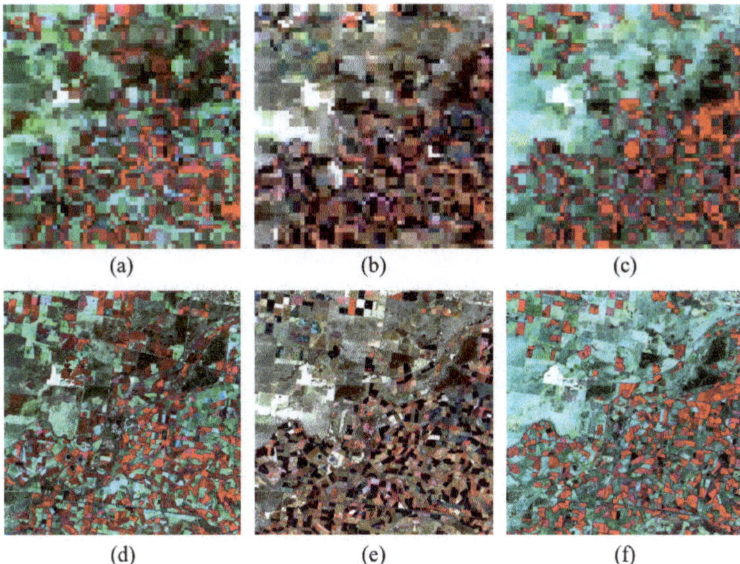

**Figure 6.** Comparison of CIA image pairs on 7 October 2001, 24 November 2001, and 9 March 2002. (**a**,**d**) are the MODIS and Landsat images on 7 October 2001, respectively. (**b**,**e**) are the MODIS and Landsat images on 24 November 2001, respectively. (**c**,**f**) are the MODIS and Landsat images on 9 March 2002, respectively.

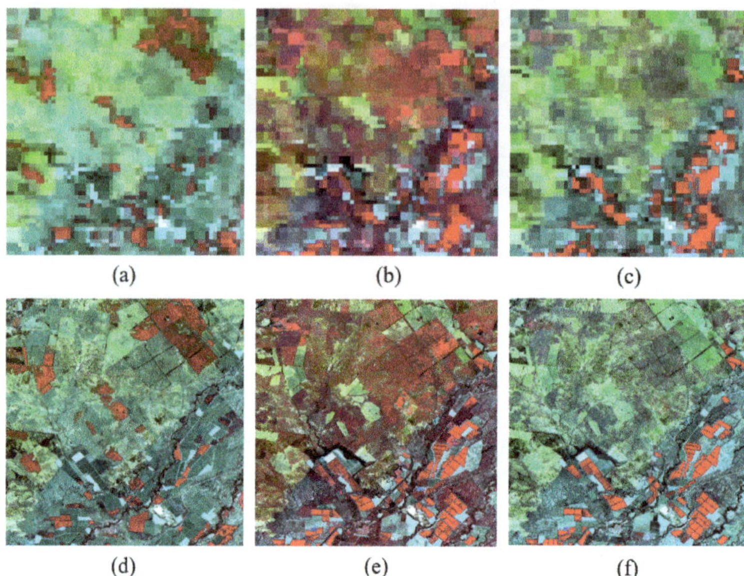

**Figure 7.** Comparison of LGC image pairs on 22 August 2004, 28 December 2004 and 13 January 2005. (**a,d**) are the MODIS and Landsat images on 22 August 2004, respectively. (**b,e**) are the MODIS and Landsat images on 28 December 2004, respectively. (**c,f**) are the MODIS and Landsat images on 13 January 2005, respectively.

### 4.2. Evaluation Indicators

To make quantitative evaluations of our proposed STF model, we compared MANet with STARFM [13], FSDAF [21], DCSTFN [27], and DMNet [18] under the same conditions. We performed the same experiment on both datasets for all methods because these methods all use two low-spatial resolution images and one high-spatial resolution image for STF.

Firstly, we used the structural similarity (SSIM) index [50] to evaluate the similarity of two images from multiple perspectives. It can be defined as:

$$\text{SSIM}(x,y) = \frac{(2\mu_x\mu_y + c_1)(2\sigma_{xy} + c_2)}{(\mu_x^2 + \mu_y^2 + c_1)(\sigma_x^2 + \sigma_y^2 + c_2)} \tag{6}$$

where $\mu_x$ and $\mu_y$ represent the mean of the predicted image $x$ and the real image $y$, respectively. $\sigma_x^2$ and $\sigma_y^2$ represent the variance of the predicted image $x$ and the real image $y$, respectively. $\sigma_{xy}$ represents the covariance of the predicted image $x$ and the true image $y$. $c_1$ and $c_2$ are two constants to avoid system errors. The range of SSIM value is $[-1, 1]$. The larger the value of SSIM is, the smaller the difference between the predicted image and the real image; that is, the predicted image quality is better.

The second indicator is the peak signal-to-noise ratio (PSNR) [51], which is used to assess the loss of signal recovery. It can be indirectly defined by the mean square error (MSE), which refers to the mean of the sum of the squared differences between the predicted and the real image pixel values. MSE can be defined as:

$$\text{MSE} = \frac{1}{mn} \sum_{i=0}^{m-1} \sum_{j=0}^{n-1} ||y(i,j) - x(i,j)||^2 \tag{7}$$

where $m$ and $n$ represent the height and width of the image, respectively. $y$ and $x$ are the real observed image and the predicted image. PSNR can be defined as:

$$\text{PSNR} = 20 \times log_{10}\left(\frac{MAX_y}{\sqrt{MSE}}\right) \tag{8}$$

where $MAX_y$ represents the maximum possible pixel value of the real image $y$. The higher the value of PSNR is, the less distortion between the predicted image and the real image; that is, the predicted image quality is better.

The third index we used is the spatial correlation coefficient (CC) [52], which measures the spatial information similarity between the predicted image $x$ and the real observed image $y$. It can be defined as:

$$CC = \frac{\sum_{i=1}^{n}(x_i - \bar{x})(y_i - \bar{y})}{\sqrt{\sum_{i=1}^{n}(x_i - \bar{x})^2 \sum_{i=1}^{n}(y_i - \bar{y})^2}} \tag{9}$$

The range of CC value is $[-1, 1]$. The closer the CC is to 1, the larger the positive correlation between the real observed image and the predicted image.

Finally, we used the root mean square error (RMSE) [27] index to measure the deviation between the predicted value $x$ and the real observed value $y$. Specifically, it is the square root of MSE. It can be defined as:

$$RMSE = \sqrt{\frac{1}{mn} \sum_{i=1}^{m} \sum_{j=1}^{n} (y(i,j) - x(i,j))^2} \tag{10}$$

where $m$ and $n$ represent the height and width of the image, respectively. $y$ and $x$ are the real observed value and the predicted value. The closer the RMSE is to 0, the closer the predicted image is to the real image.

### 4.3. Parameter Setting

STARFM [13] and FSDAF [21] are machine learning-based models that use 20% of the datasets to test directly in experiments without training. DCSTFN [27], DMNet [18] and MANet are all deep learning-based frameworks. MANet is a PyTorch-based framework that uses the Adam optimizer to optimize network training parameters. The weight attenuation is set to $1 \times 10^{-6}$, the initial learning rate is set to 0.0008, and the training epoch is set to 30. We trained MANet for 6 h in a Windows 10 professional environment, equipped with 16 GB RAM, an Intel Core I5-10400 CPU @2.90 GHz, and a NVIDIA GeForce RTX 3060 GPU.

### 4.4. Experiment Results

#### 4.4.1. Subjective Evaluation

Figure 8 shows the prediction results of various fusion methods on the CIA dataset on 26 April 2002. "GT" represents the real image, and "Proposed" is our MANet method. As Figure 8 shows, the field of the CIA dataset is relatively small, and it has strong spatial heterogeneity. For better visual comparison, we extracted and enlarged the sharp contrast part. The figure shows that all the fusion methods can improve the spatial resolution of the predicted images to a certain extent, indicating that these fusion methods can roughly recover the temporal changes, spatial variations, and spectral change of the predicted images. However, in some heterogeneous regions, the fusion results of different fusion methods are different. As shown in the figure, the fusion results of the STARFM fusion method and FSDAF fusion method have been seriously distorted in spectral details. The "GT" image shows a white area, while the STARFM predicted image shows obvious purple patches and loses texture details. This may be because the STF method is heavily affected by the search window during the process of image pixel prediction and performs poorly when the image has high-spatial heterogeneity. In the FSDAF predicted image, there are also some purple patches, and the edge of the farmland is fuzzy. This may be because the fusion method uses a TPS algorithm to predict high-spatial resolution images from low-spatial resolution images. As the figure shows, the spectral information of the white area of the DCSTFN predicted result experienced an error, and the fuzzy effect also appeared at the edge of farmland, which may be caused by the loss of spatial information after using multiple convolution layers. Although the results predicted by the DMNet fusion method show good texture details and the spatial information was retained relatively completely, the spectral distortion was relatively serious, which might be related to the use of a simple addition method for fusion. For our proposed method, the farmland edge information is well processed. Although the spectral information is not accurately reflected, the color difference is relatively small, and the white area is partially restored, making it relatively similar to that of the real image. This shows that our proposed method has a better effect on the high-spatial heterogeneity dataset than the other fusion methods. This is because we paid more attention to extracting spatial and temporal details by introducing a multiscale mechanism.

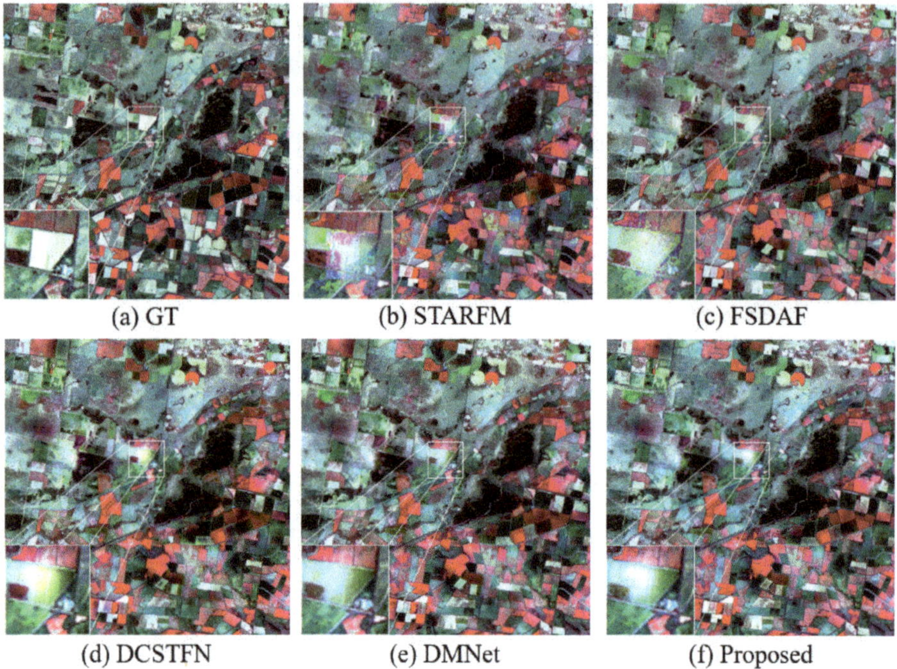

<table>
<tr><td align="center">(a) GT</td><td align="center">(b) STARFM</td><td align="center">(c) FSDAF</td></tr>
<tr><td align="center">(d) DCSTFN</td><td align="center">(e) DMNet</td><td align="center">(f) Proposed</td></tr>
</table>

**Figure 8.** Predicted results of the high-spatial resolution image (26 April 2002) on the CIA [46] dataset. Additionally, the comparison methods include STARFM [13], FSDAF [21], DCSTFN [27] and DMNet [18], which were represented by (**b–e**) in the figure, respectively. Moreover, the GT is the ground truth represented by (**a**), and (**f**) is our proposed STF method.

Figure 9 shows the prediction results of various fusion methods on the LGC dataset on 2 March 2005. "GT" represents the real image, and "Proposed" is our MANet method. Since the variation of spectral information on the LGC dataset is large, we mainly compared the spectral changes and boundary information of the fusion results. For visual comparison, we also extracted and enlarged the sharp-contrast part. As the figure shows, all fusion methods can achieve good prediction of spatial details in most areas. However, in some regions where the spectral information changes greatly, the prediction results of each fusion method are different. As shown in the figure, the predicted images of the STARFM fusion method and FSDAF fusion method exhibit spectral distortion. A red line is shown in the "GT" image, but there are red patches in the STARFM predicted image, which is a serious spectral distortion. This is because STARFM uses surrounding pixels to reconstruct the central pixel, which results in spectral distortion because it is not conducive to the restoration of boundary details. Some black patches in the red area of the "GT" image disappeared in the STARFM predicted image, indicating that spectral changes and boundary information of the STARFM fusion method were lost, which may be caused by the settings of the search window. As shown by the FSDAF prediction results, although the red patches are reduced, there is still spectral distortion, which may also be due to partial information lost in the prediction process and the TPS interpolation operation. The methods based on machine learning performed poorly in processing boundary details in the region where spectral information varies greatly. The DCSTFN and DMNet STF approaches still have some fuzzy phenomena in processing boundary information. In the DCSTFN prediction results, the red line is not smooth enough, and the texture details are not well processed. This may be caused by the loss of detailed information during the process of using multiple convolutional layers in this method. DCSTFN and DMNet can recover the spectral information of the image to some extent. The predicted result of our method is smoother than that of others methods in processing the red line, and the spectral information and boundary information can be well predicted. In general, compared with other fusion methods, our proposed method not only achieves accurate prediction of texture details, but also processes the spectral details well.

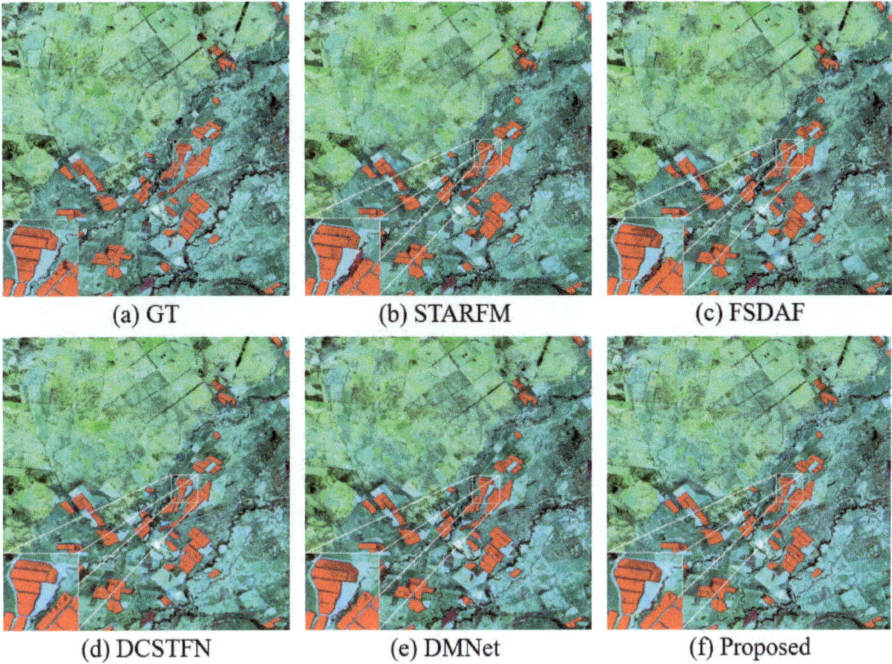

(a) GT      (b) STARFM      (c) FSDAF

(d) DCSTFN      (e) DMNet      (f) Proposed

**Figure 9.** Predicted results of the high-spatial resolution image (2 March 2005) on the LGC [46] dataset. Additionally, comparison methods include STARFM [13], FSDAF [21], DCSTFN [27] and DMNet [18], which were represented by (**b**–**e**) in the figure, respectively. Moreover, the GT is the ground truth represented by (**a**), and (**f**) is our proposed STF method.

### 4.4.2. Objective Evaluation

Table 1 shows the quantitative evaluation results of various fusion methods on the CIA dataset with high-spatial heterogeneity. The best values of the index are marked in bold. As the table shows, the prediction results of our proposed MANet fusion method are improved in terms of most indicators compared with those of other algorithms. For example, in terms of the SSIM index related to spatial information, the result of our proposed method is approximately 2.9% higher than that of the FSDAF fusion method based on machine learning. Compared with the DMNet method based on deep learning, the SSIM values of our method are improved by about 1% on multiple bands. These show that our proposed method can handle spatial variation information of the dataset with high-spatial heterogeneity well. The quantitative evaluation results obtained by the STARFM fusion method are the worst, which may be because the surrounding pixels are used for pixel reconstruction, which is not applicable in a region where spatial information changes greatly. The poor quantitative evaluation result of the FSDAF fusion method may be due to the limitation of the TPS interpolation algorithm. The spectral information is related to RMSE and CC values, and the value of RMSE represents the pixel-level error between the predicted image and the real image in particular. In the quantitative evaluation results of the DCSTFN STF method, the indices of some bands are the best, which shows that DC-STFN can predict the spectral information of these bands well. The SSIM value of DMNet method is better than that of DCSTFN method, which indicates that DMNet can better handle spatial variation information. The values of CC and RMSE of DMNet method are both worse than those of the DCSTFN method, which indicates that DCSTFN method can predict spectral information well. This may be because the DMNet method uses a simple addition strategy for fusion and ignores some useful information. The MANet method acquired the best results on other indexes, such as RMSE and CC values, which indicates that our proposed method can better predict spectral change information. The experimental results indicate that the spatial details and spectral change information of remote sensing images can be better captured by adding multiscale and attentional mechanisms to the network structure.

**Table 1.** Quantitative assessment of different STF methods on the CIA [46] dataset.

| Evaluation | Band | Method | | | | |
|---|---|---|---|---|---|---|
| | | STARFM | FSDAF | DCSTFN | DMNet | Proposed |
| SSIM | Band1 | 0.8731 | 0.9037 | 0.9355 | 0.9368 | **0.9455** |
| | Band2 | 0.8527 | 0.9172 | 0.9304 | 0.9304 | **0.9351** |
| | Band3 | 0.7938 | 0.8578 | 0.8915 | 0.8905 | **0.8989** |
| | Band4 | 0.7329 | 0.8210 | 0.8231 | 0.8271 | **0.8319** |
| | Band5 | 0.7197 | 0.8109 | 0.8165 | 0.8187 | **0.8274** |
| | Band6 | 0.7260 | 0.8194 | 0.8383 | 0.8379 | **0.8432** |
| | Average | 0.7830 | 0.8550 | 0.8726 | 0.8736 | **0.8803** |
| PSNR | Band1 | 27.4332 | 37.2104 | 38.3779 | 38.3696 | **39.2152** |
| | Band2 | 24.3359 | 36.0368 | 36.4337 | 36.3136 | **36.8910** |
| | Band3 | 24.5396 | 31.3339 | 33.2257 | 32.8116 | **33.2862** |
| | Band4 | 19.6533 | 26.9470 | 28.7492 | 28.5944 | **28.8370** |
| | Band5 | 20.8408 | 28.0493 | 28.4029 | 28.1894 | **28.5474** |
| | Band6 | 22.1580 | 25.0635 | 29.8863 | 29.7228 | **29.9921** |
| | Average | 23.1601 | 30.7735 | 32.5126 | 32.3336 | **32.7948** |
| CC | Band1 | 0.3898 | 0.8014 | 0.8374 | 0.8382 | **0.8547** |
| | Badn2 | 0.3965 | 0.7988 | 0.8603 | 0.8581 | **0.8658** |
| | Band3 | 0.5883 | 0.8302 | **0.8912** | 0.8854 | 0.8882 |
| | Band4 | 0.5039 | 0.8161 | 0.8265 | 0.8195 | **0.8272** |
| | Band5 | 0.6855 | 0.8977 | 0.9015 | 0.8989 | **0.9060** |
| | Band6 | 0.6927 | 0.9060 | 0.9153 | 0.9126 | **0.9162** |
| | Average | 0.5428 | 0.8417 | 0.8720 | 0.8688 | **0.8764** |
| RMSE | Band1 | 0.0124 | 0.0124 | 0.0123 | 0.0122 | **0.0112** |
| | Band2 | 0.0156 | 0.0162 | 0.0156 | 0.0158 | **0.0149** |
| | Band3 | 0.0227 | 0.0234 | **0.0226** | 0.0239 | 0.0229 |
| | Band4 | 0.0387 | 0.0408 | 0.0387 | 0.0395 | **0.0385** |
| | Band5 | 0.0386 | 0.0399 | 0.0386 | 0.0394 | **0.0382** |
| | Band6 | 0.0330 | 0.0329 | **0.0324** | 0.0330 | **0.0324** |
| | Average | 0.0268 | 0.0276 | 0.0267 | 0.0273 | **0.0264** |

Table 2 shows the quantitative evaluation results of various fusion methods on the LGC dataset with large spectral changes. The best values of the index are marked in bold. As the table shows, the prediction results of our proposed MANet fusion method are improved in terms of most indicators compared with those of other algorithms. For example, the result of our proposed method is approximately 1% higher than those of other methods in terms of the SSIM index, which indicates that our proposed method can handle spatial variation information. Spectral variation is related to RMSE and CC indexes, the result of our proposed method is improved to a certain degree compared with other methods, which indicates that our proposed method can better predict spectral change information. The quantitative evaluation results of the STARFM fusion method are the worst and with serious spectral distortion, because the method uses the surrounding pixels to predict center pixels with the limits of the search window, so it cannot be applied to the area with great spectral changes. The quantitative evaluation results of the FSDAF fusion method are poor compared with those of the STF methods, which may be because this method uses the TPS interpolation algorithm to predict high-resolution images and finally uses the information of adjacent regions to obtain the predicted images, which leads to spectral distortion due to information loss. In the quantitative evaluation results of the DCSTFN fusion method, the RMSE index values of some bands are optimal, which indicates that DCSTFN method can predict the spectral change information to some extent. The quantitative evaluation results of the DMNet fusion method are inferior to those of DCSTFN because it loses information through an additive fusion strategy. Table 2 shows that our method achieves the best quantitative evaluation results on the SSIM, RMSE, PSNR, and CC indexes. This is because we use high-spatial resolution image features to help restore the spectral information and spatial details of the predicted image. Finally, a non-local attention mechanism is used to pay more attention to the spatial and spectral relations between pixels. This shows that our method can be better applied to regions with large spectral changes.

**Table 2.** Quantitative assessment of different STF methods on the LGC [46] dataset.

| Evaluation | Band | Method | | | | |
|---|---|---|---|---|---|---|
| | | **STARFM** | **FSDAF** | **DCSTFN** | **DMNet** | **Proposed** |
| SSIM | Band1 | 0.8846 | 0.9264 | 0.9361 | 0.9368 | **0.9384** |
| | Band2 | 0.8837 | 0.9300 | **0.9489** | 0.9304 | 0.9488 |
| | Band3 | 0.8401 | 0.9241 | 0.9262 | 0.8905 | **0.9303** |
| | Band4 | 0.8071 | 0.8803 | 0.8901 | 0.8971 | **0.8975** |
| | Band5 | 0.7860 | 0.8693 | 0.8706 | 0.8687 | **0.8842** |
| | Band6 | 0.7908 | 0.8615 | 0.8714 | 0.8779 | **0.8804** |
| | Average | 0.8321 | 0.8986 | 0.9072 | 0.9002 | **0.9133** |
| PSNR | Band1 | 30.4687 | 38.5891 | 39.0567 | 39.5980 | **39.6168** |
| | Band2 | 23.3251 | 37.1057 | 38.0523 | 38.1447 | **38.2195** |
| | Band3 | 23.6144 | 35.0483 | 35.9674 | 35.7742 | **36.0948** |
| | Band4 | 17.4570 | 31.2650 | 31.5236 | 31.4327 | **31.8561** |
| | Band5 | 20.3062 | 30.2034 | 30.9916 | 30.8822 | **31.2151** |
| | Band6 | 21.9842 | 31.0435 | 32.1594 | 31.9054 | **32.2980** |
| | Average | 22.8593 | 33.8758 | 34.6252 | 34.6229 | **34.8834** |
| CC | Band1 | 0.7697 | 0.8802 | 0.8973 | 0.9012 | **0.9090** |
| | Band2 | 0.8775 | 0.8901 | 0.8943 | 0.8939 | **0.9003** |
| | Band3 | 0.8272 | 0.8969 | 0.9052 | 0.9067 | **0.9079** |
| | Band4 | 0.8993 | 0.9090 | 0.9198 | 0.9183 | **0.9209** |
| | Band5 | 0.7816 | 0.9216 | 0.9263 | 0.9242 | **0.9298** |
| | Band6 | 0.7270 | 0.9203 | 0.9228 | 0.9252 | **0.9264** |
| | Average | 0.8137 | 0.9030 | 0.9110 | 0.9116 | **0.9157** |
| RMSE | Band1 | 0.0122 | 0.0139 | 0.0122 | 0.0119 | **0.0117** |
| | Band2 | 0.0134 | 0.0132 | **0.0130** | **0.0130** | 0.0131 |
| | Band3 | 0.0164 | 0.0167 | **0.0162** | 0.0166 | 0.0163 |
| | Band4 | 0.0268 | 0.0276 | 0.0268 | 0.0271 | **0.0259** |
| | Band5 | 0.0291 | 0.0297 | 0.0291 | 0.0298 | **0.0286** |
| | Band6 | 0.0277 | 0.0271 | 0.0257 | 0.0266 | **0.0254** |
| | Average | 0.0209 | 0.0214 | 0.0205 | 0.0208 | **0.0202** |

## 5. Discussion

The experimental results obtained on the CIA dataset show that our method acquired the best result by introducing multiscale and attention mechanisms and a compound loss function in heterogeneous regions. The subjective evaluation shows that the prediction results of the STARFM fusion method and FSDAF fusion method both exhibit serious spectral distortion, while the image predicted by our proposed STF method is relatively closer to the real image. This shows that our method can predict the spectral variation, temporal variation, and spatial features of images in heterogeneous regions. Second, the experimental results obtained on the LGC dataset show that our method can better predict the spectral changes in regions with great spectral changes because our method pays more attention to extracting details and incorporates a new fusion method to retain more detailed features. The following was achieved with the MANet method: (1) feature extraction of low-spatial resolution remote sensing images is realized by using a multiscale mechanism; (2) the upsampling of low-spatial resolution images is performed by using the RCAU module; and (3) a new fusion strategy is introduced to further learn the global temporal and spatial change information of the fused image, which can obtain a more accurate fused image. We use the RCAU module to upsample low-spatial resolution images, in which the channel attention mechanism captures the spatial and spectral details during the upsampling process. Similarly, after the initial fusion image is generated, we send it to the APNB module so that we can capture global information of the predicted image according to the indexes of time and space. Thus, we can obtain more accurate prediction results.

### 5.1. Ablation Experiments

Three experiments were designed to further describe the importance of the multiscale mechanism, the RCAU module, and the APNB module. In the first experiment, we replaced the multiscale mechanism with an ordinary convolution and retained the RCAU module and the APNB module. In the second experiment, we removed the RCAU module and retained the multiscale mechanism and the APNB module. In the third experiment, we removed the APNB module and retained the multiscale mechanism and the RCAU module Table 3 shows the results of Experiment 1, Experiment 2, and Experiment 3, in which "ANet" refers to the network structure with the multiscale mechanism removed, "MAPNet" refers to the network structure with the RCAU module removed, and "MRNet" refers to the network structure with the APNB module removed. The best values of the index are marked in bold.

**Table 3.** The results of comparative experiments.

| Dataset | Index | ANet | MAPNet | MRNet | MANet |
|---|---|---|---|---|---|
| CIA | SSIM | 0.8794 | 0.8791 | 0.8788 | **0.8803** |
| | RMSE | 0.0266 | 0.0267 | 0.0267 | **0.0264** |
| LGC | SSIM | 0.9132 | 0.9131 | **0.9133** | **0.9133** |
| | RMSE | 0.0203 | 0.0204 | 0.0203 | **0.0202** |

As the above table shows, on the CIA dataset, the SSIM value of ANet is greater than that of MRNet and that of MAPNet, which indicates that ANet is better than MRNet and MAPNet in predicting spatial change information. The RMSE value of ANet is less than that of MRNet and that of MAPNet, which indicates that the predicted result of ANet is more accurate than that of MRNet and that of MAPNet in predicting spectral change information. These show that adding attention mechanisms is beneficial to feature extraction of the spectral and spatial variation information. On the LGC dataset, the SSIM value of MRNet is larger than that of MAPNet and that of ANet, which indicates that the predicted result of MRNet is better than that of MAPNet and that of ANet in predicting the spatial change information. The RMSE values of MRNet and ANet are smaller than that of MAPNet, which indicates that MRNet and ANet are better than MAPNet in predicting spectral change information. These show that adding multiscale and attention mechanisms is beneficial to feature extraction of the spectral and spatial variation information. The SSIM and RMSE values of MANet on the CIA dataset and LGC dataset are optimal, which indicates the MANet method can better extract spatial and spectral information compared with other STF methods. Figure 10 shows the results on the CIA dataset of these three comparative experimental methods and our proposed method with band 4 on 26 April 2002. Figure 11 shows the results on the LGC dataset of these three comparative experimental methods and our proposed method with band 4 on 28 December 2004.

    (a) GT      (b) ANet      (c) MAPNet      (d) MRNet      (e) Proposed

**Figure 10.** The results on the CIA dataset of these comparative experimental methods.

    (a) GT      (b) ANet      (c) MAPNet      (d) MRNet      (e) Proposed

**Figure 11.** The results on the LGC dataset of these comparative experimental methods.

In Figures 10 and 11, (a) represents the real image, (b) represents the ANet predicted image, (c) represents the MAPNet predicted image, (d) represents the MRNet predicted image, and (e) represents the MANet predicted image. Figure 10 shows that the ANet predicted image has obvious spectral distortion. The predicted images of MAPNet, MRNet, and MANet are more similar to the real observed images, which indicates that adding a multiscale mechanism can effectively extract the temporal changes and spectral details of images. The ANet method performs well in terms of the quantitative evaluation results, possibly because texture details are lost in the MAPNet and MRNet fusion methods. MANet performs best in terms of quantitative evaluation results, which shows that adding a multiscale mechanism can effectively extract the temporal changes and spectral details and adding attention modules can effectively extract spatial details. As Figure 11 shows, the predicted images of MAPNet and MRNet exhibit spatial and spectrum detail loss, which shows that using the attention mechanisms to extract temporal and spatial details for subsequent image recovery is important in regions with large spectrum variation. Comparatively, the MANet predicted image is more similar to the real image, which indicates that our method can deal well with spectral and spatial details. Although we improved the method of extracting spatial information and spectral details, our study still has deficiencies, such as the prediction accuracy of our method for areas with large topographic variations. Once we have collected enough qualified datasets, we can design a more suitable network structure for more advanced analysis.

*5.2. Loss Curves and the Number of Training Parameters*

Table 4 shows the number of training parameters for various fusion methods. STARFM and FSDAF are fusion methods based on machine learning, so they have no training process. As the table shows, our fusion method has fewer training parameters than other deep learning-based fusion methods. In training the network, the whole dataset is trained in each epoch. As the number of training epochs increases, the accuracy of model training increases. We input the dataset into the MANet structure according to the number of bands to optimize the weights of the network. Figure 12 shows the evolution of the loss curves at the training stage and validation stage for 30 epochs, where each color represents a different band and the solid line and dotted line represent the loss curves at the training stage and validation stage, respectively. Since the loss function is composed of content loss and vision loss, the closer it is to zero, the better the training effect. We can see from Figure 12 that the training loss value decreases rapidly at first and then stabilizes and no longer decreases after 20 epochs, while the validation loss is not stable and fluctuates greatly in the early stage. After more than 25 epochs, all the loss function curves show a relatively stable trend. Therefore, the network tends to converge when the number of epochs is greater than or equal to 30.

**Table 4.** The number of training parameters for various fusion methods.

| Method | STARFM | FSDAF | DCSTFN | DMNet | MANet |
|---|---|---|---|---|---|
| Training parameters | - | - | 298,177 | 327,061 | 77,171 |

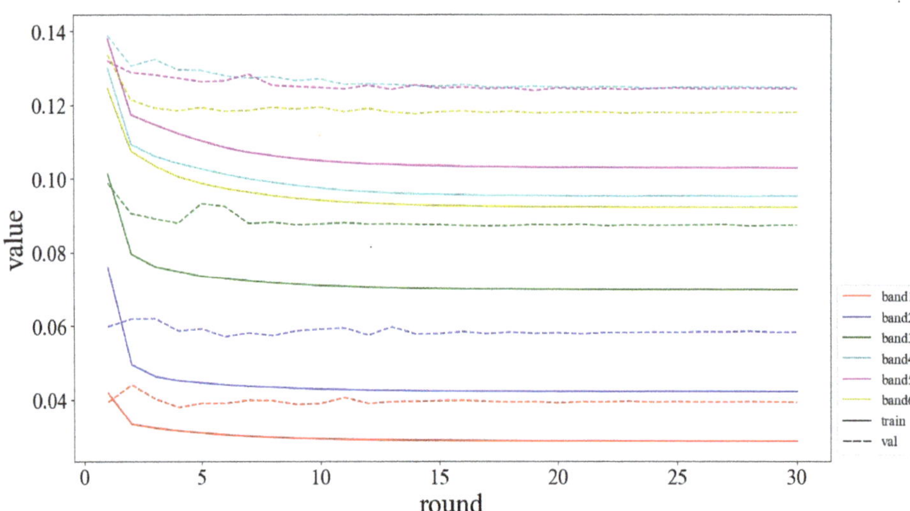


**Figure 12.** The loss curves of MANet for multiple bands on the training and test datasets.

## 6. Conclusions

We evaluated the effectiveness of our proposed STF method MANet by using two datasets with different characteristics and acquired the best final experimental results. The main contributions of our research are introducing a new STF architecture, which includes the following:

1.  The multiscale mechanism is used to extract the temporal and spatial variation of a low-spatial resolution image. The final experimental results indicated that the extraction of detail features at different scales can make the network retain more useful temporal and spatial details, and the prediction result is closer to the real result.

2. By designing the RCAU module, we not only realize the upsampling of feature maps with low-spatial resolution, but also reduce the loss of detail information by the weighting operation, which is more conducive to the reconstruction of low-spatial resolution image pixels.

3. In the fusion process, we have designed a new fusion strategy. The APNB module was added after the initial fusion image, which can effectively extract global spatial and temporal information. Experimental results show that our method can better capture the spatial details and spectral information of the predicted image.

The experimental results show that our method achieves the best prediction results on both the CIA dataset with complex spatial information and the LGC dataset with variable spectral information. From the perspective of the whole fusion framework, the feature information of low-spatial resolution images and the rich spatial information of high-spatial resolution images are both important for predicting HTHS resolution images. The low-spatial resolution image easily loses details in the upsampling process, so we introduce attention mechanisms to restore its spatial resolution and spectral information with the help of channel weights, which is significant in solving temporal and spatial problems. In the STF problems, due to the limitation of fewer available datasets, the predicted accuracy is difficult to greatly improve. Therefore, future research must map low-spatial resolution images to high-spatial resolution images without reference in the prediction stage. These problems can be further discussed.

**Author Contributions:** Data curation, H.C.; formal analysis, H.C.; methodology, H.C. and X.L.; validation, H.C. and X.L.; visualization, X.L. and Y.P.; writing—original draft, H.C.; writing—review and editing, H.C., Y.P. and T.X. The released version of the manuscript has been read and agreed by all authors. All authors have read and agreed to the published version of the manuscript.

**Funding:** This work was supported in part by the National Natural Science Foundation of China under Grant 41871226; in part by the Major Industrial Technology Research and Development Projects of high-tech industry in Chongqing under Grant D2018-82; in part by the Intergovernmental International Scientific and Technological Innovation Cooperation Project of the National key R & D Program Grant 2021YFE0194700; the key cooperation project of Chongqing Municipal Education Commission: HZ2021008.

**Data Availability Statement:** The data that support the findings of this study are openly available in MANet at https://github.com/caohuimin/MANet (accessed on 6 September 2022).

**Acknowledgments:** The authors would like to thank all of the reviewers for their valuable contributions to our article.

**Conflicts of Interest:** The authors declare no conflict of interest.

# References

1. Saah, D.; Tenneson, K.; Matin, M.; Uddin, K.; Cutter, P.; Poortinga, A.; Nguyen, Q.H.; Patterson, M.; Johnson, G.; Markert, K.; et al. Land Cover Mapping in Data Scarce Environments: Challenges and Opportunities. *Front. Environ. Sci.* **2019**, *7*, 150. [CrossRef]

2. Li, M.; Sun, D.; Goldberg, M.; Stefanidis, A. Derivation of 30-m-resolution water maps from TERRA/MODIS and SRTM. *Remote Sens. Environ.* **2013**, *134*, 417–430. [CrossRef]

3. Lv, Z.; Liu, T.F.; Zhang, P.; Benediktsson, J.A.; Lei, T.; Zhang, X. Novel adaptive histogram trend similarity approach for land cover change detection by using bitemporal very-high-resolution remote sensing images. *IEEE Trans. Geosci. Remote Sens.* **2019**, *57*, 9554–9574. [CrossRef]

4. Ma, Y.; Chen, F.; Liu, J.; He, Y.; Duan, J.; Li, X. An Automatic Procedure for Early Disaster Change Mapping Based on Optical remote sensing. *Remote Sens.* **2016**, *8*, 272. [CrossRef]

5. Huang, B.; Wang, J.; Song, H.; Fu, D.; Wong, K. Generating High Spatiotemporal Resolution Land Surface Temperature for Urban Heat Island Monitoring. *IEEE Geosci. Remote Sens. Lett.* **2013**, *10*, 1011–1015. [CrossRef]

6. Dai, P.; Zhang, H.; Zhang, L.; Shen, H. A remote sensing Spatiotemporal Fusion Model of Landsat and Modis Data via Deep Learning. In Proceedings of the IGARSS 2018—2018 IEEE International Geoscience and Remote Sensing Symposium, Valencia, Spain, 22–27 July 2018; pp. 7030–7033.

7. Song, H.; Huang, B. Spatiotemporal satellite image fusion through one-pair image learning. *IEEE Trans. Geosci. Remote Sens.* **2013**, *51*, 1883–1896. [CrossRef]

8.   Li, W.; Cao, D.; Peng, Y.; Yang, C. MSNet: A Multi-Stream Fusion Network for remote sensing Spatiotemporal Fusion Based on Transformer and Convolution. *Remote Sens.* **2021**, *13*, 3724. [CrossRef]
9.   Wu, M.; Wang, C. Spatial and Temporal Fusion of remote sensing Data using wavelet transform. In Proceedings of the 2011 International Conference on Remote Sensing, Environment and Transportation Engineering, Nanjing, China, 24–26 June 2011; pp. 1581–1584.
10.  Gu, X.; Han, L.; Wang, J.; Huang, W.; He, X. Estimation of maize planting area based on wavelet fusion of multi-resolution images. *Trans. Chin. Soc. Agric. Eng.* **2012**, *28*, 203–209. [CrossRef]
11.  Acerbi-Junior, F.W.; Clevers, J.G.P.W.; Schaepman, M.E. The assessment of multi-sensor image fusion using wavelet transforms for mapping the Brazilian Savanna. *Int. J. Appl. Earth Obs. Geoinform.* **2006**, *8*, 278–288. [CrossRef]
12.  Shevyrnogov, A.; Trefois, P.; Vysotskaya, G. Multi-satellite data merge to combine NOAA AVHRR efficiency with Landsat-6 MSS spatial resolution to study vegetation dynamics. *Adv. Space Res.* **2000**, *26*, 1131–1133. [CrossRef]
13.  Gao, F.; Masek, J.; Schwaller, M.; Hall, F. On the blending of the Landsat and MODIS surface reflectance: Predicting daily Landsat surface reflectance. *IEEE Trans. Geosci. Remote Sens.* **2006**, *44*, 2207–2218. [CrossRef]
14.  Zhu, X.; Chen, J.; Gao, F.; Chen, X.; Masek, J.G. An enhanced spatial and temporal adaptive reflectance fusion model for complex heterogeneous regions. *Remote Sens. Environ.* **2010**, *114*, 2610–2623. [CrossRef]
15.  Hilker, T.; Wulder, M.A.; Coops, N.C.; Linke, J.; McDermid, G.; Masek, J.G.; Gao, F.; White, J.C. A new data fusion model for high-spatial-and temporal-resolution mapping of forest disturbance based on Landsat and MODIS. *Remote Sens. Environ.* **2009**, *113*, 1613–1627. [CrossRef]
16.  Crist, E.P.; Kauth, R.J. The tasseled cap de-mystified. *Photogramm. Eng. Remote Sens.* **1986**, *52*, 81–86.
17.  Healey, S.P.; Cohen, W.B.; Yang, Z.; Krankina, O.N. Comparison of Tasseled Cap-based Landsat data structures for use in forest disturbance detection. *Remote Sens. Environ.* **2005**, *97*, 301–310. [CrossRef]
18.  Li, W.; Zhang X.; Peng, Y.; Dong, M. DMNet: A Network Architecture Using Dilated Convolution and Multiscale Mechanisms for Spatiotemporal Fusion of remote sensing Images. *IEEE Sens. J.* **2020**, *20*, 12190–12202. [CrossRef]
19.  Zhukov, B.; Oertel, D.; Lanzl, F.; Reinhackel, G. Unmixing-based multisensor multiresolution image fusion. *IEEE Trans. Geosci. Remote Sens.* **1999**, *37*, 1212–1226. [CrossRef]
20.  Wu, M.; Niu, Z.; Wang, C.; Wu, C.; Wang, L. Use of MODIS and Landsat time series data to generate high-resolution temporal synthetic Landsat data using a spatial and temporal reflectance fusion model. *J. Appl. Remote Sens.* **2012**, *6*, 063507. [CrossRef]
21.  Zhu, X.; Helmer, E.H.; Gao, F.; Liu, D.; Chen, J.; Lefsky, M.A. A flexible spatiotemporal method for fusing satellite images with different resolutions. *Remote Sens. Environ.* **2016**, *172*, 165–177. [CrossRef]
22.  Huang, B.; Song, H. Spatiotemporal reflectance fusion via sparse representation. *IEEE Trans. Geosci. Remote Sens.* **2012**, *50*, 3707–3716. [CrossRef]
23.  Wei, J.; Wang, L.; Liu, P.; Song, W. Spatiotemporal Fusion of remote sensing Images with Structural Sparsity and Semi-Coupled Dictionary Learning. *Remote Sens.* **2017**, *9*, 21. [CrossRef]
24.  Wu, B.; Huang, B.; Zhang, L. An error-bound-regularized sparse coding for spatiotemporal reflectance fusion. *IEEE Trans. Geosci. Remote Sens.* **2015**, *53*, 6791–6803. [CrossRef]
25.  Peng, Y.; Li, W.; Luo, X.; Du, J.; Zhang, X.; Gan, Y.; Gao, X. Spatiotemporal Reflectance Fusion via Tensor Sparse Representation. *IEEE Trans. Geosci. Remote Sens.* **2021**, *60*, 1–18. [CrossRef]
26.  Dong, C.; Loy, C.; He, K.; Tang, X. Image Super-Resolution Using Deep Convolutional Networks. *IEEE Trans. Pattern Anal. Mach. Intell.* **2015**, *38*, 295–307. [CrossRef]
27.  Tan, Z.; Yue, P.; Di, L.; Tang, J. Deriving High Spatiotemporal remote sensing Images Using Deep Convolutional Network. *Remote Sens.* **2018**, *10*, 1066. [CrossRef]
28.  Tan, Z.; Di, L.; Zhang, M.; Guo, L.; Gao, M. An enhanced deep convolutional model for spatiotemporal image fusion. *Remote Sens.* **2019**, *11*, 2898. [CrossRef]
29.  Song, H.; Liu, Q.; Wang, G.; Hang, R.; Huang, B. Spatiotemporal satellite image fusion using deep convolutional neural networks. *IEEE J. Sel. Top. Appl. Earth Obs. Remote Sens.* **2018**, *11*, 821–829. [CrossRef]
30.  Liu, X.; Deng, C.; Chanussot, J.; Hong, D.; Zhao, B. StfNet: A two-stream convolutional neural network for spatiotemporal image fusion. *IEEE Trans. Geosci. Remote Sens.* **2019**, *57*, 6552–6564. [CrossRef]
31.  Tan, Z.; Gao, M.; Li, X.; Jiang, L. A Flexible Reference-Insensitive Spatiotemporal Fusion Model for remote sensing Images Using Conditional Generative Adversarial Network. *IEEE Trans. Geosci. Remote Sens.* **2021**, *60*, 1–13. [CrossRef]
32.  Li, W.; Yang, C.; Peng, Y.; Zhang, X. A Multi-Cooperative Deep Convolutional Neural Network for Spatiotemporal Satellite Image Fusion. *IEEE J. Sel. Top. Appl. Earth Obs. Remote Sens.* **2021**, *14*, 10174–10188. [CrossRef]
33.  Yang, G.; Liu, H.; Zhong, X.; Chen, L.; Qian, Y. Temporal and Spatial Fusion of Remote Sensing Images: A Review. *Comput. Eng. Appl.* **2022**, *58*, 27–40. [CrossRef]
34.  Huang, G.; Liu, Z.; Maaten, L.V.; Weinberger, K.Q. Densely Connected Convolutional Networks. In Proceedings of the 2017 IEEE Conference on Computer Vision and Pattern Recognition (CVPR), Honolulu, HI, USA, 21–26 July 2017; pp. 2261–2269.
35.  Lin, T.; Dollár, P.; Girshick, R.; He, K.; Hariharan, B.; Belongie, S. Feature Pyramid Networks for Object Detection. In Proceedings of the 2017 IEEE Conference on Computer Vision and Pattern Recognition (CVPR), Honolulu, HI, USA, 21–26 July 2017; pp. 936–944.

36. Zhu, Z.; Xu, M.; Bai, S.; Huang, T.; Bai, X. Asymmetric Non-Local Neural Networks for Semantic Segmentation. In Proceedings of the 2019 IEEE/CVF International Conference on Computer Vision (ICCV), Seoul, Korea, 27 October–2 November 2019; pp. 593–602.
37. Yu, F.; Koltun, V. Multi-Scale Context Aggregation by Dilated Convolutions. In Proceedings of the 4th International Conference on Learning Representations, ICLR, San Juan, Puerto Rico, 2–4 May 2016.
38. Zhang, Y.; Li, K.; Li, K.; Wang, L.; Zhong, B.; Fu, Y. Image Super-Resolution Using Very Deep Residual Channel Attention Networks. In Proceedings of the European Conference on Computer Vision, Munich, Germany, 8–14 September 2018; pp. 294–310.
39. Hu, J.; Shen, L.; Albanie, S.; Sun, G.; Wu, E. Squeeze-and-excitation networks. *IEEE Trans. Pattern Anal. Mach. Intell.* **2020**, *42*, 2011–2023. [CrossRef] [PubMed]
40. Fu, J.; Liu, J.; Tian, H.; Li, Y.; Bao, Y.; Fang, Z.; Lu, H. Dual Attention Network for Scene Segmentation. In Proceedings of the 2019 IEEE/CVF Conference on Computer Vision and Pattern Recognition (CVPR), Long Beach, CA, USA, 15–20 June 2019; pp. 3146–3154.
41. Wang, X.; Girshick, R.; Gupta, A.; He, K. Non-local neural networks. In Proceedings of the IEEE Conference on Computer Vision and Pattern Recognition, Los Alamitos, CA, USA, 18–23 June 2018; pp. 7794–7803.
42. Wang, S.; Hou, X.; Zhao,X. Automatic Building Extraction From High-Resolution Aerial Imagery via Fully Convolutional Encoder-Decoder Network with Non-Local Block. *IEEE Access* **2020**, *8*, 7313–7322. [CrossRef]
43. Lai, W.S.; Huang, J.B.; Ahuja, N.; Yang, M.H. Fast and Accurate Image Super-Resolution with Deep Laplacian Pyramid Networks. *IEEE Trans. Pattern Anal. Mach. Intell.* **2019**, *41*, 2599–2613. [CrossRef]
44. Tan, Z.; Gao, M.; Yuan, J.; Jiang, L.; Duan, H. A Robust Model for MODIS and Landsat Image Fusion Considering Input Noise. *IEEE Trans. Geosci. Remote Sens.* **2022**, *60*, 5407217. [CrossRef]
45. Zhao, H.; Gallo, O.; Frosio, I.; Kautz, J. Loss Functions for Image Restoration with Neural Networks. *IEEE Trans. Comput. Imaging* **2017**, *3*, 47–57. [CrossRef]
46. Emelyanova, I.V.; McVicar, T.R.; Van Niel, T.G.; Li, L.T.; Van Dijk, A.I. Assessing the accuracy of blending Landsat–MODIS surface reflectances in two landscapes with contrasting spatial and temporal dynamics: A framework for algorithm selection. *Remote Sens. Environ.* **2013**, *133*, 193–209. [CrossRef]
47. Li, F.; Jupp, D.L.B.; Reddy, S.; Lymburner, L.; Mueller, N.; Tan, P.; Islam, A. An Evaluation of the Use of Atmospheric and BRDF Correction to Standardize Landsat Data. *IEEE J. Sel. Top. Appl. Earth Obs. Remote Sens.* **2010**, *3*, 257–270. [CrossRef]
48. Berk, A.; Anderson, G.P.; Bernstein, L.S.; Acharya, P.K.; Dothe, H.; Matthew, M.; Adler-Golden, S.; Chetwynd, J.; Richtsmeier, S.; Pukall, B.; et al. MODTRAN4 radiative transfer modeling for atmospheric correction. In Proceedings of the SPIE, Optical Spectroscopic Techniques and Instrumentation for Atmospheric and Space Research III, Denver, CO, USA, 20 October 1999.
49. Van Niel, T.G.; McVicar, T.R. Determining temporal windows for crop discrimination with remote sensing: A case study in south-eastern Australia. *Comput. Electron. Agric.* **2004**, *45*, 91–108. [CrossRef]
50. Wang, Z.; Bovik, A.C.; Sheikh, H.R.; Simoncelli, E.P. Image quality assessment: From error visibility to structural similarity. *IEEE Trans. Image Process.* **2004**, *13*, 600–612. [CrossRef]
51. Ponomarenko, N.; Ieremeiev, O.; Lukin, V.; Egiazarian, K.; Carli, M. Modified image visual quality metrics for contrast change and mean shift accounting. In Proceedings of the 2011 11th International Conference the Experience of Designing and Application of CAD Systems in Microelectronics (CADSM), Polyana, Ukraine, 23–25 February 2011; pp. 305–311.
52. Alparone, L.; Wald, L.; Chanussot, J.; Thomas, C.; Gamba, P.; Bruce, L. Comparison of Pansharpening Algorithms: Outcome of the 2006 GRS-S Data-Fusion Contest. *IEEE Trans. Geosci. Remote Sens.* **2007**, *45*, 3012–3021. [CrossRef]

*remote sensing*

Article

# Fast Seismic Landslide Detection Based on Improved Mask R-CNN

Rao Fu [1], Jing He [1,*], Gang Liu [1,2], Weile Li [2], Jiaqi Mao [1], Minhui He [1] and Yuanyang Lin [1]

[1]  School of Earth Sciences, Chengdu University of Technology, Chengdu 610059, China
[2]  State Key Laboratory of Geological Hazard Prevention and Geological Environment Protection, Chengdu 610059, China
*  Correspondence: hejing13@cdut.edu.cn

**Abstract:** For emergency rescue and damage assessment after an earthquake, quick detection of seismic landslides in the affected areas is crucial. The purpose of this study is to quickly determine the extent and size of post-earthquake seismic landslides using a small amount of post-earthquake seismic landslide imagery data. This information will serve as a foundation for emergency rescue efforts, disaster estimation, and other actions. In this study, Wenchuan County, Sichuan Province, China's 2008 post-quake Unmanned Air Vehicle (UAV) remote sensing images are used as the data source. ResNet-50, ResNet-101, and Swin Transformer are used as the backbone networks of Mask R-CNN to train and identify seismic landslides in post-quake UAV images. The training samples are then augmented by data augmentation methods, and transfer learning methods are used to reduce the training time required and enhance the generalization of the model. Finally, transfer learning was used to apply the model to seismic landslide imagery from Haiti after the earthquake that was not calibrated. With Precision and F1 scores of 0.9328 and 0.9025, respectively, the results demonstrate that Swin Transformer performs better as a backbone network than the original Mask R-CNN, YOLOv5, and Faster R-CNN. In Haiti's post-earthquake images, the improved model performs significantly better than the original model in terms of accuracy and recognition. The model for identifying post-earthquake seismic landslides developed in this paper has good generalizability and transferability as well as good application potential in emergency responses to earthquake disasters, which can offer strong support for post-earthquake emergency rescue and disaster assessment.

**Keywords:** mask R-CNN; Swin Transformer; landslide detection; UAV image; transfer learning

**Citation:** Fu, R.; He, J.; Liu, G.; Li, W.; Mao, J.; He, M.; Lin, Y. Fast Seismic Landslide Detection Based on Improved Mask R-CNN. *Remote Sens.* **2022**, *14*, 3928. https://doi.org/10.3390/rs14163928

Academic Editors: Yue Wu, Kai Qin, Qiguang Miao and Maoguo Gong

Received: 5 July 2022
Accepted: 11 August 2022
Published: 12 August 2022

**Publisher's Note:** MDPI stays neutral with regard to jurisdictional claims in published maps and institutional affiliations.

## 1. Introduction

Landslides are one of the most common natural disasters in mountainous areas, frequently resulting in significant property damage and casualties, particularly the thousands of landslide disasters caused by major earthquakes, which are more severe [1,2]. In Wenchuan County, Sichuan Province, China, on 12 May 2008, a powerful 8.0 magnitude earthquake devastated Yingxiu town. The epicenter of this strong earthquake was situated in the middle and high mountains of the western Sichuan basin, where the geological environment is quite fragile, resulting in the occurrence of numerous geological hazards, such as seismic landslides, mudslides, and hillside collapses [1]. A large number of landslides were caused by the Wenchuan earthquake, and these landslides directly caused the deaths of nearly 20,000 people [3]. Because of the serious threat posed by seismic landslides to people's lives and properties, as well as public safety, the rapid and automatic extraction of sudden landslides has become a hot topic in landslide research around the world [4–6]. The area, scale, and distribution of seismic landslides are determined by analyzing the morphology and characteristics of seismic landslide areas. It is crucial for disaster relief, mitigation, planning, and construction in the affected areas to quickly and accurately identify the location information of seismic landslides and implement targeted relevant measures in order to effectively reduce the damage caused by seismic landslides.

In early research, the majority of landslide detection and boundary extraction relied on the manual interpretation method [7–9], which has high accuracy. However, when the treated area is large or the disaster is urgent, the manual interpretation of landslides has issues such as a large workload, a long time to complete, and low efficiency, which is not conducive to the rapid extraction of large-scale landslide hazards after the disaster [10]. Additionally, because test results are subject to individual subjectivity, they will not be of the same standard if different persons interpret different areas [11].

Numerous automatic picture recognition approaches have been used to automatically detect landslides in the context of the quick development of information extraction technology. Many researchers have begun to use machine learning and deep learning algorithms for landslide detection due to the rapid development of these techniques. These algorithms include support vector machines (SVM), random forests (RF), artificial neural networks (ANN), decision trees (DT), convolutional neural networks (CNN), region-CNN (R-CNN), faster R-CNN, and others [12–15]. Gaelle Danneels et al. [16] used maximum likelihood classification and ANN classification methods to detect landslides from ASTER imagery automatically. Omid Ghorbanzadeh et al. [17] combined the ResU-Net model and the Object-Based Image Analysis (OBIA) method for landslide detection and compared the classification results with ResU-Net alone, and the proposed method improved the average intersection-bonding of maps obtained by ResU-Net by more than 22%. Faster R-CNN and the U-Net algorithm were employed by HuajinLi et al. [18] to locate landslides in large-scale satellite pictures, and they demonstrated that the suggested framework provided more precise segmentation of loess landslides than frameworks like Fully Convolutional Networks (FCN) and U-Net. ANN, SVM, RF, and CNN were utilized by Omid Ghorbanzadeh et al. [19] to perform landslide detection using optical data from the Rapid Eye satellite, and the results of these algorithms were assessed.

Deep learning applied to landslide detection has the advantages of fast detection, high automation, and low cost [20,21]. However, this kind of technology needs a lot of image data, and obtaining high-resolution data for natural hazard studies is costly and inconvenient, which makes it difficult to detect earthquakes quickly after their occurrence [15]. Transfer learning can help the learning process in new domains by using the "knowledge" gained from earlier tasks, such as data features and model parameters, which lowers the cost of gathering training data and boosts the effectiveness of model applications [22].

In conclusion, this study employs an improved Mask R-CNN algorithm, transfer learning for model training, and approaches for data augmentation to increase the sample size and automatically detect landslides from a small sample of post-earthquake UAV footage. Then, using transfer learning, the trained model is used to identify the landslide caused by the Haiti earthquake.

The main objectives of this paper are as follows:

1. Develop an earthquake landslide remote sensing recognition model with some generalizability;
2. To test the generalizability of the model, the trained model is used to extract data on seismic landslide hazards in untrained areas.

The following are this paper's significant innovations and contributions:

1. The Mask R-CNN technique is improved to increase model generalization on post-earthquake photos as well as the precision of landslide recognition;
2. The training model finished on Wenchuan UAV images is applied to seismic landslide recognition on post-earthquake satellite imagery of Haiti using transfer learning.

The remaining portions of the paper are structured as follows: Section 2 provides details on the experimental process, the improved Mask R-CNN model's framework, the experimental parameter settings, and the accuracy metrics. Section 3 describes the experimental results, comparing and analyzing the recognition results, performance and transferability of the different models. Section 4 compares the paper's major works and innovations to other researchers' discoveries. Finally, Section 5 summarizes the work and main results of the study, analyses the shortcomings of the study, and provides an outlook for future work.

## 2. Methods

### 2.1. Data

#### 2.1.1. Study Area

Because the Wenchuan earthquake caused a huge number of landslides, and we now have the results of the manual interpretation of the seismic landslides. The data sample is also rich and simple to collect. We chose UAV images of Wenchuan County taken after the 12 May 2008 Wenchuan earthquake to evaluate the efficacy of the proposed approach. These images include a large number of seismic landslides with a data resolution of 0.25 m. The location of the study area is shown in Figure 1.

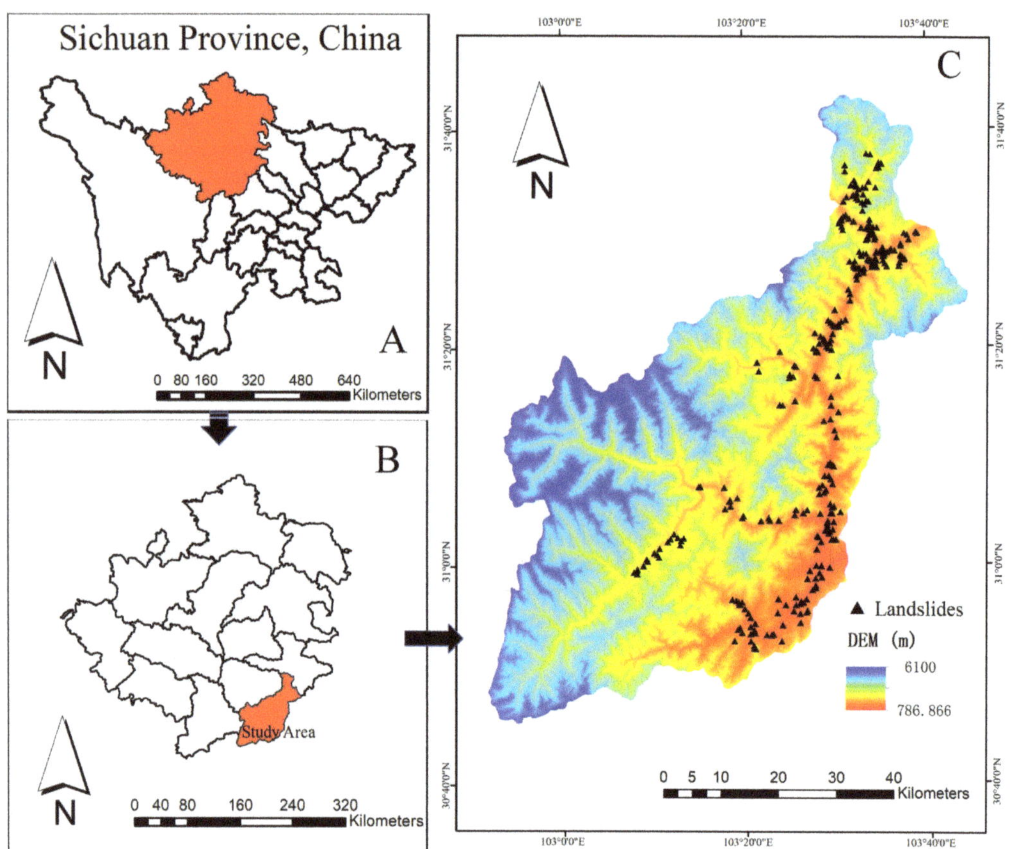


**Figure 1.** (**A**) the location of Aba Tibetan and Qiang Autonomous Prefecture in Sichuan, (**B**) the location of Wenchuan in Aba Tibetan and Qiang Autonomous Prefecture, and (**C**) the distribution of landslides in Wenchuan.

The study area has a total area of 4084 km$^2$ and is situated in Wenchuan County, Aba Tibetan and Qiang Autonomous Prefecture, Sichuan Province. It is located within 102°51′ E~103°44′ E and 30°45′ N~31°43′ N. This study area is situated in a valley between high and low mountains, with an overall undulating topography and an elevation trend that is high in the north and low in the south, as well as high in the west and low in the east, in which the highest elevation in the study area is 6100 m while the lowest elevation is only 786 m. The study area's stratigraphic lithology is primarily composed of granite, syenite, and amphibolite. The environment and seismic hazards have an impact on the surface rocks, causing the structure to break down and a large number of collapses and landslides to occur one after another. This results in the formation of a lot of loose solid material in the study area, which creates an ideal environment for the development of geological hazards such as landslides. With an average annual rainfall of 826 mm to 1049 mm, the study area's temperate monsoon climate and abundant rainfall during the rainy season create ideal conditions for the emergence of landslides and other geological hazards. Meanwhile, this region is situated in the Beichuan to Yingxiu fault zone, which has been proven to be the seismogenic fault of the Wenchuan earthquake, which is part of the Longmenshan active fault zone. The Wenchuan earthquake was caused by the sudden release of the accumulated energy in the Beichuan to Yingxiu zone of the Longmenshan thrust tectonic zone, which was brought on by the continuous Northeast compression of the Indian plate, the long-term accumulation of tectonic stresses on the eastern edge of the Tibetan Plateau, the East compression along the Longmenshan tectonic zone, and the blockage of the Sichuan Basin [23].

### 2.1.2. Dataset Production

This study used data from an SF-300 UAV equipped with a Canon EOS 5D Mark II camera that was flown over Wenchuan County in Sichuan Province, China, on 15 August 2010 at an average altitude of 2000 m, three RGB channels, a spatial resolution of 0.25 m, and an image size of 5616 × 3744 pixels.

This study pre-processed the training data and developed a dataset of UAV seismic landslide photos in COCO format. First, to ensure that the training dataset is roughly balanced, images are filtered based on image sharpness and the number of landslides on the graph. Since there were not many images, they were chosen by hand. We made an effort to choose images that had a good balance of pixels from landslides and non-landslides (foreground and background), with an average pixel ratio of roughly 55:45. There were two more pre-processing stages carried out after the selection of the photographs:

(1) Resizing the image to reduce complexity: the resized image is 512 × 512 pixels;
(2) Data annotation: this paper uses the Labelme annotator (from the Python library) to define the seismic landslides in the image and add textual descriptions to these seismic landslides, as shown in Figure 2.

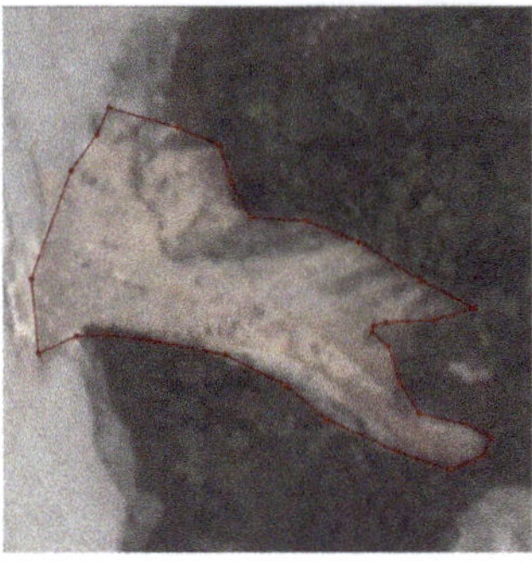

**Figure 2.** Image annotation.

### 2.1.3. Dataset Augmentation

Due to the short number of image samples used in this experiment, image data augmentation was necessary to increase the number of training photos, avoid overfitting by changing the tiny dataset to include features from large data, and optimize the deep learning algorithm's training adaption [24]. For image data augmentation in this study, image rotation and image flip are used (as shown in Figure 3). The three basic rotational processing techniques for images are 90, 180, and 270 degrees, whereas image flip involves flipping the images horizontally and up and down.

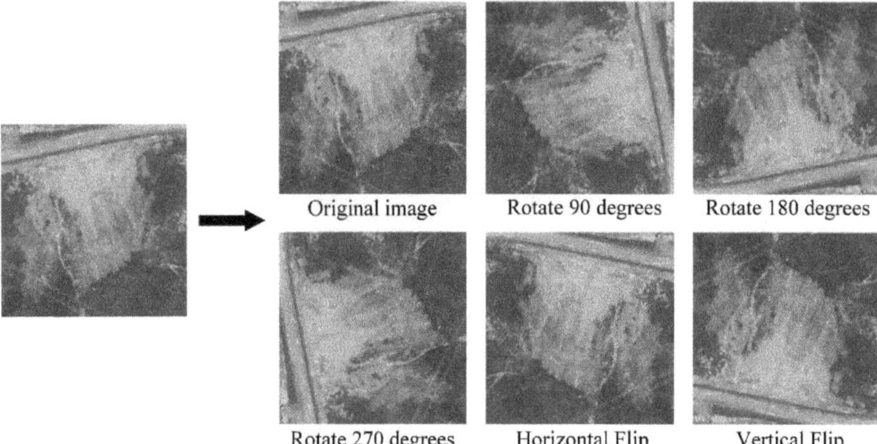


**Figure 3.** Image data augmentation.

After data augmentation, 852 landslide-containing images were obtained and split into three sets: a training set, a validation set, and a test set, with the ratio being 7:2:1. The training set is used to train the model, the validation set is used to validate the model during training, and the test set is used to assess the model. The specific values of the dataset division are shown in Table 1.

**Table 1.** Dataset division situation.

|  | Number of Images | Number of Landslides Included in the Image |
| --- | --- | --- |
| training set | 596 | 3560 |
| validation set | 170 | 898 |
| testing set | 86 | 476 |

### 2.2. Methodology Flow

In this study, we employ transfer learning to enhance the generalization and robustness of the Mask R-CNN model, which is the principal model for landslide identification based on seismic landslide photos captured by UAVs.

The following steps are primarily involved in the seismic landslide detection process: data gathering and processing, dataset production and augmentation, landslide detection, and accuracy evaluation. In Figure 4, the methodology flow is displayed.

### 2.3. Transfer Learning

Both landslide identification and landslide prediction have been successful when using deep learning. However, gathering the necessary training data is frequently challenging in real-world situations, and insufficient datasets frequently cause experimental results to be overfitting. The amount of training data required in such circumstances can be decreased by using transfer learning.

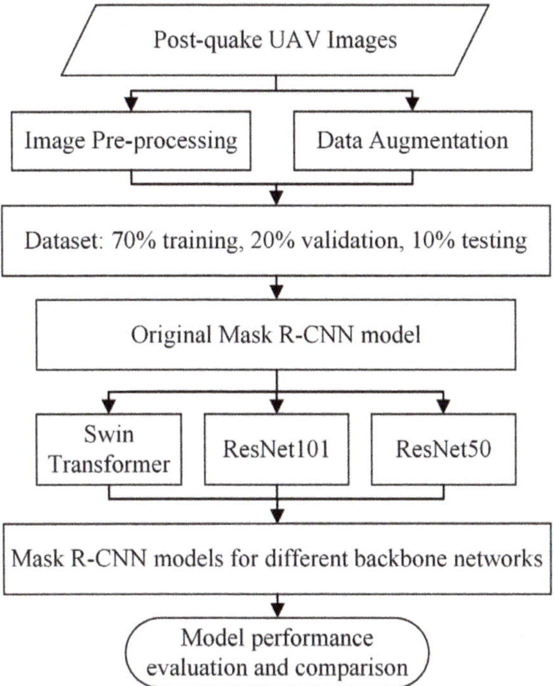


**Figure 4.** Methodology Flow.

When there is a lack of training data for the target task, transfer learning approaches can transfer information from some prior tasks to the target task [25]. As illustrated in Figure 5, the primary goal of employing transfer learning in this research is to increase experiment accuracy by transferring information from the Microsoft Common Objects in Context (MS COCO) dataset [26], which has a vast quantity of data, to a smaller landslide dataset. The model files for this experiment were pre-trained with the MSCOCO dataset and can be downloaded at this URL https://github.com/facebookresearch/detectron2/blob/main/MODEL_ZOO.md (accessed on 4 July 2022) to further reduce training time.

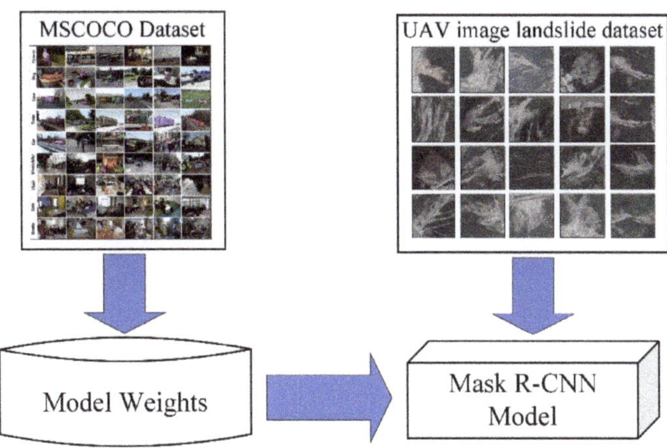


**Figure 5.** Model-based transfer learning.

### 2.4. ResNet

Kaiming He et al. [27] from Microsoft Research introduced ResNet (Residual Neural Network), successfully training a 152-layer neural network by using the ResNet Unit and taking first place in the ILSVRC 2015 competition despite using fewer parameters than VGGNet [28].

ResNet is made up of a residual structure, the basic concept of which is to expand the network by adding directly connected channels, or the Highway Network concept [29]. The performance input was transformed nonlinearly in the prior network structure, whereas the Highway Network permitted some of the output from the earlier network layers to be kept.

ResNet can be built using a variety of layer counts; the most popular ones are 50, 101, and 152 layers. All of these layer counts are achieved by stacking the aforementioned residual modules together. For this experiment, ResNet50 and ResNet101 were employed. Convolutional neural networks in ResNet50 and ResNet101 feature 50 and 101 layers, respectively.

### 2.5. Swin Transformer

Han Hu et al. from Microsoft Research made the Swin Transformer network proposal in 2021 [30], and their research got the best paper award at the 2021 ICCV. The Swin Transformer network has supplanted the traditional CNN architecture as the standard backbone in computer vision, outperforming backbone networks such as DeiT [31], ViT [32], and EfficientNet [33]. Based on the concept of the ViT model, the Swin Transformer ingeniously offers a sliding window technique that enables the model to learn data across windows. The model can handle super-resolution images thanks to the down-sampling layer, which also reduces computing work and frees it up to concentrate on global and local information. A hierarchical feature structure and linear computational complexity to image size are two characteristics of the Swin Transformer. Due to these characteristics, the model can be applied to a wide range of vision tasks. In vision tasks including target detection and picture segmentation, the Swin Transformer has achieved SOTA (state-of-the-art) results.

### 2.6. Mask R-CNN

The Mask R-CNN [34] framework consists of two stages: the first stage scans the image and produces proposals (regions that are likely to contain a target), and the second stage categorizes the proposals and produces bounding boxes and masks. The Mask R-CNN is expanded by the Faster R-CNN. The target detection framework Faster R-CNN is widely used [35], and Mask R-CNN expands it to include instance segmentation. The Mask R-CNN network structure is shown in Figure 6.

The Mask R-CNN extends the Faster R-CNN by adding a parallel branch to the existing boundary box recognition to predict the target's mask. Using the FCN to combine segmentation and classification undermines the effectiveness of instance segmentation according to the original Mask R-CNN paper. Therefore, Mask R-CNN uses the FCN to predict a Concrete Boundary for each category independently and relies on a different branch of the network to obtain the category and Boundary Box, as opposed to deriving the Boundary Box from the Concrete Boundary.

### 2.7. The Landslide Detection Method Used in This Paper

In this experiment, ResNet50, ResNet101, and Swin Transformer were employed as the backbone networks to extract image features, while Mask R-CNN was used as the primary landslide recognition model. Faster R-CNN with the semantic segmentation algorithm FCN [36] makes up the Mask R-CNN algorithm. The main network structure is shown in Figure 7.

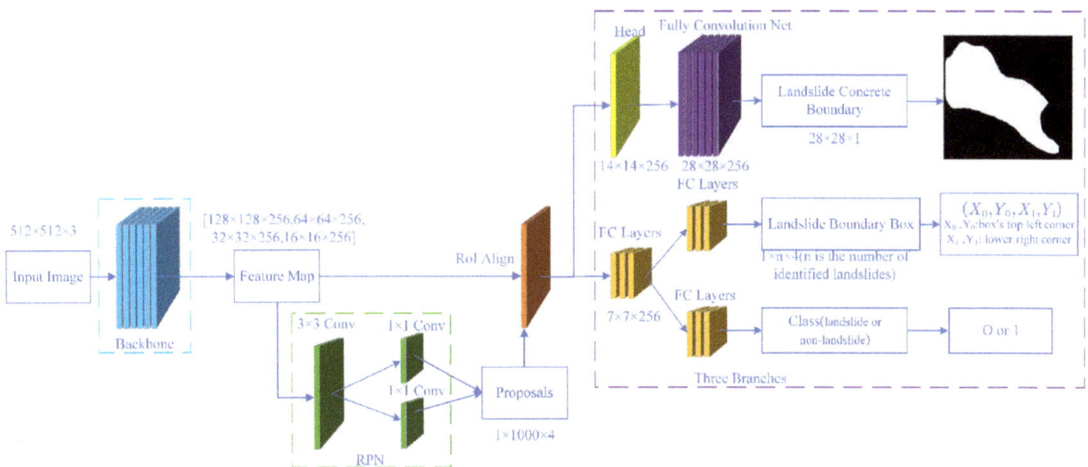

**Figure 6.** Mask R-CNN network structure.

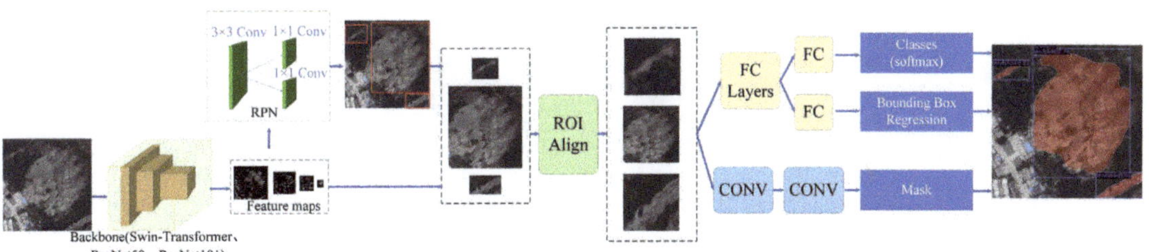


**Figure 7.** Network organization in this research.

Following the input of the seismic landslide picture into the network, the backbone first extracts the associated feature map, and then an ROI is set at each place in this feature map, yielding numerous candidate ROIs. After that, these candidate ROIs are sent into the Region Proposal Network (RPN) for regression and binary classification (slippery slope or non-slippery slope), with some of the non-slippery slope ROIs being passed off. For each anchor, RPN produces two outputs: a border accuracy to better fit the target and an anchor category to differentiate between landslides and background. The anchor that best contains the target can be chosen using RPN's predictions, and its size and position can be adjusted. If numerous anchors overlap each other, the anchor with the greatest score is kept by non-maximal suppression. The ROI Align procedure is then applied to these remaining ROIs, which first maps the original picture to the pixel of the feature map before mapping the feature map to the fixed feature. Finally, classification, Bounding box regression, and mask generation (FCN operation in each ROI) are applied to these ROIs. The network's primary modules are made up of and operate as follows.

(1)  RPN

Mask R-CNN does away with the conventional sliding window in favor of directly using RPN to create detection frames. Figure 8 depicts the precise organization of the RPN.

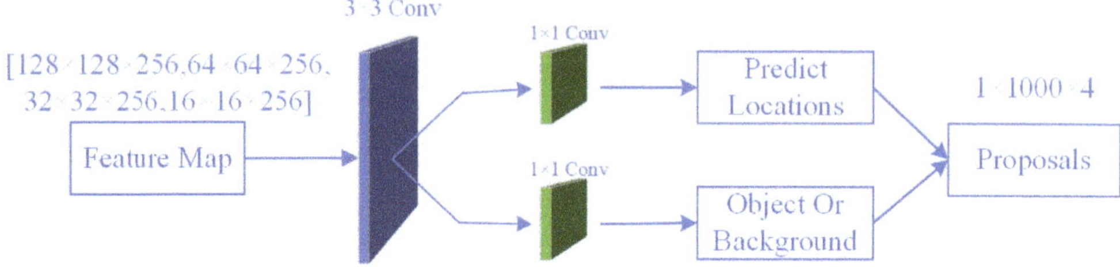



**Figure 8.** RPN architecture.

The original image is downsampled to produce feature maps. The final layer of the feature map is directly used by the general network because it has strong semantics. However, the last layer's feature map's positioning and resolution are quite poor, making it simple to miss relatively small objects. The backbone used in this paper uses multiple feature maps from the bottom to the top level for fusion, fully utilizing the extracted features at each stage in order to achieve better feature integration. Simply put, the higher-level features are transmitted to the lower-level semantics to complement them, resulting in high-resolution, strongly semantic features that make it easier to detect small targets.

With the use of sliding windows, the RPN, a lightweight neural network, scans the image and locates areas with targets. The anchors are the rectangular areas that the RPN scans, and they overlap one another to cover as much of the image as they can. The sliding window is implemented by the convolution process of RPN, which allows all regions to be scanned in parallel using the graphics processing unit (GPU). Furthermore, RPN does not scan the image directly; instead, it uses the backbone feature map, which enables RPN to utilize the extracted features effectively and prevent double counting. RPN generates two outputs for each anchor: an anchor class to distinguish foreground from background and a border to better fit the target. By using RPN's predictions, the anchor that best contains the target can be selected and its position and size fine-tuned, and if multiple anchors are overlapping each other, the anchor with the highest foreground score is retained through non-maximal suppression.

The $1 \times 1$ convolutional layer is used to output a specified number of channels of feature maps. Proposals are areas where the algorithm finds possible objects after scanning the image through a sliding window. The top 1000 proposal boxes are kept after the Proposals Layer sorts the resulting proposal boxes in descending order of score. Four coordinates are contained in each box, resulting in the final matrix, which has the dimensions $1 \times 1000 \times 4$. The $1 \times 1000 \times 4$ dimension represents the 1000 areas of the image where the target is likely to be located. Here in the Proposals Layer, the network completes its equivalent of targeting.

(2)  ROI Align

Mask R-CNN proposes the ROI Align approach in place of ROI Pooling to address the issue of region mismatch (misalignment) brought on by two quantization processes in Faster R-CNN. The ROI Align operation is shown in Figure 9, with the dashed part representing the feature map and the solid line representing the ROI, where the ROI is sliced into $2 \times 2$ cells. If there are four points to be sampled, first, each cell is divided into four small squares (represented by red lines), with the center of each serving as the sampling point. The values of these sampled pixel points are then determined because the coordinates of these sample points are typically floating-point numbers, necessitating a bilinear interpolation of the sampled pixel (as indicated by the four arrows). The last step is to max pool the four sampled points inside each cell, which results in the ROI Align result. The purpose of ROI Align is to pool the corresponding areas in the feature map to a fixed size based on the position coordinates of the proposed boxes obtained from RPN for subsequent classification and boundary box regression operations.

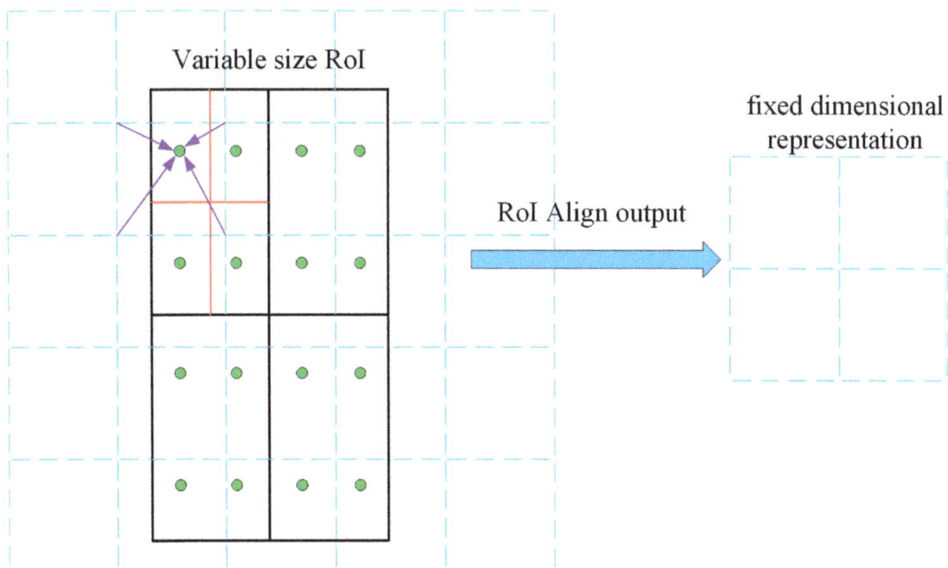


**Figure 9.** ROI Align sampling and pooling implementation process.

(3)     Fully Convolution Nets

The FCN convolves and pools the image, decreasing its feature map, then performs a deconvolution operation, which means an interpolation operation to increase its feature map, and finally classifies each pixel value. To generate the mask of the identified target in the input image, that is, the boundary of the identified target, the FCN operation is applied to each ROI of the image.

*2.8. Experimental Setup*

For our trials in this study, we employed an RTX3090 graphics processor, an Intel i9-10900k processor, and 64 GB of RAM. For the model software environment, both the original and improved Mask R-CNN models are implemented in PyTorch, the python version of Torch, a neural network framework open-sourced by Facebook, specifically for GPU-accelerated neural network programming. Torch is a traditional tensor library that is popular in machine learning and other applications that need a lot of arithmetic. It is used to manipulate multidimensional matrix data.

The AdamW algorithm [37] was selected as the gradient decent optimizer algorithm for setting model parameters because it uses less memory, trains more quickly, converges more quickly, and reduces computational costs. The remaining hyperparameters are all those that perform better on the validation set after multiple iterations. The batch size was set to 8, the number of threads was set to 4, and the learning rate was set to $10^{-3}$ as the model parameter settings. After that, 50 epochs of training were performed on all three models.

The network structure code used in the article can be downloaded at these two URLs: https://github.com/open-mmlab/mmdetection (accessed on 4 July 2022) and https://github.com/SwinTransformer/Swin-Transformer-Object-Detection (accessed on 4 July 2022).

*2.9. Indicators for Accuracy Evaluation*

In this experiment, the performance of the three seismic landslide detection models was quantitatively assessed using Precision, Recall, F1 score, Accuracy, and intersection over union (IoU) metrics [38]. Precision numbers primarily show how accurately landslides were detected on the image. The number of landslides in the image that have been successfully recognized is represented by the Recall metric. The F1 score is used to calculate the equilibrium between accuracy and recall. The F1 score is a combined indicator of the model's accuracy and is the harmonic mean of precision and recall. The accuracy rate represents the proportion of accurately predicted data among all data. The confusion matrix in Table 2 includes True-Positive (TP), False-Positive (FP), and False-Negative (FN), where TP is the number of samples that were correctly identified as landslides, FP is the number of samples that were incorrectly identified as non-landslides, and FN is the number of samples that were not identified as landslides. If the IoU is more than or equal to 0.5, a prediction for a landslide is deemed to be true, and if it is less than 0.5, it is deemed to be false.

**Table 2.** Confusion matrix of predicted result and ground truth.

| Ground Truth | Predicted Result | |
| --- | --- | --- |
| | Landslide | Non-Landslide |
| Landslide | TP (True-Positive) | FN (False-Negative) |
| Non-landslide | FP (False-Positive) | TN (True-Negative) |

False negatives play a crucial role in managing the risk of landslides. The small number of FNs guarantees that the model misses fewer landslides and identifies all affected structures and settlements, allowing for an accurate assessment of the extent and severity of damage to the landslide hazard area and prompt action to be taken to prevent and mitigate the disaster in the affected area.

IoU is a metric used to assess how accurately corresponding object boundaries are found in a set of data. IoU is a straightforward calculation criterion that may be applied to any task that produces an output with a predicted range (bounding boxes). The correlation between the true and predicted values is calculated by using this criterion, and the stronger the correlation, the higher the value. The IoU value measures how closely the system's anticipated box and the image's ground truth box overlap. The accuracy of a single detection is represented by the intersection of the detection result and the ground truth over their concatenation.

For this experiment, the test set was also labeled. After the test set's images were recognized, the program counted and filtered the IoU for each identified landslide. A TP is defined as an IoU value greater than 0.5, and the total number of TP is calculated in this way. The number of FPs is equal to the number of detected landslides less the number of TP, whereas the number of FN is equal to the number of true landslides in the tag minus the number of TP.

## 3. Results

The three models in this experiment were trained using the same landslide dataset, and since all three models were fitted before the 30th epoch, only the first 30 epochs were chosen for illustration. Figure 10 displays, following 30 training epochs, the overall validation loss curves generated by the various models. The three models' loss curves generally follow a similar trend, with faster learning and a notable decline in loss values during the initial phases of training. The model gradually converges in the middle and later phases of training, with smaller changes in loss values and a sluggish rate of decrease. The accuracy of the model's landslide detection keeps improving as the total loss value drops, and in this study, the epoch with the lowest total loss value is chosen for identification on the test set.

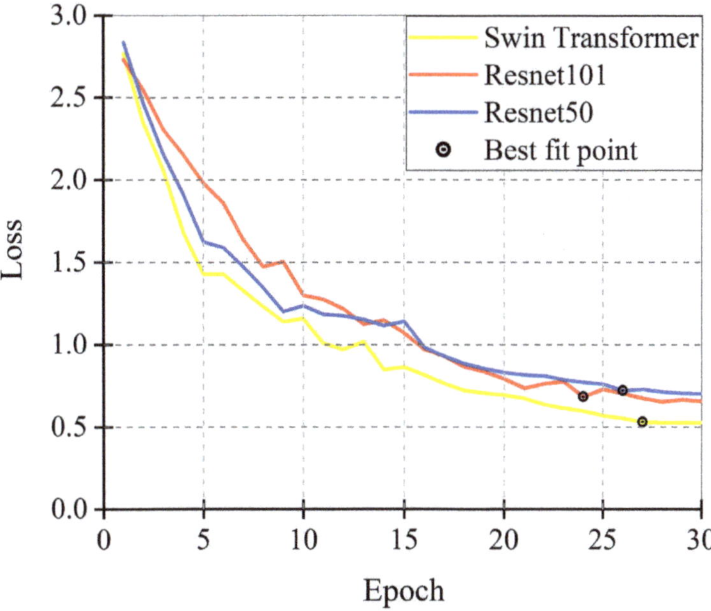

**Figure 10.** Validation Loss Curve.

The same landslide dataset was also used to train the YOLOv5, Faster R-CNN. Following training, the three models' recognition accuracies were compared to those of the classic YOLOv5 and Faster R-CNN models on the test set. The primary architect of the original YOLOv5 project methodology was Glenn Jocher of Ultralytics. YOLOv5 is a one-stage detection model that, after a single inspection, generates the class probability and position coordinate values of the object directly, without the aid of a region proposals stage. While being slower but typically more accurate, Mask R-CNN is a two-stage detection model. The results are given in Table 3.

**Table 3.** Comparison of network accuracy.

| Model | Precision (%) | Recall (%) | F1 Score (%) | Accuracy (%) |
|---|---|---|---|---|
| Mask R-CNN+Swin Transformer | 93.28 | 87.41 | 90.25 | 82.2 |
| Mask R-CNN+ResNet101 | 89.55 | 83.92 | 86.64 | 76.43 |
| Mask R-CNN+ResNet50 | 86.15 | 80.32 | 83.13 | 73.91 |
| YOLOv5 | 88.64 | 83.78 | 86.16 | 75.94 |
| Faster R-CNN | 84.47 | 78.36 | 81.30 | 69.13 |

Table 3 shows that the model with the Swin Transformer performs well on the test set, with Precision values of 0.9328, Recall values of 0.8741, F1 scores of 0.9025, and Accuracy values of 0.822. The improved algorithm outperformed the previous algorithms in all indexes when compared to the original Mask R-CNN, YOLOv5, and Faster R-CNN. According to a study of the test results, the method described in this work for detecting seismic landslides has a greater detection accuracy than the original Mask R-CNN algorithm. Figure 11 displays the outcomes of the UAV image recognition in the test set. In the diagram, the blue box represents the landslide boundary box determined by the model, and the red area represents the landslide boundary determined by model identification.

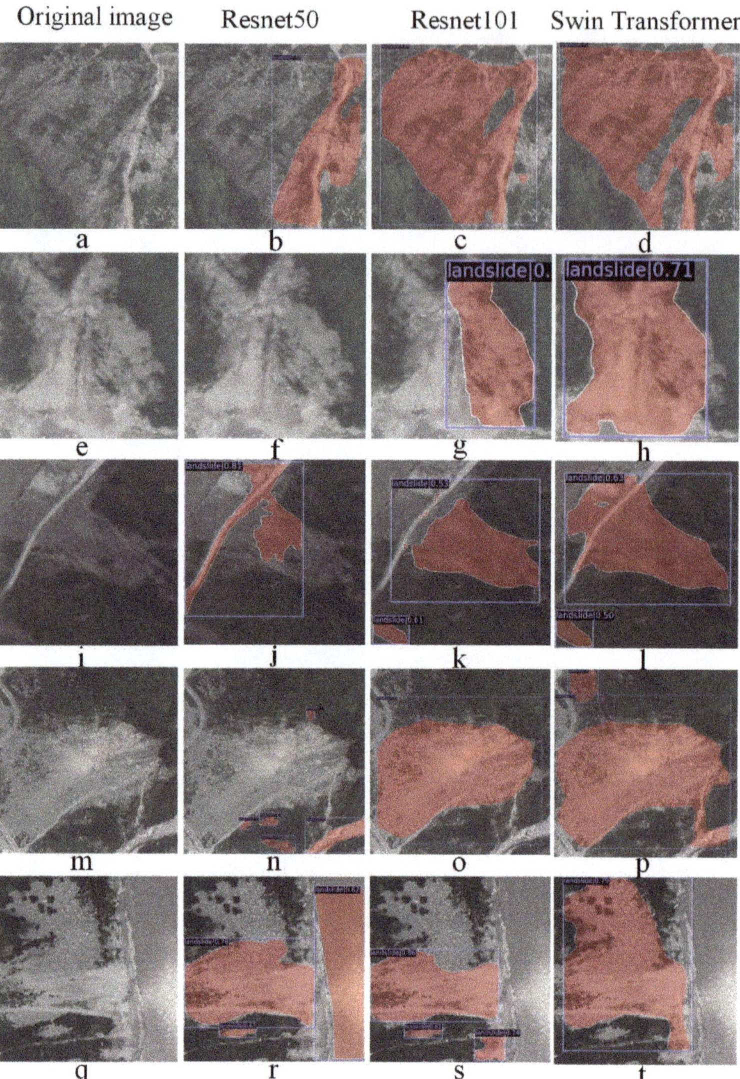



**Figure 11.** Results of UAV image recognition in the test set ((**a,e,i,m,q**) are ground truth images; (**b,f,j,n,r**) are recognition images of Resnet 50 for the backbone; (**c,g,k,o,s**) are recognition images of Resnet 101 for the backbone and (**d,h,l,p,t**) are recognition images of Swin Trasnformer for the backbone).

As seen in Figure 11a–h, the improved model is capable of correctly identifying individual landslides when there is a significant area of individual landslides on the image. The resulting landslide boundaries are also more precise than those produced by ResNet50 and ResNet101.

While the recognition results of ResNet50 and ResNet101 suffer from serious under-detection when the natural environment is complex, as shown in Figure 11i–p, the improved model can accurately detect them even if there are two or more landslides on the image at the same time. Although the landslide area below the image is only partially visible in the UAV image, the model correctly identifies it, demonstrating the improved model's increased robustness. In this way, the model can maintain high recognition accuracy even in landslide regions on the image that is partially or completely hidden by surface structures, such as buildings or forests.

As seen in Figure 11q–t, there is a river on the right side of the landslide area, and the visual characteristics of the river are very similar to those of the landslide. ResNet50 exhibits the phenomenon of mistaking the river for a landslide, but the improved model correctly ignores these occurrences and detects the landslide area on the left side, demonstrating its anti-interference ability.

The recognition results, however, demonstrate that the improved model continues to miss and misidentify locations on the map. The exposed rock above is mistakenly identified by the improved model in Figure 11p as a landslide area due to the visual similarities between the two. Figure 11t demonstrates how the improved model overlooked a minor landslide below. This suggests that there is still room for improvement in the improved model's capacity to identify landslides in challenging environments, especially when it comes to distinguishing them from exposed rock and exposed soil. In the future, steps could be taken to lessen the influence of bare rock and bare soil on the model, such as increasing the number of landslide samples and adding satellite and drone imagery of various resolutions to the dataset.

This study put the seismic landslide photos from Haiti into three trained models for identification to examine the generalizability and transferability of the models. After image correction, fusion, and other pre-processing, the Haiti seismic landslide image, which is segmented into $512 \times 512$ pixels size for identification in this paper, is created with a 1 m resolution true-color image of the Haiti post-earthquake GF-2 satellite image from 2021. Figure 12 depicts the outcomes of the identification of the Haiti satellite imagery, with the Ground Truth being the landslide boundary determined by the geohazard interpreters based on the outcomes of surface changes between the Haiti satellite images taken prior to the earthquake and those taken following it. In the diagram, the blue box represents the landslide boundary box determined by the model, and the red area represents the landslide boundary determined by model identification.

Figure 12a–d,q–t demonstrate the improved model's superior feature extraction capabilities for seismic landslides. ResNet50 and ResNet101 exhibit missed and false detections when identifying landslides, while the improved model can still recognize landslides that are untrained and have different colors. The improved model performs well in identifying small landslides, even when they are small, as shown in Figure 12e–h.

ResNet50 and ResNet101 could only identify the larger, more noticeable landslides among them, and the detection results are shown in Figure 12i–p. Despite the large number of landslides present on the image at the same time, the improved model was still able to identify the vast majority of them.

In conclusion, both ResNet-50 and ResNet-101 performed poorly in their recognition of the Haiti images, but the improved model's detection results on those same images still maintained high accuracy, demonstrating the improved model's superior robustness and transferability.

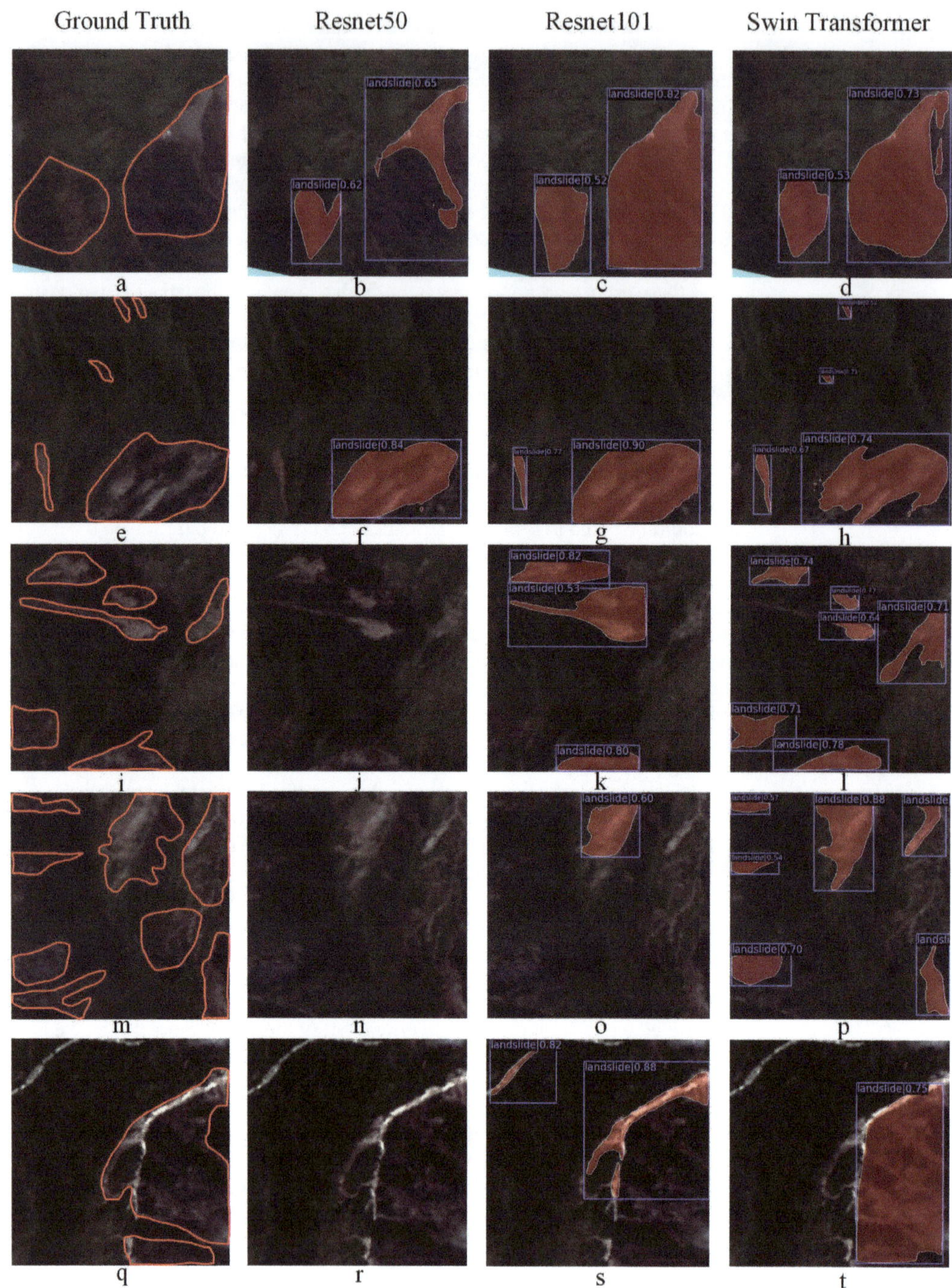


**Figure 12.** Comparison of Haiti's satellite image identification outcomes ((**a,e,i,m,q**) are ground truth images; (**b,f,j,n,r**) are recognition images of Resnet 50 for the backbone; (**c,g,k,o,s**) are recognition images of Resnet 101 for the backbone and (**d,h,l,p,t**) are recognition images of Swin Trasnformer for the backbone).

## 4. Discussion

In recent years, deep learning has been employed by many academics to identify landslides. A DLWC model for landslide detection in Hue–Saturation–Intensity (HSI) data was proposed by C. Ye et al. [39] DLWC combines the extracted features and susceptibility factors for landslide detection after using deep confidence networks to extract spatial features and spectral characteristics of landslides at high levels on hyperspectral images. To determine if it was a landslide, a logistic regression classifier with constraints was finally employed. The experimental outcomes demonstrate that the detection accuracy of landslides on remote sensing images reaches 97.91 percent, which is more accurate than the conventional hyperspectral image classification method. Utilizing contour data and the vegetation index, Bo Yu et al. [40] developed an end-to-end deep learning framework for landslide detection. The framework is divided into two sections: one for identifying areas at risk of landslides using vegetation indices and DEM and the other for accurately identifying those areas through the use of a semantic segmentation deep learning model. With a recall of 65% and an accuracy of 55.35%, the proposed methodology performed 44 percent more accurately than comparable published works when used to identify landslides in Nepal on images from Landsat 2015. In order to detect and map earthquake-induced landslides in single RapidEye satellite images, Yi Y et al. [41] proposed a new end-to-end deep learning network, LandsNet, to learn various features of landslides. To address the lack of training samples, specific training samples were first generated and a data augmentation strategy was put into place. A cascaded end-to-end deep learning network called LandsNet was subsequently built. By using morphological processing, the identified landslide maps have finally been further optimized. In two spatially distinct earthquake-affected areas, LandsNet achieved the best F1 value of about 86.89 percent, which is almost 7 and 8 percentage points higher than those of ResUNet and DeepUNet, respectively. An improved U-Net model for seismically generated landslide extraction was developed by Liu, P. et al. [42] using post-earthquake aerial remote sensing imagery to annotate a landslide dataset. The article increases the amount of feature parameters for the training samples by first adding three new bands with spatial information to the three RGB bands: DSM, slope, and aspect. In order to rebuild the U-Net model structure, a residual learning unit was then added to the conventional U-Net model. Finally, the new technique was used to identify seismic landslides in Jiuzhaigou County, Sichuan Province, China. According to the findings, the new method's accuracy is 91.3 percent, which is 13.8 percent greater than that of the conventional U-Net model.

All of the aforementioned studies have produced positive results, but there are clear drawbacks in the quick identification of post-earthquake seismic landslides, such as the challenge of quickly obtaining hyperspectral image data after an earthquake, the low identification accuracy, and the poor transferability. Automatic seismic landslide extraction's main goal is to take all necessary measures to meet seismic emergency needs and to offer technical assistance for disaster mitigation and relief efforts such as opening up lifelines and life rescue. The issue of applying a trained model to landslides in untrained areas has received less attention in recent research on landslide identification. The algorithm used in this paper can extract corresponding patterns using multi-layer learning in a neural network using spatial and spectral features of seismic landslides from remote sensing imagery. In the early stages of the study, numerous sets of labelled data are used as samples to train landslide identification models, which can be used to quickly extract data about disasters such as seismic landslides in the post-disaster period. In this study, data from the 2008 Wenchuan earthquake were used to train a recognition model with some generalizability to seismic landslides, and seismic landslide data from Haiti were used to validate the model. The method used in this study produced comparable F1 values and accuracy to the studies by Yi Y et al. [41] and Liu, P et al. [42], but the model can be used for identification immediately after an earthquake without collecting data, improving time efficiency, indicating that the model has better potential for use in emergency response to earthquake disasters.

With good results and an accuracy of 0.9328 and an F1 score of 0.9025, Resnet-50, Resnet-101, and Swin Transformer were utilized in this study as the backbone networks of Mask R-CNN for the extraction of seismic landslides in Wenchuan. On untrained post-earthquake satellite photos of Haiti, the improved model continues to produce good recognition results, and its accuracy and transferability have both increased. Compared to earlier examined methods, the one used in this study is more automated and necessitates fewer data.

This study employs some techniques to improve model accuracy and avoid overfitting while working with tiny samples of data. To obtain a larger dataset, data augmentation is first applied to the photos, rotating and flipping each image. Second, transfer learning is utilized to shorten the training time, improve the generalizability of the model, and reduce the amount of data gathering needed. Ultimately, the Swin Transformer was employed as the model's backbone network to improve its adaptability and accuracy.

The findings of this study demonstrate that seismic landslides can be successfully identified in UAV imagery by using deep learning techniques. It is anticipated that satellite and UAV imagery data of various resolutions will be added to the dataset for the study's next phase to increase data diversity and boost the precision of landslide identification. Other data, such as Digital Elevation Model (DEM) data, can also be incorporated into the model, in addition to remotely sensed imagery.

## 5. Conclusions

In this study, we created a seismic landslide sample dataset by labeling the landslides that appeared on post-quake UAV images from Wenchuan County, Sichuan Province, in 2008. To identify landslides in post-earthquake drone images of Wenchuan, this study used ResNet-50, ResNet-101, and Swin Transformer as the backbone networks. Data augmentation and transfer learning methods were also used, and the generalizability and transferability of the models were compared using seismic landslide images of Haiti. The results demonstrate that the Swin Transformer outperforms ResNet-101 and ResNet-50, obtaining a Precision value of 0.9328 and an F1 score of 0.9025 on the dataset and having greater robustness and generalization for landslide detection. In this study, a remote sensing model for identifying seismic landslides with some degree of universality was developed and successfully used to identify seismic landslides in Haiti. This indicates the accuracy of using the improved Mask R-CNN algorithm to detect landslides in post-earthquake UAV imagery. The landslide identification model developed in this paper has made some advances in terms of generalizability and transferability, and it can deliver accurate landslide data for post-earthquake emergency rescue and disaster assessment. This study still has a few flaws in it. The next step will be to streamline the model in order to reduce training time because the model parameters used in this study are numerous and demand high computer performance. In order to improve the model's accuracy and dependability, the dataset will be expanded in the future to include satellite imagery and drone imagery at various resolutions.

**Author Contributions:** Data curation, formal analysis, methodology, writing—original draft, writing—review and editing, visualization, R.F.; resources, writing—review and editing, J.H. and W.L.; formal analysis, writing—review and editing, G.L.; data curation, validation, J.M., M.H. and Y.L. All authors have read and agreed to the published version of the manuscript.

**Funding:** This research was funded by the National Key Research and Development Program of China (Grant No. 2021YFC3000401), the Chengdu Technology Innovation R&D Project (2022-YF05-01090-SN), the State Key Laboratory of Geohazard Prevention and Geoenvironment Protection Independent Research Project (Grant No. SKLGP2018Z010), the National Natural Science Foundation of China (NSFC) (Grant No. 41871303), the Sichuan Provincial Science and Technology Support Project (Grant No. 2021YFG0365), and the Department of Natural Resources of Sichuan Province (Grant No. kj-2021-3).

**Data Availability Statement:** Not applicable.

**Conflicts of Interest:** The authors declare no conflict of interest.

**Abbreviations**

| | |
|---|---|
| UAV | unmanned air vehicle |
| COCO | common objects in context |
| SVM | support vector machines |
| RF | random forests |
| ANN | artificial neural networks |
| DT | decision trees |
| CNN | convolutional neural networks |
| R-CNN | region-CNN |
| OBIA | object-based image analysis |
| MS COCO | Microsoft common objects in context |
| ROI | region of interest |
| FCN | fully convolutional networks |
| RPN | region proposal network |
| GPU | graphics processing unit |
| IoU | intersection over union |
| TP | true-positive |
| FP | false-positive |
| FN | false-negative |
| DEM | digital elevation model |

## References


1. Huang, R.; Li, W. Analysis of the Geo-Hazards Triggered by the 12 May 2008 Wenchuan Earthquake, China. *Bull. Eng. Geol. Environ.* **2009**, *68*, 363–371. [CrossRef]
2. Dai, F.C.; Lee, C.F.; Ngai, Y.Y. Landslide Risk Assessment and Management: An Overview. *Eng. Geol.* **2002**, *64*, 65–87. [CrossRef]
3. Yin, Y.; Wang, F.; Sun, P. Landslide Hazards Triggered by the 2008 Wenchuan Earthquake, Sichuan, China. *Landslides* **2009**, *6*, 139–152. [CrossRef]
4. Yang, R.; Zhang, F.; Xia, J.; Wu, C. Landslide Extraction Using Mask R-CNN with Background-Enhancement Method. *Remote Sens.* **2022**, *14*, 2206. [CrossRef]
5. Hacıefendioğlu, K.; Demir, G.; Başağa, H.B. Landslide Detection Using Visualization Techniques for Deep Convolutional Neural Network Models. *Nat. Hazards* **2021**, *109*, 329–350. [CrossRef]
6. Tavakkoli Piralilou, S.; Shahabi, H.; Jarihani, B.; Ghorbanzadeh, O.; Blaschke, T.; Gholamnia, K.; Meena, S.R.; Aryal, J. Landslide Detection Using Multi-Scale Image Segmentation and Different Machine Learning Models in the Higher Himalayas. *Remote Sens.* **2019**, *11*, 2575. [CrossRef]
7. Gorum, T.; Fan, X.; van Westen, C.J.; Huang, R.Q.; Xu, Q.; Tang, C.; Wang, G. Distribution Pattern of Earthquake-Induced Landslides Triggered by the 12 May 2008 Wenchuan Earthquake. *Geomorphology* **2011**, *133*, 152–167. [CrossRef]
8. Sato, H.P.; Hasegawa, H.; Fujiwara, S.; Tobita, M.; Koarai, M.; Une, H.; Iwahashi, J. Interpretation of Landslide Distribution Triggered by the 2005 Northern Pakistan Earthquake Using SPOT 5 Imagery. *Landslides* **2007**, *4*, 113–122. [CrossRef]
9. Keefer, D.K. Statistical Analysis of an Earthquake-Induced Landslide Distribution—The 1989 Loma Prieta, California Event. *Eng. Geol.* **2000**, *58*, 231–249. [CrossRef]
10. Galli, M.; Ardizzone, F.; Cardinali, M.; Guzzetti, F.; Reichenbach, P. Comparing Landslide Inventory Maps. *Geomorphology* **2008**, *94*, 268–289. [CrossRef]
11. Hölbling, D.; Füreder, P.; Antolini, F.; Cigna, F.; Casagli, N.; Lang, S. A Semi-Automated Object-Based Approach for Landslide Detection Validated by Persistent Scatterer Interferometry Measures and Landslide Inventories. *Remote Sens.* **2012**, *4*, 1310–1336. [CrossRef]
12. Arabameri, A.; Pradhan, B.; Rezaei, K.; Lee, C.-W. Assessment of Landslide Susceptibility Using Statistical- and Artificial Intelligence-Based FR–RF Integrated Model and Multiresolution DEMs. *Remote Sens.* **2019**, *11*, 999. [CrossRef]
13. Chang, Z.; Du, Z.; Zhang, F.; Huang, F.; Chen, J.; Li, W.; Guo, Z. Landslide Susceptibility Prediction Based on Remote Sensing Images and GIS: Comparisons of Supervised and Unsupervised Machine Learning Models. *Remote Sens.* **2020**, *12*, 502. [CrossRef]
14. Micheletti, N.; Foresti, L.; Robert, S.; Leuenberger, M.; Pedrazzini, A.; Jaboyedoff, M.; Kanevski, M. Machine Learning Feature Selection Methods for Landslide Susceptibility Mapping. *Math Geosci* **2014**, *46*, 33–57. [CrossRef]
15. Mohan, A.; Singh, A.K.; Kumar, B.; Dwivedi, R. Review on Remote Sensing Methods for Landslide Detection Using Machine and Deep Learning. *Trans. Emerg. Telecommun. Technol.* **2021**, *32*, e3998. [CrossRef]

16. Danneels, G.; Pirard, E.; Havenith, H.-B. Automatic Landslide Detection from Remote Sensing Images Using Supervised Classification Methods. In Proceedings of the 2007 IEEE International Geoscience and Remote Sensing Symposium, Barcelona, Spain, 23–28 July 2007; pp. 3014–3017.

17. Ghorbanzadeh, O.; Gholamnia, K.; Ghamisi, P. The Application of ResU-Net and OBIA for Landslide Detection from Multi-Temporal Sentinel-2 Images. *Big Earth Data* **2022**, 1–26. [CrossRef]

18. Li, H.; He, Y.; Xu, Q.; Deng, J.; Li, W.; Wei, Y. Detection and Segmentation of Loess Landslides via Satellite Images: A Two-Phase Framework. *Landslides* **2022**, *19*, 673–686. [CrossRef]

19. Ghorbanzadeh, O.; Blaschke, T.; Gholamnia, K.; Meena, S.R.; Tiede, D.; Aryal, J. Evaluation of Different Machine Learning Methods and Deep-Learning Convolutional Neural Networks for Landslide Detection. *Remote Sens.* **2019**, *11*, 196. [CrossRef]

20. Cheng, L.; Li, J.; Duan, P.; Wang, M. A Small Attentional YOLO Model for Landslide Detection from Satellite Remote Sensing Images. *Landslides* **2021**, *18*, 2751–2765. [CrossRef]

21. Ullo, S.L.; Langenkamp, M.S.; Oikarinen, T.P.; Del Rosso, M.P.; Sebastianelli, A.; Piccirillo, F.; Sica, S. Landslide Geohazard Assessment with Convolutional Neural Networks Using Sentinel-2 Imagery Data. In Proceedings of the IGARSS 2019—2019 IEEE International Geoscience and Remote Sensing Symposium, Yokohama, Japan, 28 July–2 August 2019; pp. 9646–9649.

22. Pan, S.J.; Yang, Q. A Survey on Transfer Learning. *IEEE Trans. Knowl. Data Eng.* **2010**, *22*, 1345–1359. [CrossRef]

23. Xu, X.; Wen, X.-Z.; Ye, J.-Q.; Ma, B.-Q.; Chen, J.; Zhou, R.-J.; He, H.-L.; Tian, Q.-J.; He, Y.-L.; Wang, Z.C.; et al. The Ms8.0 Wenchuan Earthquake Surface Ruptures and its Seismogenic Structure. *Seismol. Egol.* **2008**, *30*, 597.

24. Shorten, C.; Khoshgoftaar, T.M. A Survey on Image Data Augmentation for Deep Learning. *J. Big Data* **2019**, *6*, 60. [CrossRef]

25. Weiss, K.; Khoshgoftaar, T.M.; Wang, D. A Survey of Transfer Learning. *J. Big Data* **2016**, *3*, 9. [CrossRef]

26. Lin, T.-Y.; Maire, M.; Belongie, S.; Bourdev, L.; Girshick, R.; Hays, J.; Perona, P.; Ramanan, D.; Zitnick, C.L.; Dollár, P. Microsoft COCO: Common Objects in Context. *arXiv* **2015**, arXiv:1405.0312.

27. He, K.; Zhang, X.; Ren, S.; Sun, J. Deep Residual Learning for Image Recognition. In Proceedings of the 2016 IEEE Conference on Computer Vision and Pattern Recognition (CVPR), Las Vegas, NV, USA, 27–30 June 2016; pp. 770–778.

28. Simonyan, K.; Zisserman, A. Very Deep Convolutional Networks for Large-Scale Image Recognition. *arXiv* **2021**, arXiv:1409.1556.

29. Srivastava, R.K.; Greff, K.; Schmidhuber, J. Highway Networks. *arXiv* **2015**, arXiv:1505.00387.

30. Liu, Z.; Lin, Y.; Cao, Y.; Hu, H.; Wei, Y.; Zhang, Z.; Lin, S.; Guo, B. Swin Transformer: Hierarchical Vision Transformer Using Shifted Windows. In Proceedings of the 2021 IEEE/CVF International Conference on Computer Vision (ICCV), Montreal, QC, Canada, 10–17 October 2021; pp. 9992–10002.

31. Touvron, H.; Cord, M.; Douze, M.; Massa, F.; Sablayrolles, A.; Jégou, H. Training Data-Efficient Image Transformers & Distillation through Attention. *arXiv* **2021**, arXiv:2012.12877.

32. Dosovitskiy, A.; Beyer, L.; Kolesnikov, A.; Weissenborn, D.; Zhai, X.; Unterthiner, T.; Dehghani, M.; Minderer, M.; Heigold, G.; Gelly, S.; et al. An Image Is Worth 16x16 Words: Transformers for Image Recognition at Scale. *arXiv* **2021**, arXiv:2010.11929.

33. Tan, M.; Le, Q.V. EfficientNet: Rethinking Model Scaling for Convolutional Neural Networks. *arXiv* **2020**, arXiv:1905.11946.

34. He, K.; Gkioxari, G.; Dollar, P.; Girshick, R. Mask R-CNN. In Proceedings of the IEEE International Conference on Computer Vision, Venice, Italy, 22–29 October 2017; pp. 2961–2969.

35. Ren, S.; He, K.; Girshick, R.; Sun, J. Faster R-CNN: Towards Real-Time Object Detection with Region Proposal Networks. In *Advances in Neural Information Processing Systems*; Curran Associates, Inc.: New York, NY, USA, 2015; Volume 28.

36. Long, J.; Shelhamer, E.; Darrell, T. Fully Convolutional Networks for Semantic Segmentation. *arXiv* **2015**, arXiv:1411.4038.

37. Loshchilov, I.; Hutter, F. Decoupled Weight Decay Regularization. *arXiv* **2015**, arXiv:1711.05101.

38. Padilla, R.; Passos, W.L.; Dias, T.L.B.; Netto, S.L.; da Silva, E.A.B. A Comparative Analysis of Object Detection Metrics with a Companion Open-Source Toolkit. *Electronics* **2021**, *10*, 279. [CrossRef]

39. Ye, C.; Li, Y.; Cui, P.; Liang, L.; Pirasteh, S.; Marcato, J.; Gonçalves, W.N.; Li, J. Landslide Detection of Hyperspectral Remote Sensing Data Based on Deep Learning with Constrains. *IEEE J. Sel. Top. Appl. Earth Obs. Remote Sens.* **2019**, *12*, 5047–5060. [CrossRef]

40. Yu, B.; Chen, F.; Xu, C. Landslide Detection Based on Contour-Based Deep Learning Framework in Case of National Scale of Nepal in 2015. *Comput. Geosci.* **2020**, *135*, 104388. [CrossRef]

41. Yi, Y.; Zhang, W. A New Deep-Learning-Based Approach for Earthquake-Triggered Landslide Detection from Single-Temporal RapidEye Satellite Imagery. *IEEE J. Sel. Top. Appl. Earth Obs. Remote Sens.* **2020**, *13*, 6166–6176. [CrossRef]

42. Liu, P.; Wei, Y.; Wang, Q.; Chen, Y.; Xie, J. Research on Post-Earthquake Landslide Extraction Algorithm Based on Improved U-Net Model. *Remote Sens.* **2020**, *12*, 894. [CrossRef]

*remote sensing*

*Article*

# Evolutionary Computational Intelligence-Based Multi-Objective Sensor Management for Multi-Target Tracking

Shuang Liang [1], Yun Zhu [2,*], Hao Li [3] and Junkun Yan [4]

[1]  Academy of Advanced Interdisciplinary Research, Xidian University, Xi'an 710071, China; sliang@xidian.edu.cn
[2]  School of Computer Science, Shaanxi Normal University, Xi'an 710119, China
[3]  School of Electronic Engineering, Xidian University, Xi'an 710071, China; haoli@xidian.edu.cn
[4]  National Laboratory of Radar Signal Processing, Xidian University, Xi'an 710071, China; jkyan@xidian.edu.cn
*   Correspondence: yunzhu@snnu.edu.cn

**Abstract:** In multi-sensor systems (MSSs), sensor selection is a critical technique for obtaining high-quality sensing data. However, when the number of sensors to be selected is unknown in advance, sensor selection is essentially non-deterministic polynomial-hard (NP-hard), and finding the optimal solution is computationally unacceptable. To alleviate these issues, we propose a novel sensor selection approach based on evolutionary computational intelligence for tracking multiple targets in the MSSs. The sensor selection problem is formulated in a partially observed Markov decision process framework by modeling multi-target states as labeled multi-Bernoulli random finite sets. Two conflicting task-driven objectives are considered: minimization of the uncertainty in posterior cardinality estimates and minimization of the number of selected sensors. By modeling sensor selection as a multi-objective optimization problem, we develop a binary constrained evolutionary multi-objective algorithm based on non-dominating sorting and dynamically select a subset of sensors at each time step. Numerical studies are used to evaluate the performance of the proposed approach, where the MSS tracks multiple moving targets with nonlinear/linear dynamic models and nonlinear measurements. The results show that our method not only significantly reduces the number of selected sensors but also provides superior tracking accuracy compared to generic sensor selection methods.

**Keywords:** computational intelligence; intelligent sensing technique; multi-sensor systems; multi-target tracking; random finite set; sensor selection

**Citation:** Liang, S.; Zhu, Y.; Li, H.; Yan, J. Evolutionary Computational Intelligence-Based Multi-Objective Sensor Management for Multi-Target Tracking. *Remote Sens.* **2022**, *14*, 3624. https://doi.org/10.3390/rs14153624

Academic Editor: Danilo Orlando

Received: 6 July 2022
Accepted: 26 July 2022
Published: 28 July 2022

**Publisher's Note:** MDPI stays neutral with regard to jurisdictional claims in published maps and institutional affiliations.

## 1. Introduction

With the rapid development of sensing techniques, sensing systems with multi-sensor configurations have attracted lots of attention in numerous fields, such as scene analysis, military defense, habitat monitoring, and other surveillance scenarios [1–4]. As one of the most important techniques, multi-target tracking (MTT) in multi-sensor systems (MSSs) is challenging for two reasons. On the one hand, MTT itself is difficult due to target birth, target death, false alarm, miss detection, and data association uncertainty. On the other hand, due to communication and real-time constraints, intelligent sensor management is required to balance the constraints and the tracking accuracy. Under the complex, dynamic and variable circumstances, sensor control can be regarded as an optimal nonlinear control issue, and standard optimal control schemes are not directly applicable [5].

Conventional MTT approaches used in the literature can be regarded as combinations of single-target trackers. Examples of such approaches include multiple hypothesis tracking [6,7] and joint probabilistic data association [8]. However, they cannot be used in principled sensor management since it is difficult to formulate a management criterion that accommodates the multi-target in a mathematical description. A solution to solve the sensor management problem is to use finite set statistics (FISST) [9,10] in the Bayesian



135

paradigm. Under the framework of FISST, the multi-target probability density is used to describe the uncertainty of the multi-target system and can be systematically handled by random finite sets (RFSs). The probability hypothesis density (PHD) [11], cardinalized PHD [12], and multi-Bernoulli (MB) [13] filters are popular FISST-based approaches. The MB filter uses multiple independent Bernoulli RFSs to model the set of independent targets and propagates MB parameters over time. Different from the MB filter, the PHD and cardinalized PHD filters propagate moments of the multi-target posterior density. These filters were developed as crude approximations of the Bayes filter and cannot output the trajectory for each target. In [14,15], the labeled RFS was used to solve the problem of trajectory estimation. Following these studies, Vo et al. developed a multi-target tracker named generalized labeled MB (GLMB) [16]. The labeled MB (LMB) filter [17] proposed by Reuter et al. provides an efficient approximation of the GLMB filter. In terms of accuracy, the LMB filter outperforms the PHD, cardinalized PHD, and MB filters. What is more, it outputs target trajectories.

Several solutions have been proposed under the FISST framework to solve the sensor management problem. An objective function is generally required as a criterion for sensor management. The Rényi divergence [18–20], or alpha divergence, is widely used as the objective function for sensor management. The Kullback–Leibler divergence or Hellinger affinity are special cases of the Rényi divergence. Recently, a closed-form expression of the Cauchy–Schwartz divergence has been developed for Poisson densities [21], the GLMB filter [22], and the LMB filter [23], providing an alternative objective function for sensor management [22,24,25]. Although the information divergence is derived in a principled manner, it is unclear how to translate it directly into practical performance criterions such as state or cardinality estimation errors. To meet the task of sensor management in a direct way, the task-driven objective functions have been developed [23,26–29]. In [26], the cardinality variance was used to enable efficient sensor management. In [23], Gostar et al. proposed minimizing the posterior dispersion. To deal with multiple tasks simultaneously, ad hoc methods have been developed in [27–29] by estimating the relative importance of each task and assigning weights to the objective functions. It is necessary to estimate the relative importance of each task. In [30], Nguyen et al. studied the multi-objective path-planning problem and proposed competing objectives for searching for undiscovered moving targets while keeping track of discovered targets.

In this work, we consider the problem of selecting a subset of sensors acquiring high-quality measurements to alleviate the energy and bandwidth issues. For the sensor selection problem, it is usually assumed that the number of sensors to be selected is known in advance [31], as illustrated in Figure 1.

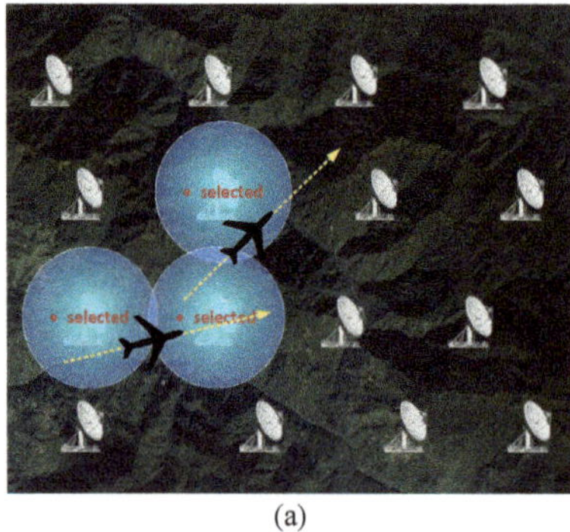


(a)

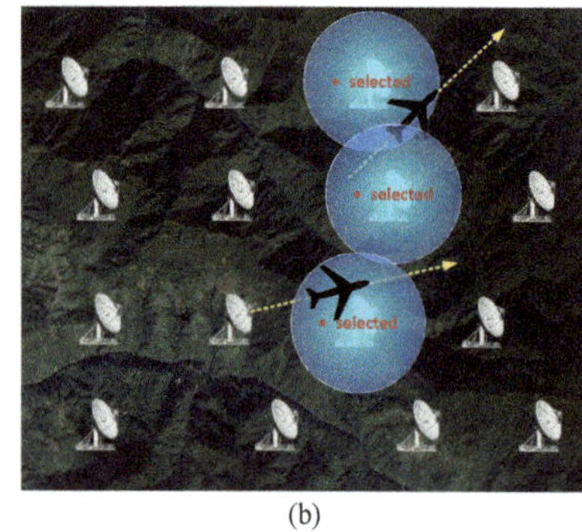


(b)

**Figure 1.** Illustration of dynamic selection of a fixed number of sensors for MTT: (**a**) At time $k$; (**b**) At time $k + 1$. It is assumed that three sensors are selected at each time step, and the blue circles show the coverage areas of the selected sensors.

However, in most practical applications, such as multi-sensor MTT, it is impossible for the system's designer to know the exact number of selected sensors before the selection operation begins. Apparently, it is necessary to study a feasible sensor selection scheme that adaptively determines the optimal number of selected sensors according to the dynamics of a multi-sensor multi-target system. In this case, sensor selection is, in fact, a global combinatorial optimization problem. When the scale of the MSS is large, sensor selection can be extremely challenging. To alleviate this issue, a spatial non-maximum suppression algorithm has been proposed in [32], but its performance is affected by a tuning parameter. The work in [33] developed an approach that decoupled the multi-sensor coordinated management into distributed management of each sensor by maximizing the local Rényi divergence. This method can be used for distributed MTT but not for sensor selection. Wang et al. [34] proposed a guided search algorithm for multi-dimensional optimization-based sensor management. It is not applicable to sensor selection applications and may become stuck at a nonstationary point because of the use of coordinate descent. Cao et al. [35] proposed a sensor selection scheme with low computational complexity based on the upper bound of the mutual information. The method is only applicable for tracking a single target.

The aim of this study is to develop a methodology that allows selection of fewer sensors while ensuring the performance of MTT. The LMB filter is used for MTT by modeling the multi-target states as LMB RFSs. In the sensor selection procedure, we develop the number of selected sensors as an objective function. The variance of the cardinality distribution is also designed as an objective function to improve the accuracy of the cardinality estimate. In addition, a constraint of the number of selected sensors is necessary to meet communication constraints while guaranteeing the performance of the filter. However, minimizing the number of selected sensors and minimizing the cardinality variance is conflicting. The problem is further compounded by the number constraint. To solve this problem, we model it as a multi-objective optimization (MOO) problem and develop a binary constrained evolutionary multi-objective algorithm to dynamically select a subset of sensors. For each selection command, the generalized covariance intersection (GCI) scheme [36] is used for implementing multi-sensor data fusion. The main contributions are summarized as follows.

First, to the best of our knowledge, it is the first study in which an evolutionary algorithm is used in the multi-objective POMDP for MTT. In general, the ideal solution of the MOO does not exist since the objective functions are conflicting. We find the Pareto solutions using an evolutionary multi-objective algorithm via non-dominated sorting and dynamically select a subset of sensors at each time step.

Second, we develop a novel binary constrained crossover and binary constrained mutation operators within the evolutionary algorithm to handle the constraint for the number of selected sensors and obtain feasible solutions.

Third, we compare the proposed evolutionary MOO (EMOO)-based sensor selection approach with several other sensor selection solutions. Simulation results prove that the proposed approach has satisfactory state estimation performances and effectively reduces the number of selected sensors.

The paper is organized as follows. Section 2 presents the existing literature on the RFS and the LMB recursion. The motivation and implementation of the EMOO-based sensor selection approach are presented in Section 3. Section 4 presents numerical simulations and results. Conclusions are given in Section 5.

## 2. Background

### 2.1. Labeled RFS

In the stochastic multi-target system, the target state is modeled as an RFS. The single-target state and the multi-target state are denoted by $x$ and $X$, respectively. It is difficult to output the trajectories of multiple targets only by using the representation of RFS, and we can only estimate the set of states at different time steps, i.e., $\{X_1, \ldots, X_k\}$. To address this issue, the labeled RFS is introduced. In the labeled RFS, the target state is augmented with a label $\ell$. To distinguish between labeled and unlabeled entities, labeled entities are bold, e.g., $x$ and $X$. At time $k$, the multi-target state $X_k$ consists of $N(k)$ single-target states $x_{k,1}, \ldots, x_{k,N(k)}$ and the multi-target measurement $Z_k$ consists of $M(k)$ measurements $z_{k,1}, \ldots, z_{k,M(k)}$. Then, $X_k$ an $Z_k$ are given as

$$X_k = \{x_{k,1}, \ldots, x_{k,N(k)}\} \in \mathcal{F}(\mathbb{X} \times \mathbb{L}), \tag{1}$$

$$Z_k = \{z_{k,1}, \ldots, z_{k,M(k)}\} \in \mathcal{F}(\mathbb{Z}), \tag{2}$$

where $\mathcal{F}(\mathbb{Z})$ denotes the space of finite subsets of $\mathbb{Z}$, and $\mathbb{X}$, $\mathbb{L}$, and $\mathbb{Z}$ denote the spaces for $X$, $\ell$, and $Z$, respectively.

The multi-target posterior density $\pi_k(X_k|Z_{1:k})$ is estimated by the Bayesian prediction and update [9,10]

$$\pi_{k|k-1}(X_k|Z_{1:k-1}) = \int f_{k|k-1}(X_k|X)\pi_{k-1}(X|Z_{1:k-1})\delta X, \tag{3}$$

$$\pi_k(X_k|Z_{1:k}) = \frac{g_k(Z_k|X_k)\pi_{k|k-1}(X_k|Z_{1:k-1})}{\int g_k(Z_k|X)\pi_{k|k-1}(X|Z_{1:k-1})\delta X}, \tag{4}$$

where $Z_{1:k} = (Z_1, \ldots, Z_k)$ represents the set of measurements accumulated to the current time; $\pi_{k|k-1}(X_k|Z_{1:k-1})$ is the predicted density; $f_{k|k-1}(\cdot|\cdot)$ is the multi-target transition density, encapsulating multi-target motion, such as target birth/death and single-target motion; $g_k(\cdot|\cdot)$ is the multi-target likelihood, encapsulating system uncertainty, such as observation noise, data association uncertainty, and detection uncertainty. The integrals given in (3) and (4) are not ordinary integrals but set integrals. For a function $f : \mathcal{F}(\mathbb{X} \times \mathbb{L}) \to \mathbb{R}$ the set integral is denoted as [9,10]

$$\int f(X)\delta X = \sum_{i=0}^{\infty} \frac{1}{i!} \int f(\{x_1, \ldots, x_i\})d(x_1, \ldots, x_i). \tag{5}$$

In the following, the standard inner product notation of $f$ and $g$ is expressed as

$$\langle f, g \rangle \triangleq \int f(x)g(x)dx, \tag{6}$$

and the multi-target exponential notation is given as

$$h^X \triangleq \prod_{x \in X} h(x). \tag{7}$$

The inclusion function $1_S(X)$ and the Kronecker delta function $\delta_S(X)$ are denoted as

$$1_S(X) \triangleq \begin{cases} 1, & \text{if } X \subseteq S \\ 0, & \text{otherwise} \end{cases}, \quad \delta_S(X) \triangleq \begin{cases} 1, & \text{if } X = S \\ 0, & \text{otherwise} \end{cases}. \tag{8}$$

### 2.2. Labeled Multi-Bernoulli Filter

In the LMB filter, a target $x \in \mathbb{X}$ with label $\ell \in \mathbb{L}$ is completely characterized by the probability of existence $r^{(\ell)}$ and the probability density $p^{(\ell)}(x)$. The LMB distribution is, therefore, represented by $\pi = \{(r^{(\ell)}, p^{(\ell)}(\cdot))\}_{\ell \in \mathbb{L}}$. Let $\triangle(X) = \delta_{|X|}(|\mathcal{L}(X)|)$ denote a distinct label indicator and $\mathcal{L} : \mathbb{X} \times \mathbb{L} \to \mathbb{L}$ be the projection $\mathcal{L}(X) = \{\mathcal{L}(x) : (x \in X)\}$. The LMB RFS density is parameterized as

$$\pi(X) = \triangle(X)w(\mathcal{L}(X))[p]^X, \tag{9}$$

where

$$w(L) = \prod_{i \in \mathbb{L}}(1 - r^{(i)})\prod_{i \in \mathbb{L}}\frac{1_{\mathbb{L}}r^{(\ell)}}{(1 - r^{(\ell)})}, \tag{10}$$

$$[p]^X = \prod_{(x,\ell) \in X} p^{(\ell)}(x), \tag{11}$$

and $L$ indicates a set of labels.

If the posterior density follows the LMB distribution and is parameterized as $\pi = \{(r^{(\ell)}, p^{(\ell)}(\cdot))\}_{\ell \in L}$ and the birth model also follows the LMB distribution with the parameter set $\pi_B = \{(r_B^{(\ell)}, p_B^{(\ell)}(\cdot))\}_{\ell \in B}$, then the predicted density is given as

$$\pi_+ = \{(r_{+,S}^{(\ell)}, p_{+,S}^{(\ell)}(\cdot))\}_{\ell \in L} \cup \{(r_B^{(\ell)}, p_B^{(\ell)}(\cdot))\}_{\ell \in B}, \tag{12}$$

where

$$r_{+,S}^{(\ell)} = \eta_S(\ell) r^{(\ell)}, \tag{13}$$

$$p_{+,S}^{(\ell)}(\cdot) = \frac{\langle p_S(\cdot, \ell) f(x|\cdot, \ell), p(\cdot, \ell)\rangle}{\eta_S(\ell)}, \tag{14}$$

$$\eta_S(\ell) = \langle p_S(\cdot, \ell), p(\cdot, \ell)\rangle, \tag{15}$$

$p_S(\cdot, \ell)$ is the state-dependent survival probability and $f(x|x', \ell)$ denotes the transition density of the target with track $\ell$. For simplicity, we denote the predicted LMB RFS by

$$\pi_+ = \{(r_+^{(\ell)}, p_+^{(\ell)}(\cdot))\}_{\ell \in L_+}, \tag{16}$$

where the label space $L_+ = L \cup B$ (with $L \cap B = \oslash$).

The family of the LMB RFS is closed under the Bayesian prediction but not closed under the Bayesian update. To solve this problem, the predicted LMB distribution is converted to a $\delta$-GLMB distribution. Then, the update of the $\delta$-GLMB is implemented, and the result is approximated by an LMB. The LMB approximation of the multi-target posterior density is denoted as

$$\pi(\cdot|Z) = \{(r^{(\ell)}, p^{(\ell)}(\cdot))\}_{\ell \in L_+}, \tag{17}$$

where

$$r^{(\ell)} = \sum_{(I_+, \theta) \in \mathcal{F}(\mathbb{L}_+) \times \Theta_{I_+}} w^{(I_+, \theta)}(Z) 1_{I_+}(\ell), \tag{18}$$

$$p^{(\ell)}(x) = \frac{1}{r^{(\ell)}} \sum_{(I_+, \theta) \in \mathcal{F}(\mathbb{L}_+) \times \Theta_{I_+}} w^{(I_+, \theta)}(Z) \times 1_{I_+}(\ell) p^{(\theta)}(x, \ell), \tag{19}$$

$$w^{(I_+, \theta)}(Z) \propto w_+(I_+)[\eta_Z^{(\theta)}(\ell)]^{I_+}, \tag{20}$$

$$p^{(\theta)}(x, \ell|Z) = \frac{p_+(x, \ell)\psi_Z(x, \ell; \theta)}{\eta_Z^{(\theta)}(\ell)}, \tag{21}$$

$$\eta_Z^{(\theta)}(\ell) = \langle p_+(\cdot, \ell), \psi_Z(\cdot, \ell; \theta)\rangle, \tag{22}$$

$$\psi_Z(x, \ell; \theta) = \begin{cases} \dfrac{p_D(x, \ell) g(z_{\theta(\ell)}|x; \ell)}{\kappa(z_{\theta(\ell)})}, & \text{if } \theta(\ell) > 0, \\ 1 - p_D(x, \ell), & \text{if } \theta(\ell) = 0, \end{cases} \tag{23}$$

and $\Theta_{I_+}$ is the space of mappings $\theta : I_+ \rightarrow \{0, 1, \ldots, |Z|\}$, such that $\theta(i) = \theta(i') > 0$ implies $i = i'$; $\kappa(\cdot)$ is the intensity of the clutter measurements; $g(z|x; \ell)$ is the likelihood of measurement $z$ given $(x, \ell)$.

In the sequential Monte Carlo (SMC) implementation, the density for each target with label $(\ell)$ is approximated by a weighted sum of particles, as follows

$$p^{(\ell)}(x) \simeq \sum_{j=1}^{J^{(\ell)}} \omega_j^{(\ell)} \delta_{x_j^{(\ell)}}(x), \tag{24}$$

where $\omega_j^{(\ell)}$ is the weight of particle $j$, and $J^{(\ell)}$ denotes the number of particles. For more details on the SMC implementation, please refer to [17].

## 3. Method

### 3.1. Objective Functions Proposal

Using sensor networks with communication constraints, sensor selection for MTT applications is usually employed to acquire the best set of measurements. As sensor management solutions, the Markov decision process and partially observable Markov decision process (POMDP) have received great attention over the last few decades [24]. The POMDP framework enables direct generalization to multiple targets by using the RFS model [9,10,24]. We model the sensor selection problem as the following discrete-time POMDP:

$$\Psi = \{X_k, \mathbb{S}, f_{k|k-1}(X_k|X_{k-1},), g_k(Z_k|X_k), \vartheta(s_k)\}, \tag{25}$$

where $\mathbb{S}$ denotes a finite set of candidate sensors and $\vartheta(s_k)$ is the objective (reward or cost) function. In stochastic filtering, the aim is to find a selection command that optimizes $\vartheta(s_k)$.

In our work, two objective functions are considered: the number of selected sensors and the variance of the cardinality distribution. Both of the objective functions are dependent on binary decision variables. Let

$$s_k = [s_{1,k}, s_{2,k}, \ldots, s_{N_s,k}], \tag{26}$$

be the selection command at time $k$, and $N_s$ is the number of all candidate sensors in the MSS. The elements of $s_k$ are binary variables, i.e., $s_{i,k} = 1$, if sensor $s_i$ is selected and $s_{i,k} = 0$ otherwise. For example, if there are ten sensors in the system, $s_k = [0,1,0,1,0,0,0,0,1,0]$ indicates the command that the sensors $s_2$, $s_4$ and $s_9$ are selected at time $k$.

In many practical applications, the number of sensors to be selected is unknown to the system designer. To control the number of selected sensors at time $k$, the following objective function is considered

$$f_1(s_k) = \sum_{i=1}^{N_s} s_{i,k}. \tag{27}$$

The other objective function is the variance of the cardinality distribution, aiming at minimizing the error for the estimated number of targets. At time $k$, the cardinality variance corresponding to the selection command $s_k$ is given by

$$f_2(s_k) = \sum_{\ell \in \mathbb{L}_+} r^{(\ell)}(s_k)[1 - r^{(\ell)}(s_k)]. \tag{28}$$

The objective function defined in (28) is computed using parameters of the updated LMB distribution. However, sensors have not been selected and it is impossible to update the LMB RFS density using measurements collected by the selected sensors. The predicted ideal measurement set (PIMS) strategy [37] is utilized to address this issue, which is dependent on the predicted LMB distribution and ideal assumption of perfect detection, no clutter, and no measurement noise. First, the predicted LMB distribution is used to estimate the number of targets and the target states. The maximum a posteriori estimate of the target number is computed as follows,

$$\hat{n} = \arg\max_n \rho(n) = \arg\max_n \rho(0) \sum_{L \subseteq \mathbb{L}, |L|=n} \left( \prod_{\ell \in L} \frac{r_+^{(\ell)}}{1 - r_+^{(\ell)}} \right), \tag{29}$$

where $\rho(0) = \prod_{\ell \in \mathbb{L}} (1 - r_+^{(\ell)})$. Then, we obtain $\hat{n}$ labels with the highest existence probabilities from the predicted LMB distribution. The a posteriori estimate of the target state with label $\ell$ is given as

$$\hat{x}^{(\ell)} = \sum_{j=1}^{J_+^{(\ell)}} \omega_{j+}^{(\ell)} x_{j+}^{(\ell)}. \tag{30}$$

A predicted ideal measurement is estimated for each $\hat{x}^{(\ell)}$ under the assumed ideal conditions, and the pseudo-update of the LMB distribution is implemented with the PIMS. Then, the objective function (28) is computed using the generated pseudo LMB distribution.

### 3.2. Evolutionary Multi-Objective Optimization

Although the number of sensors to be selected is kept unknown, the number of selected sensors should be limited to a range $N_{\min}$ and $N_{\max}$. This limit not only guarantees the performance of the filter but also meets the communication requirement. At time $k$, the constrained MOO is mathematically described as follows

$$\text{Minimize } F(s_k) = [f_1(s_k), f_2(s_k)]^{\mathrm{T}} \tag{31}$$

$$\text{Subject to } N_{\min} \leq f_1(s_k) \leq N_{\max} \tag{32}$$

where $F(s_k)$ is the objective vector.

For the MOO problem, the solutions satisfying the constraint of (32) form the feasible set. The ideal solution is the one that is optimal for all the objective functions. In general, the ideal solution does not exist since the objective functions are conflicting. Several methods have been proposed to handle the problem [38–41]. Among them, the scalarization and Pareto methods do not need complicated numerical derivations and are widely used. The scalarization method is easy to implement, but it needs to assign relative weight to each objective based on prior information. Worse, unless the search space is convex, the solution may not be found [42]. In the Pareto method, the goodness of a solution is determined by the dominance, and a compromise solution can be found along the Pareto optimal front. We solve the above MOO problem and find optimal Pareto solutions using an evolutionary multi-objective algorithm via non-dominated sorting. First, the initial population of size $N^{pop}$ is generated in which each solution is a feasible solution, represented by a vector of $N_s$ binary elements. Then, the offspring solutions are obtained by binary tournament selection, crossover and mutation operators.

In our problem with binary decision variables, a simple crossover operator called one-point crossover is used. Two parent chromosomes and a random/given point are selected. After the given/selected point, genes of parent chromosomes are interchanged. An example is given in Figure 2, in which point four is selected, and the genes of two-parent chromosomes P1 and P2 are interchanged. Assuming that the number of selected sensors is limited to the range $N_{\min} = 1$ and $N_{\max} = 3$, we can observe that the offspring solutions in Figure 2 meet the constraints of the number of selected sensors. However, there are some cases where the offspring solutions need to be modified. An example is given in Figure 3, where the parent chromosomes P1 and P2 are different from those in Figure 2. In Figure 3, we also select point four and interchange the genes of P1 and P2. The sum of all the bits in solutions C1 and C2 are $N = 0$ and $N = 4$, respectively. Apparently, these solutions cannot meet the constraints of $N_{\min} = 1$ and $N_{\max} = 3$ and, hence, are infeasible for sensor selection. To solve this problem, we develop a binary constrained crossover procedure, as shown in Algorithm 1.

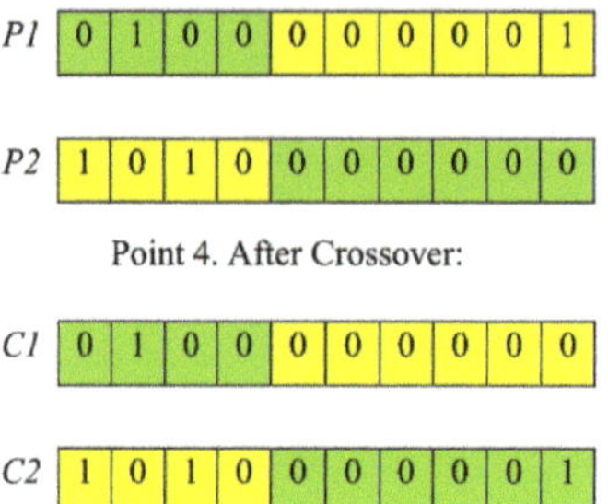

Point 4. After Crossover:

**Figure 2.** Example of the effective one-point crossover.

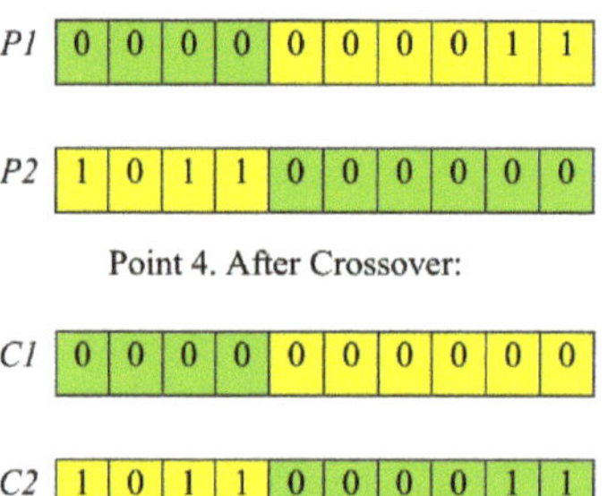

Point 4. After Crossover:

**Figure 3.** Example of the ineffective one-point crossover.

---

**Algorithm 1** Binary constrained crossover.

1. Select one crossover point.
2. Copy the binary string from the beginning to the crossover point of the first parent and the rest from the other parent.
3. Compute the sum $N$ of all the bits of the child solution.
4. If $N_{\min} \leq N \leq N_{\max}$, the child solution is reserved; otherwise, go to line 5.
5. Select and flip a point of the child solution, and go back to line 3.

---

Along with the binary constrained crossover, the mutation is also performed. For the binary issue, the bit flip mutation is one of the most commonly used mutation operators. In the bit flip mutation, one or more random bits are selected and then flipped. Figure 4 illustrates an example in which point four is selected from the parent chromosome P and flipped. We assume that the number of selected sensors is limited to the range $N_{\min} = 1$ and $N_{\max} = 3$. Then, the offspring solution in Figure 4 meets the constraints. There are some cases where the offspring solutions of the mutation need to be modified. Figure 5 shows an example, where we also select point four and flip it. The sum of all the bits in solution C is $N = 4$, which cannot meet the constraint of $N_{\max} = 3$. To solve this problem, we develop the binary constrained mutation, as shown in Algorithm 2.

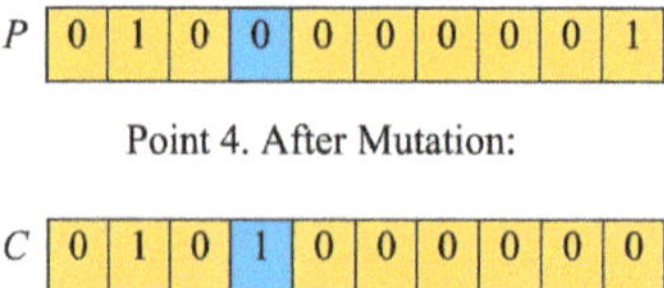

Point 4. After Mutation:

**Figure 4.** Example of the effective bit flip mutation.

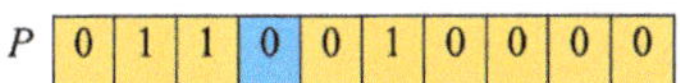

Point 4. After Mutation:

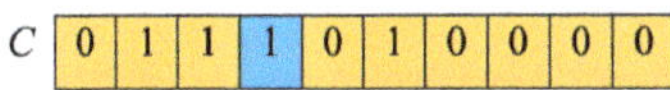

**Figure 5.** Example of the ineffective bit flip mutation.

---

**Algorithm 2** Binary constrained mutation.

---

1.    Randomly select one mutation point.
2.    Flip the selected mutation point.
3.    Compute the sum $N$ of all the bits of the child solution.
4.    If $N_{\min} \le N \le N_{\max}$, the child solution is reserved; otherwise, go back to line 1.

---

After the variants (crossover and mutation), the offspring for the next generation are generated. The new population formed by the parents and offspring is sorted according to the non-dominant relationships. The size of the population is decreased to $N^{pop}$ by eliminating the solutions with lower ranks. For the next generation, the new population is generated using binary tournament selection, binary constrained crossover, and binary constrained mutation. After several generations $G$, the Pareto-front is obtained.

The Pareto front is formed from non-dominated solutions, and it is necessary to choose one compromise solution from them. We use the gray relational analysis (GRA) strategy [43] to find the compromise solution. GRA does not require the weight of each objective function or other prior information. The gray relational coefficient (GRC) approach is used to estimate the similarity between the candidate network (formed by values of the objective functions for the Pareto solutions) and the optimal reference network (formed by the optimal value of each objective). Assuming that there are $m$ Pareto solutions obtained by the evolutionary algorithm, $f_{ij}$ is the $i$th value of the $j$th objective in the objective matrix, $\overline{f}_{ij}$ is the value of $f_{ij}$ after normalization. The main steps involved in GRA are summarized as follows.

i: Normalizethe objective function values of Pareto solutions, as follows

$$\overline{f}_{ij} = \frac{\max_{i \in m} f_{ij} - f_{ij}}{\max_{i \in m} f_{ij} - \min_{i \in m} f_{ij}}. \tag{33}$$

ii: Find the reference network points

$$f_j^* = \max_{i \in m} \overline{f}_{ij}. \tag{34}$$

iii: Estimate the difference between $f_j^*$ and $\overline{f}_{ij}$

$$\triangle I_{ij} = \left| f_j^* - \overline{f}_{ij} \right|. \tag{35}$$

iv: Find the value of GRC for each optimal solution:

$$\mathrm{GRC}_i = \frac{1}{m} \sum_{j=1}^{n} \frac{\triangle \min + \triangle \max}{\triangle I_{ij} + \triangle \max}, \tag{36}$$

where $\triangle \max = \max_{i \in m, j \in n}(\triangle I_{ij})$ and $\triangle \min = \min_{i \in m, j \in n}(\triangle I_{ij})$. v: Find the largest $\mathrm{GRC}_i$, and the corresponding solution is recommended.

*3.3. Multi-Sensor Fusion*

For each selection command candidate $s \subseteq \mathbb{S}$, the posteriors are LMB RFSs with parameters $\pi(\cdot|Z^{(s)}) = \{\{(r_{i,s_i}^{(\ell)}, p_{i,s_i}^{(\ell)}(\cdot))\}_{\ell \in \mathbb{L}_+}\}_{i=1}^{|s|}$. The posterior density of each selected sensor is approximated by

$$p_{i,s_i}^{(\ell)}(x) = \sum_{j=1}^{J_+^{(\ell)}} \omega_{i,s_i,j}^{(\ell)} \delta_{x_{j+}^{(\ell)}}(x).$$ (37)

During the update step of the LMB filter, the weights of particles are updated but the particles themselves are not changed. Therefore, the particles in (37) are the same particles used in the prediction.

We use the GCI scheme [36] to fuse those posterior LMB densities, which returns the following existence probabilities and densities,

$$r_s^{(\ell)} = \frac{\int \prod_{i=1}^{|s|} (r_{i,s_i}^{(\ell)} p_{i,s_i}^{(\ell)}(x))^{\omega_i} dx}{\prod_{i=1}^{|s|} (1 - r_{i,s_i}^{(\ell)})^{\omega_i} + \int \prod_{i=1}^{|s|} (r_{i,s_i}^{(\ell)} p_{i,s_i}^{(\ell)}(x))^{\omega_i} dx},$$ (38)

$$p_s^{(\ell)}(x) = \frac{\prod_{i=1}^{|s|} (p_{i,s_i}^{(\ell)}(x))^{\omega_i}}{\int \prod_{i=1}^{|s|} (p_{i,s_i}^{(\ell)}(x))^{\omega_i} dx},$$ (39)

where $\omega_i$ is a weight indicating the importance of sensor $s_i$ in the fusion process. The sum of all the weights is equal to 1, i.e., $\sum_{i=1}^{|s|} \omega_i = 1$. We assume that all the sensors have equal importance in the simulation studies, i.e., $\omega_i = 1/|s|$. When using the particle approximation (37) to represent each LMB density, the integrals in (38) and (39) turn to weighted sums over the particles.

*3.4. Step-by-Step Implementation*

We introduce a sensor selection solution for MTT in this paper. The framework consists of four main steps: prediction, estimation of PIMS, EMOO-based sensor selection, and fusion of local posteriors. The schematic diagram is shown in Figure 6.

Algorithm 3 shows a complete step-by-step pseudocode for a single run of the proposed algorithm that outputs a fused LMB posterior. Assume that the following parameters are always available:

- Sensor model parameters: the number of candidate sensors $N_s$ and their positions $s^{(j)} = [s_x, s_y]^T$, detection probabilities $p_D^{(j)}(\cdot)$, and clutter intensities $\kappa^{(j)}(\cdot)$ with $j = 1, 2, \ldots, N_s$;
- Birth model parameters: $\{r_B^{(\ell)}, \{\omega_{j,B}^{(\ell)}, x_{j,B}^{(\ell)}\}_{j=1}^{J_B^{(\ell)}}\}_{\ell \in \mathbb{B}}$;
- Likelihood $g(z|x, \ell)$ and transition density $f(x|\cdot, \ell)$;
- Survival probability function: $p_S(x, \ell)$;
- Constraints on the number of selected sensors: $N_{\min}$ and $N_{\max}$.

Similar to the standard particle filter, particle degeneracy is inevitable [44]. To alleviate the particle degradation problem, the particles for each hypothesized track are resampled in line 12. In a typical particle filtering implementation, Markov chain Monte Carlo steps are performed after resampling to improve the diversity of particles [44]. In line 13, multi-target states are extracted from the posterior LMB distribution and are used for error performance evaluation. The pseudocode of the algorithm for the EMOO-based sensor selection in line 6 is given in Algorithm 4.

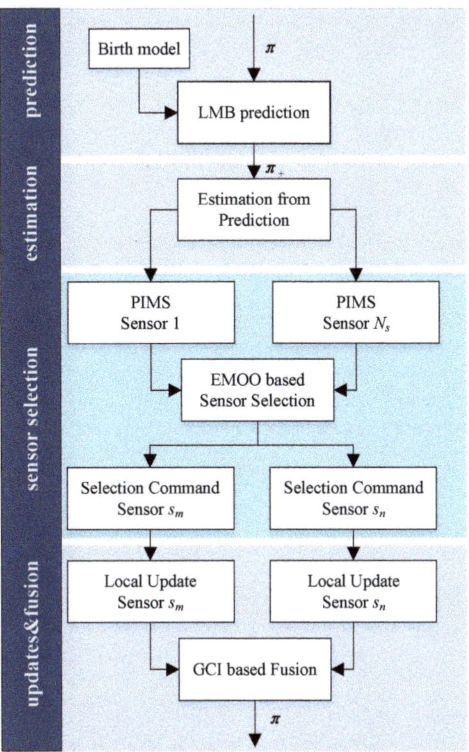



**Figure 6.** Schematic diagram of sensor selection with LMB filtering.

---

**Algorithm 3** Step-by-step pseudocode for the proposed approach with LMB filtering, sensor selection, and fusion.

---

**INPUTS:**

$\rightarrow$ LMB distribution $\pi = \{r^{(\ell)}, \{\omega_j^{(\ell)}, x_j^{(\ell)}\}_{j=1}^{J^{(\ell)}}\}_{\ell \in \mathbb{L}}$ from previous time step

**OUTPUTS:**

$\rightarrow$ The posterior parameters $\pi = \{r^{(\ell)}, \{\omega_j^{(\ell)}, x_j^{(\ell)}\}_{j=1}^{J^{(\ell)}}\}_{\ell \in \mathbb{L}}$ to be propagated to the next time step

$\rightarrow$ Estimated multi-target states at the current time

1.  Predict the LMB distribution $\pi_+ = \{(r_+^{(\ell)}, p_+^{(\ell)}(\cdot))\}_{\ell \in \mathbb{L}_+}$ using (12)–(14)
2.  Estimate the target states $\hat{X}$ using $\pi_+ = \{(r_+^{(\ell)}, p_+^{(\ell)}(\cdot))\}_{\ell \in \mathbb{L}_+}$ based on (29) and (30)
3.  **for** every sensor $s_i \in N_s$ **do**
4.      Compute the PIMS $Z^{(i)}$ of sensor $s_i$
5.  **end for**
6.  EMOO-based sensor selection
7.  Collect $Z^{(s^*)}$ from the selected sensors $s^*$
8.  **for** every sensor $s_i \in s^*$ **do**
9.      Update the local LMB distribution $\pi(\cdot|Z^{(s_i)}) = \{(r_{i,s_i}^{(\ell)}, p_{i,s_i}^{(\ell)}(\cdot))\}_{\ell \in \mathbb{L}_+}$ using (17)–(23)
10. **end for**
11. Obtain the posterior distribution $\hat{\pi}(\cdot|Z^{(s)}) = \{r^{(\ell)}, \{\omega_j^{(\ell)}, x_j^{(\ell)}\}_{j=1}^{J^{(\ell)}}\}_{\ell \in \mathbb{L}}$ based on the GCI method using (37) and (38)
12. Pruning and resampling to obtain the posterior LMB distribution
13. Extract multi-target states using (38) and (39)

---

---

**Algorithm 4** Step-by-step pseudocode for the EMOO-based sensor selection.

---

**INPUTS:**
$\rightarrow$ The predicted LMB distribution $\pi_+ = \{(r_+^{(\ell)}, p_+^{(\ell)}(\cdot))\}_{\ell \in \mathbb{L}_+}$
$\rightarrow$ PIMS from each sensor $s_i \in N_s$
$\rightarrow$ The population size $N^{pop}$
$\rightarrow$ The maximum number $G$ of generations
**OUTPUTS:**
$\rightarrow$ The sensors $s^*$ selected at current time

1.     Initialize population of size $N^{pop}$, which meet the constraint in (32)
2.     Set the generation $t = 0$
3.     **while** $t < G$
4.         Evaluate individual fitness using (27) and (28)
5.         Create a new population of offspring with the implementation of the tournament selection operator, the proposed binary constrained crossover (Algorithm 1) and binary constrained mutation (Algorithm 2)
6.         Combine the parents and offspring to create the next population
7.         Set $t = t + 1$
8.     **end**
9.     A set of non-dominated solutions is obtained
10.    Select the compromise solution using the GRA strategy (33)–(36)

---

## 4. Experiments

The performance of the proposed sensor selection approach is demonstrated within a multistatic sensor system. Compared with the traditional monostatic sensor, the multistatic sensor has many advantages [45]; for example, the information on target signatures is enhanced because of the multi-perspective and differences in the clutter properties. What is more, the receive-only multistatic sensor is passive, which provides obvious advantages in military applications. However, measurements collected by the multistatic sensor system are generally affected by noise corruption, missed detections, and false alarms, since its transmit and receive antennas are located in different places.

We use a multistatic sensor system whose structure is borrowed from [46]. As shown in Figure 7a, there is one transmitter and ten receivers within the surveillance system. The receivers are selected adaptively during the tracking of targets. The probability of detection for each receiver $j = 1, 2, \ldots, 10$ is modeled as follows [46]

$$p_D^{(j)}(x_k) = 1 - \phi(\left\|p_k - r^{(j)}\right\|; \alpha, \beta), \tag{40}$$

where $p_k$ and $r^{(j)}$ denote the target position and the position of receiver $j$, respectively. $\phi(d; \alpha, \beta) = \int_{-\infty}^{d} \mathcal{N}(v; \alpha, \beta)dv$ is the Gaussian cumulative distribution function with $\alpha = 12$ km and $\beta = (3 \text{ km})^2$; $\left\|p_k - r^{(j)}\right\|$ is the distance between the receiver and the target. Figure 7b plots the contour lines of the detection probability for each sensor in the $x - y$ plane. It can be observed from Figure 7b that the probability of detection for the multistatic sensor system decreases with the increase in the distance [47].

The sampling interval of the system is fixed as $T = 10$ s, and all the receivers have identical measurement noise. The measurement vector consists of a bearing and bistatic range, as follows

$$z_k^j = \begin{bmatrix} \varphi \\ \rho \end{bmatrix} = \begin{bmatrix} \arctan\left(\dfrac{p_{y,k} - r_y^{(j)}}{p_{x,k} - r_x^{(j)}}\right) \\ \left\|p_k - r^{(j)}\right\| + \left\|p_k - t\right\| \end{bmatrix} + \varepsilon_k^j, \tag{41}$$

where $\varepsilon_k^j \sim N(\cdot; 0, R_k)$, with $R_k = \mathrm{diag}([\sigma_\varphi^2, \sigma_\rho^2])$ and $\sigma_\varphi = (\pi/180)$ rad, $r^{(j)} = [r_x^{(j)}, r_y^{(j)}]^{\mathrm{T}}$, and $\sigma_\rho = 5$ m. The clutter measurements are uniformly distributed in $[-\pi, \pi]$ rad $\times$ $[0, 15, 000]$ m with $\kappa = 2 \times 10^{-5} (\mathrm{radm})^{-1}$.

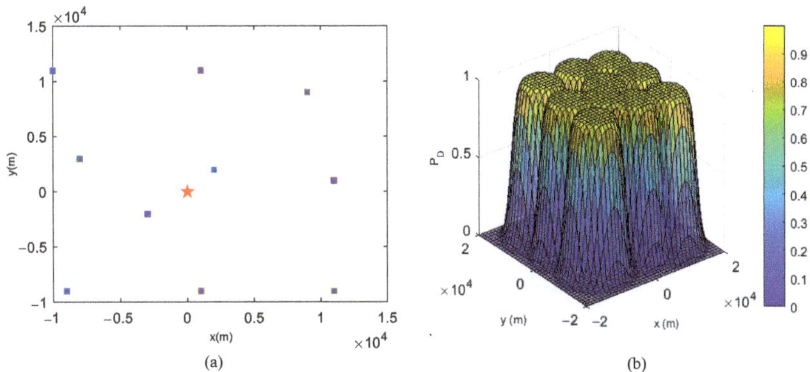

**Figure 7.** Simulation setup: (**a**) The locations of the transmitter (star) and receivers (squares); (**b**) contour plot of the probability of detection.

We use two MTT scenarios to study the performance of the EMOO approach. The first scenario has a time-varying number of targets moving with nearly constant turn (NCT) motion. The second scenario consists of three targets moving with nearly constant velocity (NCV) motion. To evaluate the performance of the EMOO approach, we compare it with three sensor selection solutions: (i) the heuristic random selection method, in which each sensor has an equal probability of being chosen; (ii) the variance-based approach using the cardinality variance defined in (28) as the cost function; (iii) the Cauchy–Schwarz divergence-based approach, which uses the Cauchy–Schwarz divergence between the predicted and updated LMB densities as the reward function. The traditional exhaustive search scheme is used to find the selection command in methods (ii) and (iii), in which the objective function is computed for all possible combinations of a fixed number of sensors in the MSS. In the following, a fixed number $N = 3$ of sensors are selected in these comparative algorithms.

The average tracking performances are obtained using 100 Monte Carlo (MC) runs. The optimal sub-pattern assignment (OSPA) [48] and OSPA$^{(2)}$ [49,50] distances are used to evaluate the tracking errors. By measuring the distance between two sets of states, the OSPA metric [48] can estimate errors in both cardinality and localization. As an adaptation of the OSPA metric, the OSPA$^{(2)}$ metric [49,50] considers sets of tracks and carries the interpretation of a per-track per-time error. All experiments are tested in Matlab R2010a and implemented on a computer with a 3.40 GHz processor.

*4.1. Scenario 1*

In this scenario, the tracking of two targets with NCT motion is studied. The target state vector is $x_k := [p_{x,k}, \dot{p}_{x,k}, p_{y,k}, \dot{p}_{y,k}, \omega_k]^{\mathrm{T}}$, in which $\omega_k$ is the turn rate. The transition model is

$$x_k = f(x_{k-1}) + Gw_{k-1}, \tag{42}$$

where

$$f(x_{k-1}) = F(\omega_{k-1})x_{k-1}, \tag{43}$$

$$F(\omega_{k-1}) = \begin{bmatrix} 1 & \frac{\sin \omega_{k-1} T}{\omega_{k-1}} & 0 & -\frac{1-\cos \omega_{k-1} T}{\omega_{k-1}} & 0 \\ 0 & \cos \omega_{k-1} T & 0 & -\sin \omega_{k-1} T & 0 \\ 0 & \frac{1-\cos \omega_{k-1} T}{\omega_{k-1}} & 1 & \frac{\sin \omega_{k-1} T}{\omega_{k-1}} & 0 \\ 0 & \sin \omega_{k-1} T & 0 & \cos \omega_{k-1} T & 0 \\ 0 & 0 & 0 & 0 & 1 \end{bmatrix}, \tag{44}$$

$$G = \begin{bmatrix} \frac{T^2}{2} & 0 & 0 \\ T & 0 & 0 \\ 0 & \frac{T^2}{2} & 0 \\ 0 & T & 0 \\ 0 & 0 & T \end{bmatrix}, \tag{45}$$

$$w_{k-1} := [w_{x,k-1}, w_{y,k-1}, w_{\omega,k-1}]^{\mathrm{T}}, \tag{46}$$

and $w_{k-1} \sim \mathcal{N}(w_{k-1}; 0, Q_{k-1})$ is white Gaussian process noise with covariance $Q_{k-1} = \mathrm{diag}(\sigma_x^2, \sigma_y^2, \sigma_\omega^2)$, where $\sigma_x = \sigma_y = 1.0 \times 10^{-4}$ m/s$^2$ and $\sigma_\omega = 1.0 \times 10^{-9}$ rad/s$^2$. The covariance of the additive process noise $Gw_{k-1}$ is $GQ_{k-1}G^{\mathrm{T}}$.

The birth process follows the LMB distribution $\{(r_B, p_B^{(i)})\}_{i=1}^2$, where $r_B = 0.02$ and $p_B^{(i)} = \mathcal{N}(x; m_B^{(i)}, P_B)$ with the mean $m_B^{(1)} = [2500, 0, -1000, 0, 0]^{\mathrm{T}}$, $m_B^{(1)} = [1750, 0, 1000, 0, 0]^{\mathrm{T}}$, and the covariance $P_B = \mathrm{diag}([50, 50, 50, 50, 6(\pi/180)]^{\mathrm{T}})^2$. The units are meters for $x$ and $y$ and meters per second for $\dot{x}$ and $\dot{y}$. The maximum and minimum numbers of particles for each hypothesized track are $L_{\max} = 1000$ and $L_{\min} = 300$, respectively. For each hypothesized track, the number of particles is proportional to its probability of existence. The probability of survival is fixed as $p_S = 0.99$. The number of components for each forward propagation is set to 100. The ground truth and estimated tracks for a single MC run with $N_{\min} = 1$ and $N_{\max} = 3$ is illustrated in Figure 8, showing the true and estimated tracks in $x$ and $y$ coordinates versus time. The plots indicate that the EMOO approach is able to identify target births and successfully accommodate nonlinearities.

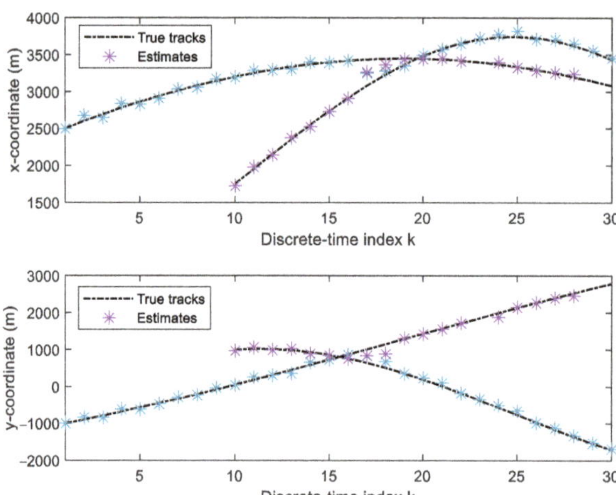

**Figure 8.** True and estimated tracks versus time in Scenario 1.

The average OSPA error (with parameters $p = 1$ and $c = 300$ m) and OSPA$^{(2)}$ error (with the same $c$, $p$, and window length $w = 10$) are given in Figure 9a,b, respectively. The average number of selected sensors is shown in Figure 9c. We observe that both the variance-based approach and the proposed EMOO approach outperform the Cauchy–Schwartz divergence-based approach in terms of OSPA and OSPA$^{(2)}$ errors. This is mainly because the objective functions of the variance-based approach and the EMOO approach are derived from the cardinality distribution, which is strongly related to the error terms computed in

OSPA and OSPA$^{(2)}$ metrics. In addition, the detection probability is unsatisfactory in the considered scenario. This underlines the importance of cardinality estimation, which is the focus of the objective function developed for the variance-based approach and the EMOO approach. Compared with the variance-based approach, the EMOO approach uses fewer sensors at each time step (as shown in Figure 9c) but provides better tracking accuracy (as shown in Figure 9a,b). For the variance-based approach, a fixed number of $N = 3$ sensors are selected at each time step. However, using more sensors does not indicate a better tracking performance. When the uncertainty of the multi-sensor tracking system is high, such as the scenario we consider, using more sensors for tracking may reduce the tracking performance.

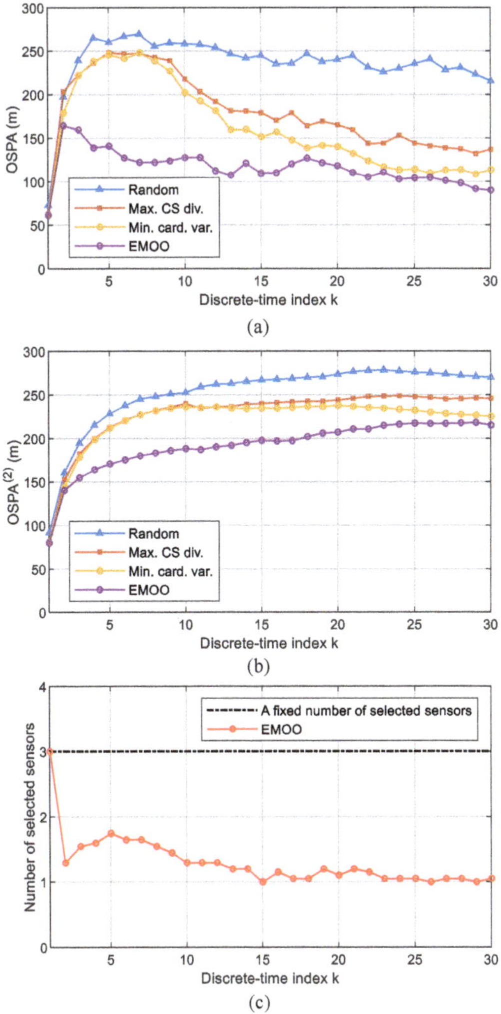

**Figure 9.** Average performance comparison in Scenario 1: (**a**) OSPA error; (**b**) OSPA$^{(2)}$ error; (**c**) the number of selected sensors.

The average computing times for the random selection approach, the CS divergence-based approach, the cardinality variance-based approach, and the EMOO approach to execute a complete MC simulation are 2.14, 153.28, 194.76, and 87.02 s, respectively. Compared with other methods, the random selection method requires less computing time because it does not use any technical method. The EMOO approach runs faster than the CS divergence-based approach and the cardinality variance-based approach.

### 4.2. Scenario 2

In this scenario, three targets with NCV motion move into the surveillance area. The state of the moving target at time $k$ is denoted as $x_k = [p_{x,k}, \dot{p}_{x,k}, p_{y,k}, \dot{p}_{y,k}]^{\mathrm{T}}$. The NCV motion of each target is modeled as

$$x_k = F_{k-1}x_{k-1} + w_{k-1},\tag{47}$$

where

$$F_{k-1} = \begin{bmatrix} 1 & T & 0 & 0 \\ 0 & 1 & 0 & 0 \\ 0 & 0 & 1 & T \\ 0 & 0 & 0 & 1 \end{bmatrix},\tag{48}$$

and $w_{k-1}$ is white Gaussian process noise with covariance $Q_{k-1}$ denoted as

$$Q_{k-1} = \sigma_w^2 \begin{bmatrix} \frac{T^4}{4} & \frac{T^3}{2} & 0 & 0 \\ \frac{T^3}{2} & T^2 & 0 & 0 \\ 0 & 0 & \frac{T^4}{4} & \frac{T^3}{2} \\ 0 & 0 & \frac{T^3}{2} & T^2 \end{bmatrix},\tag{49}$$

and $\sigma_w = 0.01$ m/s$^2$ is the standard deviation of the acceleration noise.

The birth process is an LMB RFS with parameters $\{(r_B, p_B^{(i)})\}_{i=1}^3$, where $r_B = 0.02$ and $p_B^{(i)} = \mathcal{N}(x; m_B^{(i)}, P_B)$ with $m_B^{(1)} = [3000, 0, 0, 0]^{\mathrm{T}}$, $m_B^{(2)} = [2250, 0, 2000, 0]^{\mathrm{T}}$, $m_B^{(3)} = [3000, 0, 2500, 0]^{\mathrm{T}}$, and $P_B = \mathrm{diag}([50, 50, 50, 50]^{\mathrm{T}})^2$. The units of these elements are the same as those in Scenario 1. The position estimates for a single run of the EMOO approach, assuming $N_{\min} = 1$ and $N_{\max} = 3$, are illustrated in Figure 10. It can be observed that the trajectory estimates of the EMOO approach are close to the true trajectories.

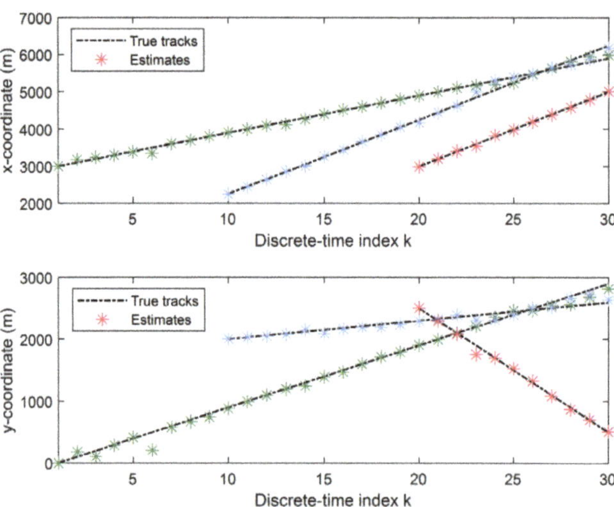


**Figure 10.** True and estimated tracks versus time in Scenario 2.

The average OSPA and OSPA$^{(2)}$ errors (with the same parameters as used in Scenario 1) are given in Figure 11a,b, respectively. The average number of selected sensors is shown in Figure 11c. It can be observed that the tracking errors of the EMOO approach are less than those of other methods in terms of OSPA and OSPA$^{(2)}$. Although the variance-based approach and the proposed EMOO method converge to similar error values, the error of the latter arrives there much earlier. Figure 11c shows that the number of selected sensors for the EMOO method is always less than that of other methods. The average computing times for the random selection approach, the CS divergence-based approach, the cardinality variance-based approach, and the EMOO approach to execute a complete MC simulation are 3.29, 279.85, 343.79, and 157.66 s, respectively. Referring to the tracking accuracy, computing time, and the number of selected sensors, the EMOO approach provides an alternative solution for sensor selection.

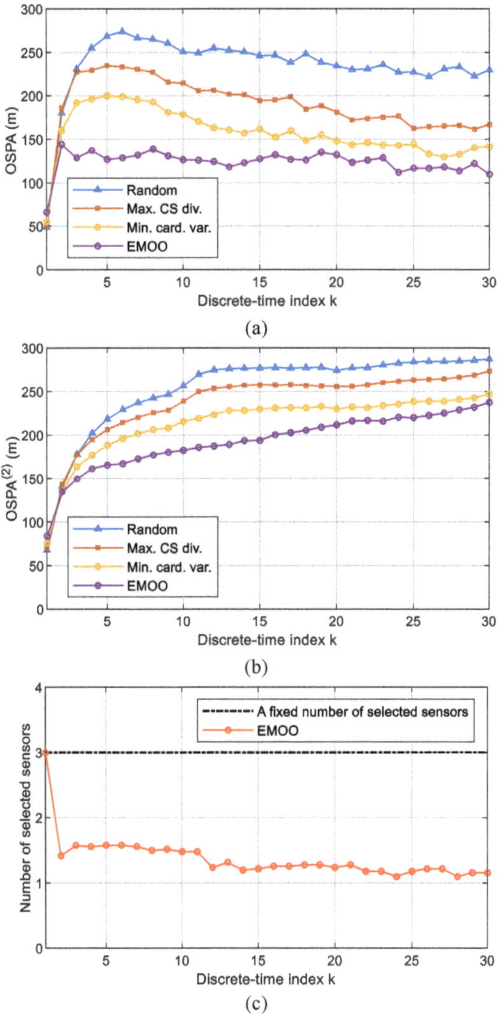

**Figure 11.** Average performance comparison in Scenario 2: (**a**) OSPA error; (**b**) OSPA$^{(2)}$ error; (**c**) the number of selected sensors.

## 5. Discussion

In the above experiments, we use two different MTT scenarios with the NCV and NCT target motions to demonstrate the performances of the proposed EMOO approach. The OSPA and OSPA$^{(2)}$ errors are used to measure the tracking accuracy, which is widely used in the RFS-based tracking field. The average tracking results obtained over 100 MC runs show that the EMOO approach performs better than the existing methods in terms of the OSPA and OSPA$^{(2)}$ errors. What is more, the EMOO approach can significantly reduce the number of selected sensors at each time step. Therefore, the energy and bandwidth problems can be effectively alleviated. The experimental results are well consistent with previous theoretical analysis.

## 6. Conclusions

A novel sensor selection approach based on evolutionary computational intelligence has been proposed under the FISST framework. The multi-target state is modeled by the LMB RFS, and the posterior density is propagated using the LMB filtering. We model the sensor selection problem as an EMOO problem with two conflicting objective functions, i.e., the number of selected sensors and the cardinality variance. The selection command is determined by optimizing the MOO problem using a novel binary constrained evolutionary algorithm. The performance of the proposed EMOO approach was verified using two scenarios in which a multistatic sensor system with poor detection ability is used for MTT. Simulation results demonstrate that the EMOO approach performs better than existing methods in terms of OSPA and OSPA$^{(2)}$ errors and significantly reduces the number of selected sensors. Our future work will consider integrating data from multiple scans to improve the tracking performance. Furthermore, the proposed EMOO methodology also applies to other advanced RFS filters such as GLMB, and this is another direction for future work.

**Author Contributions:** Conceptualization, S.L. and Y.Z.; methodology, S.L. and Y.Z.; software, Y.Z. and S.L.; validation, Y.Z., S.L., H.L. and J.Y.; formal analysis, H.L.; investigation, Y.Z. and S.L.; resources, Y.Z. and S.L.; data curation, Y.Z. and S.L.; writing—original draft preparation, Y.Z., S.L. and H.L.; writing—review and editing, H.L. and J.Y.; visualization, Y.Z.; supervision, Y.Z. and S.L.; project administration, Y.Z. and S.L.; funding acquisition, Y.Z., S.L. and H.L. All authors have read and agreed to the published version of the manuscript.

**Funding:** This work was supported in part by the National Natural Science Foundation of China under grant numbers 62007022 and 61906146, the Natural Science Foundation of Shaanxi Province under grant number 2021JQ-209, and the Fundamental Research Funds for the Central Universities under grant number GK202103082 and JB210210.

**Data Availability Statement:** In this work, we have used the free RFS MATLAB code provided by Prof. Ba-Ngu Vo and Prof. Ba-Tuong Vo at http://ba-tuong.vo-au.com/codes.html (accessed on 5 July 2022).

**Conflicts of Interest:** The authors declare no conflict of interest.

# References

1. Gao, J.; Zhang, Q.; Sun, H.; Wang, W. A Multi-Sensor Interacted Vehicle-Tracking Algorithm with Time-Varying Observation Error. *Remote Sens.* **2022**, *14*, 2176. [CrossRef]
2. Memon, S.A.; Ullah, I.; Khan, U.; Song, T.L. Smoothing Linear Multi-Target Tracking Using Integrated Track Splitting Filter. *Remote Sens.* **2022**, *14*, 1289. [CrossRef]
3. Mallick, M.; Krishnamurthy, V.; Vo, B.N. *Integrated Tracking, Classification, and Sensor Management: Theory and Applications*; Wiley Press: Hoboken, NJ, USA, 2012.
4. Bar-Shalom, Y.; Willett, P.; Tian, X. *Tracking and Data Fusion: A Handbook of Algorithms*; YBS Publishing: Storrs, CT, USA, 2011.
5. Mahler, R. Global Posterior Densities for Sensor Management. In *Acquisition, Tracking, and Pointing XII*; SPIE: Bellingham, WA, USA, 1998; pp. 252–263.
6. Reid, D. An Algorithm for Tracking Multiple Targets. *IEEE Trans. Autom. Control* **1979**, *24*, 843–854. [CrossRef]
7. Kurien, T. Issues in The Design of Practical Multitarget Tracking Algorithms. In *Multitarget-Multisensor Tracking: Advanced Applications*; Bar-Shalom, Y., Ed.; Artech House: Norwood, MA, USA, 1990; pp. 43–83.
8. Fortmann, T.; Bar-Shalom, Y.; Scheffe, M. Sonar Tracking of Multiple Targets Using Joint Probabilistic Data Association. *IEEE J. Ocean. Eng.* **2003**, *8*, 173–184. [CrossRef]
9. Mahler, R. *Statistical Multisource-Multitarget Information Fusion*; Artech House: Norwood, MA, USA, 2007.
10. Mahler, R. *Advances in Statistical Multisource-Multitarget Information Fusion*; Artech House: Norwood, MA, USA, 2014.

11. Mahler, R. Multitarget Bayes Filtering via First-order Multitarget Moments. *IEEE Trans. Aerosp. Electron. Syst.* **2003**, *39*, 1152–1178. [CrossRef]

12. Mahler, R. PHD Filters of Higher Order in Target Number. *IEEE Trans. Aerosp. Electron. Syst.* **2007**, *43*, 1523–1543. [CrossRef]

13. Vo, B.T.; Vo, B.N.; Cantoni, A. The Cardinality Balanced Multi-Target Multi-Bernoulli Filter and Its Implementations. *IEEE Trans. Signal Process.* **2009**, *57*, 409–423.

14. Vo, B.T.; Vo, B.N. Labeled Random Finite Sets and Multi-Object Conjugate Priors. *IEEE Trans. Signal Process.* **2013**, *61*, 3460–3475. [CrossRef]

15. Vo, B.T.; Vo, B.N. A Random Finite Set Conjugate Prior and Application to Multi-target Tracking. In Proceedings of the 2011 7th International Conference on Intelligent Sensors, Sensor Networks and Information Processing, Adelaide, SA, Australia, 6–9 December 2011; pp. 431–436.

16. Vo, B.N.; Vo, B.T.; Phung, D. Labeled Random Finite Sets and the Bayes Multi-Target Tracking Filter. *IEEE Trans. Signal Process.* **2014**, *62*, 6554–6567. [CrossRef]

17. Reuter, S.; Vo, B.T.; Vo, B.N.; Dietmayer, K. The Labeled Multi-Bernoulli Filter. *IEEE Trans. Signal Process.* **2014**, *62*, 3246–3260.

18. Hero, A.O.; Kreucher, C.M.; Blatt, D. Information Theoretic Approaches to Sensor Management. In *Foundations and Applications of Sensor Management*; Hero, A.O., Castanon, D., Cochran, D., Kastella, K., Eds.; Springer: Berlin/Heidelberg, Germany, 2008; Chapter 3, pp. 33–57.

19. Ristic, B.; Vo, B. Sensor Control for Multi-object State-space Estimation Using Random Finite Sets. *Automatica* **2010**, *46*, 1812–1818. [CrossRef]

20. Cai, H.; Gehly, S.; Yang, Y.; Hoseinnezhad, R.; Norman, R.; Zhang, K. Multisensor Tasking Using Analytical Renyi Divergence in Labeled Multi-Bernoulli Filtering. *J. Guid. Control Dyn.* **2019**, *42*, 2078–2085. [CrossRef]

21. Hoang, H.G.; Vo, B.N.; Vo, B.T.; Mahler, R. The Cauchy-Schwarz Divergence for Poisson Point Processes. *IEEE Trans. Inf. Theory* **2015**, *61*, 4475–4485. [CrossRef]

22. Beard, M.; Vo, B.T.; Vo, B.N.; Arulampalam, S. Void Probabilities and Cauchy-Schwarz Divergence for Generalized Labeled Multi-Bernoulli Models. *IEEE Trans. Signal Process.* **2017**, *65*, 5047–5061. [CrossRef]

23. Gostar, A.K.; Hoseinnezhad, R.; Bab-Hadiashar, A.; Liu, W. Sensor-Management for Multitarget Filters via Minimization of Posterior Dispersion. *IEEE Trans. Aerosp. Electron. Syst.* **2017**, *53*, 2877–2884. [CrossRef]

24. Nguyen, H.V.; Rezatofighi, H.; Vo, B.N.; Ranasinghe, D.C. Online UAV Path Planning for Joint Detection and Tracking of Multiple Radio-Tagged Objects. *IEEE Trans. Signal Process.* **2019**, *67*, 5365–5379. [CrossRef]

25. Jiang, M.; Yi, W.; Kong, L. Multi-sensor Control for Multi-target Tracking Using Cauchy-Schwarz Divergence. In Proceedings of the 2016 19th International Conference on Information Fusion (FUSION), Heidelberg, Germany, 5–8 July 2016; pp. 2059–2066.

26. Hoang, H.G.; Vo, B.T. Sensor Management for Multi-target Tracking via Multi-Bernoulli Filtering. *Automatica* **2014**, *50*, 1135–1142. [CrossRef]

27. Gostar, A.K.; Hoseinnezhad, R.; Bab-Hadiashar, A. Multi-Bernoulli Sensor Control via Minimization of Expected Estimation Errors. *IEEE Trans. Aerosp. Electron. Syst.* **2015**, *51*, 1762–1773. [CrossRef]

28. Panicker, S.; Gostar, A.K.; Bab-Haidashar, A.; Hoseinnezhad, R. Sensor Control for Selective Object Tracking Using Labeled Multi-Bernoulli Filter. In Proceedings of the 2018 21st International Conference on Information Fusion (FUSION), Cambridge, UK, 10–13 July 2018; pp. 2218–2224.

29. Panicker, S.; Gostar, A.K.; Bab-Hadiashar, A.; Hoseinnezhad, R. Tracking of Targets of Interest Using Labeled Multi-Bernoulli Filter with Multi-Sensor Control. *Signal Process.* **2020**, *171*, 107451. [CrossRef]

30. Nguyen, H.V.; Rezatofighi, H.; Vo, B.N.; Ranasinghe, D. Multi-Objective Multi-Agent Planning for Jointly Discovering and Tracking Mobile Object. In Proceedings of the AAAI Conference on Artificial Intelligence, New York, NY, USA, 7–12 February 2020; pp. 7227–7235.

31. Zhu, Y.; Wang, J.; Liang, S. Multi-Objective Optimization Based Multi-Bernoulli Sensor Selection for Multi-Target Tracking. *Sensors* **2019**, *19*, 980. [CrossRef]

32. Ma, L.; Xue, K.; Wang, P. Multitarget Tracking with Spatial Nonmaximum Suppressed Sensor Selection. *Math. Probl. Eng.* **2015**, *2015*, 148081. [CrossRef]

33. Ma, L.; Xue, K.; Wang, P. Distributed Multiagent Control Approach for Multitarget Tracking. *Math. Probl. Eng.* **2015**, *2015*, 903682. [CrossRef]

34. Wang, X.; Hoseinnezhad, R.; Gostar, A.K.; Rathnayake, T.; Xu, B.; Bab-Hadiashar, A. Multi-sensor Control for Multi-object Bayes Filters. *Signal Process.* **2018**, *142*, 260–270. [CrossRef]

35. Cao, N.; Choi, S.; Masazade, E.; Varshney, P.K. Sensor Selection for Target Tracking in Wireless Sensor Networks with Uncertainty. *IEEE Trans. Signal Process.* **2016**, *64*, 5191–5204. [CrossRef]

36. Fantacci, C.; Vo, B.N.; Vo, B.T.; Battistelli, G.; Chisci, L. Consensus Labeled Random Finite Set Filtering for Distributed Multi-Object Tracking. *arXiv* **2015**, arXiv:1501.01579.

37. Mahler, R. Multitarget Sensor Management of Dispersed Mobile Sensors. In *Theory and Algorithms for Cooperative Systems*; Grundel, D., Murphey, R., Pardalos, P.M., Eds.; World Scientific: Singapore, 2004; pp. 239–310.

38. Li, H.; Gong, M.; Wang, C.; Miao, Q. Pareto Self-Paced Learning Based on Differential Evolution. *IEEE Trans. Cybern.* **2021**, *51*, 4187–4200. [CrossRef]

39. Gong, M.; Li, H.; Luo, E.; Liu, J.; Liu, J. A Multiobjective Cooperative Coevolutionary Algorithm for Hyperspectral Sparse Unmixing. *IEEE Trans. Evol. Comput.* **2017**, *21*, 234–248. [CrossRef]
40. Gong, M.; Li, H.; Meng, D.; Miao, Q.; Liu, J. Decomposition-Based Evolutionary Multiobjective Optimization to Self-Paced Learning. *IEEE Trans. Evol. Comput.* **2019**, *23*, 288–302. [CrossRef]
41. Ma, L.; Gong, M.; Yan, J.; Yuan, F. A Decomposition-based Multiobjective Evolutionary Algorithm for Analyzing Network Structural Balance. *Inf. Sci.* **2017**, *378*, 144–160. [CrossRef]
42. Ngatchou, P.; Zarei, A.; El-Sharkawi, A. Pareto Multi-Objective Optimization. In Proceedings of the 2005 13th International Conference on, Intelligent Systems Application to Power Systems, Arlington, VA, USA, 6–10 November 2005; pp. 84–91.
43. Deng, J.L. Control Problems of Grey Systems. *Syst. Control Lett.* **1982**, *1*, 288–294.
44. Ristic, B.; Arulampalam, S.; Gordon, N. *Beyond the Kalman Filter-Particle Filters for Tracking Applications*; Artech House: Norwood, MA, USA, 2004.
45. Willis, N.J.; Griffiths, H.D. *Advances in Bistatic Radar*; SciTech Publishing Inc.: Raleigh, NC, USA, 2007.
46. Ristic, B.; Farina, A. Target Tracking via Multi-static Doppler Shifts. *IET Radar Sonar Navig.* **2013**, *7*, 508–516.
47. Mahafza, B. *Radar Systems Analysis and Design Using MATLAB*, 3rd ed.; Chapman and Hall/CRC Press: Boca Raton, FL, USA, 2013.
48. Schuhmacher, D.; Vo, B.T.; Vo, B.N. A Consistent Metric for Performance Evaluation of Multi-Object Filters. *IEEE Trans. Signal Process.* **2008**, *56*, 3447–3457. [CrossRef]
49. Beard, M.; Vo, B.T.; Vo, B.N. A Solution for Large-Scale Multi-Object Tracking. *IEEE Trans. Signal Process.* **2020**, *68*, 2754–2769. [CrossRef]
50. Beard, M.; Vo, B.T.; Vo, B.N. Performance Evaluation for Large-Scale Multi-Target Tracking Algorithms. In Proceedings of the 2018 21st International Conference on Information Fusion (FUSION), Cambridge, UK, 10–13 July 2018; pp. 1–5.

*Article*

# Monitoring of Iron Ore Quality through Ultra-Spectral Data and Machine Learning Methods

Ana Cristina Pinto Silva, Keyla Thayrinne Zoppi Coimbra, Levi Wellington Rezende Filho, Gustavo Pessin and Rosa Elvira Correa-Pabón *

Vale S.A., Programa de Pós-Graduação em Instrumentação, Controle e Automação de Processos de Mineração, Universidade Federal de Ouro Preto e Instituto Tecnológico Vale, Ouro Preto 35400-000, MG, Brazil; ana.cristina.silva@vale.com (A.C.P.S.); keyla.thayrinne@pq.itv.org (K.T.Z.C.); levi_wrf@yahoo.com.br (L.W.R.F.); gustavo.pessin@itv.org (G.P.)
* Correspondence: rosa.correa@itv.org

**Abstract:** Currently, most mining companies conduct chemical analyses by X-ray fluorescence performed in the laboratory to evaluate the quality of Fe ore, where the focus is mainly on the Fe content and the presence of impurities. However, this type of analysis requires the investment of time and money, and the results are often available only after the ore has already been sent by the processing plant. Reflectance spectroscopy is an alternative method that can significantly contribute to this type of application as it consists of a nondestructive analysis technique that does not require sample preparation, in addition to making the analyses available in more active ways. Among the challenges of working with reflectance spectroscopy is the large volume of data produced. However, one way to optimize this type of approach is to use machine learning techniques. Thus, the main objective of this study was the calibration and evaluation of models to analyze the quality of Fe from Sinter Feed collected from deposits in the Carajás Mineral Province, Brazil. To achieve this goal, machine learning models were tested using spectral libraries and X-ray fluorescence data from Sinter Feed samples. The most efficient models for estimating Fe were the Adaboost and support vector machine and our results highlight the possibility of application in the samples without the need for preparation and optimization of the analysis time, providing results in a timely manner to contribute to decision-making in the production chain.

**Keywords:** iron content; machine learning; reflectance spectroscopy

**Citation:** Silva, A.C.P.; Coimbra, K.T.Z.; Filho, L.W.R.; Pessin, G.; Correa-Pabón, R.E. Monitoring of Iron Ore Quality through Ultra-Spectral Data and Machine Learning Methods. *AI* 2022, 3, 554–570. https://doi.org/10.3390/ai3020032

Academic Editors: Yue Wu, Kai Qin, Maoguo Gong and Qiguang Miao

Received: 24 March 2022
Accepted: 29 April 2022
Published: 15 June 2022

**Publisher's Note:** MDPI stays neutral with regard to jurisdictional claims in published maps and institutional affiliations.

## 1. Introduction

Given the growing demand for minerals due to the increase in world population, the decline in iron ore (Fe) deposit quality is a topic of global concern. Several factors can interfere with the quality of this ore, including the reduction in Fe content in the deposits and the presence of impurities, such as phosphorus (P), alumina ($Al_2O_3$) and silica ($SiO_2$) [1].

The lower the ore quality is, the greater the pressure on the deposits is due to the increase in the amount of material to be removed from the mines, acceleration of the production of tailings and waste, and the demand for a greater volume of water, energy and other inputs used in the mine processing stage. Thus, a reduction in ore quality leads to increased production costs, in addition to the significant impacts on the environment.

Brazil is privileged to have geological provinces of extreme relevance for the mineral sector in its territory, including the Carajás Mineral Province (CMP), which represents a well-preserved Archean terrain located in the Amazon Craton (Figure 1). The CMP is known for being the first world producer of high iron content (>62% Fe) and the second most important for areas of copper and gold deposits of the IOCG type (iron oxide, copper sulfides and gold), with significant production of metals, such as manganese and nickel [2–4].

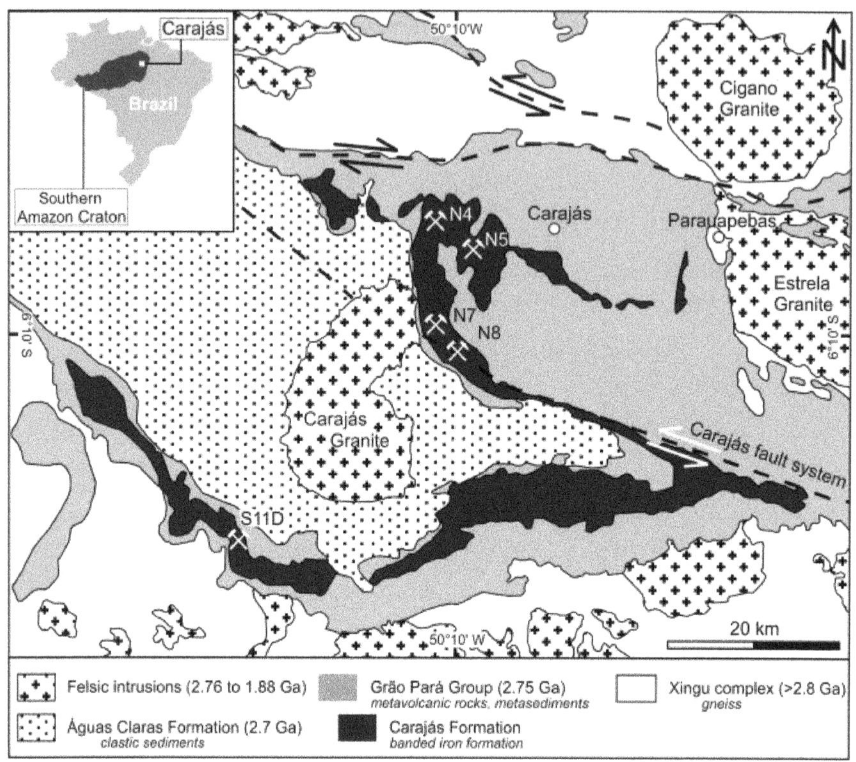



**Figure 1.** Geological map showing the Carajás Mineral Province in the context of the Amazonian Craton [5].

Currently, one of the main ways to evaluate the quality of iron ore produced at any stage of the production chain of mining companies is through chemical analysis methodologies of samples, which are analyzed by X-ray fluorescence (XRF) in the laboratory. However, these analyses are time-consuming and often used only for quality control of already-dispatched ore, considering that the results are not available in time to contribute to decision-making related to the production process.

Thus, methods, such as reflectance spectroscopy, are faster alternatives that can provide physicochemical and mineralogical information about Fe ore. In addition to optimizing the response time of the analyses, one of the great advantages of using spectroscopy is that it is a nondestructive method that does not require sample preparation.

One of the great challenges of working with reflectance spectroscopy is that in addition to the need for specialized labor, this type of analysis generates a large volume of data, which can be a hindrance for calibrating and validating the models. However, with the advancement of technology in the last decade, the use of machine learning methods has become more widespread for these purposes [6–8]. These methods contribute to optimization of the data processing time and have increased the reliability of the results, considering the possibility of developing more robust models [9].

Currently, studies using reflectance spectroscopy, together with hyperspectral imaging and machine learning, are mostly focused on soil physicochemical and mineralogical classification and environmental quality [10–19].

For the mineral sector, methodologies involving machine learning and reflectance spectroscopy are relatively recent developments and are not widespread approaches. Ref. [6] used neural networks to classify Fe ore using spectroscopy in hematite, magnetite, chlorite, phyllite and granites. The authors suggest the implementation of algorithms for the primary selection of Fe ore. In turn, ref. [7] achieved satisfactory results with hyperspectral imaging data and machine learning for the study of tin-tungsten mines in northwestern Spain. Furthermore, ref. [8] analyzed the performance of models to estimate concentrations of tin and cassiterite ores using machine learning in deposits in Germany and southern Romania.

In this context, the research presented here is directed at the mineral industry. The main objective is to analyze the feasibility of applying reflectance spectroscopy in the VSWIR range (Visible, Near Infrared and Shortwave Infrared; 400–2500 nm) in association with machine learning methods for monitoring the quality of the Sinter Feed produced in inserted mines in the CMP, focusing mainly on estimation of Fe content.

## 2. Materials and Methods

In this study, two groups of samples were used, which included the Sinter Feed Product (SFP) and Sinter Feed prepared in the laboratory (SFL), as well as their respective chemical analyses performed by XRF. The methodology used follows the flowchart in Figure 2 and is detailed in the following section.

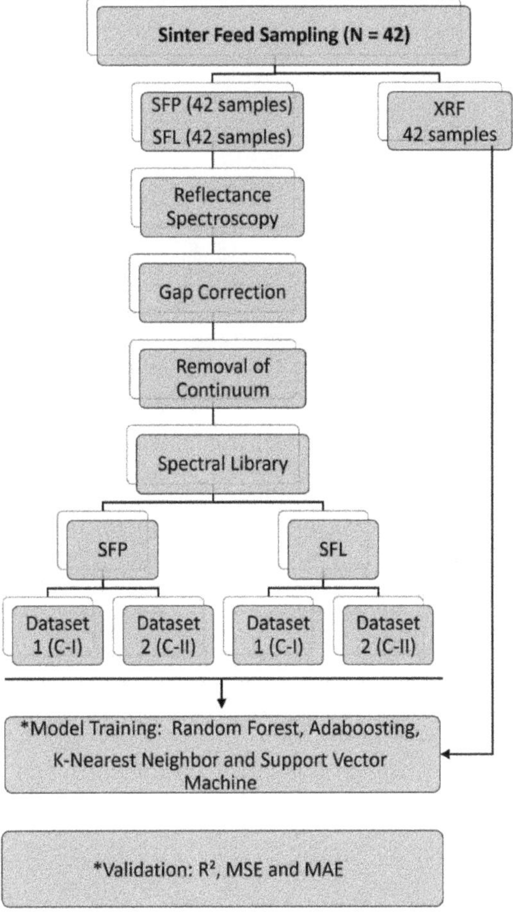


**Figure 2.** Flowchart showing the methodology used: * SFP—Sinter Feed Product; SFL—Sinter Feed Prepared in Laboratory; XRF-X-ray Fluorescence; $R^2$—coefficient of determination; MSE—mean square error; MAE—mean absolute error.

Table 1 shows the XRF analyses of the studied samples. The data were organized according to the nomenclature of the samples and the arrangement of the respective major and minor elements. In turn, in the lower rows of the table, there is the mean and the demonstration of the minimum and maximum values of each element analyzed.

**Table 1.** X-ray fluorescence (XRF) data of the 42 samples used for model validation training. Column 1 shows the nomenclature of the samples, the first row is the analyzed elements, and the lower portion of the table shows the basic statistical analyses.

| Sample | Fe | $SiO_2$ | P | $Al_2O_3$ | Mn | $TiO_2$ | CaO | MgO | $K_2O$ |
|---|---|---|---|---|---|---|---|---|---|
| CN_10551 | 61.12 | 0.39 | 0.43 | 2.82 | 0.018 | 0.402 | 0.006 | 0.030 | 0.006 |
| CN_10552 | 61.76 | 0.40 | 0.46 | 2.49 | 0.018 | 0.407 | 0.006 | 0.027 | 0.007 |
| CN_10554 | 61.99 | 0.43 | 0.41 | 2.44 | 0.019 | 0.253 | 0.006 | 0.027 | 0.006 |
| CN_10556 | 60.45 | 0.35 | 0.52 | 2.90 | 0.018 | 0.482 | 0.006 | 0.026 | 0.005 |
| CN_10558 | 60.46 | 0.32 | 0.54 | 2.93 | 0.018 | 0.497 | 0.006 | 0.025 | 0.005 |
| CN_10560 | 61.13 | 0.34 | 0.46 | 2.42 | 0.018 | 0.404 | 0.006 | 0.026 | 0.007 |
| CN_10561 | 62.56 | 0.45 | 0.32 | 1.94 | 0.020 | 0.190 | 0.006 | 0.028 | 0.007 |
| CN_10563 | 63.26 | 0.48 | 0.26 | 1.69 | 0.020 | 0.245 | 0.006 | 0.034 | 0.008 |
| CN_10564 | 64.00 | 0.67 | 0.21 | 1.47 | 0.022 | 0.176 | 0.006 | 0.035 | 0.010 |
| CN_10566 | 63.56 | 0.59 | 0.25 | 1.58 | 0.020 | 0.167 | 0.006 | 0.027 | 0.011 |
| CN_10568 | 61.93 | 0.41 | 0.41 | 2.15 | 0.018 | 0.424 | 0.006 | 0.025 | 0.007 |
| CN_10569 | 62.45 | 0.40 | 0.32 | 1.77 | 0.018 | 0.197 | 0.006 | 0.025 | 0.007 |
| CN_10573 | 63.94 | 0.89 | 0.21 | 1.52 | 0.011 | 0.200 | 0.006 | 0.025 | 0.004 |
| CN_10574 | 63.89 | 0.63 | 0.22 | 1.44 | 0.015 | 0.192 | 0.006 | 0.030 | 0.004 |
| CN_10576 | 61.61 | 0.58 | 0.37 | 1.88 | 0.008 | 0.239 | 0.006 | 0.025 | 0.004 |
| CN_10577 | 62.77 | 0.93 | 0.27 | 1.45 | 0.012 | 0.215 | 0.006 | 0.025 | 0.004 |
| CN_10578 | 63.18 | 0.46 | 0.26 | 1.43 | 0.008 | 0.219 | 0.006 | 0.025 | 0.004 |
| CN_10579 | 62.18 | 0.58 | 0.23 | 2.68 | 0.010 | 0.279 | 0.008 | 0.025 | 0.004 |
| CN_10580 | 63.54 | 0.66 | 0.21 | 1.45 | 0.015 | 0.204 | 0.006 | 0.025 | 0.004 |
| CN_10581 | 62.43 | 0.63 | 0.18 | 2.45 | 0.010 | 0.231 | 0.006 | 0.025 | 0.004 |
| CN_10582 | 61.98 | 9.65 | 0.01 | 0.41 | 0.112 | 0.042 | 0.006 | 0.100 | 0.004 |
| CN_10583 | 58.21 | 14.91 | 0.01 | 0.60 | 0.134 | 0.048 | 0.006 | 0.105 | 0.004 |
| CN_10584 | 41.41 | 39.45 | 0.01 | 0.37 | 0.023 | 0.067 | 0.006 | 0.096 | 0.004 |
| CN_10585 | 68.24 | 1.21 | 0.01 | 0.22 | 0.038 | 0.040 | 0.006 | 0.124 | 0.004 |
| CN_10759 | 62.92 | 0.58 | 0.35 | 2.27 | 0.010 | 0.190 | 0.007 | 0.050 | 0.009 |
| CN_10760 | 64.37 | 0.63 | 0.36 | 1.33 | 0.015 | 0.084 | 0.007 | 0.057 | 0.012 |
| CN_10762 | 65.66 | 0.72 | 0.14 | 0.97 | 0.013 | 0.082 | 0.008 | 0.062 | 0.016 |
| CN_10764 | 64.80 | 0.56 | 0.17 | 1.18 | 0.011 | 0.131 | 0.008 | 0.046 | 0.009 |
| CN_10765 | 66.11 | 0.71 | 0.06 | 0.60 | 0.016 | 0.068 | 0.008 | 0.064 | 0.010 |
| CN_10553 | 61.97 | 0.39 | 0.38 | 2.25 | 0.018 | 0.257 | 0.006 | 0.027 | 0.007 |
| CN_10557 | 62.15 | 0.57 | 0.36 | 2.29 | 0.018 | 0.160 | 0.006 | 0.031 | 0.007 |
| CN_10559 | 61.12 | 0.36 | 0.46 | 2.44 | 0.018 | 0.410 | 0.006 | 0.025 | 0.006 |
| CN_10562 | 63.34 | 0.47 | 0.29 | 1.61 | 0.019 | 0.190 | 0.006 | 0.028 | 0.007 |
| CN_10565 | 64.44 | 0.49 | 0.23 | 1.29 | 0.020 | 0.139 | 0.006 | 0.034 | 0.007 |
| CN_10567 | 62.06 | 0.51 | 0.35 | 2.41 | 0.019 | 0.418 | 0.006 | 0.027 | 0.008 |
| CN_10575 | 62.62 | 0.82 | 0.22 | 1.97 | 0.015 | 0.224 | 0.006 | 0.025 | 0.004 |
| CN_10757 | 62.78 | 0.56 | 0.45 | 1.90 | 0.010 | 0.126 | 0.006 | 0.043 | 0.009 |
| CN_10761 | 64.85 | 0.60 | 0.26 | 1.27 | 0.009 | 0.099 | 0.007 | 0.045 | 0.013 |
| CN_10763 | 65.35 | 0.62 | 0.13 | 0.98 | 0.010 | 0.094 | 0.007 | 0.046 | 0.010 |
| CN_10766 | 65.38 | 0.61 | 0.13 | 1.11 | 0.014 | 0.131 | 0.007 | 0.056 | 0.008 |
| CN_10550 | 60.55 | 0.38 | 0.47 | 3.03 | 0.017 | 0.436 | 0.006 | 0.028 | 0.004 |
| CN_10555 | 63.26 | 0.48 | 0.26 | 1.69 | 0.020 | 0.245 | 0.006 | 0.034 | 0.008 |
| Min | 41.41 | 0.32 | 0.01 | 0.22 | 0.008 | 0.040 | 0.006 | 0.025 | 0.004 |
| Max | 68.24 | 39.45 | 0.54 | 3.03 | 0.134 | 0.497 | 0.008 | 0.124 | 0.016 |
| Average | 62.42 | 2.04 | 0.28 | 1.74 | 0.021 | 0.222 | 0.006 | 0.040 | 0.007 |

## 2.1. Sinter Feed Samples

In this study, the database containing Sinter Feed samples and their respective XRF analyses was extracted from [20]. The samples were divided into two groups (Figure 3).

The first group comprises 42 samples of SFP, which corresponds to the material of the crushed mining front and has a particle size ranging from 6.3 to 0.150 mm. The second group includes the same 42 samples of SFL that underwent spraying and drying processes, resulting in a more homogeneous particle size.

**Figure 3.** Photographs of the Sinter Feed product samples and their corresponding samples prepared in the laboratory.

### 2.2. Acquisition and Processing of Reflectance Spectra

The instrument used to acquire the reflectance data in the VSWIR interval was the FieldSpec 4 High-Resolution Next Generation high-resolution spectroradiometer coupled with the Turntable reading device, both of which are from the Analytical Spectral Devices manufacturer [21]. FieldSpec 4 Hi-Res NG detects electromagnetic radiation in the spectral range between 350 nm and 2500 nm, with spectral resolutions of 3 nm (@700 nm) and 6 nm (@1400 nm/2100 nm). The sampling intervals are 1.4 nm between 350–1000 nm and 1.1 nm in the range of 1000–2500 nm, with 2151 channels. The reflectance data were generated from measurements of energy reflected by the target in relation to a reference material (Spectralon) in the entire VSWIR range.

The reading of the spectra was performed using RS$^3$ software in conjunction with Turntable and FieldSpec 4 equipment. The samples were placed in Petri dishes with dimensions of 150 mm × 15 mm to obtain readings through the spectroradiometer. Each sample was measured 100 times with 50 scans. Thus, each spectrum used in the spectral characterization corresponds to an average of 5000 readings.

In the preprocessing step of the spectra, gap correction was performed, considering that the FieldSpec 4 Hi-Res NG has three different sensors, and in the sequence, the average of the 100 spectral readings collected was calculated. This procedure allows the reduction in noise and artifacts that may hinder the interpretation of the absorption features.

The corrected spectra were processed using open-source software ViewSpecPro version 6.2.0 [21]. In this step, the continuum was removed, the main purpose of which was to eliminate or reduce the effects unrelated to the properties of interest for the analyzed targets and to highlight the absorption features of the spectra [22–24]. The continuum removal technique generates normalized data, in which all information is represented on the same order of magnitude, on a scale of zero to one, and it is compared with the same level of relevance.

Finally, after processing the spectra, wavelength versus reflectance factor plots were prepared to compose the spectral libraries with the SFP and SFL data.

*2.3. Datasets*

To calibrate and validate the models, two datasets were organized for SFP and two were organized for SFL, with the first considering the broadest spectrum in the range of 400 to 2500 nm (Calibration 1, or C-I) and the second considering the range of 400 to 1310 nm (Calibration 2, or C-II). The C-II data focused on the characteristic features of iron oxides and hydroxides, which are generally observed in the visible and near-infrared (VNIR) region at 670, 860 and 900 nm.

### 2.3.1. Modeling Procedures

The spectral measurements were acquired in the SFP ($n = 42$) and SFL ($n = 42$) samples. For each sample, 100 spectral readings were performed, considering an average of 25 readings, resulting in four spectra per sample and a database of 169 spectra for SFP and 169 for SFL.

The models were trained to estimate the iron contents of the sample sets from the regression relationship found by the machine learning methods, the spectral bands and the iron contents analyzed by XRF. In the composition of the database, 70% of the data were used for training (118 spectra) and 30% were used for model validation (51 spectra). The selection of samples for training and validation of the methods was performed randomly. In the case of training, the C-I and C-II datasets organized for both SFP and SFL were used.

Models that have shown satisfactory results in studies developed by different authors involving reflectance spectroscopy were tested [14,15]. Among these models, four that showed the best performance for the dataset analyzed were selected in this study: the Random Forest (RF); Adaboost (ADB); K-Nearest Neighbor (k-NN); and support vector machine (SVM).

The software used in this step was Orange Canvas, which is open source and has data visualization through streams. The choice of Orange Canvas was based on its characteristic of qualitative analysis, where the graphical interface allows a greater focus on exploratory analysis of the data instead of programming codes. It already has several open-source Python libraries, such as numpy, scipy and scikit-learn, and thus, several machine learning algorithms are available for testing and evaluating models [25].

### 2.3.2. Random Forest (RF)

An RF comprises a set of decision trees that vote together for a classification. Each tree is constructed by chance and randomly selects a subset of resources from a subset of data points. The tree is trained in these data points (only in the selected characteristics), and the rest "out of the basket" are used to evaluate the tree. RF is known to be effective in preventing overfitting [26]. After reaching a certain number of trees, the overfitting remains constant, and no superior performance is achieved.

This method works efficiently on large volumes of data as it works with the training data and thus the algorithm seeks the best conditions and where to insert each one into the flow [27].

For this study, 30 trees were used, with several attributes considered in each division of five. In the tests, it was observed that values greater than 30 did not bring performance gains in the result and thus would result only in computational overload.

### 2.3.3. Adaboosting (ADB)

The ADB is a machine learning method that uses multiple classifiers based on a combination of classifiers with lower accuracy, where the result produces a classifier with higher accuracy. Thus, for data training, the method induces the interaction of these various classifier models, and at each interaction, the boosting method generates a hypothesis. Each new hypothesis generated has the objective of correcting the errors resulting from the previously tested hypotheses. The process is repeated until the training stage is completed [28].

A total of 100 tree-type estimators were adopted, with a maximum learning rate of one. The criterion adopted was empirical, and the increase in the number of estimators above that adopted did not result in a significant improvement in performance for the model.

### 2.3.4. K-Nearest Neighbor (kNN)

The kNN algorithm is a simple model used in both classification and regression problems that predict future values based on past recorded values. In this method, the variable "K" will direct the number of neighbors; thus, the algorithm searches for the desired values closest to that point based on the distances of its "K" closest neighbors [29].

According to [30], the calculation to determine the closest neighbors can be performed using various mathematical methods to calculate the distance between two points, according to the following equations:

$$\text{Euclidean distance}: d(p,q) = \sqrt{\sum_{i=1}^{n}(\sigma - q_i)^2} \tag{1}$$

$$\text{Distance from Manhatten}: d(p,q) = \sum_{i=1}^{n}|\sigma - q_i| \tag{2}$$

$$\text{Mahalanobis distance}: d(p,q) = \sqrt{\sum_{i=1}^{n}\frac{(\sigma - q_i)^2}{\pi_i^2}} \tag{3}$$

$$\text{Chebyshev distance}: d(p,q) = \max_k|\sigma_k - q_{ik}| \tag{4}$$

In determining the values to be assigned to "K", if they are very low, they will be affected by noise in the data to influence the final result obtained, which makes it more sensitive to very close regions, which may result in overfitting. However, very high values, with the generation of several neighbors, can generate more robust models [31].

The metric adopted for the method was the Euclidean distance, with uniform weights, which showed the best performance based on empirical tests. The number of neighborhoods used was 10, specified by calculating the approximate square root of the total number of data applied to train the model.

### 2.3.5. Support Vector Machine (SVM)

The SVM was initially developed with a focus on the solution to binary classification problems. Subsequently, the technique was improved and used for multiclass classification and regression problems [32].

The concept of SVM is the separation of data into classes; for this purpose, the algorithm creates hyperplanes, where each data point belongs to the training base and is plotted as a point in n-dimensional space and with reference to a coordinate. Thus, classification is performed with the objective of finding the hyperplane that will differentiate classes [11].

The parameters adopted for the SVM method were the standard configurations of the estimator present in the Orange Canvas software, with a limit interaction number of 50.

### *2.4. Validation of Predication Models*

Different statistical metrics were used to evaluate the prediction accuracy of the models, including the coefficient of determination ($R^2$), mean absolute error (MAE), and mean square error (MSE) [33,34]. These metrics made it possible to evaluate the accuracy of the machine learning models, comparing the estimated Fe values for each method with the respective levels analyzed by XRF.

## 3. Results

This section may be divided by subheadings. It should provide a concise and precise description of the experimental results, their interpretation, as well as the experimental conclusions that can be drawn.

### 3.1. Physical and Spectral Characterization of the Sinter Feed

Figure 4 shows the spectral library used for the calibration and validation datasets of the Fe content estimation models. Figure 4A shows the SFP spectra used for C-I and C-II. Following the same logic, Figure 4B shows the corresponding SFL samples. The graphs show the reflectance values normalized by removal of the continuum on the Y axis and the wavelength on the X axis.

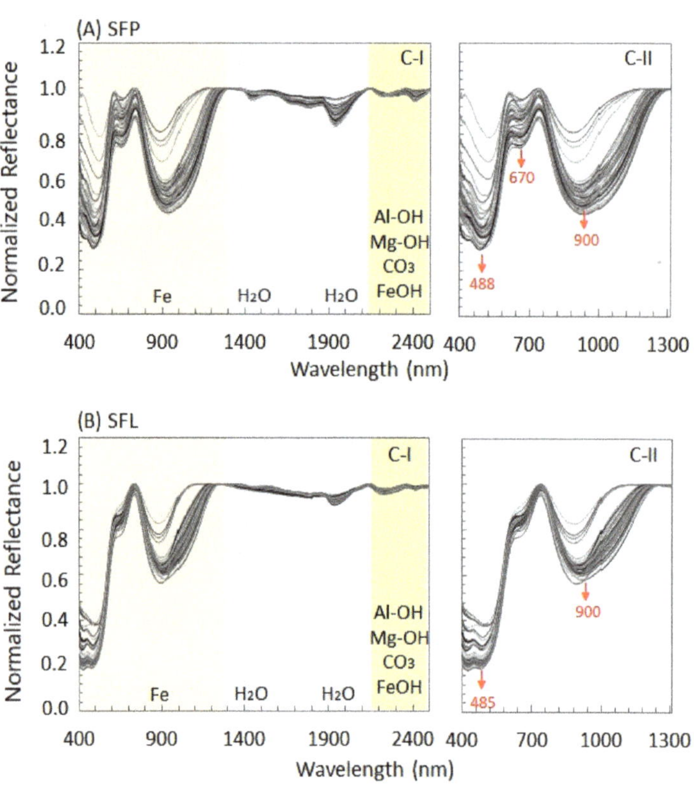

**Figure 4.** Reflectance spectra normalized by removal of the continuum from the samples used in model training: (**A**) Spectra obtained in the SFP samples, showing the database used for C-I and C-II. (**B**) Mean of the SFL spectra used for C-I and C-II. The absorption bands relevant to the study are highlighted in the spectra. SFP—Sinter Feed Product; SFL—Sinter Feed prepared in the laboratory; C-I—Calibration 1; C-II—Calibration 2.

The spectra of the SFP and SFL samples show some similarities; for example, both have absorption features in the VNIR region at approximately 860 and 900 nm. However, when dealing with the sprayed samples, some features were well attenuated, both for VNIR and SWIR. In the case of VNIR, this attenuation was observed at 670 nm, and in SWIR, it was observed at 1400, 1950 and between 2135 and 2500 nm.

The reflectance spectra, as well as photographs of the representative samples of the Sinter Feed, with higher and lower Fe contents, are shown in Figures 5 and 6. In analyzing the spectra, the VNIR region for both SFP and for SFL at 860 nm, has a deep absorption fea ture for sample CN_10585 and is relatively more discrete for sample CN_10584. Conversely, at 900 nm, marked features are observed for samples CN_10564, CN_10563, CN_10556 and CN_10550 (Figure 5A,B). In the region of 670 nm, for SFP, all samples showed absorption features, which were less pronounced for CN_10585 and CN_10584. For samples prepared in the laboratory, the region's feature was attenuated (Figure 5B).

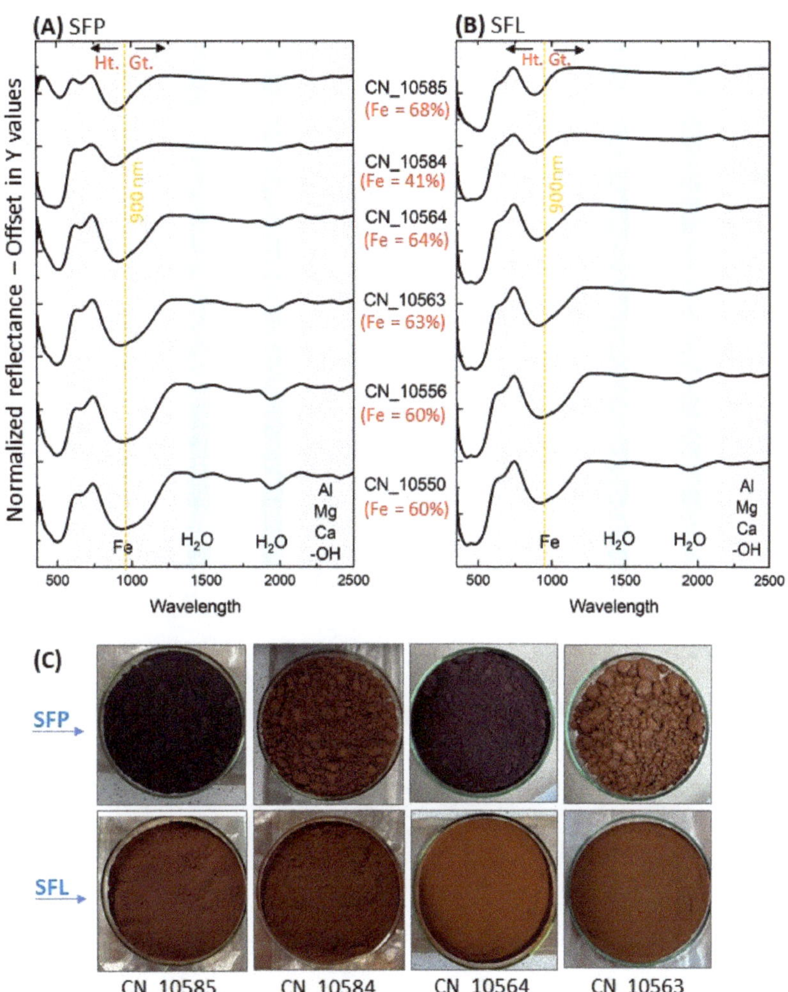



**Figure 5.** (**A,B**) Reflectance spectra normalized by removing the continuum of sinter feed product (SFP) samples and their equivalent prepared in the laboratory (SFL). The diagnostic features of hematite (Ht) samples, which can be observed at 860 nm (CN_10585, CN_10584, and CN_10564), and goethiteitic (Gt) samples, which occur close to 900 nm, stand out in both the SFP and SFL spectra (samples CN_10563, CN_10556 and CN_10550). The absorption features that occur at 670 nm, which indicates $F^{3+}$, and at 1380 and 1950 nm, which indicate the presence of $H_2O$, occur only in the SFP spectra and are attenuated in the SFL spectra. (**C**) Comparison of the SFP and SFL samples showing the physical characteristics of representative samples of higher and lower Fe contents.

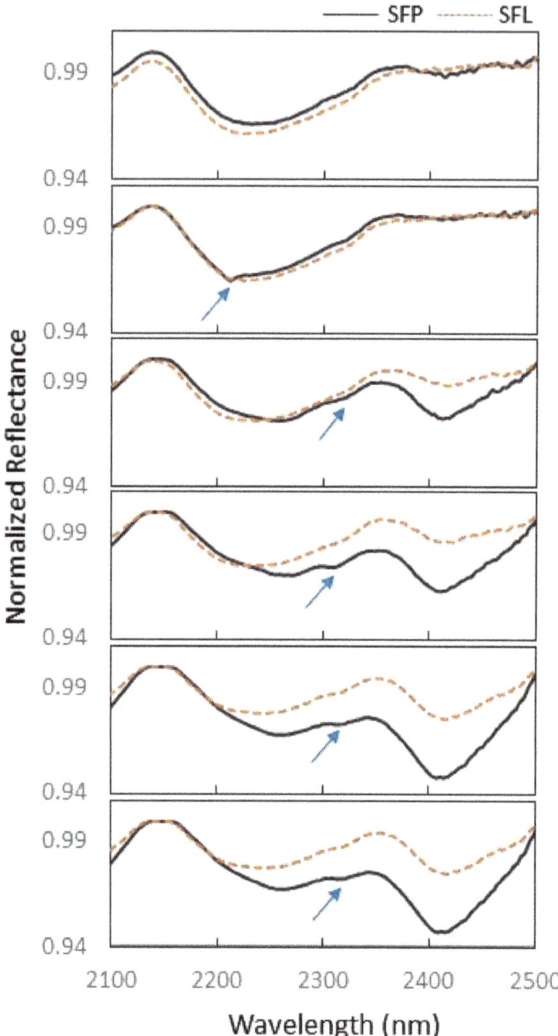

Figure 6. Reflectance spectra normalized by removing the continuum of sinter feed product (SFP) samples and their equivalent prepared in the laboratory (SFL). The figure shows in detail the absorption features in the SWIR region that occur mainly for SFP.

Regarding the physical characteristics, the Sinter Feed samples can be divided into two groups: (i) Sinter Feed with particle size ranging from fine to medium, with dark red to black tones. This group is represented by samples CN_10585, CN_10584 and CN_10564, and when subjected to the process of spraying and drying in the laboratory, the black samples tend to result in reddish tones (Figure 5C); (ii) Sinter Feed with coarse particle size, light red tones and cohesive texture. The CN_10563 sample represents this group, and after spraying, it maintained its reddish color (Figure 5C).

In the SWIR region, for SFP at 1380 and 1950 nm, there are features related to $H_2O$ and OH (Figure 5A). Conversely, between 2200 and 2500 nm, the features are commonly linked to contaminants, such as Al-OH, Mg-OH, $CO_3$ and OH (Figure 6). These features are observed for nearly all samples in Figure 6, with the exception of CN_10585. For the SFL spectra, these features were masked in all spectra shown in Figure 6.

### 3.2. Iron Estimation Models

The models tested here, RF, ADB, kNN, and SVM, provided different forecasting precision for the Fe concentrations of the Sinter Feed samples (Table 2). The table with the evaluation statistics of the models was divided into two parts, one with the evaluation of the models calibrated with the SFP samples (A) and another for the SFL samples (B). In both cases, the $R^2$, MSE and MAE statistics are presented for the models calibrated with the spectra of the C-I and C-II libraries.

**Table 2.** Results of the model statistics for the estimation of Fe contents using the reference spectra (C-I and C-II) and the XRF analyses.

| (A) SFP | | | |
|---|---|---|---|
| ** C-I: 400 to 2500 nm* | | | |
| Model | $R^2$ | MSE | MAE |
| RF | 0.768 | 0.559 | 0.638 |
| ADB | 0.801 | 0.479 | 0.554 |
| kNN | 0.680 | 0.771 | 0.717 |
| SVM | 0.865 | 0.325 | 0.450 |
| ** C-II: 400 to 1310 nm* | | | |
| RF | 0.670 | 0.795 | 0.728 |
| ADB | 0.725 | 0.662 | 0.685 |
| kNN | 0.627 | 0.898 | 0.760 |
| SVM | 0.758 | 0.584 | 0.676 |
| (B) SFL | | | |
| ** C-I: 400 to 2500 nm* | | | |
| Model | $R^2$ | MSE | MAE |
| RF | 0.777 | 0.536 | 0.567 |
| ADB | 0.837 | 0.392 | 0.507 |
| kNN | 0.724 | 0.664 | 0.677 |
| SVM | 0.878 | 0.295 | 0.451 |
| ** C-II: 400 to 1310 nm* | | | |
| RF | 0.884 | 0.280 | 0.452 |
| ADB | 0.872 | 0.308 | 0.935 |
| kNN | 0.732 | 0.647 | 0.650 |
| SVM | 0.795 | 0.493 | 0.569 |

The models tested with the spectra of the C-I dataset for SFP with the best performances were ADB and SVM (Table 2). In the case of ADB, $R^2 = 0.801$, MSE = 0.479, and MAE = 0.554. For SVM, $R^2 = 0.865$, MSE = 0.325, and MAE = 0.450. On the other hand, the RF and kNN showed lower performances, with $R^2 < 0.768$, MSE and MAE > 0.559. All models trained with the C-II dataset showed lower performance compared to C-I (Table 2), with $R^2$ ranging from 0.627 to 0.758, MSE between 0.584 and 0.898 and MAE between 0.676 and 0.760.

The results of the SFP Fe estimates performed by the tested model dataset C-I can be seen in Figure 7A and with dataset C-II in Figure 7B. In the graphs, the levels of Fe analyzed by XRF are on the X axis, and the levels estimated by the models calibrated in this study are on the Y axis.

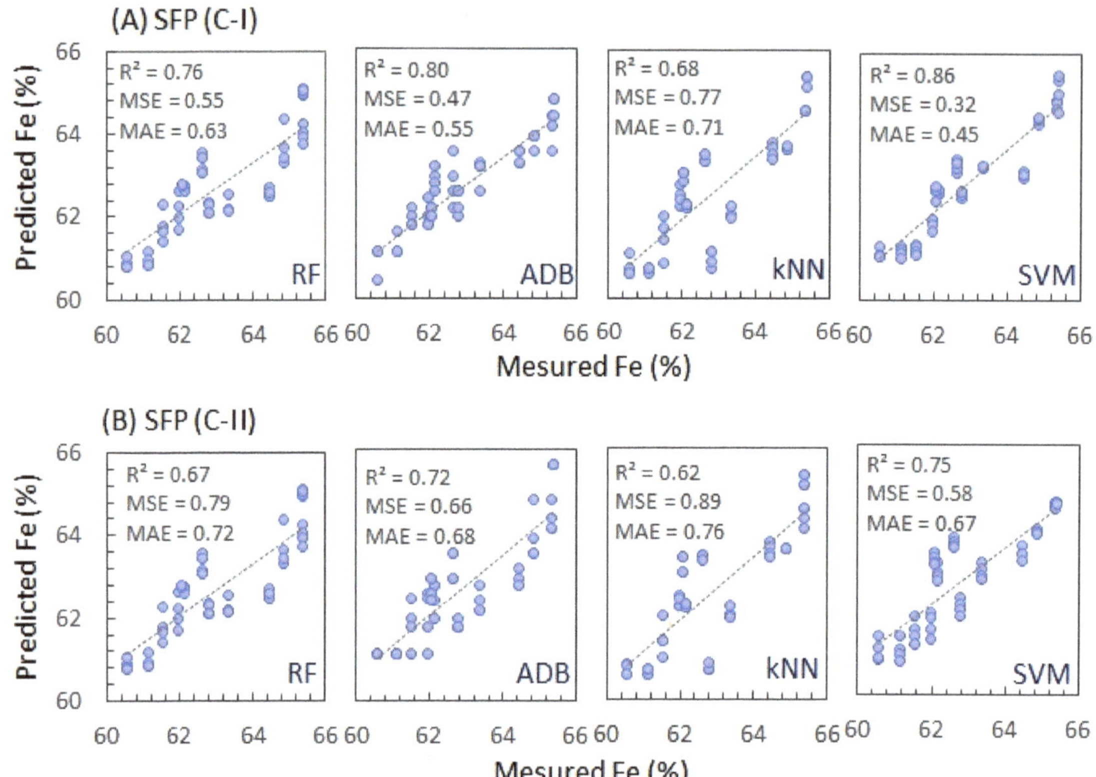


**Figure 7.** (**A**) Results of the models for estimating Fe tested with the SFP samples with the C-I spectra showing better ADB and SVM performances. (**B**) Results of the models for estimating Fe tested with the SFP samples with the C-II spectra, showing that the models were not efficient in estimating the Fe content when reducing the spectrum to 400–1310 nm.

Corroborating the results presented in the statistical table, the graphs show that the predictions made with ADB and SVM were more efficient in estimating the Fe content in SFP and C-I (Figure 7A). In this case, the samples are better fitted with the trend lines plotted in the graph. On the other hand, for both RF and kNN, the samples are more scattered and more distant from the line. Furthermore, the graphs of the Fe contents estimated with the C-II dataset also show a large dispersion between samples for the four models tested (Figure 7B).

The models calibrated with the C-I and C-II datasets of the samples of SFL showed very similar performances (Table 2). The best performances are observed in the RF, ADB and SVM models, with $R^2$ ranging from 0.777 to 0.884, MSE between 0.280 and 0.536, and MAE between 0.452 and 0.935. The kNN model had the worst performance, with $R^2$ values between 0.724 and 0.732, MSE values between 0.647 and 0.664, and MAE values between 0.650 and 0.677.

The results of the SFL Fe estimates performed by the models tested with the C-I and C-II datasets were plotted together with the Fe contents analyzed by XRF in the graphs of Figure 8A,B, respectively. In general, the arrangement of the samples in the graphs shows good alignment with the trend line for the RF, ADB and SVM models for both C-I and C-II. In contrast, the samples in the kNN plot are more dispersed for the entire analyzed dataset, thus corroborating the evaluation of the statistical data in Table 2.

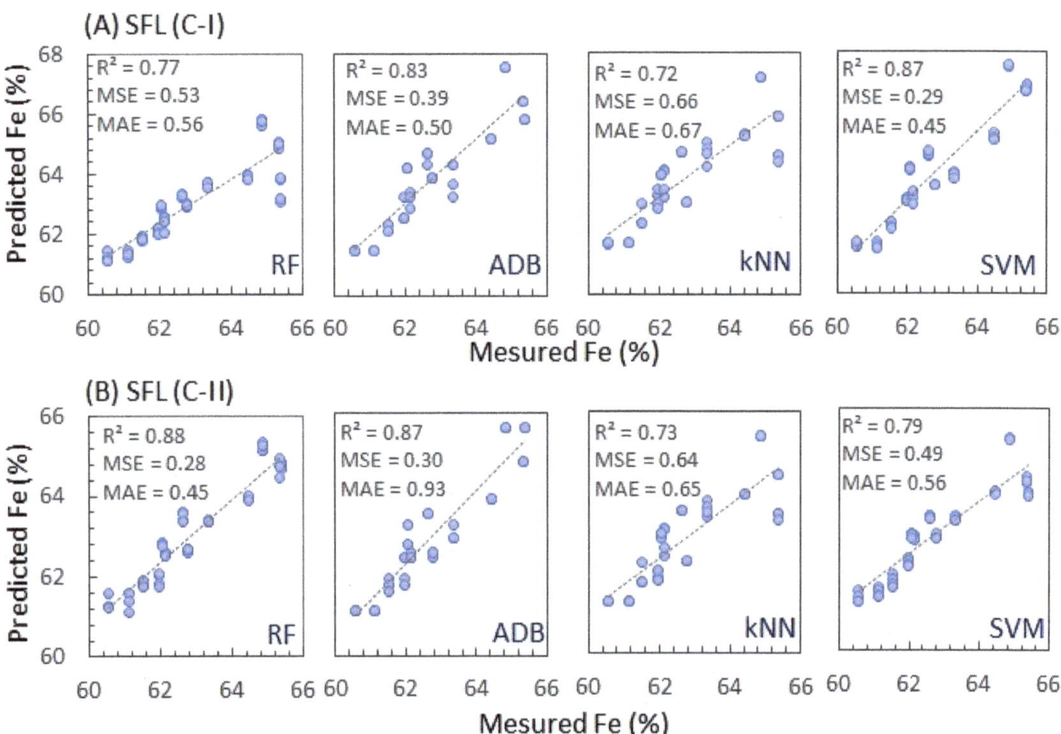

**Figure 8.** (**A**,**B**) Results of the models for the estimation of Fe tested with the Sinter Feed samples prepared in the laboratory with the C-I and C-II spectra, respectively. The graphs show good alignment of the samples for RF, ADB and SVM. For kNN, the samples are more dispersed.

## 4. Discussion

The CMP, together with the Iron Quadrangle (in southeastern Brazil) and Hamersley Province (in western Australia), hosts the largest deposits of high-content Fe in the world [35]. Pioneer studies conducted in the Carajás Province showed that the high-grade ore in the region is commonly associated with paragenesis containing hematite as an essential mineral. However, lower mineralization levels are also associated with intermediate-grade magnetite-carbonate or hematite-carbonate paragenesis, in addition to the formation of banded and goethite formations [36]. In general, high-grade ore occurs in the form of tabular bodies of friable to soft hematite containing smaller lenses of hard hematite [35].

The average percentage of Fe observed in the set of samples studied here is 62%, with the lowest value occurring for sample CN_10584 (41%) and the highest for sample CN_10585 (68%). Regarding the other elements, the means are 2.0% for $SiO_2$, 0.28% for P, 1.74% for $Al_2O_3$, 0.021% for Mn, 0.022% for $TiO_2$, 0.040% for MgO, 0.006% for CaO and 0.007% $K_2O$ (Table 1).

Samples with lower Fe contents have higher $SiO_2$ contents, for example, as samples CN_10583 (Fe = 58%) and CN_10584 (Fe = 41%), with silica contents of 14 and 39%, respectively, which may represent the reflection of lithological variation and possible associations with the Cangas [20]. In addition to $SiO_2$, it was not possible to observe a very clear relationship between Fe and the other elements analyzed.

Regarding contaminants, especially $SiO_2$, most of the samples are within the expected limit ($SiO_2$ < 0.6%) to maintain the quality of the ore for steel production, using [37] as a reference, except for samples CN_10582, CN_10583 and CN_10584, which have $SiO_2$ contents varying from 9–39%.

Regarding phosphorus (P), the mean content observed in the sample set was approximately 0.2% (Table 1). According to [1], this element is commonly associated with secondary iron oxides, such as limonite, goethite, secondary hematite, and minerals rich in alumina, including clay and gibbsite, and apatite/hydroxyapatite in magnetite ores. Such minerals will influence different regions of the spectrum.

The evaluation of the XRF chemical analyses contributed to interpretation of the spectral curves of the dataset as the physicochemical and mineralogical characteristics of a given material directly influence its reflectance spectra. Such information can be obtained by analyzing the depth and location of the absorption features of each material [38].

In the spectra evaluated here, the most developed absorption features are in the VNIR, which is the region with the best response for the characterization of iron oxides and hydroxides [38]. The features observed at 485 nm (SFL) and 488 nm (SFP) can be explained by the transfer of charges, which are common phenomena in oxides [39].

Pioneering reflectance spectroscopy studies revealed that the position of the absorption feature between 850–1000 nm is a function of the composition of Fe oxides and hydroxides, in which pure hematite is characterized in the range of ~860 nm and pure goethite in ~920 nm [40]. Thus, the comparison of the spectra of the samples illustrated in Figure 5 suggests that samples CN_10585, CN_10584 and CN_10564 are more hematinic, while the others have spectral characteristics of goethite. Nevertheless, in the VNIR region, it is possible to observe that the hematite samples have more subtle absorption features at 670 nm compared to the goethite samples. These features are commonly associated with $F^{3+}$ [40].

Even if hematite samples are commonly associated with high Fe content [35], it is important to evaluate the presence and influence of contaminants (Figure 6). For example, sample CN_10585 shows a very characteristic spectral curve of hematite, low values of contaminants, such as silica, alumina and P ($SiO_2$ = 1.21; P = 0.01; $Al_2 O_3$ = 0.22), and high iron content (Fe = 68%). In turn, between 2200 and 2500 nm, it was not possible to observe significant absorption features for the spectrum of this sample (Figure 6).

In contrast, the CN_10584 sample, which also has a diagnostic spectral curve of hematite (Figure 5), showed an absorption feature close to 2200 nm (Figure 6), which may reflect the presence of the silica observed in Table 1, where $SiO_2$ = 39%. Thus, even if the spectrum suggests the presence of hematite, this sample has the lowest iron content of the studied set (Fe = 41%).

Comparing the spectra of the three goethite samples, it is possible to observe absorption features between 2100 and 2500 nm (Figure 6). These features may be related to the presence of contaminants or even FeOH. The CN_10563 sample exhibits a very subtle feature at approximately 2300 nm, and the XRF result shows a relatively high iron content (Fe = 63%) and no significant contaminant content ($SiO_2$ = 0.62; P = 0.13; $Al_2O_3$ = 0.98). Conversely, samples CN_10556 and CN_10550, which have relatively lower iron contents, at approximately 60%, exhibit slightly deeper absorption features in the 2310 nm range, which can be explained by the high phosphorus contents (P = 0.47 a 0.52%) and alumina ($Al_2 O_3$ = 2.9 to 3.0%).

The SFP spectra have absorption features at 1380 and 1950 (Figure 5A). These features do not occur for SFL (Figure 5B), and according to Clark (1999), they are commonly associated with the presence of $H_2O$ or OH. Regarding the spraying and drying process, within the scope of spectral characterization, the results discussed here showed that there were no significant gains in making the samples homogeneous and without moisture, considering that several absorption features were attenuated for SFL, which made it a factor that hinders qualitative analysis.

Identification of the typical absorption features of hematite and goethite corroborates the information about the Sinter Feed mineralogy extracted from the physical descriptions of the samples. In this case, CN_10585, CN_10584 and CN_10564, which showed features in the 860 nm region (characteristic of hematite), were physically described as friable hematites with fine to medium particle size and dark red to black color when produced, which changed to dark red tones when prepared in the laboratory (Figure 5C). Conversely, the CN_10563 sample selected to represent the group of goethite samples, which showed an absorption feature at approximately 900 nm, has a coarse particle size with a cohesive texture, characteristic of cangas, typical of the Carajás Formation [20]. The sample is also light red when used as a product and darker red when prepared in the laboratory (Figure 5C). Despite its relatively high Fe content, due to its physical characteristics and the presence of impurities, processing the canga is more laborious than processing friable hematite [20].

Comparing the performance of the tested algorithms with the C-I and C-II spectra used to estimate the SF content of Fe, it is not possible to notice very significant differences between them. In both cases, the best estimates were made using the RF, ADB, and SVM models, in which the statistical metrics showed lower MSE and MAE values and the highest $R^2$ values. In contrast, the kNN method showed lower efficiency for estimating Fe contents and was thus less suitable for this purpose.

Regarding the algorithms trained with the SFP spectral library, the best results were achieved using the C-I spectra, with the ADB and SVM being more efficient in estimating the Fe contents. Already restricting the spectrum region, in the case of C-II, none of the algorithms showed satisfactory results, with low $R^2$ values and relatively high MSE and MAE values.

Comparing the best results of the calibrated SFP models with the SFL results, it was found that there is no significant difference between the performance of the models. In the case of the Sinter Feed product (C-I), the ADB and SVM algorithms showed $R^2$ values between 0.801 and 0.865, MSE values between 0.325 and 0.479 and MAE values between 0.450 and 0.554. For SFL (C-I and C-II), the RF, ADB and SVM showed $R^2$ values between 0.777 and 0.884, MSE values between 0.295 and 0.493, and MAE values between 0.451 and 0.935.

## 5. Conclusions

The main contributions of this study are listed as follows:

- Spectrally, there were no gains with the preparation of the Sinter Feed samples in the laboratory as with the drying and spraying procedures, the particle size of these samples became more homogeneous, thus attenuating the VSWIR absorption features used for qualitative and quantitative analyses of the physicochemical and mineralogical properties of the Sinter Feed.
- The absorption features located in the VNIR region (~860 nm) enabled the identification of more hematitic (CN_10585, CN_10584 and CN_10564) and goethite samples, starting at 900 nm (CN_10563, CN_10556 and CN_10550). This information corroborated the physical characterization of the Sinter Feed, in which the hematitic samples were described as the most friable material with fine to medium particle sizes and colors between red to black (when product) and dark red (when sprayed). On the other hand, the goethite samples had coarser particle sizes and colors varying in shades of red, both for the Sinter Feed product samples and for the samples prepared in the laboratory.
- The best Fe estimates for SFP were made with the ADB and SVM models, using only the C-I dataset, which is in the spectral range of 400 to 2500 nm.
- For SFL, the RF, ADB and SVM models were more efficient for estimating Fe using both the C-I and C-II libraries. Conversely, kNN is the least recommended for this application.
- The possibility of calibrating models, such as SVM and ADB, using only the Sinter Feed spectra without sample preparation opens space to discuss the operationalization of these methods in the processing plant routine.

- Finally, we suggest calibration and evaluation of models using reflectance spectroscopy and XRF to estimate contaminants, such as phosphorus, silica, manganese and alumina. We also suggest that other algorithms could be tested to improve the results presented here, such as decision trees and artificial neural networks.

**Author Contributions:** A.C.P.S.: Conceptualization, methodology, software, data curation, validation, formal analysis, writing—original draft preparation; K.T.Z.C.: Conceptualization, methodology, formal analysis, Writing—review and editing, Supervision, L.W.R.F.: Software, validation, formal analysis, G.P.: Conceptualization, methodology, formal analysis, Writing—review and editing, Supervision, R.E.C.-P.: Conceptualization, methodology, formal analysis, Writing—review and editing, Supervision, project administration. All authors have read and agreed to the published version of the manuscript.

**Funding:** This research was in part by the Coordenação de Aperfeiçoamento de Pessoal de Nível Superior—Brasil (CAPES)—Finance Code 001, the Conselho Nacional de Desenvolvimento Científico e Tecnológico (CNPQ), the Instituto Tecnológico Vale (ITV) and the Universidade Federal de Ouro Preto (UFOP).

**Data Availability Statement:** Not applicable.

**Acknowledgments:** We Thank Vale—Serra Sul Physical Chemical Laboratory for providing the sinter feed iron ore samples, as laboratory tests results. We also would like to acknowledge The ITV for supporting the research, as well as providing all the equipment that made it possible to accomplish the spectral readings.

**Conflicts of Interest:** The authors declare no conflict of interest. The funders had no role in the design of the study; in the collection, analyses, or interpretation of data; in the writing of the manuscript, or in the decision to publish the results.

# References


1. Yellishetty, M.; Werner, T.T.; Weng, Z. Iron Ore in Australia and the World: Resources, Production, Sustainability, and Future Prospects. In *Iron Ore*; Woodhead Publishing: Cambridge, UK, 2022; pp. 711–750.
2. Banco de Desenvolvimento de Minas Gerais. Minas Gerais Do Século XXI. Vol.IX. Belo Horizonte: Rona Editora. 2002. Available online: https://silo.tips/download/minas-gerais-do-seculo-xxi-2 (accessed on 23 March 2022).
3. Grainger, C.J.; Groves, D.I.; Tallarico, F.H.; Fletcher, I.R. Metallogenesis of the Carajás mineral province, southern Amazon craton Brazil: Varying styles of Archean through Paleoproterozoic to Neoproterozoic base-and precious-metal mineralisation. *Ore Geol. Rev.* **2008**, *33*, 451–489. [CrossRef]
4. Juliani, C.; Monteiro, L.V.S.; Fernandes, C.M.D. Potencial Mineral: Cobre. In *Recursos Minerais no Brasil: Problemas e Desafios*; Melfi, A.J., Misi, A., Campos, D.d.A., Cordani, U.G., Eds.; Academia Brasileira de Ciências: Rio de Janeiro, Brazil, 2006; pp. 134–149.
5. Cabral, A.; Creaser, R.; Nägler, T.; Lehmann, B.; Voegelin, A.; Belyatsky, B.; Pašava, J.; Gomes, A.S.; Galbiatti, H.; Böttcher, M.E.; et al. Trace-element and multi-isotope geochemistry of Late-Archean black shales in the Carajás iron-ore district, Brazil. *Chem. Geol.* **2013**, *362*, 91–104. [CrossRef]
6. Xiao, D.; Le, B.T.; Ha, T.T.L. Iron ore identification method using reflectance spectrometer and a deep neural network framework. *Spectrochim. Acta Part A Mol. Biomol. Spectrosc.* **2021**, *248*, 119168. [CrossRef] [PubMed]
7. Lobo, A.; Garcia, E.; Barroso, G.; Martí, D.; Fernandez-Turiel, J.-L.; Ibáñez-Insa, J. Machine Learning for Mineral Identification and Ore Estimation from Hyperspectral Imagery in Tin–Tungsten Deposits: Simulation under Indoor Conditions. *Remote Sens.* **2021**, *13*, 3258. [CrossRef]
8. Tuşa, L.; Kern, M.; Khodadadzadeh, M.; Blannin, R.; Gloaguen, R.; Gutzmer, J. Evaluating the performance of hyperspectral short-wave infrared sensors for the pre-sorting of complex ores using machine learning methods. *Miner. Eng.* **2020**, *146*, 106150. [CrossRef]
9. Barra, I.; Haefele, S.M.; Sakrabani, R.; Kebede, F. Soil spectroscopy with the use of chemometrics, machine learning and pre-processing techniques in soil diagnosis: Recent advances—A review. *TrAC Trends Anal. Chem.* **2021**, *135*, 116166. [CrossRef]
10. Brown, D.J.; Shepherd, K.D.; Walsh, M.G.; Mays, M.D.; Reinsch, T.G. Global soil characterization with VNIR diffuse reflectance spectroscopy. *Geoderma* **2006**, *132*, 273–290. [CrossRef]
11. Chen, S.H.; Jakeman, A.J.; Norton, J.P. Artificial Intelligence techniques: An introduction to their use for modelling environmental systems. *Math. Comput. Simul.* **2008**, *78*, 379–400. [CrossRef]
12. Richter, N.; Jarmer, T.; Chabrillat, S.; Oyonarte, C.; Hostert, P.; Kaufmann, H. Free Iron Oxide Determination in Mediterranean Soils using Diffuse Reflectance Spectroscopy. *Soil Sci. Soc. Am. J.* **2009**, *73*, 72–81. [CrossRef]
13. Rossel, R.V.; Behrens, T. Using data mining to model and interpret soil diffuse reflectance spectra. *Geoderma* **2010**, *158*, 46–54. [CrossRef]

14. Khosravi, V.; Ardejani, F.D.; Yousefi, S.; Aryafar, A. Monitoring soil lead and zinc contents via combination of spectroscopy with extreme learning machine and other data mining methods. *Geoderma* **2018**, *318*, 29–41. [CrossRef]

15. Hu, P.; Liu, X.; Cai, Y.; Cai, Z. Band Selection of Hyperspectral Images Using Multiobjective Optimization-Based Sparse Self-Representation. *IEEE Geosci. Remote Sens. Lett.* **2018**, *16*, 452–456. [CrossRef]

16. Pabón, R.E.C.; Filho, C.R.D.S.; de Oliveira, W.J. Reflectance and imaging spectroscopy applied to detection of petroleum hydrocarbon pollution in bare soils. *Sci. Total Environ.* **2019**, *649*, 1224–1236. [CrossRef] [PubMed]

17. Cardoso-Fernandes, J.; Silva, J.; Lima, A.; Teodoro, A.C.; Perrotta, M.; Cauzid, J.; Roda-Robles, E.; Ribeiro, M.D.A. Reflectance spectroscopy to validate remote sensing data/algorithms for satellite-based lithium (Li) exploration (Central East Portugal). In *Earth Resources and Environmental Remote Sensing/GIS Applications XI*; International Society for Optics and Photonics: Bellingham, WA, USA, 2020; Volume 11534.

18. Jia, X.; O'Connor, D.; Shi, Z.; Hou, D. VIRS based detection in combination with machine learning for mapping soil pollution. *Environ. Pollut.* **2021**, *268*, 115845. [CrossRef]

19. Parent, E.J.; Parent, S.É.; Parent, L.E. Machine learning prediction of particle-size distribution from infrared spectra, methodologies and soil features. *bioRxiv* **2020**. [CrossRef]

20. Silva, A.C.P. Monitoramento da Qualidade de Sinter Feed Através de Dados Espectrais Associados a Aprendizado de Máquina– estudo de Caso: Mina De Carajás Serra Sul (S11D). Master's Thesis, UFOP, ITV, Ouro Preto, Minas Gerais, Brazil, 2021. Available online: https://www.itv.org/wp-content/uploads/2022/01 (accessed on 20 January 2022).

21. ASD FieldSpec®4 Hi-Res: Espectrorradiômetro de Alta Resolução, Malvern Panalytical. Available online: https://www.malvernpanalytical.com/br/products/product-range/asd-range/fieldspec-range (accessed on 17 January 2022).

22. Clark, R.N.; Roush, T.L. Reflectance spectroscopy: Quantitative analysis techniques for remote sensing applications. *J. Geophys. Res. Earth Surf.* **1984**, *89*, 6329–6340. [CrossRef]

23. Kokaly, R.F. Investigating a Physical Basis for Spectroscopic Estimates of Leaf Nitrogen Concentration. *Remote Sens. Environ.* **2001**, *75*, 153–161. [CrossRef]

24. Ozaki, Y.; McClure, W.F.; Christy, A.A. (Eds.) *Near-Infrared Spectroscopy in Food Science and Technolog*; John Wiley & Sons: Hoboken, NJ, USA, 2006.

25. Orange Data Mining. Available online: https://github.com/biolab/orange3 (accessed on 15 January 2022).

26. Cutler, A.; Cutler, D.R.; Stevens, J.R. Random Forests. In *Ensemble Machine Learning*; Springer: Boston, MA, USA, 2012; pp. 157–175.

27. Biau, G.; Scornet, E. A random forest guided tour. *Test* **2016**, *25*, 197–227. [CrossRef]

28. Borges, F.A.; Rabelo, R.A.; Araujo, M.A.; Fernandes, R.A. Metodologia baseada no algoritmo adaboost combinado com rede reural Para localização do distúrbio de afundamento de tensão. In *Congresso Brasileiro de Automática-CBA*; Cidade Universitária Zeferino Vaz: Campinas, Brazil, 2019; Volume 1.

29. Alpaydin, E. *Introduction to Machine Learning*; MIT Press: Cambridge, MA, USA, 2020.

30. Linden, R. Técnicas de Agrupamento. *Rev. Sist. Inf. FSMA* **2009**, *4*, 18–36.

31. Lantz, B. *Machine Learning with R: Expert Techniques for Predictive Modeling*; Packt Publishing Ltd.: Birmingham, UK, 2019.

32. Filgueiras, P.R. Regressão Por Vetores de Suporte Aplicado na Determinação de Propriedades Físico-Químicas de Petróleo e Biocombustíveis. Ph.D. Thesis, Instituto de Química, Universidade Estadual de Campinas, Campinas, Brazil, 2014.

33. Gupta, H.V.; Kling, H.; Yilmaz, K.K.; Martinez, G.F. Decomposition of the mean squared error and NSE performance criteria: Implications for improving hydrological modelling. *J. Hydrol.* **2009**, *377*, 80–91. [CrossRef]

34. De Myttenaere, A.; Golden, B.; Le Grand, B.; Rossi, F. Mean absolute percentage error for regression models. *Neurocomputing* **2016**, *192*, 38–48. [CrossRef]

35. Clout, J.; Manuel, J. Mineralogical, chemical, and physical characteristics of iron ore. In *Iron Ore*; Woodhead Publishing: Cambridge, UK, 2015; pp. 45–84. [CrossRef]

36. Dalstra, H.; Guedes, S. Giant hydrothermal hematite deposits with Mg-Fe metasomatism: A comparison of the carajas, hamersley, and other iron ores. *Econ. Geol.* **2004**, *99*, 1793–1800. [CrossRef]

37. Upadhyay, R.; Venkatesh, A.S. Current strategies and future challenges on exploration, beneficiation and value addition of iron ore resources with special emphasis on iron ores from Eastern India. *Appl. Earth Sci.* **2006**, *115*, 187–195. [CrossRef]

38. Van Der Meer, F. Analysis of spectral absorption features in hyperspectral imagery. *Int. J. Appl. Earth Obs. Geoinf.* **2004**, *5*, 55–68. [CrossRef]

39. Pontual, S.; Merry, N.; Gamson, P. *Spectral Interpretation-Field Manual. GMEX. Spectral Analysis Guides for Mineral Exploration*; AusSpec International Pty. Ltd.: Warrnambool, VIC, Australia, 2008; p. 189.

40. Townsend, T.E. Discrimination of iron alteration minerals in visible and near-infrared reflectance data. *J. Geophys. Res. Earth Surf.* **1987**, *92*, 1441–1454. [CrossRef]

 **remote sensing**

*Communication*

# Self-Supervised Pre-Training with Bridge Neural Network for SAR-Optical Matching

**Lixin Qian** [1,2,†]**, Xiaochun Liu** [1]**, Meiyu Huang** [2,†] **and Xueshuang Xiang** [2,*]

[1] School of Mathematics and Statistics, Wuhan University, Wuchang District, Wuhan 430072, China; qianlixin@whu.edu.cn (L.Q.); xcliu@whu.edu.cn (X.L.)

[2] Qian Xuesen Laboratory of Space Technology, China Academy of Space Technology, Haidian District, Beijing 100086, China; huangmeiyu@qxslab.cn

[*] Correspondence: xiangxueshuang@qxslab.cn

[†] These authors contributed equally to this work.

**Abstract:** Due to the vast geometric and radiometric differences between SAR and optical images, SAR-optical image matching remains an intractable challenge. Despite the fact that the deep learning-based matching model has achieved great success, SAR feature embedding ability is not fully explored yet because of the lack of well-designed pre-training techniques. In this paper, we propose to employ the self-supervised learning method in the SAR-optical matching framework, in order to serve as a pre-training strategy for improving the representation learning ability of SAR images as well as optical images. We first use a state-of-the-art self-supervised learning method, Momentum Contrast (MoCo), to pre-train an optical feature encoder and an SAR feature encoder separately. Then, the pre-trained encoders are transferred to an advanced common representation learning model, Bridge Neural Network (BNN), to project the SAR and optical images into a more distinguishable common feature representation subspace, which leads to a high multi-modal image matching result. Experimental results on three SAR-optical matching benchmark datasets show that our proposed MoCo pre-training method achieves a high matching accuracy up to 0.873 even for the complex QXS-SAROPT SAR-optical matching dataset. BNN pre-trained with MoCo outperforms BNN with the most commonly used ImageNet pre-training , and achieves at most 4.4% gains in matching accuracy.

**Keywords:** SAR-optical fusion; image matching; self-supervised learning; representation learning

**Citation:** Qian, L.; Liu, X.; Huang, M.; Xiang, X. Self-Supervised Pre-Training with Bridge Neural Network for SAR-Optical Matching. *Remote Sens.* **2022**, *14*, 2749. https://doi.org/10.3390/rs14122749

Academic Editors: Yue Wu, Kai Qin, Qiguang Miao and Maoguo Gong

Received: 1 April 2022
Accepted: 1 June 2022
Published: 8 June 2022

**Publisher's Note:** MDPI stays neutral with regard to jurisdictional claims in published maps and institutional affiliations.



## 1. Introduction

Synthetic Aperture Radar (SAR) and optical imagery are two of the most commonly used modalities in remote sensing since they provide highly complementary content to each other. While optical imagery with good interpretability is easily affected by atmospheric conditions, SAR data can collect information all the time but suffered from serious intrinsic speckle noise. Therefore, fusion information of SAR and optical images can give rise to a better interpretation of the imaged area. For accurate SAR-optical data fusion, identifying corresponding image patches plays a crucial role as a pre-procedure. It remains a widely unsolved challenge to match SAR-optical remote sensing data due to the vast geometric and radiometric differences as shown in Figure 1.

Over the past few decades, the traditional SAR-optical image matching methods can be generally divided into area-based and feature-based approaches. Area-based methods utilize the intensity of pixel values in some regions of the image and the corresponding regional similarity evaluation is calculated, such as cross correlation (CC) [1], structural similarity (SSIM), mutual information (MI) [2,3], and so on. However, owing to the low flexibility and lack of local structure information, the area-based methods fail to avoid information loss in the measure of multimodal image similarity. Therefore, more attempts at SAR and optical image matching have been placed on the feature-based methods. Since feature-based methods rely on the invariant feature points and handcrafted descriptors can handle the geometric changes well, the feature-based methods generally outperform the area-based methods, for example, scale-invariant feature transform (SIFT) [4], SAR-SIFT [5], HOPC [6,7], etc. However, the handcrafted descriptors based on low-level semantic features are not capable of dealing with highly divergent changes between SAR and optical images. More recently, deep learning-based SAR-optical image matching approaches have achieved great success. Two-tower architecture is the most commonly exploited in multimodal image matching, such as a Siamese or pseudo-Siamese network [8–11], which consists of two convolutional neural networks (CNN) to extract the deep characteristic features—not only with the Siamese network but a novel method called Bridge Neural Network (BNN) [12] to project the multimodal data into a common representation subspace where features can be measured with Euclidean distance.

**Figure 1.** Comparison between SAR (**bottom**) and optical (**top**) imagery of the same scene.

Meanwhile, the ImageNet [13,14] supervised pre-training technique that contains prior knowledge is the most widely adopted in the SAR-related field for the scarcity of labeled SAR images in the past. However, there remain some limitations to using ImageNet pre-trained models for SAR-optical fusion tasks. Images in ImageNet are all optical images, which do not contain SAR information and characteristics; as a result, taking ImageNet supervised pre-training directly can hardly improve the learning ability of SAR images and benefit the SAR-optical fusion task.

As ImageNet pre-training can hardly improve the learning ability for SAR images, the representation learning ability to match models plays a vital role in SAR-optical matching tasks. Recent advances in self-supervised learning for computer vision present competitive results with supervised learning. Without manual annotations, useful representations can be obtained only with the help of some pretext tasks, which is probably achieved by maximizing the mutual information of learned representations. Representations pre-trained by contrastive learning that transferred to downstream tasks: classification, segmentation, and detection tasks lead to a competitive performance with supervised learning [15–21]. A contrastive learning method called Momentum Contrast (MoCo) [19] uses a momentum-update encoder to generate a dynamic dictionary query to save more negative samples with less memory. The MoCo pre-training method has achieved promising results in a variety of downstream tasks.

As discussed above, self-supervised learning can pre-train representations that can be transferred to downstream tasks by fine-tuning. In an attempt to improve multimodal SAR-optical image matching performance, we exploit the self-supervised learning technique to improve the feature learning ability of SAR and optical imagery, respectively. Then, the model is transferred to the SAR-optical matching task. The overall process is illustrated in Figure 2. More specifically, we take MoCo [19] as a pre-training strategy and BNN [12] for matching. By this method, self-supervised pre-training enhances the embedding of SAR and optical images and benefits SAR-optical matching tasks. Experimental results on three datasets show that self-supervised pre-training leads to a better matching performance. Our main contributions are summarized as follows:

- We propose a framework applying self-supervised learning to SAR-optical image matching, improving the feature learning ability of SAR and optical images.
- We take MoCo and BNN as one of the most representative works in self-supervised learning and multi-modal image matching to make the framework truly implemented.
- For the proposed framework, we conduct lots of experiments to confirm the feasibility and the effectiveness of the self-supervised learning transferred to optical-SAR image matching task, which would encourage further research in this field.

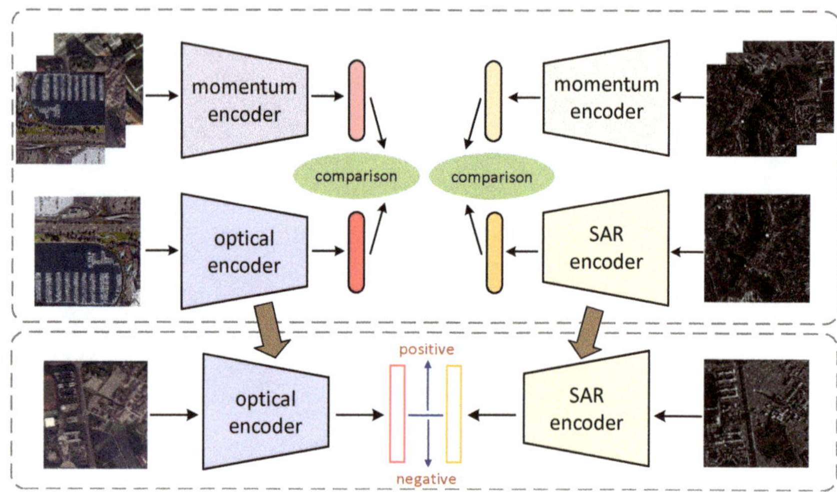



**Figure 2.** The overall process of models via MoCo pre-training transferred to BNN is described. The first box shows the process of applying MoCo to improve the representation learning ability of SAR and optical images separately. Then, the model is transferred to BNN to deal with the SAR-optical matching task in the second box.

The rest of the paper is organized as follows: Section 2 introduces MoCo, BNN, and the way we combine them in detail. The settings and results of the experiments are shown in Section 3. The discussion and conclusions are drawn in Section 4.

## 2. Method

The method in this paper consists of two steps: MoCo [19] pre-training and transferring to BNN [12] for matching tasks. In this section, we will introduce the MoCo, BNN, and the way we combine them in detail.

### 2.1. Momentum Contrast

In [19], they regard contrastive learning as training an encoder for a dictionary look-up task. Momentum Contrast (MoCo) is for building a dynamic large and consistent dictionary on-the-fly. The core of MoCo is maintaining the dictionary as a queue of data samples; therefore, the dictionary size can be decoupled from the mini-batch size for the utilization of the queue to be larger and contain more negative samples.

Given a dataset $X = \{x_i\}_{i=1}^{N}$. , where $x_i$ can be considered as a query sample. Then we can randomly select the other $k$ samples from the dataset $X$ to form a dictionary $\{d_1, d_2, \ldots, d_k\}$. There is a single sample $d_i$ in the dictionary that matches $x_i$. Therefore, the sample $x_i$ with the dictionary can be combined into one positive pair $\{x_i, d_i\}$ and $k$ negative pairs $\{x_i, d_j\}(j \neq i)$. To extract the representations, two feature encoders $f(\cdot; \theta_x), f(\cdot; \theta_d)$

are employed with parameters $\theta_x, \theta_d$ respectively to project the images into the query representations $z_i = f(x_i; \theta_x)$ and the keys of the dictionary $y_i = f(d_i; \theta_d)$. Regarding InfoNCE [16] as the contrastive loss:

$$\mathcal{L}_{contrast} = -\underset{X}{\mathbb{E}}\left[\log \frac{s(x, d_+)}{\sum_{x \in X} \sum_{j=1}^{k+1} s(x, d_j)}\right],\tag{1}$$

where $s(\cdot, \cdot)$ is the metric function and $d_+$ is the unique positive sample. Here, dot product is applied to measure the similarity between latent representations with a temperature hyper-parameter $\tau$:

$$s(x, d) = \exp\left(\frac{f(x; \theta_x) \cdot f(d; \theta_d)}{\|f(x; \theta_x)\| \cdot \|f(d; \theta_d)\|} \cdot \frac{1}{\tau}\right),\tag{2}$$

which achieves high values for positives and low scores for negatives.

**Momentum update.** The parameters $\theta_x$ of query encoder are updated by back-propagation while the parameters $\theta_d$ of key encoder are updated by:

$$\theta_d \leftarrow m\theta_d + (1 - m)\theta_x,\tag{3}$$

where $m \in [0, 1)$ is the momentum coefficient. The momentum update makes $\theta_d$ evolve more smoothly than $\theta_x$. Consequently, the keys in dictionary are encoded by slightly different encoders.

**Compared with memory bank.** The memory bank proposed in [15] is composed of the representations of samples. For every mini-batch, the keys are randomly sampled from the memory bank with no back-propagation. It can support a large mini-batch size, but the representations in the memory bank can not be consistent since it is only updated when it has been chosen. In contrast, the momentum update is more memory-efficient and guarantees the consistency of dictionary keys. Comparison is shown in Figure 3.

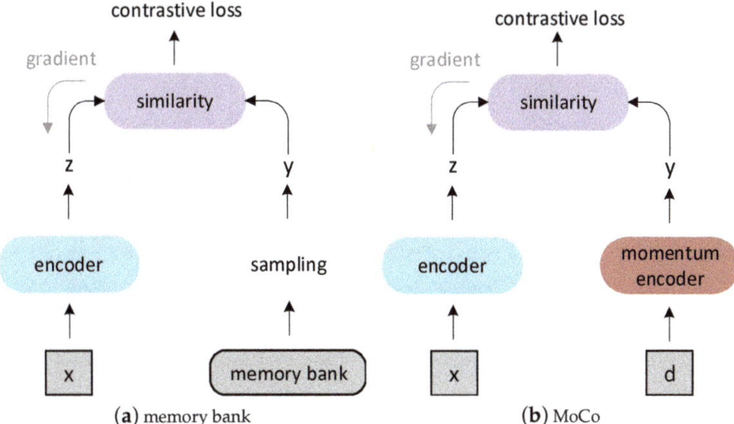


**Figure 3.** Comparison between memory bank [15] and MoCo [19]. (**a**) Forming a memory bank and sample from it as the key representations; (**b**) MoCo encodes the dictionary on-the-fly by momentum updating an encoder slowly.

**Pretext Task.** Pretext tasks act as an important strategy to learn representations of data. The pretext tasks in this paper follows [15,19]: random grayscale, random color jittering, and random horizontal flip. In addition, as depicted in [19,22], Batch Normalization (BN) [23] prevents the model from learning good representations. Thus, shuffling BN [19] is employed to solve this problem.

### 2.2. Bridge Neural Network

Despite the different modalities of SAR and optical images, BNN [12] works like a bridge and is capable of projecting the multimodal images into a common representation subspace. As depicted in Figure 4, given a SAR-optical image dataset $\{X_s, X_o\}$. $X_s = \{x_s^i\}_{i=1}^N$, $X_o = \{x_o^i\}_{i=1}^N$ are sets of SAR and optical image patches. We construct the positive sample set $S_p = \{x_s^i, x_o^i\}$, where corresponding image pairs are from the same region. Image pairs from $S_n = \{x_s^i, x_o^j\}(i \neq j)$ are from different areas, which we call negative samples. The dual networks respectively extract features from SAR and optical images: $f(\cdot)$ is for features of SAR images while $g(\cdot)$ for optical images. The separate networks reduce the images into the $n$-dimension latent vectors: $z_s = f(x_s), z_o = g(x_o)$. Then, Euclidean distance is designed to bring the latent representations of positive samples together while pushing negative samples apart in the common representation subspace. The Euclidean distance between $z_s$ and $z_o$ is defined as:

$$h(x_s, x_o) = \frac{1}{\sqrt{n}}\|(f(x_s) - g(x_o))\|_2,\tag{4}$$

which indicates whether the input data pairs $\{x_s, x_o\}$ have a potential relationship. The distance of positive samples is regressed to 0 while the distance between negative samples converges to 1. Therefore, the regression loss on positive samples and negative samples are as follows:

$$l_p(S_p) = \frac{1}{|S_p|} \sum_{(x_s, x_o) \in S_p} (h(x_s, x_o) - 0)^2,\tag{5}$$

$$l_n(S_n) = \frac{1}{|S_n|} \sum_{(x_s, x_o) \in S_n} (h(x_s, x_o) - 1)^2.\tag{6}$$

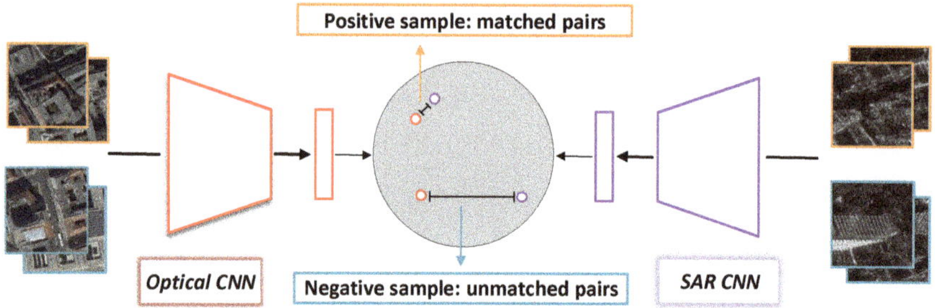


**Figure 4.** Schematic illustration of the BNN architecture for the SAR-optical image matching task. Optical network and SAR network are used in BNN to project the images from different modalities into a common subspace. The Euclidean distance of the representations of SAR and optical images is pulled close for positive samples and pulled apart for negative samples.

Hence, we add the loss on positive samples and negative samples up as the BNN loss:

$$l_{BNN}(S_p, S_n) = \frac{l_p(S_p) + \alpha \cdot l_n(S_n)}{1 + \alpha},\tag{7}$$

where $\alpha$ balances the weights of positive loss and negative loss. Thus, optimizing the loss of BNN can lead to the best networks $(f^*, g^*)$:

$$(f^*, g^*) = \mathrm{argmin}_{f,g}\, l_{BNN}(S_p, S_n)\tag{8}$$

### 2.3. Transfer MoCo Pre-Trained Model to BNN

As depicted in Figure 2, at the pre-training stage, we first pre-train MoCo to obtain an optical encoder and an SAR encoder to separately learn feature representations and measure the similarity in the latent space. At the matching stage, the pre-trained encoders transferred to BNN serve as the initialization of the optical network and SAR network for fine-tuning in the SAR-optical matching task. The encoders project the SAR and optical images into a common feature representation subspace, where it is easy to measure similarity, to determine whether they match or not.

## 3. Experiments

### 3.1. Dataset

We conduct experiments on three SAR-optical image datasets: SARptical [24], QXS-SAROPT [25], SEN1-2 [26]. **SARptical** is a dataset of over 10,000 pairs of corresponding SAR and optical image pairs, which are from TerraSAR-X and aerial UltraCAM optical images. The images are of $112 \times 112$ pixels with $100$ m $\times$ $100$ m ground coverage. We use 7577 image pairs for training and 1263 patch pairs for testing. **QXS-SAROPT** contains 20,000 pairs of SAR-optical image patches from GaoFen-3 satellite and Google Earth, covering a variety of land types. The images have a size of $256 \times 256$ pixels at a pixel spacing of $1$ m $\times$ $1$ m. The dataset is randomly divided into training and testing at a ratio of 7:3. We select the spring subset from four sub-groups in **SEN1-2** dataset, which consists of registered patch-pairs from Sentinel-2 and Sentinel-1—a total of 75,724 patch pairs of size $256 \times 256$ pixels with spatial distance of $10$ m $\times$ $10$ m, of which 52,799 are for training and 22,925 for testing.

### 3.2. Implementation

Considering the complexity and difficulty of the matching task, we only take Vgg11 [27] as the feature extraction network for SARptical while Vgg11 [27], ResNet50 [28], and Darknet53 [29] as the backbone for QXS-SAROPT and SEN1-2. More specifically, we encode the SAR and optical images into a 50-dimensional feature representation subspace. It is noted that every mini-batch input contains the same number of positive and negative samples to prevent the model from mode collapse for the unbalanced data distribution. All images are normalized w.r.t. mean and variance in preparation.

**MoCo.** We take $N - 1$ negative samples ($N$ is the number of training datasets) and set temperature parameter $\tau = 0.07$ and momentum coefficient for updating encoder is set as $m = 0.999$. We use SGD as an optimizer and a mini-batch size of 20. The weight decay is 0.001 and the SGD momentum is 0.9. For SARptical, the learning rate is set as 0.05 for 300 epochs. While training on QXS-SAROPT, the learning rate is 0.001 for the first 250 epochs and 0.0005 for the last 250 epochs. As for SEN1-2, the models are trained with a learning rate of 0.05 for 70 epochs.

**BNN.** The ratio of positive and negative samples is 1:1 and adjusting factor $\alpha = 1$. For SARptical, the learning rate is set as 0.1 to fine-tune the model for 100 epochs. Meanwhile, BNN with MoCo pre-training on QXS-SAROPT and SEN1-2 are fine-tuned with a learning rate of 0.05 for 10 epochs.

### 3.3. Results Analysis

We fine-tune BNN to SAR-optical image matching task with MoCo pre-training (MoCo-BNN) on three datasets: SARptical, QXS-SAROPT, and SEN1-2. To demonstrate the superiority of our method, we compare our method with two other initialization methods: no pre-training (NP) and ImageNet pre-training (IP).. Accuracy, precision, and recall score are employed as evaluation metrics. The matching results on three datasets can be seen in Tables 1–3. It is noted that the pair matching results of IP-BNN benchmark can be directly obtained from article [25,30].

**Table 1.** Results for BNN patch-matching on SARptical with different pre-training methods.

| Methods | Accuracy | Precision | Recall |
| --- | --- | --- | --- |
| NP-BNN | 0.887 | 0.825 | 0.993 |
| IP-BNN | **0.913** | 0.855 | **0.999** |
| MoCo-BNN(ours) | **0.913** | **0.859** | 0.991 |

**Table 2.** Results for BNN patch-matching on QXS-SAROPT with different pre-training methods.

| Backbone | Methods | Accuracy | Precision | Recall |
| --- | --- | --- | --- | --- |
| Vgg11 | NP-BNN | 0.844 | 0.781 | 0.982 |
| | IP-BNN | 0.817 | 0.744 | **0.999** |
| | MoCo-BNN(ours) | **0.858** | **0.795** | 0.990 |
| ResNet50 | NP-BNN | 0.831 | 0.750 | 0.990 |
| | IP-BNN [25,30] | 0.829 | 0.748 | 0.993 |
| | MoCo-BNN(ours) | **0.873** | **0.808** | **0.995** |
| Darknet53 | NP-BNN | 0.826 | 0.761 | 0.980 |
| | IP-BNN [25,30] | 0.828 | 0.746 | 0.995 |
| | MoCo-BNN(ours) | **0.871** | **0.809** | **0.997** |

**Table 3.** Results for BNN patch-matching on SEN1-2 with different pre-training methods.

| Backbone | Methods | Accuracy | Precision | Recall |
| --- | --- | --- | --- | --- |
| Vgg11 | NP-BNN | 0.832 | **0.800** | 0.916 |
| | IP-BNN | 0.828 | 0.760 | **0.993** |
| | MoCo-BNN(ours) | **0.841** | 0.787 | 0.960 |
| ResNet50 | NP-BNN | 0.775 | 0.721 | 0.931 |
| | IP-BNN | 0.783 | 0.722 | 0.954 |
| | MoCo-BNN(ours) | **0.800** | **0.753** | **0.968** |
| Darknet53 | NP-BNN | 0.834 | 0.775 | 0.970 |
| | IP-BNN | 0.853 | 0.788 | 0.993 |
| | MoCo-BNN(ours) | **0.862** | **0.796** | **0.998** |

**Accuracy performance.** Tables 2 and 3 suggest that our MoCo pre-trained models lead to a better matching performance on QXS-SAROPT and SEN1-2.(The bolded number represents the highest score on the backbone.) Especially on QXS-SAROPT, the accuracy of BNN taking MoCo as a pre-training strategy achieves 87.3% and 87.1% and makes a 4.4% improvement, which surpasses the other two methods by large margins. MoCo pre-training also makes an obvious improvement on SEN1-2, indicating that MoCo has a powerful representation learning ability for both SAR and optical images. Furthermore, the results of NP and IP with ResNet50 and Darknet53 as backbone are almost the same, which illustrates that the ImageNet supervised pre-trained models have no capacity for SAR information and characteristics and it hardly makes sense in the SAR-optical image matching problem. It is worth noting that IP-BNN is even worse than NP-BNN with Vgg11 as a backbone on QXS-SAROPT and SEN1-2, which is in line with the claims made in [14], i.e., shallow models can be trained from scratch as long as a proper initialization is used, whereas only when the network is large enough can the ImageNet pre-training model, which contains prior knowledge provide a good initialization to fine-tune. Besides, in Table 1, MoCo does not make any progress on SARptical. The reason may stem from the single building scenario in this dataset; the IP method learns better optical features and makes a great performance. When confronted with more complex SAR-optical datasets, optical feature embedding ability is weakened and SAR feature learning ability plays a major role. We visualize the accuracy results in Figure 5 to better show the comparison of the different pre-training strategies.

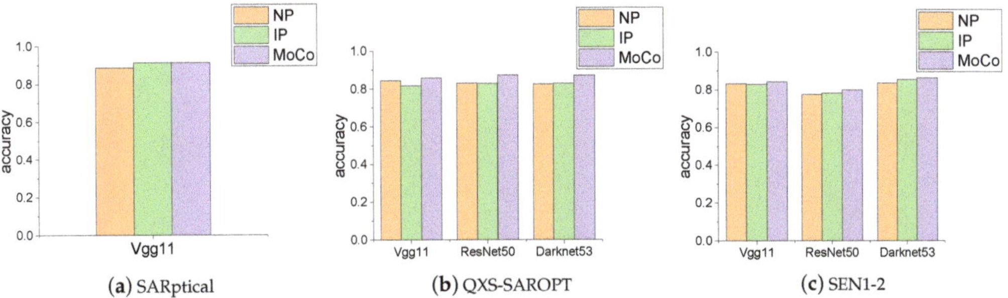

**(a)** SARptical    **(b)** QXS-SAROPT    **(c)** SEN1-2

**Figure 5.** Visualization of the accuracy results of BNN with NP, IP, and MoCo pre-training on three datasets.

We show matching results of some images pairs in Figure 6. In particular, image pairs in red box are classified correctly by our MoCo-BNN and classified incorrectly by IP-BNN. As shown in the figure, MoCo-BNN not only can distinguish the similar negative image pairs, but also can correctly discriminate among the non-obvious positive sample pairs.

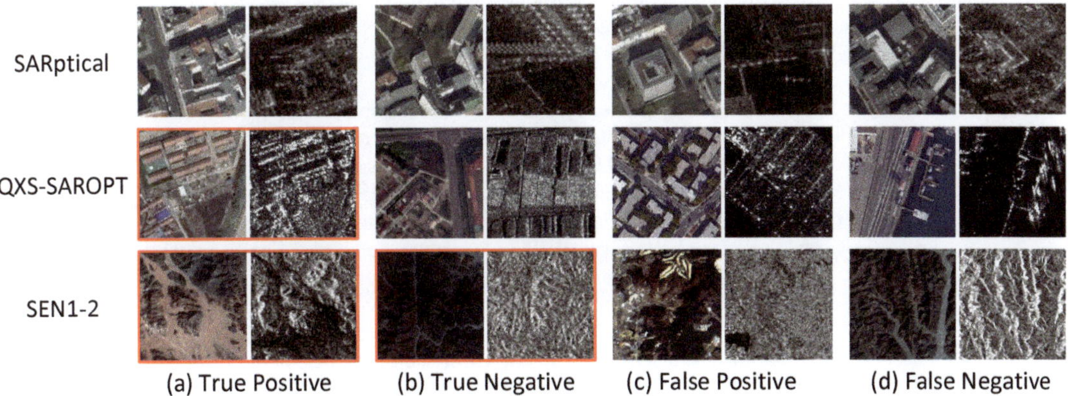

(a) True Positive    (b) True Negative    (c) False Positive    (d) False Negative

**Figure 6.** Exhibition of images pairs in different matching results. The red boxes frame the image pairs that MoCo-BNN classifies correctly and IP-BNN classifies incorrectly.

**Embedding learning performance.** To intuitively compare the feature representations learn by BNN with NP, IP, and MoCo pre-training, we visualize the embedding learning results of the test set of QXS-SAROPT. We use t-distributed Stochastic Neighbor Embedding(t-SNE) to visualize the features extracted by BNN with NP, IP, and MoCo pre-training. We first concatenate the SAR and optical 50-dimensional features to the 100-dimensional features. T-SNE projects the features to two dimensions so that the high-dimensional features are convenient to visualize. As shown in Figure 7, the features of positive samples and negative samples learned by BNN with NP and IP are mixed together while the positive features (class 1) and negative features (class 0) learned by BNN with MoCo pre-training are more gathered in each class and more separate between different classes. Therefore, the MoCo pre-trained BNN can generate a more distinguishable embedding space, leading to a better multi-modal image matching result. The reason may stem from MoCo pre-training leading to a better representation learning ability for SAR and optical images.

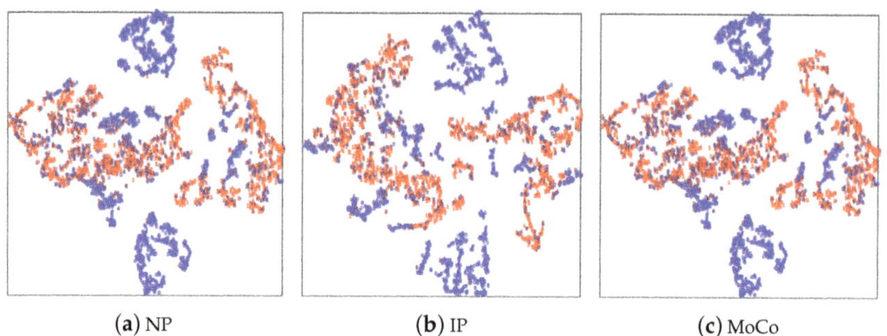

(**a**) NP        (**b**) IP        (**c**) MoCo

**Figure 7.** Visualization of positive and negative features extracted by BNN with NP, IP, and MoCo pre-training using t-SNE. The red dots represent the positive features, and the blue dots represent the negative features.

## 4. Discussion and Conclusions

Aiming at improving SAR-optical image matching performance, considering directly fine-tuning on an ImageNet supervised pre-training model as commonly used can hardly benefit for improving the learning ability for SAR images; this paper exploits a self-supervised pre-training to improve the feature learning ability of SAR and optical images respectively. Then, the pre-trained model is transferred to the SAR-optical matching tasks. The experiments demonstrate that self-supervised pre-training leads to a significant improvement.

Furthermore, we exploit a self-supervised pre-training paradigm to improve the feature learning ability of multi-modal images. However, this paper only did experiments on SAR-optical images and only for matching tasks. It is believed that our method not only can be adaptive to different kinds of remote sensing images, such as multi-spectral images, hyperspectral images and so on, but also can be transferred to different tasks, such as objective detection.

However, the experiments were only conducted on one self-supervised method MoCo, and one matching network BNN is a major deficiency. Furthermore, it has been verified in [15–21] that a large mini-batch size is necessary for self-supervised learning to learn a good representation. Nonetheless, it still works well when we train the MoCo with mini-batch size 20, which is much smaller than the commonly used mini-batch size in self-supervised learning. In the future, more experiments on a Siamese network and other self-supervised learning methods will be carried out to confirm the effectiveness of self-supervised pre-training in SAR-optical image matching.

**Author Contributions:** L.Q. wrote the manuscript and performed all the experiments; M.H. performed the experiment analysis and revised the manuscript. L.Q. and M.H. contributed equally to this work. M.H., X.L. and X.X. supervised the study. All authors have read and agreed to the published version of the manuscript.

**Funding:** This research is funded by Beijing Nova Program of Science and Technology (Grant Number: Z191100001119129).

**Data Availability Statement:** Publicly available datasets were analyzed in this study. SARptical [24] can be downloaded at http://www.sipeo.bgu.tum.de/downloads/SARptical_data.zip. QXS-SAROPT [25] can be found at https://github.com/yaoxu008/QXS-SAROPT. SEN1-2 [26] is available at https://mediatum.ub.tum.de/1436631.

**Conflicts of Interest:** The authors declare no conflict of interest.

## References

1. Burger, W.; Burge, M.J. *Principles of Digital Image Processing: Core Algorithms*; Springer Science & Business Media: Berlin/Heidelberg, Germany, 2010.
2. Walters-Williams, J.; Li, Y. Estimation of mutual information: A survey. In Proceedings of the International Conference on Rough Sets and Knowledge Technology, Gold Coast, Australia, 14–16 July 2009; Springer: Berlin/Heidelberg, Germany, 2009; pp. 389–396.
3. Suri, S.; Reinartz, P. Mutual-information-based registration of TerraSAR-X and Ikonos imagery in urban areas. *IEEE Trans. Geosci. Remote Sens.* **2009**, *48*, 939–949. [CrossRef]
4. Lowe, D.G. Distinctive image features from scale-invariant keypoints. *Int. J. Comput. Vis.* **2004**, *60*, 91–110. [CrossRef]
5. Dellinger, F.; Delon, J.; Gousseau, Y.; Michel, J.; Tupin, F. SAR-SIFT: A SIFT-like algorithm for SAR images. *IEEE Trans. Geosci. Remote Sens.* **2014**, *53*, 453–466. [CrossRef]
6. Ye, Y.; Shen, L. Hopc: A novel similarity metric based on geometric structural properties for multi-modal remote sensing image matching. *ISPRS Ann. Photogramm. Remote Sens. Spat. Inf. Sci.* **2016**, *3*, 9. [CrossRef]
7. Ye, Y.; Shan, J.; Bruzzone, L.; Shen, L. Robust registration of multimodal remote sensing images based on structural similarity. *IEEE Trans. Geosci. Remote Sens.* **2017**, *55*, 2941–2958. [CrossRef]
8. Zagoruyko, S.; Komodakis, N. Learning to compare image patches via convolutional neural networks. In Proceedings of the IEEE Conference on Computer Vision and Pattern Recognition, Boston, MA, USA, 7–12 June 2015; pp. 4353–4361.
9. Merkle, N.; Luo, W.; Auer, S.; Müller, R.; Urtasun, R. Exploiting deep matching and SAR data for the geo-localization accuracy improvement of optical satellite images. *Remote Sens.* **2017**, *9*, 586. [CrossRef]
10. Mou, L.; Schmitt, M.; Wang, Y.; Zhu, X.X. A CNN for the identification of corresponding patches in SAR and optical imagery of urban scenes. In Proceedings of the 2017 Joint Urban Remote Sensing Event (JURSE), Dubai, United Arab Emirates, 6–8 March 2017; IEEE: Piscataway, NJ, USA, 2017; pp. 1–4.
11. Hughes, L.H.; Schmitt, M.; Mou, L.; Wang, Y.; Zhu, X.X. Identifying corresponding patches in SAR and optical images with a pseudo-Siamese CNN. *IEEE Geosci. Remote Sens. Lett.* **2018**, *15*, 784–788. [CrossRef]
12. Xu, Y.; Xiang, X.; Huang, M. Task-Driven Common Representation Learning via Bridge Neural Network. In Proceedings of the AAAI Conference on Artificial Intelligence, Honolulu, HI, USA, 27 January–1 February 2019; Volume 33, pp. 5573–5580.
13. Russakovsky, O.; Deng, J.; Su, H.; Krause, J.; Satheesh, S.; Ma, S.; Huang, Z.; Karpathy, A.; Khosla, A.; Bernstein, M.; et al. Imagenet large scale visual recognition challenge. *Int. J. Comput. Vis.* **2015**, *115*, 211–252. [CrossRef]
14. He, K.; Girshick, R.; Dollár, P. Rethinking imagenet pre-training. In Proceedings of the IEEE/CVF International Conference on Computer Vision, Seoul, Korea, 27 October–2 November 2019; pp. 4918–4927.
15. Wu, Z.; Xiong, Y.; Yu, S.X.; Lin, D. Unsupervised feature learning via non-parametric instance discrimination. In Proceedings of the IEEE Conference on Computer Vision and Pattern Recognition, Salt Lake City, UT, USA, 18–22 June 2018; pp. 3733–3742.
16. Oord, A.v.d.; Li, Y.; Vinyals, O. Representation learning with contrastive predictive coding. *arXiv* **2018**, arXiv:1807.03748.
17. Tian, Y.; Krishnan, D.; Isola, P. Contrastive multiview coding. *arXiv* **2019**, arXiv:1906.05849.
18. Chen, T.; Kornblith, S.; Norouzi, M.; Hinton, G. A simple framework for contrastive learning of visual representations. In Proceedings of the International Conference on Machine Learning, PMLR, Vienna, Austria, 12–18 July 2020; pp. 1597–1607.
19. He, K.; Fan, H.; Wu, Y.; Xie, S.; Girshick, R. Momentum contrast for unsupervised visual representation learning. In Proceedings of the IEEE/CVF Conference on Computer Vision and Pattern Recognition, Seattle, WA, USA, 13–19 June 2020; pp. 9729–9738.
20. Caron, M.; Misra, I.; Mairal, J.; Goyal, P.; Bojanowski, P.; Joulin, A. Unsupervised learning of visual features by contrasting cluster assignments. *arXiv* **2020**, arXiv:2006.09882.
21. Zbontar, J.; Jing, L.; Misra, I.; LeCun, Y.; Deny, S. Barlow Twins: Self-Supervised Learning via Redundancy Reduction. *arXiv* **2021**, arXiv:2103.03230.
22. Henaff, O. Data-efficient image recognition with contrastive predictive coding. In Proceedings of the International Conference on Machine Learning, Vienna, Austria, 12–18 July 2020; pp. 4182–4192.
23. Ioffe, S.; Szegedy, C. Batch normalization: Accelerating deep network training by reducing internal covariate shift. In Proceedings of the International Conference on Machine Learning, PMLR, Lille, France, 6–11 July 2015; pp. 448–456.
24. Wang, Y.; Zhu, X.X. The sarptical dataset for joint analysis of sar and optical image in dense urban area. In Proceedings of the IGARSS 2018—2018 IEEE International Geoscience and Remote Sensing Symposium, Valencia, Spain, 22–27 July 2018; IEEE: Piscataway, NJ, USA, 2018; pp. 6840–6843.
25. Huang, M.; Xu, Y.; Qian, L.; Shi, W.; Zhang, Y.; Bao, W.; Wang, N.; Liu, X.; Xiang, X. The QXS-SAROPT Dataset for Deep Learning in SAR-Optical Data Fusion. *arXiv* **2021**, arXiv:2103.08259.
26. Schmitt, M.; Hughes, L.H.; Zhu, X.X. The SEN1-2 dataset for deep learning in SAR-optical data fusion. *arXiv* **2018**, arXiv:1807.01569.
27. Simonyan, K.; Zisserman, A. Very deep convolutional networks for large-scale image recognition. *arXiv* **2014**, arXiv:1409.1556.
28. He, K.; Zhang, X.; Ren, S.; Sun, J. Deep residual learning for image recognition. In Proceedings of the IEEE/CVF International Conference on Computer Vision, Las Vegas, NV, USA, 27–30 June 2016; pp. 770–778.
29. Redmon, J.; Farhadi, A. Yolov3: An incremental improvement. *arXiv* **2018**, arXiv:1804.02767.
30. Bao, W.; Huang, M.; Zhang, Y.; Xu, Y.; Liu, X.; Xiang, X. Boosting ship detection in SAR images with complementary pretraining techniques. *arXiv* **2021**, arXiv:2103.08251.

*Article*

# MSGATN: A Superpixel-Based Multi-Scale Siamese Graph Attention Network for Change Detection in Remote Sensing Images

Wenjing Shuai [1,*], Fenlong Jiang [2], Hanhong Zheng [2] and Jianzhao Li [2]

1   School of Electronic Engineering, Xidian University, Xi'an 710121, China
2   Key Laboratory of Intelligent Perception and Image Understanding of Ministry of Education, School of Electronic Engineering, Xidian University, Xi'an 710071, China; jiangfenlong@outlook.com (F.J.); hanhong_zheng@163.com (H.Z.); 19jzli@stu.xidian.edu.cn (J.L.)
*   Correspondence: wjshuai@xidian.edu.cn; Tel.: +86-182-2903-4065



**Abstract:** With the rapid development of Earth observation technology, how to effectively and efficiently detect changes in multi-temporal images has become an important but challenging problem. Relying on the advantages of high performance and robustness, object-based change detection (CD) has become increasingly popular. By analyzing the similarity of local pixels, object-based CD aggregates similar pixels into one object and takes it as the basic processing unit. However, object-based approaches often have difficulty capturing discriminative features, as irregular objects make processing difficult. To address this problem, in this paper, we propose a novel superpixel-based multi-scale Siamese graph attention network (MSGATN) which can process unstructured data natively and extract valuable features. First, a difference image (DI) is generated by Euclidean distance between bitemporal images. Second, superpixel segmentation is employed based on DI to divide each image into many homogeneous regions. Then, these superpixels are used to model the problem by graph theory to construct a series of nodes with the adjacency between them. Subsequently, the multi-scale neighborhood features of the nodes are extracted through applying a graph convolutional network and concatenated by an attention mechanism. Finally, the binary change map can be obtained by classifying each node by some fully connected layers. The novel features of MSGATN can be summarized as follows: (1) Training in multi-scale constructed graphs improves the recognition over changed land cover of varied sizes and shapes. (2) Spectral and spatial self-attention mechanisms are exploited for a better change detection performance. The experimental results on several real datasets show the effectiveness and superiority of the proposed method. In addition, compared to other recent methods, the proposed can demonstrate very high processing efficiency and greatly reduce the dependence on labeled training samples in a semisupervised training fashion.

**Keywords:** change detection; superpixel segmentation; graph attention network; remote sensing images



**Citation:** Shuai, W.; Jiang, F.; Zheng, H.; Li, J. MSGATN: A Superpixel-Based Multi-Scale Siamese Graph Attention Network for Change Detection in Remote Sensing Images. *Appl. Sci.* **2022**, *12*, 5158. https://doi.org/10.3390/app12105158

Academic Editor: Junseop Lee

Received: 23 April 2022
Accepted: 16 May 2022
Published: 20 May 2022

**Publisher's Note:** MDPI stays neutral with regard to jurisdictional claims in published maps and institutional affiliations.

## 1. Introduction

With the continuous collection of massive multi-temporal remote sensing images, such as multi-spectral [1,2], synthetic aperture radar (SAR) [3], hyperspectral [4], and unmanned aerial vehicle (UAV) images [5], these multi-temporal remote sensing images have been promoted in practical applications. In this data context, change detection (CD) is one of the most meaningful technologies, which aims to quantitatively and qualitatively obtain the change information of ground objects by analyzing bitemporal remote sensing images. In many practical situations, these changes have potential significance, such as urban development planning, natural disaster assessment, dynamic monitoring of ecological environment, and natural resource management [6–8].

In the early stages, in order to obtain land cover change information, the traditional CD technology usually includes the following steps. First, the bitemporal image needs to be preprocessed, including radiation correction, ensemble correction, and spatial registration [9,10]. Second, a difference image (DI) between bitemporal images can be acquired by image ratio [11], image difference [12], change vector analysis [13,14], etc. Finally, a threshold or a clustering algorithm is applied to segment DI into binary change map (BCM), such as Otsu [15,16], double-window flexible pace search [17,18], K-means [19], fuzzy c-means [20,21], and so on. Since bitemporal images are usually collected under different imaging conditions, such as illumination, season, etc., the different images may contain a large number of spurious differences [22]. Moreover, these methods usually use pixels as processing units, and either thresholding or clustering can cause a lot of noise in the results of these methods.

To address the limitations of pixel-level methods, many scholars have made great efforts in CD and proposed various object-based CD methods [23,24]. In general, object-based methods first need to segment the image to obtain multi-scale objects. Universal image segmentation techniques include fractal net evolution segmentation approach [25], simple linear iterative clustering (SLIC) superpixel segmentation [26], etc. These approaches frequently generate multi-scale objects or superpixels through region growing, i.e., objects or superpixels are obtained by gradual pixel binning with similar spectral values. Therefore, each superpixel or object is composed of a homogeneous set of pixels. The CD can then be achieved by the object for the image analysis and processing unit. For example, in the early stages, Jungho et al. proposed an object-based CD based on correlation image analysis and image segmentation [27]. An object-based approach is based on multiple classifiers and multi-scale uncertainty analysis for CD with high-resolution (HR) remote sensing images [28]. Recently, some novel object-based approaches have made some efforts. For instance, Lv et al. promoted an object-oriented key point vector distance to obtain binary CD [29]. This method can significantly improve the performance of the difference image, as it measures the difference between the key-points vectors of two objects in bitemporal images. Similar methods are available in [30–32].

Although the aforementioned approaches have made remarkable progress, some limitations are still unavoidable. These limitations mainly include the following three aspects:

- Traditional methods are difficult to deal with and analyze irregular objects effectively as the multi-scale objects or the superpixels represent unstructured data. Therefore, there is still a lack of effective representative feature extraction approach for unstructured data.
- Image segmentation itself is a challenging task, and usually some parameters need to be adjusted to obtain better segmentation results. Moreover, the error of image segmentation may accumulate in the change detection task to some extent. Therefore, object-based change detection is severely limited by the performance of image segmentation.
- Object-based CD approaches generally require more complex frameworks. This results in a lower degree of automation of the entire CD framework due to the need to individually perform image segmentation algorithms and select appropriate segmentation parameters.

With the popularization of deep learning technology, the methods based on deep neural networks have been widely used in change detection. In particular, the graph neural networks (GNNs) have been noticed due to their excellent performance in unstructured data classification. Recently, GNNs have been successfully applied to image classification [33,34] and change detection [35,36]. Specifically, in [37], graph convolutional networks (GCNs) are utilized to extract the features of different types for hyperspectral image classification. Saha et al. proposed a semisupervised CD approach based on GCNs [38], which adopts multi-scale parcel segmentation to encode multi-temporal images as a graph. However, there are still few studies on GCN-based CD at present. Therefore, GCNs-based CD still needs continuous and further research.

Considering the excellent performance of GCNs in solving image classification, we are able to model the change detection task as a graph node classification task for improving the performance of CD. With this motivation, the paper proposes a novel multi-scale superpixel graph attention network (MSGATN), which can process unstructured data natively and extract valuable features. In the proposed method, a difference image (DI) is firstly obtained by Euclidean distance between bitemporal images. Then, an SLIC algorithm is employed to divide the DI into many homogeneous superpixels. Subsequently, these superpixels are exploited to model the problem by graph theory to build a series of nodes with Based on this, the multi-scale features of each node are captured by a graph attention network (GATN). Finally, the binary change map (BCM) is generated by classifying each node using some fully connected layers.

The contributions of the proposed MSGATN approach are summarized as follows:

(1) We propose a network model based on graph theory, which can process the unstructured data of objects with irregular boundaries in OBCD and consider the adjacency relationship between objects.

(2) The proposed method is inductive, which can simultaneously adapt to graphs of different scales. Therefore, our proposed MSGATN can exploit the constructed graphs of various scales, thus improving the abilities of representation and generalization.

(3) Experiments on several real datasets obtained from different sensors demonstrate that the proposed MSGATN has high efficiency and performance, as well as certain generalization.

The rest of this paper is organized as follows. Section 2 briefly introduces some related works. In Section 3, our method is described in detail. Section 4 provides the experimental settings and results. Finally, the conclusions and future works are given in Section 5.

## 2. Related Work

### 2.1. Deep-Learning-Based CD Methods

In recent years, deep learning technology has become a new favorite in the field of CD [39,40], especially convolutional neural networks (CNNs). These deep-learning-based methods can be roughly summarized into two categories, i.e., image-level methods and patch-level methods.

(1) Image-level methods: This category of method is to acquire semantic change information by analyzing a complete bitemporal image at a time [41]. Hence, that usually requires a large number of pairs of manually labeled training image pairs. For example, Ji et al. proposed a Siamese U-Net with shared weights to acquire a building change map in an end-to-end manner [42]. Liu et al. devised a local-global pyramid network for building CD in [43]. In [44], a spatial–temporal attention-based network based on self-attention mechanism was applied to mine deep robust features for large image-to-image CD datasets. Although these approaches can achieve competitive performance, they often not only require a large number of manually labeled paired images to train the network, but also cost more storage space and computational resources.

(2) Patch-level methods: Different from image-level methods, this type of method indicates using local pixel patches or superpixels as analysis units, and capturing feature representations through convolution or fully connection to achieve CD. In the early stages, Gong et al. proposed a novel CD method based on deep learning [45], which can avoid the effect of the DI to provide a better change detection performance. In [46], a Gabor-based PCANet (GaborPCANet) was promoted for CD in SAR images, which utilizes PCA filters as convolutional filters to capture the image features. A convolutional-wavelet neural network (CWNN) was devised to detect sea ice change detection from SAR images in [47]. Jiang et al. developed a semisupervised multiple CD approach, which can detect multiple changes using only a very limited samples by training a generative adversarial network [48]. This approach introduces dual-tree complex wavelet transform into CNNs to reduce the effect of the speckle noise, thus improving detection performance. However, these methods based on local pixel patches are still limited by the selection of regular windows. To alleviate this limitation, superpixels-based CD methods have received attention, which aim to use superpixels as analysis units to capture more representative features through CNNs. To achieve this, recent methods have made further efforts. For instance, Gong et al.

presented a superpixel-based difference representation learning to extract semantic change information between bitemporal images [49]. In [50], an end-to-end superpixel-enhanced CD network was designed, which combines an adaptive superpixel merging module to mine difference information for CD. Other methods refer to [51–53].

*2.2. Graph Neural Networks*

In the early stage, among the works concerning CD approach, graph theory-based approaches have been extensively used for CD [54,55]. For instance, in [56], a weighted graph was built to measure changes for CD with SAR images. Sun et al. proposed an iterative robust graph for unsupervised heterogenous images CD [57]. This method constructs a robust K-nearest neighbor graph of bitemporal images, and calculates the difference image by comparing the graphs.

With the development of deep learning, a variant of CNNs, graph neural networks (GNNs), has received sustained attention in many applications [58–60]. In particular, graph convolutional networks (GCNs) have been successfully applied in the fields of remote sensing, such as remote sensing image retrieval [61], remote sensing image semantic segmentation [62], and hyperspectral image classification [63]. Specifically, GCNs are able to efficiently process graph-structured data by modeling the relationships between samples (or vertices). Therefore, GCNs can be naturally used to model remote spatial relationships in remote sensing images, which is not considered in CNNs. Recently, considering the previous GCNs-based research in the field of remote sensing, these methods have been developed and applied to CD tasks. For example, Wu et al. promoted a multi-scale GCN to detect land cover changes for CD in homogeneous and heterogeneous remote sensing images [64]. This approach constructs graph representations through object-wise high-level features generated by a pretrained U-Net. In [65], a multi-scale dynamic GCN was employed to mine the short-range and long-range contextual information. These GNNs-based methods have been initially applied to solve remote sensing image CD. However, there are still few CD methods for GNNs, and a large number of systematic theoretical studies and applications are still lacking. Therefore, further development of GNN-based CD methods has potential value.

## 3. Proposed Superpixel-Based MSGATN

*3.1. Overview of the Proposed MSGATN*

In this subsection, an overview of the proposed MSGATN is given briefly in Figure 1. Firstly, the difference intensity of bitemporal remote sensed images is obtained by Euclidean distance. Based on the pixel-wise similarity, the difference intensity map can be segmented to massive unstructured multi-scale superpixels of varied shapes and boundaries by simple linear iterative clustering (SLIC). With the segmented DI acquired, a region adjacency graph (RAG) can be constructed based on the mutual consistency of neighbor superpixels. The spatial–temporal relationships between these superpixels can be well modeled by the edges of constructed DI RAG. Then, the bitemporal remote sensing images are also segmented into superpixels with the guidance of the segmentation information extracted in DI superpixel segmentation. Several significant statistical characteristics, i.e., minimum, maximum, mean, standard deviation, skewness, and kurtosiscan further represent the features of multi-scale bitemporal superpixels. As a result, the input graph of graph attention network (GATN) can be constructed by the nodes obtained from features of bitemporal superpixels and the edges acquired in the RAG of DI superpixels. Finally, superpixel-level prediction is obtained by GATN and remapped to form the pixel-level change map. The detailed inference process of MSGATN can be illustrated in Algorithm 1.

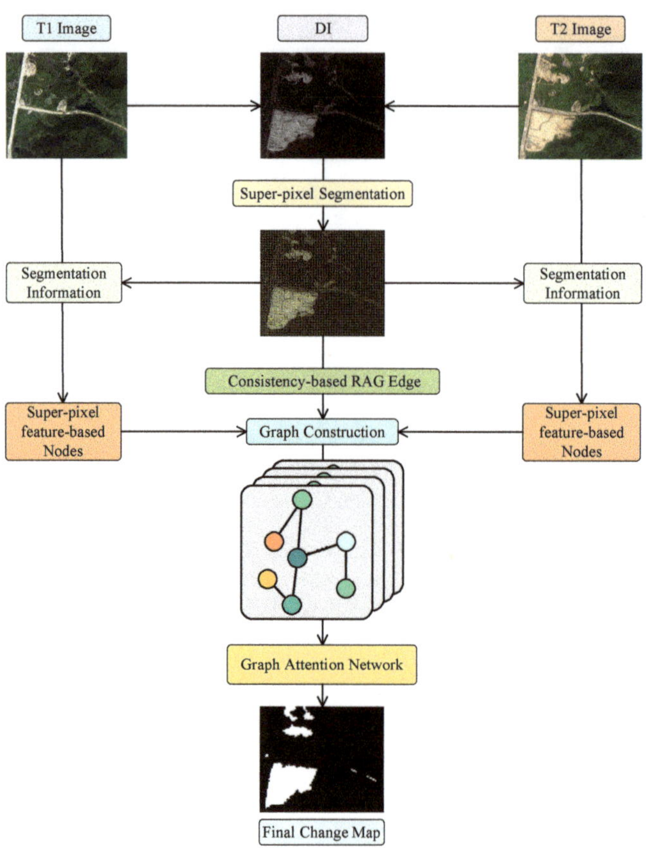

**Figure 1.** The framework and procedure of MSGATN.

---

**Algorithm 1** Inference process of MSGATN

---

**Input:** $T^1$, $T^2$: the bitemporal images.

1: Begin

2: $DI \leftarrow Euclidean\_Distance(T1, T2)$; // obtain the difference intensity

3: $S^{DI} \leftarrow SLIC(DI)$; // conduct superpixel segmentation over $DI$

4: $\mathcal{G}_{sp}(\mathcal{V}, \mathcal{E}_{sp}) \leftarrow RAG(S^{DI})$; // acquire the region adjacency graph of $S^{DI}$

5: $S^{T1} \leftarrow superpixel\_segmentation(T1)$; // segment T1 according to $S^{DI}$

6: $S^{T2} \leftarrow superpixel\_segmentation(T2)$; // segment T2 according to $S^{DI}$

7: $F^1 \leftarrow feature\_analyse(S^{T1})$; // represent the significant features of $S^{T1}$

8: $F^2 \leftarrow feature\_analyse(S^{T2})$; // represent the significant features of $S^{T2}$

9: $\mathcal{V}_f \leftarrow concatenate(F^1, F^2)$; // collect the superpixel-level bitemporal features

10: $\mathcal{G}_{input} \leftarrow (\mathcal{V}_f, \mathcal{E}_{sp})$; // construct the input graph for GATN

11: $\mathcal{F}_{output} \leftarrow GATN(\mathcal{G}_{input})$; // obtain superpixel-wise change map

12: $CM \leftarrow remap(\mathcal{F}_{output})$; // remap the superpixel prediction to acquire final $CM$

**Output:** $CM$: binary change map.

---

As shown in the procedure above, the proposed MSGATN firstly obtains the multi-scale unstructured features of bitemporal remote sensed images through superpixel segmentation, which further promotes the fine-grained CD prediction. Then, the mutual relationships inside these bitemporal superpixels are well represented and modeled by GATN. Given the overall framework and inference process of the proposed MSGATN, the detailed information of the proposed graph construction mechanism can be given in the following section.

### 3.2. Graph Construction

To acquire credible prior information for GATN, a preliminary but representative graph construction is indispensable. In the proposed graph construction method, difference intensity and bitemporal remote sensing images are integrated to obtain comprehensive non-local change information, which advances the changed region detection in GATN. The overall graph construction can be further illustrated by the following steps. Initially, the pixel-wise difference intensity $DI \in \mathbb{R}^{H \times W}$ can be represented as

$$distance = \sqrt[2]{(T1 - T2)^2} \tag{1}$$

$$DI = \frac{distance - min(distance)}{max(distance) - min(distance)} \tag{2}$$

After the difference intensity is acquired, the multi-scale superpixel segmentation over $DI$ can be given as

$$S^{DI} = \{p_1, p_2, \cdots, p_{N\_seg}\} = \text{SLIC}_{N\_seg}(DI) \tag{3}$$

where $\text{SLIC}_{N\_seg}(\cdot)$ denotes the simple linear iterative clustering with different numbers of segmented superpixels, and $N\_seg$ represents the number of superpixels. Generally, the more superpixels, the smaller they are. In this case, we can obtain multi-scale superpixels for the multi-scale feature recognition of the proposed MSGATN. Then, the region adjacency co-relations $\mathcal{G}_{sp}$ of $S^{DI}$ can be modeled as

$$\mathcal{G}_{sp}(\mathcal{V}, \mathcal{E}_{sp}) = RAG\left(S^{DI}\right) \tag{4}$$

in which $RAG(\cdot)$ represents the region adjacency graph construction operation. The edges $\mathcal{E}_{sp}$ are exploited to model the local and non-local relations inside bitemporal superpixels. To achieve this, the bitemporal superpixels need to be firstly acquired as $S^{T1} = \left\{p_1^1, p_2^1, \cdots, p_{N\_seg}^1\right\}$ and $S^{T2} = \left\{p_1^2, p_2^2, \cdots, p_{N\_seg}^2\right\}$. Then, the bitemporal features of these superpixels can be denoted as

$$F^1 = concat\left[min\left(S^{T1}\right), max\left(S^{T1}\right), mean\left(S^{T1}\right), std\left(S^{T1}\right), skew\left(S^{T1}\right), kur\left(S^{T1}\right)\right] \tag{5}$$

$$F^2 = concat\left[min\left(S^{T2}\right), max\left(S^{T2}\right), mean\left(S^{T2}\right), std\left(S^{T2}\right), skew\left(S^{T2}\right), kur\left(S^{T2}\right)\right] \tag{6}$$

where $concat(\cdot)$ indicates a feature-level integration, and $min(\cdot)$, $max(\cdot)$, $mean(\cdot)$, $std(\cdot)$, $skew(\cdot)$, and $kur(\cdot)$ represent the superpixel-wise minimum, maximum, mean, standard deviation, skewness, and kurtosis, respectively. Given these dependable and discriminative features of bitemporal superpixels, the nodes can be obtained as follows:

$$\mathcal{V}_f = concat\left(F^1, F^2\right) \tag{7}$$

at which $concat(\cdot)$ denotes a feature-wise concatenation. With the nodes and edges obtained, the input multi-scale graphs for MSGATN can be constructed as

$$\mathcal{G}_{input}^{N\_seg} = \left(\mathcal{V}_f, \mathcal{E}_{sp}\right) \tag{8}$$

With different $N\_seg$, input graphs of varied scales can be provided for the proposed MSGATN; thus, the multi-scale change objects obtain a finer cognition.

### 3.3. Multi-Scale Siamese Graph Attention Network

In the proposed MSGATN, a GATN is employed to better reveal the relationships between multi-scale unstructured superpixels, and the graph attention mechanism is the linchpin of GATN. To further facilitate the understanding for the proposed MSGATN, the mathematical style of graph attention is given as follows: Let $f_i \in \mathbb{R}^{C_I}$ and $f_j \in \mathbb{R}^{C_I}$ be the feature vectors of current node $i$ and its neighbor node $j$, respectively. Then, the edge score $e_{ij}$ can be obtained by

$$e_{ij} = \left( concat\left( f_i W, f_j W \right) \right) A \tag{9}$$

where $W \in \mathbb{R}^{C_I \times C_O}$ and $A \in \mathbb{R}^{2C_O \times 1}$ are the learnable supervised parameters, $concat(\cdot)$ represents a feature-wise integration, and $C_I$ and $C_O$ denote the input and output feature lengths, respectively. With each $e_{ij}$ acquired, the attention score $a_{ij}$ can be given as follows:

$$a_{ij} = \frac{exp\left( LeakyReLU\left( e_{ij} \right) \right)}{\sum_k exp\left( LeakyReLU(e_{ik}) \right)} \tag{10}$$

where $LeakyReLU(\cdot)$ represents a nonlinear activation, and $k$ denotes all the neighbor nodes of $i$. In the proposed MSGATN, the graph attention mechanism is widely used to refine the graph feature representation. To improve the recognition for multi-scale objects, the proposed network is trained over multi-scale graphs from superpixel segmentation of different superpixel numbers. In our method, the $N\_seg$ is set to 2000, 4000, and 6000 to obtain input graphs of different scales. Basically, GATN can tackle inductive tasks with graphs of varied scales. Based on this fact, the proposed MSGATN can learn multi-scale feature representation through training over several multi-scale constructed graphs in a Siamese framework.

## 4. Experiments

### 4.1. Dataset Descriptions

To further test and verify the ability of the proposed method, two extensively used remote sensing CD multi-spectral datasets, i.e., the Guangzhou city dataset and the Hongqi canal dataset, are exploited, which are shown in Figures 2 and 3.

#### 4.1.1. Guangzhou City Dataset

This dataset is composed of a bitemporal multispectral image pair with the spatial resolution of 2.5 m, captured by the Systeme Probatoire d'Observation de la Terre 5 (SPOT-5) satellite. It depicts the land cover change over urban areas of Guangzhou City between October 2006 and October 2007, respectively, as shown in Figure 2. The bitemporal images are the size of $877 \times 738$ pixels, including red, green, and near-infrared bands. Its annotation focuses on vegetation change.

**Figure 2.** Guangzhou City dataset: (a) $T_1$-time image, (b) $T_2$-time image, (c) ground truth image.

### 4.1.2. Hongqi Canal Dataset

The second dataset, Hongqi Canal dataset, contains two high-resolution multispectral remote sensed images which focus on the region of Yellow River Estuary near the city of Dongying in China, as shown in Figure 3. The bitemporal images, which have the spatial resolution of 2m and the size of 539 × 543, were acquired by GF-1 satellite on 9 December 2013 and 16 October 2015, respectively. It mainly describes the river changes of the Hongqi Canal settled in Xijiu village.

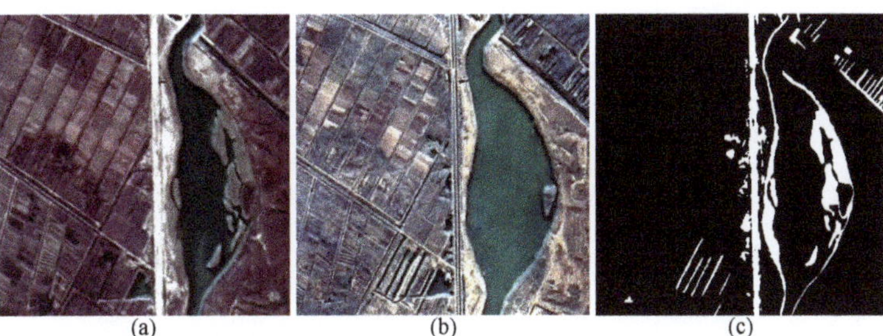


**Figure 3.** Hongqi Canal dataset: (**a**) $T_1$-time image, (**b**) $T_2$-time image, (**c**) ground truth image.

### 4.2. Comparative Methods and Related Settings

In the experiments, to evaluate the performance of the proposed MSGATN, we selected five related CD approaches for comparison with our MSGATN. All methods are described as follows:

(1) PCA_K-means [19]: This approach is one of the popular unsupervised CD methods, which adopts principal component analysis (PCA) and k-means clustering to acquire binary change map. In this method, two parameters ($h$ and $s$) should be set. For the Guangzhou City dataset, $h$ and $s$ are set to 9 and 3, respectively. For the Hongqi Canal dataset, $h$ and $s$ are set to 5 and 3, respectively.

(2) ASEA [66]: It is a state-of-the-art method that exploits the contextual information around a pixel to improve detection accuracy. This method requires no parameter setting.

(3) GaborPCANet [46]: This was proposed in [46]. It utilizes PCA filters as convolution kernels to obtain representative neighborhood features. In this approach, a parameter, patch size, is set to 5 for both experimental datasets.

(4) DBN [49]: This is a superpixel-based method, which can acquire a better detection result by difference representation learning. For our experimental datasets, the parameter patch size is fixed to 5 in this method.

(5) CWNN [47]: It devises a convolutional-wavelet neural network in SAR images. In the experiments, the parameter patch size is fixed to 7 for our datasets.

(6) Proposed MSGATN: In our MSGATN, the number of superpixels is a hyperparameter. Specifically, in our method, we selected six scales of superpixel segmentation, which, respectively, include 1000, 2000, 3000, 4000, 5000, and 6000 superpixels, to train our MSGATN in a Siamese manner. For both experimental results, the results of 6000 superpixels are chosen to be compared with other methods.

### 4.3. Evaluation Criteria

To further evaluate the performance for CD, several widely used evaluation metrics, which are precision, recall, $F1$, and overall accuracy ($OA$), are employed. Their definitions are given as follows:

$$Precision = \frac{TP}{TP + FP} \tag{11}$$

$$Recall = \frac{TP}{TP + FN} \tag{12}$$

$$F1 = \frac{2 \times Precision \times Recall}{Precision + Recall} \tag{13}$$

$$OA = \frac{TP + TN}{TP + TN + FP + FN} \tag{14}$$

where $TP$, $TN$, $FP$, and $FN$ are the numbers of true positive, true negative, false positive, false negative pixels, respectively. Based on these well-acknowledged evaluation metrics, the performance of different CD methods can be better revealed.

### 4.4. Comparative Results

In this subsection, the comparative results on two widely used CD datasets are given in detail. To further illustrate the proposed method, corresponding analysis will be given in detail. Detailed visual and quantized results and analysis are presented as follows.

### 4.4.1. Results on the Guangzhou City Dataset

The visualized and quantitative comparison results over the Guangzhou City dataset are shown in Figure 4 and Table 1, respectively. From the quantitative comparison, our MSGATN can achieve the best $F1$ and $OA$ (90.54% and 97.19%). However, DBN and GaborPCANet achieved the best precision and recall, respectively. Although the proposed MSGATN does not achieve the best precision and recall, our method still provides relatively reliable performance in terms of precision and recall. For example, compared with the DBN, despite DBN reaching the best precision (98.05%), it obtained the second-worst performance in recall (78.51%). Therefore, our MSGATN can acquire more balanced performance for the four metrics. Different from other approaches, the proposed MSGATN adopts a multi-scale graph attention network to effectively capture the representative features of unstructured data, thereby improving the detection accuracy. Regarding the visual results, the proposed MSGATN exhibits the fewest false detections compared to the other five methods. Specifically, the GaborPCANet presents many false alarms compared to the proposed MSGATN. Moreover, although the visual results of the DBN show fewer false alarms, a large number of missed pixels are unavoidable. Compared with these methods, our proposed MSGATN can obtain a more balanced performance in terms of false detections and missed detections. Moreover, the proposed MSGATN can provide more complete change information compared with other methods, except GaborPCANet. Overall, the visual results also yielded similar conclusions to the quantitative comparisons.

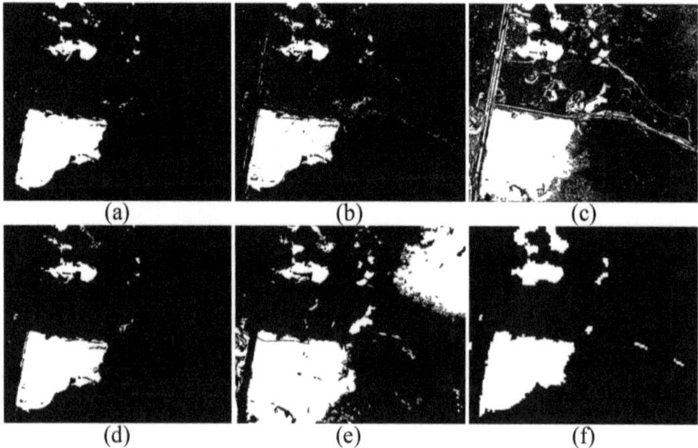

**Figure 4.** The results of different methods on the Guangzhou City dataset: (**a**) PCA_K-means, (**b**) ASEA, (**c**) GaborPCANet, (**d**) DBN, (**e**) CWNN, and (**f**) proposed MSGATN.

**Table 1.** Accuracy comparison (in %) of different methods on the Guangzhou City dataset. The best evaluation value are presented in bold for different metrics.

| Methods | Precision | Recall | F1 | OA |
|---|---|---|---|---|
| PCA_K-means | 97.33 | 78.22 | 86.74 | 96.43 |
| ASEA | 95.45 | 79.73 | 86.88 | 96.40 |
| GaborPCANet | 51.61 | **94.98** | 66.88 | 85.94 |
| DBN | **98.05** | 78.51 | 87.20 | 96.55 |
| CWNN | 42.19 | 87.17 | 56.86 | 80.23 |
| Proposed MSGATN | 91.04 | 90.04 | **90.54** | **97.19** |

### 4.4.2. Results on the Hongqi Canal Dataset

As presented in Table 2, the proposed MSGATN achieve the overall superiority over the Hongqi Canal dataset compared to other selected CD methods. That is, our method outperforms the other methods in all evaluation indicators, i.e., precision, recall, $F1$, and $OA$, with a great gap. More exactly, the best precision (80.96%) and recall (57.17%) are achieved by the proposed MSGATN, which leads to the best F1 (67.02%) for our method. It indicates that the proposed MSGATN can capture finer complete land cover and acquire better mapping for changed multi-scale objects with the help of input graphs with varied scales, and similar conclusions can be discovered in the visualized CD results depicted in Figure 5. Given the fact that the annotation of the Hongqi Canal dataset mainly focuses on the river change, massive false alarms can be found in the CMs generated by other methods. These false alarms are basically caused by the unchanged farmland around the canal. However, they are well filtered out in the proposed MSGATN, which can be attributed to the finer feature representation of our method. As a result, the river course change in the Hongqi Canal dataset is well denoted by the proposed MSGATN, which suggests the advantage of our method.

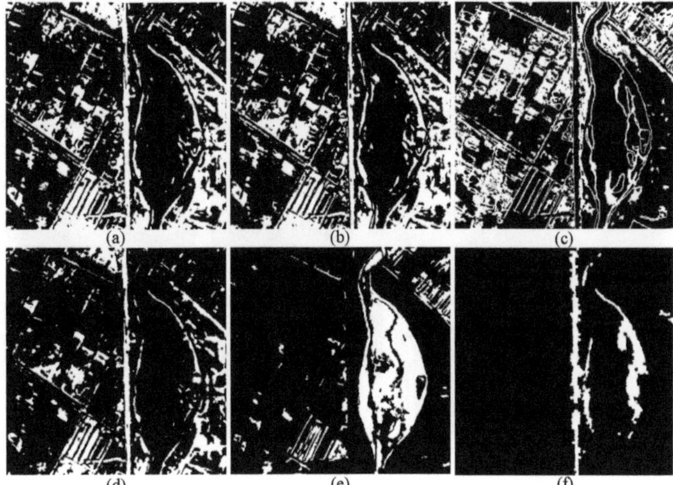

**Figure 5.** The results of different methods on the Hongqi Canal dataset: (**a**) PCA_K-means, (**b**) ASEA, (**c**) GaborPCANet, (**d**) DBN, (**e**) CWNN, and (**f**) proposed MSGATN.

**Table 2.** Accuracy comparison (in %) of different methods on the Hongqi Canal dataset. The best evaluation value are presented in bold for different metrics.

| Methods | Precision | Recall | *F1* | *OA* |
|---|---|---|---|---|
| PCA_K-means | 15.67 | 47.28 | 23.53 | 71.57 |
| ASEA | 16.26 | 50.69 | 24.62 | 71.28 |
| GaborPCANet | 3.71 | 12.89 | 5.76 | 60.99 |
| DBN | 19.17 | 34.53 | 24.66 | 80.47 |
| CWNN | 31.91 | 56.12 | 40.68 | 84.86 |
| Proposed MSGATN | **80.96** | **57.17** | **67.02** | **94.80** |

### 4.5. Parameters Analysis of the Proposed MSGATN on the Guangzhou Dataset

To further investigate the effectiveness of the proposed MSGATN, parameters analyses are performed on the Guangzhou dataset in this section. In our MSGATN, the number of superpixels is a hyperparameter. Furthermore, we selected six scales of superpixel segmentation, which, respectively, include 1000, 2000, 3000, 4000, 5000, and 6000 superpixels (as shown in Figure 6), to train our MSGATN in a Siamese manner. Generally, a larger number of superpixels indicates a smaller segmentation scale. Conversely, a smaller number of superpixels indicates a larger segmentation scale. Thanks to the characteristics of the inductive GATN, our MSGATN can easily exploit multi-scale superpixel features. By this way, features of different scales can be considered in our method. In this context, different BCMs can be generated by the proposed MSGATN at each scale, as presented in Figure 6. As the number of superpixels increases, the scale of superpixels also becomes finer. Similarly, the BCM of each scale is also finer as the number of superpixels enlarges for our MSGATN.

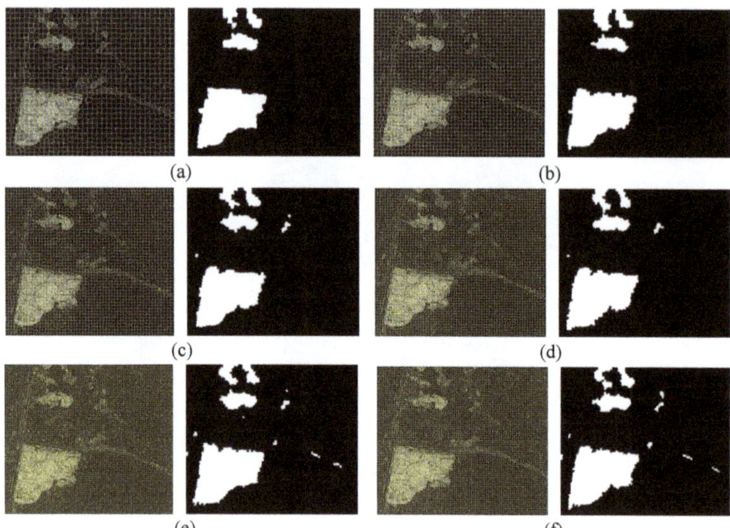

**Figure 6.** Segmentation results and the corresponding change detection results of different superpixel numbers in the proposed MSGATN on the Guangzhou dataset: (**a**) 1000 superpixels, (**b**) 2000 superpixels, (**c**) 3000 superpixels, (**d**) 4000 superpixels, (**e**) 5000 superpixels, (**f**) 6000 superpixels.

Figure 7 more intuitively demonstrates the relationship between the number of superpixels and detection accuracy. Concretely, as the number of superpixels increases, all metrics show an upward trend. However, if the number of superpixels exceeds 3000, the accuracy gradually decreases. Hence, the performance of the proposed MSGATN may not continue to increase as the number of superpixels increases. Moreover, more superpixels can lead to larger graph structures, which can significantly increase the computational cost. According to the above analysis, the number of superpixels cannot be continuously increased in our method.

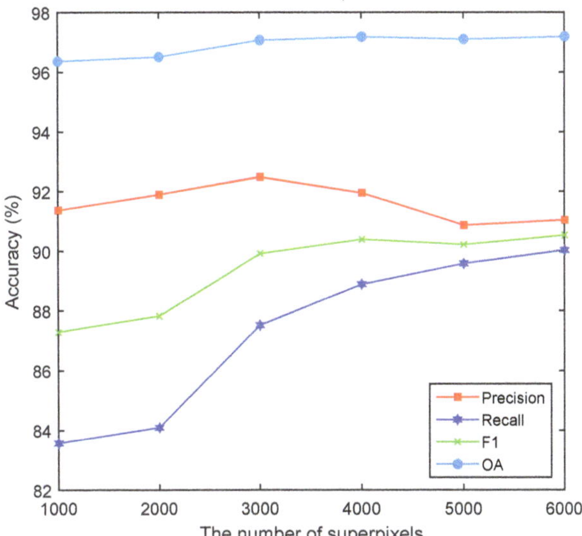

**Figure 7.** Relationship between change detection accuracy and superpixels numbers for the proposed MSGATN on the Guangzhou dataset.

## 5. Conclusions

In this work, a novel superpixel-based multi-scale Siamese graph attention network (MSGATN) is proposed for change detection in high-resolution remote sensed imagery. In the proposed method, superpixel segmentation is exploited to aggregate homogeneous difference information to construct heterogeneous change information for a better recognition of multi-scale changed land cover. In addition, multi-scale superpixel-constructed graphs are introduced to a graph attention network (GATN) in a Siamese framework, which further facilitates the cognition of multi-scale objects for the GATN, thus improving the performance. The proposed MSGATN is validated over two widely used change detection datasets, and compared to several comparative change detection methods. Corresponding results indicate that the proposed method outperforms other methods over all selected evaluation metrics.

In the future work, efforts can be made to achieve a more fine-grained changed land cover annotation in an unsupervised framework, which can be less time-consuming and laboring in practical applications.

**Author Contributions:** Conceptualization, W.S. and F.J.; methodology, F.J. and H.Z.; validation, H.Z.; investigation, J.L.; writing—original draft preparation, W.S.; writing—review and editing, W.S. and F.J. All authors have read and agreed to the published version of the manuscript.

**Funding:** This work was supported by the National Natural Science Foundation of China (61906148).

**Conflicts of Interest:** The authors declare no conflict of interest.

## References

1. Lv, Z.; Li, G.; Yan, J.; Benediktsson, J.A.; You, Z. Training Samples Enriching Approach for Classification Improvement of VHR Remote Sensing Image. *IEEE Geosci. Remote Sens. Lett.* **2021**, *19*, 1–5. [CrossRef]
2. Wu, Y.; Xiao, Z.; Liu, S.; Miao, Q.; Ma, W.; Gong, M.; Xie, F.; Zhang, Y. A Two-Step Method for Remote Sensing Images Registration Based on Local and Global Constraints. *IEEE J. Sel. Top. Appl. Earth Obs. Remote Sens.* **2021**, *14*, 5194–5206. [CrossRef]

3. Wu, Y.; Ma, W.; Gong, M.; Su, L.; Jiao, L. A novel point-matching algorithm based on fast sample consensus for image registration. *IEEE Geosci. Remote Sens. Lett.* **2014**, *12*, 43–47. [CrossRef]

4. Lv, Z.; Li, G.; Jin, Z.; Benediktsson, J.A.; Foody, G.M. Iterative training sample expansion to increase and balance the accuracy of land classification from VHR imagery. *IEEE Trans. Geosci. Remote Sens.* **2020**, *59*, 139–150. [CrossRef]

5. Liu, T.; Gong, M.; Jiang, F.; Zhang, Y.; Li, H. Landslide Inventory Mapping Method Based on Adaptive Histogram-Mean Distance With Bitemporal VHR Aerial Images. *IEEE Geosci. Remote Sens. Lett.* **2021**, *19*, 1–5. [CrossRef]

6. Zhu, Z. Change detection using landsat time series: A review of frequencies, preprocessing, algorithms, and applications. *ISPRS J. Photogramm. Remote Sens.* **2017**, *130*, 370–384. [CrossRef]

7. Shi, W.; Zhang, M.; Zhang, R.; Chen, S.; Zhan, Z. Change detection based on artificial intelligence: State-of-the-art and challenges. *Remote Sens.* **2020**, *12*, 1688. [CrossRef]

8. Tewkesbury, A.P.; Comber, A.J.; Tate, N.J.; Lamb, A.; Fisher, P.F. A critical synthesis of remotely sensed optical image change detection techniques. *Remote Sens. Environ.* **2015**, *160*, 1–14. [CrossRef]

9. Wu, Y.; Miao, Q.; Ma, W.; Gong, M.; Wang, S. PSOSAC: Particle swarm optimization sample consensus algorithm for remote sensing image registration. *IEEE Geosci. Remote Sens. Lett.* **2017**, *15*, 242–246. [CrossRef]

10. Wu, Y.; Liu, J.W.; Zhu, C.Z.; Bai, Z.F.; Miao, Q.G.; Ma, W.P.; Gong, M.G. Computational intelligence in remote sensing image registration: A survey. *Int. J. Autom. Comput.* **2021**, *18*, 1–17. [CrossRef]

11. Gong, M.; Cao, Y.; Wu, Q. A neighborhood-based ratio approach for change detection in SAR images. *IEEE Geosci. Remote Sens. Lett.* **2011**, *9*, 307–311. [CrossRef]

12. Lv, Z.; Liu, T.; Zhang, P.; Atli Benediktsson, J.; Chen, Y. Land cover change detection based on adaptive contextual information using bitemporal remote sensing images. *Remote Sens.* **2018**, *10*, 901. [CrossRef]

13. Liu, S.; Bruzzone, L.; Bovolo, F.; Zanetti, M.; Du, P. Sequential spectral change vector analysis for iteratively discovering and detecting multiple changes in hyperspectral images. *IEEE Trans. Geosci. Remote Sens.* **2015**, *53*, 4363–4378. [CrossRef]

14. Thonfeld, F.; Feilhauer, H.; Braun, M.; Menz, G. Robust Change Vector Analysis (RCVA) for multi-sensor very high resolution optical satellite data. *Int. J. Appl. Earth Obs. Geoinf.* **2016**, *50*, 131–140. [CrossRef]

15. Otsu, N. A threshold selection method from gray-level histograms. *IEEE Trans. Syst. Man Cybern.* **1979**, *9*, 62–66. [CrossRef]

16. Lv, Z.; Liu, T.; Atli Benediktsson, J.; Lei, T.; Wan, Y. Multi-scale object histogram distance for LCCD using bitemporal very-high resolution remote sensing images. *Remote Sens.* **2018**, *10*, 1809. [CrossRef]

17. Lv, Z.Y.; Liu, T.F.; Zhang, P.; Benediktsson, J.A.; Lei, T.; Zhang, X. Novel adaptive histogram trend similarity approach for land cover change detection by using bitemporal very-high-resolution remote sensing images. *IEEE Trans. Geosci. Remote Sens.* **2019**, *57*, 9554–9574. [CrossRef]

18. Lv, Z.; Liu, T.; Shi, C.; Benediktsson, J.A. Local histogram-based analysis for detecting land cover change using VHR remote sensing images. *IEEE Geosci. Remote Sens. Lett.* **2020**, *18*, 1284–1287. [CrossRef]

19. Celik, T. Unsupervised change detection in satellite images using principal component analysis and $k$-means clustering. *IEEE Geosci. Remote Sens. Lett.* **2009**, *6*, 772–776. [CrossRef]

20. Ghosh, A.; Mishra, N.S.; Ghosh, S. Fuzzy clustering algorithms for unsupervised change detection in remote sensing images. *Inf. Sci.* **2011**, *181*, 699–715. [CrossRef]

21. Gong, M.; Zhou, Z.; Ma, J. Change detection in synthetic aperture radar images based on image fusion and fuzzy clustering. *IEEE Trans. Image Process.* **2011**, *21*, 2141–2151. [CrossRef] [PubMed]

22. Wu, Y.; Li, J.; Yuan, Y.; Qin, A.; Miao, Q.G.; Gong, M.G. Commonality autoencoder: Learning common features for change detection from heterogeneous images. *IEEE Trans. Neural Netw. Learn. Syst.* **2021**. [CrossRef] [PubMed]

23. Chen, G.; Hay, G.J.; Carvalho, L.M.; Wulder, M.A. Object-based change detection. *Int. J. Remote Sens.* **2012**, *33*, 4434–4457. [CrossRef]

24. Hussain, M.; Chen, D.; Cheng, A.; Wei, H.; Stanley, D. Change detection from remotely sensed images: From pixel-based to object-based approaches. *ISPRS J. Photogramm. Remote Sens.* **2013**, *80*, 91–106. [CrossRef]

25. Chen, Q.; Li, L.; Xu, Q.; Yang, S.; Shi, X.; Liu, X. Multi-feature segmentation for high-resolution polarimetric SAR data based on fractal net evolution approach. *Remote Sens.* **2017**, *9*, 570. [CrossRef]

26. Achanta, R.; Shaji, A.; Smith, K.; Lucchi, A.; Fua, P.; Süsstrunk, S. SLIC superpixels compared to state-of-the-art superpixel methods. *IEEE Trans. Pattern Anal. Mach. Intell.* **2012**, *34*, 2274–2282. [CrossRef]

27. Im, J.; Jensen, J.; Tullis, J. Object-based change detection using correlation image analysis and image segmentation. *Int. J. Remote Sens.* **2008**, *29*, 399–423. [CrossRef]

28. Tan, K.; Zhang, Y.; Wang, X.; Chen, Y. Object-based change detection using multiple classifiers and multi-scale uncertainty analysis. *Remote Sens.* **2019**, *11*, 359. [CrossRef]

29. Lv, Z.; Liu, T.; Benediktsson, J.A. Object-oriented key point vector distance for binary land cover change detection using VHR remote sensing images. *IEEE Trans. Geosci. Remote Sens.* **2020**, *58*, 6524–6533. [CrossRef]

30. Hao, M.; Zhou, M.; Jin, J.; Shi, W. An advanced superpixel-based Markov random field model for unsupervised change detection. *IEEE Geosci. Remote Sens. Lett.* **2019**, *17*, 1401–1405. [CrossRef]

31. Zhu, L.; Zhang, J.; Sun, Y. Remote Sensing Image Change Detection Using Superpixel Cosegmentation. *Information* **2021**, *12*, 94. [CrossRef]

32. Pang, S.; Hu, X.; Zhang, M.; Cai, Z.; Liu, F. Co-segmentation and superpixel-based graph cuts for building change detection from bitemporal digital surface models and aerial images. *Remote Sens.* **2019**, *11*, 729. [CrossRef]
33. Cai, W.; Wei, Z. Remote sensing image classification based on a cross-attention mechanism and graph convolution. *IEEE Geosci. Remote Sens. Lett.* **2020**, *19*, 1–5. [CrossRef]
34. Du, X.; Zheng, X.; Lu, X.; Doudkin, A.A. Multisource remote sensing data classification with graph fusion network. *IEEE Trans. Geosci. Remote Sens.* **2021**, *59*, 10062–10072. [CrossRef]
35. Wang, R.; Wang, L.; Dong, P.; Jiao, L.; Chen, J.W. Graph-Level Neural Network for SAR Image Change Detection. In Proceedings of the 2021 IEEE International Geoscience and Remote Sensing Symposium IGARSS, Brussels, Belgium, 11–16 July 2021; pp. 3785–3788.
36. Kalinicheva, E.; Ienco, D.; Sublime, J.; Trocan, M. Unsupervised change detection analysis in satellite image time series using deep learning combined with graph-based approaches. *IEEE J. Sel. Top. Appl. Earth Obs. Remote Sens.* **2020**, *13*, 1450–1466. [CrossRef]
37. Hong, D.; Gao, L.; Yao, J.; Zhang, B.; Plaza, A.; Chanussot, J. Graph convolutional networks for hyperspectral image classification. *IEEE Trans. Geosci. Remote Sens.* **2020**, *59*, 5966–5978. [CrossRef]
38. Saha, S.; Mou, L.; Zhu, X.X.; Bovolo, F.; Bruzzone, L. Semisupervised change detection using graph convolutional network. *IEEE Geosci. Remote Sens. Lett.* **2020**, *18*, 607–611. [CrossRef]
39. Liu, S.; Marinelli, D.; Bruzzone, L.; Bovolo, F. A review of change detection in multitemporal hyperspectral images: Current techniques, applications, and challenges. *IEEE Geosci. Remote Sens. Mag.* **2019**, *7*, 140–158. [CrossRef]
40. ZhiYong, L.; Liu, T.; Benediktsson, J.A.; Falco, N. Land cover change detection techniques: Very-high-resolution optical images: A review. *IEEE Geosci. Remote Sens. Mag.* **2021**, *10*, 44–63.
41. Zheng, H.; Gong, M.; Liu, T.; Jiang, F.; Zhan, T.; Lu, D.; Zhang, M. HFA-Net: High Frequency Attention Siamese Network for Building Change Detection in VHR Remote Sensing Images. *Pattern Recognit.* **2022**, *129*, 108717. [CrossRef]
42. Ji, S.; Wei, S.; Lu, M. Fully convolutional networks for multisource building extraction from an open aerial and satellite imagery data set. *IEEE Trans. Geosci. Remote Sens.* **2018**, *57*, 574–586. [CrossRef]
43. Liu, T.; Gong, M.; Lu, D.; Zhang, Q.; Zheng, H.; Jiang, F.; Zhang, M. Building change detection for vhr remote sensing images via local-global pyramid network and cross-task transfer learning strategy. *IEEE Trans. Geosci. Remote Sens.* **2021**, *60*, 1–17. [CrossRef]
44. Chen, H.; Shi, Z. A spatial-temporal attention-based method and a new dataset for remote sensing image change detection. *Remote Sens.* **2020**, *12*, 1662. [CrossRef]
45. Gong, M.; Zhao, J.; Liu, J.; Miao, Q.; Jiao, L. Change detection in synthetic aperture radar images based on deep neural networks. *IEEE Trans. Neural Netw. Learn. Syst.* **2015**, *27*, 125–138. [CrossRef]
46. Gao, F.; Dong, J.; Li, B.; Xu, Q. Automatic change detection in synthetic aperture radar images based on PCANet. *IEEE Geosci. Remote Sens. Lett.* **2016**, *13*, 1792–1796. [CrossRef]
47. Gao, F.; Wang, X.; Gao, Y.; Dong, J.; Wang, S. Sea ice change detection in SAR images based on convolutional-wavelet neural networks. *IEEE Geosci. Remote Sens. Lett.* **2019**, *16*, 1240–1244. [CrossRef]
48. Jiang, F.; Gong, M.; Zhan, T.; Fan, X. A semisupervised GAN-based multiple change detection framework in multi-spectral images. *IEEE Geosci. Remote Sens. Lett.* **2019**, *17*, 1223–1227. [CrossRef]
49. Gong, M.; Zhan, T.; Zhang, P.; Miao, Q. Superpixel-based difference representation learning for change detection in multispectral remote sensing images. *IEEE Trans. Geosci. Remote Sens.* **2017**, *55*, 2658–2673. [CrossRef]
50. Zhang, H.; Lin, M.; Yang, G.; Zhang, L. ESCNet: An End-to-End Superpixel-Enhanced Change Detection Network for Very-High-Resolution Remote Sensing Images. *IEEE Trans. Neural Netw. Learn. Syst.* **2021**, 1–15. [CrossRef]
51. Gong, M.; Jiang, F.; Qin, A.; Liu, T.; Zhan, T.; Lu, D.; Zheng, H.; Zhang, M. A Spectral and Spatial Attention Network for Change Detection in Hyperspectral Images. *IEEE Trans. Geosci. Remote Sens.* **2021**, *60*, 1–15. [CrossRef]
52. Saha, S.; Bovolo, F.; Bruzzone, L. Unsupervised deep change vector analysis for multiple-change detection in VHR images. *IEEE Trans. Geosci. Remote Sens.* **2019**, *57*, 3677–3693. [CrossRef]
53. Du, B.; Ru, L.; Wu, C.; Zhang, L. Unsupervised deep slow feature analysis for change detection in multi-temporal remote sensing images. *IEEE Trans. Geosci. Remote Sens.* **2019**, *57*, 9976–9992. [CrossRef]
54. Vakalopoulou, M.; Karantzalos, K.; Komodakis, N.; Paragios, N. Graph-based registration, change detection, and classification in very high resolution multitemporal remote sensing data. *IEEE J. Sel. Top. Appl. Earth Obs. Remote Sens.* **2016**, *9*, 2940–2951. [CrossRef]
55. Wang, T.; Lu, G.; Liu, J.; Yan, P. Graph-based change detection for condition monitoring of rotating machines: Techniques for graph similarity. *IEEE Trans. Reliab.* **2018**, *68*, 1034–1049. [CrossRef]
56. Pham, M.T.; Mercier, G.; Michel, J. Change detection between SAR images using a pointwise approach and graph theory. *IEEE Trans. Geosci. Remote Sens.* **2015**, *54*, 2020–2032. [CrossRef]
57. Sun, Y.; Lei, L.; Guan, D.; Kuang, G. Iterative Robust Graph for Unsupervised Change Detection of Heterogeneous Remote Sensing Images. *IEEE Trans. Image Process.* **2021**, *30*, 6277–6291. [CrossRef]
58. Fan, X.; Gong, M.; Wu, Y.; Qin, A.; Xie, Y. Propagation Enhanced Neural Message Passing for Graph Representation Learning. *IEEE Trans. Knowl. Data Eng.* **2021**. [CrossRef]
59. Gong, M.; Zhou, H.; Qin, A.; Liu, W.; Zhao, Z. Self-Paced Co-Training of Graph Neural Networks for Semi-Supervised Node Classification. *IEEE Trans. Neural Netw. Learn. Syst.* **2022**. [CrossRef] [PubMed]
60. Fan, X.; Gong, M.; Xie, Y.; Jiang, F.; Li, H. Structured self-attention architecture for graph-level representation learning. *Pattern Recognit.* **2020**, *100*, 107084. [CrossRef]

61. Chaudhuri, U.; Banerjee, B.; Bhattacharya, A. Siamese graph convolutional network for content based remote sensing image retrieval. *Comput. Vis. Image Underst.* **2019**, *184*, 22–30. [CrossRef]
62. Ouyang, S.; Li, Y. Combining deep semantic segmentation network and graph convolutional neural network for semantic segmentation of remote sensing imagery. *Remote Sens.* **2020**, *13*, 119. [CrossRef]
63. Qin, A.; Shang, Z.; Tian, J.; Wang, Y.; Zhang, T.; Tang, Y.Y. Spectral–spatial graph convolutional networks for semisupervised hyperspectral image classification. *IEEE Geosci. Remote Sens. Lett.* **2018**, *16*, 241–245. [CrossRef]
64. Wu, J.; Li, B.; Qin, Y.; Ni, W.; Zhang, H.; Fu, R.; Sun, Y. A multiscale graph convolutional network for change detection in homogeneous and heterogeneous remote sensing images. *Int. J. Appl. Earth Obs. Geoinf.* **2021**, *105*, 102615. [CrossRef]
65. Tang, X.; Zhang, H.; Mou, L.; Liu, F.; Zhang, X.; Zhu, X.X.; Jiao, L. An Unsupervised Remote Sensing Change Detection Method Based on Multiscale Graph Convolutional Network and Metric Learning. *IEEE Trans. Geosci. Remote Sens.* **2021**, *60*, 1–15. [CrossRef]
66. Lv, Z.; Wang, F.; Liu, T.; Kong, X.; Benediktsson, J.A. Novel Automatic Approach for Land Cover Change Detection by Using VHR Remote Sensing Images. *IEEE Geosci. Remote Sens. Lett.* **2021**, *19*, 1–5. [CrossRef]

*remote sensing*

*Article*

# EfficientUNet+: A Building Extraction Method for Emergency Shelters Based on Deep Learning

Di You [1,2,†], Shixin Wang [1,2,†], Futao Wang [1,2,3,*], Yi Zhou [1,2], Zhenqing Wang [1,2], Jingming Wang [1,2] and Yibing Xiong [1,2]

1   Aerospace Information Research Institute, Chinese Academy of Sciences, Beijing 100094, China; youdi@aircas.ac.cn (D.Y.); wangsx@radi.ac.cn (S.W.); zhouyi@radi.ac.cn (Y.Z.); wangzhenqing19@mails.ucas.ac.cn (Z.W.); wangjingming19@mails.ucas.ac.cn (J.W.); xiongyibing19@mails.ucas.ac.cn (Y.X.)
2   University of Chinese Academy of Sciences, Beijing 100049, China
3   Key Laboratory of Earth Observation of Hainan Province, Hainan Aerospace Information Research Institute, Sanya 572029, China
*   Correspondence: wangft@aircas.ac.cn; Tel.: +86-134-264-025-82
†   These authors contributed equally to this work.

**Abstract:** Quickly and accurately extracting buildings from remote sensing images is essential for urban planning, change detection, and disaster management applications. In particular, extracting buildings that cannot be sheltered in emergency shelters can help establish and improve a city's overall disaster prevention system. However, small building extraction often involves problems, such as integrity, missed and false detection, and blurred boundaries. In this study, EfficientUNet+, an improved building extraction method from remote sensing images based on the UNet model, is proposed. This method uses EfficientNet-b0 as the encoder and embeds the spatial and channel squeeze and excitation (scSE) in the decoder to realize forward correction of features and improve the accuracy and speed of model extraction. Next, for the problem of blurred boundaries, we propose a joint loss function of building boundary-weighted cross-entropy and Dice loss to enforce constraints on building boundaries. Finally, model pretraining is performed using the WHU aerial building dataset with a large amount of data. The transfer learning method is used to complete the high-precision extraction of buildings with few training samples in specific scenarios. We created a Google building image dataset of emergency shelters within the Fifth Ring Road of Beijing and conducted experiments to verify the effectiveness of the method in this study. The proposed method is compared with the state-of-the-art methods, namely, DeepLabv3+, PSPNet, ResUNet, and HRNet. The results show that the EfficientUNet+ method is superior in terms of Precision, Recall, F1-Score, and mean intersection over union (mIoU). The accuracy of the EfficientUNet+ method for each index is the highest, reaching 93.01%, 89.17%, 91.05%, and 90.97%, respectively. This indicates that the method proposed in this study can effectively extract buildings in emergency shelters and has an important reference value for guiding urban emergency evacuation.

**Keywords:** deep learning; emergency shelter; building extraction; Google Image; transfer learning; EfficientUNet+

**Citation:** You, D.; Wang, S.; Wang, F.; Zhou, Y.; Wang, Z.; Wang, J.; Xiong, Y. EfficientUNet+: A Building Extraction Method for Emergency Shelters Based on Deep Learning. *Remote Sens.* **2022**, *14*, 2207. https://doi.org/10.3390/rs14092207

Academic Editors: Yue Wu, Kai Qin, Qiguang Miao and Maoguo Gong

Received: 8 April 2022
Accepted: 2 May 2022
Published: 5 May 2022

**Publisher's Note:** MDPI stays neutral with regard to jurisdictional claims in published maps and institutional affiliations.

## 1. Introduction

Extracting buildings is of great significance for applications such as urban planning, land use change, and environmental monitoring [1,2], particularly for buildings in emergency shelters. This process helps improve disaster prevention and mitigation and other management capabilities [3]. An emergency shelter is a safe place for emergency evacuation and temporary dwelling for residents in response to sudden disasters such as earthquakes [4]. These temporary facilities mainly include open spaces, such as parks, green spaces, stadiums, playgrounds, and squares [5]. When disasters occur, buildings are prone to collapse and can injure people [6]. Some areas cannot be used for evacuation. Therefore, extracting buildings from emergency shelters has important guiding relevance in evaluating the emergency evacuation capabilities of shelters.

In the early days, the building footprint in emergency shelters was mainly obtained by manual measurement. The spatial resolution of satellite remote sensing images has reached the submeter level with the development of Earth observation technology. High-resolution remote sensing images have the advantages of rich ground object information, multiple imaging spectral bands, and short revisit time [7–9]. Thus, these images can accurately show the details of urban areas, providing critical support for extracting buildings. Despite the detailed information that these images provide, spectral errors, such as "intra-class spectral heterogeneity" and "inter-class spectral homogeneity", exist [10]. These errors increase the difficulty of building extraction. Moreover, buildings have various features, such as shapes, materials, and colors, complicating the quick and accurate extraction of buildings from high-resolution remote sensing images [11,12].

The traditional methods of extracting buildings based on remote sensing images mainly include image classification based on pixel features and object-oriented classification. The extraction methods based on pixel features mainly rely on the information of a single pixel for classification; these methods include support vector machine and morphological building index, which are relatively simple and efficient to use [13]. However, they ignore the relationship between adjacent pixels and lack the use of spatial information of ground objects. They are prone to "salt and pepper noise", resulting in the blurred boundaries of the extracted buildings [14]. Based on object-oriented extraction methods, pixels are clustered according to relative homogeneity to form objects for classification, utilizing spatial relationships or context information to obtain good classification accuracy [15]. However, classification accuracy largely depends on image segmentation results, and the segmentation scale is difficult to determine; thus, problems such as oversegmentation or undersegmentation are prone to occur [16], resulting in complex object-oriented classification methods.

Deep learning has a strong generalization ability and efficient feature expression ability [17]. It bridges the semantic gap, integrates feature extraction and image classification, and avoids preprocessing, such as image segmentation, through the hierarchical end-to-end construction method. It can also automatically perform hierarchical feature extraction on massive raw data, reduce the definition of feature rules by humans, lessen labor costs and solve problems such as the inaccurate representation of ground objects caused by artificially designed features [18,19]. With the rapid development of artificial intelligence technology in recent years, deep learning has played a prominent role in image processing, change detection, and information extraction. It has been widely used in building extraction, and the extraction method has been continuously improved.

Convolutional neural network (CNN) is the most widely used method for structural image classification and change detection [20]. CNN can solve the problems caused by inaccurate empirically designed features by eliminating the gap between different semantics; it can also learn feature representations from the data in the hierarchical structure itself [21], improving the accuracy of building extraction. Tang et al. [22] proposed to use the vector "capsule" to store building features. The encoder extracts the "capsule" from the remote sensing image, and the decoder calculates the target building, which not only realizes the effective extraction of buildings, but also has good generalization. Li et al. [23] used the improved faster regions with a convolutional neural network (R-CNN) detector; the spectral residual method is embedded into the deep learning network model to extract the rural built-up area. Chen et al. [24] used a multi-scale feature learning module in CNN to achieve better results in extracting buildings from remote sensing images. However, CNN requires ample storage space, and repeated calculations lead to low computational efficiency. Moreover, only some local features can be extracted, limiting the classification performance.

Fully convolutional neural network (FCN) is an improvement based on CNN. It uses a convolutional layer to replace the fully connected layer after CNN; it also realizes end-to-end semantic segmentation for the first time [25]. FCN fuses deep and shallow features of the same resolution to recover the spatial information lost during feature extraction [26]. It is widely used in image semantic segmentation. Bittner et al. [27] proposed an end-to-end FCN method based on the automatic extraction of relevant features and dense image classification. Their proposed method effectively combines spectral and height information from different data sources (high-resolution imagery and digital surface model, DSM). Moreover, the network increases additional connections, providing access to high-frequency information for the top-level classification layers and improving the spatial resolution of building outline outputs. Xu et al. [28] pointed out that the FCN model can detect different classes of objects on the ground, such as buildings, curves of roads, and trees, and predict their shapes. Wei et al. [29] introduced multiscale aggregation and two postprocessing strategies in FCN to achieve accurate binary segmentation. They also proposed a specific, robust, and effective polygon regularization algorithm to convert segmented building boundaries into structured footprints for high building extraction accuracy. Although FCN has achieved good results in building extraction, it does not consider the relationship between pixels. It also focuses mainly on global features and ignores local features, resulting in poor prediction results and a lack of edge information. However, FCN is symbolic in the field of image semantic segmentation, and most of the later deep learning network models are improved and innovated based on it.

The UNet network model belongs to one of the FCN variants. It adds skip connections between the encoding and decoding of FCN. The decoder can receive low-level features from the encoder, form outputs, retain boundary information, fuse high- and low-level semantic features of the network, and achieve good extraction results through skip connections [30]. In recent years, many image segmentation algorithms have used the UNet network as the original segmentation network model, and these algorithms have been fine-tuned and optimized on this basis. Ye et al. [31] proposed RFN-UNet, which considers the semantic gap between features at different stages. It also uses an attention mechanism to bridge the gap between feature fusions and achieves good building extraction results in public datasets. Qin et al. [32] proposed a network structure $U^2$Net with a two-layer nested UNet. This model can capture a large amount of context information and has a remarkable effect on change detection. Peng et al. [33] used UNet++ as the backbone extraction network and proposed a differentially enhanced dense attention CNN for detecting changes in bitemporal optical remote sensing images. In order to improve the spatial information perception ability of the network, Wang et al. [34] proposed a building method, B-FGC-Net, with prominent features, global perception, and cross-level information fusion. Wang et al. [35] combined UNet, residual learning, atrous spatial pyramid pooling, and focal loss, and the ResUNet model was proposed to extract buildings. Based on refined attention pyramid networks (RAPNets), Tian et al. [36] embedded salient multi-scale features into a convolutional block attention module to improve the accuracy of building extraction.

Most of the above methods of extracting buildings are performed on standard public datasets or large-scale building scenarios. They rarely involve buildings in special scenarios, such as emergency shelters. The volume and footprint of buildings in emergency shelters are generally small. For such small buildings, UNet [30] structure can integrate high- and low-level features effectively and restore fine edges, thereby reducing the problems of missed and false detection and blurred edges during building extraction. We use UNet as the overall framework to design a fully convolutional neural network, namely, the EfficientUNet+ method. We verify this method by taking an emergency shelter within the Fifth Ring Road of Beijing as an example. The innovations of the EfficientUNet+ method are as follows:

(1)   We use EfficientNet-b0 as the encoder to trade off model accuracy and speed. The features extracted by the model are crucial to the segmentation results; we also embed the spatial and channel squeeze and excitation (scSE) in the decoder to achieve positive correction of features.

(2)   The accurate boundary segmentation of positive samples in the segmentation results has always been a challenge. We weight the building boundary area with the cross-entropy function and combine the Dice loss to alleviate this problem from the perspective of the loss function.

(3)   Producing a large number of samples for emergency shelters within the Fifth Ring Road of Beijing is time-consuming and labor-intensive. We use the existing public WHU aerial building dataset for transfer learning to achieve high extraction accuracy using a few samples. It can improve the computational efficiency and robustness of the model.

This paper is organized as follows: Section 2 "Methods" introduces the EfficientUNet+ model overview, which includes EfficientNet-b0, scSE module, loss function, and transfer learning; Section 3 "Experimental Results" presents the study area and data, experimental environment and parameter settings, and accuracy evaluation and experimental results of the EfficientUNet+ method; Section 4 "Discussion" validates the effectiveness of the proposed method through comparative experiments and ablation experiments; Section 5 "Conclusion" presents the main findings of this study.

## 2. Methods

In this study, the method of deep learning was used to extract buildings in a special scene, emergency shelters. Given that the buildings in the emergency shelters are generally small, the use of high-resolution remote sensing images to extract buildings is prone to the problems of missed mentions and false and blurred boundaries. Based on the UNet model, EfficientUNet+, an improved building extraction method from high-resolution remote sensing images, was proposed. Beijing's Fifth Ring Road emergency shelters comprised the research area. Figure 1 shows the technical route of this study.

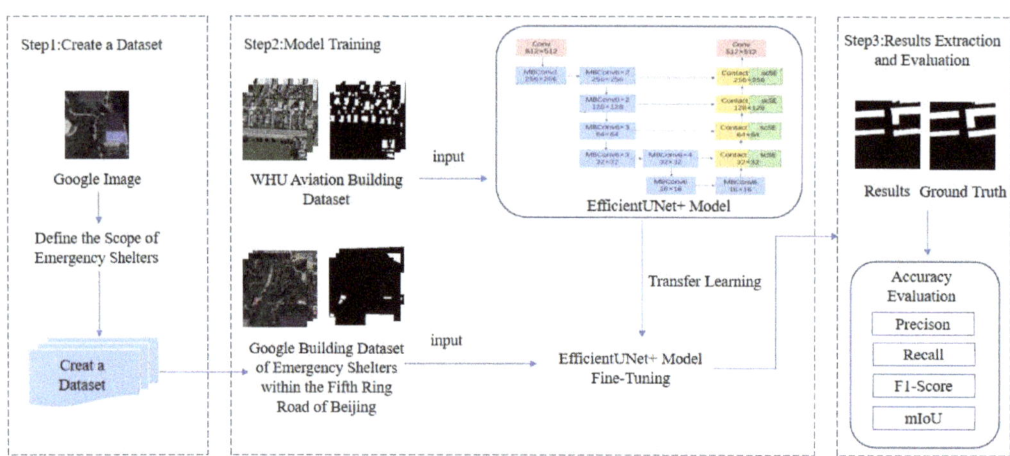


**Figure 1.** The technical route of this study.

### 2.1. EfficientUNet+ Module Overview

The UNet model is an encoder–decoder architecture, which consists of a compressed path for capturing context and a symmetric expansion path for precise localization. It uses skip connections to fuse the high- and low-level semantic information of the network [37]. Good segmentation results can be obtained when the training set is small. However, the original UNet model uses VGG-16 as the encoder, which has many model parameters, and the feature learning ability is weak. This study follows the model framework of UNet, applies EfficientNet in the UNet encoder, and proposes a deep learning-based method for extracting buildings in emergency shelters, namely, EfficientUNet+. Figure 2 shows the

EfficientUNet+ module structure. The emergency shelters within the Fifth Ring Road of Beijing were taken as the research area to verify the effectiveness of the method in this study. The method is improved as follows. (1) The deep learning model used by the encoder is EfficientNet-b0, which is a new model developed using composite coefficients to scale the three dimensions of width/depth/resolution and achieves satisfactory classification accuracy with few model parameters and fast inference [38,39]. (2) The scSE is embedded in the decoder. Embedding spatial squeeze and excitation (sSE) into low-level features can emphasize salient location information and suppress background information; combining channel squeeze and excitation (cSE) with high-level features extracts salient meaningful information [40], thereby reducing false lifts of buildings. (3) The cross-entropy function is used to weigh the boundary area, improving the accuracy of building boundary extraction. The Dice loss is combined to solve the problem of blurred boundary extraction. (4) Given the small number of samples in the study area, a transfer learning method is introduced to transfer the features of the existing WHU aerial building dataset to the current Beijing Fifth Ring Road emergency shelter building extraction task, thereby reducing the labor cost of acquiring new samples and further improving the accuracy of building extraction.

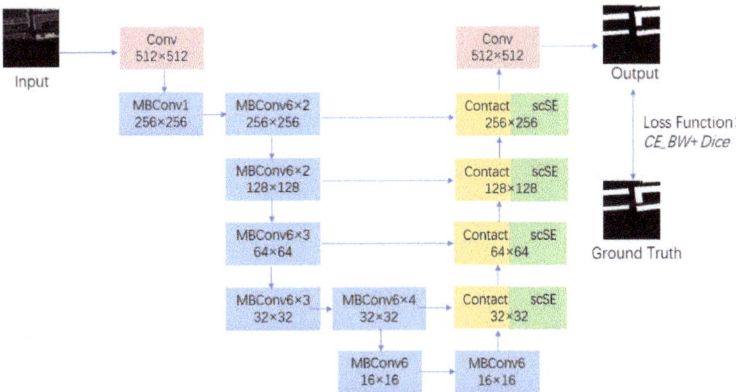


**Figure 2.** EfficientUNet+ module structure.

### 2.2. EfficientNet-b0

In 2019, the EfficientNet model proposed by Google made a major breakthrough in the field of image classification. The network model was applied to the ImageNet dataset and showed superior performance. The model uses compound coefficients to scale the three dimensions of network depth (depth), network width (width), and input image resolution (resolution) uniformly; thus, the optimal classification effect can be obtained by balancing each dimension [41]. Compared with traditional methods, this network model has a small number of parameters and can learn the deep semantic information of images, greatly improving the accuracy and efficiency of the model [37,39]. EfficientNet also has good transferability [42].

The EfficientNet network consists of a multiple-module mobile inversion bottleneck (MBConv) with a residual structure. Figure 3 shows the MBConv structure. The MBConv structure includes $1 \times 1$ convolution layer (including batch normalization (BN) and Swish), $k \times k$ DepthwiseConv convolution (including BN and Swish; the value of $k$ is 3 or 5), squeeze and excitation (SE) module, common $1 \times 1$ convolutional layer (including BN), and dropout layer. This structure can consider the number of network parameters while enhancing the feature extraction capability.

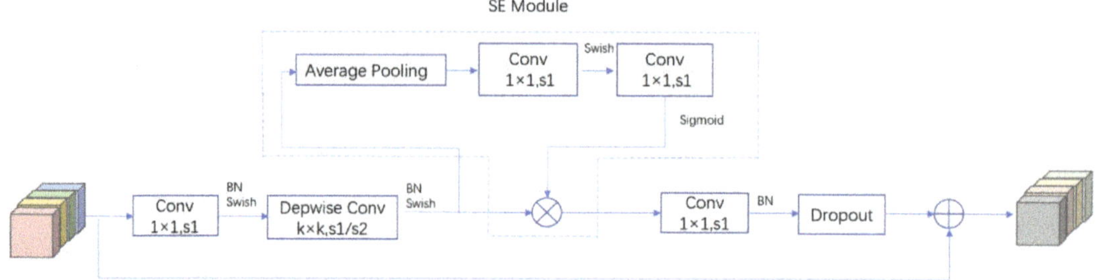

**Figure 3.** MBConv structure.

EfficientNet-b0 is a baseline architecture for lightweight networks in image classification [42]. As shown in Table 1, EfficientNet-b0 consists of nine stages. Stage 1 consists of $3 \times 3$ convolution kernels with a stride of 2. Stages 2 to 8 consist of repeated stacking of MBConv, and the column parameter layers represent the number of times the MBConv is repeated. Stage 9 consists of a $1 \times 1$ convolution kernel, average pooling, and a fully connected layer. Each MBConv in the table is followed by number 1 or number 6 These numbers are the magnification factors. In particular, the first convolutional layer in the MBConv expands the channels of the input feature map to n times the original. $k3 \times 3$ or $k5 \times 5$ represents the size of the convolution kernel in the DepthwiseConv convolutional layer in MBConv. Resolution represents the size of the feature map output by this stage.

**Table 1.** Network structure of EfficientNet-b0.

| Stage | Operator | Resolution | Layers |
|---|---|---|---|
| 1 | Conv $3 \times 3$ | $512 \times 512$ | 1 |
| 2 | MBConv1, $k3 \times 3$ | $256 \times 256$ | 1 |
| 3 | MBConv6, $k3 \times 3$ | $256 \times 256$ | 2 |
| 4 | MBConv6, $k5 \times 5$ | $128 \times 128$ | 2 |
| 5 | MBConv6, $k3 \times 3$ | $64 \times 64$ | 3 |
| 6 | MBConv6, $k5 \times 5$ | $32 \times 32$ | 3 |
| 7 | MBConv6, $k5 \times 5$ | $32 \times 32$ | 4 |
| 8 | MBConv6, $k3 \times 3$ | $16 \times 16$ | 1 |
| 9 | Conv$1 \times 1$&Pooling&FC | $8 \times 8$ | 1 |

The EfficientNetb1-b7 series of deep neural networks chooses the most suitable one in width (the number of channels of the feature map), depth (the number of convolutional layers), and resolution (the size of the feature map) according to the depth, width, and resolution of EfficientNet-b0. The basic principle is that increasing the depth of the network can obtain rich and complex features. This approach can be applied to other tasks. However the gradient disappears, the training becomes difficult, and the time consumption increases if the network depth is too deep. Given that the sample data are relatively small, we used EfficientNet-b0 as the backbone of the segmentation model.

### 2.3. scSE Module

scSE is a mechanism that combines spatial squeeze and excitation (sSE) and channel squeeze and excitation (cSE) [43]. It comprises two parallel modules, namely, the sSE and the cSE. Figure 4 shows the operation flow of the scSE module. This mechanism compresses features and generates weights on channels and spaces, respectively, and then reassigns different weights to increase the attention to the content of interest and ignore unnecessary features [44].

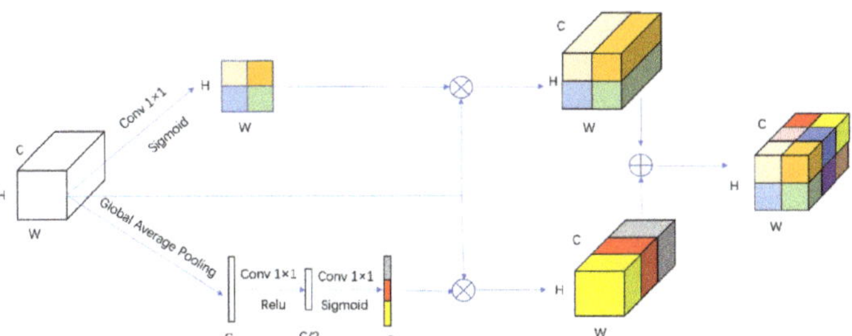

**Figure 4.** Operation flow of scSE module.

sSE is a spatial squeeze and excitation that improves the effectiveness of important features by assigning different weights to different spatial locations on the feature map. First, channel compression is performed on the feature map $(C, H, W)$ using a $1 \times 1$ convolution block with channel $C$ to transform this feature map $(1, H, W)$. Then, the spatial location weights of the features on each channel are generated by normalization by the Sigmoid function. After the reconstruction of the spatial position relationship of the original feature map, a new feature map is finally generated. Equation (1) presents the calculation formulas.

$$\mathbf{U}_{sSE} = [\sigma(q_{1,1})\mathbf{u}^{1,1}, ..., \sigma(q_{i,j})\mathbf{u}^{i,j}, ..., \sigma(q_{H,W})\mathbf{u}^{H,W}] \tag{1}$$

where $\mathbf{U}_{sSE}$ is the new feature map, $\sigma$ is the activation function, $q_{i,j}$ is the linear combination of spatial positions $(i, j)$ under channel $C$, and $\mathbf{u}^{i,j}$ is the spatial location of the feature.

cSE is a channel squeeze and excitation, which generates a channel-reweighted feature map by integrating the weight relationship between different channels. Thus, a channel-reweighted feature map is generated. First, the feature map $(C, H, W)$ is generated by a global average pooling vector $Z \in \mathrm{R}^{C \times 1 \times 1}$, where $C$, $H$, and $W$ represent the channel number, height, and width of the feature map, respectively. The vector $Z$ is operated by two fully connected layers to output a $C \times 1 \times 1$ vector. Then, a weight vector reflecting the importance of different channels is obtained through the Sigmoid function. Finally, the feature map is reweighted to generate a new feature map after feature filtering on the channel. Equations (2)–(5) present the calculation formulas [21].

$$\mathbf{u}_c = \sum_{s=1}^{C'} \mathbf{v}_c^s * \mathbf{x}^s \tag{2}$$

where $\mathbf{u}_c$ is the output feature map; $C'$ and $C$ are the number of input and output channels, respectively; $\mathbf{v}_c$ is the second two-dimensional spatial convolution kernel; $*$ means convolution operation; and $\mathbf{x}^s$ is the sth input feature map.

$$\mathbf{z}_c = \frac{1}{H \times W} \sum_{i=1}^{H} \sum_{j=1}^{W} u_c(i,j) \tag{3}$$

where $\mathbf{z}_c$ is the generated vector through $\mathbf{u}_c$ after global average pooling (squeeze operation), and $H$ and $W$ represent the height and width of the feature map, respectively.

$$\mathbf{s} = \sigma(\mathbf{W}_2 \sigma(\mathbf{W}_1 \mathbf{z})) \tag{4}$$

where $\mathbf{s}$ is the vector output through $\mathbf{z}$ after the excitation operation, $\mathbf{W}_1 \in \mathbf{R}^{\frac{C}{r} \times C}$, $\mathbf{W}_2 \in \mathbf{R}^{C \times \frac{C}{r}}$, and $r$ is the scaling factor. Through the operation, $\mathbf{z}$ converts to $\hat{\mathbf{z}}$ and generates a new feature map as follows:

$$\mathbf{U}_{cSE} = [\sigma(\hat{z}_1)\mathbf{u}_1, \sigma(\hat{z}_2)\mathbf{u}_2, ..., \sigma(\hat{z}_c)\mathbf{u}_c] \tag{5}$$

### 2.4. Loss Function

The loss function is used to calculate the difference between the predicted value and the true value. The network model parameters are updated through the backpropagation of the error. The smaller the loss function value is, the better the model fitting effect is and the more accurate the prediction is [45]. The cross-entropy loss function is the most commonly used loss function in deep learning semantic segmentation. Equation (6) presents the formula of the two-category cross-entropy function.

$$Loss_{CE} = \frac{1}{N}\sum_i -[y_i \cdot \log(p_i) + (1 - y_i) \cdot \log(1 - p_i)] \tag{6}$$

where $y$ is the prediction result and $p$ is the ground truth. The weight of each pixel is equal by considering the cross-entropy function. The boundary area of the building is difficult to segment. We weigh the area's cross-entropy loss from the perspective of the loss function. In backpropagation, the network is enhanced to learn the boundary regions. Equations (7) and (8) present the cross-entropy function formula for boundary weighting.

$$Loss_{CE\_BW} = Loss_{CE} \cdot Weight \tag{7}$$

$$Weight = \begin{cases} 1, not\ boundary \\ w, \quad boundary \end{cases} \tag{8}$$

In this study, the value of $w$ is 4. We introduce Dice loss to alleviate the imbalance in the number of positive and negative samples. Equations (9) and (10) present the final model loss function.

$$Loss = Loss_{CE\_BW} + Loss_{Dice} \tag{9}$$

$$Loss_{Dice} = 1 - \frac{\sum_i |p_i \cap y_i|}{\sum_i (|p_i| + |y_i|)} \tag{10}$$

### 2.5. Transfer Learning

Training often relies on a large amount of sample data to prevent overfitting in the process of training deep learning models. However, collecting sample data by visual interpretation requires a certain amount of experience and knowledge. It is also time-consuming and labor-intensive. In the case of a small number of samples, the existing data can be fully utilized through the transfer learning method. Transfer learning is further tuned by building a pretrained model on the source domain for feature extraction or parameter initialization and applying it to a related but different target domain [46,47]. Compared with training from scratch on a dataset with small sample size, transfer learning can improve computational efficiency and generalization of the model.

Given the complex, diverse, and changeable shapes and colors of target buildings, obtaining a large number of fine samples in the process of extracting buildings from emergency shelters within the Fifth Ring Road of Beijing is difficult even with manual visual interpretation, resulting in a small amount of sample data. Supporting the learning needs of a large number of network parameters is challenging. At present, most of the transfer learning research in the field of remote sensing uses ImageNet dataset for pretraining. However, ImageNet belongs to the field of natural images, and features such as resolution and depth of field are quite different from remote sensing data.

The WHU aerial building dataset is an open large-scale database often used for building extraction. The WHU aerial building dataset is very similar to the requirements of our task. Although the characteristics of the two building datasets are different, 8188 image data with a size of 512 × 512 pixels were obtained through WHU because of the relatively large amount of data in the WHU dataset. The characteristics of the buildings still have great versatility. Therefore, this study used the transfer learning method to pretrain the model based on the WHU aerial building dataset. The pretrained model parameters were used as the initial values of the Beijing building extraction model, effectively increasing the generalization ability of the model on the building dataset of the emergency shelters within the Fifth Ring Road of Beijing.

## 3. Experimental Results

### 3.1. Study Area and Data

#### 3.1.1. Study Area

Beijing is the capital of China, covering an area of 16.4 km$^2$, with a resident population of 21.893 million [48]. It has become a distribution center of population, economy, and resources in the country. It also has an important geographical location in the country and even the world. Beijing is located at 39°26′N–41°03′N, 115°25′E–117°30′E, in the Yinshan–Yanshan seismic zone. It is one of the only three capitals in the world located in an area with a high earthquake intensity of magnitude 8. It is a key fortified city for disaster prevention and mitigation in the country. The central urban area of Beijing has dense buildings, a concentrated population, and the coexistence of old and new buildings. Once a disaster occurs, the damage caused by casualties and economic losses in this city is far greater than that in other areas. Therefore, the emergency shelters within the Fifth Ring Road of Beijing were selected as the research area, including parks, green spaces, squares, stadiums, playgrounds, and other outdoor open spaces. Among the emergency shelter types, the park exhibits large types and numbers of buildings. Thus, only the extraction of buildings in the park's emergency shelters is considered in this study. According to the list of emergency shelters published by the Beijing Earthquake Administration and the list of registered parks published by the Beijing Municipal Affairs Resources Data Network, the Fifth Ring Road of Beijing has 118 parks that can be used as emergency shelters. Figure 5 shows the spatial distribution of park emergency shelter sites within the Fifth Ring Road of Beijing.

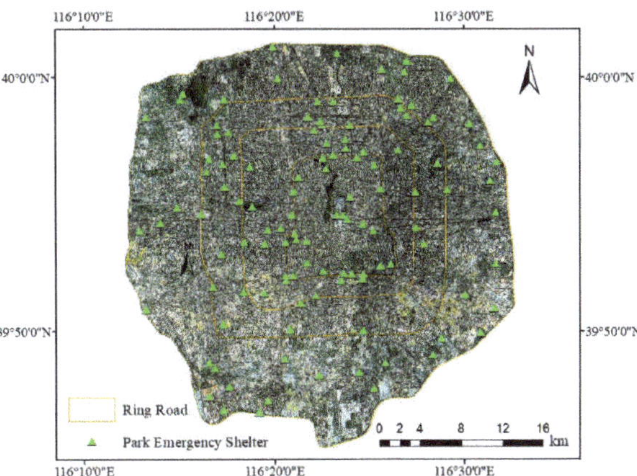


**Figure 5.** Spatial distribution of park emergency shelter sites within the Fifth Ring Road of Beijing.

#### 3.1.2. Dataset

The WHU aerial building dataset was used in this study to pretrain the model. Then, the created Google building dataset of emergency shelters within the Fifth Ring Road of

Beijing was used to verify the effectiveness of the proposed method. Partial details of the two datasets are shown in Figure 6.

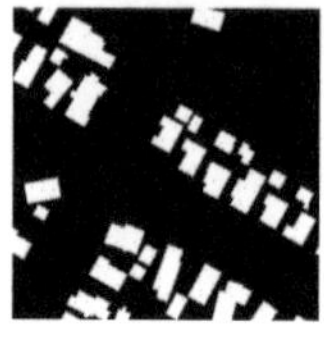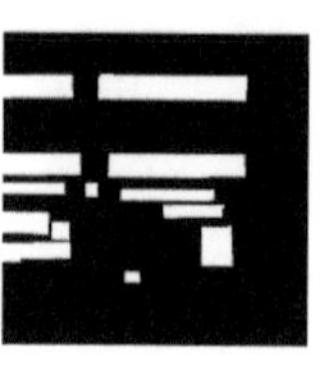

(**a**)                                          (**b**)

**Figure 6.** Part of the details of WHU aerial building dataset and Google building dataset of emergency shelters within the Fifth Ring Road of Beijing. (**a**) WHU aerial building dataset. (**b**) Google building dataset of emergency shelters within the Fifth Ring Road of Beijing.

WHU aerial building dataset: The WHU dataset is divided into an aerial building dataset and a satellite building dataset. Given that the data used in this study are Google Images, the WHU aerial building dataset is similar to Google Image features. Thus the standard open-source high-resolution WHU aerial dataset was used in this study as the training sample for transfer learning. The dataset was acquired in New Zealand, covering 220,000 buildings of different shapes, colors, and sizes, with an area of 450 km$^2$. The initial spatial resolution of the image is 0.075 m. Considering the memory and operating efficiency of the computer, Ji et al. [49] downsampled the spatial resolution of the image to 0.3 m and cropped the image in the area to a size of 512 × 512 pixels, forming an image dataset with 8188 images, including 4736 in the training set, 1036 in the validation set, and 2416 in the test set.

Google building dataset of emergency shelters within the Fifth Ring Road of Beijing The dataset uses Google's high-resolution remote sensing imagery with a spatial resolution of 0.23 m. We selected 21 typical parks with varying image sizes using expert visual interpretation to produce ground truth values for model training and evaluation. The 21 images and the corresponding ground truth values were cropped by the sliding window method to obtain 1110 image blocks with a size of 512 × 512 pixels. A total of 710 images were randomly selected as the training set for model parameter tuning, 178 images were used as the validation set for model parameter selection, and 222 images were used as the test set to evaluate the effect of the final model.

### 3.2. Experimental Environment and Parameter Settings

The experimental platform uses an Intel Core i7-8700@3.20 GHz 6-core processor equipped with 32.0 G memory and an Nvidia GeForce RTX 3090. In terms of the software environment, we used the Windows 10 Professional Edition 64-bit operating system The programming language is Python 3.7, the model building tool is PyTorch 1.7, and the graphics processing unit (GPU) computing platform is CUDA 11.0.

During model training, the batch size was set to 32, the initial learning rate was set to 0.001, the learning rate was adjusted by cosine annealing (the minimum learning rate is 0.001), the optimizer used Adam with weight decay (weight decay coefficient is 0.001) the number of iteration rounds was 120 epochs, and the model parameters corresponding to the rounds with the highest accuracy in the validation set were selected as the final model parameters. In addition, data augmentation operations of horizontal flip, vertical flip, diagonal flip, and 90-degree rotation were performed on the training data.

### 3.3. Accuracy Evaluation

This study used four indicators, Precision, Recall, F1-Score, and mean intersection over union (mIoU), to assess the building extraction accuracy and quantitatively evaluate the performance of the proposed method in extracting buildings [50,51]. Precision represents the proportion of the number of correctly predicted building pixels to the number of pixels whose prediction result is a building. Precision also focuses on evaluating whether the result is misjudged. Recall represents the proportion of the correctly predicted building pixels to the real building pixels. It focuses on evaluating whether the results have omissions. The F1-Score combines the results of Precision and Recall. It is the harmonic mean of Precision and Recall. The mIoU calculates the intersection ratio of each class and then accumulates the average. The mIoU also represents the ratio of the number of predicted building pixels to the intersection and union of the two sets of real buildings, that is, the overlap ratio of the predicted map and the label map. Equations (11)–(14) present the calculation formulas.

$$Precision = TP/(TP + FP) \tag{11}$$

$$Recall = TP/(TP + FN) \tag{12}$$

$$F1 = 2 \times Precision \times Recall/(Precision + Recall) \tag{13}$$

$$mIoU = \frac{1}{k}\sum\nolimits_{i=0}^{k}[TP/(FN + FP + TP)] \tag{14}$$

where $TP$ means that the predicted building is correctly identified as a building; $FP$ means that the predicted building is misidentified as a building; $TN$ means that the predicted non-buildings are correctly identified as non-buildings; $FN$ means real buildings are wrongly identified as non-buildings; and $k$ is the number of categories.

### 3.4. Experimental Results

The EfficientUNet+ method proposed in this study was used to pretrain the model of the public dataset WHU aerial buildings. The experiments were conducted on the park emergency shelter buildings in the study area through the transfer learning method. The emergency shelter in Chaoyang Park has a large area and complex building types, shapes, and colors. Therefore, we took the emergency shelter in Chaoyang Park as an example. Figure 7 shows the results of the buildings extracted by the EfficientUNet+ method.

(a)

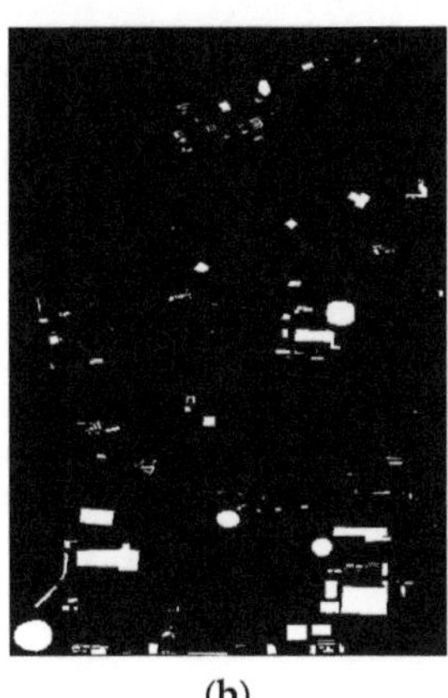
(b)

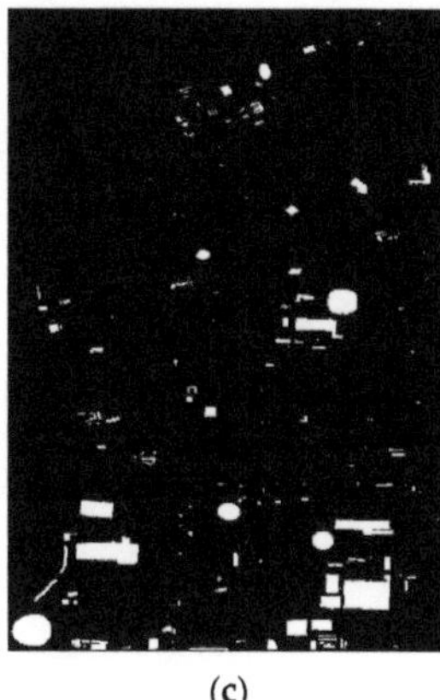
(c)

**Figure 7.** Original image, building ground truth value, and extraction results of the emergency shelter in Chaoyang Park. (**a**) Original image. (**b**) Ground truth. (**c**) Extraction results.

Five local areas of A, B, C, D, and E in the emergency shelter of Chaoyang Park were selected to see the details of the experimental results clearly. Figure 8 shows the original image, the corresponding ground truth, and extraction results.

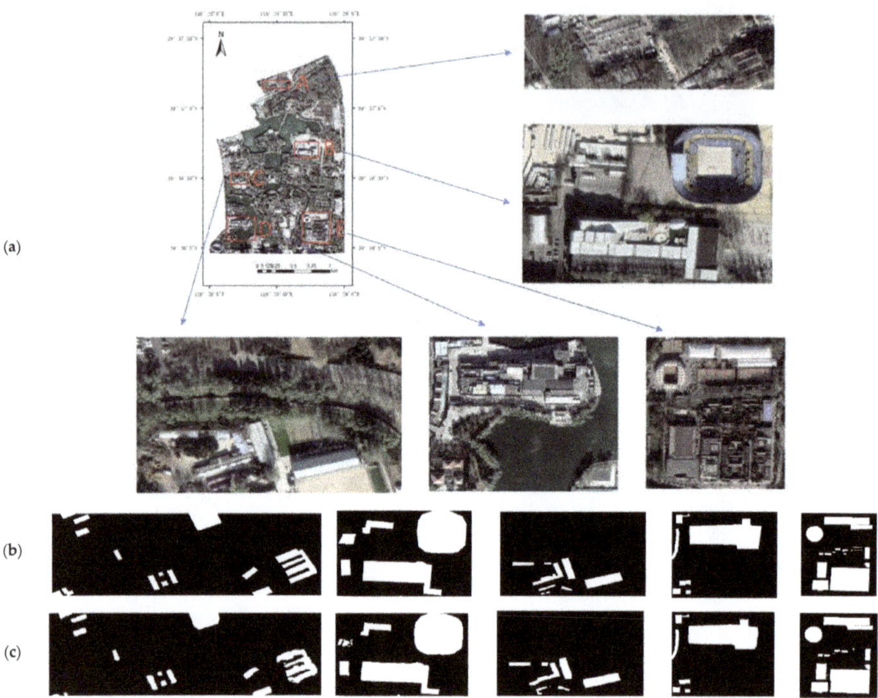

**Figure 8.** Extraction results of buildings in emergency shelters of Chaoyang Park. (**a**) Google image. (**b**) Building ground truth. (**c**) Building extraction results.

Figure 8 shows that the outlines of the buildings in the emergency shelter are all extracted, the boundaries are complete and clearly visible, and only a few occluded buildings have broken boundaries. This observation shows that the EfficientUNet+ method proposed in this study can pay attention to the details in information while obtaining deep semantic information to achieve a complete building image, effectively extracting buildings in remote sensing images.

The four indicators, namely, Precision, Recall, F1-Score, and mIoU, were selected to evaluate the building extraction accuracy by the EfficientUNet+ method proposed in this study. The evaluation results are shown in Table 2.

**Table 2.** Accuracy of EfficientUNet+ method for extracting buildings.

| Precision | Recall | F1-Score | mIoU |
| --- | --- | --- | --- |
| 93.01% | 89.17% | 91.05% | 90.97% |

Table 2 shows the quantitative results of using the EfficientUNet+ method to extract buildings from remote sensing images. The evaluation indicators reach approximately 90%; in particular, the Precision is 93.01%, the Recall is 89.17%, the F1-Score is 91.05%, and the mIoU is 90.97%. This finding indicates that the method can effectively extract buildings in high-resolution remote sensing images.

We further visualize the multi-scale architectural features extracted by the proposed model at different depths, as shown in Figure 9. From Figure 9b–f, we can see that the low-resolution architectural features are gradually refined as the feature resolution increases. The example in column (f) of Figure 9 illustrates that the semantic information of small-scale buildings cannot be captured by high-level features, because they occupy less than one pixel at low resolution.

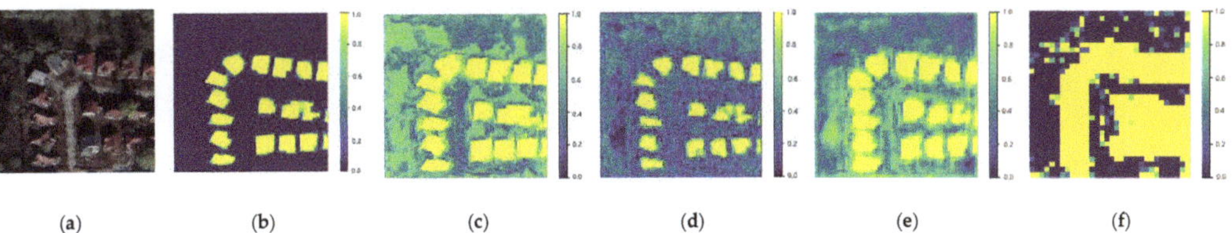

(a)   (b)   (c)   (d)   (e)   (f)

**Figure 9.** Feature map visualization. (**a**) Sample image. (**b**) Depth = 1. (**c**) Depth = 2. (**d**) Depth = 3. (**e**) Depth = 4. (**f**) Depth = 5.

## 4. Discussion

### 4.1. Comparison to State-of-the-Art Studies

To verify whether the proposed method performs better than other state-of-the-art methods, several deep learning methods commonly used in semantic segmentation and building extraction were selected as comparison methods, namely, DeepLabv3+, pyramid scene parsing network (PSPNet), deep residual UNet (ResUNet), and high-resolution Net (HRNet). Among these methods, the DeepLabv3+ method introduces a decoder, which can achieve accurate semantic segmentation and reduce the computational complexity [52]. The PSPNet method extends pixel-level features to global pyramid pooling to make predictions more reliable [53]. The ResUNet method is a variant of the UNet structure with state-of-the-art results in road image extraction [54]. The HRNet method maintains high-resolution representations through the whole process, and its effectiveness has been demonstrated in previous studies [55]. Some detailed images of emergency shelters were selected to compare the extracted accuracy and edge information clearly. Figure 10 shows the results of different methods.

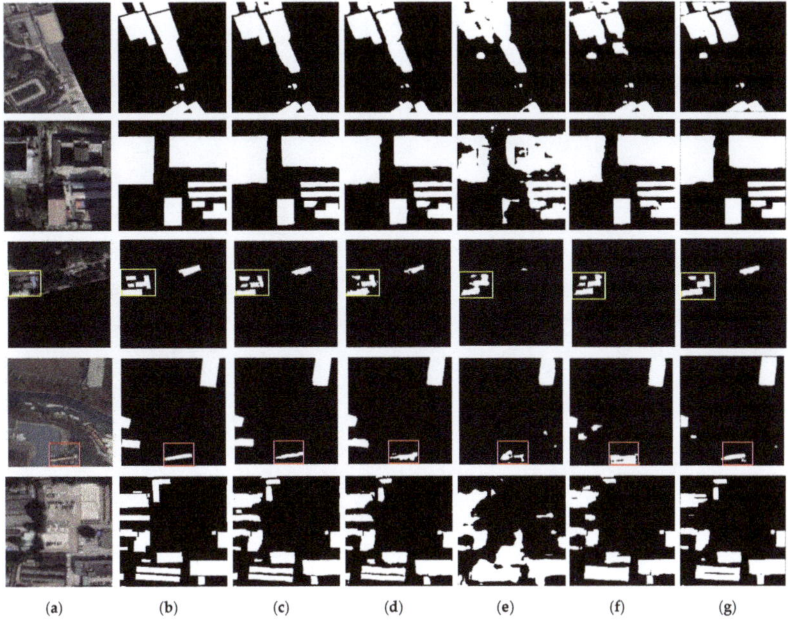

(a)   (b)   (c)   (d)   (e)   (f)   (g)

**Figure 10.** Partial details of the building in the emergency shelter through different methods. (**a**) Original image. (**b**) Ground truth. (**c**) EfficientUNet+. (**d**) DeepLabv3+. (**e**) PSPNet. (**f**) ResUNet. (**g**) HRNet.

Figure 10 shows that compared with other methods, the EfficientUNet+ method extracts almost all the buildings in the image and clearly shows the details, such as the edges and corners of the buildings, closely representing the real objects. The red box in Figure 10 shows that the above methods can extract the approximate location of the building. However, the EfficientUNet+ method can also extract the edge of the building, and its detail retention is higher than that of the other methods. The yellow box in Figure 10 shows that the results of DeepLabv3+, PSPNet, ResUNet, and HRNet methods have areas of misrepresentation and omission, whereas the EfficientUNet+ method can extract buildings more accurately than the other methods.

Four indicators were used to evaluate the extraction results of the EfficientUNet+, DeepLabv3+, PSPNet, ResUNet, and HRNet methods and to quantitatively analyze and evaluate the extraction accuracy. The results are shown in Table 3. The accuracy comparison chart of the extraction results is shown in Figure 11 to intuitively compare the extraction accuracy of each method.

**Table 3.** Accuracy comparison of the extraction results of different methods.

| Methods | Precision | Recall | F1-Score | mIoU |
| --- | --- | --- | --- | --- |
| DeepLabv3+ [52] | 90.52% | 87.15% | 88.80% | 88.92% |
| PSPNet [53] | 76.40% | 75.34% | 75.87% | 78.36% |
| ResUNet [54] | 88.51% | 80.72% | 84.44% | 85.16% |
| HRNet [55] | 89.14% | 83.43% | 86.19% | 86.63% |
| EfficientUNet+ | 93.01% | 89.17% | 91.05% | 90.97% |

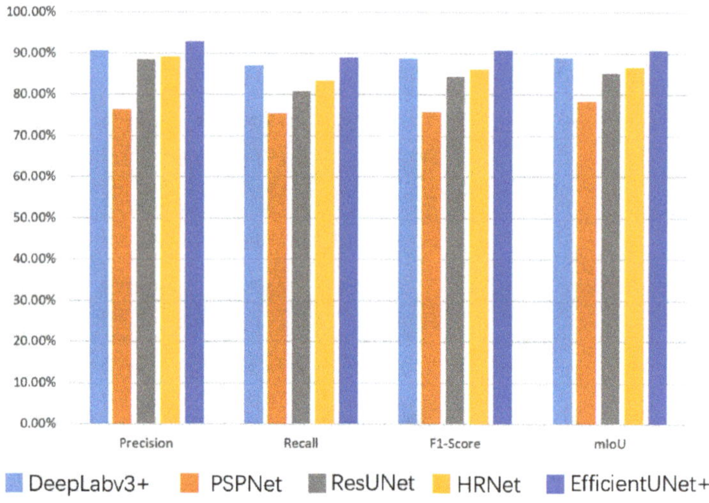


**Figure 11.** Accuracy comparison chart of different methods.

Table 3 and Figure 11 show that the accuracy of the EfficientUNet+ method for extracting buildings is 2.49%, 16.61%, 4.5%, and 3.87% higher than that of DeepLabv3+, PSPNet, ResUNet, and HRNet, respectively. The Recall of the EfficientUNet+ method is 2.02%, 13.83%, 8.45%, and 5.74% higher than that of DeepLabv3+, PSPNet, ResUNet, and HRNet, respectively. The F1-Score of the EfficientUNet+ method is 2.25%, 15.18%, 6.61%, and 4.86% higher than that of DeepLabv3+, PSPNet, ResUNet, and HRNet, respectively. The mIoU of the EfficientUNet+ method is 2.05%, 12.61%, 5.81%, and 4.34% higher than that of DeepLabv3+, PSPNet, ResUNet, and HRNet, respectively. In summary, the EfficientUNet+ method has the highest accuracy in each index, indicating that the EfficientUnet+ method proposed in this study can effectively extract the semantic information of buildings and improve the generalization ability of the model. The proposed method has certain advantages in extracting buildings from remote sensing images.

### 4.2. Ablation Experiment

### 4.2.1. scSE Module

The following ablation experiments were designed in this study to verify the effectiveness of adding the scSE module to the decoder trained by the model: (1) the network model with the scSE; (2) the network model without the scSE. Other experimental conditions are the same. The two methods were applied to the experiments on the building dataset of emergency shelters. The local details of the extraction results are shown in Figure 12. The accuracy comparison is shown in Table 4.

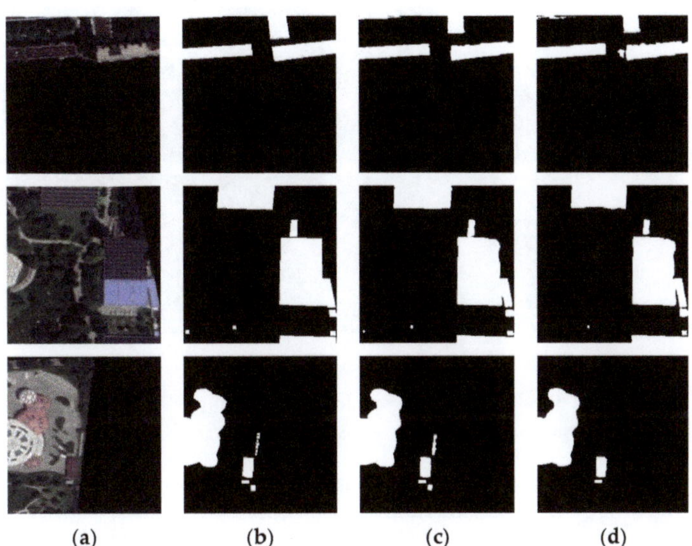

(a)    (b)    (c)    (d)

**Figure 12.** Building extraction results with or without the scSE. (**a**) Original image. (**b**) Ground truth. (**c**) EfficientUNet+. (**d**) EfficientUNet (without scSE).

**Table 4.** Accuracy comparison of extraction results of different decoders.

| Method | Decoder | Precision | Recall | F1-Score | mIoU |
|---|---|---|---|---|---|
| EfficientUNet | Without scSE | 90.81% | 88.23% | 89.50% | 89.54% |
| EfficientUNet+ | With scSE | 93.01% | 89.17% | 91.05% | 90.97% |

Figure 12 shows that the EfficientUNet+ method with the scSE can basically extract all the buildings in the image, whereas the buildings extracted by the EfficientUNet method without the scSE have missed and false detection. Table 4 shows that adding the scSE to the decoder can improve the accuracy of model extraction of buildings. The extraction result analysis shows that the accuracy of each evaluation index after adding the scSE is improved. In particular, the Precision, Recall, F1-Score, and mIoU are increased by 2.2%, 0.94%, 1.55%, and 1.43%, respectively. The scSE added to the decoder enhances the feature learning of the building area, improves the attention of the features of interest, and suppresses the feature response of similar background areas, thereby reducing the false detection of buildings and improving the classification effect.

### 4.2.2. Loss Function

The following ablation experiments were designed in this study to verify the effectiveness of the boundary weighting in the loss function: (1) the cross-entropy function is weighted on the boundary area, and the Dice loss is combined; (2) the regular cross-entropy function and joint Dice loss are used. Other experimental conditions are the same. Figure 13 shows the local details of the extraction results. Table 5 shows the accuracy comparison.

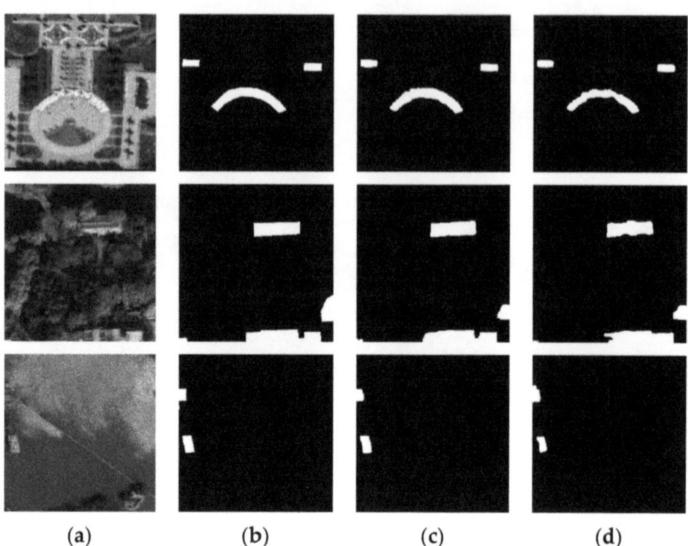

(a)        (b)        (c)        (d)

**Figure 13.** Building extraction results with different loss functions. (**a**) Original image. (**b**) Ground truth. (**c**) $\text{Loss}_{CE\_BW} + \text{Loss}_{Dice}$. (**d**) $\text{Loss}_{CE} + \text{Loss}_{Dice}$.

**Table 5.** Comparison of the accuracy of prediction results of different loss functions.

| Loss Function | Precision | Recall | F1-Score | mIoU |
|---|---|---|---|---|
| $\text{Loss}_{CE} + \text{Loss}_{Dice}$ | 92.07 | 87.39 | 89.67 | 89.71 |
| $\text{Loss}_{CE\_BW} + \text{Loss}_{Dice}$ | 93.01 | 89.17 | 91.05 | 90.97 |

Figure 13 shows the results extracted by the EfficientUNet+ method using boundary-weighted cross-entropy and Dice joint loss function. The boundary of the building is complete, and the edge is clearly visible. However, the buildings extracted by the EfficientUNet+ method without boundary weighting on the loss function have damaged and jagged boundaries. Table 5 shows that the area boundary weighting on the cross-entropy loss function improves the clarity, integrity, and accuracy of the edge details of the buildings in the result. The reason is that the boundary region has a substantial weight in backpropagation. The model also pays considerable attention, alleviating the boundary ambiguity problem of building extraction to a certain extent.

### 4.2.3. Transfer Learning

The following ablation experiments were designed in this study to verify the effectiveness of transfer learning: (1) the EfficientUNet+ method is first pretrained on the WHU aerial building dataset and then adopts transfer learning techniques; (2) the EfficientUNet+ method is directly applied to the Google emergency shelter building dataset. Other experimental conditions are the same. Figure 14 shows the local details of the extraction results. Table 6 shows the accuracy comparison, where "$\sqrt{}$" indicates that the transfer learning technology is used for the experiment and "—" indicates that the transfer learning technology is not used for building extraction.

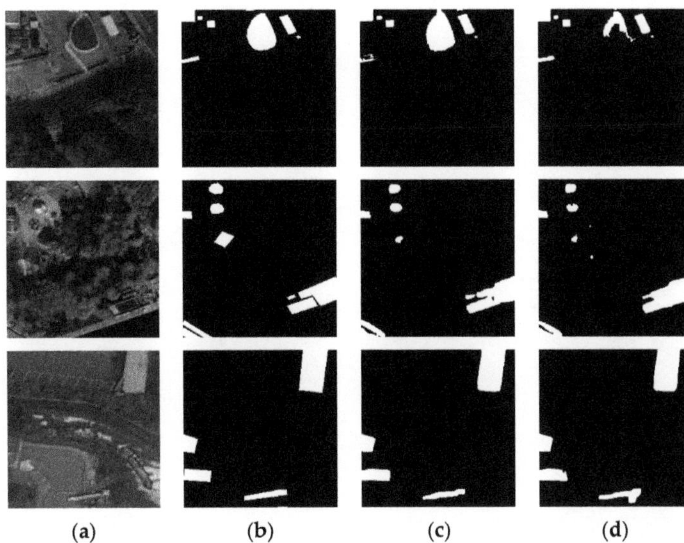

**(a)**       **(b)**       **(c)**       **(d)**

**Figure 14.** Building extraction results with and without transfer learning. (**a**) Original image. (**b**) Ground truth. (**c**) EfficientUNet+ with transfer learning. (**d**) EfficientUNet+ without transfer learning.

**Table 6.** Accuracy comparison of prediction results with and without transfer learning.

| Transfer Learning | Precision | Recall | F1-Score | mIoU |
|---|---|---|---|---|
| — | 92.75% | 88.92% | 90.79% | 90.73% |
| √ | 93.01% | 89.17% | 91.05% | 90.97% |

Figure 14 shows that the pretrained model EfficientUNet+ on the existing public WHU aerial building dataset is applied to the created Google building dataset using the transfer learning technology, thereby increasing the model's ability to extract buildings and its generalization ability. Table 6 shows that the extraction accuracy of the transfer learning technology applied to the real object dataset is high, and the performance is stable. This finding shows that transfer learning can make full use of the existing data information, effectively solve the insufficient number of samples leading to model overfitting, and improve the generalization ability of the network. Thus, it can achieve satisfactory results in information extraction.

### 4.3. Efficiency Evaluation

We visualize the training loss versus epoch in Figure 15. It can be seen that the training loss of the proposed method decreases the fastest, far exceeding other comparison methods, which verifies its efficiency in the training phase. In addition, in order to verify the extraction efficiency of the proposed method, we count the operation time of the validation set, as shown in the table. It can be seen that the inference time and training time of the proposed method are 11.61 s and 279.05 min respectively, which are the shortest and the most efficient of all the compared methods. Table 7 shows that the method proposed in this study can quickly extract the buildings in emergency shelters.

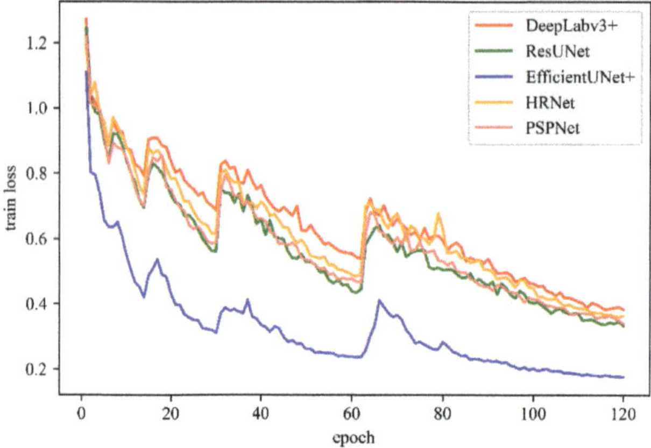

**Figure 15.** Visualization graph of training loss and epochs.

**Table 7.** Operation time of buildings extracted by different methods.

| Time | DeepLabv3+ | PSPNet | ResUnet | HRNet | EfficientUNet+ |
|---|---|---|---|---|---|
| Inference time | 16.31 s | 13.42 s | 15.96 s | 32.05 s | 11.16 s |
| Train time | 362.77 min | 312.82 min | 334.77 min | 427.98 min | 279.05 min |

## 5. Conclusions

Buildings in special scenes, such as emergency shelters, are generally small. The extraction of such small buildings is prone to problems, such as integrity, misrepresentation and omission, and blurred boundaries. An improved deep learning method, EfficientUNet+, is proposed in this study, taking the emergency shelters within the Fifth Ring Road of Beijing as the research area. The effectiveness of the proposed method to extract buildings is verified. The following are the conclusions: (1) EfficientNet-b0 is used as the encoder and the scSE is embedded in the decoder, which can accurately correct the feature map. Thus, the features extracted by the model are conducive to building extraction. (2) The joint loss function of building boundary-weighted cross-entropy and Dice loss can enforce constraints on building boundaries, making the building extraction results close to the ground truth. (3) Transfer learning technology can complete the high-precision extraction of buildings with few training samples in a specific scene background and improve the generalization ability of the model. The Precision, Recall, F1-Score, and mIoU of the EfficientUnet+ method are 93.01%, 89.17%, 91.05%, and 90.97%, respectively. Its accuracy is the highest among all evaluation indicators. This finding shows that the EfficientUnet+ method has suitable performance and advantages in extracting buildings in emergency shelters. The extraction results have guiding relevance in improving urban emergency evacuation capabilities and building livable cities.

However, the model sometimes misses extracting buildings that are obscured by trees. In the future, we will continue to optimize and improve the EfficientUNet+ method try to extract buildings under different phenological conditions in summer and winter and improve the accuracy and performance of remote sensing image building extraction. The method proposed in this study is suitable for optical remote sensing images. In the future, we will try to apply the proposed method to other datasets, such as side-scan sonar to further verify the advantages of this method in small building extraction.

**Author Contributions:** Conceptualization, D.Y., F.W. and S.W.; methodology, Z.W., D.Y., F.W. and S.W.; software, D.Y. and Z.W.; validation, Y.X., F.W. and S.W.; formal analysis, D.Y.; investigation, D.Y.; resources, F.W.; data curation, J.W. and Y.X.; writing—original draft preparation, D.Y.; writing—review and editing, S.W. and Y.Z.; visualization, F.W. and Z.W.; supervision, S.W. and Y.Z.; project administration, D.Y. and S.W.; funding acquisition, D.Y. and S.W. All authors have read and agreed to the published version of the manuscript.

**Funding:** This research was funded by the Finance Science and Technology Project of Hainan Province (no. ZDYF2021SHFZ103) and the National Key Research and Development Program of China (no. 2021YFB3901201).

**Data Availability Statement:** The data presented in this study are available on request from the corresponding author.

**Acknowledgments:** We thank Wuhan University for providing the open access and free aerial image dataset. We would also like to thank the anonymous reviewers and the editors for their insightful comments and helpful suggestions.

**Conflicts of Interest:** The authors declare no conflict of interest.

## Abbreviations

The following abbreviations are used in this manuscript:

| | |
|---|---|
| CNN | Convolutional Neural Network |
| FCN | Fully Convolutional Network |
| DSM | Digital Surface Model |
| GIS | Geographic Information System |
| scSE | Spatial and Channel Squeeze and Excitation |
| sSE | Spatial Squeeze and Excitation |
| cSE | Channel Squeeze and Excitation |
| BN | Batch Normalization |
| SE | Squeeze and Excitation |
| mIoU | Mean Intersection over Union |
| TP | True Positive |
| FP | False Positive |
| FN | False Negative |
| Adam | Adaptive Moment Estimation |
| GPU | Graphics Processing Unit |
| PSPNet | Pyramid Scene Parsing Network |
| ResNet | Residual UNet |
| HRNet | High-Resolution Net |

## References

1. Chen, Q.; Wang, L.; Waslander, S.; Liu, X. An end-to-end shape modeling framework for vectorized building outline generation from aerial images. *ISPRS J. Photogramm. Remote Sens.* **2020**, *170*, 114–126. [CrossRef]
2. Janalipour, M.; Mohammadzadeh, A. Evaluation of effectiveness of three fuzzy systems and three texture extraction methods for building damage detection from post-event LiDAR data. *Int. J. Digit. Earth* **2018**, *11*, 1241–1268. [CrossRef]
3. Melgarejo, L.; Lakes, T. Urban adaptation planning and climate-related disasters: An integrated assessment of public infrastructure serving as temporary shelter during river floods in Colombia. *Int. J. Disaster Risk Reduct.* **2014**, *9*, 147–158. [CrossRef]
4. GB21734-2008; Earthquake Emergency Shelter Site and Supporting Facilities. National Standards of People's Republic of China: Beijing, China, 2008.
5. Jing, J. Beijing Municipal Planning Commission announced the Outline of Planning for Earthquake and Emergency Refuge Places (Outdoor) in Beijing Central City. *Urban Plan. Newsl.* **2007**, *21*, 1.
6. Yu, J.; Wen, J. Multi-criteria Satisfaction Assessment of the Spatial Distribution of Urban Emergency Shelters Based on High-Precision Population Estimation. *Int. J. Disaster Risk Sci.* **2016**, *7*, 413–429. [CrossRef]
7. Hui, J.; Du, M.; Ye, X.; Qin, Q.; Sui, J. Effective Building Extraction from High-Resolution Remote Sensing Images with Multitask Driven Deep Neural Network. *IEEE Geosci. Remote Sens. Lett.* **2018**, *16*, 786–790. [CrossRef]
8. Xu, Z.; Zhou, Y.; Wang, S.; Wang, L.; Li, F.; Wang, S.; Wang, Z. A Novel Intelligent Classification Method for Urban Green Space Based on High-Resolution Remote Sensing Images. *Remote Sens.* **2020**, *12*, 3845. [CrossRef]

9. Dai, Y.; Gong, J.; Li, Y.; Feng, Q. Building segmentation and outline extraction from UAV image-derived point clouds by a line growing algorithm. *Int. J. Digit. Earth* **2017**, *10*, 1077–1097. [CrossRef]
10. Zeng, Y.; Guo, Y.; Li, J. Recognition and extraction of high-resolution satellite remote sensing image buildings based on deep learning. *Neural Comput. Appl.* **2021**, *5*, 2691–2706. [CrossRef]
11. Jing, W.; Xu, Z.; Ying, L. Texture-based segmentation for extracting image shape features. In Proceedings of the 19th International Conference on Automation and Computing (ICAC), London, UK, 13–14 September 2013; pp. 13–14.
12. Huang, Z.; Cheng, G.; Wang, H.; Li, H.; Shi, L.; Pan, C. Building extraction from multi-source remote sensing images via deep deconvo-lution neural networks. In Proceedings of the 2016 IEEE International Geoscience and Remote Sensing Symposium (IGARSS), Beijing, China, 10–15 July 2016; pp. 1835–1838.
13. Blaschke, T. Object based image analysis for remote sensing. *ISPRS J. Photogramm. Remote Sens.* **2010**, *65*, 2–16. [CrossRef]
14. Chen, G.; Zhang, X.; Wang, Q.; Dai, F.; Gong, Y.; Zhu, K. Symmetrical dense-shortcut deep fully convolutional networks for semantic segmentation of very-high-resolution remote sensing images. *IEEE J. Sel. Top. Appl. Earth Obs. Remote Sens.* **2018**, *11*, 1633–1644. [CrossRef]
15. Zhang, J.; Li, T.; Lu, X.; Cheng, Z. Semantic classification of high-resolution remote-sensing images based on mid-level features. *IEEE J. Sel. Top. Appl. Earth Obs. Remote Sens.* **2016**, *9*, 2343–2353. [CrossRef]
16. Gong, M.; Zhan, T.; Zhang, P.; Miao, Q. Superpixel-based difference representation learning for change detection in multispectral remote sensing images. *IEEE Trans. Geosci. Remote Sens.* **2017**, *55*, 2658–2673. [CrossRef]
17. Ren, S.; He, K.; Girshick, R.; Sun, J. Faster R-CNN: Towards Real-Time Object Detection with Region Proposal Networks. *IEEE Trans. Pattern Anal. Mach. Intell.* **2017**, *39*, 1137. [CrossRef]
18. Liu, H.; Luo, J.; Huang, B.; Hu, X.; Sun, Y.; Yang, Y.; Xu, N.; Zhou, N. DE-Net: Deep Encoding Network for Building Extraction from High-Resolution Remote Sensing Imagery. *Remote Sens.* **2019**, *11*, 2380. [CrossRef]
19. Zhu, Q.; Liao, C.; Han, H.; Mei, X.; Li, H. MAPGnet: Multiple attending path neural network for building footprint extraction from remote sensed imagery. *IEEE Trans. Geosci. Remote Sens.* **2021**, *59*, 6169–6181. [CrossRef]
20. Yang, X.; Sun, H.; Fu, K.; Yang, J.; Sun, X.; Yan, M.; Guo, Z. Automatic ship detection in remote sensing images from google earth of complex scenes based on multiscale rotation dense feature pyramid networks. *Remote Sens.* **2018**, *10*, 132. [CrossRef]
21. Jin, Y.; Xu, W.; Zhang, C.; Luo, X.; Jia, H. Boundary-Aware Refined Network for Automatic Building Extraction in Very High-Resolution Urban Aerial Images. *Remote Sens.* **2021**, *13*, 692. [CrossRef]
22. Tang, Z.; Chen, C.; Jiang, C.; Zhang, D.; Luo, W.; Hong, Z.; Sun, H. Capsule–Encoder–Decoder: A Method for Generalizable Building Extraction from Remote Sensing Images. *Remote Sens.* **2022**, *14*, 1235. [CrossRef]
23. Li, S.; Fu, S.; Zheng, D. Rural Built-Up Area Extraction from Remote Sensing Images Using Spectral Residual Methods with Embedded Deep Neural Network. *Sustainability* **2022**, *14*, 1272. [CrossRef]
24. Chen, X.; Qiu, C.; Guo, W.; Yu, A.; Tong, X.; Schmitt, M. Multiscale Feature Learning by Transformer for Building Extraction from Satellite Images. *IEEE Geosci. Remote Sens. Lett.* **2022**, *19*, 2503605. [CrossRef]
25. Long, J.; Shelhamer, E.; Darrell, T. Fully Convolutional Networks for Semantic Segmentation. *arXiv* **2015**, arXiv:1411.4038.
26. Maggiori, E.; Tarabalka, Y.; Charpiat, G.; Alliez, P. Convolutional Neural Networks for Large-Scale Remote-Sensing Image Classification. *IEEE Trans. Geosci. Remote Sens.* **2017**, *55*, 645–657. [CrossRef]
27. Bittner, K.; Adam, F.; Cui, S.; Korner, M.; Reinartz, P. Building footprint extraction from VHR remote sensing images combined with normalized DSMs using fused Fully Convolutional Networks. *IEEE J. Sel. Top. Appl. Earth Obs. Remote Sens.* **2018**, *11*, 2615–2629. [CrossRef]
28. Xu, Y.; Wu, L.; Xie, Z.; Chen, Z. Building Extraction in Very High Resolution Remote Sensing Imagery Using Deep Learning and Guided Filters. *Remote Sens.* **2018**, *10*, 144. [CrossRef]
29. Wei, S.; Ji, S.; Lu, M. Toward automatic building footprint delineation from aerial images using CNN and regularization. *IEEE Trans. Geosci. Remote Sens.* **2020**, *58*, 2178–2189. [CrossRef]
30. Ronneberger, O.; Fischer, P.; Brox, T. U-net: Convolutional networks for biomedical image segmentation. *arXiv* **2015**, arXiv:1505.04597.
31. Ye, Z.; Fu, Y.; Gan, M.; Deng, J.; Comber, A.; Wang, K. Building extraction from very high resolution aerial imagery using joint attention deep neural network. *Remote Sens.* **2019**, *11*, 2970. [CrossRef]
32. Qin, X.; Zhang, Z.; Huang, C.; Dehghan, M.; Zaiane, O.; Jagersand, M. $U^2$Net: Going deeper with nested U-structure for salient object detection. *Pattern Recognit.* **2020**, *106*, 107404. [CrossRef]
33. Peng, X.; Zhong, R.; Li, Z.; Li, Q. Optical remote sensing image change detection based on attention mechanism and image difference. *IEEE Trans. Geosci. Remote Sens.* **2021**, *59*, 7296–7307. [CrossRef]
34. Wang, Y.; Zeng, X.; Liao, X.; Zhuang, D. B-FGC-Net: A Building Extraction Network from High Resolution Remote Sensing Imagery. *Remote Sens.* **2022**, *14*, 269. [CrossRef]
35. Wang, H.; Miao, F. Building extraction from remote sensing images using deep residual U-Net. *Eur. J. Remote Sens.* **2022**, *55*, 71–85. [CrossRef]
36. Tian, Q.; Zhao, Y.; Li, Y.; Chen, J.; Chen, X.; Qin, K. Multiscale Building Extraction with Refined Attention Pyramid Networks. *IEEE Geosci. Remote Sens. Lett.* **2022**, *19*, 1–5. [CrossRef]
37. Cao, D.; Xing, H.; Wong, M.; Kwan, M.; Xing, H.; Meng, Y. A Stacking Ensemble Deep Learning Model for Extraction from Remote Sensing Images. *Remote Sens.* **2021**, *13*, 3898. [CrossRef]

38. Tadepalli, Y.; Kollati, M.; Kuraparthi, S.; Kora, P. EfficientNet-B0 Based Monocular Dense-Depth Map Estimation. *Trait. Signal* **2021**, *38*, 1485–1493. [CrossRef]
39. Zhao, P.; Huang, L. Multi-Aspect SAR Target Recognition Based on Efficientnet and GRU. In Proceedings of the IEEE International Geoscience and Remote Sensing Symposium (IGARSS), Electr Network, Waikoloa, HI, USA, 26 September–2 October 2020; pp. 1651–1654.
40. Alhichri, H.; Alswayed, A.; Bazi, Y.; Ammour, N.; Alajlan, N. Classification of Remote Sensing Images using EfficientNet-B3 CNN Model with Attention. *IEEE Access* **2021**, *9*, 14078–14094. [CrossRef]
41. Ferrari, L.; Dell'Acqua, F.; Zhang, P.; Du, P. Integrating EfficientNet into an HAFNet Structure for Building Mapping in High-Resolution Optical Earth Observation Data. *Remote Sens.* **2021**, *13*, 4361. [CrossRef]
42. Tan, M.; Le, Q. EfficientNet: Rethinking Model Scaling for Convolutional Neural Networks. In Proceedings of the 36th International Conference on Machine Learning(ICML), Long Beach, CA, USA, 9–15 June 2019; p. 97.
43. Roy, A.; Navab, N.; Wachinger, C. Concurrent Spatial and Channel Squeeze & Excitation in Fully Convolutional Networks. *arXiv* **2018**, arXiv:1803.02579.
44. Mondal, A.; Agarwal, A.; Dolz, Z.; Desrosiers, C. Revisiting CycleGAN for semi-supervised segmentation. *arXiv* **2019**, arXiv:1908.11569.
45. Qin, X.; He, S.; Yang, X.; Dehghan, M.; Qin, Q.; Martin, J. Accurate outline extraction of individual building from very high-resolution optical images. *IEEE Geosci. Remote Sens. Lett.* **2018**, *15*, 1775–1779. [CrossRef]
46. Das, A.; Chandran, S. Transfer Learning with Res2Net for Remote Sensing Scene Classification. In Proceedings of the 11th International Conference on Cloud Computing, Data Science and Engineering (Confluence), Amity Univ, Amity Sch Engn & Technol, Electr Network, Noida, India, 28–29 January 2021; pp. 796–801.
47. Zhu, Q.; Shen, F.; Shang, P.; Pan, Y.; Li, M. Hyperspectral Remote Sensing of Phytoplankton Species Composition Based on Transfer Learning. *Remote Sens.* **2019**, *11*, 2001. [CrossRef]
48. *Seventh National Census Communiqué*; National Bureau of Statistics: Beijing, China, 2021.
49. Ji, S.; Wei, S. Building extraction via convolution neural networks from an open remote sensing building dataset. *Arca Geod. Cartogr. Sin.* **2019**, *48*, 448–459.
50. Wang, Y.; Chen, C.; Ding, M.; Li, J. Real-time dense semantic labeling with dual-Path framework for high-resolution remote sensing image. *Remote Sens.* **2019**, *11*, 3020. [CrossRef]
51. Ji, S.; Wei, S.; Lu, M. Fully convolutional networks for multisource building extraction from an open aerial and satellite imagery data set. *IEEE T Rans. Geosci. Remote Sens.* **2018**, *57*, 574–586. [CrossRef]
52. Chen, L.C.E.; Zhu, Y.; Papandreou, G.; Schroff, F.; Adam, H. Encoder-Decoder with Atrous Separable Convolution for Semantic Image Segmentation. In Proceedings of the 15th European Conference on Computer Vision (ECCV), Munich, Germany, 8–14 September 2018; pp. 833–851.
53. Zhao, H.; Shi, J.; Qi, X.; Wang, X.; Jia, J. Pyramid Scene Parsing Network. In Proceedings of the 30th IEEE/CVF Conference on Computer Vision and Pattern Recognition (CVPR), Honolulu, HI, USA, 21–26 July 2017; pp. 6230–6239.
54. Zhang, Z.X.; Liu, Q.J.; Wang, Y.H. Road Extraction by Deep Residual U-Net. IEEE Geosci. *Remote Sens. Lett.* **2018**, *15*, 749–753. [CrossRef]
55. Wang, J.; Sun, K.; Cheng, T.; Jiang, B.; Deng, C.; Zhao, Y.; Liu, D.; Mu, Y.; Tan, M.; Wang, X.; et al. Deep High-Resolution Representation Learning for Visual Recognition. *IEEE Trans. Pattern Anal. Mach. Intell.* **2019**, *43*, 3349–3364. [CrossRef]

*Article*

# Use of a DNN-Based Image Translator with Edge Enhancement Technique to Estimate Correspondence between SAR and Optical Images

Hisatoshi Toriya [1,2,*], Ashraf Dewan [3], Hajime Ikeda [1], Narihiro Owada [1], Mahdi Saadat [1], Fumiaki Inagaki [1], Youhei Kawamura [4] and Itaru Kitahara [2]

1   Faculty of International Resource Sciences, Akita University, 1-1 Tegatagakuen-Machi, Akita-City 0100862, Akita, Japan; ikeda@gipc.akita-u.ac.jp (H.I.); owada@gipc.akita-u.ac.jp (N.O.); mahdi.saadat1@gipc.akita-u.ac.jp (M.S.); fumiaki.inagaki@gipc.akita-u.ac.jp (F.I.)
2   Center for Computational Sciences, University of Tsukuba, 1-1-1 Tennodai, Tsukuba-City 3058577, Ibaraki, Japan; kitahara@ccs.tsukuba.ac.jp
3   School of Earth and Planetary Sciences, Curtin University, Kent St. Bentley, WA 6102, Australia; a.dewan@curtin.edu.au
4   Faculty of Engineering, Hokkaido University, Kita 8, Nishi 5, Kita-Ku, Sapporo-City 0608628, Hokkaido, Japan; kawamura@eng.hokudai.ac.jp
*   Correspondence: toriya@gipc.akita-u.ac.jp

**Abstract:** In this paper, the local correspondence between synthetic aperture radar (SAR) images and optical images is proposed using an image feature-based keypoint-matching algorithm. To achieve accurate matching, common image features were obtained at the corresponding locations. Since the appearance of SAR and optical images is different, it was difficult to find similar features to account for geometric corrections. In this work, an image translator, which was built with a DNN (deep neural network) and trained by conditional generative adversarial networks (cGANs) with edge enhancement, was employed to find the corresponding locations between SAR and optical images. When using conventional cGANs, many blurs appear in the translated images and they degrade keypoint-matching accuracy. Therefore, a novel method applying an edge enhancement filter in the cGANs structure was proposed to find the corresponding points between SAR and optical images to accurately register images from different sensors. The results suggested that the proposed method could accurately estimate the corresponding points between SAR and optical images.

**Keywords:** image registration; keypoint matching; synthetic aperture radar; deep neural network; generative adversarial networks

**Citation:** Toriya, H.; Dewan, A.; Ikeda, H.; Owada, N.; Saadat, M.; Inagaki, F.; Kawamura, Y.; Kitahara, I. Use of a DNN-Based Image Translator with Edge Enhancement Technique to Estimate Correspondence between SAR and Optical Images. *Appl. Sci.* **2022**, *12*, 4159. https://doi.org/10.3390/app12094159

Academic Editor: Yue Wu

Received: 18 March 2022
Accepted: 13 April 2022
Published: 20 April 2022

**Publisher's Note:** MDPI stays neutral with regard to jurisdictional claims in published maps and institutional affiliations.

## 1. Introduction

When a natural disaster such as an earthquake or tsunami occurs, visual information can provide essential data for emergency management. Aerial images obtained from satellites, aircraft, and drones can simultaneously capture a wide range of features; thus they may be utilized for response and recovery operations after a disaster event [1,2]. The higher the shooting altitude, the wider the view; however, the visibility of optical images can deteriorate owing to the lack of a light source or the influence of clouds. Alternatively, synthetic aperture radars (SARs) can capture data over a large area without much deterioration. Therefore, SARs are often used in disaster situations, such as damage area detection [3–5], infrastructure damage assessment [6,7], etc. However, SAR has some problems in practical use, including (a) they are relatively less readable by humans and (b) landmark points may be difficult to locate, especially with coarse resolution data. To obtain geographic information, the combined use of SAR images and optical data can provide a wealth of information. Thus, the geometric registration of the two data is essential, which may be challenging. The accurate registration of SAR images and optical data can provide valuable information, which would otherwise be difficult, especially during natural disasters, such as floods of large magnitude. Hence, the precise registration of optical and microwave data can support rapid scanning of flood areas or the identification of collapsed buildings after an earthquake or tsunami for rescue and evacuation efforts.

The pixel values in SAR images represent the intensity of electromagnetic waves, and they are expressed in backscatter values. This differs from visible light, thereby resulting in varying responses of several ground features. Thus, when handling SAR images, it is common to compensate for their low readability by registering them with optical images.

Image registration is traditionally performed using a digital elevation model (DEM) [8]. However, when the spatial resolution of both the optical and SAR images is high, DEM may not be effective. Moreover, when a disaster event occurs, landforms may be drastically changed such that DEM may not achieve accurate image registration. An alternative method is image-based registration, which is independent of DEM. In this method, complementing factors associated with the acquisition and processing of optical and SAR data are considered because optical images are prone to atmospheric disturbances, especially during disasters such as floods, while SAR can acquire data in any weather. Combining the SAR data acquired during the disaster and the optical images before/after the disaster can make response and recovery operations much easier. Meanwhile, an automated method for the geometric rectification of images from different sensors is not as easy due to variations in the spectral response of ground features. Hence, two datasets can be combined by translating the appearance of a SAR image to an optical image using generative adversarial networks (GANs) [9] for subsequent keypoint matching.

We proposed a method for finding local feature correspondence between multimodal (SAR and optical) images using an image-based feature keypoint extraction, description, and matching algorithm [10], as shown in Figure 1. Image translation with a GAN is used to transform a SAR image into an optical image. Although this method has good accuracy, the blurring of features could be a significant issue (Figure 2). To obtain more corresponding points, blurring should be removed to highlight the local features. However, it is difficult to achieve this with conventional GANs.

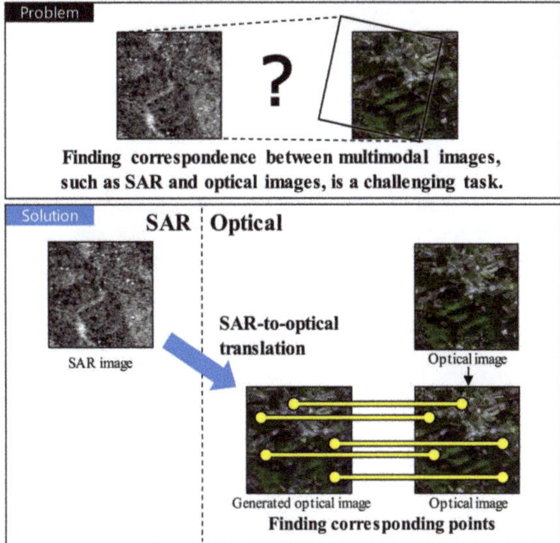


**Figure 1.** Outline of the proposed method. Using a GAN as a pre-processing step before keypoint matching, local correspondence was established for multimodal image registration.

**Figure 2.** An optical image generated by a conditional GAN ([11], **left**) and an optical image (**right**). Note the loss of local features due to blurring.

In this work we proposed a SAR and optical image registration method for overcoming the above-mentioned problem. By training the GAN with an edge-enhancement technique, we could obtain an image translator (generator) with higher quality than a conventional GAN. Furthermore, comparative evaluation was conducted to quantitatively analyze the effectiveness and performance of the proposed method. Meanwhile, local features were necessary for finding the corresponding points, and they could be obtained by applying an edge-enhancement filter. The main contributions of this work are as follows.

a    A novel method was proposed for finding the corresponding points between SAR and optical images for image registration.

b    The SAR-to-optical image translator was improved by training a GAN with edge-enhanced images to maintain local image features and improve keypoint extraction.

c    The efficiency of the proposed method was demonstrated by a comparison experiment with conventional methods and a qualitative experiment.

This paper is organized as follows. Conventional image-registration methods for multimodarl images are discussed in Section 2, as well as image translation methods with GANs. In Section 3, we describe how to train a GAN to obtain a generator that can perform SAR-to-optical-image translation. The methods for finding the corresponding points between the optical and generated optical images are described in Section 4. The experimental setting, results, and discussions are given in Section 5. Finally, the conclusion and major findings of this work are discussed in Section 6.

## 2. Related Work

An image-based (not DEM-based) registration method is expected to match multimodal images based on the similarities of their edges and corners. Thus, template-based methods with traditional metrics, such as normalized cross-correlation (NCC) or mutual information between two images, have been proposed for image registration [12,13]. Meanwhile, other suitable metrics have been proposed for more accurate template-matching of SAR and optical images [14–17]. However, these methods use image features of a relatively wide area. The template-matching accuracy decreases with small template size, whereas its robustness for occlusions and partial difference (e.g., between pre- and post-disaster) decreases with large template size. Although machine-learning-based (especially deep-learning-based) SAR and optical image-matching methods have been proposed [18–20], the range of pixels considered in these methods are whole image correspondences or some of them have a limitation in rotation robustness. The identification of local correspondence is not possible with these methods; hence, a new technique that can perform image registration with local features when applied to disaster sites is required since landform responses may be partially or fully changed after a disaster event.

Local image features are often employed to achieve the image-based registration of satellite data. Particularly, keypoint-based methods [21–23] can estimate the correspondence between two images using uncorrelated local features, similar to template-based methods. Methods that adopt a keypoint detector and a feature descriptor [24–27] are frequently used to match (i.e., estimate the correspondence of) two images. These methods describe features that are robust to geometric fluctuation (e.g., rotation, scaling) and changes in environmental (e.g., lighting) conditions. They find keypoint pairs that have similar features as corresponding points, thereby achieving image matching. Since these methods use local features, they can achieve partial correspondence, even when a part of the captured area has collapsed due to disaster.

Meanwhile, machine learning methods based on deep neural networks (DNNs) have been widely used for image modal translation [28–32]. In this work, we mainly focus on GANs. A GAN trains a generator that generates data and a discriminator that determines the authenticity of the data to produce a generator that can generate data similar to the original features through comparison. The loss function of a GAN, $\mathcal{L}_{\text{GAN}}$, is given as follows:

$$\mathcal{L}_{\text{GAN}}(G, D) = \mathbb{E}_y[logD(y)] + \mathbb{E}_z[log(1 - D(G(z)))], \tag{1}$$

where $G()$ indicates the generated data based on the input data by the generator ($G$), $D()$ indicates the probability that a discriminator ($D$) can correctly discriminate between a real input and a fake (artificial) input, $y$ and $z$ indicate the answer and a random value, respectively. The purpose of training a GAN is to obtain a well-trained generator that can produce sufficient fake data to deceive the discriminator. The generator $G^*$ is given as follows:

$$G^* = arg \min_G \max_D \mathcal{L}_{\text{GAN}}(G, D). \tag{2}$$

An application of a GAN is a conditional GAN (cGAN), whose generator obtains inputs rather than random values. It is a common multipurpose image interpretation method, and its loss function $\mathcal{L}_{\text{cGAN}}$ is given as follows:

$$\mathcal{L}_{\text{cGAN}}(G, D) = \mathbb{E}_y[logD(y)] + \mathbb{E}_{x,z}[log(1 - D(x, G(x, z)))], \tag{3}$$

where $x$ indicates the input. A well-trained generator for the cGAN can be obtained in the same way as Equation (2). It has been demonstrated that cGAN permits multimodal image-to-image translation, e.g., from an artificial room image to a real photo [33], or a sketched image to a real photo [11].

Another advantage of a GAN is that fewer training datasets are required [11]. Although a large number of training datasets is necessary for conventional machine learning methods, GANs can achieve high performance with fewer datasets owing to their generator and discriminator models. This was a crucial advantage for this study because it was difficult to prepare several datasets of aligned SAR and optical images under a disaster situation.

Therefore, we proposed a method for performing SAR-to-optical image translation using a GAN so that a keypoint-matching algorithm could be applied to multimodal images. By applying cGAN to SAR-to-optical image translation, the challenging task of multimodal image registration was reduced to the conventional task of monomodal image registration, which could be solved using a feature-based image-matching algorithm.

Blurring, as shown in Figure 2, is one of the factors that reduces the accuracy of matching. Edge enhancement for super-resolution [34] and that for small object detection [35] have been proposed. Although these methods include edge enhancement in the neural network structures, they increase the complexity of implementation, such as parameter tuning from applying the methods to problems. Therefore, in this paper, we used a neural network structure, which was used in previous studies and its performance had been established. The edge enhancement was used to pre-process the image, which simplified the implementation and improved the performance.

## 3. Training a GAN for SAR-to-Optical Image Translation with Edge Enhancement

In this section, the process of training a GAN to generate optical images (generated optical images) from SAR images is described. The generated optical images were obtained by machine-learning-based prediction using SAR images as input. Figure 3 shows the cGAN training model, in which the generator $G$ and discriminator $D$ are combined to obtain higher-quality generated optical images. In preparing SAR and optical image pairs that were already co-registered, we set the SAR images as input $x$ to the model and $y$ as the correct answer for training $G$ and $D$. With this SAR-to-optical image translation process the generated and original optical images had the same modality, and it was possible to perform image-registration processing with keypoint matching.

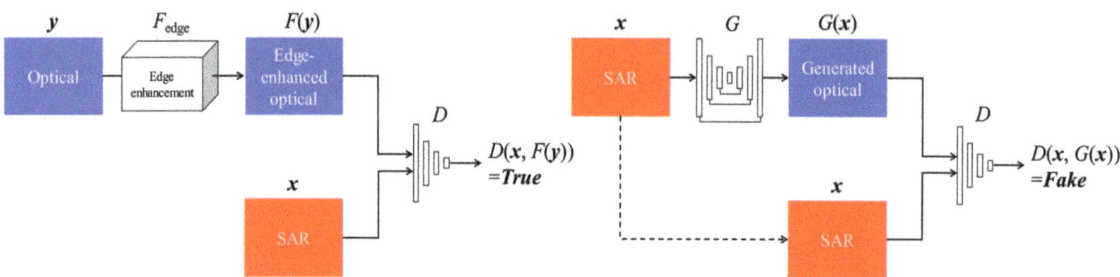


**Figure 3.** Structure of the training generator and discriminator model with an edge enhancement filter of the proposed method.

Since the objective of image translation is to find the corresponding points, important pieces of information in the generated image are the local image features (edges and corners). However, as shown in Figure 2, there were cases where the details were blurred and local features were lost in the cGAN. We solved this problem by proposing a method that applied an edge-enhancement filter, which adjusted pixel values of pixels along edges to emphasize edges, to the training data in advance to enable the cGAN network to learn the edges and corners more actively. The discriminator training is shown in Figure 3.

## 4. Finding Corresponding Points Using the Keypoint Detector and Descriptor

The keypoint-matching process is shown in Figure 4. As mentioned previously, the cGAN training model was used to obtain $G$, which generated images that were similar to the original optical images. Afterwards, keypoint matching was performed between the optical and generated optical images.

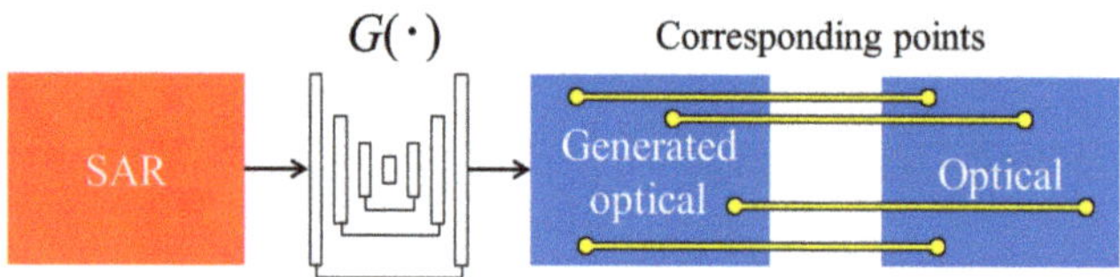


**Figure 4.** The keypoint-matching process. The trained $G$ obtained in Section 2 predicted optical images (generated optical images) from the input of SAR images.

Finding corresponding points consists of three major steps: keypoint detection, keypoint description, and keypoint matching. A typical process using SIFT [24], which is a major algorithm in keypoint matching, is outlined below. In the feature point detection step the DoG (difference of Gaussian) image is used to detect tentative keypoints and the range (scale) is used for feature description. In the subsequent localization step, the sub-pixel positions are estimated by deleting the sub-pixels from the detected tentative keypoints that are not suitable for keypoints. In the feature description step, the direction of each feature point is determined from the bright gradient to obtain rotation-invariant features and 128-dimensional feature vectors are described for each feature point according to the direction and scale. In the final keypoint-matching step, two points in two images with similar feature vectors are extracted as corresponding points.

During the process of finding the corresponding points, false correspondences were also obtained. These false correspondences might decrease the registration accuracy; therefore, a process was required to remove them. In the case of matching between map-projected images, the difference between the two images could be regarded as scale, rotational, and translational transformation. Hence, the false corresponding points were removed based on the scale value and the gradient direction information of the correspondences [36]. Figure 5 shows the results of the corresponding points after eliminating the false correspondences.

**Figure 5.** An example of a local correspondence between two images. The yellow lines are the connectors of corresponding points and the blue points are the points without corresponding points.

## 5. Evaluation of Keypoint Matching

### 5.1. Objective of the Evaluation

The objective of this experiment was to quantitatively evaluate the accuracy of the corresponding points to verify the effectiveness of the proposed method. Specifically, the accuracy of the positions where the corresponding points were extracted was evaluated. We used co-registered SAR and optical images for this experiment. In the evaluation scale, the average of the Euclidean distances $\bar{d}$ of the positions of the corresponding points in the SAR and optical image was used, which is given by the following:

$$\bar{d} = \frac{1}{n} \sum_{i}^{n} \| p_i - p_i' \|,$$ (4)

where $p_i$ and $p_i'$ are the positions of the corresponding points in the optical and generated optical image, respectively. If the registration was perfect, the corresponding points ideally had the same positions in each coordinate; hence, the closer the evaluation scale $\bar{d}$ was to 0, the higher the accuracy.

The results of the proposed method were compared with those of six other methods, which were as follows: (1) DLSC [14]—a method for finding corresponding points using dense local self-similarity; (2) HOPC [15]—a method for finding corresponding points using a histogram of orientated phase congruency (HOPC), which is based on the structural properties of images; (3) CFOG [16]—a method for finding corresponding points using CFOG, which is an extension of the pixel-wise HoG (histogram of Gaussian) descriptor; (4) Pix2pix [11]—a method that uses SIFT [24] for keypoint detection and description, and images obtained by a Pix2pix prediction; (5) Pix2pix + Edge Enhancement (EE)—a method that uses SIFT and edge-enhanced images obtained by applying an edge-enhancement filter to the Pix2pix prediction; and (6) Proposed—our method that uses SIFT for keypoint detection and description, and a discriminator trained by edge-enhanced images.

### 5.2. Environment of the Evaluation

#### 5.2.1. The GAN Structure and Loss Function of the Experiment

Figure 6 shows the network structure used for learning. U-Net [37] was used for the generator, and PatchGAN [11,38] was used for the discriminator. U-Net and PatchGAN, which were used in the original Pix2pix, showed good and stable image translation results in its paper. Therefore, our proposed method was based on the Pix2pix network structure, and we mainly evaluated the effectiveness of the edge enhancement filter.

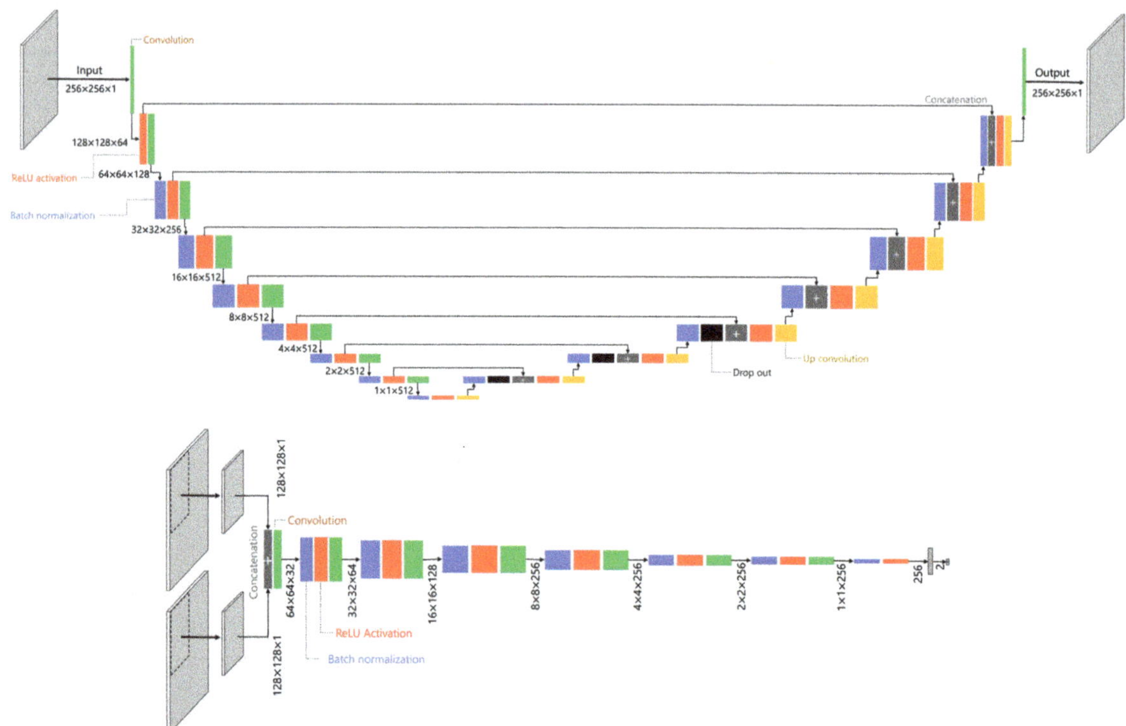


**Figure 6.** Structure of the generator and discriminator networks.

U-Net has skip structures in the layer, which makes it possible to pass information to the previous layer without loss before the convolution layer compresses the information. The number of down/up convolution layers was set to 8 ($= \log_2 256$), depending on the size of the training data, which was $256 \times 256$ pixels. To introduce randomness to the network, dropout (probability = 0.5) layers were added to three layers, as shown in Figure 6. The U-Net is considered appropriate for optical and SAR image translation because the proposed method preserves the edges and corners of the features, which are common in both optical and SAR images.

For the discriminator, the PatchGAN discriminator was used to discriminate the separated input images. The internal structure uses a general 7-layer convolutional encoder.

For the loss function, we adopted a function that uses $L_1$ norm, which is the same as that used in Pix2pix. The Pix2pix loss function is given by the following:

$$\begin{aligned}
\mathcal{L}_{\text{Pix2pix}}(G, D) &= \mathcal{L}_{\text{cGAN}}(G, D) + \lambda \mathcal{L}_{L1}(G) \\
&= \mathcal{L}_{\text{cGAN}}(G, D) + \lambda \mathbb{E}_{x,\,y,z}[\| \, y - G(x,\, z) \, \|_1].
\end{aligned} \tag{5}$$

Referring to Equations (3) and (5), our proposed loss function is given by the following:

$$\mathcal{L}_{\text{Proposed}}(G, D) = \mathbb{E}_y\left[logD\left(F_{\text{edge}}(\boldsymbol{y})\right)\right] + \mathbb{E}_{x,z}[log(1 - D(\boldsymbol{x},\ G(\boldsymbol{x}, \boldsymbol{z})))] + \lambda\mathbb{E}_{x,\ y,z}\left[\|\ F_{\text{edge}}(\boldsymbol{y}) - G(\boldsymbol{x},\ \boldsymbol{z})\ \|_1\right], \quad (6)$$

where $\lambda$ indicates a constant value, which was set to 10 [11] in this experiment, and $F_{\text{edge}}()$ indicates edge enhancement. Similar to Equation (2), the objective generator $G^{**}$ is given by the following:

$$G^{**} = arg \min_G \max_D \mathcal{L}_{\text{Proposed}}(G, D). \quad (7)$$

### 5.2.2. Dataset

The SEN1-2 dataset [39] was prepared from Sentinel-1 (SAR satellite) and Sentinel-2 (optical satellite) of the European Space Agency [40]. The dataset contains co-registered Sentinel-1 and Sentinel-2 image patches. Each item of the Sentinel-1 data has one 8-bit and $256 \times 256$-pixel channel (C-band, VV polarization), while that of Sentinel-2 data has three 8-bit and $256 \times 256$-pixel channels (red, green, and blue band). Their spatial resolution is 10 m per pixel. The "Urban", "Farm", and "Hill" areas were selected from the "spring" data in the SEN1-2 dataset due to their importance during a disaster event, and 2936, 3320, and 3224 image pairs were extracted, respectively. From these pairs, 300 sets each were selected as the test and validation data, respectively, and the remaining sets were used as training data. Figure 7 shows examples of the training, validation, and test image data for the experiment.

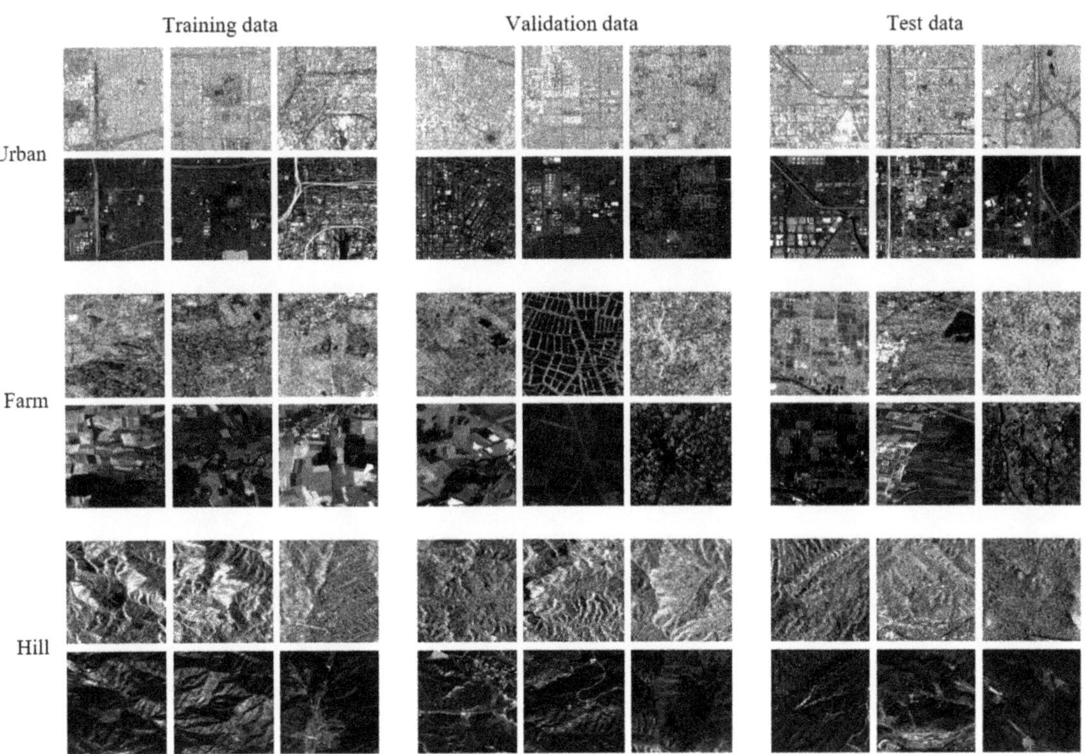


**Figure 7.** Examples of training, validation, and test images for the experiment. The upper half of each dataset represents SAR images, and the others represent optical images.

5.2.3. Implementation

The PatchGAN patch size was set to $128 \times 128$ pixels, with a batch size of 32. NVIDIA Tesla V100 GPU was used for parallel processing, and Adam [41] optimization was used to train the generator and discriminator, with a learning rate of $10^{-3}$.

Each training time was 36 h. The training time represented the time when the $L_1$ losses of the generators were low enough. The loss curves for each dataset are shown in Figure 8

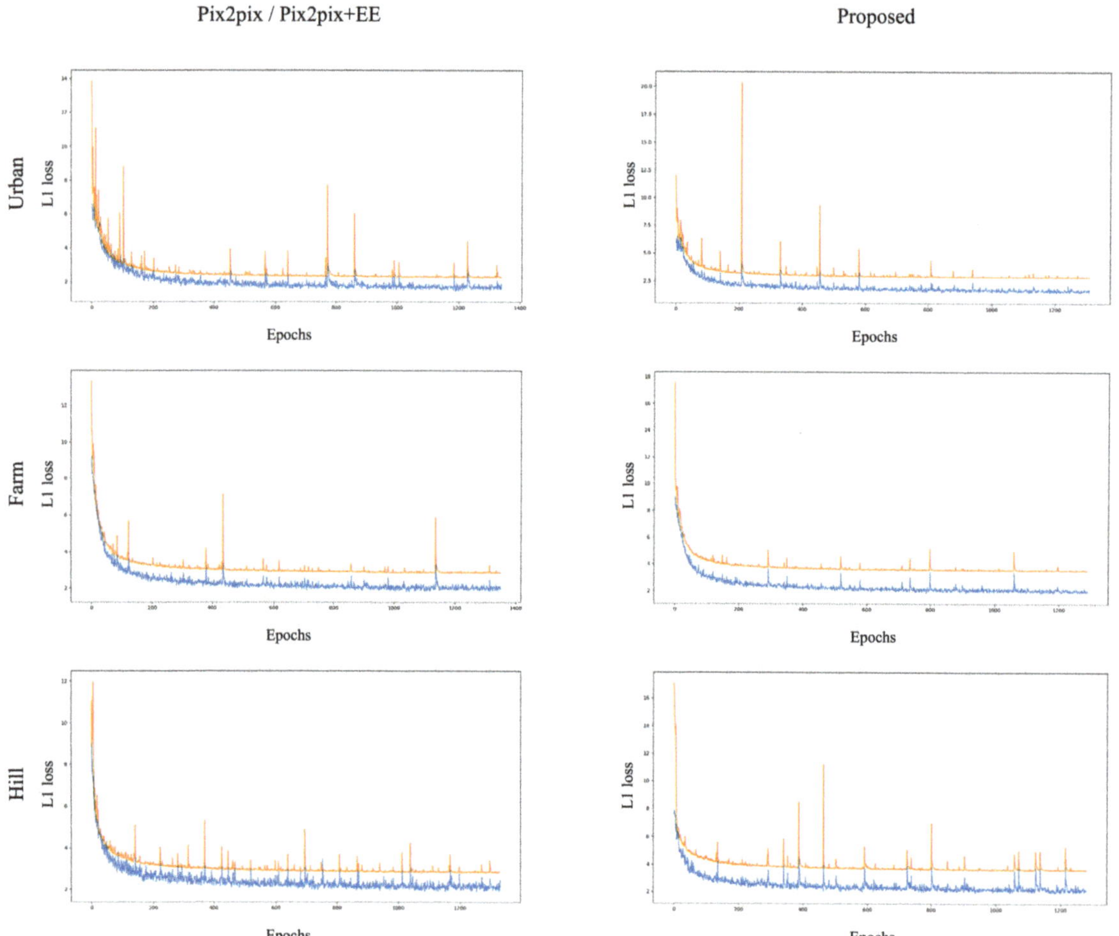



**Figure 8.** Loss curves of each data set. Blue lines show training, and orange lines show validation.

SIFT [24] was used as the keypoint detector and descriptor, and the parameters were the default values of OpenCV [42] version 3.4.3. To remove the false corresponding points [36], the threshold values of the scale and gradient direction were set to two octave layers and 5 degrees, respectively.

The template window sizes of the DLSC, HOPC, and CFOG were set to $100 \times 100$ pixels and the distance threshold of the corresponding points was set to 1.5 pixels. This indicated that corresponding points with distances greater than 1.5 pixels we re considered outliers. The same threshold was applied to all the other methods.

An edge-enhancement filter based on the Laplacian filter was used, which is given by the following:

$$F_{edge}(I) = \begin{bmatrix} -v & -v & -v \\ -v & 1+8v & -v \\ -v & -v & -v \end{bmatrix} I, \tag{8}$$

where $v$ is a parameter for setting the strength of the edge enhancement. In this experiment, $v = 0.1$ was used because it revealed good results in preliminary experiments.

### 5.3. Result and Discussion

The generated optical images were properly translated when Pix2pix or the proposed method—which used GANs in their structure—were applied. The average, standard deviation and median values of the peak signal-to-noise ratio (PSNR) of 300 test images for Pix2pix were 22.70, 3.08, and 22.50 dB, respectively, whereas those of the proposed method were 23.01, 3.14, and 22.87 dB, respectively. PSNR is calculated as

$$PSNR(I_1,\ I_2) = 10 \times log_{10} \frac{MAX(I_1)^2}{MSE(I_1,\ I_2)}, \tag{9}$$

where $MAX(I_1)$ is the possible maximum value of image $I_1$, and $MSE(I_1,\ I_2)$ is the mean squared error between $I_1$ and $I_2$. $MAX(I_1)$ needs to equal $MAX(I_2)$.

Figure 9 shows the results of the SAR-to-optical image translation. It shows the input, output, and ground truth sets with the best PSNR values. Afterward, we evaluated the probability of these images for keypoint matching. Figure 10 shows an example of improvement by the proposed method. The proposed method improved the local blur and low reproducibility that were problems in the conventional cGANs.

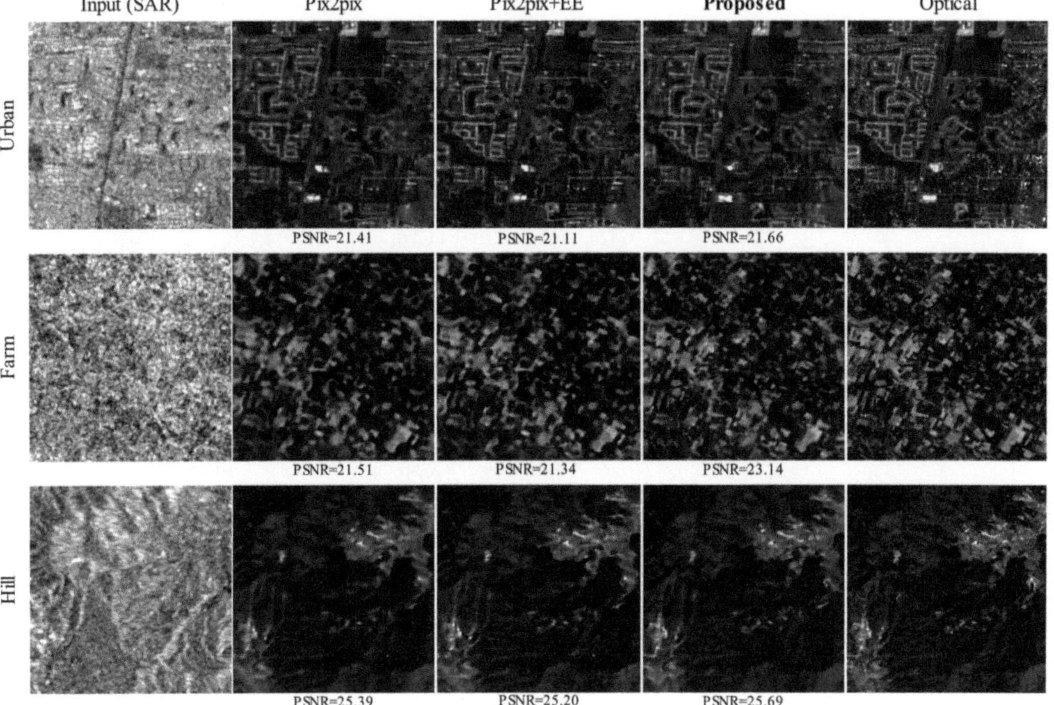


**Figure 9.** Results of SAR-to-optical image translation.

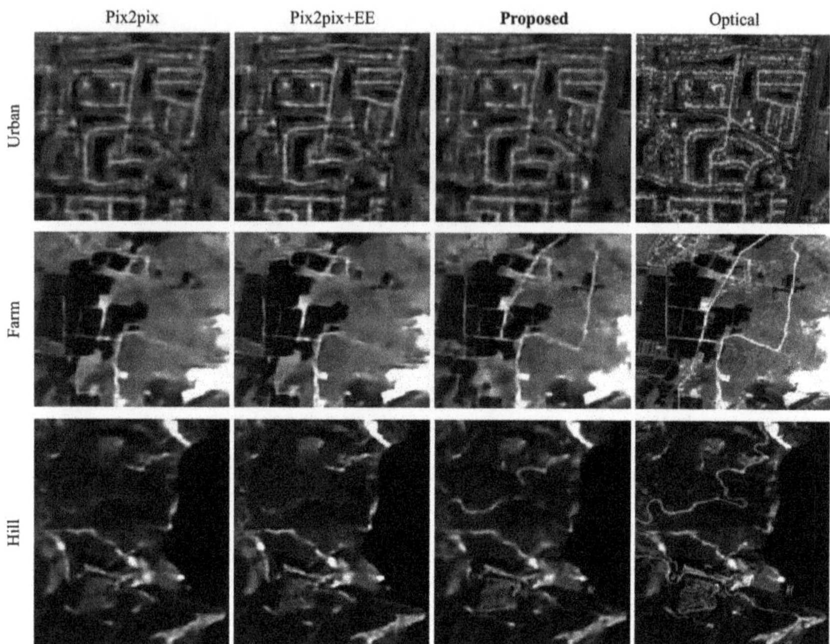



**Figure 10.** Improvement examples of image translation with our proposed method.

Tables 1 and 2 summarize the results of the quantitative evaluation experiments.

**Table 1.** Mean average (pixels) of the $\bar{d}$ in Equation (4).

|  | DLSC | HOPC | CFOG | Pix2pix | Pix2pix + EE | Proposed |
|---|---|---|---|---|---|---|
| Urban | 0.965 | 0.863 (10.6%)[3] | 0.838 (13.2%) | 0.564 [2] (41.6%) | 0.566 (41.3%) | **0.538** [1] **(44.2%)** |
| Farm | 0.847 | 0.847 (0%) | 0.700 (17.4%) | 0.675 (20.3%) | 0.676 (20.2%) | **0.569** **(32.8%)** |
| Hill | 1.128 | 1.052 (6.7%) | 1.114 (1.2%) | 0.668 (40.8%) | 0.667 (40.9%) | **0.577** **(48.9%)** |

[1] A bold number represents the best value in the row. [2] An underlined number represents the second best value in the row. [3] A value on each second row shows improving rate (%) compared to DLSC. This was calculated as $Improve\_rate_X = 100 \times (1 - Result_X / Result_{DLSC})$.

**Table 2.** Mean number of corresponding points (correct matches). The improvement score was larger in the "Farm" and "Hill" datasets.

|  | DLSC | HOPC | CFOG | Pix2pix | Pix2pix + EE | Proposed |
|---|---|---|---|---|---|---|
| Urban | 49.8 | **80.7** [1] **(62.0%)** | 63.0 (26.5%) | 67.6 (35.7%) | 63.5 (27.5%) | 78.5 [2] (57.6%) |
| Farm | 62.5 | 65.9 (5.4%) | 62.1 (−0.6%) | 57.4 (−8.2%) | 57.2 (−8.5%) | **101.1** **(61.8%)** |
| Hill | 17.2 | 45.2 (162.8%) | 22.6 (31.4%) | 57.5 (234.3%) | 56.4 (227.9%) | **85.0** **(394.2%)** |

[1] A bold number represents the best value in the row. [2] An underlined number represents the second best value in the row. [3] A value on each second row shows improving rate (%) compared to DLSC. This was calculated as $Improve\_rate_X = 100 \times (Result_X - Result_{DLSC}) / Result_{DLSC}$.

Table 1 shows the accuracy of the corresponding point detection. In the test dataset for each "Urban", "Farm", and "Hill" area, the proposed method had high precision with an average accuracy of 0.538 pixels, 0.569 pixels, and 0.577 pixels, respectively. The result of the proposed method was more accurate than those of DLSC, HOPC, CFOG, Pix2pix, and Pix2pix + EE; the absolute accuracy was the highest in the "Urban" area. However, the improved accuracy of the "Farm" and "Hill" was higher than "Urban". This was because several objects had strong local features with sharp edges and corners, such as artificial structures, in urban areas. On the other hand, the local features of the "Farm" and "Hill" areas are weaker than those of the "Urban" area.

When edge enhancement was applied to the prediction result of Pix2pix in the Pix2pix + EE method, the accuracy was not as high as that of the proposed method, probably because the enhancement was equally applied to the artifacts generated in the prediction result. Hence, it was difficult to selectively enhance the effective local features.

Table 2 shows the number of detected corresponding points (correct matches). In the test dataset for each of the "Urban", "Farm", and "Hill" areas, the average scores of the proposed method were 78.5, 101.0, and 85.0 points, respectively. As the results in Table 1 indicate, the improvement score was larger in the "Farm" and "Hill" datasets. Although the number of corresponding points in the "Urban" dataset for HOPC was more than that of the proposed method, only the proposed method achieved more accuracy and several corresponding points.

Table 3 shows the inlier (correct matches) rates, in which our proposed method achieved a high inlier rate. For Pix2pix, Pix2pix + EE, and the proposed method, some outliers were already removed based on the scale value and the gradient direction of each keypoint [36]. This was the reason why the three methods achieved higher inlier rates than DLSC, HOPC, and CFOG.

**Table 3.** Mean inlier (correct matches) rates.

| | DLSC | HOPC | CFOG | Pix2pix | Pix2pix + EE | Proposed |
|---|---|---|---|---|---|---|
| Urban | 0.437 | 0.703 (60.8%) [3] | 0.575 (31.6%) | 0.900 (105.9%) | 0.900 [2] (105.9%) | **0.913** [1] **(108.9%)** |
| Farm | 0.588 | 0.615 (4.6%) | 0.606 (3.1%) | 0.805 (36.9%) | 0.815 (38.6%) | **0.896** **(52.4%)** |
| Hill | 0.155 | 0.396 (155.5%) | 0.200 (29.0%) | 0.809 (421.9%) | 0.808 (421.3%) | **0.880** **(467.7%)** |

[1] A bold number represents the best value in the row. [2] An underlined number represents the second best value in the row. [3] A value on each second row shows improving rate (%) compared to DLSC. This was calculated as $Improve\_rate_X = 100 \times (Result_X - Result_{DLSC})/Result_{DLSC}$.

Figure 11 shows the results of the optical image generation and the subsequent feature-point matching. A sufficient number of correspondences were estimated, and it was confirmed that Pix2pix was effective for SAR-to-optical image translation. Although false correspondences remained, they could be removed using a robust estimation method, such as RANSAC [43].

As Figures 9 and 10 show, the Pix2pix + EE results indeed gave sharper images than those of Pix2pix and Proposed. However, the quantitative evaluation result showed better results than the others. The reason was that Pix2pix + EE enhanced not only effect image features but also artifacts. After all, the Pix2pix + EE results seemed sharper, but did not yield as good results.

Considering the "Urban" dataset in Figure 11 more closely, the center part of the generated optical image failed to translate in both methods. Even if fewer corresponding points were found in some parts of the images, it was possible to correctly calculate the total correspondences in a case where partial correspondence was correctly achieved in other areas since SIFT or other keypoint detection/description methods could calculate local features. Through this experiment, we confirmed that precise feature matching between the optical and generated optical images was possible.

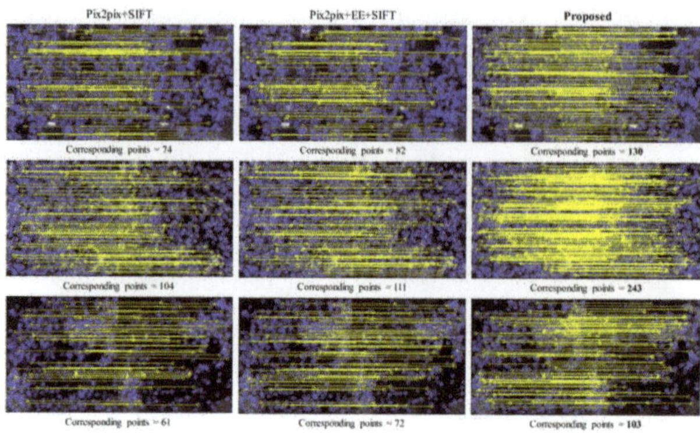



**Figure 11.** Results of keypoint matching. Yellow points connected by yellow lines mean corresponding points, and blue points mean other keypoints. No local feature correspondences were found in a collapsed area, but many corresponding points could be found in the remaining area.

A practical example of SAR-to-optical registration is shown in Figures 12 and 13. A SAR image was projected onto an optical image that had a rotation and translation formation. The projection was estimated by finding the corresponding points between the optical and generated optical images. We confirmed that high-precision image registration was achieved, even on the assumption of planarity, since this dataset covered an area that was not very wide. Alternatively, we could estimate a 3D projection of the corresponding keypoint if necessary [22].

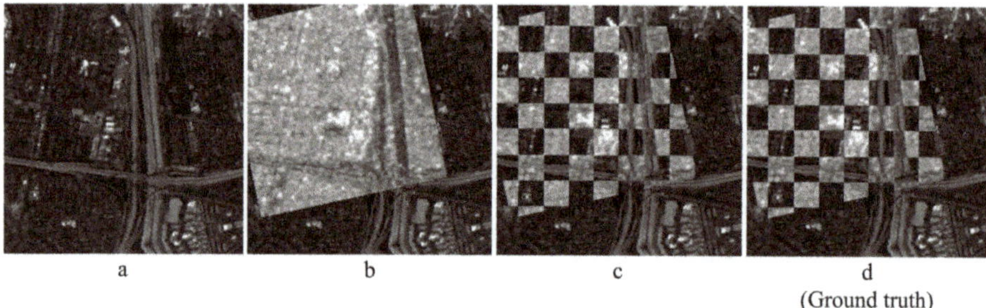



**Figure 12.** Example of SAR and optical image registration assuming planarity. To overlay a SAR image on the target optical image (**a**), after finding the local feature correspondences between the optical and generated optical images, the original SAR image was projected onto the optical image using an estimated homography transform (**b**). Half of the area was transparent, similar to a checkerboard pattern (**c**), and ground truth, which was registered by accurate DEM (**d**).

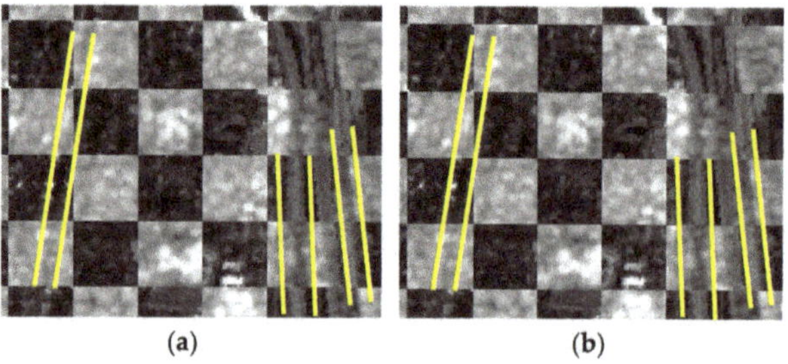



**Figure 13.** (**a,b**) are Close-up of a part of Figures c and d in Figure 12. Yellow lines were drawn along the roads to demonstrate the registration accuracy. It could be confirmed that image registration between SAR and optical images was performed with high accuracy.

A limitation of the proposed method was that it was hard to apply this method to very high spatial resolution images; we confirmed this with 3-m/pixel imagery by Cosmo-SkyMed [44]. That was because layovers in SAR images prevented training datasets from accurate image registration. We confirmed that the proposed method worked well in middle resolution images (about 10 m/pixel). Considering the balance between the size of the observation area and its detail, the dataset of 10 m/pixel used in this evaluation was considered to be an appropriate spatial resolution dataset.

## 6. Conclusions

In this paper, we proposed a method for translating SAR images to optical images using a GAN so that a keypoint-matching algorithm could be applied to multimodal images. By applying this method for SAR-to-optical image translation, we performed keypoint-matching on the monomodal images. Through quantitative evaluations of the keypoint-matching accuracy, we confirmed that the proposed method could achieve accurate keypoint matching between the optical and generated optical images using a cGAN. In the translation from optical to generated optical images using the cGAN, the local features could not be obtained, and the corresponding points could not be established due to blurring. Therefore, an improvement was achieved in our proposed method by applying the edge-enhancement filter to the training data for the discriminator and training the generator to actively learn the local features. Furthermore, we conducted a comparative experiment and confirmed that our proposed method was effective for finding local corresponding points.

**Author Contributions:** Conceptualization, H.T., A.D. and I.K.; methodology, H.T., M.S. and I.K.; software, H.T.; validation, H.T., N.O. and F.I.; formal analysis, H.T.; investigation, H.T.; resources, A.D. and I.K.; data curation, H.T. and H.I.; writing—original draft preparation, H.T., H.I. and N.O.; writing—review and editing, H.T., A.D. and I.K.; visualization, H.T. and N.O.; supervision, A.D., Y.K. and I.K.; project administration, I.K.; funding acquisition, I.K. All authors have read and agreed to the published version of the manuscript.

**Funding:** This work was supported by JST CREST under Grant JPMJCR16E3; and Grant-in-Aid 456 for JSPS Fellows under Grant JP19J11514.

**Institutional Review Board Statement:** Not applicable.

**Informed Consent Statement:** Not applicable.

**Data Availability Statement:** Not applicable.

**Acknowledgments:** Drone photography was conducted in cooperation with DRONEBIRD by Crisis Mappers Japan (NPO), a disaster drone rescue team. (http://dronebird.org/, accessed on 14 March 2022.)

**Conflicts of Interest:** The authors declare no conflict of interest.

## References

1. Voigt, S.; Giulio-Tonolo, F.; Lyons, J.; Kučera, J.; Jones, B.; Schneiderhan, T.; Platzeck, G.; Kaku, K.; Hazarika, M.K.; Czaran, L.; et al. Global Trends in Satellite-Based Emergency Mapping. *Science* **2016**, *353*, 247–252. [CrossRef] [PubMed]
2. Ogata, T. Disaster Management in Japan. *Jpn. Med. Assoc. J.* **2016**, *59*, 27–30.
3. Gokon, H.; Koshimura, S.; Matsuoka, M. Object-Based Method for Estimating Tsunami-Induced Damage Using TerraSAR-X Data. *J. Disaster Res.* **2016**, *11*, 225–235. [CrossRef]
4. Liu, W.; Yamazaki, F.; Sasagawa, T. Monitoring of the Recovery Process of the Fukushima Daiichi Nuclear Power Plant from VHR SAR Images. *J. Disaster Res.* **2016**, *11*, 236–245. [CrossRef]
5. Bai, Y.; Adriano, B.; Mas, E.; Koshimura, S. Machine Learning Based Building Damage Mapping from the ALOS-2/PALSAR-2 SAR Imagery: Case Study of 2016 Kumamoto Earthquake. *J. Disaster Res.* **2017**, *12*, 646–655. [CrossRef]
6. Shiraishi, M.; Ashiya, H.; Konno, A.; Morita, K.; Noro, T.; Nomura, Y.; Kataoka, S. Development of Real-Time Collection, Integration, and Sharing Technology for Infrastructure Damage Information. *J. Disaster Res.* **2019**, *14*, 333–347. [CrossRef]
7. Ge, P.; Gokon, H.; Meguro, K. Building Damage Assessment Using Intensity SAR Data with Different Incidence Angles and Longtime Interval. *J. Disaster Res.* **2019**, *14*, 456–465. [CrossRef]
8. Jensen, J.R.; Lulla, K. Introductory Digital Image Processing: A Remote Sensing Perspective. *Geocarto Int.* **1987**, *2*, 65. [CrossRef]

9.  Goodfellow, I.; Pouget-Abadie, J.; Mirza, M.; Xu, B.; Warde-Farley, D.; Ozair, S.; Courville, A.; Bengio, Y. Generative Adversarial Nets. In *Advances in Neural Information Processing Systems*; Ghahramani, Z., Welling, M., Cortes, C., Lawrence, N., Weinberger, K.Q., Eds.; Curran Associates, Inc.: Red Hook, NY, USA, 2014; Volume 27, pp. 2672–2680.

10. Toriya, H.; Dewan, A.; Kitahara, I. SAR2OPT: Image Alignment between Multi-Modal Images Using Generative Adversarial Networks. In Proceedings of the International Geoscience and Remote Sensing Symposium, Yokohama, Japan, 28 July–2 August 2019; pp. 923–926.

11. Isola, P.; Zhu, J.Y.; Zhou, T.; Efros, A.A. Image-to-Image Translation with Conditional Adversarial Networks. In Proceedings of the IEEE Conference on Computer Vision and Pattern Recognition, Honolulu, HI, USA, 21 November 2017; Volume 2017, pp. 5967–5976.

12. Dame, A.; Marchand, E. Accurate Real-Time Tracking Using Mutual Information. In Proceedings of the IEEE International Symposium on Mixed and Augmented Reality, Seoul, Korea, 13–16 October 2010.

13. Xiong, B.; Li, W.; Zhao, L.; Lu, J.; Zhang, X.; Kuang, G. Registration for SAR and Optical Images Based on Straight Line Features and Mutual Information. In Proceedings of the International Geoscience and Remote Sensing Symposium, Beijing, China, 10–15 July 2016.

14. Ye, Y.; Shen, L.; Hao, M.; Wang, J.; Xu, Z. Robust Optical-to-SAR Image Matching Based on Shape Properties. *IEEE Geosci. Remote Sens. Lett.* **2017**, *14*, 564–568. [CrossRef]

15. Ye, Y.; Shan, J.; Bruzzone, L.; Shen, L. Robust Registration of Multimodal Remote Sensing Images Based on Structural Similarity. *IEEE Trans. Geosci. Remote Sens.* **2017**, *55*, 2941–2958. [CrossRef]

16. Ye, Y.; Bruzzone, L.; Shan, J.; Bovolo, F.; Zhu, Q. Fast and Robust Matching for Multimodal Remote Sensing Image Registration. *IEEE Trans. Geosci. Remote Sens.* **2019**, *57*, 9059–9070. [CrossRef]

17. Xiong, X.; Xu, Q.; Jin, G.; Zhang, H.; Gao, X. Rank-Based Local Self-Similarity Descriptor for Optical-to-SAR Image Matching. *IEEE Geosci. Remote Sens. Lett.* **2019**, *17*, 1742–1746. [CrossRef]

18. Lenc, K.; Vedaldi, A. Large Scale Evaluation of Local Image Feature Detectors on Homography Datasets. In Proceedings of the British Machine Vision Conference, BMVC, Cardiff, UK, 9–12 September 2019.

19. Bürgmann, T.; Koppe, W.; Schmitt, M. Matching of TerraSAR-X Derived Ground Control Points to Optical Image Patches Using Deep Learning. *ISPRS J. Photogramm. Remote Sens.* **2019**, *158*, 241–248. [CrossRef]

20. Merkle, N.; Auer, S.; Muller, R.; Reinartz, P. Exploring the Potential of Conditional Adversarial Networks for Optical and SAR Image Matching. *IEEE J. Sel. Top. Appl. Earth Obs. Remote Sens.* **2018**, *11*, 1811–1820. [CrossRef]

21. Zhou, H.; Pan, Y.; Zhang, Z. A Speeded-up Affine Invariant Detector. In Proceedings of the International Congress on Image and Signal Processing, CISP, Sichuan, China, 16–18 October 2012; pp. 401–406.

22. Szeliski, R. *Computer Vision: Algorithms and Applications*; Springer: Berlin/Heidelberg, Germany, 2009.

23. Patel, M.I.; Thakar, V.K.; Shah, S.K. Image Registration of Satellite Images with Varying Illumination Level Using HOG Descriptor Based SURF. In *Procedia Computer Science*; Elsevier B.V.: Amsterdam, The Netherlands, 2016; Volume 93, pp. 382–388.

24. Lowe, D.G. Object Recognition from Local Scale-Invariant Features. In Proceedings of the IEEE Conference on Computer Vision and Pattern Recognition, Fort Collins, CO, USA, 23–25 June 1999.

25. Bay, H.; Ess, A.; Tuytelaars, T.; Van Gool, L. Speeded-Up Robust Features (SURF). *Comput. Vis. Image Underst.* **2008**, *110*, 346–359. [CrossRef]

26. Rublee, E.; Rabaud, V.; Konolige, K.; Bradski, G. ORB: An Efficient Alternative to SIFT or SURF. In Proceedings of the IEEE Conference on Computer Vision and Pattern Recognition, Washington, DC, USA, 20–25 June 2011; pp. 2564–2571.

27. Alcantarilla, P.F.; Nuevo, J.; Bartoli, A. Fast Explicit Diffusion for Accelerated Features in Nonlinear Scale Spaces. In Proceedings of the British Machine Vision Conference, Bristol, UK, 9–13 September 2013.

28. Simo-serra, E. Learning to Simplify: Fully Convolutional Networks for Rough Sketch Cleanup. *ACM Trans. Graph. (TOG)* **2016**, *35*, 1–11. [CrossRef]

29. Enomoto, K.; Sakurada, K.; Wang, W.; Kawaguchi, N.; Matsuoka, M.; Nakamura, R. Image Translation between SAR and Optical Imagery with Generative Adversarial Nets. In Proceedings of the International Geoscience and Remote Sensing Symposium, Valencia, Spain, 22–27 July 2018.

30. Hughes, L.H.; Marcos, D.; Lobry, S.; Tuia, D.; Schmitt, M. A Deep Learning Framework for Matching of SAR and Optical Imagery. *ISPRS J. Photogramm. Remote Sens.* **2020**, *169*, 166–179. [CrossRef]

31. Reyes, M.F.; Auer, S.; Merkle, N.; Henry, C.; Schmitt, M. SAR-to-Optical Image Translation Based on Conditional Generative Adversarial Networks-Optimization, Opportunities and Limits. *Remote Sens.* **2019**, *11*, 2067. [CrossRef]

32. Wang, L.; Xu, X.; Yu, Y.; Yang, R.; Gui, R.; Xu, Z.; Pu, F. SAR-to-Optical Image Translation Using Supervised Cycle-Consistent Adversarial Networks. *IEEE Access* **2019**, *7*, 129136–129149. [CrossRef]

33. Wang, X.; Gupta, A. Generative Image Modeling Using Style and Structure Adversarial Networks. In *Lecture Notes in Computer Science (Including Subseries Lecture Notes in Artificial Intelligence and Lecture Notes in Bioinformatics)*; Springer: Berlin/Heidelberg, Germany, 2016; Volume 9908, pp. 318–335.

34. Jiang, K.; Wang, Z.; Yi, P.; Wang, G.; Lu, T.; Jiang, J. Edge-Enhanced GAN for Remote Sensing Image Superresolution. *IEEE Trans. Geosci. Remote Sens.* **2019**, *57*, 5799–5812. [CrossRef]

35. Rabbi, J.; Ray, N.; Schubert, M.; Chowdhury, S.; Chao, D. Small-Object Detection in Remote Sensing Images with End-to-End Edge-Enhanced GAN and Object Detector Network. *Remote Sens.* **2020**, *12*, 1432. [CrossRef]

36. Toriya, H.; Kitahara, I.; Ohta, Y. Mobile Camera Localization Using Aerial-View Images. *IPSJ Trans. Comput. Vis. Appl.* **2014**, *6*, 111–119. [CrossRef]

37. Ronneberger, O.; Fischer, P.; Brox, T. U-Net: Convolutional Networks for Biomedical Image Segmentation. In Proceedings of the International Conference on Medical Image Computing and Computer-Assisted Intervention, Munich, Germany, 5–9 October 2015; Volume 9351, pp. 234–241.

38. Li, C.; Wand, M. Precomputed Real-Time Texture Synthesis with Markovian Generative Adversarial Networks. In Proceedings of the European Conference on Computer Vision, Amsterdam, The Netherlands, 11–14 October 2016; Volume 9907, pp. 702–716.

39. Schmitt, M.; Hughes, L.H.; Zhu, X.X. The Sen1-2 Dataset for Deep Learning in Sar-Optical Data Fusion. In Proceedings of the ISPRS Annals of the Photogrammetry, Remote Sensing and Spatial Information Sciences, Delft, The Netherlands, 4 July 2018; Volume 4, pp. 141–146.

40. Berger, M.; Moreno, J.; Johannessen, J.A.; Levelt, P.F.; Hanssen, R.F. ESA's Sentinel Missions in Support of Earth System Science. *Remote Sens. Environ.* **2012**, *120*, 84–90. [CrossRef]

41. Kingma, D.P.; Ba, J. Adam: A Method for Stochastic Optimization. *arXiv* **2014**, arXiv:1412.6980.

42. Bradski, G. The OpenCV Library. *Dr. Dobb's J. Softw. Tools Prof. Program.* **2000**, *25*, 120–123.

43. Fischler, M.A.; Bolles, R.C. Random Sample Consensus: A Paradigm for Model Fitting with Applications to Image Analysis and Automated Cartography. *Commun. ACM* **1981**, *24*, 381–395. [CrossRef]

44. Covello, F.; Battazza, F.; Coletta, A.; Lopinto, E.; Fiorentino, C.; Pietranera, L.; Valentini, G.; Zoffoli, S. COSMO-SkyMed an Existing Opportunity for Observing the Earth. *J. Geodyn.* **2010**, *49*, 171–180. [CrossRef]

*Article*

# Performance Evaluation of Feature Matching Techniques for Detecting Reinforced Soil Retaining Wall Displacement

Yong-Soo Ha, Jeongki Lee and Yun-Tae Kim *

Department of Ocean Engineering, Pukyong National University, Busan 48513, Korea;
hys90@pukyong.ac.kr (Y.-S.H.); lee545@wisc.edu (J.L.)
* Correspondence: yuntkim@pknu.ac.kr; Tel.: +82-51-629-6587

**Abstract:** Image registration technology is widely applied in various matching methods. In this study, we aim to evaluate the feature matching performance and to find an optimal technique for detecting three types of behaviors—facing displacement, settlement, and combined displacement—in reinforced soil retaining walls (RSWs). For a single block with an artificial target and a multiblock structure with artificial and natural targets, five popular detectors and descriptors—KAZE, SURF, MinEigen, ORB, and BRISK—were used to evaluate the resolution performance. For comparison, the repeatability, matching score, and inlier matching features were analyzed based on the number of extracted and matched features. The axial registration error (ARE) was used to verify the accuracy of the methods by comparing the position between the estimated and real features. The results showed that the KAZE method was the best detector and descriptor for RSWs (block shape target), with the highest probability of successfully matching features. In the multiblock experiment, the block used as a natural target showed similar matching performance to that of the block with an artificial target attached. Therefore, the behaviors of RSW blocks can be analyzed using the KAZE method without installing an artificial target.

**Keywords:** feature matching; image registration; natural target; RSW

**Citation:** Ha, Y.-S.; Lee, J.; Kim, Y.-T. Performance Evaluation of Feature Matching Techniques for Detecting Reinforced Soil Retaining Wall Displacement. *Remote Sens.* **2022**, *14*, 1697. https://doi.org/10.3390/rs14071697

Academic Editors: Yue Wu, Kai Qin, Maoguo Gong and Qiguang Miao

Received: 22 February 2022
Accepted: 30 March 2022
Published: 31 March 2022

**Publisher's Note:** MDPI stays neutral with regard to jurisdictional claims in published maps and institutional affiliations.



## 1. Introduction

Since the concept of reinforced earth was proposed by Vidal in the late 1950s [1], reinforced soil retaining walls (RSWs) have been widely used, owing to their low cost and rapid construction. A general safety inspection is carried out at least two times a year, and a precision safety inspection is conducted once every 1–3 years, according to the Special Acts On Safety Control And Maintenance Of Establishments [2], to inspect the physical condition of RSWs. However, RSWs frequently collapse, owing to heavy rainfall or dynamic loads, such as those produced during earthquakes. It is very difficult to detect the risk of an unexpected collapse of RSWs through a periodic safety inspection. Several RSWs that were diagnosed as safe (Grade B—no risk of collapse, management required) collapsed within 6 months after safety inspections (e.g., one in Gwangju in 2015 and one in Busan in 2020 [3,4]). Therefore, a real-time or continuous safety monitoring system is necessary for detecting RSW collapse, which is hard to detect with only periodic safety inspections [5,6].

Various studies have been conducted to analyze the behavior of RSWs using a strain gauge, displacement sensor, inclinometer, and pore water pressure sensor [7–10]. However, these experimental approaches only monitor a specific location in RSWs where sensors were installed. In addition, numerous sensors and considerable effort are required to measure the overall behavior of RSWs, resulting in inefficiency in terms of cost and labor. Therefore, this study was carried out as a pilot-phase experiment to analyze the overall behavior of structures based on images and to overcome the abovementioned disadvantages of current monitoring methods. The analysis of the RSW behavior can be divided into the matching procedure of the target and the calculation procedure of the displacement based on the matching result. The matching procedure finds and matches identical points between image pairs, and the calculation procedure calculates displacement from the changes in the pixels. In this study, we focused on the matching procedure. A feature matching technique was applied to accurately match the target before and after the behavior.

With the continuous development of image processing technologies, feature matching techniques have been widely applied across various fields, such as biomedical image registration [11–14], unmanned aerial vehicle (UAV) image registration [15], and geographic information systems (GISs) [16–18]). Additionally, a fusion study of feature and machine learning technology was performed in the field of vehicle matching [19]. Feature matching techniques show high-resolution image pairing performance between different scales, transformations, rotations, and three-dimensional (3D) projections of transformed objects in images. Tareen and Saleem [20] matched buildings and background images with overlapping parts using this technique. Image matching results were evaluated based on the changes in size, rotation, and viewpoint. SIFT, SURF, and BRISK are scale-invariant feature detectors. ORB and BRISK were invariant to affine change, and SIFT had the highest overall accuracy. Pieropan et al. [21] applied several methods to track moving elements in images based on specificity, tracking accuracy, and tracking performance. Here, AKAZE and SIFT showed high performance when detecting the deformations of small objects. Moreover, ORB and BRISK showed high performance when sequences with significant motion blur were matched. Mikolajczyk and Schmid [22] evaluated the performance of descriptors based on various image states, targets, and transforms. Their results showed that GLOH and SIFT had excellent performance. Chien et al. [23] evaluated feature performance on KITTI benchmark datasets produced using monocular visual odometry. SURF, AKAZE, and SIFT showed excellent performances in each condition of the interframe and accumulated drift errors and the segmented motion error on the translational and rotational components. Previous studies indicated that different feature matching methods should be applied according to the feature extraction and matching conditions (i.e., type of target, change of scale, and rotation). In addition, various studies have been conducted to select optimal feature matching techniques with different types of targets and transforms. Therefore, it is necessary to select an optimal feature detector and descriptor with the best performance to accurately evaluate RSW behavior through image processing.

The use of remote sensing methods in combination with high-resolution image recording technology could allow for the continuous evaluation of structure movement and displacement behavior. Feng et al. [24] used a template matching method to analyze the time-lapse displacement of attached arbitrary targets on railway and pedestrian bridges. Apparent targets (i.e., patterns, features, and textures) of the surrounding features enable easier comparison of structure images in different states and times. Lee and Shinozuka [25] measured the dynamic displacement of target panels attached to bridges and piers through texture recognition techniques from motion pictures. Choi et al. [26] carried out deformation interpretation of the artificial target (AT) to reference two-story steel frames taken by a dynamic displacement vision system. However, a critical limitation of these techniques was that only a local movement near the AT could be detected.

Therefore, in this study, the RSW behavior was simulated through single- and multiblock laboratory experiments to determine the most efficient feature matching method for the RSW structure. A facing of a block with a sheet target was defined as the AT, and a block face without an additional pattern was defined as the natural target (NT). Five feature matching methods were applied to the single-block experiment with ATs and the multiblock experiment with ATs and NTs. Furthermore, the usability of NTs in the multiblock experiment was evaluated by comparing the performance of four NTs with that of ATs. Through this analysis, we determined a feature matching technique that analyzes NTs similar to the performance of analyzing ATs.

## 2. Background and Objectives

### 2.1. Behavior of RSW

A single point existing in the 3D space can be moved in all directions. However, a specific behavior only occurs predominantly in the case of a geotechnical structure, such as an RSW, that was constructed continuously in the lateral direction. Berg et al. [27] specified the measurement of the facing displacement and settlement of RSWs to evaluate both the internal and external stabilities during the design and construction stage. Koerner and Koerner [28] investigated the collapse of 320 geosynthetically reinforced, mechanically stabilized earth (MSE) walls worldwide due to facing displacement and settlement caused by surcharge and infiltration. Therefore, facing displacement (corresponding to the bulging type), settlement, and their combined displacement were considered in this study. Horizontal displacements in the lateral direction were excluded because they hardly occurred based on the nature of the RSW.

### 2.2. Feature Detection and Matching

Feature detection and matching techniques were used to detect the features of target regions in image pairs and to match the features presumed to be identical to the detected features, respectively. These technologies are widely used in various computer machine vision fields, such as object detection, object tracking, and augmented reality [20]. As applications expand, various methods for detecting and matching features have been developed and improved to provide higher accuracy. Feature detection and matching techniques consist of a detector and a descriptor. The detector is used to discover and locate areas of interest in a given image, such as edge and junction. The areas should contain strong signal changes and were used to identify the same area in images captured with different view angles and movement of objects in the image. By contrast, the descriptor provides robust characterization of the detected features. It provides high matching performance through high invariance, even for changes in scale, rotation, and partial affine image transformation of each feature in the image pairs [29].

In this study, we used five methods—MinEigen [30], SURF [31], BRISK [32], ORB [33], and KAZE [34]—provided in MATLAB to evaluate the performance of each feature detector and description algorithm. Feature detection and matching for images before and after the behavior was performed in the following order: (1) set a target for detection and matching analysis in the image, (2) detect target features using each detector and express each feature as a feature vector through each descriptor, and (3) calculate pairwise distances for each feature vector in image pairs. Each feature was matched when the distance between two feature vectors was less than the matching threshold (a matching threshold of 10 was applied to binary feature vectors, such as MinEigen, ORB, and BRISK, and a matching threshold of 1 was employed for KAZE and SURF). Moreover, the sum of squared differences (SSD) method was used to evaluate the feature matching metric for KAZE and SURF, and the Hamming distance was used for binary features, such as MinEigen, ORB, and BRISK. The MSAC algorithm with 100,000 iterations and 99% confidence was used to exclude the outliers and determine the $3 \times 3$ transformation matrix. The inlier-matched features had to exist within two pixels of the position of the point projected through the transformation matrix.

### 2.3. Feature Performance Evaluation

Various evaluation methods have been suggested to evaluate the performance of feature detection and matching techniques. We analyzed the repeatability and matching score, which were widely used for quantitative evaluation based on features. In addition, the number of inlier features was evaluated to avoid image distortion. Image distortion may occur when the matching technology is not suitable for the image features and matching conditions. Even if the extracted and actual feature vectors have high similarity, distortion may occur if the number of exact matching features is insufficient. Therefore, for successful feature matching, two solutions are used to reduce the matching error and distortion:

(1) matching with more precisely matching feature vectors that result in higher similarity and (2) selecting proper feature detectors and descriptors to obtain highly consistent transformation matrices with a large number of inlier matching features. However, the feature vectors of the first solution are inherent properties that do not change unless the image changes. Therefore, in this study, it was not possible to arbitrarily add more feature vectors with higher similarity because specific targets were already selected as the facing of the block. Instead, the second solution could be used by selecting the optimal feature detector and descriptor to obtain a highly consistent transformation matrix with many inlier matching features for the given target.

The repeatability for a pair of images is the number of feature correspondences found between the pair of images divided by a minimum number of features detected in the image pair [35,36]. In this study, feature matching was performed on the target image extracted from the initial state and the entire image in which the behavior occurs. Because the minimum number of features were always detected in the former, the number of features in the target image at initial was used as the denominator.

$$\text{Repeatability} = \frac{\text{No. of matching features}}{\text{No. of detected features( in target image at initial)}} \tag{1}$$

The matching score is the average ratio between ground truth correspondences and the number of detected features in a shared viewpoint region [36–38]. Moreover, repeatability focuses on finding the same feature in two image pairs. Further, the matching score indicates how accurate matching is performed by excluding outliers. Matching scores of 1 and 0 indicate that the matched features are perfectly inlier and perfectly outlier, respectively.

$$\text{Matching score} = \frac{\text{No. of inlier matching features}}{\text{No. of detected features( in target image at initial)}} \tag{2}$$

The number of inlier matching features represents the number of features from which the outlier matching features have been removed among the matching features. More accurate matching can be performed when estimating the transformation using more inlier matching features.

Registration error is used to quantify the error when images (e.g., CT and MRI images) are matched [39,40]. It is used as a criterion for evaluating the matching performance in various forms, such as target registration error (TRE), landmark registration error (LRE), and mean of target registration error (mTRE). All registration errors calculate the error for each feature in the image plane based on their locations. However, it is difficult to express the error in a shape change of a block occurring in 3D space. Therefore, in this study, we proposed ARE to quantify the error of the x-axis (upper and lower edges of block) and y-axis (left and right edges of the block) during the matching between the images before and after the behavior. Figure 1 shows an example of the features and vertices for calculating TRE and ARE. The inlier matching features (in the initial image$_{t=\text{initial}}$) at the front of the block were defined as $A_i$, $B_i$, $C_i$, $D_i$, and $E_i$. Meanwhile, the inlier matching features (in the image$_{t=n}$) at the front of the block were defined as $A_n$, $B_n$, $C_n$, $D_n$, and $E_n$. The transformation matrix ($T_{OD}$) was calculated using the pairs of the inlier matching features in the image pair, and the transformed target was derived by applying the transformation matrix to the target in the initial image. The outlier matching features were defined as $F_i$, $Gi$, $H_n$, and $J_n$ in the images. $A_{i,n}$, $B_{i,n}$, $C_{i,n}$, $D_{i,n}$, and $E_{i,n}$ in the image pair are correctly estimated and matched inlier features by feature matching, and $F_i$, $G_i$, $H_n$, and $J_n$ are the mismatched outlier features. The TRE calculates the registration error for each matching point as follows:

$$\left.\begin{array}{l} \text{TRE}(A) = |T_{OD}A_i - A_n| \\ \text{TRE}(B) = |T_{OD}B_i - B_n| \\ \text{TRE}(C) = |T_{OD}C_i - C_n| \\ \text{TRE}(D) = |T_{OD}D_i - D_n| \\ \text{TRE}(E) = |T_{OD}E_i - E_n| \end{array}\right] \tag{3}$$

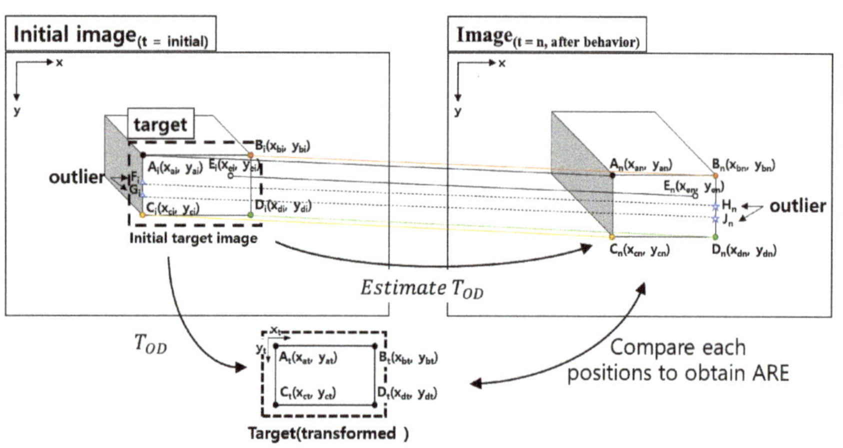


**Figure 1.** Example of features and vertices to determine TRE and ARE.

Different features may be extracted and matched depending on the type of feature detector and descriptor. Thus, the features may not be extracted and matched at the desired location, such as vertices, and it is difficult to quantify the registration error for the block shape with TRE. Therefore, ARE must be used to calculate the registration error for the block shape based on locations of vertices. All vertices were manually selected to exactly represent the block shape. The derivation of the ARE is given by Equations (4) and (5).

$$\left.\begin{array}{l} \text{ARE}_{h1} = \quad |T_{OD}A_i - T_{OD}B_i| - |A_n - B_n| = |A_t - B_t| - |A_n - B_n| \\ \text{ARE}_{v1} = |A_t - C_t| - |A_n - C_n| \\ \text{ARE}_{h2} = |C_t - D_t| - |C_n - D_n| \\ \text{ARE}_{v2} = |B_t - D_t| - |B_n - D_n| \end{array}\right] \tag{4}$$

$$\text{ARE}_h = \frac{\text{ARE}_{h1} + \text{ARE}_{h2}}{2}, \quad \text{ARE}_v = \frac{\text{ARE}_{v1} + \text{ARE}_{v2}}{2}, \quad \text{ARE} = \frac{\text{ARE}_h + \text{ARE}_v}{2}. \tag{5}$$

where $T_{OD}$ is the transformation matrix; $A$, $B$, $C$, and $D$ are the target block vertices of top left, top right, bottom left, and bottom right, respectively; subscripts $i$, $n$, and $t$ represent initial image, subsequent image, and transformed image, respectively, as shown in Figure 1. $E_i$ and $E_n$ are the detected features of the block image at the initial condition and after behavior, respectively; and $T_{OD}$ is estimated through the relationship between image$_{t=initial}$ and image$_{t=n}$ of the target. In addition, the transformed target image is obtained by applying $T_{OD}$ to the initial target image, as shown in Figure 1. $\text{ARE}_h$, $\text{ARE}_v$, and ARE were calculated by applying Equations (4) and (5) to four vertices in the transformed target and target after the behavior. $\text{ARE}_{h1}$ and $\text{ARE}_{h2}$ were calculated at the top and bottom sides of the block, respectively, and $\text{ARE}_{v1}$ and $\text{ARE}_{v2}$ were calculated at the left and right sides of the block, respectively. Based on these values, we quantitatively evaluated the horizontal and vertical errors of the target block. The conversion registration error of the block type can be analyzed intuitively. ARE was calculated by comparing the positions where the transformation matrix was applied and the position of the block after the behavior

The location of each vertex was extracted based on the pixel information of the block in each image.

## 3. Laboratory Experiment

In this study, two experiments were performed to evaluate the performance of feature detection and matching based on the behavior of blocks. Figure 2a,b show the single-block experimental setup for constant displacement on a linear stage and multiblock experiments on a moving table, respectively. In the experiments, a block with a height of 56.5 mm, a width of 89 mm, and a length of 190 mm was used to simulate the facing of the RSW.

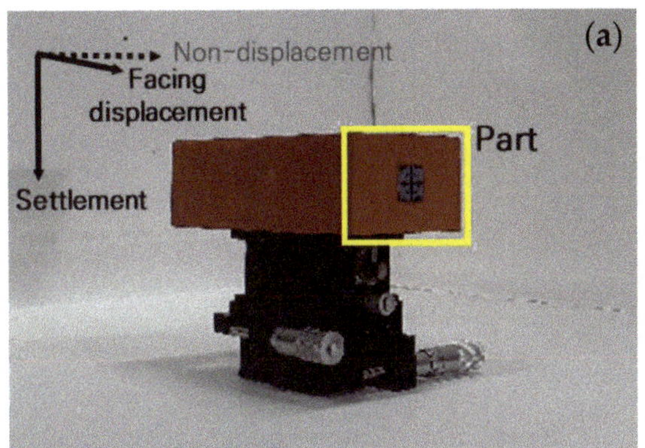


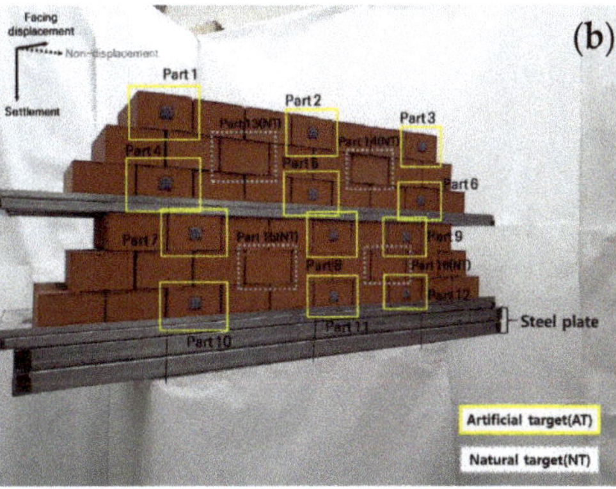


**Figure 2.** Experimental setups and target parts for single block and multiple blocks. (**a**) Single-block experiment on linear stage; (**b**) Multiblock experiment on steel plate.

The three types of behavior were generated in a single-block experiment, as shown in Figure 2a. Images were taken and analyzed before and after each displacement occurred. Subsequently, the matching performance was analyzed for 12 incident angles between 5° and 85°. The analyses were repeated 10 times under identical conditions to quantify the error. In the multiblock experiment presented in Figure 2b, three types of behaviors were generated similar to that in the single-block experiment. Images were taken and analyzed at the incidence angle that indicated excellent performance in the single-block experiment. To analyze the matching performance of blocks evenly distributed in the image among 51 blocks, 12 block facings with artificial targets were evaluated as AT (parts 1–12). In the case of the RSW structure, the features of a specific target could be confused with those of other blocks in the RSW structure because similar feature vectors are detected in the same blocks. Therefore, AT could be applied for reliable point identification and matching in the image matching method [41]. In addition, several studies reported that specific types of ATs (i.e., metal plate, black circle centered at a cross, roundel, concentric circles, cross, and speckle patterns) should be used to obtain sufficient intensity variations when a region of interest (ROI) is not sufficient [25,26,42,43]. However, because an object with sufficient strength change can be used as an NT without installing a sheet target, four NTs were assigned in the multiblock experiment to verify the usability, as shown in Figure 2b. The matching performance of four NTs was evaluated by comparing the 12 ATs. The best detector and descriptor and the usability of the block as NTs were evaluated when matching based on the behavior of the SRW structure.

## 4. Experimental Results and Discussion

### 4.1. Single-Block Experiment

In the single-block experiment, images were taken and analyzed for 12 incidence angles distributed from 5° to 85° to evaluate the performance of five feature matching methods. Figure 3 shows the repeatability of each method for a single block with an AT.

MinEigen, ORB, and BRISK have repeatabilities of less than 0.2 to the block facing. KAZE and SURF have relatively high repeatabilities of 0.288–0.875 based on the results of 10 repetitions. Moreover, KAZE and SURF showed better matching performance compared to other methods for a single block, although the repeatabilities fluctuated as the incidence angle changed. The repeatability of three different behaviors was relatively high in the section with low incidence angles, such as 5–15°, indicating that it was effective in matching because high-intensity variations were expressed in the facing of the blocks captured at low incidence angles.

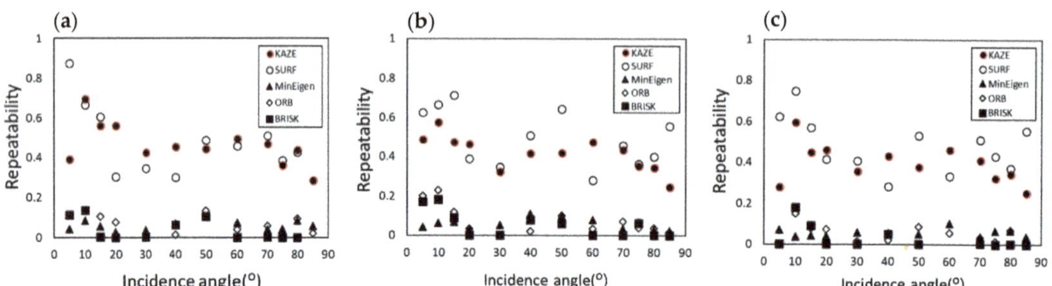

**Figure 3.** Repeatability of five feature matching methods. (**a**) Facing displacement; (**b**) Settlement; (**c**) Combined displacement.

The accurately detected and matched features were used to calculate the matching score, as shown in Figure 4. For all types of displacement, KAZE and SURF have better matching scores than MinEigen, ORB, and BRISK. We obtained more inlier matching features on the facing displacement than the settlement and combined displacements because the facing displacement in the image causes a relatively small shape deformation. In addition, the average matching scores for the overall incidence angle gradually decrease in the order of facing, settlement, and combined displacements. The changes in the image of the facing of the block, settlement, and combined displacements were compared with the normal state, as shown in Figure 5. The displacement in reality shows the actual behavior in the target area (dotted line) when the displacements (facing displacement, settlement, and combined displacement) of the block, which is located in the center of the target area, were generated. The displacement in the image shows what types of deformation appear in the image plane when actual behavior occurs from the normal state. Then, we described the scale transformation in the facing displacement, shearing transformation in the settlement, and scale and shearing transformation in the combined displacement. Compared to the shearing transformation, the scale transformation shows relatively better matching performance because it appears more similar to the existing feature vectors of the block in a normal state. Furthermore, the feature vector appears differently when both types of transformations occur together. Therefore, higher repeatability and matching scores were obtained in the order of facing, settlement, and combined displacement. Both KAZE and SURF have excellent repeatabilities and matching scores. Furthermore, KAZE shows less deviation of the matching score at all incidence angles, making it more appropriate than SURF for the images before and after the behavior of a single block.

Figure 6 presents the number of inlier features with respect to the incidence angle for five feature matching methods. KAZE detects a remarkably large number of inlier features compared to other methods. The number of inlier matching features tends to increase with the number of matching features. However, it does not increase significantly compared to the increase in the number of detected features when the incidence angle increases. For better matching performance, more inlier matching features are required, as described previously. Therefore, images taken and used at incident angles of 50–80° were recommended to extract features with sufficient resolution in the registration process.

There must be at least four 2D inlier matching features to be converted to a 3D projection transformation geometry (x-, y-, and z-axis with perspective) [37].

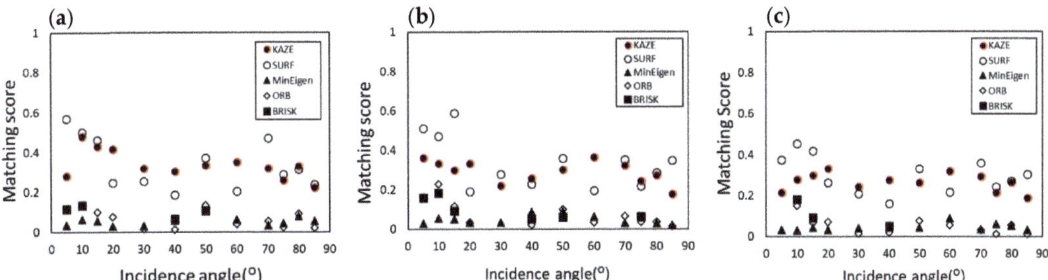

**Figure 4.** Matching scores of five feature matching methods. (**a**) Facing displacement; (**b**) Settlement; (**c**) Combined displacement.

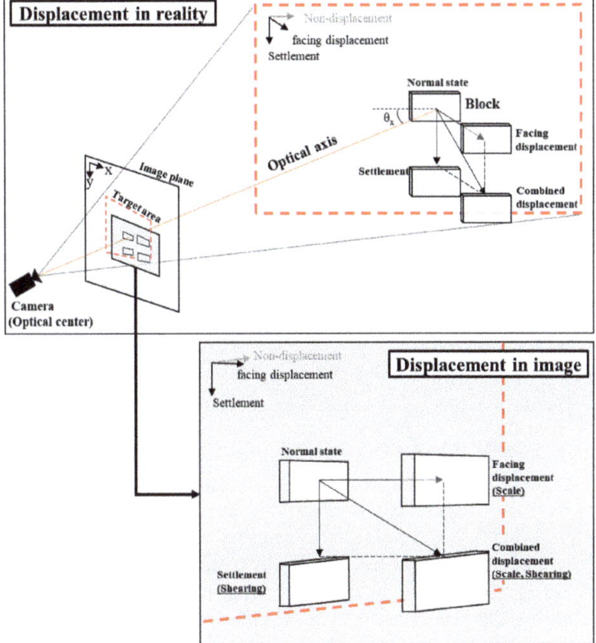



**Figure 5.** Block shape in reality and in the image according to the occurrence of three types of behaviors.

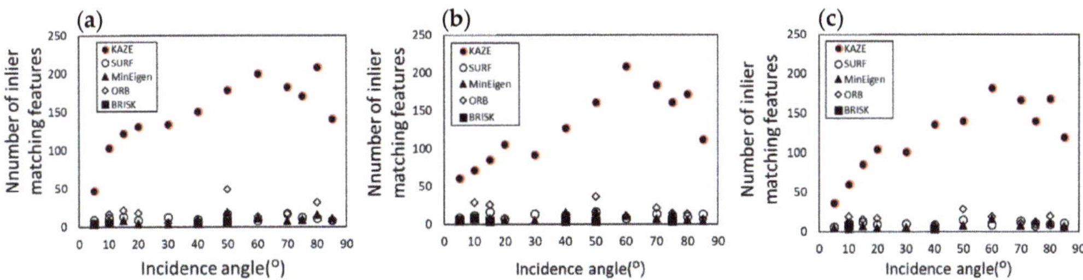

**Figure 6.** Number of inlier matching features of five feature matching methods. (**a**) Facing displacement; (**b**) Settlement; (**c**) Combined displacement.

Although all methods performed matching based on four or more matching features, the matching results of images appeared in various forms, as shown in Figure 7. The KAZE and SURF methods were successfully transformed, as shown in Figure 7a,b. The MinEigen, ORB, and BRISK methods were transformed with distortion, even if more than four matched features were used. Figure 7c–e shows examples of each image transformation. This problem occurred when the transformation matrix was estimated and matched through insufficient evidence in pairs of features. The KAZE and SURF methods did not result in distorted matching cases.

**Figure 7.** Successful and failed transformed image results. (**a**) Successful result with KAZE (incidence angle = 60°, combined displacement); (**b**) successful result with SURF (incidence angle = 60°, combined displacement); (**c**) failed result with MinEigen (incidence angle = 70°, facing displacement); (**d**) failed result with ORB (incidence angle = 30°, facing displacement); (**e**) failed result with BRISK (incidence angle = 40°, combined displacement).

For KAZE, SURF, and MinEigen, the x-axis registration error ($ARE_h$), y-axis registration error ($ARE_v$), and combined axial registration error (ARE) were evaluated for different incidence angles and each method in facing displacement (Figure 8), settlement (Figure 9), and combined displacement (Figure 10). In ORB, features were not detected for several angles of incidence, and BRISK matched less than four inlier matching features, even though features were extracted at most incident angles. Therefore, we found that BRISK and ORB could not perform ARE analysis and were inappropriate for detection and matching based on the behavior of blocks. Therefore, the results for BRISK and ORB were excluded from subsequent analysis. Moreover, KAZE has a registration error lower than 2 at an incident angle of 50–80° and has less deviation compared to other methods, as shown in Figures 8–10. To calculate the ARE, the x-y coordinates of the vertices of the block where the actual behavior occurred and the block where the behavior is estimated should be extracted and compared. However, the vertices of the block appear in various shapes as the behavior occurs. In addition, it is difficult to equally define the vertices with different shapes of blocks according to the matching methods. Moreover, vertices are not clearly present in one pixel and are distributed within two or more pixels, owing to the characteristics of an image composed of pixel units. Therefore, each vertex must be manually selected for the same behavioral shape, and ARE contains minor errors by default. Even if minor errors are included, the performance of each method could be relatively compared because the coordinates of the vertices after the actual behavior occurs were similar in each method. Table 1 lists the average and standard deviation of $ARE_h$, $ARE_v$, and ARE estimated in the behavior of each method. Specifically, KAZE showed better performance than SURF and MinEigen, and the matching performance of KAZE was consistently superior at all incidence angles, as shown in the repeated results. The KAZE method resulted in an ARE of less than 2 pixels between 50° and 80° ($ARE_h$: 0.3–1.86 pixels, $ARE_v$: 0.22–1.69 pixels, and ARE: 0.4–1.4 pixels). Based on the above results, the optical feature matching method for three types of behaviors for RSW was determined as KAZE, and the subsequent analyses were performed.

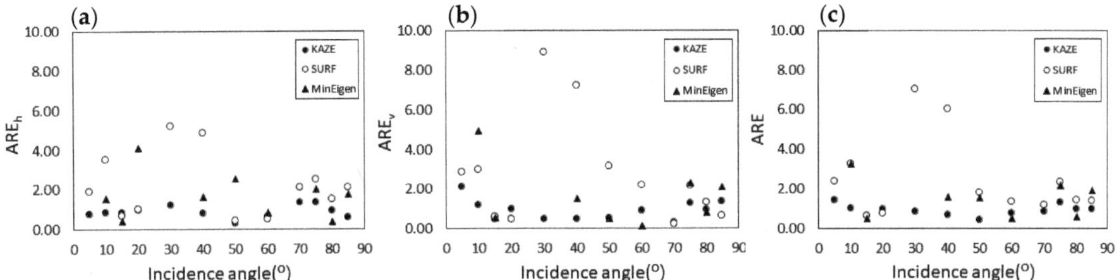

**Figure 8.** Distribution of $ARE_h$, $ARE_v$, and ARE with different feature matching methods at facing displacement. (**a**) $ARE_h$; (**b**) $ARE_v$; (**c**) ARE.

**Table 1.** Average and standard deviation of $ARE_h$, $ARE_v$, and ARE with different feature matching methods.

| | Method | Facing Displacement | | Settlement | | Combined Displacement | |
|---|---|---|---|---|---|---|---|
| | | Average | Standard Deviation | Average | Standard Deviation | Average | Standard Deviation |
| $ARE_h$ | KAZE | 0.87 | 0.31 | 1.14 | 0.57 | 0.80 | 0.43 |
| | SURF | 2.19 | 1.59 | 2.16 | 2.13 | 3.35 | 4.21 |
| | MinEigen | 24.45 | 42.26 | 71.77 | 226.51 | 9.34 | 17.52 |
| $ARE_v$ | KAZE | 0.89 | 0.51 | 1.17 | 0.66 | 1.36 | 0.88 |
| | SURF | 2.71 | 2.72 | 2.35 | 3.13 | 4.55 | 8.35 |
| | MinEigen | 24.92 | 50.25 | 67.76 | 195.74 | 11.59 | 22.77 |
| ARE | KAZE | 0.88 | 0.28 | 1.15 | 0.40 | 1.08 | 0.52 |
| | SURF | 2.45 | 2.05 | 2.25 | 2.60 | 3.95 | 6.17 |
| | MinEigen | 24.68 | 42.76 | 69.76 | 210.89 | 10.47 | 19.59 |

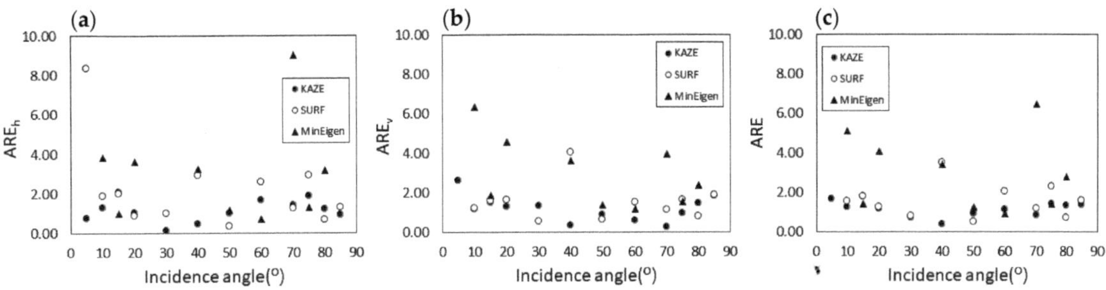

**Figure 9.** Distribution of $ARE_h$, $ARE_v$, and ARE with different feature matching methods at settlement. (**a**) $ARE_h$; (**b**) $ARE_v$; (**c**) ARE.

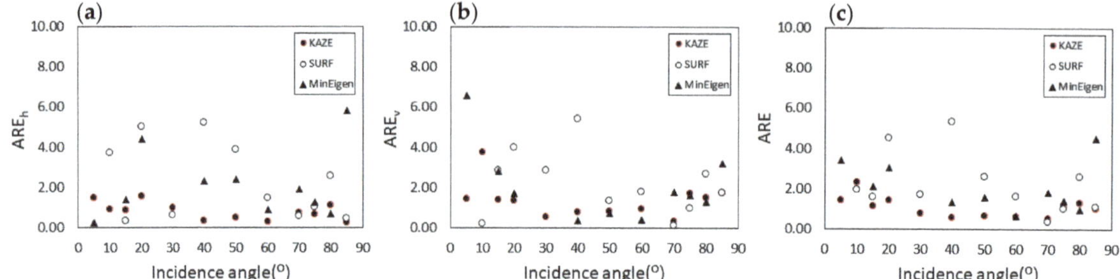

**Figure 10.** Distribution of $ARE_h$, $ARE_v$, and ARE with different feature matching methods at combined displacement. (**a**) $ARE_h$; (**b**) $ARE_v$; (**c**) ARE.

### 4.2. Multiblock Experiment

Figure 11 presents an example of images of the same size (800 × 640 pixels) for the multiblock experiment and graffiti images taken from well-known datasets from the University of Oxford [44]. In addition, 3138 and 8638 features were extracted from the laboratory experiment images and graffiti images, respectively. In the laboratory experiment images, relatively fewer features were detected because the facing of blocks was smooth and simple. Specific patterns (edges, corners, and blobs) were not included in the multiblock structure in which blocks with low-feature-intensity variation were repeatedly arranged. Therefore, repeated feature vectors with low-intensity variations may be disadvantageous in the image matching procedure. Therefore, the KAZE method, which was verified as the technique with the best performance in the single-block experiment, was applied and validated in this experiment. We analyzed how the aforementioned phenomenon affected the feature matching procedure and whether it exhibited excellent performance for RSW structures simulated with multiple blocks.

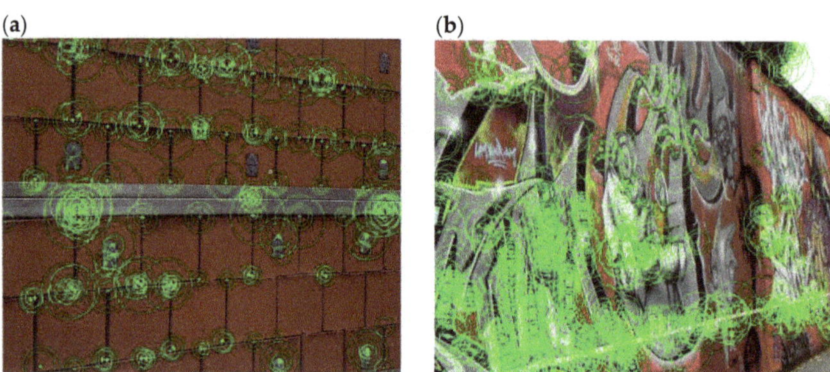

**Figure 11.** Detected features with different types of images. (**a**) Laboratory experiment; (**b**) Graffiti.

An experiment was performed to verify the matching performance of KAZE for a multiblock structure, where same-sized blocks, including 12 ATs and 4 NTs, were repeatedly arranged, as shown in Figure 2b. The multiblock experiment was analyzed by taking images at an incident angle of 60° (targeting the center of the multiple blocks) included in the range of 50–80° with the highest resolution in the single-block experiment. Figure 12 shows the repeatability, matching score, and number of inlier matching features in the multiblock experiment. The results of ATs showed that the repeatability, matching score, and inlier matching features were 0.36–0.83, 0.15–0.47, and 90–440, respectively. This was a sufficient result for high-resolution feature matching compared with the result of the single-block experiment. Moreover, relatively high feature matching performance was detected in blocks collinear in the gravitational direction of the optical axis of the camera under all types of displacement (i.e., parts 2, 5, 8, and 11). Furthermore, the resolution decreased slightly as the angle decreased or increased. In particular, parts 3, 6, 9, and 12 exhibited low inlier matching performances with incidence angles lower than 60°. However, these parts also had 90–218 inlier matching features, which is sufficient for 3D projected geometry transformations. Based on the results of NTs, the repeatability, matching score, and inlier matching features were 0.38–0.65, 0.22–0.43, and 126–327, respectively, with a similar distribution to that reported for ATs at similar incidence angles. Furthermore, there was an adequate number of inlier matching features for 3D transformation. This shows a sufficient potential for NTs to replace ATs when analyzed using the KAZE method as a feature matching technique in RSW structures.

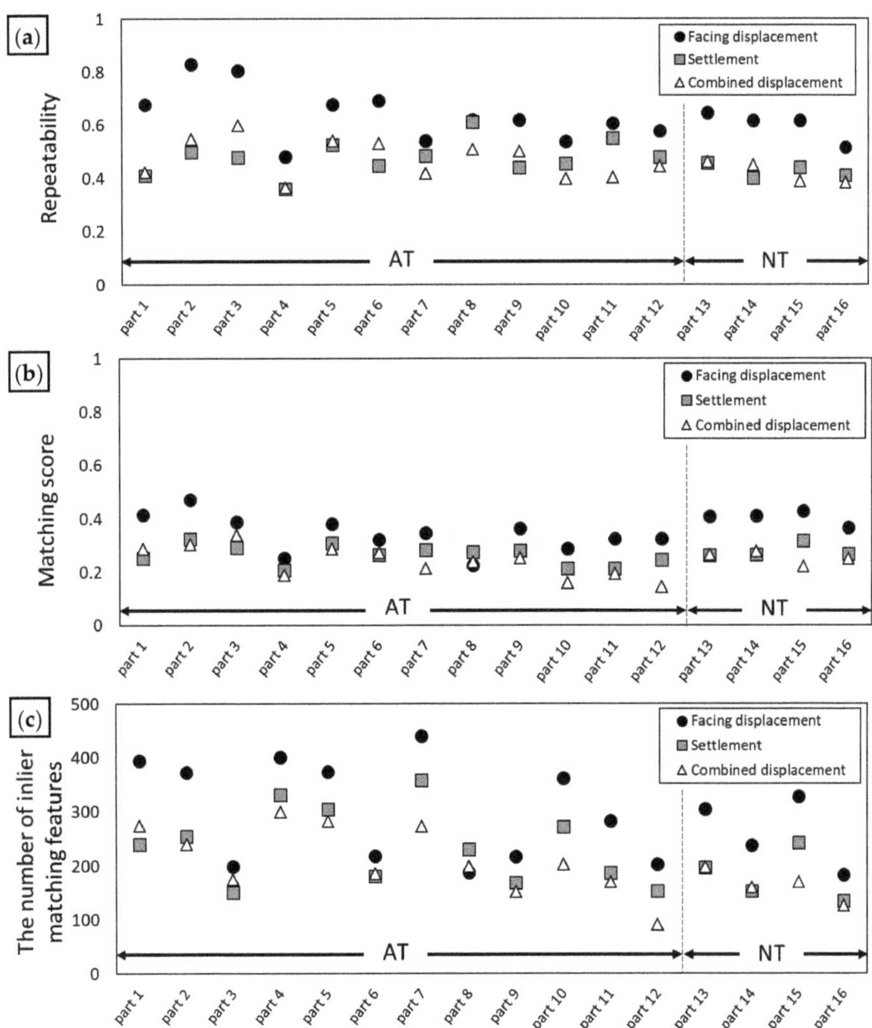

**Figure 12.** Comparison metrics in multiblock experiment with KAZE method (incidence angle = 60°). (**a**) Repeatability; (**b**) Matching score; (**c**) The number of inlier matching features.

Table 2 lists the minimum and maximum values for the feature performance evaluation measured from the single- and multiblock experiments, including ATs and NTs. Each value of the multiblock experiment was distributed similarly to the results of the single-block experiment. Therefore, the repetitive arrangement of blocks in the image is not a disadvantage if the ROI region is properly set in the process of extracting and matching features. Further, light casts a shadow on the block boundaries when the blocks are repeatedly arranged, allowing for more feature extraction than the edges of the single block. This phenomenon allows for more successful feature matching in the image after the behavior occurs. Figure 13 reveals the feature extraction results used as targets in single- and multiblock experiments. The single-block features were not extracted at the left edge of the facing, owing to the low color contrast, as shown in Figure 13a. However, the features of continuous blocks in the RSW structure were extracted from all corners by the color contrast between block and shadow, as shown in Figure 13b. As a result, 570 and 789 features were extracted for a single and multiple blocks, respectively, showing a significant difference. The ratio of inlier matching features and total extracted features was 35.09% (200/570) for the single block and 49.68% (392/789) for multiple blocks, indicating the relatively higher resolution of the matching performance of the latter compared to that of the former. Therefore, multiple repeatedly arranged blocks constituted an advantage rather than a disadvantage in the feature detection and matching procedure. In the multiblock experiment, multiple blocks were variously distributed in the image. Therefore, size and feature vector characteristics appeared differently depending on the location of the target. Therefore the performances of feature matching were slightly different.

**Table 2.** Minimum and maximum values for the feature performance evaluation measured from the single- and multiblock experiments.

| Comparison Matrix | Single-Block Experiment (at Incidence Angle = 50–80°) | | Multiblock Experiment | |
|---|---|---|---|---|
| | Min. | Max. | Min. | Max. |
| Repeatability | 0.3247 | 0.4965 | 0.3599 | 0.8289 |
| Matching score | 0.2131 | 0.3654 | 0.1449 | 0.4715 |
| Number of inlier matching features | 139.9 | 208.4 | 90 | 440 |

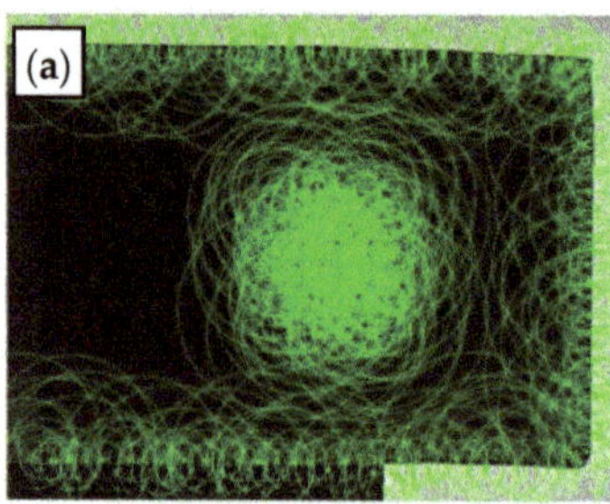

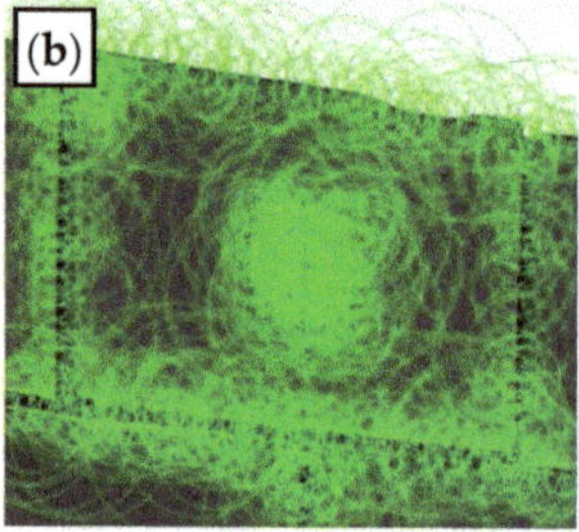


**Figure 13.** Examples of extracting features at facing displacement. (**a**) Single block (pixel size 350 × 268); (**b**) Block (part 2) in multiple blocks (pixel size: 330 × 298).

Figure 14 shows the distributions of $ARE_h$, $ARE_v$, and ARE for each displacement type of RSW structure. The distributions of $ARE_h$, $ARE_v$, and ARE for ATs in each behavior type were approximately 0.18–1.11, 0.12–1.32, and 0.15–1.27 pixels, respectively, for facing displacement; 0.10–1.55, 0.40–1.70, and 0.32–1.27 pixels, respectively, for settlement and 0.39–1.84, 0.16–1.68, and 0.54–1.29 pixels, respectively, for combined displacement. Moreover, the distributions of $ARE_h$, $ARE_v$, and ARE for NTs in each behavior type were 0.38–1.58, 0.29–1.31, and 0.49–1.27 pixels, respectively, for facing displacement; 0.58–1.03, 0.40–1.70, and 0.49–1.16 pixels, respectively, for settlement; and 0.78–1.56, 0.31–1.01, and

0.54–1.29 pixels, respectively, for combined displacement. The averages of $ARE_h$, $ARE_v$, and ARE were 0.86, 0.74, and 0.80 for the ATs and 0.92, 0.83, and 0.87, respectively, for the NTs. Although the results of NTs were slightly higher, they were similar to those of single-block experiments at incidence angles between 50° and 80° (0.99, 0.88, and 0.94). In addition, maximum $ARE_h$, $ARE_v$ and ARE values were obtained in AT (1.84), NT (1.68), and AT (1.29), respectively. The behavior of the entire section in the image could be analyzed at ATs and NTs by using KAZE; these results are compared in Table 3. The repeatability, matching score, number of inlier matching features, and ARE between ATs and NTs show similar performance. Therefore, we confirmed that NTs are a suitable target for feature matching and behavior analysis based on the feature performance evaluation, even if it does not attach a sheet target to the center of the block like AT. In addition, NTs can overcome the limitation of ATs that a specific target must be manually installed and detected.

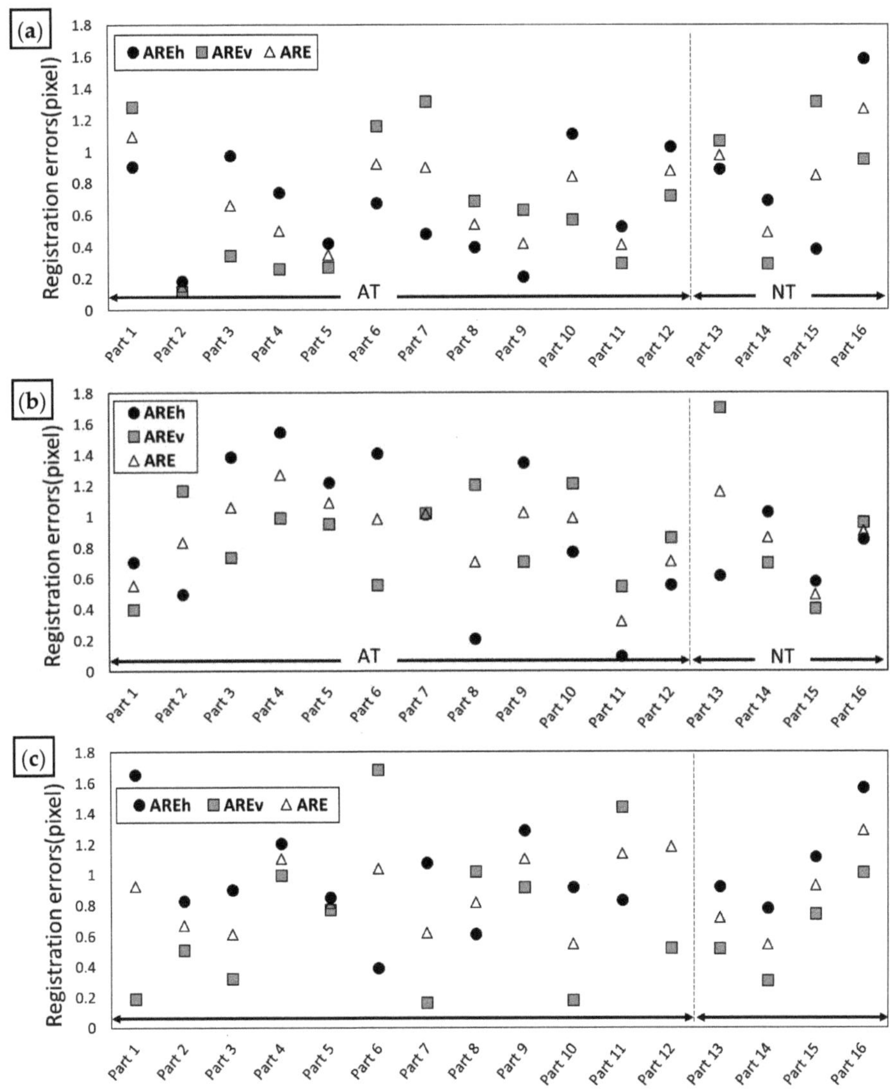

**Figure 14.** Distribution of $ARE_h$, $ARE_v$, and ARE for three types of behaviors. (**a**) Facing displacement; (**b**) Settelemt; (**c**) Combined displacement.

**Table 3.** Average values for feature performance evaluation of KAZE method with ATs and NTs.

| Average of All Displacement Types | Repeatability | Matching Score | No. of Inlier Matching Points | ARE |
|---|---|---|---|---|
| AT (Parts 1–12) | 0.53 | 0.28 | 250 | 0.8 |
| NT (Parts 13–16) | 0.48 | 0.31 | 203 | 0.87 |

## 5. Conclusions

Experiments were conducted to find the optimal feature matching method (detector and descriptor) to detect behaviors of blocks. Two laboratory experiments (i.e., single- and multiblock experiments) were analyzed. In the single-block experiment, the best feature matching method was selected by analyzing the values for feature performance evaluation. In contrast, in the multiblock experiment, the applicability and performance of the primarily simplified feature matching method were evaluated. Subsequently, the feature matching performance of NTs and ATs were compared to confirm the applicability of NTs for RSW structures. The main findings of this study are as follows:

1. Feature matching technology was applied to detect and match target changes in image pairs according to block behavior. Both the KAZE and SURF methods showed excellent performance in repeatability and matching score, which were based on the number of features. In particular, the KAZE method showed a remarkably large number of inlier matching features and obtained stable results at all incidence angles compared to other methods. In addition, ARE based on the position of the vertices of the block in the image pair (original image, transformed image) was the best in KAZE. Therefore, the KAZE method was selected as the best feature matching method, owing to its great ability (among the compared methods) to detect and match image changes based on the behavior of the block type.

2. The feature matching performance of the KAZE method was evaluated according to the behavior of multiple blocks where blocks were consecutively arranged. The repeatability, matching score, number of inlier matching features, and ARE showed excellent performance. All these results are similar to the single-block experiment results listed in Table 2. The KAZE results show that the matching performance of NTs (parts 13–16) was similar to that of ATs (parts 1–12). Therefore, the KAZE method could be applied to perform feature matching when evaluating the behavior of blocks in RSWs without installing ATs.

The ability to accurately match the 3D behavior of multiple blocks can be used as a basic step to quantitatively analyze the behavior of RSW structures in three-dimensional space. Therefore, if the KAZE technique is applied to RSWs, it can be used to accurately analyze the behavior of the retaining wall through high-accuracy behavior matching performance.

**Author Contributions:** Conceptualization, Y.-S.H.; methodology, Y.-S.H.; formal analysis, Y.-S.H.; investigation, Y.-S.H.; data curation, Y.-S.H.; writing—original draft preparation, Y.-S.H.; writing—review and editing, J.L. and Y.-T.K.; supervision, Y.-T.K.; project administration, Y.-S.H. and Y.-T.K.; funding acquisition, Y.-T.K. All authors have read and agreed to the published version of the manuscript.

**Funding:** This work is supported by a Korea Agency for Infrastructure Technology Advancement(KAIA) grant funded by the Ministry of Land, Infrastructure and Transport (Grant 22TSRD-C151228-04).

**Data Availability Statement:** Not applicable.

**Conflicts of Interest:** The authors declare no conflict of interest.

## References


1. Anastasopoulos, I.; Georgarakos, T.; Georgiannou, V.; Drosos, V.; Kourkoulis, R. Seismic Performance of Bar-Mat Reinforced-Soil Retaining Wall: Shaking Table Testing Versus Numerical Analysis with Modified Kinematic Hardening Constitutive Model. *Soil Dyn. Earthquake Eng.* **2010**, *30*, 1089–1105. [CrossRef]
2. KISTEC. *Safety Inspection Manual for National Living Facilities*; Korea Infrastructure Safety Corporation: Seoul, Korea, 2019; p. 6.
3. Jang, D.J. Gwangju Apartment Retaining Wall Collapse, Dozens of Vehicles Sunk and Damaged, Yonhap News. 2015. Available online: https://www.yna.co.kr/view/AKR20150205012552054 (accessed on 5 February 2015).
4. Kim, Y.R. Retaining Wall With 'Good' Safety Rating Also Collapsed, Korean Broadcasting System News. 2020. Available online: https://news.kbs.co.kr/news/view.do?ncd=5009697 (accessed on 22 September 2020).
5. Laefer, D.; Lennon, D. Viability Assessment of Terrestrial LiDAR for Retaining Wall Monitoring. In Proceedings of the GeoCongress 2008, Orleans, LA, USA, 9–12 March 2008; pp. 247–254. [CrossRef]
6. Hain, A.; Zaghi, A.E. Applicability of Photogrammetry for Inspection and Monitoring of Dry-Stone Masonry Retaining Walls. *Transp. Res. Rec.* **2020**, *2674*, 287–297. [CrossRef]
7. Bathurst, R.J. Case study of a monitored propped panel wall. In Proceedings of the International Symposium on Geosynthetic-Reinforced Soil Retaining Walls, Denver, CO, USA, 8–9 August 1992; pp. 214–227.
8. Sadrekarimi, A.; Ghalandarzadeh, A.; Sadrekarimi, J. Static and Dynamic Behavior of Hunchbacked Gravity Quay Walls. *Soil Dyn. Earthq. Eng.* **2008**, *28*, 2564–2571. [CrossRef]
9. Yoo, C.S.; Kim, S.B. Performance of a Two-Tier Geosynthetic Reinforced Segmental Retaining Wall Under a Surcharge Load: Full-Scale Load Test and 3D Finite Element Analysis. *Geotext. Geomembr.* **2008**, *26*, 460–472. [CrossRef]
10. Yang, G.; Zhang, B.; Lv, P.; Zhou, Q. Behaviour of Geogrid Reinforced Soil Retaining Wall With Concrete-Rigid Facing. *Geotext. Geomembr.* **2009**, *27*, 350–356. [CrossRef]
11. Datteri, R.D.; Liu, Y.; D'Haese, P.F.; Dawant, B.M. Validation of a Nonrigid Registration Error Detection Algorithm Using Clinical MRI Brain Data. *IEEE Trans. Med. Imaging* **2015**, *34*, 86–96. [CrossRef]
12. Gupta, S.; Chakarvarti, S.K.; Zaheeruddin, N.A. Medical Image Registration Based on Fuzzy c-means Clustering Segmentation Approach Using SURF. *Int. J. Biomed. Eng. Technol.* **2016**, *20*, 33–50. [CrossRef]
13. Kaucha, D.P.; Prasad, P.W.C.; Alsadoon, A.; Elchouemi, A.; Sreedharan, S. Early Detection of Lung Cancer Using SVM Classifier in Biomedical Image Processing. In Proceedings of the IEEE International Conference on Power, Control, Signals and Instrumentation Engineering (ICPCSI), Chennai, India, 21–22 September 2017; pp. 3143–3148. [CrossRef]
14. Yang, W.; Zhong, L.; Chen, Y.; Lin, L.; Lu, Z.; Liu, S.; Wu, Y.; Feng, Q.; Chen, W. Predicting CT Image From MRI Data Through Feature Matching With Learned Nonlinear Local Descriptors. *IEEE Trans. Med. Imaging* **2018**, *37*, 977–987. [CrossRef]
15. Wei, C.; Xia, H.; Qiao, Y. Fast Unmanned Aerial Vehicle Image Matching Combining Geometric Information and Feature Similarity. *IEEE Geosci. Remote Sens. Lett.* **2020**, *18*, 1731–1735. [CrossRef]
16. Regmi, K.; Shah, M. Bridging the Domain Gap for Ground-to-Aerial Image Matching. In Proceedings of the IEEE International Conference on Computer Vision (ICCV), Seoul, Korea, 27 October–2 November 2019; pp. 470–479. [CrossRef]
17. Song, F.; Dan, T.T.; Yu, R.; Yang, K.; Yang, Y.; Chen, W.Y.; Gao, X.Y.; Ong, S.H. Small UAV-Based Multi-Temporal Change Detection for Monitoring Cultivated Land Cover Changes in Mountainous Terrain. *Remote Sens. Lett.* **2019**, *10*, 573–582. [CrossRef]
18. Shao, Z.; Li, C.; Li, D.; Altan, O.; Zhang, L.; Ding, L. An Accurate Matching Method for Projecting Vector Data Into Surveillance Video to Monitor and Protect Cultivated Land. *ISPRS Int. J. Geo Inf.* **2020**, *9*, 448. [CrossRef]
19. Thornton, S.; Dey, S. Machine Learning Techniques for Vehicle Matching with Non-Overlapping Visual Features. In Proceedings of the 2020 IEEE 3rd Connected and Automated Vehicles Symposium (CAVS), Victoria, BC, Canada, 18 November-16 December 2020; pp. 1–6. [CrossRef]
20. Tareen, S.A.K.; Saleem, Z. A Comparative Analysis of SIFT, SURF, KAZE, AKAZE, ORB, and BRISK. In Proceedings of the 2018 IEEE International Conference on Computing, Mathematics and Engineering Technologies (iCoMET), Sukkur, Pakistan, 3–4 March 2018; pp. 1–10. [CrossRef]
21. Pieropan, A.; Bjorkman, M.; Bergstrom, N.; Kragic, D. Feature Descriptors for Tracking by Detection: A Benchmark. *arXiv* **2016**, arXiv:1607.06178.
22. Mikolajczyk, K.; Schmid, C. A Performance Evaluation of Local Descriptors. *IEEE Trans. Pattern Anal. Mach. Intell.* **2005**, *27*, 1615–1630. [CrossRef] [PubMed]
23. Chien, H.J.; Chuang, C.C.; Chen, C.Y.; Klette, R. When to Use What Feature? SIFT, SURF, ORB, or A-KAZE Features for Monocular Visual Odometry. In Proceedings of the IEEE International Conference on Image and Vision Computing, Palmerston North, New Zealand, 21–22 November 2016; pp. 1–6. [CrossRef]
24. Feng, D.; Feng, M.Q.; Ozer, E.; Fukuda, Y. A Vision-Based Sensor for Noncontact Structural Displacement Measurement. *Sensors* **2015**, *15*, 16557–16575. [CrossRef] [PubMed]
25. Lee, J.J.; Shinozuka, M. Real-Time Displacement Measurement of a Flexible Bridge Using Digital Image Processing Techniques. *Exp. Mech.* **2006**, *46*, 105–114. [CrossRef]
26. Choi, H.S.; Cheung, J.H.; Kim, S.H.; Ahn, J.H. Structural Dynamic Displacement Vision System Using Digital Image Processing. *NDT E Int.* **2011**, *44*, 597–608. [CrossRef]

27. Berg, R.R.; Christopher, B.R.; Samtani, N.C. Design of MSE walls. In *Design and Construction of Mechanically Stabilized Earth Walls and Reinforced Soil Slopes*; US Department of Transportation Federal Highway Administration: New Jersey, DC, USA, 2009; Volume 1, pp. 4.1–4.80.

28. Koerner, R.M.; Koerner, G.R. An Extended Data Base and Recommendations Regarding 320 Failed Geosynthetic Reinforced Mechanically Stabilized Earth (MSE) Walls. *Geotext. Geomembr.* **2018**, *46*, 904–912. [CrossRef]

29. Ji, R.; Gao, Y.; Duan, L.Y.; Hongxun, Y.; Dai, Q. *Learning-Based Local Visual Representation and Indexing*; Morgan Kaufmann: Burlington, MA, USA, 2015; pp. 17–40.

30. Shi, J.; Tomasi, C. Good Features to Track. In Proceedings of the IEEE Computer Society Conference on Computer Vision and Pattern Recognition (CVPR'94), Seattle, WA, USA, 21–23 June 1994; pp. 593–600. [CrossRef]

31. Bay, H.; Tuytelaars, T.; Van Gool, L. SURF: Speeded up Robust Features. In Proceedings of the 9th European Conference on Computer Vision (ECCV 2006), Graz, Austria, 7–13 May 2006; Volume 3951, pp. 404–417. [CrossRef]

32. Leutenegger, S.; Chli, M.; Siegwart, R.Y. BRISK: Binary Robust Invariant Scalable Keypoints. In Proceedings of the 2011 IEEE International Conference on Computer Vision, Barcelona, Spain, 6–13 November 2011; pp. 2548–2555. [CrossRef]

33. Rublee, E.; Rabaud, V.; Konolige, K.; Bradski, G. ORB: An Efficient Alternative to SIFT or SURF. In Proceedings of the IEEE International Conference on Computer Vision, Barcelona, Spain, 6–13 November 2011; pp. 2564–2571. [CrossRef]

34. Alcantarilla, P.F.; Bartoli, A.; Davison, A.J. KAZE Features. In Proceedings of the European Conference on Computer Vision, Florence, Italy, 7–13 October 2012; pp. 214–227. [CrossRef]

35. Mikolajczyk, K.; Mikolajczyk, K. Scale & Affine Invariant Interest Point Detectors. *Int. J. Comput. Vis.* **2004**, *60*, 63–86. [CrossRef]

36. Revaud, J.; Weinzaepfel, P.; de Souza, C.R.; Pion, N.; Csurka, G.; Cabon, Y.; Humenberger, M. R2D2: Repeatable and Reliable Detector and Descriptor. *arXiv* **2019**, arXiv:1906.06195.

37. Yi, K.M.; Trulls, E.; Lepetit, V.; Fua, P. Lift: Learned Invariant Feature Transform. In *European Conference on Computer Vision*; Springer: Cham, Switzerland, 2016; pp. 467–483. [CrossRef]

38. DeTone, D.; Malisiewicz, T.; Rabinovich, A. Superpoint: Self-Supervised Interest Point Detection and Description. In Proceedings of the IEEE Conference on Computer Vision and Pattern Recognition Workshops, Salt Lake City, UT, USA, 18–22 June 2018; pp. 224–236. [CrossRef]

39. Forsberg, D.; Farnebäck, G.; Knutsson, H.; Westin, C.F. Multi-Modal Image Registration Using Polynomial Expansion and Mutual Information. In Proceedings of the International Workshop on Biomedical Image Registration, Nashville, TN, USA, 7–8 July 2012; pp. 40–49. [CrossRef]

40. Schmidt-Richberg, A.; Ehrhardt, J.; Werner, R.; Handels, H. Fast Explicit Diffusion for Registration With Direction-Dependent Regularization. In Proceedings of the International Workshop on Biomedical Image Registration, Nashville, TN, USA, 7–8 July 2012; pp. 220–228. [CrossRef]

41. Wang, Z.; Kieu, H.; Nguyen, H.; Le, M. Digital Image Correlation in Experimental Mechanics and Image Registration in Computer Vision: Similarities, Differences and Complements. *Opt. Lasers Eng.* **2015**, *65*, 18–27. [CrossRef]

42. Esmaeili, F.; Varshosaz, M.; Ebadi, H. Displacement Measurement of the Soil Nail Walls by Using Close Range Photogrammetry and Introduction of CPDA Method. *Measurement* **2013**, *46*, 3449–3459. [CrossRef]

43. Zhao, S.; Kang, F.; Li, J. Displacement Monitoring for Slope Stability Evaluation Based on Binocular Vision Systems. *Optik* **2018**, *171*, 658–671. [CrossRef]

44. Visual Geometry Group. Affine Covariant Regions Datasets. 2004. Available online: http://www.robots.ox.ac.uk/~{}vgg/data (accessed on 15 July 2007).

*Article*

# SAR Image Segmentation by Efficient Fuzzy *C*-Means Framework with Adaptive Generalized Likelihood Ratio Nonlocal Spatial Information Embedded

**Jingxing Zhu** [1,2,3]**, Feng Wang** [1,2,*] **and Hongjian You** [1,2,3]

[1]  Aerospace Information Research Institute, Chinese Academy of Sciences, Beijing 100094, China; zhujingxing20@mails.ucas.ac.cn (J.Z.); hjyou@mail.ie.ac.cn (H.Y.)
[2]  Key Laboratory of Technology in Geo-Spatial Information Processing and Application System, Chinese Academy of Sciences, Beijing 100190, China
[3]  School of Electronic, Electrical and Communication Engineering, University of Chinese Academy of Sciences, Beijing 101408, China
*  Correspondence: wangfeng003020@aircas.ac.cn

**Abstract:** The existence of multiplicative noise in synthetic aperture radar (SAR) images makes SAR segmentation by fuzzy *c*-means (FCM) a challenging task. To cope with speckle noise, we first propose an unsupervised FCM with embedding log-transformed Bayesian non-local spatial information (LBNL_FCM). This non-local information is measured by a modified Bayesian similarity metric which is derived by applying the log-transformed SAR distribution to Bayesian theory. After, we construct the similarity metric of patches as the continued product of corresponding pixel similarity measured by generalized likelihood ratio (GLR) to avoid the undesirable characteristics of log-transformed Bayesian similarity metric. An alternative unsupervised FCM framework named GLR_FCM is then proposed. In both frameworks, an adaptive factor based on the local intensity entropy is employed to balance the original and non-local spatial information. Additionally, the membership degree smoothing and the majority voting idea are integrated as supplementary local information to optimize segmentation. Concerning experiments on simulated SAR images, both frameworks can achieve segmentation accuracy of over 97%. On real SAR images, both unsupervised FCM segmentation frameworks work well on SAR homogeneous segmentation in terms of region consistency and edge preservation.

**Keywords:** image segmentation; synthetic aperture radar (SAR); fuzzy *c*-means (FCM); speckle noise; non-local means

**Citation:** Zhu, J.; Wang, F.; You, H. SAR Image Segmentation by Efficient Fuzzy C-Means Framework with Adaptive Generalized Likelihood Ratio Nonlocal Spatial Information Embedded . *Remote Sens.* **2022**, *14*, 1621. https://doi.org/10.3390/rs14071621

Academic Editors: Yue Wu, Kai Qin, Maoguo Gong and Qiguang Miao

Received: 4 March 2022
Accepted: 25 March 2022
Published: 28 March 2022

**Publisher's Note:** MDPI stays neutral with regard to jurisdictional claims in published maps and institutional affiliations.



## 1. Introduction

Segmentation is a fundamental problem in SAR image analysis and applications. The primary purpose of segmentation is to segment the image into non-intersecting and consistent regions that are homogeneous [1]. Due to coherent speckle noise, which can be modeled as a powerful multiplicative noise, SAR image segmentation is recognized as a complex task. So far, many SAR image segmentation methods have been proposed to cope with the effect of speckle noise on image segmentation, such as threshold-based method [2], edge-based methods [3], region-based methods [4–9], cluster methods [10–13], Markov random field methods [3,14], Level set methods [15], graph-based methods [16,17], and deep learning based methods [18–21]. Among these methods, clustering is a commonly used method in segmentation tasks due to its effectiveness and stability. The fuzzy *s*-means (FCM) [22] is a classical clustering algorithm and has been extensively used to segment images. Unlike the hard clustering strategy, FCM is a soft clustering algorithm that allocates membership degrees to every category for each pixel. The FCM can achieve a good result for noise-free images. However, the standard FCM is noise-sensitive and lacks robustness without considering any spatial information. Thus, many modified algorithms have been proposed to enhance the effectiveness and robustness of standard FCM against noise.

Ref. Ahmed et al. [23] incorporated the spatial neighborhood term into the objective function of FCM, named BCFCM. BCFCM can modify the label of the center pixel by neighborhood weight distance and enhance the robustness to noise. However, it is time consuming. To reduce the complexity, Ref. Chen and Zhang [24] replaced the spatial neighborhood term with a mean-filtered and median-filtered image, respectively, called FCM_S1 and FCM_S2. Because of the availability of these two images in advance, the time complexity is greatly reduced. Besides, kernel methods were embedded into FCM_S1 and FCM_S2 to explore the non-Euclidean structure of data. Then two kernelized versions, KFCM_S1 and KFCM_S2, were derived. Ref. Szilagyi et al. [25] proposed the enhanced FCM, named EnFCM, which executed clustering on a gray level histogram rather than pixels to reduce the computation cost considerably. Afterwards, the fast generalized FCM (FGFCM) was proposed by Cai et al. [26]. In FGFCM, a new factor $S_{ij}$ was used to measure the local (both spatial and gray) similarity instead of $\alpha$ in EnFCM. The original image and its local spatial and gray level neighborhood are used to construct a non-linear weighted sum image, and then the clustering process is executed on the gray level histogram of the summed image. Thus, the computational load is very light. It is noteworthy that in all the aforementioned algorithms, the parameters for balancing noise immunity and edge preservation are needed. To avoid the parameter selection, Ref. Krinidis and Chatzis [27] introduced a new factor, $G_{ki}$, incorporating local spatial and gray information into the objective function in a fuzzy way and proposed a new FCM named FLICM. This algorithm completely avoids the selection of parameters and is relatively independent of the type of noise.

However, when an image is contaminated with powerful noise, the local information may also be contaminated and unreliable. Actually, for a pixel, plenty of pixels with a similar neighborhood structural configuration exist on the image [28]. Exploring a larger space and incorporating nonlocal spatial information is necessary. Ref. Wang et al. [29] proposed a modified FCM with incorporating both local and non-local spatial information. Ref. Zhu et al. [30] introduced a novel membership constraint and a new objective function was constructed, named GIFP_FCM. Afterwards, Ref. Zhao et al. [31] incorporated non-local information into the objective function of the standard FCM and GIFP_FCM, respectively, and proposed two improved FCMs: An FCM with non-local spatial information (FCM_NLS) [31] and a novel FCM with a non-local adaptive spatial constraint term (FCM_NLASC) [32].

While the improved FCM listed above works well on simulated, nature, and MR images, none of them consider the statistical characteristics of SAR images. Consequently, the above-mentioned methods cannot assure a segmentation result on SAR images. To solve this problem, Ref. Feng et al. [33] proposed a robust non-local FCM with edge preservation (NLEP_FCM). In this algorithm, a modified ratio distance to measure patch similarity for SAR images was defined, and a sum image was constructed. The edge was rectified on the summed image. Ref. Ji and Wang [34] defined an adaptive binary weight NL-means and adopted an adaptive filter degree parameter to balance noise removed and detail preservation. Besides, a fuzzy between-cluster variation term was embedded into the objective function. Eventually, a new FCM named NS_FCM was proposed. However, the NS_FCM applied Euclidean distance, which is not suitable for SAR. Ref. Wan et al. [35] directly considered the statistical distribution of SAR image and derived a patch-similarity metric for SAR image based on Bayes theory. However, the assumption of additive Gaussian noise in the Bayes equation is not considered. Therefore, it is still a challenge to segment SAR images effectively.

In this paper, we incorporate the non-local spatial information into the objective function of FCM and propose two improved FCMs for segmenting SAR images effectively. In [36], an implicit assumption that the NL-means can emerge from the Bayesian approach is that the image is affected by additive Gaussian noise. Hence, we first apply the logarithmic transformation to convert the SAR multiplicative model into an additive model and then apply the Bayesian formula to derive a modified patch similarity metric. We then incorporate the non-local spatial information obtained by this new similarity metric into FCM and propose a more robust FCM named LBNL_FCM. Afterward, this Bayesian theory-based similarity metric is analyzed. Three undesirable properties that are incompatible with human intuition are determined, even if LBNL_FCM yields a satisfactory regional consistency. In order to avoid these undesirable distance characteristics, a statistical test method called generalized likelihood ratio (GLR) is introduced. The generalized likelihood ratio was applied to SAR images in the study of Deledalle et al. [37] and was proven to possess good distance properties. However, unlike the logarithm summation form in [37], we construct the patch similarity as the continued products of the similarity of corresponding pixels by combining the SAR statistical distribution. This continued product GLR-based similarity metric is used to generate an additional image that is insensitive to speckle noise. The additional auxiliary image is then added into the objective function of FCM as the non-local spatial information term and we propose GLR_FCM. Besides, an adaptive factor based on local intensity entropy is utilized to balance the original image and the nonlocal spatial information. Eventually, a simple membership degree smoothing and majority voting are adopted in LBNL_FCM and GLR_FCM to compensate for local spatial information. The basic idea is that the membership degree of a pixel should be influenced by neighborhood pixels. Experiments will demonstrate that LBNL_FCM can achieve a better result in region consistency than previous algorithms. GLR_FCM avoids the decay parameter selection and achieves a good balance between region consistency and edge preservation.

The main contributions are as follows:

(1) A robust unsupervised FCM framework incorporating adaptive Bayesian non-local spatial information is proposed. This non-local spatial information is measured by the log-transformed Bayesian metric which is induced by applying the log-transformed SAR distribution into the Bayesian theory.

(2) To avoid undesirable properties of the log-transformed Bayesian metric, we construct the similarity between patches as the continued product of corresponding pixel similarity measured by the generalized likelihood ratio. An alternative unsupervised FCM framework is then proposed, named GLR_FCM.

(3) An adaptive factor is employed to balance the original and non-local spatial information. Besides, a sample membership degree smoothing is adopted to provide the local spatial information iteratively.

The rest of this paper is organized as follows. In Section 2, the relevant theories are described in detail. Section 3 presents the experimental results and parameters analysis. In Section 4, the qualitative evaluations of results are discussed. The conclusion is provided in Section 5.

## 2. Materials and Methods

### 2.1. Theoretical Background

#### 2.1.1. The Standard FCM

Fuzzy C-Means Clustering is based on fuzzy set theory, proposed by Bezdek [38]. The standard FCM segments the image $X$ into $c$ clusters by iteratively minimizing the objective function. The objective function of the FCM algorithm is

$$\min J_m(U, V) = \sum_{k=1}^{c} \sum_{i=1}^{N} u_{ki}^m \|x_i - v_k\|^2 \tag{1}$$

where $X = \{x_1, x_2, ..., x_N\}$ denotes an image with $N$ pixels, $m$ is the fuzzy weighing exponent, usually set as 2, $c$ is the number of clusters, and $v_k$ is the $k$th cluster center. $u_{ki}^m$ represents the membership degree of the $i$th pixel belonging to the $k$th cluster, satisfying $u_{ki} \in [0, 1]$ and $\sum_{k=1}^{c} u_{ki} = 1$.

We can minimize Equation (1) by the Lagrange multiplier method. The $u_{ki}$ and $v_k$ can be update by

$$u_{ki} = \frac{1}{\sum_{j=1}^{c}\left(\frac{\|x_i-v_k\|^2}{\|x_i-v_j\|^2}\right)^{\frac{1}{m-1}}} \tag{2}$$

$$v_k = \frac{\sum_{i=1}^{N} u_{ki}^m x_i}{\sum_{i=1}^{N} u_{ki}^m} \tag{3}$$

When the objective function reaches the minimum, we can convert the membership degree $U$ into a segmentation result by assigning each pixel a class possessing the largest membership degree.

### 2.1.2. Nonlocal Means Method

Many algorithms have demonstrated the effectiveness of local information for the segmentation of low-noise images. However, the local information may be disturbed and unreliable when the noise is severe. In addition to local information, for a particular pixel, many pixels with a similar neighborhood configuration [28] exist over the entire image. We call this nonlocal spatial information. More specifically, for the $i$th pixel in image $X$, its non-local spatial information $\tilde{x}_i$ can be calculated by the following formula

$$\tilde{x}_i = \sum_{j \in W_i^r} w_{ij} x_j \tag{4}$$

where $W_i^r$ denotes the non-local search window of radius $r$ centered at the $i$th pixel, $w_{ij}(j \in W_i^r)$ represents the normalized weight coefficient depending on the similarity of patches centered at the $i$th and $j$th pixel, i.e., $v_s(N_i)$ and $v_s(N_j)$. The similarity $w_{ij}$ can be defined as

$$w_{ij} = \frac{1}{Z_i} \exp\left(-\frac{\|v_s(N_i) - v_s(N_j)\|_{2,\sigma}^2}{h^2}\right) \tag{5}$$

where $h$ controls the smoothing degree, $Z_i = \sum_{j \in W_i^r} \exp\left(-\frac{\|v_s(N_i)-v_s(N_j)\|_{2,\sigma}^2}{h^2}\right)$ is the normalized constant, $v(N_i) = \{x_k, k \in N_i\}$ indicates the vectorized patch at pixel $i$, $N_i$ is the local neighborhood with size $s \times s$ at pixel $i$, and $\|v_s(N_i) - v_s(N_j)\|_{2,\sigma}^2$ denotes the Euclidean distance between patches $v_s(N_i)$ and $v_s(N_j)$.

### 2.1.3. Nonlocal Spatial Information Based on Bayesian Approach

Kervrann et al. [36] claims that the NL-means filter can also emerges from the Bayesian formulation and the Bayesian estimator $\hat{u}_s(N_i)$ of vectorized patch centered at the $i$th pixel can be written as

$$\hat{u}_s(N_i) \approx = \frac{\sum_{j \in W_i^r} v_s(N_j) p(v_s(N_i)|v_s(N_j))}{\sum_{j \in W_i^r} p(v_s(N_i)|v_s(N_j))} \tag{6}$$

where $W_i^r$ denotes the non-local spatial information search window centered at pixel $i$ with size $r \times r$, $v_s(N_i)$ is the observed vectorized patch centered at pixel $i$, the set $\{v_s(N_1), ..., v_s(N_{r2})\}$ is the observed patch samples in $W_i^r$. Once we know $p(v_s(N_i)|v_s(N_j))$, we can calculate the Bayesian estimator $\hat{u}_s(N_i)$.

In [36], a usual additive noise model is considered, i.e., $v(x_i) = u(x_i) + n(x_i)$, $v(x_i)$ is the grayscale value of pixel $i$ in the observed image, $u(x_i)$ is the grayscale value of pixel $i$ in the noise-free image, $n(x_i)$ is the additive Gaussian white noise. The likelihood can be factorized as

$$p(v_s(N_i)|v_s(N_j)) = \prod_{k=1}^{s^2} p((x_i^k)|(x_j^k)) \tag{7}$$

Due to the additive Gaussian noise model being considered, the $v_s(N_i)|v_s(N_j)$ follows a multivariate normal distribution. Thus, the Bayesian estimator $\hat{u}_s(N_i)$ is analogous to NL-means (Equation (4)) in form, and we can get

$$p(v_s(N_i)|v_s(N_j)) = \prod_{k=1}^{s^2} p((x_i^k)|(x_j^k)) \propto \exp\left(-\frac{\|v_s(N_i) - v_s(N_j)\|^2}{h^2}\right) \tag{8}$$

*2.2. The Modified FCM Based on Log-Transformed Bayesian Nonlocal Spatial Information*

The initial NL-means can emerge from the Bayesian approach on the premise that the image is disturbed by additive Gaussian noise. Different from the work in Wan et al. [35] that directly considers Nakagami–Rayleigh distribution, we first utilize the logarithmic transformation to convert the multiplicative speckle noise model into the additive model. Then the Bayesian approach (Equation (8)) is used on log-transformed distribution to derive a new similarity metric for SAR images. We note that this is a reasonable treatment. Actually, Ref. Xie et al. [39] has proved that, for the amplitude concerning the SAR image, the PDF of the log-transformed distribution is statistically very close to the Gaussian PDF. Therefore, the image analysis methods based on the Gaussian noise image can work equally well on the log-transformed amplitude SAR image.

Considering the multiplicative noise model, which can be described as

$$X = R_X * n_X \tag{9}$$

where X represents the observed image, $R_X$ is the noise-free amplitude image and is equal to $R^{\frac{1}{2}}$, R is the radar cross section, $n_X$ is the speckle noise. Under the assumption of fully developed speckle [40], the PDF of $L$-look amplitude of SAR images obeys the Nakagami–Rayleigh distribution [41], represented as

$$p(X|R) = \frac{2L^L}{\Gamma(L)R^L} X^{2L-1} \exp(-\frac{LX^2}{R}) \tag{10}$$

where $\Gamma(\cdot)$ is the Gamma function; then the log transformation converts Equation (10) into

$$\bar{X} = \bar{R}_X + \bar{n}_X \tag{11}$$

where $\bar{X} = \ln X$, $\bar{R}_X = \ln R_X$, $\bar{n}_X = \ln n_X$. Since the logarithmic transformation is monotonic, the PDF of $\bar{X}$ is

$$p_{\bar{X}}(\bar{X}|R) = \frac{2}{\Gamma(L)}\left(\frac{L}{R}\right)^L \exp(-\frac{L\exp(2\bar{X})}{R})\exp(2L\bar{X}) \tag{12}$$

Then, applying Equation (12) to the Bayesian formulation, we obtain

$$P(v_s(N_i)|v_s(N_j)) = \prod_{k=1}^{s^2} p(x_i^k|x_j^k)$$

$$= \prod_{k=1}^{s^2} \frac{2}{\Gamma(L)} \left(\frac{L}{x_j^k}\right)^L \exp\left(-\frac{Le^{2x_i^k}}{x_j^k}\right) \exp(2Lx_i^k)$$

$$= \left(\frac{2}{\Gamma(L)}\right)^{s^2} L^{Ls^2} \prod_{k=1}^{s^2} \exp\left(-L\ln x_j^k - \frac{L\exp(2x_i^k)}{x_j^k} + 2Lx_j^k\right)$$

$$= \left(\frac{2}{\Gamma(L)}\right)^{s^2} L^{Ls^2} \exp\left(-L\sum_{k=1}^{s^2} \ln x_j^k + \frac{\exp(2x_i^k)}{x_j^k} - 2x_i^k\right) \tag{13}$$

$$\propto \exp\left[-L\sum_{k=1}^{s^2}\left(\ln x_j^k + \frac{\exp(2x_i^k)}{x_j^k} - 2x_i^k\right)\right]$$

$$\propto \exp\left[-\frac{\sum_{k=1}^{s^2}\left(\ln x_j^k + \frac{\exp(2x_i^k)}{x_j^k} - 2x_i^k\right)}{h^2}\right]$$

where $s^2$ denotes the number of pixels in patch $v_s(N_i)$ and $v_s(N_j)$, $x_i^k$ is the $k$th pixel in the patch centered at the $i$th pixel, $h^2 = \frac{1}{L}$ is the decay parameter of the filter. Then, a new patch similarity metric based on the Bayesian approach and log-transformed statistical distribution of SAR is derived. So far, $\|v_s(N_i) - v_s(N_j)\|^2$ in Equation (5) can be replaced by

$$\bar{D}_s(v_s(N_i), v_s(N_j)) = \sum_{k=1}^{s^2} \ln x_j^k + \frac{\exp(2x_i^k)}{x_j^k} - 2x_i^k \tag{14}$$

Hence, the weight $w_{ij}$ between patches $v_s(N_i)$ and $v_s(N_j)$ can be calculated by

$$w_{ij} = \frac{1}{Z_i} \exp\left(\frac{\bar{D}_s(v_s(N_i), v_s(N_j))}{h^2}\right) \tag{15}$$

Equation (15) can be applied to Equation (4). Thus an additional auxiliary image $\tilde{I}$, which is speckle noise insensitive, can be obtained. With $\tilde{I}$ as the non-local spatial information term, incorporating into the standard FCM, a new robust FCM based on the log-transformed Bayesian non-local information (LBNL_FCM) can be obtained. The objective function is as follows

$$\min J_m(U, V) = \sum_{k=1}^{c}\sum_{i=1}^{N} u_{ki}^m \|x_i - v_k\|^2 + \sum_{k=1}^{c}\sum_{i=1}^{N} \eta_i u_{ki}^m \|\tilde{x}_i - v_k\|^2 \tag{16}$$

$$s.t. \sum_{k=1}^{c} u_{ki} = 1, \quad 0 \leq u_{ki} \leq 1, \quad 0 \leq \sum_{i=1}^{N} u_{ki} \leq N$$

Minimizing Equation (16) by using the Lagrange multiplier method, the membership degree $u_{ki}$ and cluster $v_k$ can be updated by

$$u_{ki} = \frac{1}{\sum_{j=1}^{c}\left(\frac{\|x_i - v_k\|^2 + \eta_i \|\tilde{x}_i - v_k\|^2}{\|x_i - v_j\|^2 + \eta_i \|\tilde{x}_i - v_j\|^2}\right)^{\frac{1}{m-1}}} \tag{17}$$

$$v_k = \frac{\sum_{i=1}^{N}\left(u_{ki}^m x_i + \eta_i u_{ki}^m \tilde{x}_i\right)}{\sum_{i=1}^{N}\left(u_{ki}^m + \eta_i u_{ki}^m\right)} \tag{18}$$

### 2.3. Some Problems on Patch Similarity Metric by Bayesian Theory

In the last section, we made the amplitude SAR image log transformed and combined the Bayesian equation to derive a new similarity metric. This new metric for patch satisfies the assumptions in [36] and the non-local spatial information can be appropriately measured. However, there are still three problems that bother us.

Problem 1: In Equation (15), a decay parameter $h$ is always needed to calculate the weights of the non-local spatial information. In most cases, it is difficult to obtain a satisfactory value.

Problem 2: The logarithmic transformation is homoerotic transformation (nonlinear transformation), which converts multiplier noise into additive noise while reducing the contrast of the SAR image. The original statistical distribution is changed.

Problem 3: In experiments, the LBNL_FCM effectively suppresses speckle noise and achieves the best region consistency. However, this similarity metric has three distance characteristics that do not match the characteristics one would intuitively expect. Here, we list three properties that Deledalle [37] used for the assessment of a similarity metric.

**Property 1** (Symmetry). *A good similarity metric should be invariant to changes in position.*

$$\ell(z_1, z_2) = \ell(z_2, z_1) \quad for \quad \forall z_1, z_2 \tag{19}$$

**Property 2** (Self-Similarity Maximum). *A good similarity measurement should have the property of being the maximum similarity between itself.*

$$\ell(z_1, z_1) >= \ell(z_1, z_2) \quad for \quad \forall z_1, z_2 \tag{20}$$

**Property 3** (Self-Similarity Equal). *For a good similarity measurement, the maximum similarity should not depend on the variation of variables .*

$$\ell(z_1, z_1) = \ell(z_2, z_2) \quad for \quad \forall z_1, z_2 \tag{21}$$

To further illustrate, we consider $\mathbf{x} = 1, 2, 3, 4, 5$ and $\mathbf{y} = 1, 2, 3, 4, 5$. We set $x_{i,k} = x$ and $x_{j,k} = y$. Then we put $x_{i,k}$ and $x_{j,k}$ into Equation (14) and get the similarity matrix.

Figure 1 shows the similarity matrix. From the green square we can see Property 1 is not satisfied; from the orange square we can see Property 2 is not be satisfied; from the purple square we can see Property 3 is not be satisfied. The problems discussed above encourage us to find other better similarity metrics, even if the Bayesian similarity metric is good at keeping region consistency in segmentation. Fortunately, Deledalle [37] proposed that the similarity of patches can be measured by statistical test. He proved the generalized likelihood ratio satisfied properties used in evaluating the similarity metric.

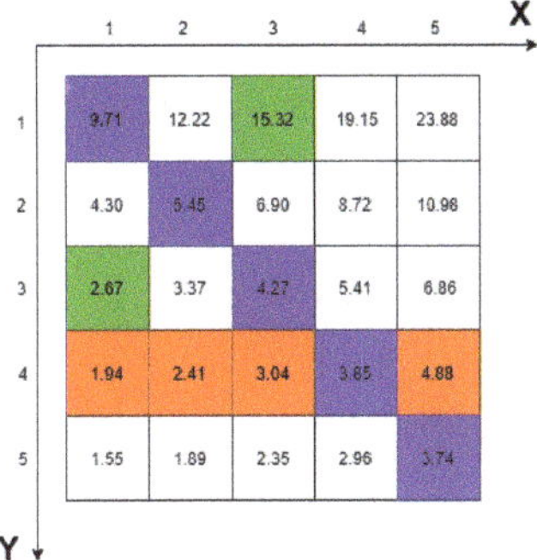

**Figure 1.** The similarity matrix between **x** and **y**. The elements marked green, orange and purple are sampled to illustrate the unsatisfied properties of the log-transformed Bayesian distance.

### 2.4. The New FCM Based on Generalized Likelihood Ratio

Generalized likelihood testing is defined as the ratio between the maximum value of the likelihood function with constraints to the maximum value of the likelihood function without constraints. The basic idea is that, if the parameters imposed on the model are valid, adding such a constraint should not lead to a significant decrease in the maximum value of the likelihood function. Considering Nakagami–Rayleigh distribution, for a pair of patches $(v_s(N_i), v_s(N_j))$ on a SAR image, we can define its likelihood ratio (LR)

$$\psi_{LR}(v_s(N_i), v_s(N_j)) = \frac{p(v_s(N_i), v_s(N_j), R_i = R_0, R_j = R_0; \varkappa_0)}{p(v_s(N_i), v_s(N_j), R_i = R_1, R_j = R_2; \varkappa_1)} \tag{22}$$

where $\varkappa_0$ and $\varkappa_1$ represent two hypotheses, defined as

$$\begin{aligned} \varkappa_0 &: R_i = R_j = R_0 (\textit{Null Hypothesis}) \\ \varkappa_1 &: R_i = R_1; R_j = R_2; R_1 \neq R_2 (\textit{Alternative Hypothesis}) \end{aligned} \tag{23}$$

$v_s(N_i)$ is the patch centered at pixel $i$, and $v_s(N_j)$ denotes the non-local patch centered at pixel $j$. $R_i$ and $R_j$ as the hypothesis parameters denote the noise-free backscatter value of center pixel $i$. Hypothesis $\varkappa_0$ means a parametric constraint on statistical distribution that the two patches $(v_s(N_i), v_s(N_j))$ come from the same distribution. Thus, they have the same backscatter value, formalized as $R_i = R_j = R_0$. Hypothesis $\varkappa_1$ means no constraint on the statistical distribution of $v_s(N_i)$ and $v_s(N_j)$, formalized as $R_i \neq R_j$. For the sake of mathematical simplicity, we choose parameters in this way

$$\begin{aligned} R_0 &= \max_{\Theta} p(v_s(N_i), v_s(N_j), R_i = R_j = R_0; \varkappa_0) \\ R_1 \text{ or } R_2 &= \max_{R_1, R_2 \in \Theta} p(z_s(N_i), z_s(N_j), R_i = R_1, R_j = R_2; \varkappa_1) \end{aligned} \tag{24}$$

Thus, Equation (22) becomes the generalized likelihood ratio (GLR), defined as

$$\psi_{GLR}(v_s(N_i), v_s(N_j)) = \frac{\sup_{R_0} p(v_s(N_i), v_s(N_j), R_i = R_j = R_0)}{\sup_{R_1, R_2} p(v_s(N_i), v_s(N_j), R_i = R_1, R_j = R_2, R_1 \neq R_2)} \tag{25}$$

where $0 < \psi_{GLR}(v_s(N_i), v_s(N_j)) < 1$; the larger the $\psi_{GLR}(z_s(N_i), z_s(N_j))$, the larger the probability that hypothesis $\varkappa_0$ holds, and the more inclined to accept $\varkappa_0$. This also means that there is a higher probability of two patches $v_s(N_i)$ and $v_s(N_j)$ coming from the same distribution. Thus, we can use GLR to measure the similarity between two patches.

Unlike the Deledalle [37] approach, we construct the patch similarity as the continued product of corresponding pixel similarity. Next, we will give a detailed derivation.

Now, we assume $v_s(N_i)$ and $v_s(N_j)$ are irrelevant, and the corresponding pixel within the patch is independent. Thus, the similarity between $v_s(N_i)$ and $v_s(N_i)$ can be calculated by

$$\psi_{GLR}(v_s(N_i), v_s(N_j)) = \prod_{k=1}^{N} \xi_{GLR}(x_i^k, x_j^k) \tag{26}$$

where $N = s^2$ is the number of pixels in the patch, and $\xi_{GLR}(x_i^k, x_j^k)$ is defined as

$$\begin{aligned} \xi_{GLR}(x_i^k, x_j^k) &= \frac{\sup_{R_0} p(x_i^k, x_j^k; R_1 = R_2 = R_0)}{\sup_{R_1, R_2} p(x_i^k, x_j^k; R_i = R_1, R_j = R_2, R_1 \neq R_2)} \\ &= \frac{\sup_{R_0} [p(x_i^k, x_j^k; R_1 = R_2 = R_0)]}{[\sup_{R_1} p(x_i^k; R_i = R_1)] * [\sup_{R_2} p(x_j^k; R_j = R_2)]} \end{aligned} \tag{27}$$

$x_i^k$ and $x_j^k$ denote the $k$th pixel in patch; $R_0$, $R_1$, $R_2$ denote noise-free backscatter value. To obtain the maximum likelihood value $\sup_{R_0} p(x_i^k, x_j^k, R_i = R_j = R_0)$, we need get joint probability

$$
\begin{aligned}
p(x_i^k, x_j^k; R_i = R_j = R_0) &= p(x_i^k; R_0) * p(x_j^k; R_0) \\
&= \left(\frac{2}{\Gamma(L)}\right)^2 * \left(\frac{L}{R_0}\right)^{2L} * \left(x_i^k x_j^k\right)^{2L-1} * \exp\left\{-\frac{L}{R_0}\left[\left(x_i^k\right)^2 + \left(x_j^k\right)^2\right]\right\}
\end{aligned}
\tag{28}
$$

To obtain the maximum likelihood estimator $\hat{R}_0$ of $R_0$, we construct the maximum likelihood function

$$
\begin{aligned}
\mathcal{L}(R_0) &= \prod_{m=1}^{M} p(x_i^{k_m}; R_0) * p(x_j^{k_m}; R_0) \\
&= \prod_{m=1}^{M} \left(\frac{2}{\Gamma(L)}\right)^2 * \left(\frac{L}{R_0}\right)^{2L} * \left(x_i^{k_m} x_j^{k_m}\right)^{2L-1} \\
&\quad * \exp\left\{-\frac{L}{R_0}\left[\left(x_i^{k_m}\right)^2 + \left(x_j^{k_m}\right)^2\right]\right\}
\end{aligned}
\tag{29}
$$

Then, making the logarithm on $\mathcal{L}(R_0)$ and differentiating

$$
\begin{aligned}
\frac{\partial \ln \mathcal{L}(R_0)}{\partial R_0} &= \frac{\partial}{\partial R_0}\left\{\sum_{m=1}^{M} \ln \frac{4L^{2L}}{\Gamma^2(L)} - 2L \ln R_0 + (2L-1)\ln\left(x_i^{k_m} z_j^{k_m}\right) - \frac{L}{R_0}\left[\left(x_i^{k_m}\right)^2 + \left(x_j^{k_m}\right)^2\right]\right\} \\
&= -\frac{2LM}{R_0} + \frac{L}{R_0^2}\sum_{m=1}^{M}\left[\left(x_i^{k_m}\right)^2 + \left(x_j^{k_m}\right)^2\right]
\end{aligned}
\tag{30}
$$

Let $\frac{\partial \ln \mathcal{L}(R_0)}{\partial R_0} = 0$; then, we get

$$
\hat{R}_0 = \frac{1}{2M}\sum_{m=1}^{M}\left[\left(x_i^{k_m}\right)^2 + \left(x_j^{k_m}\right)^2\right]
\tag{31}
$$

considering that there is only one available observation for each pixel in the patch, that is to say $M = 1$; thus, we can get

$$
\hat{R}_0 = \frac{1}{2}\left[\left(x_i^k\right)^2 + \left(x_j^k\right)^2\right]
\tag{32}
$$

With the same derivation process as above, we can obtain the maximum likelihood estimator $\hat{R}_1$ and $\hat{R}_2$ for $R_1$ and $R_2$

$$
\begin{aligned}
\hat{R}_1 &= \left(x_i^k\right)^2 \\
\hat{R}_2 &= \left(x_j^k\right)^2
\end{aligned}
\tag{33}
$$

Now, we replace $R_0$, $R_1$, $R_2$ with maximum likelihood estimators $\hat{R}_0$, $\hat{R}_1$, and $\hat{R}_2$ in Equation (27); then, we get the similarity between corresponding pixels

$$
\begin{aligned}
\xi_{GLR}(x_i^k, x_j^k) &= \frac{\sup_{R_0} p(x_i^k, x_j^k; R_1 = R_2 = R_0)}{\sup_{R_1,R_2} p(x_i^k, x_j^k; R_i = R_1, R_j = R_2, R_1 \neq R_2)} \\
&= \frac{\frac{4L^{2L}}{\Gamma(L)} * \left\{\frac{1}{2}\left[\left(x_i^k\right)^2 + \left(x_j^k\right)^2\right]\right\}^{-2L} * \left(x_i^k x_j^k\right)^{2L-1} * \exp(-2L)}{\left\{\frac{2L^L}{\Gamma(L)}(x_i^k)^{-2L} * (x_i^k)^{2L-1} * \exp(-L)\right\} * \left\{\frac{2L^L}{\Gamma(L)}(x_j^k)^{-2L} * (x_j^k)^{2L-1} * \exp(-L)\right\}}
\end{aligned}
\tag{34}
$$

After simplifying Equation (34), we get

$$\xi_{GLR}(x_i^k, x_j^k) = \left[\frac{2x_i^k x_j^k}{(x_i^k)^2 + (x_j^k)^2}\right]^{2L} \tag{35}$$

Equation (35) can measure the similarity between corresponding pixels within two patches. Figure 2a shows the similarity $\xi_{GLR}(x_i^k, x_j^k)$, where $x_i^k = [1, 2, 3, 4, 5]$ and $x_j^k = [1, 2, 3, 4, 5]$. From Figure 2a we can see that the Properties 1 and 3 mentioned earlier can be satisfied. Figure 2b is the change curve of similarity $\xi_{GLR}(x_i^k, x_j^k)$ when $x_i^k$ is fixed at 1 and $x_j^k = [1, 2, ..., 10]$. The maximum $\xi_{GLR}(x_i^k, x_j^k)$ can be obtained when $x_i^k = x_j^k = 1$. Besides, $\xi_{GLR}(x_i^k, x_j^k)$ gradually decreases with increasing distance. Thus, Property 2 can be proved.

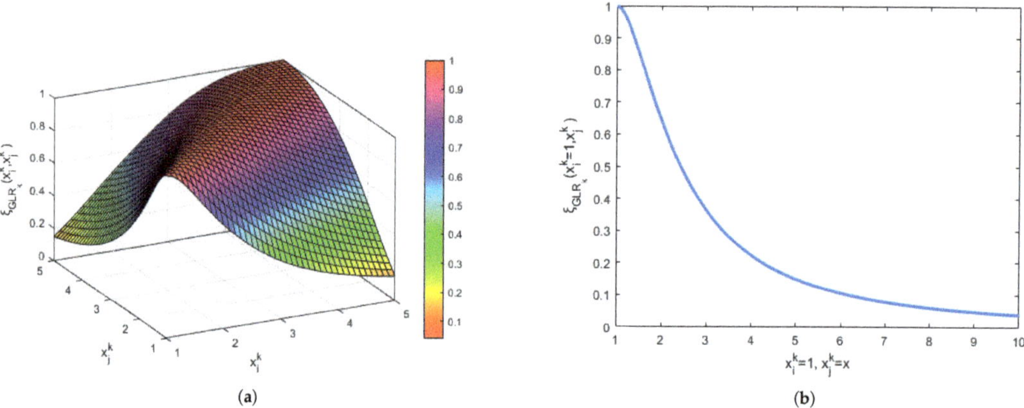

(a)    (b)

**Figure 2.** The similarity value $\xi_{GLR}(x_i^k, x_j^k)$ based on GLR. (**a**) The X and Y axes indicate values of $x_i^k$ and $x_j^k$. (**b**) The similarity when $x_i^k = 1$ and $x_j^k$ are taken from 1 to 10.

Therefore, by putting Equation (35) into Equation (26), a patch similarity metric based on GLR can be derived as follows

$$\psi_{GLR}(v_s(N_i), v_s(N_j)) = \prod_{k=1}^{N} \xi_{GLR}(x_i^k, x_j^k) = \prod_{k=1}^{N} \left[\frac{2z_i^k x_j^k}{(x_i^k)^2 + (x_j^k)^2}\right] \tag{36}$$

We then can use this similarity metric based on GLR (Equation (36)) to obtain the weight of each patch in a non-local search space centered at pixel $i$. Then the recovered amplitude of pixel $i$ in in SAR image can be calculated as follows

$$\tilde{x}_i = \sum_{j \in W_i^r} \psi_{GLR}(v_s(N_i), v_s(N_j)) * x_j \tag{37}$$

where $\tilde{x}_i$ is the estimator of the $i$th pixel, $\psi_{GLR}(v_s(N_i), v_s(N_j))$ is the weight between patch $v_s(N_i)$ and $v_s(N_j)$. After visiting all pixels in SAR image, we can construct an auxiliary image $\tilde{I} = \{\tilde{x}_1, \tilde{x}_2, ...\tilde{x}_i, ..., \tilde{x}_N\}$. Then $\tilde{I}$ is added into the objective function of standard FCM as non-local spatial information term and we can obtain GLR_FCM

$$\min J_m(U, V) = \sum_{k=1}^{c} \sum_{i=1}^{N} u_{ki}^m ||x_i - v_k||^2 + \sum_{k=1}^{c} \sum_{i=1}^{N} \eta_i u_{ki}^m ||\tilde{x}_i - v_k||^2 \tag{38}$$

$$s.t. \sum_{k=1}^{c} u_{ki} = 1, \quad 0 \le u_{ki} \le 1, \quad 0 \le \sum_{i=1}^{N} u_{ki} \le N$$

By minimizing Equation (38) using Lagrange multiplier method, the membership degree $u_{ki}$ and cluster $v_k$ can be updated by

$$u_{ki} = \frac{1}{\sum_{j=1}^{c}\left(\frac{\|x_i-v_k\|^2+\eta_i\|\tilde{x}_i-v_k\|^2}{\|x_i-v_j\|^2+\eta_i\|\tilde{x}_i-v_j\|^2}\right)^{\frac{1}{m-1}}} \tag{39}$$

$$v_k = \frac{\sum_{i=1}^{N}\left(u_{ki}^m x_i + \eta_i u_{ki}^m \tilde{x}_i\right)}{\sum_{i=1}^{N}\left(u_{ki}^m + \eta_i u_{ki}^m\right)} \tag{40}$$

In the objective function of LBNL_FCM and GLR_FCM, an adaptive factor based on local intensity entropy $\eta_i$ is introduced to balance the original detail information and non-local spatial information. $\eta_i$ is defined as

$$\eta_i = \alpha \times \frac{exp(\max E_i) - exp(E_i)}{exp(\max E_i) - 1} \tag{41}$$

$$\alpha = Med\{\sigma_1, \sigma_2, ..., \sigma_i, ..., \sigma_{N-1}, \sigma_N\}$$

where $E_i = -\sum_{j=1}^{k} p_i \log(p_i)$ denotes the information entropy of the local area histogram at the $i$th pixel. $k$ is the number of quantized gray levels. $\sigma_i$ denotes the local variance at the $i$th pixel, $Med$ indicates a median operation, and $N$ is the total number of pixels.

In Equation (41), $\eta_i$ is determined by the local intensity entropy $E_i$. In the homogeneous region, the amplitude values tend to be the same, and $E_i$ is small; hence, a large weight $\eta_i$ will be assigned for non-local spatial information. Conversely, at the edges, where the local entropy $E_i$ is relatively large, and $\eta_i$ receives a small value, the original SAR information is given more consideration.

Figure 3a–e are original SAR image slices and Figure 3f–j are the $\eta_i$ maps for Figure 3a,b, respectively. We can see a black color near the edge, which indicates that the intensity value of $\eta_i$ at the edge is small and relatively large in the homogeneous regions. Thus, the original image information and non-local spatial information can be dynamically balanced and adjusted.

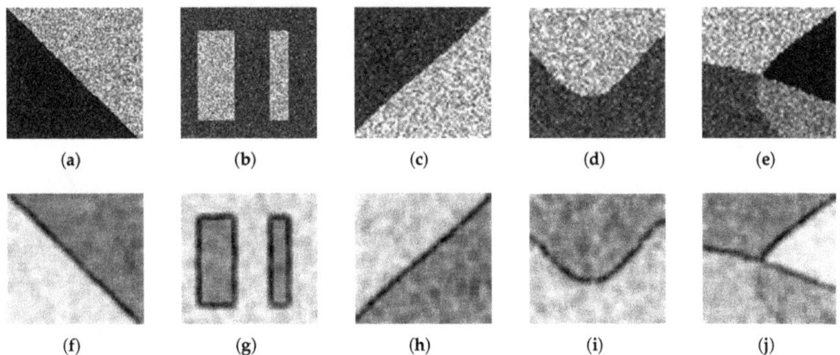

Figure 3 captions overline (a)–(e) and (f)–(j)

**Figure 3.** The results of dynamically balanced factor $\eta_i$. (**a**–**e**) Five sample SAR image slices. (**f**–**j**) $\eta_i$ maps of (**a**–**e**), respectively.

*2.5. The Membership Degree Smoothing and Label Correction*

In addition to non-local spatial information, local spatial information is also useful. For a pixel, its class should be influenced by the surrounding pixels. Thus, we add membership degree smoothing into the iteration process. For the $i$th pixel in the SAR image, we sum the membership vector of the neighborhood pixels to obtain a weight vector

$\phi_i(\phi_i = [\phi_{1i} \quad \phi_{2i} \quad \dots \quad \phi_{ci}])$, and $\phi_i$ is weighted to the membership vector of the $i$th pixel. Then we can get the new membership degree $u_i'$ for the $i$th pixel.

$$\phi_{ki} = \sum_{j \in N_i} u_{kj}$$

$$u_i' = u_i \bullet \phi_i \tag{42}$$

where $N_i$ is the neighborhood pixels of the $i$th pixel, $u_i$ is the membership before smoothing and $u_i'$ is the weighted membership degree. Figure 4 shows the calculation process.

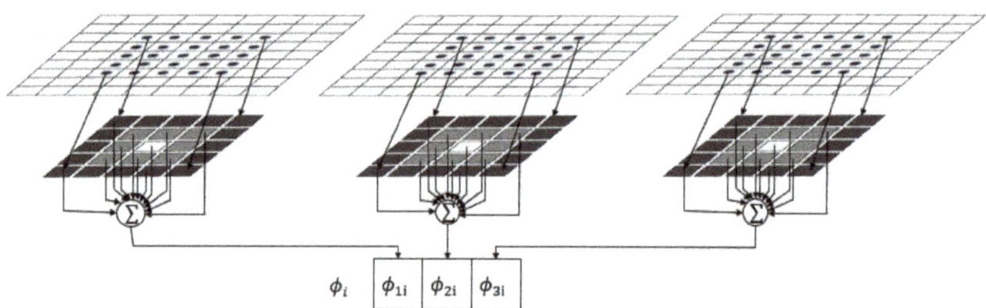


**Figure 4.** The calculation process of weight vector $\phi_i$ for three classes. In this example, the neighborhood size is specified as $5 \times 5$.

Besides, label correction is used as a homogeneous region smoothing technique in SAR segmentation in [42]. It has been shown to be effective in the correction of error class labels. Hence, we will adopt a simple method to correct the error pixel class. This framework uses the majority voting strategy to revise the error pixel label upon completion of the iteration. Specifically, a fixed-scale window is utilized to slide over the image. The class label with the largest number in the slid window is the final class of the central pixel. Figure 5 shows that the framework of GLR_FCM and LBNL_FCM is alike.

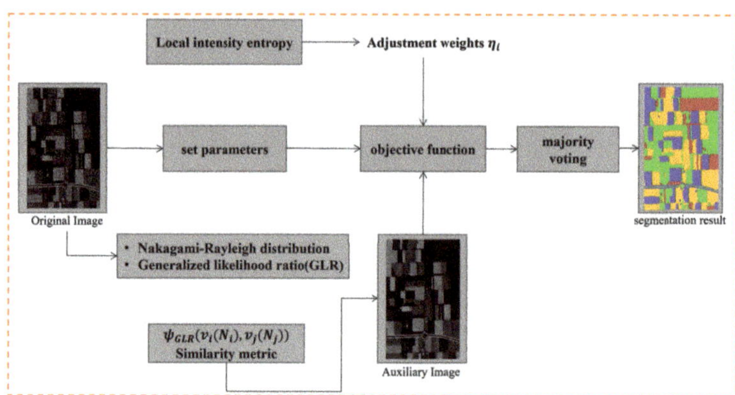


**Figure 5.** The framework of proposed segmentation algorithm GLR_FCM, and the LBNL_FCM is similar.

## 3. Experiments and Results

In this section, we perform LBNL_FCM and GLR_FCM on simulated SAR images and real SAR images to illustrate the effectiveness of our proposed algorithms. The segmentation results are evaluated qualitatively and quantitatively. Several popular improved FCM algorithms are used as baselines to illustrate the advantages of the proposed algorithms in edge preservation and region consistency. These methods are FCM [22], FCM_S1 and FCM_S2 [24], KFCM_S1 and KFCM_S2 [24], EnFCM [25], FGFCM [26],

FCM_NLS [31], NS_FCM [34], and RFCM_BNL [35]. Note that, for real SAR images, we focus more on visual inspection because it is difficult to obtain its ground truth. Experiment images are selected from four different satellites, including AIRSAR, ALOS PolSAR, TerraSAR-X, and GF3.

### 3.1. Experimental Setting

For all algorithms, the parameters are selected as follows: The stopping threshold $\delta = 10^{-5}$, Maximum iterations $T = 200$, membership exponent $m = 2$. We set $\alpha = 5$ for FCM_S1, FCM_S2, KFCM_S1, KFCM_S2, EnFCM, and FCM_NLS. According to [34], we set $\alpha = 6$ for NS_FCM. $\lambda_s$ and $\lambda_g$ in FGFCM are set to 2 and 7, respectively. For NS_FCM and FCM_NLS, the local neighbor size is $5 \times 5$, and the non-local search window is set to $11 \times 11$ and $15 \times 15$, respectively. For RFCM_BNL, LBNL_FCM and GLR_FCM, the local neighbor window is set to $3 \times 3$ and the non-local search window is set to $15 \times 15, 9 \times 9$, and $23 \times 23$. For LBNL_FCM and GLR_FCM, the membership degree smoothing and label correction window is set to $5 \times 5$. In LBNL_FCM and GLR_FCM, when calculating $\eta_i$, the gray level is quantized into 16 bins, i.e., $k = 16$.

### 3.2. Evaluation Indicators

Evaluating results is a key step in measuring the effectiveness of the algorithms. In this paper, the effectiveness of the proposed and reference algorithms is assessed from both objective and subjective aspects. Moreover, we concentrate on two crucial aspects of the segmentation results: Compactness and separation. Whether it is a visual inspection by human eyes or a quantitative evaluation, a good segmentation algorithm should make the intra-class dissimilarity as small as possible and the inter-class variability as large as possible, i.e., corresponding to compactness and separation, respectively. Table 1 shows several assessment indicators that we intend to use to quantitatively evaluate these two properties, whose efficacy was proved in [43].

**Table 1.** The quantitative evaluation indicators used in simulation SAR image experiments for results.

| Indicator | Formulation | Description |
|---|---|---|
| PC (Partition Coefficient) [44] | $PC = \frac{1}{N} \sum_{c=1}^{c} \sum_{i=1}^{N} u_{ci}^2$ | The larger the PC value, the better the partition result |
| PE (Partition Entropy) [45] | $PE = -\frac{1}{N} \sum_{c=1}^{c} \sum_{i=1}^{N} u_{ci} \log(u_{ci})$ | The smaller the PE value, the better the partition result |
| MPC (Modified PC) [46] | $MPC = \frac{C \times PC - 1}{C - 1}$ <br> $PC = \frac{1}{N} \sum_{c=1}^{c} \sum_{i=1}^{N} u_{ci}^2$ | The MPC eliminates the dependency on c, the large the MPC is, the better the partition result |
| MPE (Modified PE) [46] | $MPE = \frac{N \times PE}{N - C}$ <br> $PE = -\frac{1}{N} \sum_{c=1}^{c} \sum_{i=1}^{N} u_{ci} \log(u_{ci})$ | Similar to above that the smaller the MPE is, the better the partition result |
| FS(Fukuyama-Sugeno Index) [47] | $FS = J_m(U, V) - K_m(U, V)$ <br> $= J_m - \sum_{i=1}^{N} \sum_{c=1}^{c} u_{ci}^m \|v_c - \tilde{v}\|^2$ <br> where $\tilde{v} = \frac{1}{N} \sum_{i=1}^{N} x_i$ | The first term indicates the compactness and the second term indicates the separation. And the minimum FS implies the optimal partition |

### 3.3. Segmentation Results on Simulated SAR Images

We can obtain accurate ground truth for simulated SAR images, so we use segmentation accuracy to evaluate the segmentation performance. In addition, five numerical evaluation indexes are computed. The segmentation accuracy is defined as the number of correctly segmented pixels divided by the total number of pixels, and the formula is as follows:

$$SA = \frac{\sum_{k=1}^{c} A_k \cap C_k}{\sum_{j=1}^{c} C_j} \tag{43}$$

where $c$ represents the number of segmentation objects, $C_k$ denotes the number of pixels within the $k$th class in the real SAR image, $A_k$ indicates the number of pixels belonging to the $k$th class in the segmentation result, and $\sum_{j=1}^{c} C_j$ corresponds to the total number of pixels.

### 3.3.1. Experiment 1: Testing on the First Simulated SAR Image

The first experiments are carried out on a one-look simulated SAR image with $250 \times 200$ pixels as shown in Figure 6a. This simulated SAR image includes five classes with intensity value taken as 10, 50, 100, 150, 200. Its gray and color ground truth are shown in Figure 6b,c.

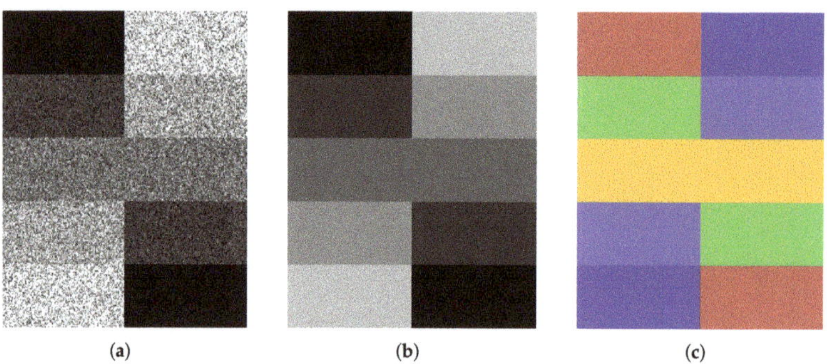

(a)      (b)      (c)

**Figure 6.** The simulated SAR image and ground truth. (**a**) Simulated SAR image; (**b**) ground truth with gray; (**c**) ground truth with color

The experiment results of the proposed algorithms and comparative algorithms are shown in Figure 7. It can be seen that the original FCM has the worst result in regional consistency and many noise points are present. FCM_S1 and FCM_S2 enhance the segmentation result by adding local information. The kernel distance versions of KFCM_S1 and KFCM_S2 obtain further enhancement results. Nevertheless, there are still plenty of noise pixels. The reason is that the local neighborhood information on SAR images is contaminated by noise. The reliability of local spatial information is severely weakened, which ultimately leads to the failure of segmentation.

FCM_NLS and NS_FCM in Figure 7h,i consider the non-local information. However, the non-local spatial information is measured by Euclidean distance, which is inappropriate for SAR images. So they still have significant misclassification problems. The RFCM_BNL takes into account the characteristics of SAR images and therefore achieves a relatively good result in terms of the regional coherence. However, there is still a large number of isolated pixels near the edges. The result of LBNL_FCM presents a better continuity of edges and homogeneous regions cleaner than that of RFCM_BNL. However, the Bayesian-based FCM algorithm is not the best in terms of edge preservation in Figure 7j,k. There is a serious misclassification phenomenon at the edges, i.e., the region between the green region and the blue region is divided into yellow class. In Figure 7l, GLR_FCM achieves the best visual result for maintaining regional consistency and edge preservation. Effectively eliminating the false class of RFCM_BNL and LBNL_FCM at the edges and almost no isolated noise pixels.

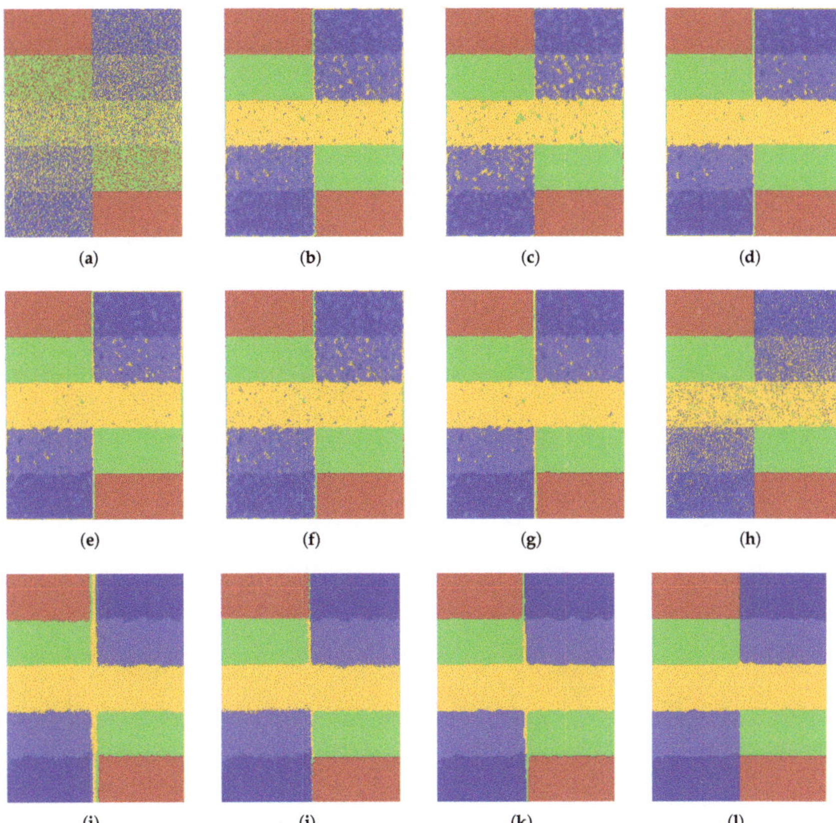

**Figure 7.** The segmentation results on simulated SAR image. (**a**) FCM. (**b**) FCM_S1. (**c**) FCM_S2. (**d**) KFCM_S1. (**e**) KFCM_S2. (**f**) EnFCM. (**g**) FGFCM. (**h**) FCM_NLS. (**i**) NS_FCM. (**j**) RFCM_BNL. (**k**) LBNL_FCM. (**l**) GLR_FCM.

Table 2 displays the SA (%) and executed time of each algorithm. We see that the kernel method is valid for results. The non-local information is more useful for SAR image segmentation compared to local information. Because of the statistical property of SAR images, higher segmentation accuracy is obtained by FCM_RBNL, LBNL_FCM and GLR_FCM. Besides, GLR_FCM obtains the best segmentation accuracy of 99.16%, consistent with the visualization in Figure 7. The algorithms based on the non-local information have higher time consumption because each pixel is visited in computing auxiliary.

**Table 2.** SA (%) and executed time(s) on the first simulated SAR image.

| Method | SA (%) | Time (s) | Method | SA (%) | Time (s) |
|---|---|---|---|---|---|
| FCM | 60.58 | 2.16 | FGFCM | 94.65 | 5.64 |
| FCM_S1 | 90.49 | 1.11 | FCM_NLS | 83.61 | 7.27 |
| FCM_S2 | 90.49 | 1.46 | NS_FCM | 95.03 | 7.77 |
| KFCM_S1 | 92.66 | 1.27 | RFCM_BNL | 97.29 | 10.88 |
| KFCM_S2 | 91.42 | 1.20 | LBNL_FCM | 97.64 | 12.11 |
| EnFCM | 90.63 | 1.85 | GLR_FCM | 99.16 | 17.73 |

Table 3 shows the quantitative evaluation for the first simulated SAR image. $V_{PC}$ and $V_{MPC}$ express the fuzziness of the partition result. The larger the value, the better the partition result. In contrast, the minimums of $V_{PE}$ and $V_{MPE}$ imply the optimal result. The

$V_{FS}$ describes the compactness and separation. The best partition can be obtained with the minimum $V_{FS}$. In addition to the optimal value obtained by the NS_FCM on $V_{FS}$, the LBNL_FCM and GLR_FCM obtain the best value in the other criteria.

Figure 8 provides the change curve of the objective function. We can see that the objective function of LBNL_FCM descends fastest and obtains the minimum value. The objective function of GLR_FCM decreases at a similar speed to that of LBNL_FCM. Moreover, a relatively small value of the objective function is obtained.

**Table 3.** Quantitative evaluation on the first simulated SAR image.

| Method | $V_{PC}$ | $V_{PE}$ | $V_{MPC}$ | $V_{MPE}$ | $V_{FS}$ |
|---|---|---|---|---|---|
| FCM | 0.7994 | 0.3995 | 0.7492 | 0.3995 | $-3.12 \times 10^8$ |
| FCM_S1 | 0.7203 | 0.5581 | 0.6504 | 0.5582 | $-1.36 \times 10^8$ |
| FCM_S2 | 0.7350 | 0.5347 | 0.6688 | 0.5347 | $-1.78 \times 10^8$ |
| KFCM_S1 | 0.6783 | 0.6623 | 0.5978 | 0.6624 | $-1.01 \times 10^8$ |
| KFCM_S2 | 0.6861 | 0.6537 | 0.6076 | 0.6537 | $-1.39 \times 10^8$ |
| EnFCM | 0.8518 | 0.3031 | 0.8147 | 0.3031 | $-1.56 \times 10^8$ |
| FGFCM | 0.8750 | 0.2595 | 0.8438 | 0.2595 | $-2.33 \times 10^8$ |
| FCM_NLS | 0.7175 | 0.5892 | 0.6469 | 0.5893 | $-1.70 \times 10^8$ |
| NS_FCM | 0.6932 | 0.6342 | 0.6165 | 0.6342 | $-9.03 \times 10^7$ |
| RFCM_BNL | 0.8069 | 0.4165 | 0.7587 | 0.4165 | $-1.34 \times 10^8$ |
| LBNL_FCM | 0.9609 | 0.0792 | 0.9511 | 0.0792 | $-1.54 \times 10^8$ |
| GLR_FCM | 0.9855 | 0.0260 | 0.9819 | 0.0260 | $-1.78 \times 10^8$ |

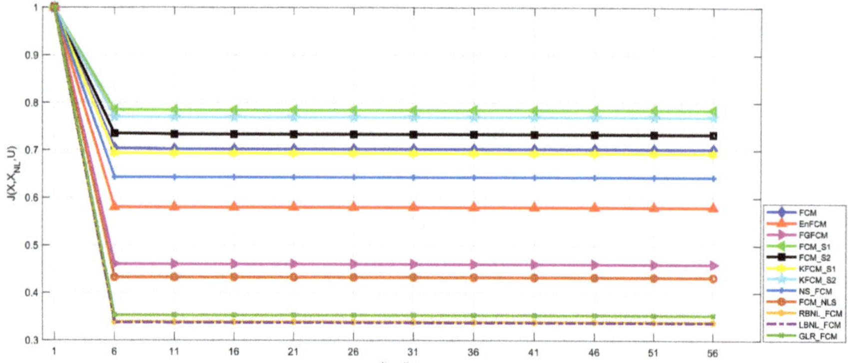


**Figure 8.** The objective function change curve of each algorithm on the first simulated SAR image.

### 3.3.2. Experiment 2: Testing on the Second Simulated SAR Image

The second simulated SAR image is composed of 283*283 pixels, and includes five classes with amplitude values settled as (0, 64, 128, 192, 255). Figure 9a–c show the original simulated SAR image and the ground truth. Figure 9d–o show the segmentation results of each algorithm.

Visually, the result of FCM (Figure 9d) has plenty of noise points. In Figure 9e–j, due to integration of the local spatial information, the isolated speckle pixels are significantly suppressed. The result of FCM_NLS obtains a better region consistency in red and green classes. However, there are still some blocks that are not properly classified under other categories. The results of NS_FCM and RFCM_BNL yield good regional coherence and smoothed edges. However, there are still serious classification mistakes on the periphery of different regions. In contrast, LBNL_FCM and GLR_FCM obtain relatively satisfactory segmentation results. Isolated pixels and blocks of speckle noise are practically non-existent there in homogeneous regions. In terms of structural information, GLR_FCM protects the continuity and smoothness of the edges, even when crossing regions with similar magnitude values. The edge can be well discriminated as shown in Figure 9o. Only slightly blurred edges exist at the nodes adjacent to the three regions.

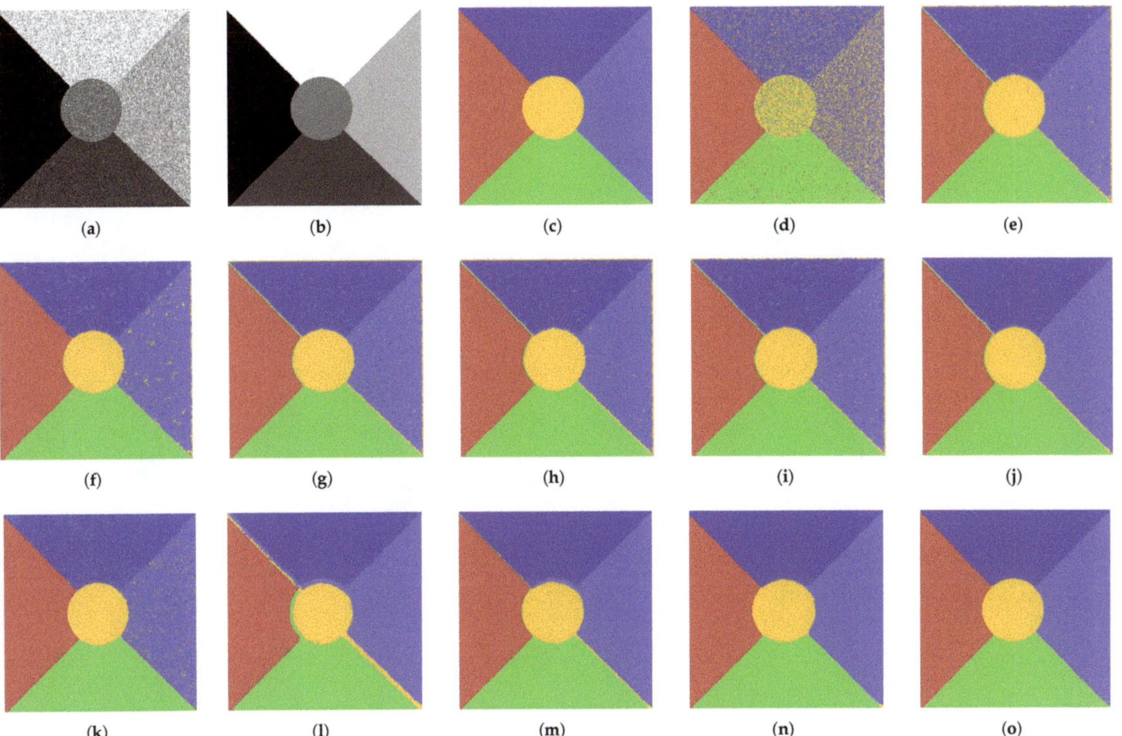

**Figure 9.** The segmentation results on the second simulated SAR image. (**a**) Original Image (**b**) Ground Truth (Gray) (**c**) Ground Truth (Color) (**d**) FCM (**e**) FCM_S1 (**f**) FCM_S2 (**g**) KFCM_S1 (**h**) KFCM_S2 (**i**) EnFCM (**j**) FGFCM (**k**) FCM_NLS (**l**) NS_FCM (**m**) RFCM_BNL (**n**) LBNL_FCM (**o**) GLR_FCM.

A conclusion similar to the first experiment can be obtained from Table 4. In SAR image, non-local spatial information is more robust to speckle noise compared to local information. Thus, the FCMs with the non-local information terms obtain relatively good segmentation accuracy above 96%. However, they are time consuming because of the auxiliary image calculated in advance.

**Table 4.** SA (%) and executed time(s) on the second simulated SAR image.

| Method | SA (%) | Time (s) | Method | SA (%) | Time (s) |
|---|---|---|---|---|---|
| FCM | 73.82 | 3.47 | FGFCM | 97.88 | 9.36 |
| FCM_S1 | 95.83 | 1.16 | FCM_NLS | 95.03 | 8.85 |
| FCM_S2 | 96.55 | 1.27 | NS_FCM | 96.10 | 9.59 |
| KFCM_S1 | 96.36 | 1.02 | RFCM_BNL | 98.66 | 16.58 |
| KFCM_S2 | 96.94 | 1.22 | LBNL_FCM | 98.82 | 16.83 |
| EnFCM | 95.88 | 2.03 | GLR_FCM | 99.86 | 18.45 |

The quantitative evaluation indicators of each algorithm are recorded in Table 5. GLR_FCM obtains the optimal value on $V_{PC}$, $V_{PE}$, $V_{MPC}$, and $V_{MPE}$ and significantly outperforms other algorithms. The LBNL_FCM has relatively optimal indicators. On $V_{FS}$, EnFCM obtains the minimum value of $-6.27 \times 10^9$.

**Table 5.** Quantitative evaluation of the second simulated SAR image.

| Method | $V_{PC}$ | $V_{PE}$ | $V_{MPC}$ | $V_{MPE}$ | $V_{FS}$ |
|---|---|---|---|---|---|
| FCM | 0.8354 | 0.3298 | 0.7943 | 0.3298 | $-6.28 \times 10^8$ |
| FCM_S1 | 0.8204 | 0.3667 | 0.7755 | 0.3667 | $-4.76 \times 10^8$ |
| FCM_S2 | 0.8298 | 0.3524 | 0.7872 | 0.3524 | $-5.34 \times 10^8$ |
| KFCM_S1 | 0.7880 | 0.4492 | 0.7351 | 0.4492 | $-4.34 \times 10^8$ |
| KFCM_S2 | 0.7923 | 0.4448 | 0.7404 | 0.4448 | $-4.86 \times 10^8$ |
| EnFCM | 0.9060 | 0.1971 | 0.8825 | 0.1971 | $-6.27 \times 10^9$ |
| FGFCM | 0.9307 | 0.1511 | 0.9134 | 0.1511 | $-5.66 \times 10^8$ |
| FCM_NLS | 0.8171 | 0.3890 | 0.7714 | 0.3890 | $-4.90 \times 10^8$ |
| NS_FCM | 0.8085 | 0.4102 | 0.7607 | 0.4103 | $-4.26 \times 10^8$ |
| RFCM_BNL | 0.8939 | 0.2414 | 0.8674 | 0.2414 | $-4.79 \times 10^8$ |
| LBNL_FCM | 0.9882 | 0.0208 | 0.9852 | 0.0208 | $-4.99 \times 10^8$ |
| GLR_FCM | 0.9972 | 0.0051 | 0.9965 | 0.0051 | $-5.24 \times 10^8$ |

Figure 10 shows the curve of objective function on the second simulated SAR image. It can be seen that FCM_RBNL, LBNL_FCM, and GLR_FCM consider the statistical properties of SAR images, so their objective function decreases fastest and only needs two iterations to converge. In addition, at the convergence, GLR_FCM has the minimum loss value of the objective function. This also implies best segmentation performance.

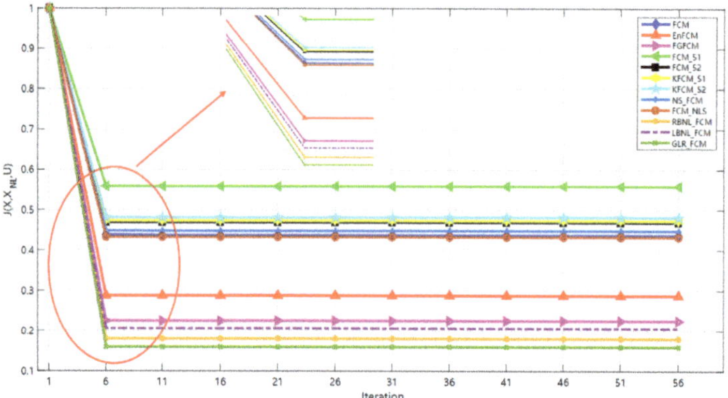

**Figure 10.** The objective function curve of each algorithm on the second simulated SAR image.

### 3.4. Segmentation Results on Real SAR Images

Experiments on simulated SAR images only illustrate the validity and feasibility of algorithms. Therefore, we will test the practicality of proposed algorithms on real SAR images taken from different satellites. It is difficult to get ground truths for real SAR images; thus, segmentation results are accessed mainly by visual inspection.

#### 3.4.1. Experiment 1: Experiment on the First Real SAR Image

The first experiment was carried out on an L-band, HH-polarized, SAR image with 2 m spatial resolution taken by AIRSAR in the Flevoland area of the Netherlands, as shown in Figure 11a. This area includes roughly four crop types, and the amplitudes are bright, dark, darker, and black. Figure 12 shows the segmentation results.

The auxiliary images of LBNL_FCM and GLR_FCM are shown in Figure 11b,c. The auxiliary image used by LBNL_FCM (see Figure 11b) strongly suppresses the noise and has a strong smoothing ability. The auxiliary image used in GLR_FCM (see Figure 11c reduces speckle noise while retaining the structural information. However, a slight texture noise remains inside the homogeneous region, which can be easily attenuated or removed by local information such as membership smoothing.

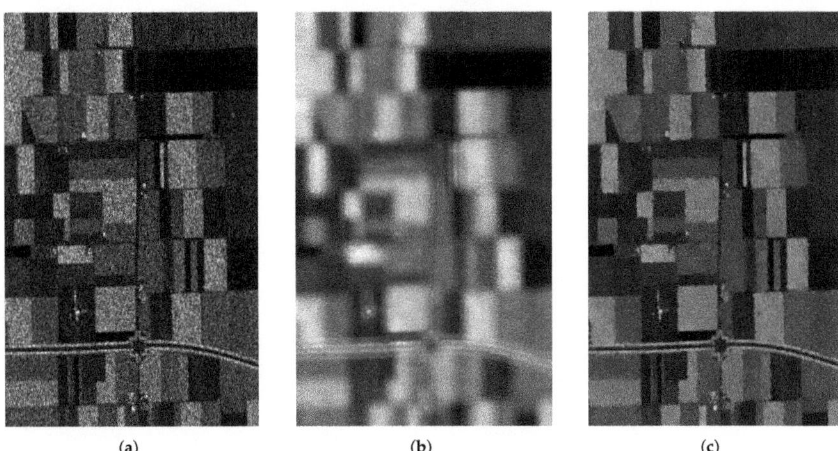

**Figure 11.** The first real SAR image and auxiliary images. (**a**) Original real SAR image; (**b**) The auxiliary image of LBNL_FCM; (**c**) The auxiliary image of GLR_FCM.

The most terrible result is provided by FCM in Figure 12a and almost fails when processing SAR images. The FCM_S1, FCM_S2, and the kernel methods suppress the noise to some extent. However, the results are still not very desirable. The EnFCM and FGFCM enhance the consistency of segmented regions compared with the previous methods by incorporating local information and using the histogram as the segmentation object. However, the darker region is misclassified to dark class from Figure 12f,g. FCM_NS has a better region coherence than FCM_NLS, but there is still severe misclassification in the region with similar intensity. Among these methods, RFCM_BNL, LBNL_FCM, and GLR_FCM obtain relatively satisfactory results. Visually, the segmentation results almost correctly reflect the region information of the original image. The large regions which are misclassed in other algorithms are correctly classified. However, the edges in RFCM_BNL and LBNL_FCM are not satisfactory enough, as shown in Figure 12j,k. A third class may appear in the middle of two adjacent regions. The result of GLR_FCM effectively overcomes this problem with the suitable similarity properties. Besides, most of the structure information is preserved in Figure 12l. A balance between regional homogeneity and edge preservation can be achieved well.

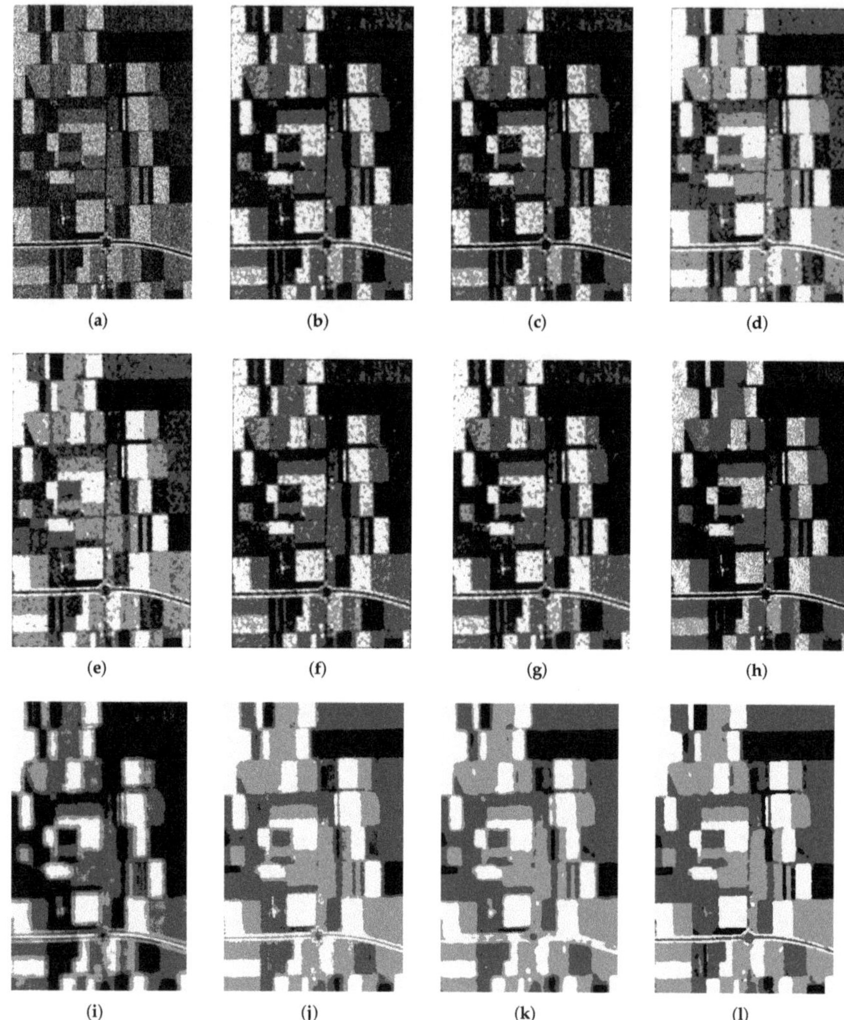

**Figure 12.** The segmentation result of each algorithm on the real SAR image. (**a**) FCM. (**b**) FCM_S1 (**c**) FCM_S2. (**d**) KFCM_S1. (**e**) KFCM_S2. (**f**) EnFCM. (**g**) FGFCM. (**h**) FCM_NLS. (**i**) NS_FCM (**j**) RFCM_BNL. (**k**) LBNL_FCM. (**l**) GLR_FCM.

3.4.2. Experiment 2: Experiment on the Second Real SAR Image

An L-band, HH-polarized SAR image taken by AIRSAR is selected in this experiment Figure 13a presents the original image. This area contains four kinds of crops shown as bright, gray, dark, and black. Figure 13a shows that the region with the brightest magnitude suffers from speckle noise. There is a gradual change in amplitude value.

The segmentation results of each algorithm are shown in Figure 13. The results of NS_FCM and FCM_NLS (Figure 13i,j) are relatively clean and accurate. However, there are many misclassified categories at the intersection of different regions. The segmentation results of RFCM_BNL and LBNL_FCM eliminate the isolated pixels and obtain good region conformity. However, RFCM_BNL and LBNL_FCM are prone to misclassification at the edge. The segmentation results of GLR_FCM are cleaner. The serious misclassification at the edge is weakened in GLR_FCM. Some small scale regions can also be segmented such as roads that appear black being correctly segmented. However, with the noise enhancement, GLR_FCM tends to produce isolated patches when combined with label correction. Figure 14 shows the local detail map of four non-local spatial information FCMs. There is a significant reduction in misclassification at the edge of GLR_FCM.

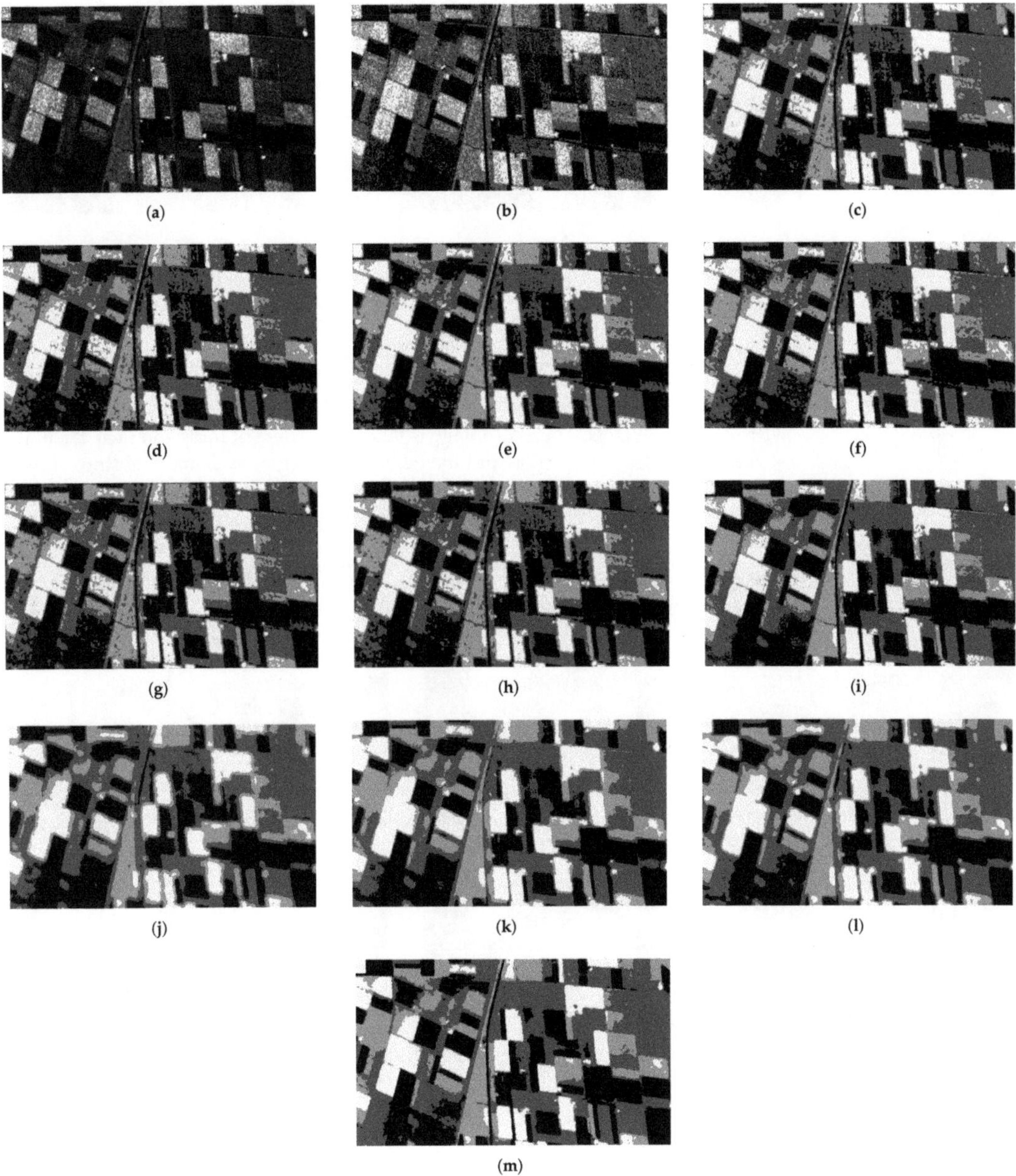

**Figure 13.** The segmentation results of each algorithms on the real SAR image. (**a**) Original image. (**b**) FCM. (**c**) FCM_S1. (**d**) FCM_S2. (**e**) KFCM_S1. (**f**) KFCM_S2. (**g**) EnFCM. (**h**) FGFCM. (**i**) FCM_NLS. (**j**) NS_FCM. (**k**) RFCM_BNL. (**l**) LBNL_FCM. (**m**) GLR_FCM.

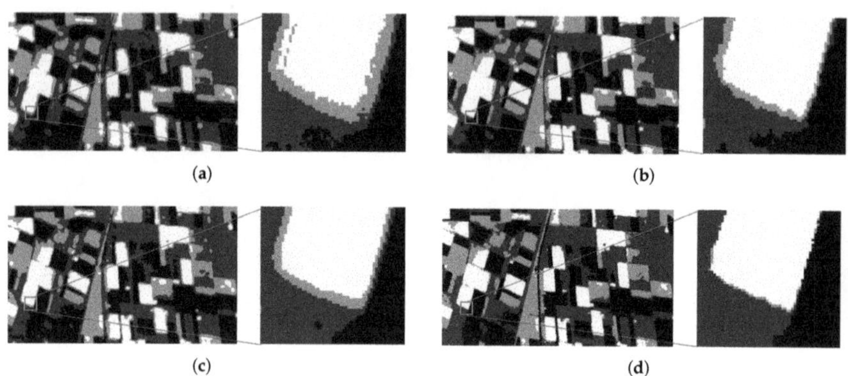

(a)

(b)

(c)

(d)

**Figure 14.** Local detail maps of segmentation results. (**a**) The result of NS_FCM. (**b**) The result of RFCM_BNL. (**c**) The result of LBNL_FCM. (**d**) The result of GLR_FCM.

### 3.4.3. Experiment 3: Experiment on the Third Real SAR Image

The fourth experiment is performed on a TerraSAR image shown in Figure 15a, which has 5 m spatial resolution and HH polarization in X-band strip imaging mode with 402 × 381 pixels. The SAR image is taken of an area of farmland near the border of Saxony in the German region and includes four categories. Some buildings show high amplitude values, and roads show low amplitude values. These unfavorable factors make it difficult to segment SAR images.

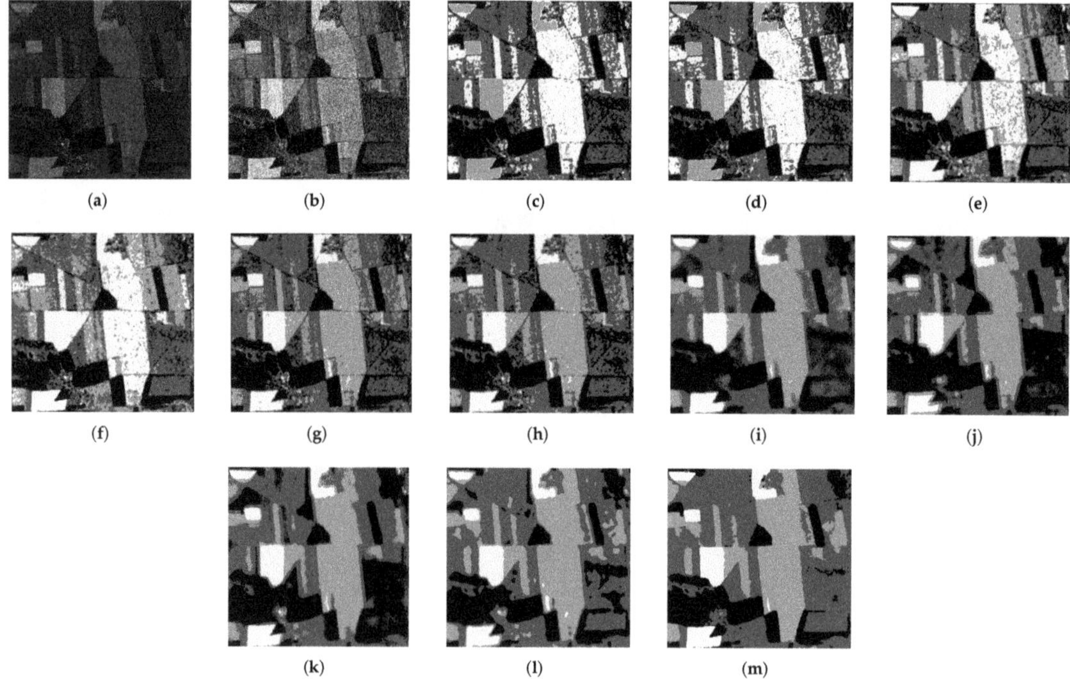

(a)

(b)

(c)

(d)

(e)

(f)

(g)

(h)

(i)

(j)

(k)

(l)

(m)

**Figure 15.** The segmentation results on real image 4. (**a**) Original image. (**b**) FCM. (**c**) FCM_S1. (**d**) FCM_S2. (**e**) KFCM_S1. (**f**) KFCM_S2. (**g**) EnFCM. (**h**) FGFCM. (**i**) FCM_NLS. (**j**) NS_FCM. (**k**) RFCM_BNL. (**l**) LBNL_FCM. (**m**) GLR_FCM.

The partition results of each algorithm are provided in Figure 15b–m. Obviously, the results of FCM, FCM_S1, FCM_S2 and kernel editions are not satisfactory. Because of the effect of speckle noise, many misclassified pixels, blocks and regions exist. The EnFCM and FGFCM correct the middle area label that is misclassified into highlighted categories in Figure 15c–f. However, the pixels in gray and darker are substantially confused. The addition of NLS_FCM and NS_FCM with non-local information reduces the misclassification, but there is still some isolated noise due to unsuitable Euclidean distance.

Moreover, RFCM_BNL (Figure 15k) obtains good region conformity, but a tiny portion of darker areas is still segmented into black classes. The result of LBNL_FCM significantly weakens the influence of speckle noise, and the best smoothing effect is obtained. GLR_FCM (Figure 15m) is enabled to balance the speckle noise suppression and edge preservation. The region consistency is guaranteed without damaging structure information.

### 3.4.4. Experiment 4: Experiment on the Fourth Real SAR Image

The fifth experiment is a 3 m spatial resolution, 222 × 516 pixels, HH-polarized SAR image taken from GF-3 with the imaging mode of the strip, and this area is located near the Daxing Airport in Beijing. The original image is shown in Figure 16a. The buildings, land, and runways are included in this SAR image; they show in magnitude as highlighted, dark, and black, respectively. Some small areas, such as lakes, also appear black.

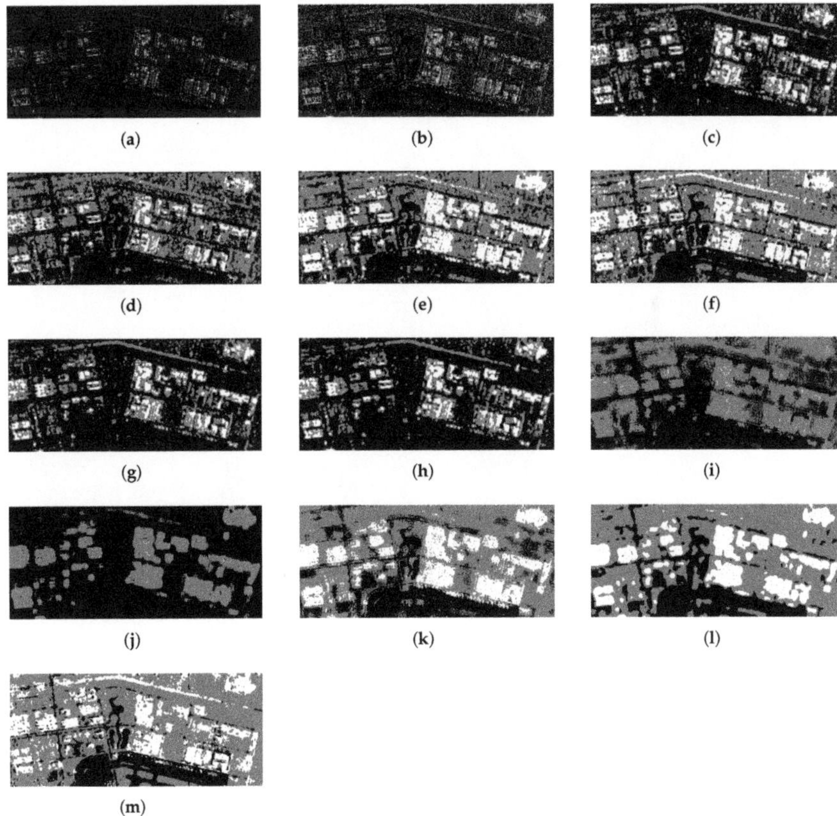

**Figure 16.** Segmentation results of each algorithm on GaoFen-3 SAR Image. (**a**) Original image. (**b**) FCM. (**c**) FCM_S1. (**d**) FCM_S2. (**e**) KFCM_S1. (**f**) KFCM_S2. (**g**) EnFCM. (**h**) FGFCM. (**i**) FCM_NLS. (**j**) NS_FCM. (**k**) RFCM_BNL. (**l**) LBNL_FCM. (**m**) GLR_FCM.

Figure 16b–m display the experimental results. As can be seen from Figure 16b the FCM is sensitive to the speckle noise. FCM_S1 and FCM_S2 slightly improve the results. The kernel versions further enhance the separability and homogeneity. However, some speckle blocks are not removed. Due to the complexity of this SAR image, EnFCM, FGFCM, FCM_NLS, and NS_FCM can barely segment correctly. Specifically, EnFCM and FGFCM cannot distinguish the ground and lake. In the results of FCM_NLS and NS_FCM, the building area and ground mix into the same category. This illustrates that the Euclidean distance is unreliable concerning SAR images. Due to the distribution of SAR being considered, RFCM_BNL, LBNL_FCM, and GLR_FCM obtain relatively satisfactory results. The two Bayesian-based FCMs slightly outperform the GLR_FCM in terms of region consistency. However, they are poor in edge localization. Additionally, in terms of structure information preserving, the GLR_FCM surpasses all the algorithms. In Figure 16m, we notice the contour of the lake can be segmented explicitly.

### 3.5. Sensitivity Analysis to Speckle Noise

In this section, we evaluate the sensitivity of proposed frameworks to noise intensity by adding different levels of speckle noise to Figure 6a. The SA (%) of different algorithms on images with eight speckle look is shown in Figure 17a. The SA (%) of most methods improves with the weakening of speckle noise. GLR_FCM obtains the best SA (%), which always exceeds 97%. Besides, the stability to different intensity of noise can be observed. LBNL_FCM obtains relatively good SA (%) and is stable for speckle look. The SA (%) of some algorithms fluctuates significantly to the number of speckle look. The variation between the best SA (%) and worst SA (%) exceeds 60%. The partial enlarged view can be seen in Figure 17b.

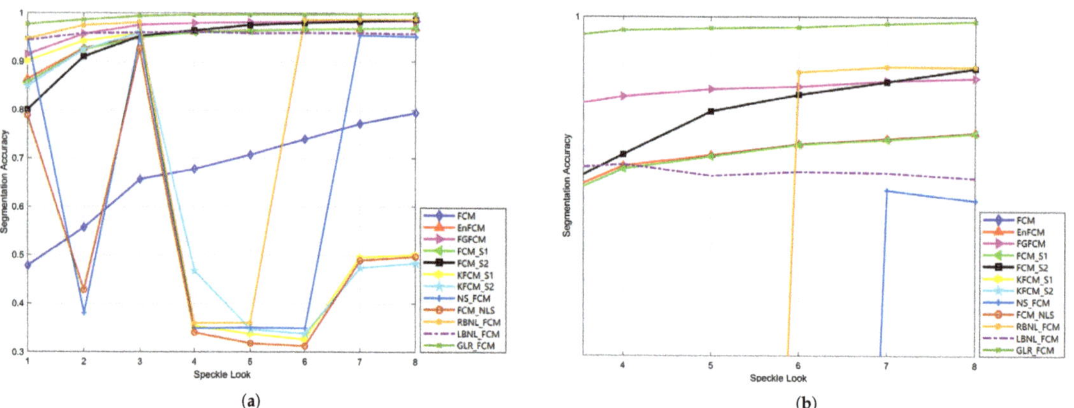

(a)  (b)

**Figure 17.** The segmentation accuracy of different algorithms testing on the first simulated SAR image with adding speckle noise of different looks. (**a**) SA curves of different methods. (**b**) The partial enlarged view of (**a**).

### 3.6. Parameters Analysis and Selection

The non-local search window size $w \times w$ and the square neighborhood size $r \times r$ are two crucial parameters related to the non-local spatial information. In this section, we investigate the optimal parameters on two simulated SAR images (Figures 6a and 9a) for LBNL_FCM and GLR_FCM.

On the first simulated SAR image (Figure 6a), we set the non-local spatial information search window $w = [5, 7, 9, 11, 13, 15, 17, 19, 21, 23]$ and local neighborhood patch $r = [3, 5, 7, 9, 11]$. The SA (%) of LBNL_FCM and GLR_FCM on the first simulated SAR image is shown in Figure 18a,b, respectively. From Figure 18a, the SA curve of LBNL_FCM decreases rapidly for ab arbitrary $r$ value when $w$ exceeds 9. One reason for this is that the logarithm transformation reduces the contrast of image amplitude. As the window $w$ expands, more pixels are included to calculate non-local information. Hence, the weight of reliable pixels decreases. The SA (%) curve of GLR_FCM can be seen from Figure 18b. The SA curve of $r = 3$ is always higher than others and the accuracy achieves the optimal with $w = 23$. Therefore, in the parameter range above, the optimal value for $r$ is 3. On the first simulated SAR image, we set $r = 3$, $w = 9$ for LBNL_FCM and $r = 3$, $w = 23$ for GLR_FCM. Figure 19 shows the SA curve of LBNL_FCM and GLR_FCM on the second simulated SAR image. Some similar phenomena can be observed. The curve with local neighborhood size $r = 3$ is always more accurate than others.

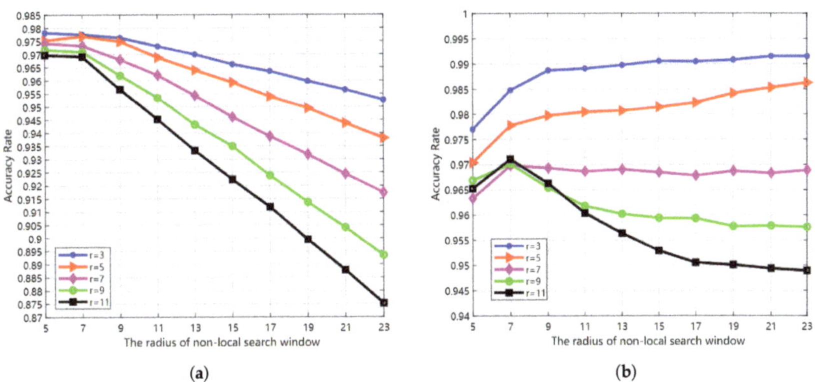

**Figure 18.** The SA (%) of LBNL_FCM and GLR_FCM carried out on the first simulated SAR image with different sizes of search window $w \times w$ and different sizes of local neighborhood $r \times r$. (**a**) LBNL_FCM; (**b**) GLR_FCM.

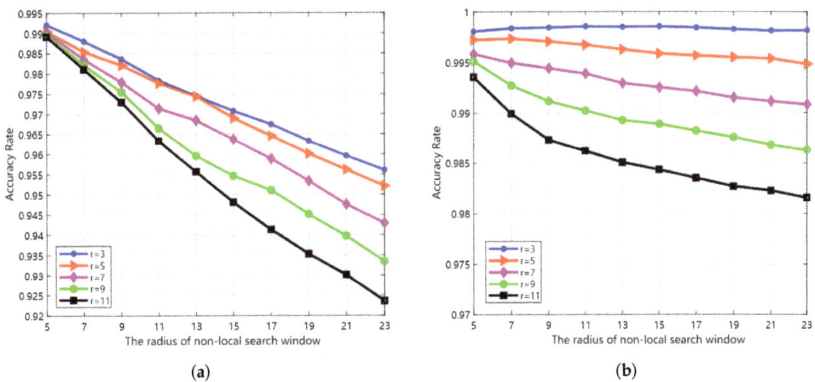

**Figure 19.** The SA (%) of LBNL_FCM and GLR_FCM carried out on the second simulated SAR image with different sizes of search window $w \times w$ and different sizes of local neighborhood $r \times r$. (**a**) LBNL_FCM; (**b**) GLR_FCM.

### 3.7. Computational Complexity Analysis

The computational complexity of the aforementioned algorithms is given in Table 6. Where $N$ is total pixels, $c$ denotes the number of clustering centers, $T$ represents the iterations, $w$ is the size of the window, $r$ is the size of the non-local search window, $s$ is the size of the neighborhood, $W$ denotes the sliding window for calculating the factor $\eta_i$, $Q$ corresponds to the number of gray levels.

The computational complexity of proposed frameworks LBNL_FCM and GLR_FCM consists of three parts. The first part $O(N \times r^2 \times s^2)$ is contributed by the calculation of the non-local spatial information. It is calculated before the iterative process. The second part $O(N \times W^2)$ comes from the calculation of the factor $\eta_i$. The third part $O(N \times c \times T)$ is from the iteration process. To sum up, the total computational complexity of LBNL_FCM and GLR_FCM is $O(N \times r^2 \times s^2 + N \times W^2 + N \times c \times T)$.

**Table 6.** The computational complexity of algorithms used in this study.

| Method | Computational Complexity | Method | Computational Complexity |
|---|---|---|---|
| FCM | $O(N \times c \times T)$ | FGFCM | $O(N \times w^2 + Q \times c \times T)$ |
| FCM_S1 | $O(N \times w^2 + N \times c \times T)$ | FCM_NLS | $O(N \times r^2 \times s^2 + N \times c \times T)$ |
| FCM_S2 | $O(N \times w^2 + N \times c \times T)$ | NS_FCM | $O(N \times r^2 \times s^2 + N \times c \times T)$ |
| KFCM_S1 | $O(N \times w^2 + N \times c \times T)$ | RFCM_BNL | $O(N \times r^2 \times s^2 + N \times W^2 + N \times c \times T)$ |
| KFCM_S2 | $O(N \times w^2 + N \times c \times T)$ | LBNL_FCM | $O(N \times r^2 \times s^2 + N \times W^2 + N \times c \times T)$ |
| EnFCM | $O(N \times w^2 + Q \times c \times T)$ | GLR_FCM | $O(N \times r^2 \times s^2 + N \times W^2 + N \times c \times T)$ |

## 4. Discussion

In the previous experiments, the effectiveness and robustness of both frameworks are verified. On the simulated SAR images, both algorithms obtain high segmentation accuracy (always exceeding 97%), and some unsupervised assessment indicators, such as $v_{PC}$, $v_{PE}$, also state that the fuzziness of clustering centers in results is reduced. On the real SAR images, LBNL_FCM shows a best region consistency in results compared with the previous algorithms. However, like FCM_NLS, NS_FCM, and RFCM_BNL, artifacts appear at the edge. Except for the factor that the amplitude value is prone to blur near the edge, it is also related to the characteristic of the log-transformed Bayesian metric reducing image contrast. Compared with FCM_NLS and NS_FCM, the results of GLR_FCM show satisfactory region uniformity; no isolated pixels exist. Compared with RFCM_FCM and LBNL_FCM, GLR_FCM can preserve the image details and the edges can be properly defined. The main reason is that the similarity metric constructed by the continued product of the generalized likelihood ratio is a ratio form in mathematical expression. That makes it easy to give a small contribution weight to the patches possessing dissimilar amplitude values with the central pixel, which implies the patches involved in reconstructing the real amplitude of central pixel in Equation (37) are trustworthy. Another feature of the proposed unsupervised FCM frameworks is that the non-local spatial information can be adaptively adjusted. Remarkably, in most previous methods, the relevant parameter is empirically set to a constant. Consequently, edge blurred artefacts are greatly reduced in GLR_FCM.

In addition to the methods involved in this article, there are many methods combining FCM with machine learning. For instance, MFCCM, proposed by Balakrishnan et al. [48], fused the characteristics of deep learning to clustering, and produces a satisfactory fuzzy clustering result. However, the disadvantage of its high computational complexity is also significant. Besides, a semi-supervised method combining CNN and IFCM [49] provided a more in-depth understanding and representation of the data features, although it requires a lot of training data. Compared to the advanced deep learning models, our proposed unsupervised FCM frameworks can quickly and efficiently deliver segmentation results. However, due to the lack of feature extraction and feature expression, the image data cannot be understood in depth.

In the parameter analysis, we confirm that $r = 3$ is an optimal value for neighborhood size when measuring patches. However, we found the optimal size of non-local search window of GLR_FCM is $w = 23$, which is different from the optimal value $w = 15$ explored by other algorithms, such as FCM_NLSL, NS_FCM, and RFCM_BNL. We speculate that, because of the strong inhibition of the GLR_FCM on dissimilar patches, more reliable patches can be obtained by expanding the scope of the search window.

In this paper, an empirical statistical distribution (Nakagami–Reigh) is utilized to describe SAR images. The dedicated model is appropriate for the homogeneous region of the SAR. In other scenarios, such as mountainous areas, urban areas, etc., statistical properties may not be expressed correctly. Besides, The relatively high computational complexity is a limitation of the proposed method. In Section 3.7, the computational complexity was listed ($O(N \times r^2 \times s^2 + N \times W^2 + N \times c \times T)$). From the loss function curve shown in Figures 8 and 10, we can observe that the iterative speed is very fast. Therefore, in the practical application, the computational cost mainly comes from the calculation of the non-local spatial information. In addition, appropriately reducing the number of iterations can also improve the efficiency without reducing the accuracy.

## 5. Conclusions

To suppress the effect of speckle noise on SAR image segmentation by clustering algorithms, we propose two unsupervised FCM frameworks incorporating non-local spatial information term, named LBNL_FCM and GLR_FCM, respectively. The non-local spatial information in LBNL_FCM and GLR_FCM is obtained by combining the statistical properties of SAR images with Bayesian methods and generalized likelihood ratio methods. Therefore, speckle noise can be suppressed. In both frameworks, a simple membership smoothing strategy complements the local information, allowing the membership of the pixel to be iteratively adjusted towards the most probable class in the local neighborhood. Besides, we add a balance factor to adaptively control the effect of non-local spatial information on the edges, so as to reduce the artifact caused by blurred edges. On the synthetic SAR images, both unsupervised FCM frameworks can obtain 99% segmentation accuracy. Several unsupervised evaluation indicators also indicate LBNL_FCM and GLR_FCM can reduce the fuzziness of the divided clusters in results ($v_{PC} = 0.9855, v_{PE} = 0.0260$). Experiments on the real SAR images show that LBNL_FCM can achieve best region consistency, and GLR_FCM can balance noise removal while preserve image detail and reduce edge blur artifacts.

In future research, we will consider combining unsupervised FCM with the characteristic of deep learning to explore intelligent clustering computing.

**Author Contributions:** Conceptualization, J.Z.; methodology, J.Z.; software, J.Z.; validation, J.Z.; formal analysis, J.Z.; investigation, J.Z.; resources, J.Z.; data curation, J.Z.; writing—original draft preparation, J.Z.; writing—review and editing, J.Z., F.W. and H.Y.; visualization, J.Z.; supervision, J.Z.; project administration, F.W. and H.Y.; funding acquisition, F.W. and H.Y. All authors have read and agreed to the published version of the manuscript.

**Funding:** This research received no external funding.

**Acknowledgments:** The author would like to thank the reviewers for their valuable suggestions and comments. We also would like to thank the production team for revising the format of the manuscript.

**Conflicts of Interest:** The authors declare no conflict of interest.

## References

1. Rahmani, M.; Akbarizadeh, G. Unsupervised feature learning based on sparse coding and spectral clustering for segmentation of synthetic aperture radar images. *IET Comput. Vis.* **2015**, *9*, 629–638. [CrossRef]
2. Jiao, S.; Li, X.; Lu, X. An Improved Ostu Method for Image Segmentation. In Proceedings of the 2006 8th international Conference on Signal Processing, Guilin, China, 16–20 November 2006; Volume 2. [CrossRef]
3. Yu, Q.; Clausi, D.A. IRGS: Image Segmentation Using Edge Penalties and Region Growing. *IEEE Trans. Pattern Anal. Mach. Intell.* **2008**, *30*, 2126–2139. [CrossRef] [PubMed]
4. Carvalho, E.A.; Ushizima, D.M.; Medeiros, F.N.; Martins, C.I.O.; Marques, R.C.; Oliveira, I.N. SAR imagery segmentation by statistical region growing and hierarchical merging. *Digit. Signal Process.* **2010**, *20*, 1365–1378. [CrossRef]
5. Xiang, D.; Zhang, F.; Zhang, W.; Tang, T.; Guan, D.; Zhang, L.; Su, Y. Fast Pixel-Superpixel Region Merging for SAR Image Segmentation. *IEEE Trans. Geosci. Remote Sens.* **2021**, *59*, 9319–9335. [CrossRef]
6. Yu, H.; Zhang, X.; Wang, S.; Hou, B. Context-Based Hierarchical Unequal Merging for SAR Image Segmentation. *IEEE Trans. Geosci. Remote Sens.* **2013**, *51*, 995–1009. [CrossRef]
7. Wang, M.; Dong, Z.; Cheng, Y.; Li, D. Optimal segmentation of high-resolution remote sensing image by combining superpixels with the minimum spanning tree. *IEEE Trans. Geosci. Remote Sens.* **2017**, *56*, 228–238. [CrossRef]
8. Ma, F.; Zhang, F.; Xiang, D.; Yin, Q.; Zhou, Y. Fast Task-Specific Region Merging for SAR Image Segmentation. *IEEE Trans. Geosci. Remote Sens.* **2022**, *60*, 1–16. [CrossRef]

9. Zhang, W.; Xiang, D.; Su, Y. Fast Multiscale Superpixel Segmentation for SAR Imagery. *IEEE Geosci. Remote Sens. Lett.* **2022**, *19*, 1–5. [CrossRef]

10. Zhang, X.; Jiao, L.; Liu, F.; Bo, L.; Gong, M. Spectral Clustering Ensemble Applied to SAR Image Segmentation. *IEEE Trans. Geosci. Remote Sens.* **2008**, *46*, 2126–2136. [CrossRef]

11. Mukhopadhaya, S.; Kumar, A.; Stein, A. FCM Approach of Similarity and Dissimilarity Measures with $\alpha$-Cut for Handling Mixed Pixels. *Remote Sens.* **2018**, *10*, 1707. [CrossRef]

12. Xu, Y.; Chen, R.; Li, Y.; Zhang, P.; Yang, J.; Zhao, X.; Liu, M.; Wu, D. Multispectral image segmentation based on a fuzzy clustering algorithm combined with Tsallis entropy and a gaussian mixture model. *Remote Sens.* **2019**, *11*, 2772. [CrossRef]

13. Madhu, A.; Kumar, A.; Jia, P. Exploring Fuzzy Local Spatial Information Algorithms for Remote Sensing Image Classification. *Remote Sens.* **2021**, *13*, 4163. [CrossRef]

14. Xia, G.S.; He, C.; Sun, H. Integration of synthetic aperture radar image segmentation method using Markov random field on region adjacency graph. *IET Radar Sonar Navig.* **2007**, *1*, 348–353. [CrossRef]

15. Shuai, Y.; Sun, H.; Xu, G. SAR Image Segmentation Based on Level Set With Stationary Global Minimum. *IEEE Geosci. Remote Sens. Lett.* **2008**, *5*, 644–648. [CrossRef]

16. Bao, L.; Lv, X.; Yao, J. Water extraction in SAR Images using features analysis and dual-threshold graph cut model. *Remote Sens.* **2021**, *13*, 3465. [CrossRef]

17. Luo, F.; Zou, Z.; Liu, J.; Lin, Z. Dimensionality reduction and classification of hyperspectral image via multi-structure unified discriminative embedding. *IEEE Trans. Geosci. Remote Sens.* **2021**, *60*, 5517916. [CrossRef]

18. Ma, F.; Gao, F.; Sun, J.; Zhou, H.; Hussain, A. Weakly supervised segmentation of SAR imagery using superpixel and hierarchically adversarial CRF. *Remote Sens.* **2019**, *11*, 512. [CrossRef]

19. Wang, C.; Pei, J.; Wang, Z.; Huang, Y.; Wu, J.; Yang, H.; Yang, J. When Deep Learning Meets Multi-Task Learning in SAR ATR Simultaneous Target Recognition and Segmentation. *Remote Sens.* **2020**, *12*, 3863. [CrossRef]

20. Colin, A.; Fablet, R.; Tandeo, P.; Husson, R.; Peureux, C.; Longépé, N.; Mouche, A. Semantic Segmentation of Metoceanic Processes Using SAR Observations and Deep Learning. *Remote Sens.* **2022**, *14*, 851. [CrossRef]

21. Zhang, R.; Chen, J.; Feng, L.; Li, S.; Yang, W.; Guo, D. A Refined Pyramid Scene Parsing Network for Polarimetric SAR Image Semantic Segmentation in Agricultural Areas. *IEEE Geosci. Remote Sens. Lett.* **2022**, *19*, 1–5. [CrossRef]

22. Bezdek, J.C.; Ehrlich, R.; Full, W. FCM: The fuzzy c-means clustering algorithm. *Comput. Geosci.* **1984**, *10*, 191–203. [CrossRef]

23. Ahmed, M.N.; Yamany, S.M.; Mohamed, N.; Farag, A.A.; Moriarty, T. A modified fuzzy c-means algorithm for bias field estimation and segmentation of MRI data. *IEEE Trans. Med. Imaging* **2002**, *21*, 193–199. [CrossRef] [PubMed]

24. Chen, S.; Zhang, D. Robust image segmentation using FCM with spatial constraints based on new kernel-induced distance measure. *IEEE Trans. Syst. Man Cybern. Part Cybern.* **2004**, *34*, 1907–1916. [CrossRef]

25. Szilagyi, L.; Benyo, Z.; Szilágyi, S.M.; Adam, H. MR brain image segmentation using an enhanced fuzzy c-means algorithm. In Proceedings of the 25th Annual International Conference of the IEEE Engineering in Medicine and Biology Society (IEEE Cat. No 03CH37439), Cancun, Mexico, 17–21 September 2003; Volume 1, pp. 724–726.

26. Cai, W.; Chen, S.; Zhang, D. Fast and robust fuzzy c-means clustering algorithms incorporating local information for image segmentation. *Pattern Recognit.* **2007**, *40*, 825–838. [CrossRef]

27. Krinidis, S.; Chatzis, V. A robust fuzzy local information C-means clustering algorithm. *IEEE Trans. Image Process.* **2010**, *19*, 1328–1337. [CrossRef] [PubMed]

28. Buades, A.; Coll, B.; Morel, J.M. A non-local algorithm for image denoising. In Proceedings of the 2005 IEEE Computer Society Conference on Computer Vision and Pattern Recognition (CVPR'05), San Diego, CA, USA, 20–25 June 2005; Volume 2, pp. 60–65.

29. Wang, J.; Kong, J.; Lu, Y.; Qi, M.; Zhang, B. A modified FCM algorithm for MRI brain image segmentation using both local and non-local spatial constraints. *Comput. Med. Imaging Graph.* **2008**, *32*, 685–698. [CrossRef]

30. Zhu, L.; Chung, F.L.; Wang, S. Generalized fuzzy c-means clustering algorithm with improved fuzzy partitions. *IEEE TRansactions Syst. Man Cybern. Part B Cybern.* **2009**, *39*, 578–591.

31. Zhao, F.; Jiao, L.; Liu, H. Fuzzy c-means clustering with non local spatial information for noisy image segmentation. *Front Comput. Sci. China* **2011**, *5*, 45–56. [CrossRef]

32. Zhao, F.; Jiao, L.; Liu, H.; Gao, X. A novel fuzzy clustering algorithm with non local adaptive spatial constraint for image segmentation. *Signal Process.* **2011**, *91*, 988–999. [CrossRef]

33. Feng, J.; Jiao, L.; Zhang, X.; Gong, M.; Sun, T. Robust non-local fuzzy c-means algorithm with edge preservation for SAR image segmentation. *Signal Process.* **2013**, *93*, 487–499. [CrossRef]

34. Ji, J.; Wang, K.L. A robust nonlocal fuzzy clustering algorithm with between-cluster separation measure for SAR image segmentation. *IEEE J. Sel. Top. Appl. Earth Obs. Remote Sens.* **2014**, *7*, 4929–4936. [CrossRef]

35. Wan, L.; Zhang, T.; Xiang, Y.; You, H. A robust fuzzy c-means algorithm based on Bayesian nonlocal spatial information for SAR image segmentation. *IEEE J. Sel. Top. Appl. Earth Obs. Remote Sens.* **2018**, *11*, 896–906. [CrossRef]

36. Kervrann, C.; Boulanger, J.; Coupé, P. Bayesian non-local means filter, image redundancy and adaptive dictionaries for noise removal. In *International Conference on Scale Space and Variational Methods in Computer Vision*; Springer: Berlin/Heidelberg, Germany, 2007; pp. 520–532.

37. Deledalle, C.A.; Denis, L.; Tupin, F. How to compare noisy patches? Patch similarity beyond Gaussian noise. *Int. J. Comput. Vis.* **2012**, *99*, 86–102. [CrossRef]

38. Bezdek, J.C. *Pattern Recognition with Fuzzy Objective Function Algorithms*; Springer: Berlin/Heidelberg, Germany, 1981.
39. Xie, H.; Pierce, L.; Ulaby, F. Statistical properties of logarithmically transformed speckle. *IEEE Trans. Geosci. Remote Sens.* **2002**, *40*, 721–727. [CrossRef]
40. Goodman, J.W. Some fundamental properties of speckle. *JOSA* **1976**, *66*, 1145–1150. [CrossRef]
41. Oliver, C.; Quegan, S. *Understanding Synthetic Aperture Radar Images*; SciTech Publishing: Raleigh, NC, USA, 2004.
42. Shang, R.; Lin, J.; Jiao, L.; Li, Y. SAR Image Segmentation Using Region Smoothing and Label Correction. *Remote Sens.* **2020**, *12*, 803. [CrossRef]
43. Liu, Y.; Zhang, X.; Chen, J.; Chao, H. A validity index for fuzzy clustering based on bipartite modularity. *J. Electr. Comput. Eng.* **2019**, *2019*, 2719617. [CrossRef]
44. Bezdek, J.C. Numerical taxonomy with fuzzy sets. *J. Math. Biol.* **1974**, *1*, 57–71. [CrossRef]
45. Bezdek, J.C. *Cluster Validity with Fuzzy Sets*; Taylor & Francis: Abingdon, UK, 1973.
46. Dave, R.N. Validating fuzzy partitions obtained through c-shells clustering. *Pattern Recognit. Lett.* **1996**, *17*, 613–623. [CrossRef]
47. Fukuyama, Y. A new method of choosing the number of clusters for the fuzzy c-mean method. In Proceedings of the 5th Fuzzy Systems Symposium, Kobe, Japan, 3 June 1989; pp. 247–250.
48. Balakrishnan, N.; Rajendran, A.; Palanivel, K. Meticulous fuzzy convolution C means for optimized big data analytics: adaptation towards deep learning. *Int. J. Mach. Learn. Cybern.* **2019**, *10*, 3575–3586. [CrossRef]
49. Wang, Y.; Han, M.; Wu, Y. Semi-supervised Fault Diagnosis Model Based on Improved Fuzzy C-means Clustering and Convolutional Neural Network. In Proceedings of the IOP Conference Series: Materials Science and Engineering. IOP Publishing, Shaanxi, China, 8–11 October 2020; Volume 1043, p. 052043.

*Article*

# A Supervoxel-Based Random Forest Method for Robust and Effective Airborne LiDAR Point Cloud Classification

Lingfeng Liao [1], Shengjun Tang [1,*], Jianghai Liao [1], Xiaoming Li [1], Weixi Wang [1], Yaxin Li [2] and Renzhong Guo [1]

[1] School of Architecture and Urban Planning, Research Institute for Smart Cities, Key Laboratory of Urban Land Resources Monitoring and Simulation, Ministry of Natural Resources, Shenzhen University, Shenzhen 518060, China; 2018104003@email.szu.edu.cn (L.L.); liaojh@szu.edu.cn (J.L.); lixming@szu.edu.cn (X.L.); wangwx@szu.edu.cn (W.W.); guorz@szu.edu.cn (R.G.)
[2] Shenzhen Research Institute, The Hong Kong Polytechnic University, Shenzhen 518057, China; yaxin-1.li@polyu.edu.hk
* Correspondence: shengjuntang@szu.edu.cn; Tel.: +86-18665363527; Fax: +86-755-2653-0210

**Abstract:** As an essential part of point cloud processing, autonomous classification is conventionally used in various multifaceted scenes and non-regular point distributions. State-of-the-art point cloud classification methods mostly process raw point clouds, using a single point as the basic unit and calculating point cloud features by searching local neighbors via the k-neighborhood method. Such methods tend to be computationally inefficient and have difficulty obtaining accurate feature descriptions due to inappropriate neighborhood selection. In this paper, we propose a robust and effective point cloud classification approach that integrates point cloud supervoxels and their locally convex connected patches into a random forest classifier, which effectively improves the point cloud feature calculation accuracy and reduces the computational cost. Considering the different types of point cloud feature descriptions, we divide features into three categories (point-based, eigen-based, and grid-based) and accordingly design three distinct feature calculation strategies to improve feature reliability. Two International Society of Photogrammetry and Remote Sensing benchmark tests show that the proposed method achieves state-of-the-art performance, with average F1-scores of 89.16 and 83.58, respectively. The successful classification of point clouds with great variation in elevation also demonstrates the reliability of the proposed method in challenging scenes.

**Keywords:** point cloud classification; supervoxel; random forest; feature fusion; segmentation

**Citation:** Liao, L.; Tang, S.; Liao, J.; Li, X.; Wang, W.; Li, Y.; Guo, R. A Supervoxel-Based Random Forest Method for Robust and Effective Airborne LiDAR Point Cloud Classification. *Remote Sens.* **2022**, *14*, 1516. https://doi.org/10.3390/rs14061516

Academic Editors: Yue Wu, Kai Qin, Maoguo Gong and Qiguang Miao

Received: 21 February 2022
Accepted: 11 March 2022
Published: 21 March 2022

**Publisher's Note:** MDPI stays neutral with regard to jurisdictional claims in published maps and institutional affiliations.

## 1. Introduction

With the development of photogrammetry and light detection and ranging (LiDAR) technologies, urban three-dimensional (3D) point clouds can be easily obtained. Three-dimensional point cloud data are used in many applications, such as power line inspections [1], urban 3D modeling [2,3], and unmanned vehicles [4]. However, the most basic requirement for these applications is the semantic classification of 3D point cloud data, which has been a research focus among photogrammetry and remote sensing communities.

Early classification efforts mainly focused on extracting low-level geometric primitives, such as point features, line features, and surface features, which were used for surface reconstruction or point cloud alignment. In recent years, researchers have developed methods for extracting high-level semantic features for structure model reconstruction from point cloud data through machine learning-and deep learning-based methods [5–7]. The core challenges of point cloud data classification are extracting discriminative features from neighborhoods and constructing point cloud classifiers [8,9]. Accurate classification depends on a combination of robust point cloud features and proper classifiers [8,10]. Recent works have applied deep learning networks to directly learn per-point features from raw point clouds [11,12]. Similar to traditional machine learning, these methods focus on the extraction of higher-order features from point cloud data by building a new convolutional neural network. Although remarkable performance has been achieved using these methods, large training sample sets are required to pre-train the classification models. These semantic tags require manual labeling, which is time-consuming and labor-intensive. Moreover, the training models obtained by such methods are difficult to generalize to other scenarios [13].

To solve the model generalization and incomplete label data problems, many researchers prefer traditional machine learning methods, which require only a small sample dataset to achieve fast and accurate semantic point cloud data classification [14–16]. However, original point cloud features are often highly unstable due to the influence of point cloud data accuracy and noise, especially data acquired by tilt photogrammetry. Thus, more researchers are exploiting high-order features and their contextual information for scene classification. As dimensional objects expanding upon the concept of the "superpixel" [17], "supervoxels" [18] are generated by partitioning 3D space as point clusters. Supervoxels have been increasingly applied to describe adjoining points related to the same objects [16,19]. Transferring the original point cloud to the "supervoxel cloud" propagates simple point-based classification to an object-based level. Some point cloud segmentation methods, such as locally convex connected patches (LCCP), recognize points through supervoxel-adjacent relationships. In addition to features, classifiers that can effectively deal with massive data must be considered. Machine learning methods such as random forest (RF) that are capable of handling complex data are gaining attention for this purpose [20,21].

Here, we propose a robust and effective point cloud classification approach that integrates point cloud supervoxels and their LCCP relationships into an RF classifier. The proposed method involves three strategies to effectively improve classification accuracy. (1) Features are divided into three categories based on their description types (point-based, eigen-based, and grid-based), and three unique feature calculation strategies are designed to improve feature reliability. (2) A centroid point is used to represent supervoxel geometries, and every point that belongs to the same cluster shares all properties. (3) Supervoxel local neighborhoods are segmented by LCCP to avoid the inclusion of object borders.

The rest of this paper is organized into four sections. In Section 2, we review and compare similar methods for solving classification issues in two categories. Section 3 presents the framework of the proposed supervoxel-based RF model, providing the feature descriptions and RF model process and algorithm. The statistical and visual results of the data training and validation are shown in Section 4, and our research conclusion and remarks are given in Section 5.

## 2. Related Works

Previous classification approaches can be categorized as knowledge-driven and model-driven methods predicated on the classifier type. Reviews of the logical bases for these methods are presented below.

### 2.1. Knowledge-Driven Methods

Knowledge-driven methods involve the detection of structural features consisting of points; human expert knowledge of the terrestrial surface is then used to extract various objects from the original point cloud. In some cases, correction systems are applied to fix obvious faults [16]. Typically, these approaches focus on two crucial points: what features to extract and how to build a reliable human-knowledge-based system for classification. Generally, some human-eye optical features, such as height, slope, and color, can be used in real cases. Huang et al. [22] integrated multispectral imagery and ALS data to obtain the ground truth red–green–blue (RGB) color and surface elevation values in each pixel and built a classification system based on color information and urban elevation knowledge for executing segmentation of different objects. Germaine and Hung [23] constructed two systems based on surface height and surface slope, respectively. Polygonal features can also be used in knowledge-based approaches. For instance, Zheng et al. [24] used the Fourier fitting method [25] to classify the pointcloud, in which the geometrical eigen features and basic features were integrated in their classification algorithm. Including spectral information assures reliable results, and combining various features ensures the system has high performance. Additionally, simple rules derived from gained features facilitate increased accuracy in the postprocessing stage. By regularizing objects placed at different heights and with distinctive surface slopes, a correction system can fix local classification faults in point clouds [22,26].

Knowledge-driven methods are well-acknowledged for their succinct and distinct processes based on the human recognition of ground objects [26]. However, these approaches rely on prior information, and precise airborne imagery is essential for acquiring reliable outputs. Moreover, matching the LiDAR dataset and multispectral image coordinates is time-consuming, which restricts knowledge-based processes to a small area range and can create spectral error accumulation. Furthermore, specific knowledge cannot generalize to diverse situations, such as vehicles and clusters on a small scale, which may generate errors in the final output. Thus, complex urban scenes may be challenging to classify using knowledge-driven methods.

### 2.2. Model-Driven Methods

Model-driven methods construct classification models from features extracted from or calculated based on point clouds, before segregating clouds into a training dataset and validation dataset. The training set fits the model and modifies the original parameters, and the validation set provides the current classification performance of the model. Appropriate model structures are crucial for such methods. The primary differences between knowledge and model-driven methods are the classifier types and structures.

Many approaches use convolutional neural networks [27] as the basic model structure [28–31]. The network structure is designed according to the actual composition of the point cloud dataset, and then the points are separated into clusters used for input. Through many rounds of forward and backward propagation, a relatively reliable classification model can be built. Varied features are included to increase the input complexity and optimize model performance. Wang et al. [31] developed a dynamic graph network structure that could simultaneously finish classification and segmentation to identify shape properties and include neighborhood features. Hong et al. [28] built upon this method by including a modification module to balance the performance and cost and using an optimized skip connection network for efficient training. Classic models, such as RF, conditional random fields [14] with integrated RF, and support vector machines [32], have also been used for the labeling process [21,33,34]. The supervoxel-based method representing object-based routes has been incorporated into simple classifiers [35], and the supervoxel-adjacency relationship can also be considered as a feature of the local neighborhood [36].

Most existing model-driven methods based on supervoxel extraction are prone to include real object boundaries in the local neighborhood of voxels, which decreases the homogeneity of supervoxel adjacency and polygonal feature accuracy. Combining a precise object segmentation utility with previous model-driven methods will effectively solve this problem. Object edges can be detected by particular network structures or LCCP [37]. Feng et al. [38] developed a local attention-edge convolutional network that identified objects by summarizing the features of all neighbors as a weight value learned by the network. The LCCP examined the connection between two adjacent supervoxels and determined whether they relate to one object by calculating the included angle of two normal vectors. The former method focused on whole object segmentation, whereas the latter recognized as many connected edges as possible. To better exploit supervoxel features and their contextual relationships for point cloud classification, we propose a robust and effective classification approach that integrates point cloud supervoxels and their LCCP relations into an RF classifier to improve the accuracy of feature calculation and reduce computational costs.

### 3. Methodology

#### 3.1. Overview of the Approach

The approach starts with a voxel-grid-based downsampling algorithm [39] to prevent the point cloud from becoming over-dense without impacting the original structure. Next, a noise-rejection statistical-outlier-removal filter is used to remove dynamic objects and erroneous points from the aerial laser point cloud. The threshold is calculated from the average distance between a single point and its k-neighbors referring to a certain range of standard deviation.

The technical route for our approach after data preprocessing is shown in Figure 1. The features are divided into three categories, point-based, eigen-based, and grid-based. First, the original 3D point cloud is transformed into a set of supervoxels by the supervoxel calculation method, in which points located in the same supervoxel generally have similar feature descriptions. The original point cloud is also divided using a regular grid to facilitate the extraction of grid-based elevation features in the later stage. Instead of semantic labeling of the raw points, supervoxels are used as the basic unit for semantic classification, and the centroids of the supervoxels are generated from the supervoxel structure. Three kinds of features are calculated: (1) The eigen-based features are first calculated using a principal component analysis algorithm, and the corresponding geometric shape features are generated by deformation and combination with those eigenvalues. Specifically, the adjacency relationship built by voxel cloud connectivity segmentation (VCCS) is used to determine the supervoxel neighborhood ranges. (2) The point-based features, including the local density, point feature histogram, point's normal vectors, elevation values, and RGB color properties, are obtained via neighborhood calculation or the point cloud's raw attributes. (3) We introduce a grid-based elevation feature to decrease the influence of uneven topography during point cloud classification. Based on the regularized grid of the point cloud data, the relative elevation of the horizontal location is used as the elevation feature of each supervoxel centroid. Finally, all three feature types are used to train the supervoxel-based RF model, which is used for point cloud classification.

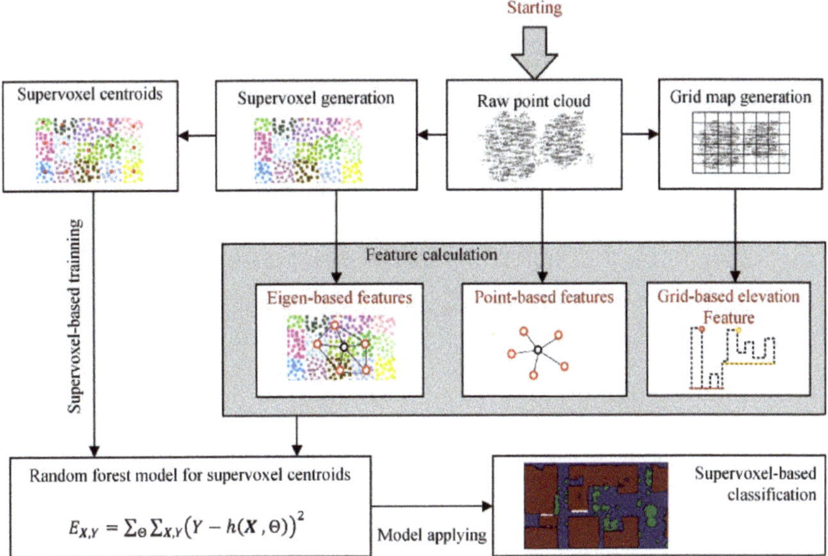



$$E_{X,Y} = \sum_{\Theta} \sum_{X,Y} (Y - h(X, \Theta))^2$$

**Figure 1.** Supervoxel-based random forests framework for point cloud classification. The equation of the random forest model located at the bottom-left refers to the least squares method applied in the model to predict unlabeled points, in which $Y$ represents the label, $X$ represents an individual centroid point, and $\Theta$ represents the coefficient matrix.

### 3.2. Two-Level Graphical Model Generation for Feature Extraction

Supervoxels are defined as groups of points that contain similar geometric features or attributes, such as location, color, and normal direction. Additionally, adjacency relationships embedded in supervoxels can provide more effective information for neighborhood searching, improving the robustness and accuracy of feature calculation. For this classification method, we use supervoxels, rather than single points, as the basic unit to construct the RF model, and the domain information is constrained via LCCP segmentation. Therefore, a two-level graphical model using supervoxel calculation and LCCP optimization is generated from the raw point cloud. Figure 2 illustrates the two-level graphical model generation process.

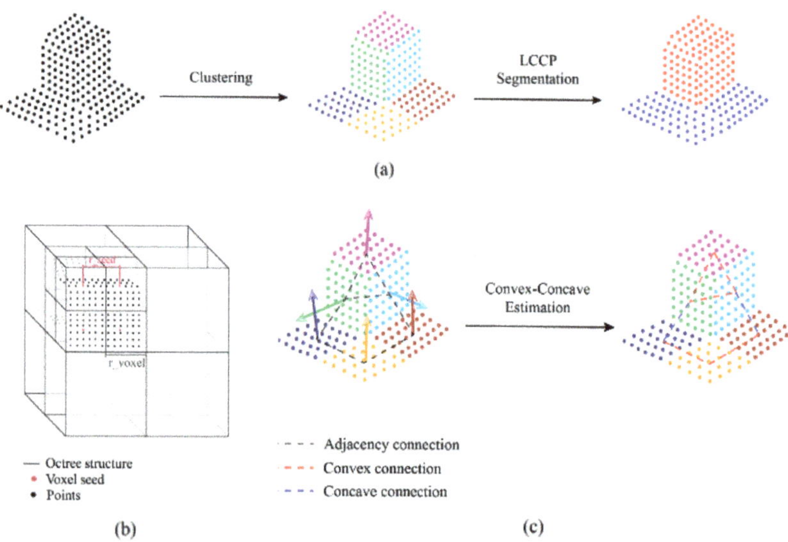


**Figure 2.** Illustration of two-level graphical model generation. (**a**) The fundamental process of supervoxel-based object segmentation. (**b**) The octree structure used for supervoxel clustering. (**c**) The locally convex connected patches (LCCP) segmentation scheme. Colored arrows show the corresponding normal vectors of supervoxels.

### 3.2.1. First-Level Graphical Model Generation by the Supervoxel and VCCS Algorithm

First, we generate the supervoxel model in two steps, namely, randomly setting down seeds within the point cloud and clustering by calculating the feature distances among neighboring points. The supervoxel clustering algorithm estimates the point homogeneity via color, space, and normal dimensions as in Equation (1).

$$d = i_{space} * d_{space} + i_{normal} * d_{normal}$$
$$d_{space} = \frac{\sqrt{\Delta x^2 + \Delta y^2 + \Delta z^2}}{r_{voxel}} \tag{1}$$
$$d_{normal} = \frac{v_1 * v_2}{|v_1| * |v_2|}$$

where $d$ represents the summarized estimation of homogeneity across all dimensions, $d_{space}$ represents the Euclidean distances between the seed points and surrounding points, and $d_{normal}$ is the normal of the fitted plane by the least squares fitting method based on the neighbor points. In this approach, the weights for distance $i_{space}$ and normal $i_{normal}$ during supervoxel clustering are set to 0.4 and 0.6, in which the higher the weight, the greater the contribution. $r_{voxel}$ is the size of each supervoxel, and $v_1$ and $v_2$ are the normal vectors of pairwise adjacent supervoxels. The entire point cloud is clustered into supervoxels using Voxel Cloud Connectivity Segmentation (VCCS) as proposed by [18]. Figure 2b shows the schemes for supervoxel generation in which the octree structure is used to define branches and separate areas. Based on the supervoxel clustering results, the centroids of each supervoxel are calculated and then used for RF point cloud classification. All points within their respective supervoxel have similar features, and the centroid points are ordered in a mesh-like shape to simplify the complex computation of plane shape features. Specifically, an adjacency map containing the adjacent connections relations among supervoxels is simultaneously generated, which presents coterminous connection information that can greatly reduce the cost of neighborhood searching and improve the robustness and accuracy during neighbor calculation [40,41].

### 3.2.2. Second-Level Graphical Model Generation via LCCP Calculation

In order to determine the neighborhood relationship more accurately, we realize the extraction of a second-level graphical model by applying the Locally Connected Convex Patches (LCCP) algorithm on the first-level supervoxel model. In this algorithm, the connection relations implicit in the supervoxels are used for the determination of the neighborhood information, and these connection relations are defined as edges. The edges between adjacent supervoxels are given concave and convex type information based on a surface convexity detection. In order to ensure the aggregation of neighboring super voxels with similar characteristics, we calculate the "robust neighbors" of each supervoxel by judging the concave–convex relationship of edges. "Robust neighbors" means that the domain information can more reasonably represent the geometric features of the current location. Figure 2 shows the convex–concave estimation method among the supervoxels. The method of determining the concave–convex relationship is shown in Equation (2). When two super voxels have a concave domain relationship, they are considered to belong to two different objects. Therefore, after LCCP-based calculations, the adjacency relations of super voxels are given concave and convex properties, which can assist in obtaining more robust domain information quickly and accurately during feature calculations.

$$\hat{d} = \frac{\vec{x_1} - \vec{x_2}}{\| \vec{x_1} - \vec{x_2} \|}$$
$$\Delta\alpha = \vec{n_1} \cdot \hat{d} - \vec{n_2} \cdot \hat{d} \tag{2}$$

where $\vec{x_1}$ and $\vec{x_2}$ indicate the centroids of these two observed two supervoxels, and $\vec{n_1}$ and $\vec{n_2}$ represent their normal vectors. The relationship is considered a convex connection when $\Delta\alpha > 0$, which indicates the angle between the normal vector of the current supervoxel and the linear vector defined by $\vec{x_1} - \vec{x_2}$ is small. Alternatively, the relationship is considered a concave connection when $\Delta\alpha < 0$.

### 3.3. Hybrid Feature Description

#### 3.3.1. Point-Based Feature Description

Considering that some features are extracted from the original point cloud with better robustness, we present five types of point cloud feature description and extraction methods. The five main types contain "Local density", "Point feature histogram (PFH)", "Direction", "Relative elevation", and "RGB color", as follows.

(1) **Local density of the point cloud:** the density feature is calculated as the average distance from one point to the nearest k-neighbors. For each centroid in the super voxel, fast retrieval of domain points is achieved by the construction of a KDTREE and the fast library for approximate nearest neighbors (FLANN) algorithm [42]. Then, the local density feature of the point is obtained by calculating the average of the Euclidean distance between two pairs of neighboring points.

(2) **Point feature histogram (PFH):** The goal of the PFH formulation is to encode a point's k-neighborhood geometrical properties by generalizing the mean curvature around

the point using a multidimensional histogram of values [43,44]. A Point Feature Histogram representation is based on the relationships between the points in the k-neighborhood and their estimated surface normals. In this work, the PFH feature of each centroid point is calculated by KDTREE searching from the original point cloud

(3)   **Direction:** The direction feature indicates the angle between the normal of the location and the horizontal plane, which is calculated as follows (Equation (3)).

$$c = \frac{n_1 \cdot n_2}{\mid n_1 \mid \cdot \mid n_2 \mid} = \frac{z_1}{\sqrt{x_1{}^2 + y_1{}^2 + z_1{}^2}} \tag{3}$$

where $c$ refers to the cosine value, $n_1$ represents the normal vector of the supervoxel and $n_2$ is the normal vector of the horizontal plane (defined as (0,0,1)), respectively. In this paper, to facilitate feature normalization, the cosine value is used to represent the directional features of the supervoxels.

(4)   **Relative elevation:** The relative elevation feature is the distance from the center point of the supervoxel to the ground in the extended z-direction. Considering the influence of ground undulation on elevation features, this paper proposes a grid-based optimization strategy for elevation feature extraction (see Section 3.3.3).

(5)   **RGB color:** RGB color information can achieve effective judgment of feature types and this paper uses color features as a basic feature of supervoxels. Considering that this paper uses supervoxels as the basic unit for feature classification experiments, their color features are determined by the average value of points inside the supervoxels.

### 3.3.2. Eigen-Based Feature Description

Eigen values illustrate the local shape characteristics of the point neighborhood, which helps distinguish objects, such as ground points which have small values in one direction and vegetation points which have similar values. The traditional method of computing Eigen-based features is implemented by K-neighborhood search of point clouds. In order to obtain more robust neighborhood information, this paper implements accurate neighborhood estimation based on the LCCP algorithm, which can accurately estimate the boundaries of different types of objects. Then these neighborhood supervoxels satisfying the LCCP conditions are used for Eigen-based feature calculation. Figure 3 shows the flow of the super voxel neighborhood calculation method in which the concave–convex relationship between supervoxels is derived from the second-level graphic model.

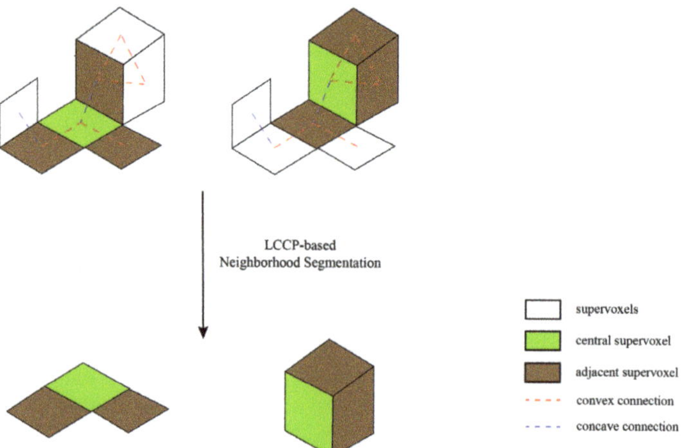


**Figure 3.** Locally convex connected patches (LCCP) neighborhood optimization. The neighborhood ranges used to calculate eigenvalues are shown at the bottom.

The three eigenvalues will be calculated by feature decomposition, and sorted in descending order ($\lambda_1 \geq \lambda_2 \geq \lambda_3 \geq 0$). Based on the mathematical meaning of eigenvalues, different combinations of eigenvalues demonstrate particular shape characteristics [10]. In this work, five types of shape features are used for the classification of supervoxels, including "Curvature", "Linearity", "Planarity", "Scattering", and "Anisotropy". The specific calculation formulas are shown in Table 1.

(1)   Curvature: Describes the extent of the curve for a point group.
(2)   Linearity: Describes the extent of the line-like shape for a point group.
(3)   Planarity: Describes the extent of the plane-like shape for a point group.
(4)   Scattering: Describes the extent of the sphere-like shape for a point group.
(5)   Anisotropy: Describes the difference between the extents of entropy in respective directions of eigenvectors for a point group.

**Table 1.** Computing method for eigenvalue-based shape features. Feature definitions on the left are described in Section 3.3.2. Three eigenvalue symbols are sorted in descending order from 1 to 3 in the formulas.

| Feature Definition | Computing Formula |
| --- | --- |
| Curvature | $\mathcal{C}_e = \frac{\lambda_3}{\lambda_1 + \lambda_2 + \lambda_3}$ |
| Linearity | $\mathcal{L}_e = \frac{\lambda_1 - \lambda_2}{\lambda_1}$ |
| Planarity | $\mathcal{P}_e = \frac{\lambda_2 - \lambda_3}{\lambda_1}$ |
| Scattering | $\mathcal{S}_e = \frac{\lambda_1}{\lambda_3}$ |
| Anisotropy | $\mathcal{A}_e = \frac{\lambda_3 - \lambda_1}{\lambda_3}$ |

### 3.3.3. Grid-Based Elevation Feature Description

When the original elevation features of point cloud data are used for point cloud data classification, it is easy to produce misclassification in areas with large topographic undulations. In particular, features with similar geometric shapes or colors can easily cause confusion in classification, such as the ground and the top surfaces of buildings. Some methods use DEM information to reduce the influence of terrain height difference on data classification, but it is often difficult to obtain accurate DEM data. Therefore, this paper proposes a grid-based method for calculating elevation features, which can accurately calculate the relative elevation information between the features and the ground. As shown in Figure 4a, we first project the original point cloud data onto a 2D plane, i.e., XOY plane and then divide the projected data into a grid according to the area size. Therefore, the relative elevation of each point can be obtained by subtracting the ground elevation from that point. In general, we take the smallest elevation value in the grid as the ground elevation of the target location. However, some hindrances, such as the absence of ground points below the building roof and large-scale clusters, are typical in 3D urban scenes due to the shortage of ray reflection, meaning that roof points, especially with a flat shape, are occasionally confused with ground points. A lattice filter kernel is used to solve the ground detection error problem, the basic principle of which is similar to image processing [45]. As illustrated in Figure 4b, each cell is checked by a $5 \times 5$ filter kernel, the outliers are first removed by the Gaussian distribution strategy. Then the algorithm corrects the ground elevation value of the current cell with the average value of the filter, when the standard deviation does not satisfy the Gaussian distribution condition [46].

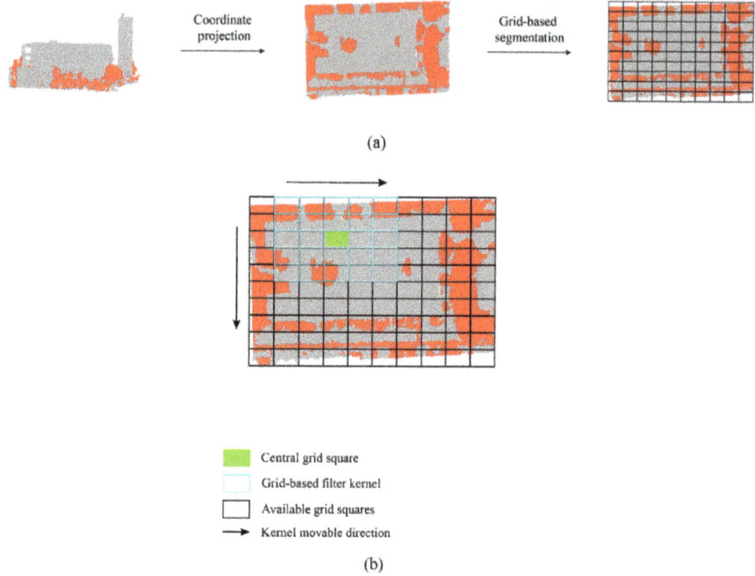



**Figure 4.** Grid-based elevation computation and filtering. (**a**) The illustrated point cloud data (left) and the 2D-projected data with grid segmentation (right). (**b**) The grid filter examining anomalies of calculated elevation values in grid squares.

### 3.4. Supervoxel-Based Random Forests (RF) Model

The RF model is an ensemble learning method for classification, regression, and other tasks that operates by constructing a multitude of decision trees at training time. For classification tasks, the output of the random forest is the class selected by the most trees. In order to integrate the above three hybrid features for point cloud data classification, a supervoxel-based RF model is constructed in this paper. In this method, the supervoxels will be used as the basic classification units, and the extracted hybrid features will be used as training information input for decision tree generation. The random forest construction process is constrained by two main parameters including the "max depth" and the "total number of decision trees". Here, the "max depth" represents the depth of each tree in the forest. The deeper the tree, the more splits it has, and it captures more information about the data. However, too large a depth value can easily cause problems such as overfitting or excess processing time. In this paper, to balance operational efficiency and classification accuracy, the max depth and the total number of decision trees are set to 25 and 10, respectively. So to obtain an optimal number, the accuracy of the output RF model is verified with the validation set. The algorithm applies the mean squared generalization error to evaluate the classification correctness, as Equation (4) shown in [20].

$$E_{X,Y} = \Sigma_{\Theta}\Sigma_{X,Y}(Y - h(X, \Theta))^2 \tag{4}$$

where $X$ refers to the random feature vector, and $Y$ refers to the corresponding label. $\Theta$ is a single tree inside the forest, appearing in tandem with one $X$.

The framework proposed by the ETH Zurich RF template library [47] is used to train the supervoxel-based random forest model. It should be noted that the framework contains three kinds of classification method, including ordinary classification, local smooth classification, and graph cut-based classification. In our approach, graph cut-based classification is employed for training purposes, since it is optimized with an energy minimization method [48] and provides the best overall classification accuracy.

## 4. Experimental Results

To verify the effectiveness of the proposed method in this paper, two sets of data were used for classification testing and accuracy analysis. The publicly available dataset from the ISPRS benchmark [49] contains data collected in Toronto, Canada, and Vaihingen, Germany, both the Toronto and German datasets were used for accuracy verification. Subsequently, a classification experiment was conducted with the airborne LiDAR dataset collected in Shenzhen City, China. In our experiments, three accuracy assessment metrics were used for accuracy evaluation according to the conventional accuracy assessment methods for point cloud classification [50,51]. We selected three indices, including the overall accuracy (OA), the mean intersection over union (mIoU), and the F1-score, which were calculated as follows.

$$OA = \frac{True\ Positive}{True\ Positive + True\ Negative + False\ Positive + False\ Negative}$$

$$mIoU = \frac{True\ Positive}{True\ Positive + 2 \times (True\ Negative + False\ Positive + False\ Negative)}$$

$$p = \frac{True\ Positive}{True\ Positive + False\ Positive} \tag{5}$$

$$r = \frac{True\ Positive}{True\ Positive + False\ Negative}$$

$$F_1 = \frac{2 * p * r}{p + r}$$

where *True Positive* (TP), *False Positive* (FP), *True Negative* (TN), and *False Negative* (FN) values are extracted from the confusion matrix of the classification result, and $p$ and $r$ are the precision and recall percentages, respectively.

### 4.1. ISPRS Benchmark Datasets

#### 4.1.1. Toronto Sites

The Toronto dataset was divided into two regions, Area 1 and Area 2 for testing purposes. The classification results are shown in Figure 5. The overall scene was divided into four types, buildings, vegetation, ground, and background. As shown in Figure 5, there was a large amount of overlap and crossover between buildings and vegetation in the Toronto data, as well as incomplete facade collection, which can easily lead to the problem of confusion between tree and building facades during classification process. Meanwhile, due to the lack of color information in Toronto's point cloud data, the classifier relied more on geometric features for semantic classification.Thanks to the grid-based elevation features and the supervoxel-optimized Eigen features, the proposed algorithm still achieved good classification results when only geometric features were used. Figure 6 shows the comparison of the classification accuracy before and after using the grid-based elevation features, in which it can be clearly seen that the ground level and the top surfaces of the buildings could be accurately distinguished after the optimization of the elevation features.

However, the method proposed in this paper still suffered from some classification errors. As illustrated in Figure 7, some misclassified areas are shown enlarged; those errors were mainly caused by similar geometric features or missing data. For example, some buildings were incorrectly classified as ground due to their low elevation values, and some buildings with missing facades were classified as ground.

In addition, the quantitative classification results were compared with those of five state-of-the-art algorithms, including MAR_2, MSR, ITCM, TICR, and TUM. The first two rely mainly on the geometric information of the original point cloud for classification, while the last three approaches fuse point cloud and image features for classification.The OA, mIoU, and F1-score are listed in Table 2. The proposed method achieved high accuracy classification results in both regions, similar to the classification accuracy of MAR_2 and MAR. It should be noted that the MSR method achieved better classification accuracy in most cases, mainly due to the use of DEM data. The proposed method achieved an OA accuracy of 93.2%, mIoU accuracy of 87.4%, and an F1-score of 92.6% in Area 1; in particular, the F1 accuracy was the best among all methods. Similarly, in Area 2, the classification method proposed in this paper achieved an OA accuracy of 93.1%, an mIoU accuracy of 87%, and an F1_score of 85.8% respectively.

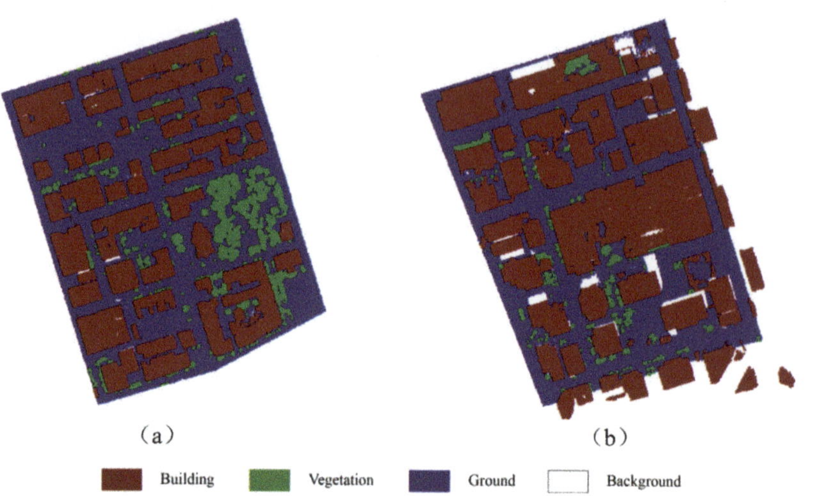


**Figure 5.** Classification results of two Toronto site areas. (**a**) The classification result of Area 1 and (**b**) the clasification result of Area 2.

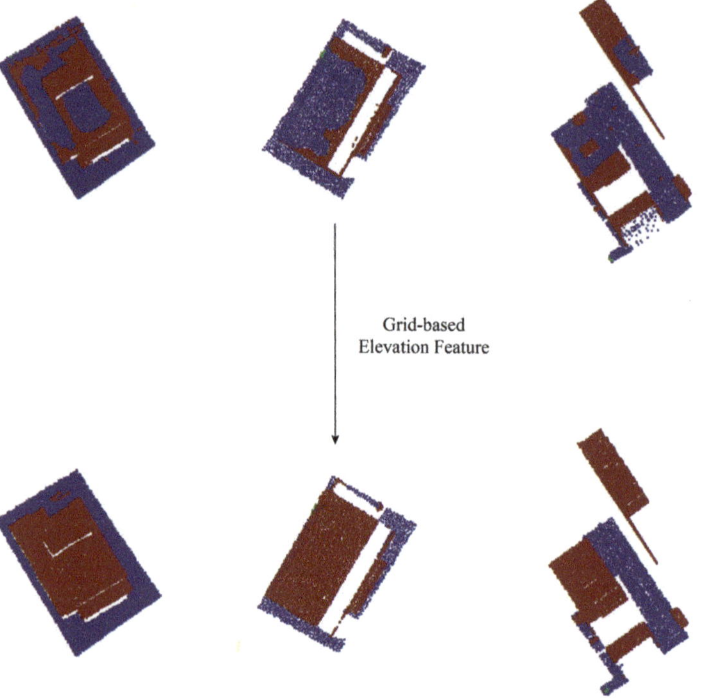


**Figure 6.** The comparison of the classification accuracy before and after using the grid-based elevation features on the Toronto sites.

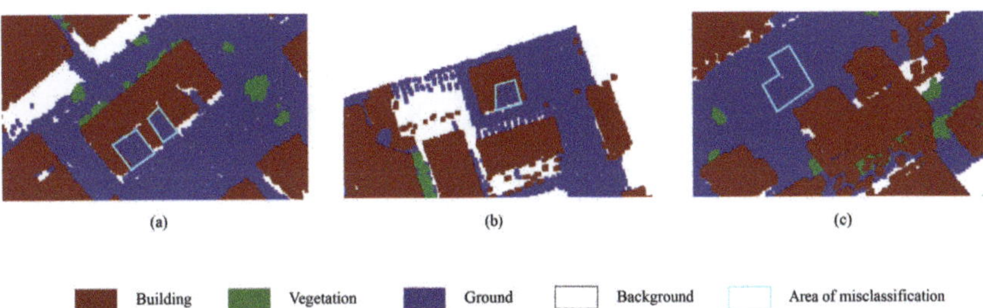



**Figure 7.** Misclassification cases in which roof points were recognized as ground points in the Toronto sites. (**a–c**) refer to different types of misclassification results from roof to ground separately.

**Table 2.** Quantitative comparison of the proposed method and previous related methods tested on the Toronto sites. Two methods, MAR_2 and MSR, used only the point cloud for classification; MSR applied terrestrial digital models. ITCM, ITCR, and TUM used the point cloud and images.

| Methods | Area 1 | | | Area 2 | | |
|---|---|---|---|---|---|---|
| | OA (%) | mIoU (%) | F1-Score (%) | OA (%) | mIoU (%) | F1-Score (%) |
| MAR_2 | 94.3 | 89.2 | 88.9 | 94.0 | 88.7 | 88.4 |
| MSR | 95.5 | 91.4 | 91.2 | 94.8 | 90.1 | 89.7 |
| ITCM | 81.3 | 68.5 | 66.1 | 83.0 | 70.9 | 67.9 |
| ITCR | 84.2 | 72.7 | 69.2 | 85.4 | 74.5 | 72.4 |
| TUM | 82.6 | 70.4 | 68.1 | 83.1 | 71.1 | 68.9 |
| Our method | 93.2 | 87.4 | 92.6 | 93.1 | 87.0 | 85.8 |

OA, overall accuracy; mIoU, mean intersection over union.

### 4.1.2. Vaihingen Sites

The height of buildings in the Vaihingen data was similar to the vegetation and did not contain color information, which was be a major challenge for point cloud data classification for the data set. Similar to the experiment of the Toronto area, the scene was divided into four categories of labels, buildings, vegetation, ground, and background. The classification results are shown in Figure 8. It can be clearly seen that the classification results were worse than those of the Toronto data, which was mainly caused by the similarity of geometric features among different types. Due to connections between supervoxels containing medium-height vegetation and building facades and some oddly curved roof surfaces, points with building groundtruth values were more likely to be partially or completely misjudged as trees. Figure 9 shows some cases of misclassification in the Vaihingen region, in which some parts of buildings were misclassified into trees.

Meanwhile, seven existing classification algorithms were used for comparative analysis of classification accuracy. The OA, mIoU, and F1-score are listed in Table 3. It can be seen that the classification algorithm proposed in this paper achieved the best classification accuracy of 85.2% OA, 74.2% mIoU accuracy, and 83.6% F1_score accuracy, respectively.

However, with the building outline explicitly extracted, the proposed method performed well in the remaining areas, achieving an overall F1-score above 83%, which surpassed some methods using heterogeneous data sources.

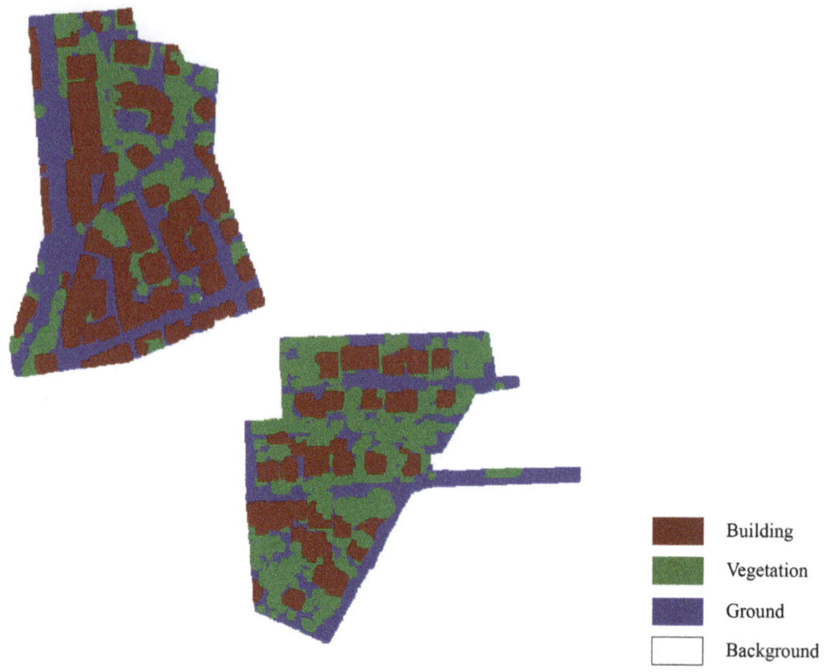



**Figure 8.** Classification results of the Vaihingen sites.

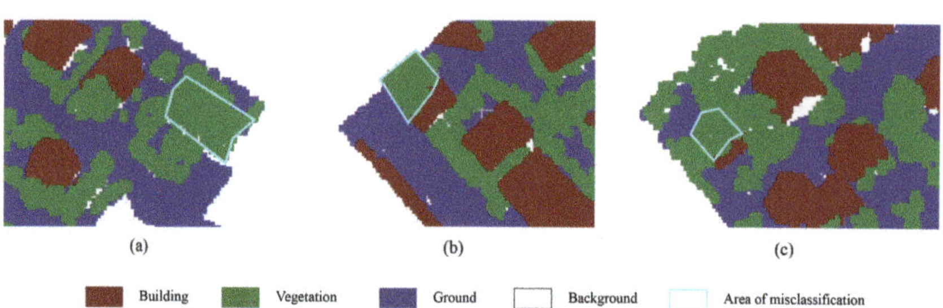



**Figure 9.** Misclassified regions in the Vaihingen site caused by unexpected connections between supervoxels of different objects. (**a–c**) mean misclassification situations in different minor scenes from roof to vegetation.

**Table 3.** Quantitative comparison of the proposed method and previous related methods tested on the Vaihingen sites sorted by overall accuracy (OA) in ascending order. The F1-score was computed based on the same categories (building, vegetation, and ground).

| Methods | OA (%) | mIoU (%) | F1-Score (%) |
|---|---|---|---|
| UM | 80.8 | 67.8 | 78.1 |
| BIJ_W | 81.5 | 68.8 | 78.6 |
| LUH | 81.6 | 68.9 | 80.4 |
| RIT_1 | 81.6 | 68.9 | 79.0 |
| D_FCN | 82.2 | 69.8 | 80.9 |
| WhuY3 | 82.3 | 69.9 | 81.0 |
| WhuY4 | 84.9 | 73.8 | 80.8 |
| Our method | 85.2 | 74.2 | 83.6 |

mIoU, mean intersection over union.

*4.2. Airborne Laser Scanner Dataset in Urban Scenes of Shenzhen*

RGB color information plays a significant role in the proposed classifier because three discriminative features are computed by RGB reflection data, and multispectral aerial images cannot be included. Furthermore, the two datasets used for testing carried little or incomplete spectral band information. Point cloud data assisted by spectral information during generation and reconstruction with complete color data and high resolution can more comprehensively prove the performance of the proposed method. Integrated reconstruction of the facade is also beneficial for the extraction of buildings.

The selected dataset included four urban regions, one for the training set and three for independent validation [marked as (a), (b), (c)]. The training area was 350 m × 200 m, and the validation areas were approximately 400 m × 300 m. The entire dataset was downsampled to a resolution of 0.3 m. The classification results are illustrated in Figure 10. Most vegetation points and ground points were accurately classified, and explicit outlines of buildings were visible in the resulting figure. In most scenes, vegetation was distinguished from adjacent buildings. Moreover, the centroid-based classification method enabled low computation costs, even though each validation area contained more than four million points after the downsampling process. This demonstrates that the proposed classifier successfully handles large datasets. The point-based classification method in CGAL library [52] was used for comparison purpose. The quantitative performance evaluations of our proposed method and the pointbased method are shown in Table 4. As expected, the super voxel-based method proposed in this paper achieved better classification accuracy in all three regions compared to the traditional point cloud-based methods. Specifically, the proposed method achieved 3.6, 5.8, and 4.4 percent, respectively, in the OA, mIoU, and F1_score in Area (a). Similar results were found in the other two regions.

The average performance of the proposed method was higher for the Shenzhen dataset than the Vaihingen and Toronto datasets. The mostly rectangular rooftop shapes and integrated facade structures prevented building points from being recognized as vegetation, whereas the uncertainty of object consistency in the Vaihingen set led to false classification. Compared with the Toronto sites, which were comparably generated except without color information, most elevated vegetation points and buildings with low height and more detailed facades were successfully distinguished using RGB color features in the Shenzhen dataset. However, some exceptional situations in the dataset affected the overall accuracy of the classification results. As shown in Figure 11a, the neighborhood information of partial rooftop points that were similar to roads, such as rises at the edge or street light posts, reduced the contextual consistency of the local region and affected the classification. Additionally, due to the intricate and uncertain shape appearances in modern urban scenes, a single training area provided limited polygonal examples. Parts of buildings with minor scale or unusual contours that were not provided in the training region were misclassified as ground pieces in the validation sets [Figure 11b], which reduced the overall classification accuracy.

Benefiting from supervoxel extraction processing, the point cloud of Shenzhen University can be rapidly aggregated into supervoxel structures, which effectively reduced the point cloud density and complexity. In turn, with supervoxels as the basic unit, the classification method proposed in this paper achieved point cloud classification with high efficiency, and the overall computation costs were about 1.5 h. Moreover, the utilization of LCCP object homogeneity segmentation in supervoxel-based neighborhoods contributed to the considerable classification precision with complete object surfaces consisting of point arrays, which advanced the object-based theory.

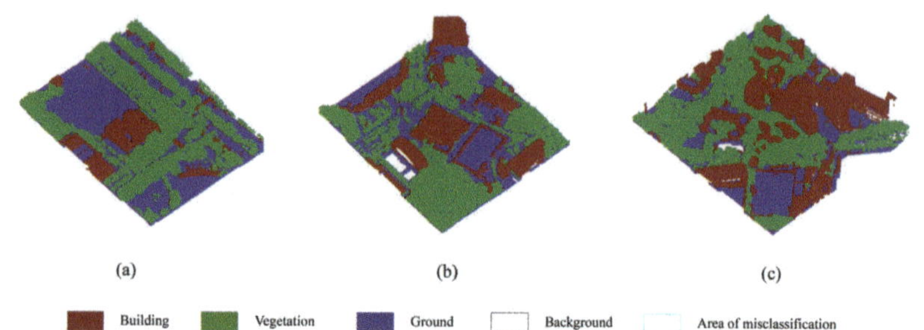

(a)        (b)        (c)



**Figure 10.** Classification results of airborne LiDAR-generated Shenzhen sites. Three selected sites have been marked as (**a–c**).

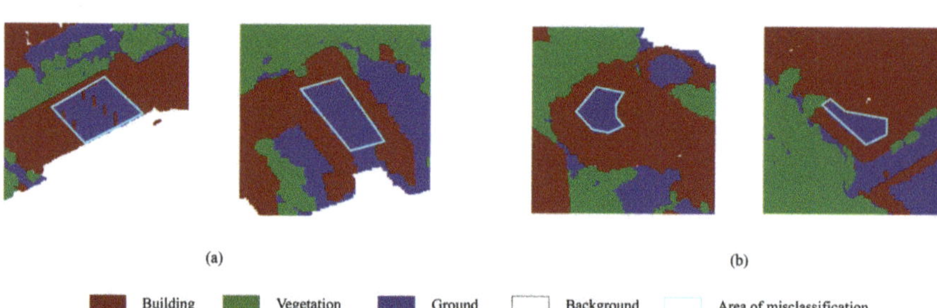

(a)        (b)



**Figure 11.** Misclassification cases in the Shenzhen dataset. (**a**) Faults due to edge interruption. (**b**) Faults due to untrained object shapes.

**Table 4.** Quantitative evaluation of the supervoxel-based results and point-based results of the proposed method on the Shenzhen airborne LiDAR dataset.

| | Area (a) | | | Area (b) | | | Area (c) | | |
|---|---|---|---|---|---|---|---|---|---|
| | OA (%) | mIoU (%) | F1-Score (%) | OA (%) | mIoU (%) | F1-Score (%) | OA (%) | mIoU (%) | F1-Score (%) |
| Our method | 94.0 | 88.7 | 90.1 | 93.5 | 87.8 | 91.8 | 93.5 | 87.8 | 91.7 |
| Point based | 90.6 | 82.9 | 85.6 | 87.6 | 78.0 | 84.2 | 86.1 | 75.6 | 79.8 |

OA, overall accuracy; mIoU, mean intersection over union.

### *4.3. Discussions of the Experimental Results*

For the classification results of the ISPRS benchmark datasets, due to missing RGB color information and some incomplete facades of buildings, the classifier lacked RGB band features, and eigen features were less discriminative. As a result, separated low roofs were classified as vegetation with a similar height. However, most of the borders dividing buildings and vegetation were successfully detected, which showed the excellent effect of applying VCCS and LCCP object-based segmentation into the classifier. For the result of the dataset of the Shenzhen urban scene, although complicated urban scenes provided multi-aspect obstacles for the classifier, the outcome of the proposed method reached our expectations. The proposed classifier achieved a high accuracy classification using only 3D point cloud data without the assistance of digital models and multispectral images, as illustrated in the ISPRS benchmark site outputs. Furthermore, benefited by the RGB information contained in this dataset, the borders between two objects in different types were more distinct, which means color information assisted the object-based classification process.

### 5. Conclusions

In this paper, we proposed a robust and effective airborne LiDAR point cloud classification method that integrated hybrid features, including point-based features, eigen-based features, and elevation-based features, into a supervoxel RF model. Three main innovations were applied to effectively improve the classification accuracy of the proposed model.

(1) Rather than single points, we used supervoxels as the basic entity to construct the RF model and constrain the domain information via LCCP segmentation.

(2) A two-level graphical model involving supervoxel calculation and LCCP optimization was generated from the raw point cloud, which significantly improved the reliability and accuracy of neighborhood searching.

(3) The features were divided into three categories based on feature descriptions (point-based, eigen-based, and grid-based), and three unique feature calculation strategies were accordingly designed to improve feature reliability. We conducted three experiments using ALS data provided by ISPRS and real scene data collected from Shenzhen, China, respectively. We compared the quantitative analysis of ALS datasets with other state-of-the-art methods, and the classification results demonstrated the robustness and effectiveness of the proposed method. Furthermore, this method achieved fine-scale classification when the point clouds had different densities.

However, the proposed method still had some limitations on scene generalizability. The algorithm may fail to recognize roof components when lacking facade information, which is caused by a loss of the connection relationship between supervoxels. In the future, we would like to integrate external constraints into the classification process to prevent the influence of over-segmentation.

**Author Contributions:** Data curation, J.L.; Formal analysis, L.L.; Funding acquisition, W.W.; Investigation, R.G.; Methodology, L.L. and S.T.; Project administration, S.T.; Supervision, S.T. and R.G.; Validation, X.L.; Visualization, Y.L.; Writing—original draft, L.L.; Writing—review & editing, S.T. All authors have read and agreed to the published version of the manuscript.

**Funding:** This research received no external funding.

**Institutional Review Board Statement:** Not applicable.

**Informed Consent Statement:** Not applicable.

**Data Availability Statement:** Not applicable.

**Acknowledgments:** This work was supported in part by the National Key Research and Development Program of China (Projects Nos. 2019YFB210310, 2019YFB2103104) and in part by a Research Program of Shenzhen S and T Innovation Committee grant (Projects Nos. JCYJ20210324093012033, JCYJ20210324093600002), the Natural Science Foundation of Guangdong Province grant (Projects No. 2121A1515012574), the Open Fund of Key Laboratory of Urban Land Resources Monitoring and Simulation, MNR (Nos. KF-2021-06-125, KF-2019-04-014), the National Natural Science Foundation of China grant (Projects Nos. 71901147, 41901329, 41971354, 41971341) and the Foshan City to promote scientific and technological achievements of universities to serve industrial development support projects (Projects No. 2020DZXX04).



# References

1. Chen, C.; Peng, X.; Song, S.; Wang, K.; Qian, J.; Yang, B. Safety Distance Diagnosis of Large Scale Transmission Line Corridor Inspection Based on LiDAR Point Cloud Collected With UAV. *Power Syst. Technol.* **2017**, *41*, 2723–2730.
2. Croce, V.; Caroti, G.; De Luca, L.; Jacquot, K.; Piemonte, A.; Véron, P. From the Semantic Point Cloud to Heritage-Building Information Modeling: A Semiautomatic Approach Exploiting Machine Learning. *Remote Sens.* **2021**, *13*, 461. [CrossRef]
3. Javernick, L.; Brasington, J.; Caruso, B. Modeling the topography of shallow braided rivers using Structure-from-Motion photogrammetry. *Geomorphology* **2014**, *213*, 166–182. [CrossRef]
4. Yue, X.; Wu, B.; Seshia, S.A.; Keutzer, K.; Sangiovanni-Vincentelli, A.L. A lidar point cloud generator: From a virtual world to autonomous driving. In Proceedings of the 2018 ACM on International Conference on Multimedia Retrieval, Yokohama, Japan, 11–14 June 2018; pp. 458–464.
5. Lafarge, F.; Mallet, C. Creating large-scale city models from 3D-point clouds: A robust approach with hybrid representation. *Int. J. Comput. Vis.* **2012**, *99*, 69–85. [CrossRef]

6.  Xiong, B.; Jancosek, M.; Elberink, S.O.; Vosselman, G. Flexible building primitives for 3D building modeling. *ISPRS J. Photogramm. Remote Sens.* **2015**, *101*, 275–290. [CrossRef]

7.  Zhou, Q.Y.; Neumann, U. 2.5 d dual contouring: A robust approach to creating building models from aerial lidar point clouds. In Proceedings of the European Conference on Computer Vision, Heraklion, Crete, Greece, 5–11 September 2010; Springer: Berlin/Heidelberg, Germany, 2010; pp. 115–128.

8.  Hackel, T.; Wegner, J.D.; Schindler, K. Fast semantic segmentation of 3D point clouds with strongly varying density. *ISPRS Ann. Photogramm. Remote Sens. Spat. Inf. Sci.* **2016**, *3*, 177–184. [CrossRef]

9.  Jie, S.; Zulong, L. Airborne LiDAR feature selection for urban classification using random forests. *Geomat. Inf. Sci. Wuhan Univ.* **2014**, *39*, 1310–1313.

10. Wang, Y.; Chen, Q.; Liu, L.; Li, X.; Sangaiah, A.K.; Li, K. Systematic comparison of power line classification methods from ALS and MLS point cloud data. *Remote Sens.* **2018**, *10*, 1222. [CrossRef]

11. Qi, C.R.; Yi, L.; Su, H.; Guibas, L.J. Pointnet++: Deep hierarchical feature learning on point sets in a metric space. *arXiv* **2017**, arXiv:1706.02413.

12. Li, X.; Wang, L.; Wang, M.; Wen, C.; Fang, Y. DANCE-NET: Density-aware convolution networks with context encoding for airborne LiDAR point cloud classification. *ISPRS J. Photogramm. Remote Sens.* **2020**, *166*, 128–139. [CrossRef]

13. Li, X.; Wen, C.; Cao, Q.; Du, Y.; Fang, Y. A novel semi-supervised method for airborne LiDAR point cloud classification. *ISPRS J. Photogramm. Remote Sens.* **2021**, *180*, 117–129. [CrossRef]

14. Niemeyer, J.; Rottensteiner, F.; Soergel, U. Contextual classification of lidar data and building object detection in urban areas. *ISPRS J. Photogramm. Remote Sens.* **2014**, *87*, 152–165. [CrossRef]

15. Niemeyer, J.; Rottensteiner, F.; Sörgel, U.; Heipke, C. Hierarchical higher order crf for the classification of airborne lidar point clouds in urban areas. *Int. Arch. Photogramm. Remote Sens. Spat. Inf. Sci.-ISPRS Arch.* **2016**, *41*, 655–662. [CrossRef]

16. Zhu, Q.; Li, Y.; Hu, H.; Wu, B. Robust point cloud classification based on multi-level semantic relationships for urban scenes. *ISPRS J. Photogramm. Remote Sens.* **2017**, *129*, 86–102. [CrossRef]

17. Achanta, R.; Shaji, A.; Smith, K.; Lucchi, A.; Fua, P.; Süsstrunk, S. SLIC superpixels compared to state-of-the-art superpixel methods. *IEEE Trans. Pattern Anal. Mach. Intell.* **2012**, *34*, 2274–2282. [CrossRef]

18. Papon, J.; Abramov, A.; Schoeler, M.; Worgotter, F. Voxel cloud connectivity segmentation-supervoxels for point clouds. In Proceedings of the IEEE Conference on Computer Vision and Pattern Recognition, Portland, OR, USA, 23–28 June 2013; pp. 2027–2034.

19. Wu, F.; Wen, C.; Guo, Y.; Wang, J.; Yu, Y.; Wang, C.; Li, J. Rapid localization and extraction of street light poles in mobile LiDAR point clouds: A supervoxel-based approach. *IEEE Trans. Intell. Transp. Syst.* **2016**, *18*, 292–305. [CrossRef]

20. Breiman, L. Random forests. *Mach. Learn.* **2001**, *45*, 5–32. [CrossRef]

21. Ni, H.; Lin, X.; Zhang, J. Classification of ALS point cloud with improved point cloud segmentation and random forests. *Remote Sens.* **2017**, *9*, 288. [CrossRef]

22. Huang, M.J.; Shyue, S.W.; Lee, L.H.; Kao, C.C. A knowledge-based approach to urban feature classification using aerial imagery with lidar data. *Photogramm. Eng. Remote Sens.* **2008**, *74*, 1473–1485. [CrossRef]

23. Germaine, K.A.; Hung, M.C. Delineation of impervious surface from multispectral imagery and lidar incorporating knowledge based expert system rules. *Photogramm. Eng. Remote Sens.* **2011**, *77*, 75–85. [CrossRef]

24. Zheng, M.; Wu, H.; Li, Y. An adaptive end-to-end classification approach for mobile laser scanning point clouds based on knowledge in urban scenes. *Remote Sens.* **2019**, *11*, 186. [CrossRef]

25. Chen, H.; Wang, C.; Chen, T.; Zhao, X. Feature selecting based on fourier series fitting. In Proceedings of the 2017 8th IEEE International Conference on Software Engineering and Service Science (ICSESS), Beijing, China, 24–26 November 2017; pp. 241–244.

26. Ponciano, J.J.; Roetner, M.; Reiterer, A.; Boochs, F. Object Semantic Segmentation in Point Clouds—Comparison of a Deep Learning and a Knowledge-Based Method. *ISPRS Int. J. Geo-Inf.* **2021**, *10*, 256. [CrossRef]

27. Gu, J.; Wang, Z.; Kuen, J.; Ma, L.; Shahroudy, A.; Shuai, B.; Liu, T.; Wang, X.; Wang, G.; Cai, J.; et al. Recent advances in convolutional neural networks. *Pattern Recognit.* **2018**, *77*, 354–377. [CrossRef]

28. Hong, J.; Kim, K.; Lee, H. Faster Dynamic Graph CNN: Faster Deep Learning on 3D Point Cloud Data. *IEEE Access* **2020**, *8*, 190529–190538. [CrossRef]

29. Li, Y.; Tong, G.; Li, X.; Zhang, L.; Peng, H. MVF-CNN: Fusion of multilevel features for large-scale point cloud classification. *IEEE Access* **2019**, *7*, 46522–46537. [CrossRef]

30. Song, W.; Zhang, L.; Tian, Y.; Fong, S.; Liu, J.; Gozho, A. CNN-based 3D object classification using Hough space of LiDAR point clouds. *Hum.-Cent. Comput. Inf. Sci.* **2020**, *10*, 19. [CrossRef]

31. Wang, Y.; Sun, Y.; Liu, Z.; Sarma, S.E.; Bronstein, M.M.; Solomon, J.M. Dynamic graph cnn for learning on point clouds. *ACM Trans. Graph. (TOG)* **2019**, *38*, 1–12. [CrossRef]

32. Mountrakis, G.; Im, J.; Ogole, C. Support vector machines in remote sensing: A review. *ISPRS J. Photogramm. Remote Sens.* **2011**, *66*, 247–259. [CrossRef]

33. Vosselman, G.; Coenen, M.; Rottensteiner, F. Contextual segment-based classification of airborne laser scanner data. *ISPRS J. Photogramm. Remote Sens.* **2017**, *128*, 354–371. [CrossRef]

34. Zhang, J.; Lin, X.; Ning, X. SVM-based classification of segmented airborne LiDAR point clouds in urban areas. *Remote Sens.* **2013**, *5*, 3749–3775. [CrossRef]
35. Chen, D.; Peethambaran, J.; Zhang, Z. A supervoxel-based vegetation classification via decomposition and modelling of full-waveform airborne laser scanning data. *Int. J. Remote Sens.* **2018**, *39*, 2937–2968. [CrossRef]
36. Wang, H.; Wang, C.; Luo, H.; Li, P.; Chen, Y.; Li, J. 3-D point cloud object detection based on supervoxel neighborhood with Hough forest framework. *IEEE J. Sel. Top. Appl. Earth Obs. Remote Sens.* **2015**, *8*, 1570–1581. [CrossRef]
37. Christoph Stein, S.; Schoeler, M.; Papon, J.; Worgotter, F. Object partitioning using local convexity. In Proceedings of the IEEE Conference on Computer Vision and Pattern Recognition, Columbus, OH, USA, 23–28 June 2014; pp. 304–311.
38. Feng, M.; Zhang, L.; Lin, X.; Gilani, S.Z.; Mian, A. Point attention network for semantic segmentation of 3D point clouds. *Pattern Recognit.* **2020**, *107*, 107446. [CrossRef]
39. Rusu, R.B.; Cousins, S. 3D is here: Point cloud library (PCL). In Proceedings of the 2011 IEEE International Conference on Robotics and Automation, Shanghai, China, 9–13 May 2011; pp. 1–4.
40. Weinmann, M.; Urban, S.; Hinz, S.; Jutzi, B.; Mallet, C. Distinctive 2D and 3D features for automated large-scale scene analysis in urban areas. *Comput. Graph.* **2015**, *49*, 47–57. [CrossRef]
41. Zhou, Y.; Yu, Y.; Lu, G.; Du, S. Super segments based classification of 3D urban street scenes. *Int. J. Adv. Robot. Syst.* **2012**, *9*, 248. [CrossRef]
42. Muja, M.; Lowe, D.G. Scalable nearest neighbor algorithms for high dimensional data. *IEEE Trans. Pattern Anal. Mach. Intell.* **2014**, *36*, 2227–2240. [CrossRef]
43. Rusu, R.B.; Blodow, N.; Marton, Z.C.; Beetz, M. Aligning point cloud views using persistent feature histograms. In Proceedings of the 2008 IEEE/RSJ International Conference on Intelligent Robots and Systems, Nice, France, 22–26 September 2008; pp. 3384–3391.
44. Rusu, R.B.; Marton, Z.C.; Blodow, N.; Beetz, M. Learning informative point classes for the acquisition of object model maps. In Proceedings of the 2008 10th International Conference on Control, Automation, Robotics and Vision, Hanoi, Vietnam, 17–20 December 2008; pp. 643–650.
45. Miao, Z.; Jiang, X. Weighted iterative truncated mean filter. *IEEE Trans. Signal Process.* **2013**, *61*, 4149–4160. [CrossRef]
46. Liang, J.M.; Shen, S.Q.; Li, M.; Li, L. Quantum anomaly detection with density estimation and multivariate Gaussian distribution. *Phys. Rev. A* **2019**, *99*, 052310. [CrossRef]
47. Walk, S. Random Forest Template Library. Available online: https://prs.igp.ethz.ch/research/Source_code_and_datasets.html (accessed on 15 July 2021).
48. Boykov, Y.; Veksler, O.; Zabih, R. Fast approximate energy minimization via graph cuts. *IEEE Trans. Pattern Anal. Mach. Intell.* **2001**, *23*, 1222–1239. [CrossRef]
49. Rottensteiner, F.; Sohn, G.; Jung, J.; Gerke, M.; Baillard, C.; Benitez, S.; Breitkopf, U. The ISPRS benchmark on urban object classification and 3D building reconstruction. *ISPRS Ann. Photogramm. Remote Sens. Spat. Inf. Sci. I-3 (2012) Nr. 1* **2012**, *1*, 293–298. [CrossRef]
50. Garcia-Garcia, A.; Orts-Escolano, S.; Oprea, S.; Villena-Martinez, V.; Garcia-Rodriguez, J. A review on deep learning techniques applied to semantic segmentation. *arXiv* **2017**, arXiv:1704.06857.
51. Wen, C.; Yang, L.; Li, X.; Peng, L.; Chi, T. Directionally constrained fully convolutional neural network for airborne LiDAR point cloud classification. *ISPRS J. Photogramm. Remote Sens.* **2020**, *162*, 50–62. [CrossRef]
52. Fabri, A.; Pion, S. CGAL: The computational geometry algorithms library. In Proceedings of the 17th ACM SIGSPATIAL International Conference on Advances in Geographic Information Systems, Seattle, WA, USA, 4–6 November 2009; pp. 538–539.

 *remote sensing*

*Article*

# MLCRNet: Multi-Level Context Refinement for Semantic Segmentation in Aerial Images

Zhifeng Huang, Qian Zhang * and Guixu Zhang

School of Computer Science and Technology, East China Normal University, Shanghai 200062, China;
51205901122@stu.ecnu.edu.cn (Z.H.); gxzhang@cs.ecnu.edu.cn (G.Z.)
* Correspondence: qzhang@cs.ecnu.edu.cn; Tel.: +86-152-2105-4245

**Abstract:** In this paper, we focus on the problem of contextual aggregation in the semantic segmentation of aerial images. Current contextual aggregation methods only aggregate contextual information within specific regions to improve feature representation, which may yield poorly robust contextual information. To address this problem, we propose a novel multi-level context refinement network (MLCRNet) that aggregates three levels of contextual information effectively and efficiently in an adaptive manner. First, we designed a local-level context aggregation module to capture local information around each pixel. Second, we integrate multiple levels of context, namely, local-level, image-level, and semantic-level, to aggregate contextual information from a comprehensive perspective dynamically. Third, we propose an efficient multi-level context transform (EMCT) module to address feature redundancy and to improve the efficiency of our multi-level contexts. Finally, based on the EMCT module and feature pyramid network (FPN) framework, we propose a multi-level context feature refinement (MLCR) module to enhance feature representation by leveraging multi-level contextual information. Extensive empirical evidence demonstrates that our MLCRNet achieves state-of-the-art performance on the ISPRS Potsdam and Vaihingen datasets.

**Keywords:** semantic segmentation; aerial imagery; feature extraction; multi-level context modeling; feature refinement

**Citation:** Huang, Z.; Zhang, Q.; Zhang, G. MLCRNet: Multi-Level Context Refinement for Semantic Segmentation in Aerial Images. *Remote Sens.* **2022**, *14*, 1498. https://doi.org/10.3390/rs14061498

Academic Editors: Yue Wu, Kai Qin, Maoguo Gong and Qiguang Miao

Received: 14 February 2022
Accepted: 17 March 2022
Published: 20 March 2022

**Publisher's Note:** MDPI stays neutral with regard to jurisdictional claims in published maps and institutional affiliations.



## 1. Introduction

Image segmentation or semantic annotation is an exceptionally significant topic in remote sensing image interpretation and plays a key role in various real-world applications, such as geohazard monitoring [1,2], urban planning [3,4], site-specific crop management [5,6], autonomous driving systems [7,8], and land change detection [9]. This task aims to segment and interpret a given image into different image regions associated with semantic categories.

Recently, deep learning methods represented by deep convolutional neural networks [10] have demonstrated powerful feature extr4action capabilities compared with traditional feature extraction methods, thereby sparking the interest of researchers and prompting a series of works [11–16]. Among these works, FCN [11] is a pioneer in deep convolutional neural networks and has made great progress in the field of image segmentation. Its encoder–decoder architecture first employs several down-sampling layers in the encoder to reduce the spatial resolution of the feature map to extract features. Then, it uses several up-sampling layers in the decoder to restore the spatial resolution, and it exhibits many improvements in semantic segmentation. However, limited by the structure of the encoder–decoder, FCN suffers from inadequate contextual and detail information. On one hand, some of the detail information is usually dropped by the down-sampling operation. On the other hand, due to the inherent nature of convolution, FCN does not provide adequate contextual information. This task leaves plenty of room for improvement. The key to improving the performance of semantic segmentation is to obtain strong semantic representation with detail information (e.g., detailed target boundaries, location, etc.) [17]

To restore detail information, several studies fuse features that come from encoder (low-level features) and decoder (high-level features) by long-range skip connections. FPN-based approaches [18–20] employ a long-range lateral path to refine feature representations across layers iteratively. SFNet [17] extracts location information from low-level features at a limited scope (e.g., $3 \times 3$ kernel size) and then applies it to calibrate the target boundaries of high-level features. Although impressive, these methods solely focus on harvesting contextual information from a local perspective (the local level) and do not aggregate contextual information from a more comprehensive perspective.

Furthermore, to improve the intra-class consistency of feature representation, some studies enhance feature representation by aggregating contextual information. Wang et al. [21] proposed the self-attention mechanism, a long-range contextual relationship modeling approach that is used by the segmentation model [22–25] to aggregate contextual information across an image adaptively. EDFT [26] designed the Depth-aware Self-attention (DSA) Module, which uses the self-attention mechanism to aggregate image-level contextual information to merge RGB features and depth features. Nevertheless, these approaches only focus on harvesting contextual information from the perspective of the whole image (the image level) without explicit guidance of prior context information [27], and they suffer from high computational complexity $\mathcal{O}((HW)^2)$, where $HW$ is the input image size [28]. In addition, OCRNet [29], ACFNet [30], and SCARF [31] model the contextual relationships within a specific category region based on coarse segmentation (the semantic level). However, in some regions, the contextual information tends to be unbalanced (e.g., pixels in the border or small-scale object regions are susceptible to interference from another category), leading to the misclassification of these pixels. Moreover, ISNet [32] models contextual information from the perspective of the image level and semantic level. HMANet [33] designed a Class Augmented Attention (CAA) module to capture semantic-level context information and a Region Shuffle Attention (RSA) module to exploit region-wise image level context information. Although these methods improve the intra-class consistency of the feature representation, they still lack local detail information, resulting in lower classification accuracy in the object boundary region.

Several works have attempted to combine local-level and image-level contextual information to enhance the detail information and intra-class consistency of feature maps. MANet [34] introduces the multi-scale context extraction module (MCM) to extract both local-level and image-level contextual information in low-resolution feature maps. Zhang et al. [35] aggregate local-level contextual information in a high-resolution branch and harvest image-level contextual information in a low-resolution branch based on HRNet. HRCNet [36] proposes a light-weight dual attention (LDA) module to obtain image-level contextual information, and then the feature enhancement feature pyramid (FEFP) module is designed to exploit the local-level and image-level contextual information in parallel structure. Although these methods harvest local-level and image-level contextual information within the single module or between different modules, they are still missing the contextual dependencies of distinct classes. This paper seeks to provide a solution to these issues by integrating different levels of contextual information efficiently to enhance feature representation.

To this end, we propose a novel network called the multi-level context refinement network (MLCRNet) to harvest contextual information from a more comprehensive perspective efficiently. The basic idea is to embed local-level and image-level contextual information into semantic-level contextual relations to obtain more comprehensive and accurate contextual information to augment feature representation. Specifically, inspired by the flow alignment module in SFNet [17], we first design a local-level context aggregation module, which discards the warp operation that demands extensive computation and enhances the feature representation with a local contextual relationship matrix directly. Then, we propose the multi-level context transform (MCT) module to integrate three levels of context, namely, local-level, image-level, and semantic-level, to capture contextual information from multiple aspects adaptively, which can improve model performance but dramatically increased GPU memory usage and inference time. Thus, an efficient MCT (EMCT) module is presented to address feature redundancy and to improve the efficiency of our MCT module. Subsequently, based on the EMCT block and FPN framework, we propose a multi-level context prior feature refinement module called the multi-level context refinement (MLCR) module to enhance feature representation by aggregating multi-level contextual information. Finally, our model refines the feature map iteratively across FPN [18] decoder layers with MLCR.

In summary, our contribution falls into three aspects:

1. We propose a MCT module, which dynamically harvests contextual information from the semantic, image, and local perspectives.
2. The EMCT module is designed to address feature redundancy and improve the efficiency of our MCT module. Furthermore, a MLCR module is proposed on the basis of EMCT and FPN to enhance feature representation by aggregating multi-level contextual information.
3. We propose a novel MLCRNet based on the feature pyramid framework for accurate semantic segmentation.

## 2. Related Work

### 2.1. Semantic Segmentation

Over the past decade, deep learning methods represented by convolutional neural networks have made substantial advances in the field of semantic segmentation. FCN is a seminal work that applies convolutional layers on the entire image to replace fully connected layers to generate pixel-by-pixel labels, and many researchers have made great improvements based on it. These improvements can be roughly divided into two categories. One is for encoders to improve the robustness of feature representation. Yu et al. [37] designed an efficient structure called STDC for the semantic segmentation task, which obtains variant scalable receptive fields with a small number of parameters. HRNet [38] obtains a strong semantic representation with detail information by parallelizing multiple branches with different spatial resolutions. The other improvement is for the decoder, which introduces richer contextual information to enhance feature representation. DeepLab [13–15] presents the ASPP module that collects multi-scale contexts by employing a series of convolutions with different dilation rates. SENet [39] harvests global contexts by using global average pooling (GAP), and GCNet [40] adopts query-independent attention to model global contexts. This work concentrates on the latter, which aggregates more robust contextual information to enhance feature representation.

### 2.2. Context Aggregation

Based on the scope of context modelling, we can roughly categorize these contextual aggregation methods into three categories, namely, local level, image level, and semantic level. OCRNet [29], ACFNet [30], and SCARF [31] model contextual relationships within a specific category region based on coarse segmentation results. FLANe [41] and DANet [22] use self-attention [21] to gather image-level contexts along channel and spatial dimensions. Li et al. [42] present a kernel attention with linear complexity to capture image-level context in the spatial dimension. ISANet [43] disentangles dense image-level contexts into the product of two sparse affinity matrices. CCNet [44] iteratively collects contextual information at a criss-cross pathway to approximate image-level contextual information. PSPNet [45] and DeepLab [13–15] harvest context at multiple scales, and SFNet [17] harvests local-level contextual information by using the flow alignment module.

### 2.3. Semantic Segmentation of Aerial Imagery

Unlike natural images, the use of semantic segmentation in aerial images is more challenging. Niu et al. [33] proposed hybrid multiple attention (HMA), which models attention in channel, spatial, and category dimensions to augment feature representation. Yang et al. [46] designed a collaborative network for image super-resolution and the segmentation of remote sensing images, which takes low-resolution images as input to obtain high-resolution semantic segmentation and super-resolution image reconstruction results, thereby effectively alleviating the constraints of inconvenient high-resolution data as well as limited computational resources. Saha et al. [47] proposed a novel unsupervised joint segmentation method, which separately feeds multi-temporal images to a deep network, and the segmentation labels are obtained from the argmax classification of the final layer. Du et al. [48] proposed an object-constrained higher-order CRF model to explore local-level and semantic-level contextual information to optimize segmentation results. EANet [49] combines aerial image segmentation with edge prediction tasks in a multi-task learning approach to improve the classification accuracy of pixels in object contour regions.

## 3. Methods

### 3.1. General Contextual Refinement Framework

As shown in Figure 1, the general contextual refinement scheme can be divided into three parts, namely, context modeling, transformation, and weighting:

$$C = f_c(X) \tag{1}$$

$$A = f_t(C) \tag{2}$$

$$X' = f_w(A, g(X)) \tag{3}$$

where $X \in R^D$ is the input feature map, $f_c$ is the contextual information aggregate function, $C$ is the context relation matrix, function $f_t$ is adopted to transform context relation into context the attention matrix $A \in R^D$, $f_w$ is the weighting function, and $X' \in R^D$ is the output feature map. The function $g$ is used to calculate a better embedding of the input feature map. In this paper, we take $g$ as part of $f_w$ and set $g$ as identity embedding: $g(x) = x$.

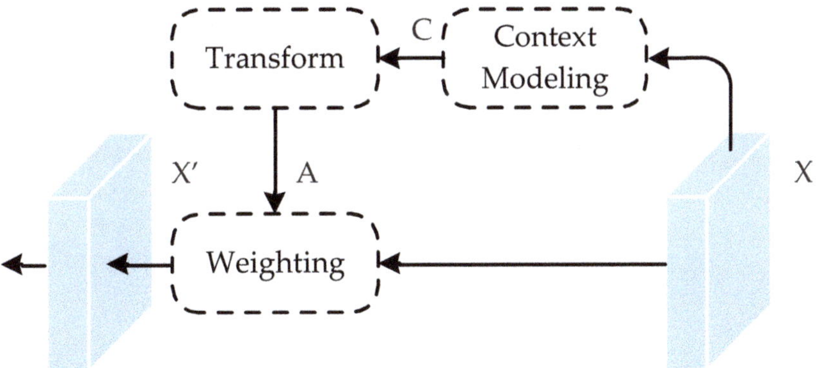


**Figure 1.** General contextual refinement framework.

According to the different context modelling methods, the generic definition can be divided into three specific examples, namely, local-level context, image-level context, and semantic-level context.

### 3.1.1. Local-Level Context

The main purpose of proposed local-level context is to calibrate misalignment pixels between fine and coarse feature maps from the encoder and decoder. Concretely, standard encoder–decoder semantic segmentation architecture relies heavily on up-sampling methods to up-sample the low spatial resolution strong semantic feature maps into high spatial resolution. However, the widely used up-sampling approaches, such as bilinear up-sampling, can not recover spatial detail information, which is lost during the down-sampling process. Therefore, the misalignment problem must be solved by utilizing the precise position information from the encoder feature map. As depicted in Figure 2, we first harvest local-level context information $C_L$:

$$C_L = \zeta(Cat(\tau(F), \beta(X))) \tag{4}$$

where $F \in R^{C' \times HW}$ is a $C'$-dimensional feature map from the encoder; $X \in R^{C \times H \times W}$ is the decoder feature map; $\tau$ and $\beta$ are used to compress the channel depth of $F$ and $X$ to be the same, respectively; $Cat$ represents the channel concatenation operation; $\zeta$ is implemented by one $3 \times 3$ convolutional layer; $C_L \in R^{K \times HW}$; and $K$ is the category number. Then, $C_L$ is transformed into the local-level context attention matrix $A_L$:

$$A_L = \varphi(C_L), \tag{5}$$

where $\varphi$ is the local-level context transformation function and implemented by one $1 \times 1$ convolutional layer, and $A_L \in R^{C \times HW}$.

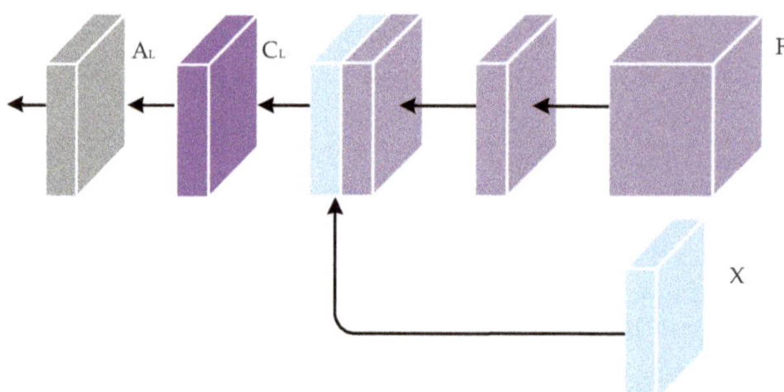


**Figure 2.** Local-level context module.

### 3.1.2. Image-Level Context

The main purpose of the image-level context is to model the contextual information from the perspective of the whole image [32]. Here, we adopt the $GAP$ operation to gather image-level prior context information $C_I$:

$$C_I = \rho(GAP(X)) \tag{6}$$

where $\rho$ is implemented by two $1 \times 1$ convolutional layer, and $C_I \in R^{C \times 1}$. Then, *repeat* is adopted to generate the image-level context attention matrix $A_I$:

$$A_I = repeat(C_I) \tag{7}$$

where $A_I \in R^{C \times HW}$ is the image-level context attention matrix.

### 3.1.3. Semantic-Level Context

The central idea of semantic-level context is to aggregate contextual information based on semantic-level prior information [29–31]. We first employ an auxiliary segmentation head $\xi$ and class dimension normalized exponential function *Softmax* to predict the category posterior probability distribution $P$:

$$P = Softmax(\xi(X)) \tag{8}$$

where $X \in R^{C \times HW}$ ($C$, $H$, and $W$ stand for the number of channels, height, and width of the feature map, respectively), and $P \in R^{K \times HW}$ ($K$ is the number of semantic categories)

Then, we aggregate the semantic prior context $C_S$ according to the category posterior probability distribution:

$$C_S = XP^T \tag{9}$$

where $C_S \in R^{C \times K}$ is the semantic-level contextual information. Finally, we apply self-attention to generate the semantic-level context attention matrix $A_S$:

$$A_S = \eta(C_S) Softmax\left(\frac{\phi(C_S^T)\psi(X)}{\sqrt{d}}\right) \tag{10}$$

where $A_S \in R^{C \times HW}$ is the semantic-level context attention matrix, $\eta$, $\phi$, and $\psi$ are embeddings implemented by two $1 \times 1$ convolutional layer, and $d$ is the number of the middle channel.

### 3.2. EMCT

The intuition of the proposed EMCT is to efficiently and dynamically extract contextual information from the category, image, and local perspectives.

#### 3.2.1. Multi-Level Context Transform

The most straightforward way to transform multi-level contextual information is to directly sum up all levels' context attention matrices. As shown in Figure 3, we propose a multi-level context transformation block, called MCT block, which first computes the local-level, image-level and semantic-level contextual attention matrices separately, and then directly sums them together to obtain the multi-level contextual attention matrix:

$$\hat{A}_{ML} = reshape(A_L + A_I + A_S) \tag{11}$$

where $A_L \in R^{C \times HW}$, $A_I \in R^{C \times HW}$, and $A_S \in R^{C \times HW}$ are the local-level, image-level and semantic-level contextual attention matrices mentioned in Section 3.1, *reshape* is adopted to switch the dimension of the multi-level context attention matrix to $R^{C \times H \times W}$, and $\hat{A}_{ML} \in R^{C \times H \times W}$ is the multi-level context attention matrix.

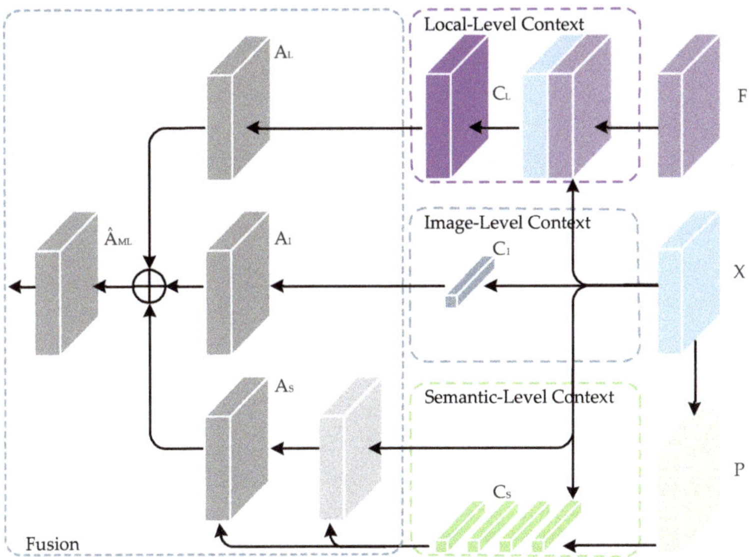


**Figure 3.** The multi-level context transform (MCT) module.

### 3.2.2. Reduction of Computational Complexity

To alleviate contextual information redundancy and reduce computational complexity, we design an EMCT module by reframing the context transform operation based on the MCT block. As illustrated in Figure 4, we construct the EMCT block as:

$$A_{ML} = (C_S \odot C_I)C_L \tag{12}$$

where $A_{ML} \in R^{C \times H \times W}$ and $\odot$ is the broadcast element-wise multiplication that we use to embed image-level contextual information into semantic level contextual information. Then, we further fuse it with the local contextual information matrix $C_L$ by matrix multiplication to generate the multi-level contextual relationship matrix $A_{ML}$. Our designed EMCT module outperforms the MCT module in terms of time complexity and space complexity. Detailed complexity comparison results are presented in Section 4.2.4.

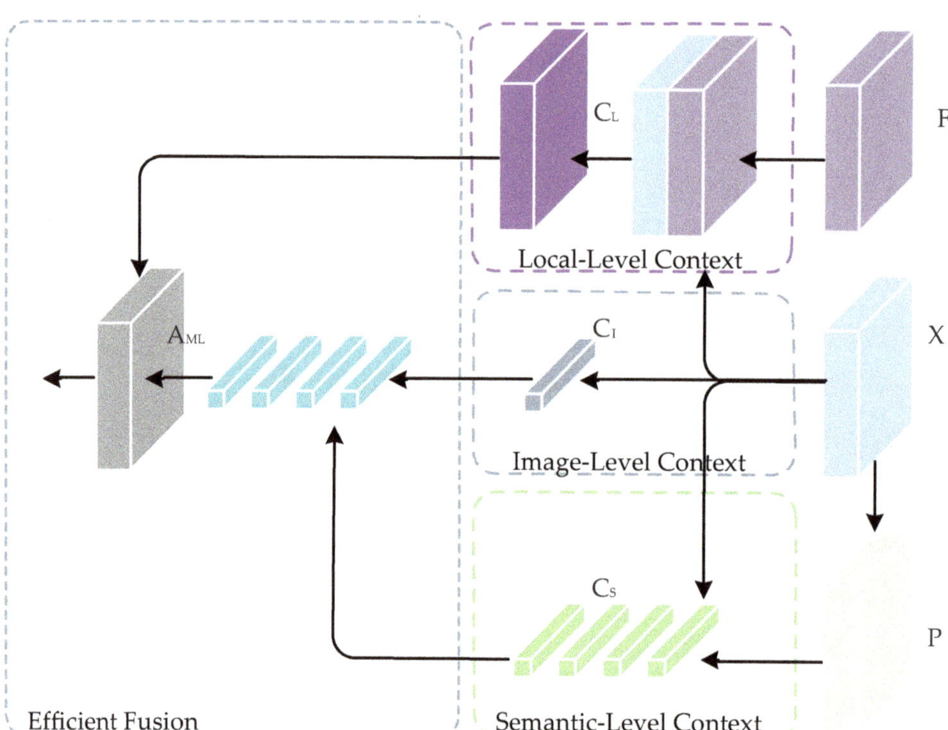


**Figure 4.** The efficient multi-level context transform (EMCT) module. The image-level contextual information $C_I$ is first embedded into semantic-level contextual information $C_S$, then we further fuse them with local contextual information matrix $C_L$ by matrix multiplication to generate multi-level contextual attention matrix $A_{ML}$.

### 3.3. Multi-Level Context Refinement Module

Based on the EMCT block, we propose a multi-level context feature refinement module called the MLCR module. According to Figure 5, we construct the MLCR block as:

$$X' = [EMCT(Upsample_{2\times}(X), F) \odot Upsample_{2\times}(X)] \oplus F \tag{13}$$

where $F \in R^{C \times H \times W}$ is the fine feature map from the encoder, $X \in R^{C \times H/2 \times W/2}$ is the prior decoder layer output, $Upsample_{2\times}$ is the bilinear up-sample operation, $\oplus$ stands for the broadcast element-wise addition, and $X'$ is the refined feature map.

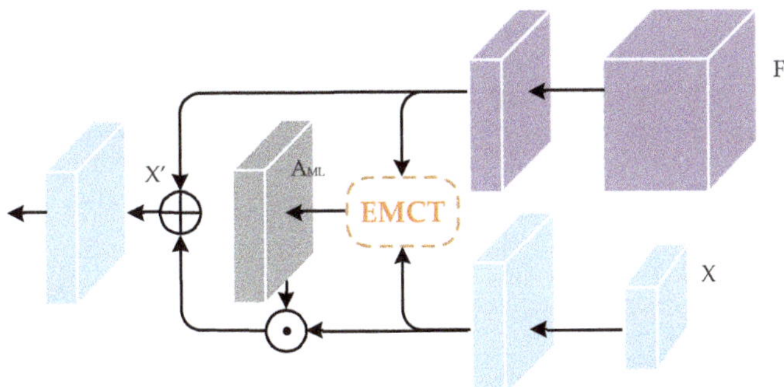

**Figure 5.** MLCR module.

### 3.4. MLCRNet

Finally, we construct a coarse-to-fine network based on the MLCR module called MLCRNet (Figure 6). MLCRNet incorporates the backbone network and FPN decoder, and any standard classification network with four stages (e.g., ResNet series [16,50,51]) can serve as the backbone network. The FPN [18] decoder progressively fuses high-level and low-level features by bilinear up-sampling to build up a hierarchical multi-scale pyramid network. As shown in Figure 6, the decoder can be seen as an FPN armed with multiple MLCRs.

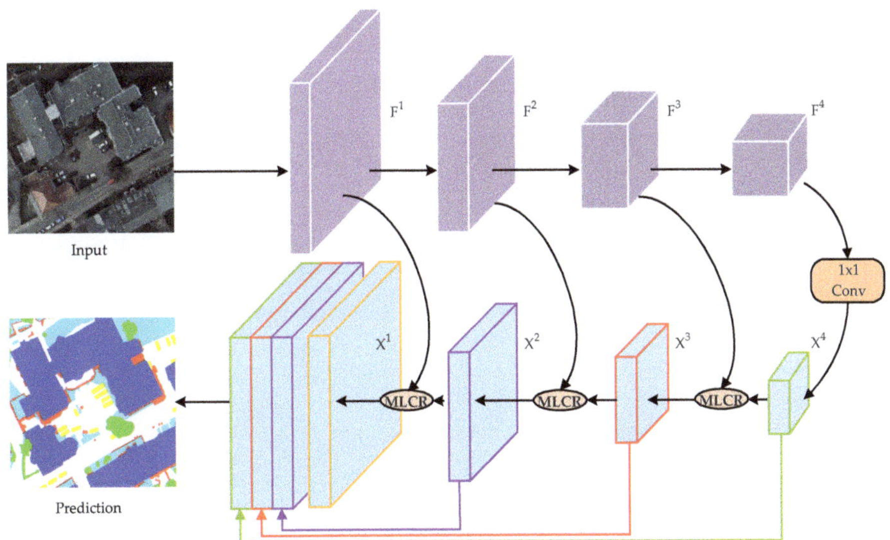

**Figure 6.** Overview of the proposed MLCRNet.

Initially, we feed the input image $I \in R^{3 \times H \times W}$ into the backbone network and projected it to a set of feature maps $\{F^s\}_{s \in [1,4]}$ from each network stage, where $F^s \in R^{C_s \times H_s \times W_s}$ denotes the $i$-th stage of the backbone output, $H_s = \frac{H}{2^{s+1}}$, and $W_s = \frac{W}{2^{s+1}}$. Then, considering the complexity of the aerial image segmentation task and the overall network computation cost, we replace the 4th stage of the FPN [18] decoder with one $1 \times 1$ convolution layer, reduce the channel dimension to $C_d$, and obtain the feature maps $X^4 \in R^{C_d \times H_4 \times W_4}$. Then, we replace all the rest of the stages of the FPN decoder with MLCR:

$$X^s = \mathrm{MLCR}\left(X^{s+1}, F^{s+1}\right) \tag{14}$$

where $X^s \in R^{C_d \times H_s \times W_s}$ is the FPN decoder output feature map of stage $s \in [1,3]$, MLCR is the MLCR module, and $F^s$ is the backbone network output feature map of stage $s$. The coarse feature map $X^s$ and the fine feature map $F^s$ are fed into the MLCR module to produce the fine feature map $X^1$. We obtain the output feature map $X^1$ by refining the feature maps iteratively. Finally, following the same setting of FPN, $\{F^s\}_{s=1,2,3,4}$ are up-sampled to the same spatial size of $F^1$ and concatenated together for prediction.

## 4. Experiments and Results

In this part, we first introduce the benchmarks, implementation, and training details of the proposed network. Next, we introduce the evaluation metric. Afterwards, we perform a string of ablation experiments on the Potsdam dataset. Finally, we compare the proposed method with the others from Potsdam and Vaihingen.

### 4.1. Experimental Setup

#### 4.1.1. Benchmarks

We conducted experiments on two challenging datasets from the challenging 2D Semantic Labeling Contest held by the International Society for Photogrammetry and Remote Sensing (ISPRS).

**Potsdam.** The ISPRS Potsdam [52] data set contains 38 orthorectified patches, each of which is composed of four wave bands, namely, red (R), green (G), blue (B), and near infrared (NIR), plus the corresponding digital surface model (DSM). All patches have a spatial resolution of 6000 × 6000 pixels and a ground sampling distance (GSD) of 5 cm. In terms of dataset partitioning, we randomly selected 17 images as the training set, 14 images as the test set, and 1 image as the validation set. It should be noted that we do not use NIR and DSM in our experiments.

**Vaihingen.** Unlike the Potsdam semantic labeling dataset, Vaihingen [52] is a relatively small dataset with only 33 patches and an average size of 2494 × 2064 pixels. Each of them contains NIR-R-G channels. Following the division method suggested by the dataset publisher, we used 16 patches for training and 17 for testing.

#### 4.1.2. Implementation Details

We utilized ResNet50 [16] pre-trained on ImageNet [53] as the backbone by dropping the last several fully connected layers and by replacing the last stage down-sampling operations by dilated convolutional layer with dilation rate 2. Aside from the backbone, we applied Kaiming initialization [54] to initialize the weights. We replaced all batch normalization (BN) [55] layers in the network with Sync-BN [56]. Given that our model adopted deep supervision [57], for fair comparison, we used deep supervision in all experiments.

#### 4.1.3. Training Settings

In the training phase, we adopted the stochastic gradient descent (SGD) optimizer with a batch size of 16, and the initial learning rate, momentum, and weight decay were set to 0.001, 0.9, and 5 $\times 10^{-4}$, respectively. As a common practice, "Poly" learning rate schedules were adopted to update the initial learning rate by a decay factor $\left(1 - \frac{cur\_iter}{total\_iter}\right)^{0.9}$ after each iteration. For Potsdam and Vaihingen, we set the training iterations as 73.6 K.

In practice, suitably enlarging the size of the input image can improve network performance. After balancing performance and memory constraints, we employed a sliding window with 25% overlap and clipped the original image into pixel 512 × 512 patches. We adopted random horizontal flip, random transpose, random scaling (scale ratio from 0.5 to 2.0), and random cropping with a crop size of 512 × 512 as our data augmentation strategy for all benchmarks.

### 4.1.4. Inference Settings

During inference, we used the same clipping method as the training phase. By default, we do not use any test time data augmentation. For the comprehensive quantitative evaluation of our proposed method, the mean intersection of union (mIoU), overall accuracy (OA), and average F1 score (F1) were used for accurate comparison. Furthermore, a number of float-point operations (FLOPs), memory cost (Memory), number of parameters (Parameter), and frames per second (FPS) were adopted for computation cost comparison.

### 4.1.5. Reproducibility

We conducted all experiments based on the PyTorch (version $\geq 1.3$) [58] framework and trained on tow NVIDIA RTX 3090 GPUs with a 24 GB memory per card. Aside from our method, all models were obtained from open sourcing code.

### 4.2. Ablation Study

#### 4.2.1. Ablation Studies of the MLCR Module to Different Layers

To demonstrate the effectiveness of the MLCR, we replaced various FPN [18] decoder stages with our MLCR. As illustrated in Table 1, from the top four rows, MLCR enhances all stages and exhibits the most progress at Stage 1, bringing an improvement of 1.3% mIoU. By replacing MLCR in all stages, we achieved 76.0% mIoU by an improvement of 1.9%.

**Table 1.** Ablation results for MLCR module to different insert positions on Potsdam test set.

| Method | 3 | 2 | 1 | mIoU (%) | $\Delta\alpha$ (%) |
|---|---|---|---|---|---|
| Baseline | | | | 74.1 | — |
| MLCR | ✓ | | | 74.8 | 0.7 ↑ |
| MLCR | | ✓ | | 75.0 | 0.9 ↑ |
| MLCR | | | ✓ | 75.4 | 1.3 ↑ |
| MLCR | | ✓ | ✓ | 75.7 | 1.6 ↑ |
| MLCR | ✓ | ✓ | ✓ | 76.0 | 1.9 ↑ |

We up-sampled and visualized the feature maps outputted from the 4th stage of FPN [18] and after MLCR enhancement, as shown in Figure 7. The features enhanced by MLCR are more structural.

#### 4.2.2. Ablation Studies of Different Level Contexts

To explore the impact of different levels of context on performance, we set the irrelevant contextual information to one and then observed how performance was affected by different levels of contextual information (e.g., set the image level context information $C_I$ and local level context information $C_L$ to one when investigating the importance of semantic level context). As shown in Table 2, the first to fourth rows suggest that improvements can come from any single level of context. Compared with the baseline, the addition of semantic-level and image-level contextual information brings 1.2% and 1.3% mIoU improvement, respectively. However, the addition of local-level context information only results in a 0.9 app mIoU improvement, most likely because local-level context improves the accuracy of object boundary areas, which occupy a comparatively small area. Meanwhile, combining semantic-level context and image-level context yields a result of 75.7% mIoU, which brings 1.4% improvement. Similarly, combining image-level context with local-level context also results in a 1.5% mIoU improvement. Finally, when we integrated local-level, image-level, and semantic-level context, it behaved superiorly compared with other methods, thereby further improving to 76.0%. In summary, our approach brings great benefit via exploiting multi-level context.

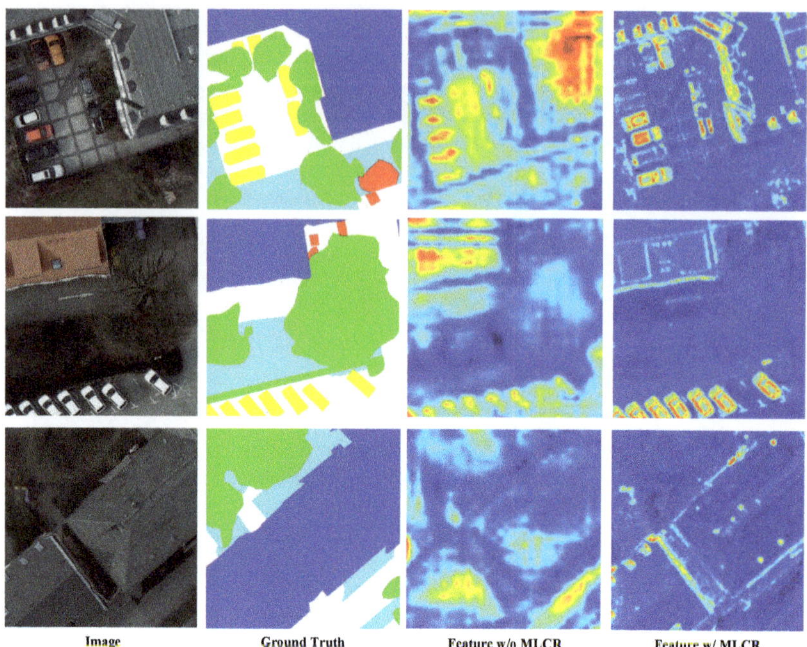



**Figure 7.** Visualization of features. Our module enhances the representation of more structural features.

**Table 2.** Ablation studies of different level context on Potsdam test set.

| Method | S | I | L | mIoU (%) | $\Delta\alpha$ (%) |
|---|---|---|---|---|---|
| Baseline | — | — | — | 74.1 | — |
| | ✓ | | | 75.3 | 1.2 ↑ |
| | | ✓ | | 75.4 | 1.3 ↑ |
| | | | ✓ | 75.0 | 0.9 ↑ |
| | | ✓ | ✓ | 75.6 | 1.5 ↑ |
| | ✓ | ✓ | | 75.5 | 1.4 ↑ |
| | ✓ | | ✓ | 75.6 | 1.5 ↑ |
| | ✓ | ✓ | ✓ | 76.0 | 1.9 ↑ |

### 4.2.3. Ablation Studies of Local-Level Context Receptive Fields

To evaluate our proposed local-level context, we varied the kernel size to investigate the effect of different harvesting scopes on local-level contextual information, and the results are reported in Table 3. Appropriate kernel sizes (e.g., $3 \times 3$) can achieve maximum accuracy (76.0% mIoU) with a small additional computational cost. However, larger convolutions (e.g., $5 \times 5$) achieve results (75.8%) similar to those of $3 \times 3$ but come with a significant additional computational expense. Notably, smaller kernel sizes (e.g., $1 \times 1$) yield results similar to those when local context information (e.g., set local contextual relation $C_L$ as one) is eliminated, with results of 75.5% and 75.5%, respectively. This finding demonstrates that our proposed local-level context is effective in harvesting local information within an appropriate scope.

**Table 3.** Ablation study on kernel size k in local-level context module.

| Method | mIoU (%) | FLOPs (G) |
|---|---|---|
| k = 1 | 75.5 | 42.7 |
| k = 3 | 76.0 | 43.3 |
| k = 5 | 75.8 | 44.3 |
| k = 7 | 75.8 | 45.9 |

### 4.2.4. Ablation Studies of Computation Cost

We further studied the efficiency of the MLCR module by applying it to the baseline model. We reported the model memory cost, parameter number, FLOPs, FPS, and performance in the inference stage with the batch of size one. As illustrated in Table 4, the performance difference between MCT and EMCT is statistically negligible. However, EMCT only incurs minimal additional computation cost overhead. Specifically, MCT increases GPU memory usage by 255 M compared with the Baseline. However, EMCT increased it by only 2 M, and the same was true for the Parameter (+2.1 vs. +0.5), GFLOPs (+8.0 vs. +0.6), and FPS ($-26.7$ vs. $-10.3$).

**Table 4.** Ablation study on computation cost.

| Method | Memory (Mb) | Parameter (M) | FLOPs (G) | FPS | mIoU (%) |
|---|---|---|---|---|---|
| Baseline | 915 | 25.2 | 42.7 | 90.3 | 74.1 |
| MCT | 1170 (+255) | 27.3 (+2.1) | 50.7 (+8.0) | 63.6 ($-26.7$) | 75.8 (+1.7) |
| Efficient MCT | 917 (+2) | 25.7 (+0.5) | 43.3 (+0.6) | 80.0 ($-10.3$) | 76.0 (+1.9) |

### 4.3. Comparison with State-of-the-Art

**Potsdam.** Given that some models (e.g., ACFNet [30], SFNet [17], and SCARF [31]) apply additional context modelling blocks, such as ASPP [13] or PPM [45], between the backbone network and the decoder, we removed these additional blocks for a fair comparison. Considering that the ASPP module is part of the decoder in DeepLabV3+ [15], we retained the ASPP module in DeepLabV3+. Likewise, we preserved the PPM module in PSPNet [45]. Tables 5 and 6 compare the quantification results on the Potsdam test set. At first glance, our method achieves the best performance (76.0% mIoU) among these approaches. In the subsequent sections, we analyze and compare these approaches in detail.

**Table 5.** Quantitative comparisons with state-of-the-arts on Potsdam test set.

| Model | Backbone | Stride | mIoU (%) | Acc (%) | F1 | Parameter (M) | FLOPs (G) |
|---|---|---|---|---|---|---|---|
| FCN [11] | ResNet50 | 16× | 72.5 | 83.0 | 83.5 | 32.9 | 33.7 |
| OCRNet [29] | ResNet50 | 16× | 73.9 | 84.0 | 84.4 | 39.0 | 47.6 |
| CCNet [44] | ResNet50 | 16× | 74.1 | 84.1 | 84.6 | 47.4 | 57.4 |
| ISANet [43] | ResNet50 | 16× | 74.5 | 84.5 | 84.8 | 40.0 | 49.5 |
| PSPNet [45] | ResNet50 | 16× | 74.5 | 84.2 | 84.8 | 46.6 | 52.0 |
| ACFNet [30] | ResNet50 | 16× | 74.7 | 84.3 | 84.9 | 30.1 | 39.3 |
| DANet [22] | ResNet50 | 16× | 74.9 | 84.4 | 85.1 | 47.4 | 198.1 |
| DepLabV3+ [15] | ResNet50 | 16× | 75.1 | 84.7 | 85.1 | 40.3 | 69.3 |
| MANet [42] | ResNet50 | 16× | 75.2 | 84.7 | 85.2 | 33.5 | 49.6 |
| AttUNet [59] | ResNet50 | 16× | 75.3 | 84.6 | 85.3 | 96.5 | 207.8 |
| SFNet [17] | ResNet50 | 16× | 75.4 | 84.9 | 85.4 | 30.6 | 100.1 |
| ISNet [32] | ResNet50 | 16× | 75.7 | 85.0 | 85.6 | 44.5 | 58.8 |
| SCARF [31] | ResNet50 | 16× | 75.7 | 85.3 | 85.6 | 25.9 | 45.0 |
| Ours | ResNet50 | 16× | 76.0 | 85.2 | 85.8 | 25.7 | 43.3 |

**Table 6.** Per-class results (mean intersection over union) on the Potsdam test set.

| Model | Imp.sur | Building | Low.veg | Tree | Car | Clutter | mIoU(%) |
|---|---|---|---|---|---|---|---|
| FCN [11] | 79.8 | 90.3 | 70.6 | 72.6 | 72.2 | 49.8 | 72.5 |
| OCRNet [29] | 80.9 | 90.9 | 71.6 | 73.5 | 74.7 | 51.9 | 73.9 |
| CCNet [44] | 81.1 | 91.5 | 71.9 | 73.3 | 75.6 | 51.3 | 74.1 |
| ISANet [43] | 81.2 | 91.5 | 72.4 | 74.1 | 74.7 | 52.8 | 74.5 |
| PSPNet [45] | 81.4 | 91.3 | 72.1 | 74.1 | 75.4 | 52.5 | 74.5 |
| ACFNet [30] | 81.3 | 91.4 | 71.5 | 73.4 | 79.4 | 51.0 | 74.7 |
| DANet [22] | 81.7 | 91.5 | 72.0 | 74.4 | 76.4 | 53.2 | 74.9 |
| DepLabV3+ [15] | 81.5 | 91.4 | 72.0 | 73.1 | 80.9 | 51.4 | 75.1 |
| MANet [42] | 81.6 | 91.1 | 72.2 | 73.8 | 81.7 | 50.6 | 75.2 |
| AttUNet [59] | 81.6 | 91.3 | 71.9 | 73.1 | 81.4 | 52.3 | 75.3 |
| SFNet [17] | 81.9 | 91.5 | 72.5 | 73.7 | 81.0 | 51.8 | 75.4 |
| ISNet [32] | 82.1 | 91.7 | 72.7 | 74.3 | 81.1 | 52.1 | 75.7 |
| SCARF [31] | 82.1 | 91.5 | 72.8 | 74.1 | 81.4 | 52.1 | 75.7 |
| Ours | 82.3 | 91.4 | 73.1 | 73.7 | 81.6 | 53.7 | 76.0 |

Table 5 shows that MLCRNet outperforms existing approaches with 76.0% mIoU 85.2% OA, and a 85.8 F1 score on the Potsdam test set. Among previous works, semantic level context methods, for instance, OCRNet [29], ACFNet [30], and SCARF [31], achieve 73.9% mIoU, 74.7% mIoU, and 75.7% mIoU, respectively. Image-level context models such as CCNet [44], ISANet [43], and DANet [22], achieve 74.1% mIoU, 74.5% mIoU and 74.9% mIoU, respectively. Local-level context approach SFNet [17] yields a result of 75.4% mIoU, 84.9% OA, and an 85.4 F1 score. Multi-level context methods, such as ISNet MANet, DeepLabV3+, and PSPNet, reach 75.7% mIoU, 75.2% mIoU, 75.1% mIoU and 74.5% mIoU, respectively. Compared with these methods, MLCRNet harvests contextual in formation from a more comprehensive perspective, thereby achieving the best performance results with the lowest number of parameters (25.7 M) and relatively modest FLOPs (43.3 G)

Table 6 summarizes the detailed per-category comparisons. Our method achieves improvements in categories such as impervious surfaces, low vegetation, cars, and clut ter. Our method effectively preserves the consistency of segmentation within objects at various scales.

Figure 8 shows the visualization results of our proposed MLCRNet and baseline model on the Potsdam datasets, which further proves the reliability of our proposed method As can be observed, by introducing multi-level contextual information, the segmentation performance of large and small objects can be well improved. For example, in the first and third rows, our method improves the consistency of segmentation within large objects. In the second rows, our MLCR improves the consistency of segmentation within large objects In the second row, our method not only enhances the consistency of the segmentation within small objects but also improves the performance of regions that are easily confused (e.g., the region sheltered by trees, buildings, or shadows). In addition, some robustness experiment results are presented in the Appendix A.

**Vaihingen.** We conducted further experiments on Vaihingen datasets, which is a challenging remote sensing image semantic labelling dataset with a total data volume (number of pixels) of roughly 8.1% of that of Potsdam. Table 7 summarizes the results, and our method achieves 68.1% mIoU, 77.5% OA, and a 79.8 F1 score, thereby significantly outperforming previous state-of-the-art methods by 1% mIoU, 1.1% OA, and a 0.8 F1 score due to the robustness of MLCRNet.

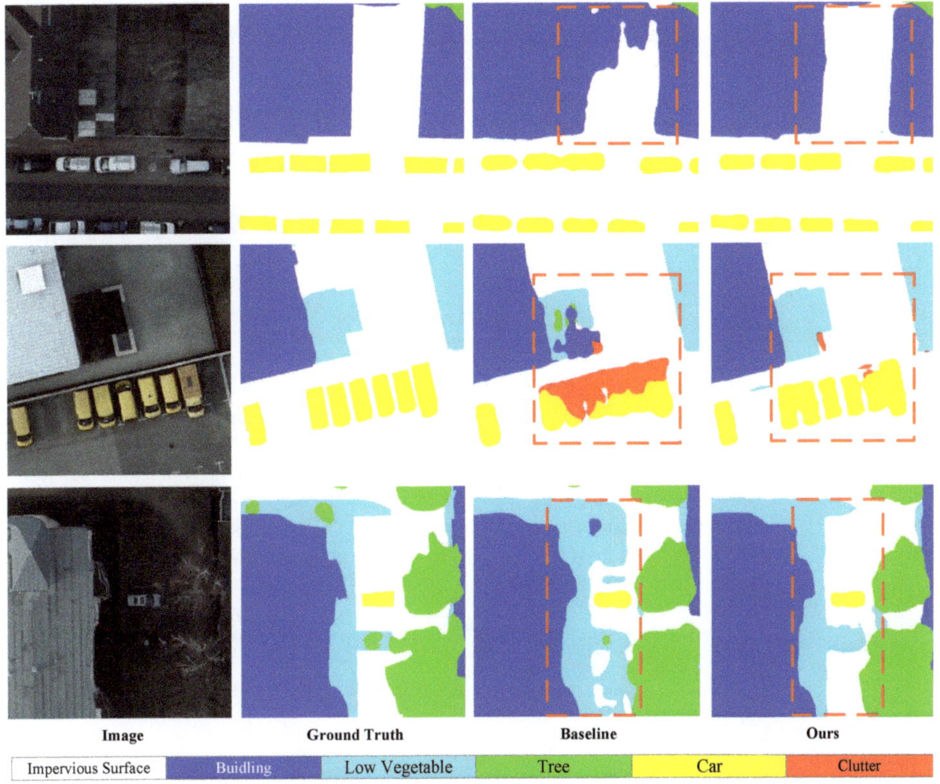



**Figure 8.** Qualitative comparisons against the Baseline on the Potsdam test set. We marked the improved regions with red dashed boxes (best viewed when colored and zoomed in).

**Table 7.** Quantitative comparisons with state-of-the-arts on Vaihingen test set.

| Model | Backbone | Stride | mIoU (%) | Acc (%) | F1 |
|---|---|---|---|---|---|
| FCN [11] | ResNet50 | 16× | 64.6 | 74.7 | 77.1 |
| CCNet [44] | ResNet50 | 16× | 65.5 | 75.2 | 77.7 |
| OCRNet [29] | ResNet50 | 16× | 66.3 | 76.5 | 78.6 |
| ISNet [32] | ResNet50 | 16× | 66.4 | 76.7 | 78.6 |
| ISANet [43] | ResNet50 | 16× | 66.6 | 76.4 | 78.7 |
| PSPNet [45] | ResNet50 | 16× | 66.6 | 76.0 | 78.6 |
| ACFNet [30] | ResNet50 | 16× | 66.7 | 76.4 | 78.7 |
| DANet [22] | ResNet50 | 16× | 66.8 | 76.4 | 78.8 |
| DepLabV3+ [15] | ResNet50 | 16× | 66.9 | 76.4 | 78.8 |
| MANet [42] | ResNet50 | 16× | 66.9 | 76.2 | 78.8 |
| AttUNet [59] | ResNet50 | 16× | 67.1 | 76.4 | 79.0 |
| Ours | ResNet50 | 16× | 68.1 | 77.5 | 79.8 |

As listed in Table 8, our proposed method achieves outstanding performance consistently in categories such as impervious surfaces, buildings, low vegetation, trees, and cars.

**Table 8.** Per-class results (mean intersection over union) on the Vaihingen test set.

| Model | Imp.sur | Buildings | Low.veg | Tree | Car | Clutter | mIoU (%) |
|---|---|---|---|---|---|---|---|
| FCN [11] | 78.9 | 86.1 | 63.8 | 72.8 | 49.9 | 36.0 | 64.6 |
| CCNet [44] | 80.1 | 86.7 | 65.0 | 73.5 | 52.5 | 35.3 | 65.5 |
| OCRNet [29] | 79.6 | 86.5 | 64.6 | 73.5 | 54.1 | 39.4 | 66.3 |
| ISNet [32] | 79.8 | 86.1 | 63.8 | 72.9 | 58.8 | 36.9 | 66.4 |
| ACFNet [30] | 80.6 | 87.1 | 65.2 | 74.1 | 57.8 | 35.3 | 66.7 |
| DANet [22] | 80.1 | 86.4 | 65.3 | 73.8 | 59.4 | 36.0 | 66.8 |
| DepLabV3+ [15] | 80.4 | 86.5 | 64.3 | 73.7 | 61.3 | 35.2 | 66.9 |
| MANet [42] | 80.3 | 86.5 | 64.1 | 73.5 | 63.4 | 33.7 | 66.9 |
| AttUNet [59] | 80.4 | 86.6 | 64.3 | 73.7 | 63.2 | 34.4 | 67.1 |
| Ours | 81.3 | 87.2 | 65.4 | 74.3 | 64.4 | 36.1 | 68.1 |

To further understand our model, we displayed the segmentation results of the Baseline and MLCRNet on the Vaihingen datasets, which can be seen in Figure 9. By integrating different levels of contextual information to reinforce feature representation, MLCRNet increases the differences among the different categories. For example, in the first and second rows, some regions suffer from local noise (e.g., occluders such as trees, buildings, or shadows) and tend to be misclassified. Our proposed MLCRNet assembles different levels of contextual information to eliminate local noise and to improve the classification accuracy in these regions.

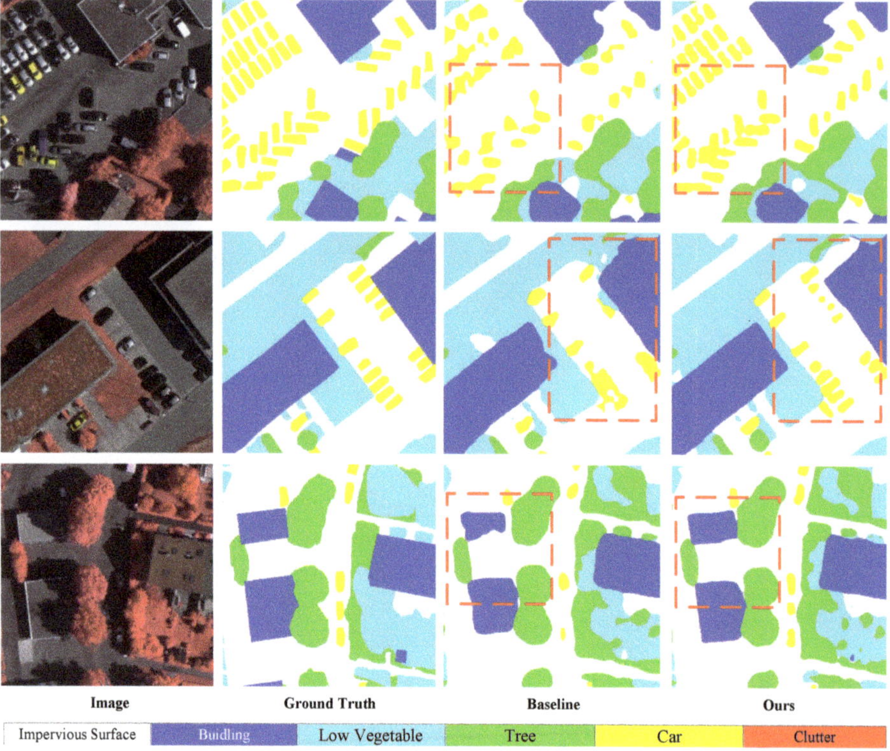


**Figure 9.** Qualitative comparisons between our method and Baseline on Vaihingen test set. We marked the improved regions with red dashed boxes (best viewed when colored and zoomed in).

## 5. Discussion

Previous studies have explored the importance of different levels of context and have made many improvements in semantic segmentation. However, these approaches tend to only focus on level-specific contextual relationships and do not harvest contextual information from a more holistic perspective. Consequently, these approaches are prone to suffer from a lack of contextual information (e.g., image-level context provides little improvement in identifying small targets). To this end, we aimed to seek an efficient and comprehensive approach that can model and transform contextual information.

Initially, we directly integrated local-level, image-level, and semantic-level contextual attention matrices, which improved model performance but dramatically increased GPU memory usage and inference time. We realize that these three levels of context are not orthogonal. Moreover, concatenating the three levels of contextual attention matrices directly suffers from the redundancy of contextual information. Hence, we designed the EMCT module to transform the three levels of contextual relationships into a contextual attention matrix effectively and efficiently. The experimental results suggest that our proposed method has three advantages over other methods. First, our proposed MLCR module has made progress in quantitative experimental results, and ablation experimental results on the Potsdam test set reveal the effectiveness of our proposed module, thereby lifting the mIoU by 1.9% compared with the Baseline and outperforming other state-of-the-art models. Second, the computational cost of our proposed MLCR module is less than those of other contextual aggregation methods. Relative to DANet, MLCRNet reduces the number of parameters by 46% and the FLOPs by 78%. Lastly, from the qualitative experimental results, our MLCR module increases the consistency of intra-class segmentation and object boundary accuracy, as shown in the first row of Figure 10. MLCNet improves the quality of the car edges while solving the problem of misclassification of disturbed areas (e.g., areas between adjacent vehicles, areas obscured by building shadows). The second and third rows of Figure 10 show the power of MLCRNet to improve the intra-class consistency of large objects (e.g., buildings, roads, grassy areas, etc.). Nevertheless, for future practical applications, we need to continue to improve accuracy.

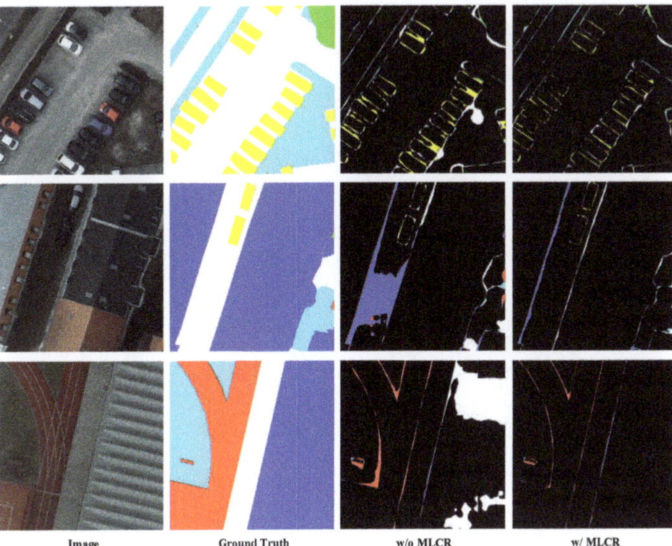



**Figure 10.** Qualitative comparison in terms of prediction errors on Potsdam test set, where correctly predicted pixels are shown with a black background and incorrectly predicted pixels are colored using the prediction results (best viewed when colored and zoomed in).

## 6. Conclusions

In this paper, we designed a novel MLCRNet that dynamically harvests contextual information from the semantic, image, and local perspectives for aerial image semantic segmentation. Concretely, we first integrated three levels of context, namely, local level, image level, and semantic level, to capture contextual information from multiple aspects adaptively. Next, an efficient fusion block is presented to address feature redundancy and improve the efficiency of our multi-level context. Finally, our model refines the feature map iteratively across FPN layers with MLCR. Extensive evaluations on Potsdam and Vaihingen challenging datasets demonstrate that our model can gather the multi-level contextual information efficiently, thereby enhancing the structure reasoning of the model.

**Author Contributions:** Z.H. and Q.Z. conceived of the presented idea and designed the study, respectively. Z.H. derived the models and performed the experiments. The manuscript was drafted by Z.H. with support from Q.Z. and G.Z. All authors discussed the results and contributed to the final manuscript. All authors have read and agreed to the published version of the manuscript.

**Funding:** This work was supported by the Natural Science Foundation of Shanghai (21ZR1421200), the National Nature Science Foundation of China (Grant Nos. 61731009 and 41301472), and the Science and Technology Commission of Shanghai Municipality (Grant Nos. 19511120600 and 18DZ2270800).

**Institutional Review Board Statement:** Not applicable.

**Informed Consent Statement:** Not applicable.

**Data Availability Statement:** The data and the code of this study are available from the corresponding author upon request.

**Conflicts of Interest:** The authors declare no conflict of interest.

## Appendix A. Robustness Evaluation

*Appendix A.1. Incorrect Labels and Rectification*

During the early experiments, we noticed that two labels in Potsdam datasets (e.g. IDs: 4_12 and 6_7) were incorrect, with all pixels of labels 4_12 and some pixels of 6_7 (approximately 6000 pixels) inconsistent with the labels defined by the dataset publisher. We randomly selected three $512 \times 512$ patches in 4_12 (Figure A1). As shown in the second column, the original labels are mixed with noise, most likely because the dataset publisher failed to remove the original image channels after the tagging was completed.

After comparing the RGB channels of the incorrect labels with normals, we found that the RGB channels of the incorrect labels were shifted to varying degrees (offset $\leq 127$). Therefore, we used the binarization operation to process the incorrect label:

$$GT_{k,i,j} = \begin{cases} 255, & if \ GT'_{k,i,j} \geq T \\ 0, & oterwise \end{cases} \tag{A1}$$

where $GT' \in R^{3 \times H \times W}$ is the original ground truth; $GT \in R^{3 \times H \times W}$ is the fixed ground truth; and $T$ is the threshold, which is set as $T = 127$. We show the modified result in the third column of Figure A1. Next, we are to present the results of quantitative experiments on a training set that includes incorrect labels. Note that we have re-implemented the experiment with corrected labels and reported the results in the main text.

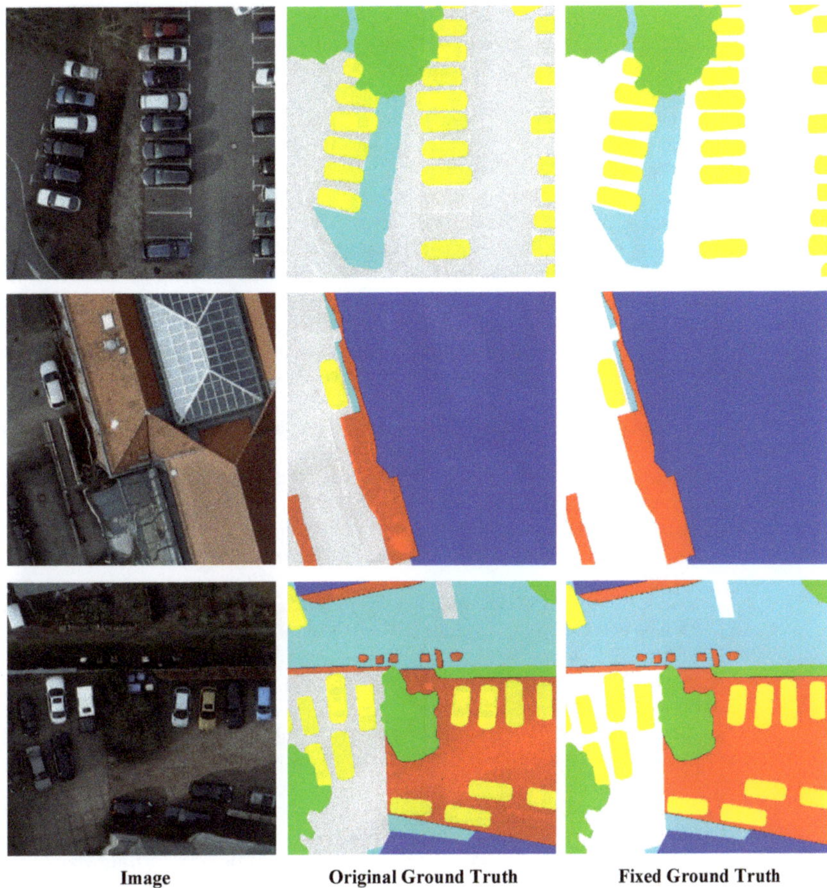

|  |  |  |
|:---:|:---:|:---:|
| **Image** | **Original Ground Truth** | **Fixed Ground Truth** |

**Figure A1.** Error and binarization-corrected labels in the Potsdam datasets (best viewed when colored and zoomed in).

*Appendix A.2. Robustness Evaluation Results*

We presented the experimental results before fixing the incorrect label to demonstrate the robustness of our proposed method. Table A1 shows that our method is less affected by the incorrect label than the other methods.

**Table A1.** Robustness evaluation results on the Potsdam test set.

| Model | Backbone | Stride | mIoU (%) | Acc (%) | F1 |
|:---:|:---:|:---:|:---:|:---:|:---:|
| ISNet [32] | ResNet50 | 16× | 70.2 | 81.3 | 81.8 |
| FCN [11] | ResNet50 | 16× | 71.5 | 81.9 | 82.8 |
| OCRNet [29] | ResNet50 | 16× | 73.6 | 83.6 | 84.2 |
| DepLabV3+ [15] | ResNet50 | 16× | 74.5 | 84.2 | 84.8 |
| SCARF [31] | ResNet50 | 16× | 74.6 | 83.9 | 84.8 |
| SFNet [17] | ResNet50 | 16× | 74.7 | 84.1 | 84.9 |
| Ours | ResNet50 | 16× | 75.3 | 84.6 | 85.4 |

## References


1.  Kang, Y.; Lu, Z.; Zhao, C.; Xu, Y.; Kim, J.-W.; Gallegos, A.J. InSAR monitoring of creeping landslides in mountainous regions: A case study in Eldorado National Forest, California. *Remote Sens. Environ.* **2021**, *258*, 112400. [CrossRef]
2.  Bianchi, F.M.; Grahn, J.; Eckerstorfer, M.; Malnes, E.; Vickers, H. Snow avalanche segmentation in SAR images with fully convolutional neural networks. *IEEE J. Sel. Top. Appl. Earth Obs. Remote Sens.* **2020**, *14*, 75–82. [CrossRef]
3.  Xu, F.; Somers, B. Unmixing-based Sentinel-2 downscaling for urban land cover mapping. *ISPRS J. Photogramm. Remote Sens.* **2021**, *171*, 133–154. [CrossRef]
4.  Luo, X.; Tong, X.; Pan, H. Integrating multiresolution and multitemporal Sentinel-2 imagery for land-cover mapping in the Xiongan New Area, China. *IEEE Trans. Geosci. Remote Sens.* **2021**, *59*, 1029–1040. [CrossRef]
5.  Azizi, A.; Abbaspour-Gilandeh, Y.; Vannier, E.; Dusséaux, R.; Mseri-Gundoshmian, T.; Moghaddam, H.A. Semantic segmentation: A modern approach for identifying soil clods in precision farming. *Biosyst. Eng.* **2020**, *196*, 172–182. [CrossRef]
6.  Anand, T.; Sinha, S.; Mandal, M.; Chamola, V.; Yu, F.R. AgriSegNet: Deep aerial semantic segmentation framework for iot-assisted precision agriculture. *IEEE Sens. J.* **2021**, *21*, 17581–17590. [CrossRef]
7.  Azimi, S.M.; Fischer, P.; Korner, M.; Reinartz, P. Aerial LaneNet: Lane-marking semantic segmentation in aerial imagery using wavelet-enhanced cost-sensitive symmetric fully convolutional neural networks. *IEEE Trans. Geosci. Remote Sens.* **2019**, *57*, 2920–2938. [CrossRef]
8.  Xu, Y.; Chen, H.; Du, C.; Li, J. MSACon: Mining spatial attention-based contextual information for road extraction. *IEEE Trans. Geosci. Remote Sens.* **2021**, *60*, 5604317. [CrossRef]
9.  Song, L.; Xia, M.; Jin, J.; Qian, M.; Zhang, Y. SUACDNet: Attentional change detection network based on siamese U-shaped structure. *Int. J. Appl. Earth Obs. Geoinf.* **2021**, *105*, 102597. [CrossRef]
10. Krizhevsky, A.; Sutskever, I.; Hinton, G.E. ImageNet classification with deep convolutional neural networks. In *Advances in Neural Information Processing Systems*; Pereira, F., Burges, C.J.C., Bottou, L., Wein-berger, K.Q., Eds.; Curran Associates, Inc.: Red Hook, NY, USA, 2012; Volume 25.
11. Long, J.; Shelhamer, E.; Darrell, T. Fully convolutional networks for semantic segmentation. In Proceedings of the IEEE Conference on Computer Vision and Pattern Recognition (CVPR), Boston, MA, USA, 7–12 June 2015; IEEE Computer Society: Los Alamitos, CA, USA, 2015; pp. 3431–3440.
12. Girshick, R. Fast R-CNN. In Proceedings of the 2015 IEEE International Conference on Computer Vision (ICCV), Santiago, Chile, 7–13 December 2015; pp. 1440–1448. [CrossRef]
13. Chen, L.-C.; Papandreou, G.; Kokkinos, I.; Murphy, K.; Yuille, A.L. DeepLab: Semantic image segmentation with deep convolutional nets, atrous convolution, and fully connected CRFs. *IEEE Trans. Pattern Anal. Mach. Intell.* **2018**, *40*, 834–848. [CrossRef]
14. Chen, L.-C.; Papandreou, G.; Schroff, F.; Adam, H. Rethinking atrous convolution for semantic image segmentation. *arXiv* **2017**, arXiv:1706.05587.
15. Chen, L.-C.; Zhu, Y.; Papandreou, G.; Schroff, F.; Adam, H. Encoder-decoder with atrous separable convolution for semantic image segmentation. In Proceedings of the European Conference on Computer Vision (ECCV), Munich, Germany, 8–14 September 2018.
16. He, K.; Zhang, X.; Ren, S.; Sun, J. Deep residual learning for image recognition. In Proceedings of the 2016 IEEE Conference on Computer Vision and Pattern Recognition (CVPR), Las Vegas, NV, USA, 27–30 June 2016; pp. 770–778.
17. Li, X.; You, A.; Zhu, Z.; Zhao, H.; Yang, M.; Yang, K.; Tan, S.; Tong, Y. Semantic flow for fast and accurate scene parsing. In Proceedings of the Computer Vision—ECCV 2020, Glasgow, UK, 23–28 August 2020; Vedaldi, A., Bischof, H., Brox, T., Frahm, J.-M., Eds.; Springer International Publishing: Cham, Switzerland, 2020; pp. 775–793.
18. Kirillov, A.; Girshick, R.; He, K.; Dollár, P. Panoptic feature pyramid networks. In Proceedings of the 2019 IEEE/CVF Conference on Computer Vision and Pattern Recognition (CVPR), Long Beach, CA, USA, 15–20 June 2019; pp. 6399–6408.
19. Ronneberger, O.; Fischer, P.; Brox, T. U-Net: Convolutional networks for biomedical image seg-mentation. In Proceedings of the Medical Image Computing and Computer-Assisted Intervention—MICCAI 2015, Munich, Germany, 5–9 October 2015; Navab, N., Hornegger, J., Wells, W.M., Frangi, A.F., Eds.; Springer International Publishing: Cham, Switzerland, 2015; pp. 234–241.
20. Lin, T.Y.; Dollár, P.; Girshick, R.; He, K.; Hariharan, B.; Belongie, S. Feature pyramid networks for object detection. In Proceedings of the 2017 IEEE Conference on Computer Vision and Pattern Recognition (CVPR), Honolulu, HI, USA, 21–26 July 2017; pp. 936–944. [CrossRef]
21. Wang, X.; Girshick, R.; Gupta, A.; He, K. Non-local neural networks. In Proceedings of the 2017 IEEE Conference on Computer Vision and Pattern Recognition (CVPR), Honolulu, HI, USA, 21–26 July 2017.
22. Fu, J.; Liu, J.; Tian, H.; Li, Y.; Bao, Y.; Fang, Z.; Lu, H. Dual attention network for scene segmentation. In Proceedings of the 2019 IEEE/CVF Conference on Computer Vision and Pattern Recognition (CVPR), Long Beach, CA, USA, 15–20 June 2019; pp. 3141–3149. [CrossRef]
23. Hu, P.; Caba, F.; Wang, O.; Lin, Z.; Sclaroff, S.; Perazzi, F. Temporally distributed networks for fast video semantic segmentation. In Proceedings of the 2020 IEEE/CVF Conference on Computer Vision and Pattern Recognition (CVPR), Seattle, WA, USA, 13–19 June 2020; pp. 8815–8824.
24. Zhang, H.; Wang, C.; Xie, J. Co-occurrent features in semantic segmentation. In Proceedings of the 2019 IEEE/CVF Conference on Computer Vision and Pattern Recognition (CVPR), Long Beach, CA, USA, 15–20 June 2019; pp. 548–557.

25. Zhu, Z.; Xu, M.; Bai, S.; Huang, T.; Bai, X. Asymmetric non-local neural networks for semantic segmentation. In Proceedings of the 2019 IEEE/CVF International Conference on Computer Vision (ICCV), Seoul, Korea, 27 October–2 November 2019; pp. 593–602.

26. Yan, L.; Huang, J.; Xie, H.; Wei, P.; Gao, Z. Efficient depth fusion transformer for aerial image semantic segmentation. *Remote Sens.* **2022**, *14*, 1294. [CrossRef]

27. Yu, C.; Wang, J.; Gao, C.; Yu, G.; Shen, C.; Sang, N. Context prior for scene segmentation. In Proceedings of the 2020 IEEE/CVF Conference on Computer Vision and Pattern Recognition (CVPR), Seattle, WA, USA, 13–19 June 2020; pp. 12413–12422.

28. Bello, I. LambdaNetworks: Modeling long-range interactions without attention. In Proceedings of the International Conference on Learning Representations, Vienna, Austria, 4 May 2021.

29. Yuan, Y.; Chen, X.; Wang, J. Object-contextual representations for semantic segmentation. In Proceedings of the Computer Vision—ECCV 2020, Glasgow, UK, 23–28 August 2020; Vedaldi, A., Bischof, H., Brox, T., Frahm, J.-M., Eds.; Springer International Publishing: Cham, Switzerland, 2020; pp. 173–190.

30. Zhang, F.; Chen, Y.; Li, Z.; Hong, Z.; Liu, J.; Ma, F.; Han, J.; Ding, E. ACFNet: Attentional class feature network for semantic segmentation. In Proceedings of the 2019 IEEE/CVF International Conference on Computer Vision (ICCV), Seoul, Korea, 27 October–2 November 2019; pp. 6797–6806.

31. Ding, X.; Shen, C.; Che, Z.; Zeng, T.; Peng, Y. SCARF: A semantic constrained attention refinement network for semantic segmentation. In Proceedings of the 2021 IEEE/CVF International Conference on Computer Vision Workshops (ICCVW), Montreal, BC, Canada, 11–17 October 2021; pp. 3002–3011.

32. Jin, Z.; Liu, B.; Chu, Q.; Yu, N. ISNet: Integrate image-level and semantic-level context for semantic segmentation. In Proceedings of the IEEE/CVF International Conference on Computer Vision, Montreal, BC, Canada, 11–17 October 2021.

33. Niu, R.; Sun, X.; Tian, Y.; Diao, W.; Chen, K.; Fu, K. Hybrid multiple attention network for semantic segmentation in aerial images. *IEEE Trans. Geosci. Remote Sens.* **2021**, *60*, 5603018. [CrossRef]

34. Shang, R.; Zhang, J.; Jiao, L.; Li, Y.; Marturi, N.; Stolkin, R. Multi-scale adaptive feature fusion network for semantic segmentation in remote sensing images. *Remote Sens.* **2020**, *12*, 872. [CrossRef]

35. Zhang, J.; Lin, S.; Ding, L.; Bruzzone, L. Multi-scale context aggregation for semantic segmentation of remote sensing images. *Remote Sens.* **2020**, *12*, 701. [CrossRef]

36. Xu, Z.; Zhang, W.; Zhang, T.; Li, J. Hrcnet: High-resolution context extraction net-work for semantic segmentation of remote sensing images. *Remote Sens.* **2020**, *13*, 71. [CrossRef]

37. Fan, M.; Lai, S.; Huang, J.; Wei, X.; Chai, Z.; Luo, J.; Wei, X. Rethinking BiSeNet for real-time semantic segmentation. In Proceedings of the 2021 IEEE/CVF Conference on Computer Vision and Pattern Recognition (CVPR), Nashville, TN, USA, 20–25 June 2021; pp. 9711–9720.

38. Wang, J.; Sun, K.; Cheng, T.; Jiang, B.; Deng, C.; Zhao, Y.; Liu, D.; Mu, Y.; Tan, M.; Wang, X.; et al. Deep high-resolution representation learning for visual recognition. *IEEE Trans. Pattern Anal. Mach. Intell.* **2021**, *43*, 3349–3364. [CrossRef]

39. Hu, J.; Shen, L.; Sun, G. Squeeze-and-excitation networks. In Proceedings of the 2018 IEEE Conference on Computer Vision and Pattern Recognition (CVPR), Salt Lake City, UT, USA, 18–23 June 2018; pp. 7132–7141.

40. Cao, Y.; Xu, J.; Lin, S.; Wei, F.; Hu, H. Gcnet: Non-local networks meet squeeze-excitation networks and beyond. In Proceedings of the 2019 IEEE/CVF International Conference on Computer Vision Workshop (ICCVW), Seoul, Korea, 27–28 October 2019; pp. 1971–1980. [CrossRef]

41. Song, Q.; Li, J.; Li, C.; Guo, H.; Huang, R. Fully attentional network for semantic segmentation. *arXiv* **2021**, arXiv:2112.04108.

42. Li, R.; Zheng, S.; Zhang, C.; Duan, C.; Su, J.; Wang, L.; Atkinson, P.M. Multiattention network for semantic segmentation of fine-resolution remote sensing images. *IEEE Trans. Geosci. Remote Sens.* **2021**, *60*, 5607713. [CrossRef]

43. Huang, L.; Yuan, Y.; Guo, J.; Zhang, C.; Chen, X.; Wang, J. Interlaced sparse self-attention for semantic segmentation. *arXiv* **2019**, arXiv:1907.12273.

44. Huang, Z.; Wang, X.; Wei, Y.; Huang, L.; Shi, H.; Liu, W.; Huang, T.S. CCNet: Criss-cross attention for semantic segmentation. *IEEE Trans. Pattern Anal. Mach. Intell.* **2020**, *1*, 9133304. [CrossRef]

45. Zhao, H.; Shi, J.; Qi, X.; Wang, X.; Jia, J. Pyramid scene parsing network. In Proceedings of the 2017 IEEE Conference on Computer Vision and Pattern Recognition (CVPR), Honolulu, HI, USA, 21–26 July 2017; pp. 6230–6239. [CrossRef]

46. Zhang, Q.; Yang, G.; Zhang, G. Collaborative network for super-resolution and semantic segmentation of remote sensing images. *IEEE Trans. Geosci. Remote Sens.* **2021**, *60*, 4404512. [CrossRef]

47. Saha, S.; Mou, L.; Qiu, C.; Zhu, X.X.; Bovolo, F.; Bruzzone, L. Unsupervised deep joint segmentation of multitemporal high-resolution images. *IEEE Trans. Geosci. Remote Sens.* **2020**, *58*, 8780–8792. [CrossRef]

48. Du, S.; Du, S.; Liu, B.; Zhang, X. Incorporating DeepLabv3+ and object-based image analysis for semantic segmentation of very high resolution remote sensing images. *Int. J. Digit. Earth* **2020**, *14*, 357–378. [CrossRef]

49. Yang, G.; Zhang, Q.; Zhang, G. EANet: Edge-aware network for the extraction of buildings from aerial images. *Remote Sens.* **2020**, *12*, 2161. [CrossRef]

50. He, K.; Zhang, X.; Ren, S.; Sun, J. Identity mappings in deep residual networks. In Proceedings of the Computer Vision—ECCV 2016, Amsterdam, The Netherlands, 11–14 October 2016; Leibe, B., Matas, J., Sebe, N., Welling, M., Eds.; Springer International Publishing: Cham, Switzerland, 2016; pp. 630–645.

51. Xie, S.; Girshick, R.; Dollár, P.; Tu, Z.; He, K. Aggregated Residual Transformations for Deep Neural Networks. In Proceedings of the 2017 IEEE Conference on Computer Vision and Pattern Recognition (CVPR), Honolulu, HI, USA, 21–26 July 2017; pp. 5987–5995. [CrossRef]

52. ISPRS 2D Semantic Labeling Contest. 2016. Available online: https://www2.isprs.org/commissions/comm2/wg4/benchmark 2d-sem-label-potsdam/ (accessed on 15 December 2020).

53. Russakovsky, O.; Deng, J.; Su, H.; Krause, J.; Satheesh, S.; Ma, S.; Huang, Z.; Karpathy, A.; Khosla, A.; Bernstein, M.; et al. ImageNet large scale visual recognition challenge. *Int. J. Comput. Vis.* **2015**, *115*, 211–252. [CrossRef]

54. He, K.; Zhang, X.; Ren, S.; Sun, J. Delving deep into rectifiers: Surpassing human-level performance on imagenet classification. In Proceedings of the International Conference on Computer Vision (CVPR), Las Condes, Chile, 11–18 December 2015; pp. 1026–1034.

55. Ioffe, S.; Szegedy, C. Batch normalization: Accelerating deep network training by reducing internal covariate shift. In Proceedings of the 32nd International Conference on International Conference on Machine Learning, Lille, France, 7–9 July 2015; Volume 37, pp. 448–456.

56. Bulo, S.R.; Porzi, L.; Kontschieder, P. In-place activated batchnorm for memory-optimized training of DNNs. In Proceedings of the 2018 IEEE/CVF Conference on Computer Vision and Pattern Recognition, Salt Lake City, UT, USA, 18–23 June 2018; pp. 5639–5647.

57. Lee, C.-Y.; Xie, S.; Gallagher, P.; Zhang, Z.; Tu, Z. Deeply-supervised nets. In Proceedings of the Eighteenth International Conference on Artificial Intelligence and Statistics, San Diego, CA, USA, 9–12 May 2015; Lebanon, G., Vishwanathan, S.V.N., Eds.; PMLR: San Diego, CA, USA, 2015; Volume 38, pp. 562–570.

58. Paszke, A.; Gross, S.; Massa, F.; Lerer, A.; Bradbury, J.; Chanan, G.; Killeen, T.; Lin, Z.; Gimelshein, N.; Antiga, L.; et al. PyTorch: An imperative style, high-performance deep learning library. In Proceedings of the Advances in Neural Information Processing Systems 32: Annual Conference on Neural Information Processing Systems 2019, NeurIPS 2019, Vancouver, BC, Canada, 8–14 December 2019; Wallach, H.M., Larochelle, H., Beygelzimer, A., d'Alché-Buc, F., Fox, E.B., Garnett, R., Eds.; Curran Associates Inc.: Red Hook, NY, USA, 2019; pp. 8024–8035.

59. Oktay, O.; Schlemper, J.; Folgoc, L.L.; Lee, M.C.H.; Heinrich, M.P.; Misawa, K.; Mori, K.; McDonagh, S.G.; Hammerla, N.Y.; Kainz, B.; et al. Attention U-Net: Learning where to look for the pancreas. *arXiv* **2018**, arXiv:1804.03999.

*Article*

# A Mutual Teaching Framework with Momentum Correction for Unsupervised Hyperspectral Image Change Detection

**Jia Sun, Jia Liu, Ling Hu, Zhihui Wei and Liang Xiao ***

School of Computer Science and Engineering, Nanjing University of Science and Technology, Nanjing 210094, China; sun_jia@njust.edu.cn (J.S.); omegaliuj@njust.edu.cn (J.L.); njusthuling@njust.edu.cn (L.H.); gswei@njust.edu.cn (Z.W.)
* Correspondence: xiaoliang@mail.njust.edu.cn

**Abstract:** Deep-learning methods rely on massive labeled data, which has become one of the main impediments in hyperspectral image change detection (HSI-CD). To resolve this problem, pseudo-labels generated by traditional methods are widely used to drive model learning. In this paper, we propose a mutual teaching approach with momentum correction for unsupervised HSI-CD to cope with noise in pseudo-labels, which is harmful for model training. First, we adopt two structurally identical models simultaneously, allowing them to select high-confidence samples for each other to suppress self-confidence bias, and continuously update pseudo-labels during iterations to fine-tune the models. Furthermore, a new group confidence-based sample filtering method is designed to obtain reliable training samples for HSI. This method considers both the quality and diversity of the selected samples by determining the confidence of each group instead of single instances. Finally, to better extract the spatial–temporal spectral features of bitemporal HSIs, a 3D convolutional neural network (3DCNN) is designed as an HSI-CD classifier and the basic network of our framework. Due to mutual teaching and dynamic label learning, pseudo-labels can be continuously updated and refined in iterations, and thus, the proposed method can achieve a better performance compared with those with fixed pseudo-labels. Experimental results on several HSI datasets demonstrate the effectiveness of our method.

**Keywords:** change detection; bitemporal hyperspectral image; pseudo-label; mutual teaching

**Citation:** Sun, J.; Liu, J.; Hu, L.; Wei, Z.; Xiao, L. A Mutual Teaching Framework with Momentum Correction for Unsupervised Hyperspectral Image Change Detection. *Remote Sens.* **2022**, *14*, 1000. https://doi.org/10.3390/rs14041000

Academic Editors: Yue Wu, Kai Qin, Maoguo Gong and Qiguang Miao

Received: 7 January 2022
Accepted: 15 February 2022
Published: 18 February 2022

**Publisher's Note:** MDPI stays neutral with regard to jurisdictional claims in published maps and institutional affiliations.

## 1. Introduction

Hyperspectral imaging techniques can obtain continuous spectral information over a wide range of spectral wavelengths. The ability to display the subtle spectral variations of different ground objects has played an important role in many land-cover monitoring applications, such as mineral exploration [1,2], land-use monitoring [3,4], and military defense [5]. Change detection (CD) is the process of identifying differences in the state of an object or phenomenon by observing it at different times [6], which has been an indispensable application in the remote sensing field for a long time. Because of the rapid increase in spectral information, hyperspectral images (HSIs) are able to help detect finer changes than other remote sensing images and observe more change details. However, due to the spectral variability and redundant information, it is still a substantial challenge to effectively mine the spectral–spatial information to complete the HSI-CD task.

For decades, a variety of unsupervised methods have been applied to HSI-CD. Change detection aims to generate an accurate binary change map. In traditional methods, the change map can be obtained by analyzing the difference image (DI), which is usually based on differencing or log-rationing function. The most typical method is the change vector analysis (CVA) method [7], which identifies the changed pixels and the type of change according to the magnitude and direction of the spectral change vector. Some techniques utilize image transformation to extract new features for better performance. Principal component analysis (PCA) [8] retains the main information of original images according to the statistical characteristics, which reduces data redundancy enormously. Both multivariate alteration detection (MAD) [9] and the improved iteratively reweighted MAD (IRMAD) [10] methods calculate the degree of change by canonical correlation analysis. These measures all assume that the characteristics of unchanged pixels are uniform, but in reality, due to atmospheric conditions, illumination, etc., completely identical features rarely exist [11]. To suppress the difference in unchanged pixels, slow feature analysis (SFA) extracts the most temporally invariant component and then converts images into a new feature space [12]. Furthermore, some methods [13,14] directly determine the changing type of pixels using the postclassification comparison (PCC), but their performances fully depend on the accuracy of the classifier.

Recently, deep-learning (DL) methods have been favored by many researchers because of their strong nonlinear representation ability. Change detection needs to process bitemporal images simultaneously, as feature fusion must be carried out to form a single feature vector, which is usually a similarity measure between those two features [15]. Conventional methods inevitably lose partial information via difference or other processing, while deep-learning methods can avoid this problem. Mou et al. [16] proposed an end-to-end network. A convolutional neural network (CNN) extracts spectral–spatial features and a recurrent neural network (RNN) analyzes the temporal dependence between images. Considering the mixed pixels in HSIs, some methods [17] utilize subpixel-level information obtained by unmixing to improve detection accuracy. Chen et al. [18] proved that the 3D convolution kernel combined with regularization can effectively extract the spectral–spatial features of HSIs for classification tasks. Based on this, a 3D convolutional neural network (3DCNN) for hyperspectral image change detection is designed as the basic model of the proposed framework.

However, the great success of existing deep learning methods in many tasks mainly benefits from a large amount of labeled data. Pixel-wise labels for bitemporal HSIs need to be annotated by experts, which is time-consuming and expensive. Thus, it is difficult to obtain in large quantities. To solve the problem, existing unsupervised HSI-CD methods usually use pseudo-labels generated by traditional algorithms [17,19–21]. One of the main challenges is that the training process of neural networks is susceptible to noise in pseudo-labels. It is difficult to deal with the high-dimensionality of hyperspectral data for traditional CD methods. Additionally, affected by atmospheric conditions, illumination, and topography changes, the spectral variability of ground objects further increases the difficulty of change detection. Due to the limitations of these traditional methods, there certainly exist some discrepancies between the pseudo-labels and the true labels. Zhang et al. [22] proved that advanced neural networks can easily fit training sets with arbitrary labels. Once the network fits inaccurate labels, it will seriously affect the classification results. Wang et al. [17] utilized subpixel information to enhance robustness of the model. Du et al. [19] designed a deep slow feature analysis (DSFA) algorithm based on SFA theory and deep network to extract invariant components. These methods largely ignore handling the noisy labels. Li et al. [20] added a noisy model with zero-mean Gaussian distribution to their loss function, yet the experimental effect was general. The authors in [21] adopted two unsupervised algorithms to jointly generate credible labels. However, the same problem still exists, where it is impossible to filter all noisy labels by only one-time sample selection.

To address the noisy labels, we propose dynamically correcting pseudo-labels instead of safely relying on labels. The momentum correction approach is based on mutual teaching, where two learning models are mutually updated to jointly learn. Dynamic learning approaches by sample selection are popular in robust learning from noisy labels [23–27]. Yao et al. [24] adjusted the number of training samples in each iteration according to the learning curve. Self-paced learning (SPL) [25,26], which reduces the confidence threshold as the number of iterations increases, automatically selects more complex samples. However, the self-training of networks is prone to self-confidence bias and cannot be corrected when errors accumulate. Co-teaching [27] trains two classifiers simultaneously and enables them to select small loss samples for each other in every mini-batch, effectively suppressing the phenomenon of overfitting. It is generally assumed that small loss samples are more likely to be correctly labeled. Nevertheless, simply using loss to select training data is not suitable for bitemporal HSIs with complex variations. The selected samples are easily concentrated in the two categories of maximum and minimum changes, which is not conducive to the generalization of models. Therefore, we divide all samples into multiple groups according to the similarity of the difference vector in advance and randomly select from the high-confidence groups to ensure that multiclass samples can be selected. In addition, although our approach uses incompletely correct labels during initialization, utilizing the newly derived more reliable results to update the pseudo-labels can further boost the classifier performance [28,29]. The main contributions are summarized as follows:

(1)  We introduce to a novel mutual teaching framework with momentum correction for resisting noisy labels generated by traditional methods in unsupervised HSI-CD. Due to mutual teaching and dynamic label learning, pseudo-labels can be continuously updated and refined in iterations, and thus the proposed method can achieve superior results.

(2)  A group confidence-based sample selection approach is proposed to avoid selecting the two most extreme types of samples, and it is used alternately with another selection mechanism in iteration to ensure that complex samples can participate in training.

(3)  An end-to-end 3DCNN is designed as a classifier for HSI-CD and the basic model of the proposed framework. Experiments on four datasets demonstrate that our framework can effectively improve model performance.

## 2. Related Work

### 2.1. Unsupervised Deep Methods for Change Detection

Remote sensing image annotation is more difficult than that of natural images, especially for pixel-level change detection. Therefore, unsupervised methods without manual labeling steps have more advantages. Currently, unsupervised deep-learning methods can be divided into two categories. As shown in Figure 1a, the network is treated as a feature extractor to transform original images into a new feature space, and the model parameters are optimized based on the analysis of current output features in each iteration. For example, Liu et al. [30] proposed a symmetric convolutional coupling network (SCCN), which was initialized by a denoising autoencoder, and then minimized the feature difference between those unchanged pixels. Zhang et al. [31] adopted clustering analysis and detected multiple types of changes. Liu et al. [32] established an energy function driven network according to the feature difference. The advantage of these methods is that the newly derived features are used in each iteration to progressively improve the accuracy of the results. However, due to the limitation of optimization, it is difficult to use more complex models without any labels and the high dimension of HSIs is not conducive to model convergence. The other is shown in Figure 1b. The results obtained by the traditional algorithms are assigned to all samples as pseudo-labels to train neural networks, which is more commonly used [17,19–21,25,26,33,34]. These methods are easy to implement and closer to end-to-end patterns, avoiding the intermediate steps of difference image analysis. The only problem is that the pseudo-labels are not completely correct, which may mislead the network training. Inspired by the first category of methods, we utilize new predictive values to update pseudo-labels in multiple iterations to gradually reduce noise labels.

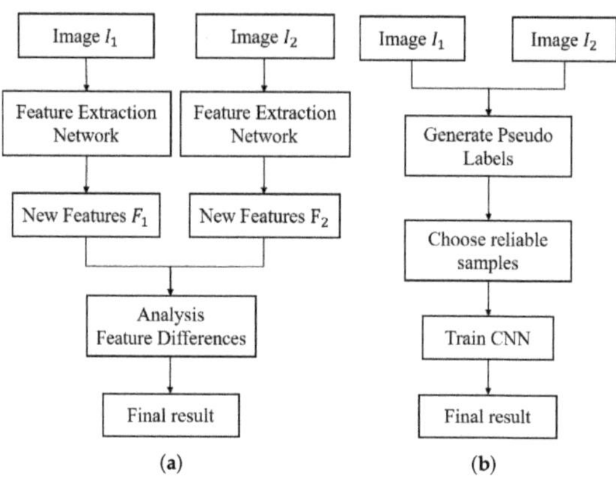

**Figure 1.** Architectures of two unsupervised deep methods for change detection. (**a**) Feature analysis driven model training. (**b**) Pseudo-labels-driven model training.

### 2.2. Deep Learning with Noisy Labels

Noisy labels are ubiquitous in deep-learning applications, such as in large-scale low quality datasets collected from the internet or crowdsourcing platforms in supervised learning, predictive pseudo-labels in semi-supervised learning and in domain adaptation learning. The overfitting of noisy labels will directly weaken the generalization of the models. Thus, learning with noisy labels still attracts researchers' attention. The noise transition matrix and robust loss function are commonly used for antinoise training. Goldberger et al. [35] added another softmax layer to capture the transitional relationship between the noisy and true labels. Ghosh et al. [36] confirmed that the loss function based on the mean-absolute error is inherently robust to noise. CleanNet [37] determined whether the sample label is correct by comparing it with a representative "class prototype". However, these methods generally require prior knowledge or rely on certain constraints. To avoid consuming additional resources or more complex networks, it is a good way to select clean parts from noisy instances to update models. The memorization effects of deep neural networks show that the samples with smaller values collected from the loss function are more likely to be correctly annotated. Therefore, some studies [24–27] allow the model to select reliable samples for itself in each iteration to improve classification accuracy, which is similar to active learning and reinforcement learning.

### 2.3. Mutual Teaching Paradigm

Although sample selection can effectively prevent noisy labels from participating in training, it is difficult to ensure that the selected labels are absolutely clean. The self-training process is sensitive to noise and outliers, and multiple iterations will accumulate the model bias caused by a few wrongly selected instances or unbalanced samples. For this purpose, MentorNet [38] is learned to compute time-varying weights for each training sample based on a predefined course, which provides meaningful supervision to help StudentNet overcome corrupted labels. However, the problem of error accumulation still exists. Inspired by co-training [39], co-teaching [27] trains two identical deep networks and lets them select small loss samples for each other in every minibatch. The difference between them is that co-training needs to establish two viewpoints to generate reliable pseudo-labels, which are generally used for semisupervised learning. Co-teaching only needs one viewpoint, which utilizes the randomness of the network training process to resist self-confidence bias, similar to finding their potential shortcomings by "peer-review" Likewise, deep mutual learning [40] enables multiple student networks to learn from each other to produce a more robust and generalized network in model distillation. This simple and effective learning paradigm is easily extended to other applications [41–44]

Unfortunately, few studies have focused on pseudo-label noise in HSI-CD. Therefore, we develop a dynamic change detection framework using the novel mutual teaching approach and an improved sample selection method.

## 3. Methodology

In this section, we detail the proposed method from three aspects: the training process of the mutual teaching framework, sample selection and class balancing, and the classifier for HSI-CD.

### 3.1. The Mutual Teaching Framework

An overview of the proposed mutual teaching framework for bitemporal HSI-CD is shown in Figure 2. First, the original pseudo-labels are obtained from a traditional method, and after screening, they are used to initialize two DL models with the same structure. After each iteration, the two DL models update pseudo-labels for each other with new predictions and alternately use two different sample selection procedures to ensure the accuracy and diversity of training instances. In the end, the final result is generated by combining the predictions of the two models.

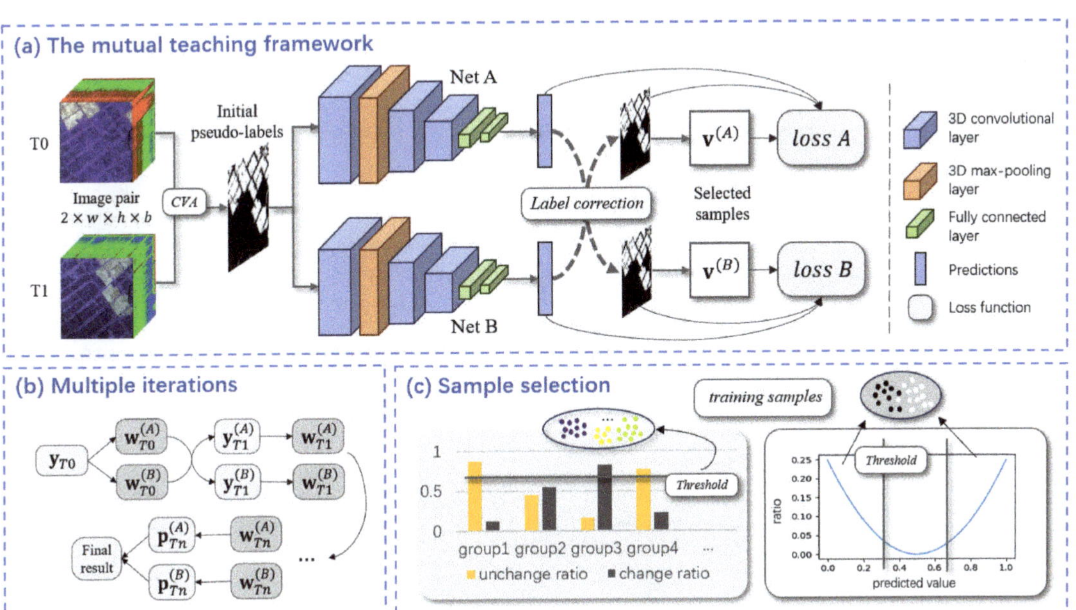


**Figure 2.** Graphical illustration of the proposed method. (**a**) Overall framework of the proposed mutual teaching based on collaborative training and label correction. Both models update their pseudo-labels with each other's predictions and select clean samples to optimize parameters. (**b**) The model parameters and pseudo-labels are alternately updated and the final result is generated by the predictions of the two models. (**c**) Two sample selection methods jointly ensure the accuracy and diversity of training samples.

In this work, the HSI-CD task is regarded as a classification problem, that is, to determine whether the sample corresponding to each pixel belongs to the changed or unchanged class. Taking a pair of pixels in the same position of bitemporal images as a training instance. With a total of $m$ samples, $x_i$ is the $i$th sample, and $m$ equals $w \times h$, where $w$ and $h$ are the width and height of the original image, respectively. We adopt the CVA algorithm to obtain initial pseudo-labels $\mathbf{y} \in \mathbb{R}^m$, $y_i$ equals 1 to represent the change sample, and 0 is unchanged. In contrast to the existing deep-learning-based change detection methods, the pseudo-labels generated by CVA are only used to initialize the parameters of the two networks $\mathbf{w}^{(A)}$ and $\mathbf{w}^{(B)}$ and serve as initial values of $\mathbf{y}^{(A)}$ and $\mathbf{y}^{(B)}$. They will be updated dynamically by mutually training the two networks.

To further improve the accuracy of classifiers, it is necessary to select samples with the label as correct as possible. After feeding all pseudo-labels into the sample selection program, two sets of training data $\mathbf{v}^{(A)}$ and $\mathbf{v}^{(B)}$ can be obtained. Here $\mathbf{v} \in \mathbb{R}^m$ indicates whether the sample $x_i$ is selected, where $v_i$ equals 1 indicating selected and 0 unselected. The parameter updating procedures of both models are described below:

$$\widehat{\mathbf{w}}^{(A)} = \underset{\mathbf{w}^{(A)}}{\arg\min} \sum_{i=0}^{m-1} v_i^{(A)} L\left(y_i^{(A)}, f\left(x_i, \mathbf{w}^{(A)}\right)\right)$$

$$\widehat{\mathbf{w}}^{(B)} = \underset{\mathbf{w}^{(B)}}{\arg\min} \sum_{i=0}^{m-1} v_i^{(B)} L\left(y_i^{(B)}, f\left(x_i, \mathbf{w}^{(B)}\right)\right) \tag{1}$$

where $L(y_i, f(x_i, \mathbf{w}))$ is the loss between the classifier's predicted value $f(x_i, \mathbf{w})$ and the pseudo-label $y_i$. Then, we can update the pseudo-labels with a new prediction:

$$\hat{y}_i^{(A)} = \alpha y_i^{(A)} + (1 - \alpha) f\left(x_i, \mathbf{w}^{(B)}\right)$$

$$\hat{y}_i^{(B)} = \alpha y_i^{(B)} + (1 - \alpha) f\left(x_i, \mathbf{w}^{(A)}\right) \tag{2}$$

where $\alpha$ is the momentum parameter. Note that both models use each other's predicted values to update their own pseudo-labels for mutual teaching purposes.

Sample selection can effectively reduce noisy labels, but it is impossible to completely screen them. Due to various factors such as unbalanced samples and noisy labels, it is inevitable for the classifier to generate confidence bias. The error will be transferred back to itself in the next iteration, and it should be increasingly accumulated in the self-training process. Benefiting from the respective training of the two models, they can filter out different types of errors by mutual teaching and effectively reduce the accumulation of these errors. Meanwhile, with the improvement of model prediction accuracy, the influence of noisy labels can also be mitigated by gradually modifying pseudo-labels. After multiple iterations, the final results are derived from the predicted values of two classifiers. When their predictions are different, we choose one with less loss.

### 3.2. Sample Selection

To reduce the impact of noisy labels, sample selection is utilized in large studies. The most common approach is to judge the credibility according to the sample loss, which can be formulated as:

$$v_i = \begin{cases} 1, & \text{if } |y_i - f(x_i, w)| < \lambda \\ 0, & \text{otherwise} \end{cases} \tag{3}$$

The threshold $\lambda$ is a critical parameter. When $\lambda$ is too large, noisy labels will increase, and when $\lambda$ is too small, the lack of complex samples is not conducive to the generalization of the classifier.

In previous studies [25,26], the above selection method was used, and it is designed for synthetic aperture radar (SAR) image data, which is comparatively simple. However, it performs poorly on HSIs. The selected samples tend to focus on the simplest regions and ignore other types. Thus, we need to design a more appropriate sample selection algorithm for such HSIs. Considering the distribution of data, we use a clustering-based method to select data in blocks. The data in the same cluster have high similarity. When most samples in the cluster have consistent prediction, it is relatively reliable. To select samples of different types evenly and ensure correct labels, a group confidence-based sample selection approach is designed, as shown in Figure 3. First, the PCA is used to reduce the feature dimension of the difference image, and k-means algorithm is applied on the results to obtain the grouping information. Then, we can obtain a grouping label vector $\mathbf{c} \in \mathbb{R}^m$, which indicates the grouping information of all samples and takes the value in $\{0, 1, \ldots, n-1\}$, while $n$ is the total number of groups. We consider the label with the highest proportion in each group as the group label:

$$g_j = \max_{l \in \{0,1\}} \left\{ \sum_{i=0}^{m-1} (c_i == j) \times (y_i == l) \right\}, \ j = 0, 1, \cdots, n-1 \tag{4}$$

where $y_i$ is the pseudo-label; $\mathbf{g} = [g_0, \ldots, g_{n-1}] \in \mathbb{R}^n$, $g_j$ is the group label. If there are more changed samples than unchanged samples within the group, $g_j$ takes 1; otherwise, it takes 0. The group confidence is determined by the proportion of group labels:

$$r_j = \frac{\sum_{i=0}^{m-1} (c_i == j) \times (y_i == g_j)}{\sum_{i=0}^{m-1} (c_i == j)}, \ j = 0, 1, \cdots, n-1 \tag{5}$$

where $\mathbf{r} \in \mathbb{R}^n$ represents a group confidence vector. When the value exceeds a certain threshold, the sample in the group is considered reliable:

$$v_i = \begin{cases} y_i == g_j, & \text{if } r_j \geq \sigma \\ 0, & \text{otherwise} \end{cases} \quad \text{s.t. } j = c_i \tag{6}$$

where $\sigma$ is the group confidence threshold. Note that we only select samples with the same label as the group. In this way, the selected training dataset contains samples of varying degrees of change and has a low proportion of noisy labels.

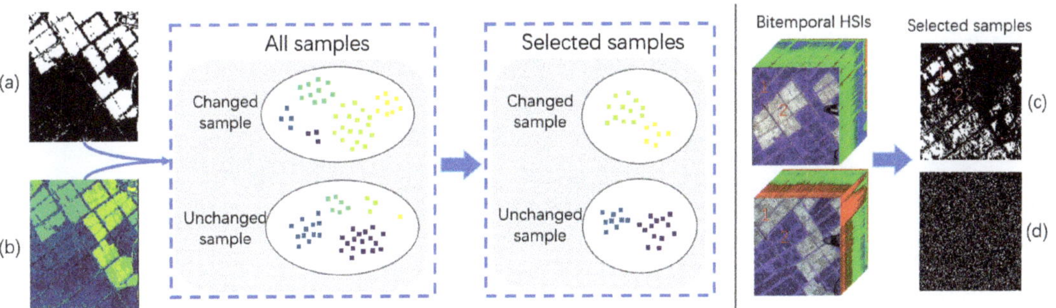


**Figure 3.** Sample selection based on group confidence. (**a**) Pseudo-labels. (**b**) Multiclass map. (**c**) Sample loss-based method. (**d**) Group confidence-based method. In (**c,d**), white is the selected samples with correct label, red is the selected noisy samples, and black is the discarded samples.

Figure 3c,d shows the samples selected in two ways respectively. It is obvious that the dataset contains two main changes, and the changed samples selected by a single sample confidence focus on one class while ignoring the other. Moreover, the sampling of our method is more even, and the noisy labels contained in both methods are negligible. Another advantage of our method is that it can be used on pseudo-labels of discrete values, such as 0 or 1. The sample loss-based algorithm can only be used on continuous values, which are usually between 0 and 1.

The group confidence-based sample selection approach can consider screening noisy labels and the diversity of the training data. There are mainly two kinds of samples to be discarded: one with low group confidence and the other has a different label from most of the samples in the group. Thus, some complex samples may never participate in training, and the two models cannot adequately exchange information. Therefore, we alternately use two selection strategies to jointly guarantee the accuracy and the stability of the final results, as shown in Figure 2c. The group confidence-based sample selection method is used to select samples which are as clean possible, to improve the accuracy of models, and the parameter $\sigma$ is set to 0.8. The loss-based method is used to select as many samples as possible to encourage these easily overlooked complex samples to participate in training, and the parameter $\lambda$ is set to 0.4.

In addition, we apply different weights for sample loss to balance class. General binary classification uses cross-entropy loss, which can be defined as follows:

$$l_{ce}(y_i, p_i) = -y_i \log p_i - (1 - y_i) \log(1 - p_i) \tag{7}$$

where $p_i$ is the predicted value. Then, the final weighted loss function is:

$$L(y_i, f(x_i, \mathbf{w})) = |y_i - f(x_i, \mathbf{w})|^{\gamma} \cdot l_{ce}(y_i, f(x_i, \mathbf{w})) \tag{8}$$

where the first item enhances the weight of large loss samples, and $\gamma$ is set to 2 according to article [45]. The weighted loss can balance the multiclass samples to avoid a large deviation of the model. The entire procedure for the proposed method is summarized in Algorithm 1.

---

**Algorithm 1:** Procedure of the proposed method.

---

**Input:** Two images $\mathbf{I}_1$ and $\mathbf{I}_2$; thresholds $\sigma$ and $\lambda$; the number of iterations $n_t$; the momentum parameter $\alpha$.

**Output:** The final result $\mathbf{p}$.

*// Initialization*
Get pseudo-labels $\mathbf{y}$ and multiclass map $\mathbf{c}$; initialize $\mathbf{y}^{(A)}$ and $\mathbf{y}^{(B)}$;
Randomly initialize $\mathbf{w}^{(A)}$ and $\mathbf{w}^{(B)}$;
**for** $i \leftarrow 1$ **to** $n_t$ **do**
    **if** $i \% 2 == 1$ **then**
        Update selected sample $\mathbf{v}^{(A)}$ and $\mathbf{v}^{(B)}$ by (6);
    **else**
        Update selected sample $\mathbf{v}^{(A)}$ and $\mathbf{v}^{(B)}$ by (3);
    **end**
    Update model parameters $\mathbf{w}^{(A)}$ and $\mathbf{w}^{(B)}$ by (1) and (8);
    Update pseudo-labels $\mathbf{y}^{(A)}$ and $\mathbf{y}^{(B)}$ by (2);
**end**
$\mathbf{p} \Leftarrow \{\mathbf{p}^{(A)} = f(\mathbf{x}, \mathbf{w}^{(A)}), \mathbf{p}^{(B)} = f(\mathbf{x}, \mathbf{w}^{(B)})\}$

---

### 3.3. A 3D Convolutional Neural Network Establishment

The bitemporal hyperspectral data have four dimensions, two spatial axes, a spectral axis and a temporal axis. To extract features using general 2D convolution kernels, most change detection methods reduce one dimension of data by stacking or with a difference operation. However, direct stacking increases the number of convolution kernel channels and network parameters, especially for HSIs with hundreds of channels, and the difference operation leads to the loss of original information. In HSI classification, the authors in [18] verified that 3D convolution can better extract spectral spatial features of HSI than 2D convolution. In some video processing applications, 3D convolution kernel has been used to extract temporal and spatial features simultaneously. Similar to change detection, these kinds of data have an additional temporal dimension relative to a single image. Therefore, 3D convolution is an appropriate feature extractor without additional operations in HSI-CD.

In convolutional layers, the calculation of new features uses convolution kernels to multiply local domain features of the previous layer, then adds a bias and passes through an activation function. For 2D convolution, the value of the feature map extracted by the $i$th convolution kernel of the $l$th layer at position $(x, y)$ is calculated as:

$$X_{l,i}^{xy} = f\left(\sum_m \sum_{p=0}^{P_i-1} \sum_{q=0}^{Q_i-1} W_{l,i}^{pq} X_{l-1,m}^{(x+p)(x+q)} + b_{l,i}\right) \tag{9}$$

where $f(\cdot)$ is the activation function, $P_i$ and $Q_i$ are the height and width of the kernel, $W_{l,i}^{pq}$ is the value of the kernel connected to the feature map at position $(p,q)$, $m$ represents the $l-1$th layer feature map connected to the current feature, and $b_{l,i}$ is the bias. For 3D convolution, the value of the feature map extracted by the $i$th convolution kernel of the $l$th layer at position $(x,y,z)$ is calculated as:

$$X_{l,i}^{xyz} = f\left(\sum_m\sum_{p=0}^{P_i-1}\sum_{q=0}^{Q_i-1}\sum_{r=0}^{R_i-1} W_{l,i}^{pqr} X_{l-1,m}^{(x+p)(y+q)(z+r)} + b_{l,i}\right) \tag{10}$$

where $R_i$ is the size of the 3D kernal along the spectral dimension. Because the adjacent spectral channels of HSIs have a strong correlation, it is reasonable to extract the spatial and spectral neighborhood information of two images simultaneously with 3D convolution.

For HSI-CD tasks, we design a 3D convolutional neural network as the basis classifier, as shown in Figure 4. A sample consists of two data blocks of a neighborhood size of 3 extracted from bitemporal images at the same location, and is filled to size 5 with 0 when input into the network. After three 3D convolution layers and a pooling layer, the fused feature vector is extracted, and finally, the change information is output through two fully connected layers. The last layer is activated by the softmax function, and the other uses the ReLU function. After all of the samples are fed into the network, a result map with the same size as the original image representing the degree of change can be obtained. To verify the effectiveness of 3D convolution, we design a similar 2D convolutional neural network for comparison in subsequent experiments.

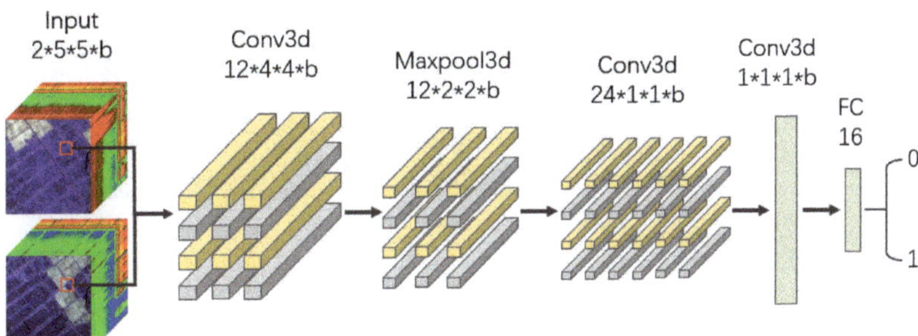


**Figure 4.** Architecture of the proposed 3DCNN.

## 4. Result

To verity the effectiveness of the proposed method, several experiments were conducted using a multispectral dataset and three popular hyperspectral datasets. This section first introduces the datasets used in the experiment. Then the evaluation measures of change detection and experimental setup are described. Finally, comparative experiments with other methods are analyzed in detail.

### 4.1. Introduction to Datasets

The first dataset "Bastrop" is shown in Figure 5a,b, which consists of two multispectral images (MSI) taken before and after a forest fire in Bastrop County, Texas, in September 2011 and October 2011 [46]. This multispectral dataset was selected from Landsat 5 Thematic Mapper (TM) multispectral images consisting of six spectral bands with a spatial resolution of 30 m for bands 1–5 and 7 and one thermal band (band 6). Their spatial size is 1534 × 808 pixels with 7 bands.

The other three HSI datasets were collected from Earth Observing-1 (EO-1) Hyperion data. EO-1 has a spectral resolution of 10 nm and a spatial resolution of approximately 30 m, with a total of 242 different bands. The second dataset "Umatilla" is irrigated farmland in Umatilla County, OR, USA, as shown in Figure 5d,e. The images contain

390 × 200 pixels and 242 bands. The third dataset, "Yancheng", was acquired on 3 May 2006, and 23 April 2007, in Yancheng, Jiangsu Province, China, as shown in Figure 5g,h. The two images both consist of 450 × 140 pixels with 155 bands after eliminating the noise. The fourth dataset, "river", was obtained on 3 May 2013, and 31 December 2013, in Jiangsu Province, China, as shown in Figure 5j,k. This dataset contains two HSIs with 463 × 241 pixels and 198 channels [17].

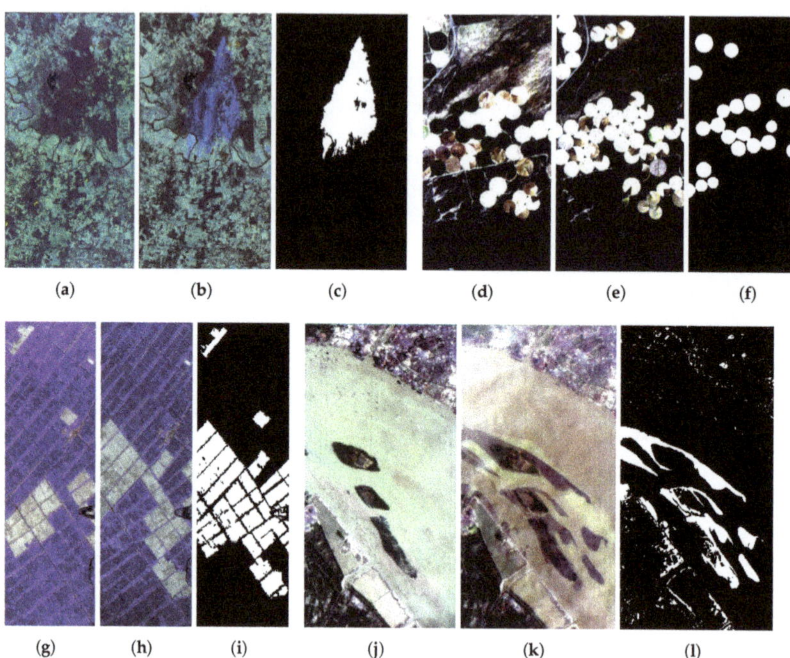

**Figure 5.** Experimental datasets. (**a**) Bastrop dataset in September 2011. (**b**) Bastrop dataset in October 2011. (**d**) Umatilla dataset on 1 May 2004. (**e**) Umatilla dataset on 8 May 2007. (**g**) Yancheng dataset on 3 May 2006. (**h**) Yancheng dataset on 23 April 2007. (**j**) River dataset on 3 May 2013. (**k**) River dataset on 31 December 2013. (**c,f,i,l**) groundtruth change map for Bastrop, Umatilla, Yancheng and River dataset, respectively.

### 4.2. Evaluation Measures and Experimental Configurations

In this paper, specific evaluation metrics are used to evaluate the change detection results of all methods on the datasets. Generally, the results of change detection use pixel-level indicators, which mainly include the following four metrics: true positives ($TP$), true negatives ($TN$), false positives ($FP$), and false negatives ($FN$). The positive sample refers to the changed samples, displayed in white in the result image, and the negative sample refers to unchanged samples, displayed in black. The correct rate of classification is represented by the overall accuracy ($OA$), and the formula is

$$OA = \frac{TP + TN}{TP + TN + FP + FN} \tag{11}$$

Compared with $OA$, the kappa coefficient and $F1$ score can better reflect the consistency between the predicted results and the actual results. It is calculated as

$$PRE = \frac{(TP + FP) \times (TP + FN) + (FN + TN) \times (FP + TN)}{(TP + TN + FP + FN)^2} \tag{12}$$

$$Kappa = \frac{OA - PRE}{1 - PRE} \tag{13}$$

$$Precision = \frac{TP}{TP + FP} \tag{14}$$

$$Recall = \frac{TP}{TP + FN} \tag{15}$$

$$F1 = \frac{2 \times Precision \times Recall}{Precision + Recall} \tag{16}$$

For the experimental setting, the proposed 3DCNN network structure is shown in Figure 4. For hyperspectral images, the size of the first two 3D convolution kernels is $2 \times 2 \times 5$, while on the Bastrop dataset, the size of the 3D convolution kernels is $2 \times 2 \times 1$ because it has only seven bands. The parameter $n$ is the total number of groups. For the Bastrop datasets, $n$ is set to 10, and for the other three datasets, $n$ is set to 20. The momentum parameter $\alpha$ is set to 0.4.

### 4.3. Comparison with Other Methods

To verity the effectiveness of the proposed method, we tested our method on three hyperspectral datasets and a multispectral dataset, then compared it with other classical methods, including the change vector analysis (CVA) [7], iteratively reweighted multivariate alteration detection (IRMAD) [10], iterative slow feature analysis (ISFA) [12], support vector machines (SVM), GETNET [17], 2DCNN, and 3DCNN. The Otsu threshold algorithm is used in CVA to generate the final change detection result, which is used as pseudo-labels for other methods that require labeled data. Among the above methods, only CVA, IRMAD and ISFA do not require labeled samples. Other classification-based methods use the same pseudo-labels for supervised training and select training samples through our proposed sample selection method.

### 4.3.1. Experiments on the Bastrop Dataset

Figure 6 shows the final binary result images of the eight methods, and Table 1 lists the results of the numerical evaluation. It can be clearly seen from the figure that there is a large amount of misclassification noise in CVA, mainly false negative samples, and the kappa coefficient is only 0.7241. IRMAD is the worst and ISFA is relatively better among the three unsupervised traditional algorithms, but they have the same problems. Other methods use the results of CVA as pseudo-labels to train their models. Although SVM significantly reduces FP values, it also leads to a huge deviation that causes the changed samples to be mistaken for unchanged. Then, the kappa coefficient was reduced by 22%. Deep-learning methods have wonderful advantages. They all increase the OA and the kappa coefficients and outperform traditional algorithms visually. Remarkably, the performance of our method is considerably better than other methods on this dataset, with kappa rising to 0.9406, which is 9% higher than the second-highest value. In particular, a large number of false negative samples have been well-corrected.

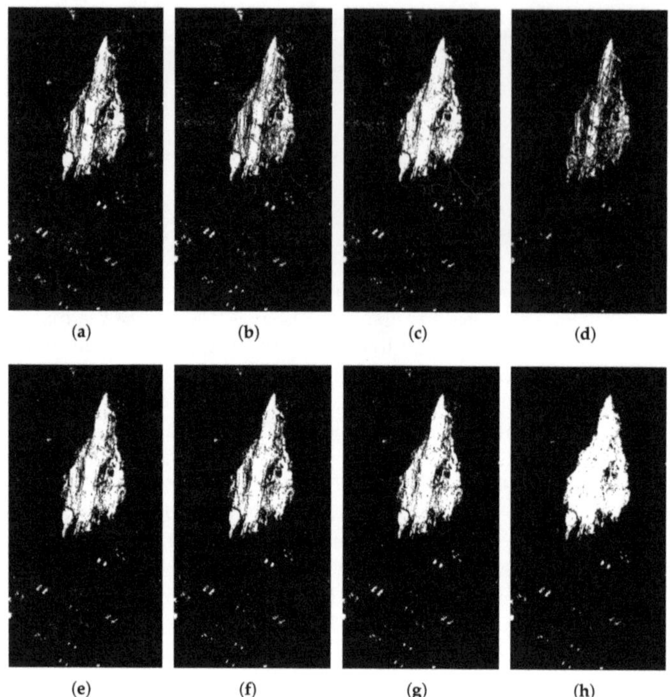

**Figure 6.** The change map on Bastrop dataset. (**a**) CVA. (**b**) IRMAD. (**c**) ISFA. (**d**) SVM. (**e**) GETNET (**f**) 2DCNN. (**g**) 3DCNN. (**h**) ours.

**Table 1.** Quantitative evaluation of CD results by different methods for Bastrop dataset.

| Methods | FP | FN | OA | Kappa | F1 |
|---|---|---|---|---|---|
| CVA | 10,272 | 46,813 | 0.9539 | 0.7241 | 0.7487 |
| IRMAD | 13,490 | 54,000 | 0.9455 | 0.6688 | 0.6977 |
| ISFA | 10,082 | 32,010 | 0.9660 | 0.8073 | 0.8259 |
| SVM | **2799** | 83,435 | 0.9304 | 0.4992 | 0.5290 |
| GETNET | 6744 | 31,319 | 0.9693 | 0.8241 | 0.8408 |
| 2DCNN | 6923 | 34,276 | 0.9668 | 0.8076 | 0.8257 |
| 3DCNN | 7420 | 26,299 | 0.9728 | 0.8473 | 0.8623 |
| ours | 7212 | **6811** | **0.9887** | **0.9406** | **0.9469** |

4.3.2. Experiments on the Umatilla Dataset

These dataset results are shown in Figure 7 and listed in Table 2. Among the three unsupervised traditional algorithms, CVA has the most serious noise and the lowest accuracy. From the visual effect, the results of IRMAD are closest to the real labels, but there is no substantial advantage compared with ISFA in quantitative analysis. Deep-learning methods can basically filter out the background noise, which also confirms the effectiveness of our sample selection method. Although the gap is small, our method has achieved the best performance in quantitative analysis.

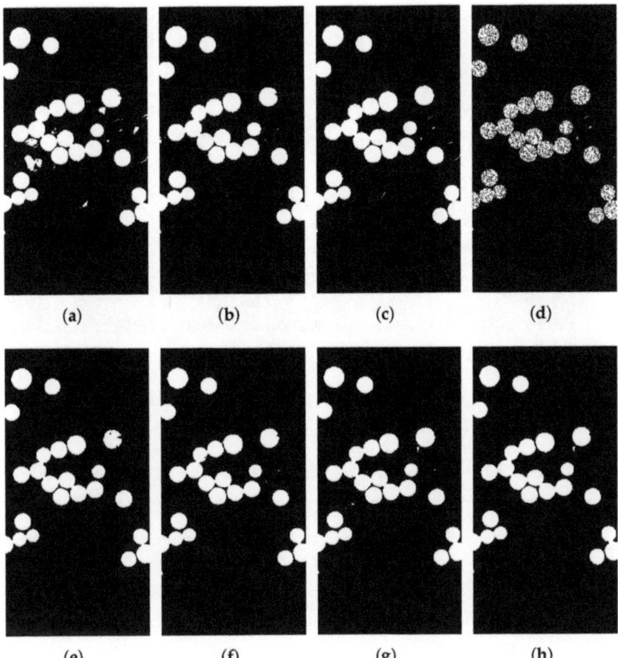

Figure 7. The change map on Umatilla dataset. (**a**) CVA. (**b**) IRMAD. (**c**) ISFA. (**d**) SVM. (**e**) GETNET. (**f**) 2DCNN. (**g**) 3DCNN. (**h**) ours.

**Table 2.** Quantitative evaluation of CD results by different methods for Umatilla dataset.

| Methods | FP | FN | OA | Kappa | F1 |
|---|---|---|---|---|---|
| CVA | 1092 | 198 | 0.9835 | 0.9258 | 0.9352 |
| IRMAD | 452 | 246 | 0.9911 | 0.9586 | 0.9637 |
| ISFA | 506 | **191** | 0.9911 | 0.9588 | 0.9639 |
| SVM | 256 | 2125 | 0.9695 | 0.8442 | 0.8612 |
| GETNET | 216 | 337 | 0.9929 | 0.9667 | 0.9707 |
| 2DCNN | 210 | 445 | 0.9916 | 0.9604 | 0.9651 |
| 3DCNN | 277 | 291 | 0.9927 | 0.9660 | 0.9701 |
| ours | **151** | 309 | **0.9941** | **0.9723** | **0.9756** |

### 4.3.3. Experiments on the Yancheng Dataset

The changes in this dataset are mainly related to farmland. The results are shown in Figure 8 and listed in Table 3. The Yancheng dataset is a relatively simple, traditional method that can also achieve a good performance, especially the performance of ISFA and deep-learning methods that are very similar. Additionally, their OAs are all over 97%. The performance of SVM is the worst and there is too much noise in the changed area. In addition, these four deep-learning methods have all performed very well, but GETNET and 2DCNN still have obvious noise in the unchanged regions, and 3DCNN performs poorly in the changed regions. Only our method eliminates the background noise and also ensures the accuracy of the changed region with multiple iterations.

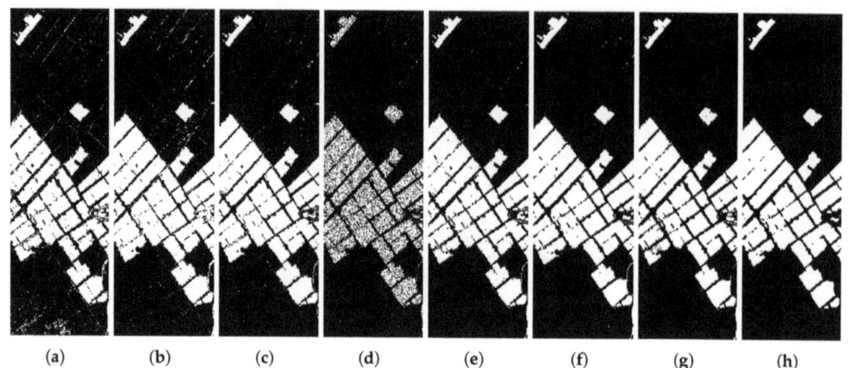

Figure 8. The change map on Yancheng dataset. (a) CVA. (b) IRMAD. (c) ISFA. (d) SVM. (e) GETNET. (f) 2DCNN. (g) 3DCNN. (h) ours.

**Table 3.** Quantitative evaluation of CD results by different methods for Yancheng dataset.

| Methods | FP | FN | OA | Kappa | F1 |
|---------|------|------|--------|--------|--------|
| CVA | 1833 | 1158 | 0.9525 | 0.8860 | 0.9197 |
| IRMAD | 2268 | 356 | 0.9583 | 0.9019 | 0.9318 |
| ISFA | 1303 | **296** | 0.9746 | 0.9394 | 0.9574 |
| SVM | **512** | 4619 | 0.9186 | 0.7882 | 0.8419 |
| GETNET | 810 | 792 | 0.9746 | 0.9383 | 0.9562 |
| 2DCNN | 1162 | 611 | 0.9719 | 0.9323 | 0.9522 |
| 3DCNN | 554 | 1059 | 0.9744 | 0.9373 | 0.9553 |
| ours | 548 | 817 | **0.9783** | **0.9472** | **0.9624** |

### 4.3.4. Experiments on the River Dataset

The River dataset is more complex than the other datasets and contains a variety of changes, mainly the disappearance of substances in rivers. Figure 9 shows the maps obtained by eight methods and the quantitative comparison is shown in Table 4. It is obvious from the numerical indicators that the results of CVA are extremely unbalanced, and the number of false-positive samples is approximately 6 times that of the false-negative samples. In addition, ISFA, which performs relatively well in the other datasets, has the worst accuracy here. There is no significant difference among the results of the three networks. The OA can grow to more than 95%, which once again proves that the deep neural network has a strong learning ability and that sample selection can effectively suppress noisy labels. Through multiple iterations and sample selection, the proposed method eliminates the huge deviation of the initial pseudo-labels and obtains the best performance.

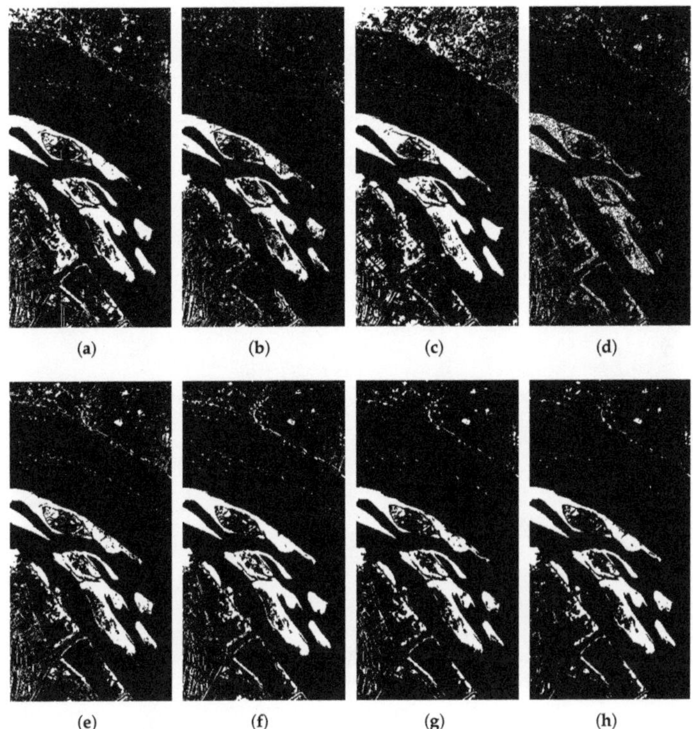

**Figure 9.** The change map on River dataset. (**a**) CVA. (**b**) IRMAD. (**c**) ISFA. (**d**) SVM. (**e**) GETNET. (**f**) 2DCNN. (**g**) 3DCNN. (**h**) ours.

**Table 4.** Quantitative evaluation of CD results by different methods for River dataset.

| Methods | FP | FN | OA | Kappa | F1 |
|---------|------|------|--------|--------|--------|
| CVA | 6196 | **1123** | 0.9344 | 0.7103 | 0.7467 |
| IRMAD | 3343 | 3089 | 0.9424 | 0.7005 | 0.7328 |
| ISFA | 10,244 | 1355 | 0.8961 | 0.5897 | 0.6453 |
| SVM | 2595 | 6007 | 0.9229 | 0.5373 | 0.5784 |
| GETNET | 4185 | 1369 | 0.9502 | 0.7636 | 0.7915 |
| 2DCNN | 3618 | 1127 | 0.9575 | 0.7958 | 0.8196 |
| 3DCNN | 2447 | 2215 | 0.9582 | 0.7827 | 0.8061 |
| ours | **1595** | 1809 | **0.9695** | **0.8387** | **0.8558** |

## 5. Discussion

### 5.1. Ablation Study

To argue the effectiveness of the mutual teaching paradigm, on the above four datasets we make the two networks perform mutual teaching and separate training under the same conditions. Figure 10 shows the OA of the results in 10 consecutive iterations. Classifiers A and B refer to each other's predicted values, while A$'$ and B$'$ only use their own results. The two sets of experiments have the same initialization. In the mutual teaching framework, high-precision classifiers are often dragged down by low-precision classifiers, undergoing raising and lowering changes. However, overall, the performances of the two models basically show an upward trend. Although this process has some fluctuations, it does not affect the overall performance. The self-training performance is relatively poor, and the Umatilla and Yancheng datasets are almost not improved. The improvement of the Bastrop and River datasets is mainly due to sample selection and label correction, but it is also inferior to the mutual teaching models.

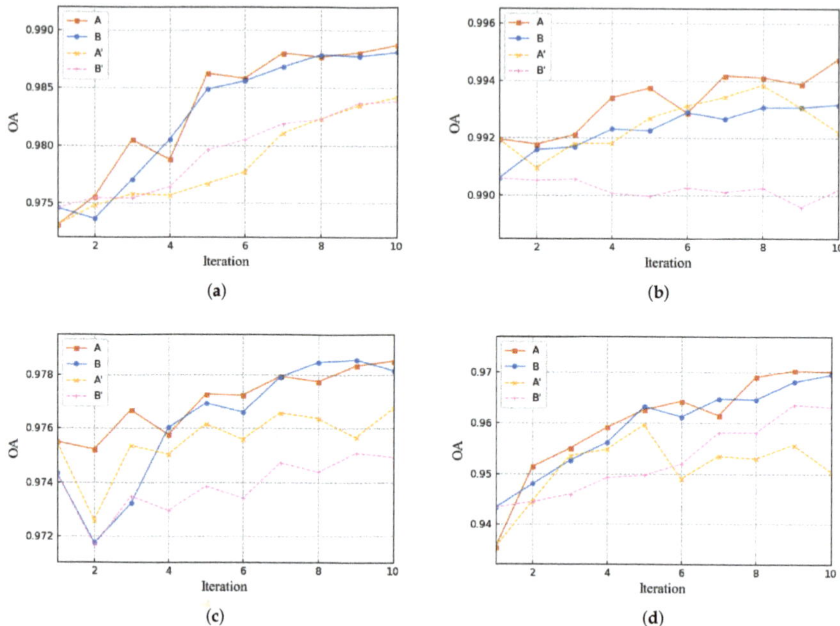

**Figure 10.** The iterative performance of the mutual teaching framework, where A and B use mutual teaching and A′ and B′ are self-training. (**a**) Bastrop dataset. (**b**) Umatilla dataset. (**c**) Yancheng dataset. (**d**) River dataset.

As shown in Figure 11, we compare the alternating training with only one sample selection method. If a dataset itself is relatively simple, the difference between these results is not large. To show the difference in performance, we only use the most complicated River dataset. Figure 11 shows the overall accuracy of the final results and the variance between two models under three settings in each iteration on the River dataset. Although the model accuracy increases faster when only the group confidence-based sample selection method is used, the accuracy no longer increases and remains stable from the sixth iteration. The overall accuracy is further improved by alternating training and significantly exceeds other settings, which proves that the participation of complex samples in training is beneficial to improving the model performance and preserving the details of the change map.

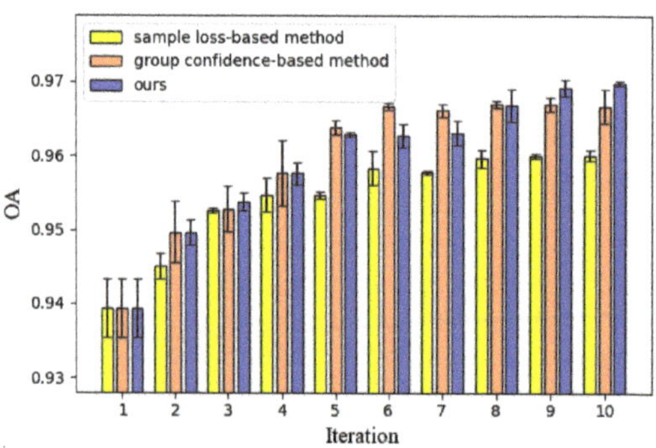


**Figure 11.** The result accuracy and the variance of two models under three settings on River dataset.

*5.2. Compatibility of the Proposed Framework with Other Models*

Figure 12 displays the change detection results of three networks during initialization and after 10 iterations in our framework. GETNET and 3DCNN have considerably more false-positive samples during initialization, which is mainly misled by the pseudo-labels generated by CVA. However, after multiple corrections, these noises have been improved to a certain extent, especially for the 3DCNN (as shown in the red box). The main error of the 2DCNN result is that some changed regions were not detected, and it had also been recovered after iteration. In other words, both false-positive and false-negative noise labels have the opportunity to be corrected under the proposed framework. The results demonstrate that the mutual teaching framework can also benefit other deep-learning methods based on pseudo-labels.

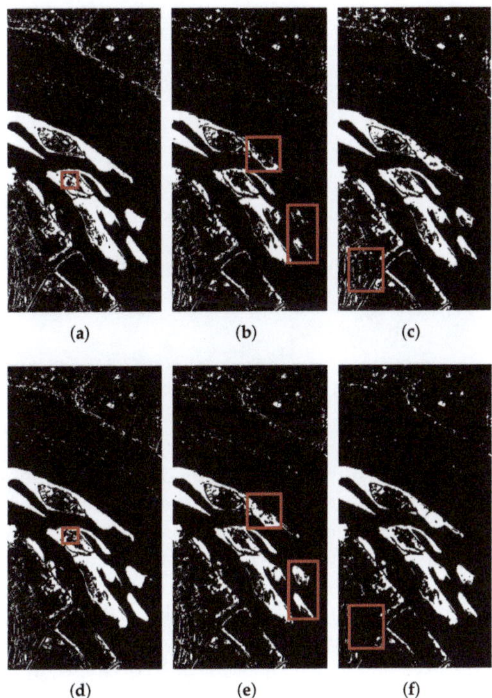

**Figure 12.** Results of different networks in the first and last iterations on River dataset. (**a**) GETNET in the first iteration. (**b**) 2DCNN in the first iteration. (**c**) 3DCNN in the first iteration. (**d**) GETNET in the last iteration. (**e**) 2DCNN in the last iteration. (**f**) 3DCNN in the last iteration.

*5.3. Hyperparametric Analysis*

5.3.1. Analysis of the Pseudo-Label Update Rate

In the process of pseudo-label correction, the momentum parameter $\alpha$ (in Equation (2)) selection is worth discussing. When $\alpha = 0$, pseudo-labels in each iteration are determined only by new predicted values; when $\alpha = 1$, our method depends entirely on the initial pseudo-label without any updates. We measure the results of different parameters on four datasets, as shown in Figure 13. Experimental results show that the update of pseudo-labels can bring a better performance. If the false labels are not corrected, they will inevitably limit the final result. With the increase in $\alpha$, the overall accuracy shows a downward trend. A value of 0.2~0.5 is a suitable range for all datasets. Therefore, we choose $\alpha = 0.4$ for our experiments.

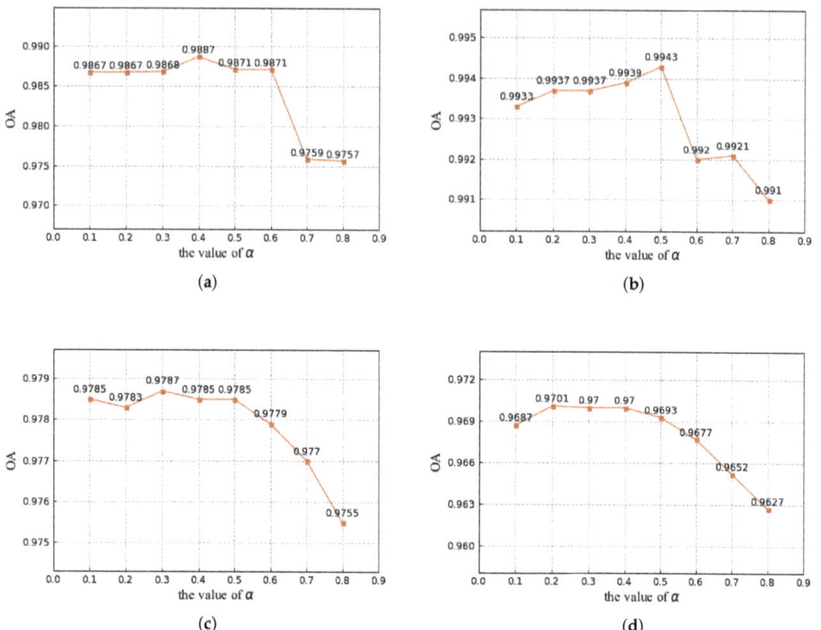

**Figure 13.** Analysis of the momentum parameter $\alpha$. (**a**) Bastrop dataset. (**b**) Umatilla dataset. (**c**) Yancheng dataset. (**d**) River dataset.

### 5.3.2. Analysis of the Number of Groups

Sample selection by clustering methods can ensure the diversity of training samples. However, too detailed a classification makes it difficult to remove noise for datasets with simple ground objects types (the sample loss-based method is more effective here). For complicated datasets, especially HSI that are sensitive to ground changes, classified sampling is crucial to class balance and model generalization. The results for different numbers of groups are shown in Figure 14. Since the Bastrop dataset has only one change type, and the spectral information of MSI is much less than that of HSI, the value of $n$ needs to be small. Based on the experiment, we choose $n = 10$ on the Bastrop dataset. For the other three HSI datasets, we choose $n = 20$. Moreover, other clustering methods that automatically determine the number of groups can be considered to avoid parameter selection.

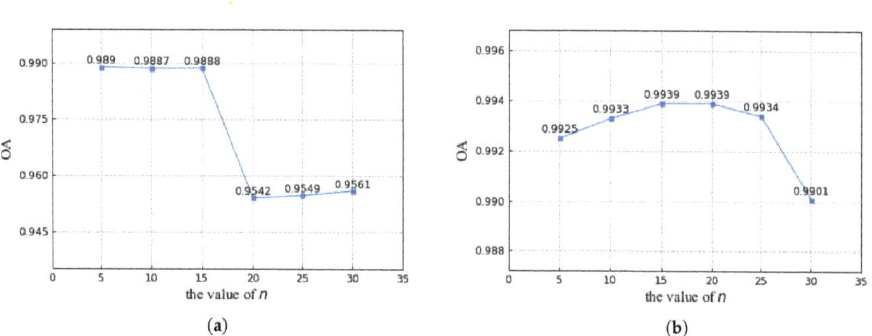

**Figure 14.** *Cont.*

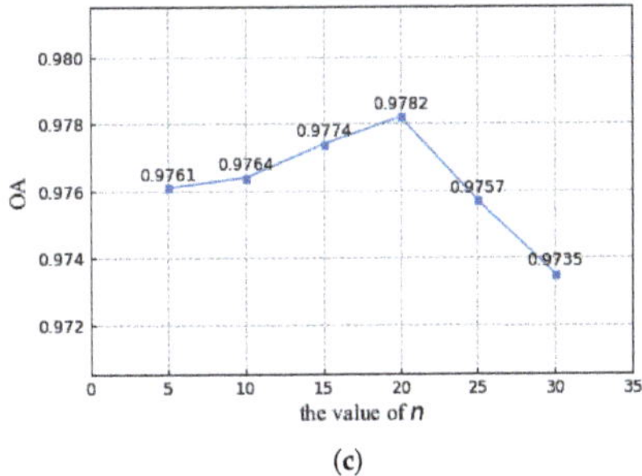
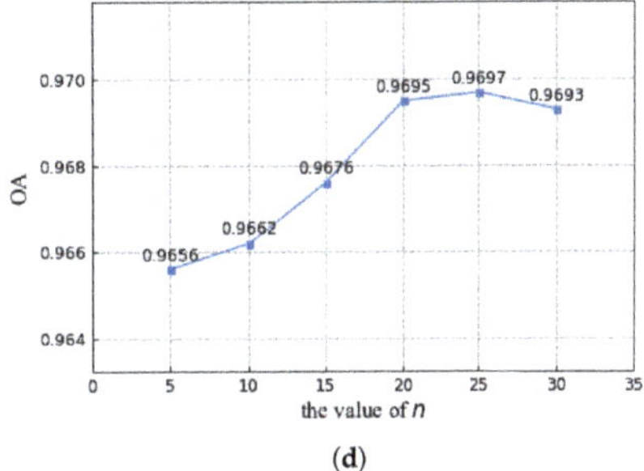

(c)         (d)

**Figure 14.** Analysis of the number of groups $n$. (**a**) Bastrop dataset. (**b**) Umatilla dataset. (**c**) Yancheng dataset. (**d**) River dataset.

### 5.4. Computing Time

The computing device is equipped with Intel i7-9700K CPU (3.6 GHz) and NVIDIA GeForce RTX2080Ti GPU. The program is written in Python via the code library of PyTorch. Here, we list the computing time for each dataset in Table 5. With the multiple models and numerous iterations for optimization, the proposed method suffers from high computational complexity. Theoretically, the optimization process of the two models is independent. Therefore, the method can be accelerated by parallel computing to reduce the computing time by 50%, which is the same as the self-training time of a single model.

**Table 5.** Time cost (seconds) of each dataset.

|        | Bastrop | Umatilla | Yancheng | River  |
|--------|---------|----------|----------|--------|
| 3DCNN  | 343.52  | 51.18    | 46.94    | 67.99  |
| ours   | 2919.93 | 414.29   | 339.44   | 496.96 |

## 6. Conclusions

In this article, a general mutual teaching framework with momentum correction is proposed for the HSI-CD task by dual-3DCNN. It aims to perform robust training for deep-learning methods using pseudo-labels generated by traditional approaches. Adopting the idea of collaborative training, the proposed framework encourages the two models to teach each other to mitigate self-confidence bias and boosts label correction in the iterative process to further improve performance. Then, focusing on the complexity of HSI change types, a new sample selection method based on group confidence is designed to extract better quality and diverse training data. Furthermore, the 3DCNN can effectively extract spatiotemporal spectral features of bitemporal HSIs, and thus, it is developed as the basic classifier of the above framework. Our approach uses pseudo-labels obtained by unsupervised algorithms, which means it can also be compatible with other networks that require labeled data.

We implemented our approach and performed experiments on a multispectral dataset, as well as on three public hyperspectral datasets. The visual and quantitative results show that our method can effectively improve the robustness and generalization of the deep neural network for the HSI-CD task.

**Author Contributions:** Conceptualization, J.S. and L.H.; methodology, L.X.; software, J.S.; validation, J.S., L.X. and J.L.; formal analysis, L.X. and J.L.; writing—original draft preparation, J.S.; writing—review and editing, L.X. and J.L.; visualization, J.S., L.H. and Z.W.; supervision, L.X., J.L. and Z.W.; project administration, L.X.; funding acquisition, L.X. All authors have read and agreed to the published version of the manuscript.

**Funding:** This research was funded by the National Natural Science Foundation of China under Grant 61571230, 61871226, and 61906093, in part by the Jiangsu Provincial Social Developing Project under Grant BE2018727, in part by the Jiangsu Provincial Nature Science Foundations of China under Grant BK20190451, in part by the Fundamental Research Funds for the Central Universities under Grant 30918011104 and 30920021134; and in part by the National Major Research Plan of China under Grant 2016YFF0103604.

**Institutional Review Board Statement:** Not applicable.

**Informed Consent Statement:** Not applicable.

**Data Availability Statement:** Not applicable.

**Conflicts of Interest:** The authors declare no conflict of interest.

## References


1. Scafutto, R.D.M.; de Souza Filho, C.R.; de Oliveira, W.J. Hyperspectral remote sensing detection of petroleum hydrocarbons in mixtures with mineral substrates: Implications for onshore exploration and monitoring. *ISPRS J. Photogramm. Remote Sens.* **2017**, *128*, 146–157. [CrossRef]
2. Carrino, T.A.; Crósta, A.P.; Toledo, C.L.B.; Silva, A.M. Hyperspectral remote sensing applied to mineral exploration in southern Peru: A multiple data integration approach in the Chapi Chiara gold prospect. *Int. J. Appl. Earth Obs. Geoinf.* **2018**, *64*, 287–300. [CrossRef]
3. Zhang, X.; Sun, Y.; Shang, K.; Zhang, L.; Wang, S. Crop classification based on feature band set construction and object-oriented approach using hyperspectral images. *IEEE J. Sel. Top. Appl. Earth Obs. Remote Sens.* **2016**, *9*, 4117–4128. [CrossRef]
4. Vali, A.; Comai, S.; Matteucci, M. Deep learning for land use and land cover classification based on hyperspectral and multispectral earth observation data: A review. *Remote Sens.* **2020**, *12*, 2495. [CrossRef]
5. Shimoni, M.; Haelterman, R.; Perneel, C. Hypersectral imaging for military and security applications: Combining myriad processing and sensing techniques. *IEEE Geosci. Remote Sens. Mag.* **2019**, *7*, 101–117. [CrossRef]
6. Singh, A. Review article digital change detection techniques using remotely-sensed data. *Int. J. Remote Sens.* **1989**, *10*, 989–1003. [CrossRef]
7. Bovolo, F.; Bruzzone, L. A theoretical framework for unsupervised change detection based on change vector analysis in the polar domain. *IEEE Trans. Geosci. Remote Sens.* **2006**, *45*, 218–236. [CrossRef]
8. Celik, T. Unsupervised change detection in satellite images using principal component analysis and *k*-means clustering. *IEEE Geosci. Remote Sens. Lett.* **2009**, *6*, 772–776. [CrossRef]
9. Nielsen, A.A.; Conradsen, K.; Simpson, J.J. Multivariate alteration detection (MAD) and MAF postprocessing in multispectral bitemporal image data: New approaches to change detection studies. *Remote Sens. Environ.* **1998**, *64*, 1–19. [CrossRef]
10. Nielsen, A.A. The regularized iteratively reweighted MAD method for change detection in multi-and hyperspectral data. *IEEE Trans. Image Process.* **2007**, *16*, 463–478. [CrossRef]
11. Coppin, P.; Jonckheere, I.; Nackaerts, K.; Muys, B.; Lambin, E. Review ArticleDigital change detection methods in ecosystem monitoring: A review. *Int. J. Remote Sens.* **2004**, *25*, 1565–1596. [CrossRef]
12. Wu, C.; Du, B.; Zhang, L. Slow feature analysis for change detection in multispectral imagery. *IEEE Trans. Geosci. Remote Sens.* **2013**, *52*, 2858–2874. [CrossRef]
13. Ahlqvist, O. Extending post-classification change detection using semantic similarity metrics to overcome class heterogeneity: A study of 1992 and 2001 US National Land Cover Database changes. *Remote Sens. Environ.* **2008**, *112*, 1226–1241. [CrossRef]
14. Wan, L.; Xiang, Y.; You, H. A post-classification comparison method for SAR and optical images change detection. *IEEE Geosci. Remote Sens. Lett.* **2019**, *16*, 1026–1030. [CrossRef]
15. Liu, T.; Yang, L.; Lunga, D. Change detection using deep learning approach with object-based image analysis. *Remote Sens. Environ.* **2021**, *256*, 112308. [CrossRef]
16. Mou, L.; Bruzzone, L.; Zhu, X.X. Learning spectral-spatial-temporal features via a recurrent convolutional neural network for change detection in multispectral imagery. *IEEE Trans. Geosci. Remote Sens.* **2018**, *57*, 924–935. [CrossRef]
17. Wang, Q.; Yuan, Z.; Du, Q.; Li, X. GETNET: A general end-to-end 2-D CNN framework for hyperspectral image change detection. *IEEE Trans. Geosci. Remote Sens.* **2018**, *57*, 3–13. [CrossRef]
18. Chen, Y.; Jiang, H.; Li, C.; Jia, X.; Ghamisi, P. Deep feature extraction and classification of hyperspectral images based on convolutional neural networks. *IEEE Trans. Geosci. Remote Sens.* **2016**, *54*, 6232–6251. [CrossRef]
19. Du, B.; Ru, L.; Wu, C.; Zhang, L. Unsupervised deep slow feature analysis for change detection in multi-temporal remote sensing images. *IEEE Trans. Geosci. Remote Sens.* **2019**, *57*, 9976–9992. [CrossRef]
20. Li, X.; Yuan, Z.; Wang, Q. Unsupervised deep noise modeling for hyperspectral image change detection. *Remote Sens.* **2019**, *11*, 258. [CrossRef]
21. Li, Q.; Gong, H.; Dai, H.; Li, C.; He, Z.; Wang, W.; Feng, Y.; Han, F.; Tuniyazi, A.; Li, H.; et al. Unsupervised Hyperspectral Image Change Detection via Deep Learning Self-generated Credible Labels. *IEEE J. Sel. Top. Appl. Earth Obs. Remote Sens.* **2021**, *14*, 9012–9024. [CrossRef]
22. Zhang, C.; Bengio, S.; Hardt, M.; Recht, B.; Vinyals, O. Understanding deep learning (still) requires rethinking generalization. *Commun. ACM* **2021**, *64*, 107–115. [CrossRef]

23. Yu, X.; Han, B.; Yao, J.; Niu, G.; Tsang, I.; Sugiyama, M. How does disagreement help generalization against label corruption? In Proceedings of the International Conference on Machine Learning, Long Beach, CA, USA, 10–15 June 2019; pp. 7164–7173.

24. Yao, Q.; Yang, H.; Han, B.; Niu, G.; Kwok, J.T.Y. Searching to exploit memorization effect in learning with noisy labels. In Proceedings of the International Conference on Machine Learning, Virtual Event, 13–18 July 2020; pp. 10789–10798.

25. Shang, R.; Yuan, Y.; Jiao, L.; Meng, Y.; Ghalamzan, A.M. A self-paced learning algorithm for change detection in synthetic aperture radar images. *Signal Proc.* **2018**, *142*, 375–387. [CrossRef]

26. Gong, M.; Duan, Y.; Li, H. Group self-paced learning with a time-varying regularizer for unsupervised change detection. *IEEE Trans. Geosci. Remote Sens.* **2019**, *58*, 2481–2493. [CrossRef]

27. Han, B.; Yao, Q.; Yu, X.; Niu, G.; Xu, M.; Hu, W.; Tsang, I.; Sugiyama, M. Co-teaching: Robust Training of Deep Neural Networks with Extremely Noisy Labels. In Proceedings of the International Conference on Neural Information Processing Systems, Stockholm, Sweden, 10–15 July 2018; pp. 8536–8546.

28. Li, P.; Xu, Y.; Wei, Y.; Yang, Y. Self-correction for human parsing. *arXiv* **2020**, arXiv:1910.09777.

29. Zheng, G.; Awadallah, A.H.; Dumais, S. Meta label correction for noisy label learning. In Proceedings of the AAAI Conference on Artificial Intelligence, Virtual Event, 2–9 February 2021.

30. Liu, J.; Gong, M.; Qin, K.; Zhang, P. A deep convolutional coupling network for change detection based on heterogeneous optical and radar images. *IEEE Trans. Neural Netw. Learn. Syst.* **2016**, *29*, 545–559. [CrossRef] [PubMed]

31. Zhang, P.; Gong, M.; Zhang, H.; Liu, J.; Ban, Y. Unsupervised difference representation learning for detecting multiple types of changes in multitemporal remote sensing images. *IEEE Trans. Geosci. Remote Sens.* **2018**, *57*, 2277–2289. [CrossRef]

32. Liu, J.; Zhang, W.; Liu, F.; Xiao, L. A Probabilistic Model Based on Bipartite Convolutional Neural Network for Unsupervised Change Detection. *IEEE Trans. Geosci. Remote Sens.* **2021**, *60*, 4701514. [CrossRef]

33. Gong, M.; Zhao, J.; Liu, J.; Miao, Q.; Jiao, L. Change detection in synthetic aperture radar images based on deep neural networks. *IEEE Trans. Neural Netw. Learn. Syst.* **2015**, *27*, 125–138. [CrossRef]

34. Li, Y.; Peng, C.; Chen, Y.; Jiao, L.; Zhou, L.; Shang, R. A deep learning method for change detection in synthetic aperture radar images. *IEEE Trans. Geosci. Remote Sens.* **2019**, *57*, 5751–5763. [CrossRef]

35. Goldberger, J.; Ben-Reuven, E. Training deep neural-networks using a noise adaptation layer. In Proceedings of the International Conference of Learning Representations (ICLR), San Juan, Puerto Rico, 2–4 May 2016.

36. Ghosh, A.; Kumar, H.; Sastry, P. Robust loss functions under label noise for deep neural networks. In Proceedings of the AAAI Conference on Artificial Intelligence, San Francisco, CA, USA, 4–9 February 2017; Volume 31.

37. Lee, K.H.; He, X.; Zhang, L.; Yang, L. Cleannet: Transfer learning for scalable image classifier training with label noise. In Proceedings of the IEEE Conference on Computer Vision and Pattern Recognition, Salt Lake City, UT, USA, 18–22 June 2018; pp. 5447–5456.

38. Jiang, L.; Zhou, Z.; Leung, T.; Li, L.J.; Li, F.-F. Mentornet: Learning data-driven curriculum for very deep neural networks on corrupted labels. In Proceedings of the International Conference on Machine Learning, PMLR, Stockholm, Sweden, 10–15 July 2018; pp. 2304–2313.

39. Blum, A.; Mitchell, T. Combining labeled and unlabeled data with co-training. In Proceedings of the Eleventh Annual Conference on Computational Learning Theory, Madison, WI, USA, 24–26 July 1998; pp. 92–100.

40. Zhang, Y.; Xiang, T.; Hospedales, T.M.; Lu, H. Deep mutual learning. In Proceedings of the IEEE Conference on Computer Vision and Pattern Recognition, Salt Lake City, UT, USA, 18–22 June 2018; pp. 4320–4328.

41. Ge, Y.; Chen, D.; Li, H. Mutual mean-teaching: Pseudo label refinery for unsupervised domain adaptation on person re-identification. In Proceedings of the International Conference on Learning Representations, Addis Ababa, Ethiopia, 26–30 April 2020.

42. Yang, F.; Li, K.; Zhong, Z.; Luo, Z.; Sun, X.; Cheng, H.; Guo, X.; Huang, F.; Ji, R.; Li, S. Asymmetric co-teaching for unsupervised cross-domain person re-identification. In Proceedings of the AAAI Conference on Artificial Intelligence, New York, NY, USA, 7–12 February 2020; Volume 34, pp. 12597–12604.

43. Liu, J.; Li, R.; Sun, C. Co-Correcting: Noise-tolerant Medical Image Classification via mutual Label Correction. *IEEE Trans. Med. Imag.* **2021**, *40*, 3580–3592 [CrossRef] [PubMed]

44. Tai, X.; Li, M.; Xiang, M.; Ren, P. A Mutual Guide Framework for Training Hyperspectral Image Classifiers with Small Data. *IEEE Trans. Geosci. Remote Sens.* **2021**, *60*, 5510417. [CrossRef]

45. Lin, T.Y.; Goyal, P.; Girshick, R.; He, K.; Dollár, P. Focal loss for dense object detection. In Proceedings of the IEEE International Conference on Computer Vision, Venice, Italy, 22–29 October 2017; pp. 2980–2988.

46. Volpi, M.; Camps-Valls, G.; Tuia, D. Spectral alignment of multi-temporal cross-sensor images with automated kernel canonical correlation analysis. *ISPRS J. Photogramm. Remote Sens.* **2015**, *107*, 50–63. [CrossRef]

*remote sensing*

MDPI

*Article*

# An Adaptive Surrogate-Assisted Endmember Extraction Framework Based on Intelligent Optimization Algorithms for Hyperspectral Remote Sensing Images

**Zhao Wang [1], Jianzhao Li [2], Yiting Liu [2], Fei Xie [3,*] and Peng Li [1]**

[1] Key Laboratory of Electronic Information Countermeasure and Simulation Technology, Ministry of Education, Xidian University, No. 2 South Taibai Road, Xi'an 710075, China; wangzhao@xidian.edu.cn (Z.W.); penglixd@xidian.edu.cn (P.L.)

[2] Key Laboratory of Intelligent Perception and Image Understanding of Ministry of Education of China, Xidian University, No. 2 South Taibai Road, Xi'an 710071, China; 19jzli@stu.xidian.edu.cn (J.L.); ytliu@stu.xidian.edu.cn (Y.L.)

[3] Academy of Advanced Interdisciplinary Research, Xidian University, No. 2 South Taibai Road, Xi'an 710068, China

* Correspondence: fxie@xidian.edu.cn

**Abstract:** As the foremost step of spectral unmixing, endmember extraction has been one of the most challenging techniques in the spectral unmixing processing due to the mixing of pixels and the complexity of hyperspectral remote sensing images. The existing geometrial-based endmember extraction algorithms have achieved the ideal results, but most of these algorithms perform poorly when they do not meet the assumption of simplex structure. Recently, many intelligent optimization algorithms have been employed to solve the problem of endmember extraction. Although they achieved the better performance than the geometrial-based algorithms in different complex scenarios, they also suffer from the time-consuming problem. In order to alleviate the above problems, balance the two key indicators of accuracy and running time, an adaptive surrogate-assisted endmember extraction (ASAEE) framework based on intelligent optimization algorithms is proposed for hyperspectral remote sensing images in this paper. In the proposed framework, the surrogate-assisted model is established to reduce the expensive time cost of the intelligent algorithms by fitting the fully constrained evaluation value with the low-cost estimated value. In more detail, three commonly used intelligent algorithms, namely genetic algorithm, particle swarm optimization algorithm and differential evolution algorithm, are specifically designed into the ASAEE framework to verify the effectiveness and robustness. In addition, an adaptive weight surrogate-assisted model selection strategy is proposed, which can automatically adjust the weights of different surrogate models according to the characteristics of different intelligent algorithms. Experimental results on three data sets (including two simulated data sets and one real data set) show the effectiveness and the excellent performance of the proposed ASAEE framework.

**Keywords:** hyperspectral remote sensing; intelligent optimization algorithms; endmember extraction; surrogate-assisted model

**Citation:** Wang, Z.; Li, J.; Liu, Y.; Xie, F.; Li, P. An Adaptive Surrogate-Assisted Endmember Extraction Framework Based on Intelligent Optimization Algorithms for Hyperspectral Remote Sensing Images. *Remote Sens.* **2022**, *14*, 892. https://doi.org/10.3390/rs14040892

Academic Editor: Junjun Jiang

Received: 10 January 2022
Accepted: 9 February 2022
Published: 13 February 2022

## 1. Introduction

Hyperspectral remote sensing image, with hundreds of continuous spectra containing rich ground object information in each pixel [1], has been used in various fields, such as terrain change detection [2], geological exploration [3] and agricultural monitoring [4]. However, limited to the low spatial resolution of the hyperspectral remote sensor, mixed pixels are inevitably appear in the hyperspectral image. The mixed pixels contain at least one ground object material, such as water, soil, and trees, etc., which interferes with the accurate analysis of the hyperspectral image to a certain extent [5,6]. Spectral unmixing, as an efficient technique to solve the problem of mixed pixels, aims to decompose the mixed pixels into a set of pure substances (also known as endmembers) and estimate the proportion of the corresponding endmembers (also called abundances) [7]. In general, endmember extraction and abundance estimation are the two main tasks of spectral unmixing. There are many mixture models in spectral unmixing, such as linear mixture model (LMM), bilinear mixture model and nonlinear mixture models [8]. However, other models have higher complexity than LMM, most endmember extraction researches are based on the LMM. In the LMM [9], it assumes that each observable pixel can be expressed as a linear combination of pure endmembers. Therefore, for a hyperspectral image consisting of $m$ endmembers, each pixel in the LMM can be written as

$$y_i = E\alpha + n_i \tag{1}$$

where $y_i = [y_1, y_2, ..., y_L]^T$ is the $i$-th mixed pixel in the hyperspectral image; $E = [e_1, e_2, ..., e_m]$ represents the set of endmembers that reconstructs the hyperspectral image $Y$, in which each endmember in $e$ contains $L$ spectral bands; $\alpha = [\alpha_1, \alpha_2, ..., \alpha_m]^T$ denotes the abundance vector of the corresponding endmember set $e$; $n_i = [n_1, n_2, ..., n_L]^T$ is the noise term for the $i$-th mixed pixel. For a hyperspectral image with $N$ observed pixels, (1) can be written as the matrix form

$$Y = EA + N \tag{2}$$

where $Y = [y_1, y_2, ..., y_N]$ is the hyperspectral image matrix, $A = [\alpha_1, \alpha_2, ..., \alpha_N]$ and $N = [n_1, n_2, ..., n_N]$ represent the abundance matrix and the noise matrix, respectively. Due to physical limitations and constraints, the abundance needs to satisfy two constraints, namely the abundance sum-to-one constraint (ASC, $\sum_{j=1}^{m} \alpha_j = 1$) and the abundance nonnegative constraint (ANC, $\alpha_j \geq 0, j = 1, 2, ..., m$) [10].

Geometrically, it is assumed that there are pure endmembers in the hyperspectral image, and all pixels can be contained in a simplex whose vertices correspond to the endmember set constituting the image [11]. Therefore, in recent decades, many geometrial-based endmember extraction methods have been proposed to obtain the vertices of the simplex, among which the classical algorithms include the pixel purity index (PPI) [12], N-FINDR [13] and the vertex component analysis (VCA) [14], etc. PPI extracts the endmember set by projecting spectral vectors into random vectors and employed the minimum noise fraction for reducing the dimension. N-FINDR selects the simplex with the largest volume, and their vertices are used as the terminal endmembers. VCA obtains the endmembers by continuously projecting the extreme values until reaching the prescribed number of endmembers. Low computational complexity and high accuracy of extraction results are the advantages of these algorithms, but there are also some unavoidable shortcomings. For example, when the data do not meet the simple structure, the extraction accuracy of geometrial-based method will be significantly reduced, and it is also vulnerable to noise and outliers [15,16].

In order to alleviate the above problems, some intelligent optimization algorithms have been applied in endmember extraction in recent years. In the literature, the intelligent-based endmember extraction algorithms can be roughly divided into three main categories, which are the based on the genetic algorithms (GA) [17–20], the particle swarm optimization (PSO) algorithms [21–27] and the differential evolution (DE) algorithms [28,29]. Zhang et al. [21] employed the discrete particle swarm optimization (DPSO) to minimize the root mean square error (RMSE) between the reconstructed image and the original image to obtain the appropriate endmember set by encoding each particle as the potential position of the active endmember in the hyperspectral image. In [19], the combination of genetic algorithm and the orthogonal projection, called genetic orthogonal projection (GOP), was proposed to solve the problem of endmember extraction. To overcome the problem of poor performance and low efficiency, Zhong et al. [28] proposed an adaptive differential evolution (ADEE) algorithm, which explore the endmember set with the adaptive crossover and mutation strategies to avoid manual setting of parameters. Liu et al. [23] explored a novel quantum-behaved particle swarm optimization (QPSO) with the row-column coding to overcome the dimension disaster of the standard PSO algorithm, and a cooperative approach is designed to expand the whole particle swarm search space. In [26], an improved QPSO (IQPSO) was designed to enhance the precision of extracted endmembers. Although the above intelligent-based endmember extraction algorithms are more effective and robust than the traditional algorithms, they suffer from a serious problem of time-consuming. It is difficult to satisfy the rapidity of endmember extraction in practical applications. In addition, if the time-consuming is reduced by decreasing the number of algorithm cycles, the intelligent based algorithms are easy to fall into the local optimum.

As mentioned earlier, the abundance is subject to the ASC and the ANC, and four abundance estimation strategies including the fully constrained least squares (FCLS), the sum-to-one constrained least squares (SCLS), the nonnegative constrained least squares (NCLS) and the unconstrained least squares (UCLS) can be employed to solve this problem. However, mathematically, it takes much less time to solve the abundance inversion problem with the other three methods than FCLS. Therefore, some intelligent-based endmember extraction algorithms [25,30] directly employed the UCLS for the abundance inversion in order to reduce the computational cost. Nevertheless, the extraction accuracy of the SCLS, NCLS and UCLS is not convincing compared with the FCLS. Therefore, it is urgent to reduce the computational cost for the intelligent-based endmember extraction algorithms. The surrogate-assisted evolutionary algorithms (SAEAs), have been widely adopted as one of the most effective methods to solve the expensive optimization problems [31–33]. As the name suggests, it aims to establish a surrogate model to approximate the expensive objective evaluation function to significantly reduce the computational cost. At present, many efficient surrogate models such as polynomial regression (PR) [34], support vector machines (SVM) [35–37], radial basis function networks (RBF) [38,39] and Gaussian processes (GP) [40,41] have been studied and developed. Taking this cue, it is a natural idea to estimate the expensive evaluation value of FCLS from the cheap estimates with the SAEAs. In summary, an efficient adaptive surrogate-assisted intelligent algorithms (ASAEE) framework for endmember extraction is proposed in this paper to overcome the costly time problem. The major contributions of this paper are threefold:

(1) This paper solves the endmember extraction problem with the proposed ASAEE framework. The overall convergence characteristics and the time-consuming issue can be significantly improved by the proposed framework.

(2) Three algorithms of ASAEE-GA, ASAEE-PSO and ASAEE-DE based on the ASAEE framework are specifically designed. The experimental results of these three algorithms have been greatly improved compared with the corresponding state-of-the-art intelligent-based endmember extraction algorithms.

(3) An adaptive weight surrogate-assisted model selection algorithm is designed, which is able to automatically adjust the weights of different surrogate-assisted models according to the characteristics of different intelligent optimization algorithms.

(4) We also transfer the ASAEE framework to other intelligent-based endmember extraction algorithms, which greatly reduces the expensive time cost while maintaining the accuracy.

The remainder of this paper is structured as follows. Section 2 briefly reviews the research related to the employing of intelligent algorithms to solve the endmember extraction problem. In Section 3, the proposed ASAEE framework and its combination with three intelligent optimization algorithms, namely ASAEE-GA, ASAEE-PSO, and ASAEE-DE are described in detail. Section 4 reports the experimental results of the proposed method compared with several state-of-the-art endmember extraction algorithms. Conclusions are drawn in Section 5.

## 2. Related Work

In this section, we first review the intelligent-based optimization algorithms for endmember extraction. Relevant researches on the surrogate-assisted models are also briefly introduced.

### 2.1. Intelligent-Based Endmember Extraction Algorithms

The intelligent-based endmember extraction methods can effectively compensate for the shortcomings of geometrial-based methods in terms of reduced accuracy when the simplex condition is not satisfied. Furthermore, they can obtain the better accuracy for endmembers with undesirable distribution in the search space, which means that the intelligent-based algorithms have higher robustness and less dependence on data. Most of the intelligent-based methods transform the endmember extraction task of hyperspectral images into a combinatorial optimization problem and solve it by intelligent optimization algorithms, such as GA, PSO and DE.

The optimization objective function is important for the final results of endmember extraction. Most of the endmember extraction methods focus on the RMSE value, and the result will have a smaller RMSE when it is closer to the ground truth. Zhang et al. [21] proposed the DPSO which represents the combination of endmembers in hyperspectral image with binary encoding for particle positions and velocities, and searches for the optimal result in the discrete feasible space with the classical discrete particle swarm optimization. In subsequent studies, QPSO [23] is designed with the quantum-behaved strategy to strengthen the robustness and the convergence rate and in the multi-dimensional search space. IQPSO [26] improved the QPSO in the global search capability and the high-dimensional difficulty. In [19], Rezaei et al. employed a GA to determinate the exact number and position of each endmember obtained by the projecting the data in an orthogonal subspace. In [28], Zhong et al. designed an adaptive differential evolution strategy to the classical DE algorithm, which solves the drawback that traditional differential evolution method requires multiple runs to find the appropriate parameters for different practical problems.

The other commonly used optimization objective is to maximize the volume. As the volume of the convex simplex with the endmembers as the vertices, the larger volume is obtained when the result is closer to the real endmember set, which is defined as follows

$$Volume(E) = \frac{\left| det \begin{bmatrix} 1 & 1 & \dots & 1 \\ e_1 & e_2 & \dots & e_m \end{bmatrix} \right|}{(m-1)!} \tag{3}$$

where $m$ is the number of endmembers. A novel mutation operator accelerated quantum-behaved particle swarm optimization (MOAQPSO) [24] proposed by Xu et al. is one of the methods based on maximization of volume. Different from the DPSO, there is no velocity vector in MOAQPSO, which explores the best combination of endmembers by the position of the particles and employs the mutation rate to avoid falling into local optimum.

Some researches also employed other intelligent-based optimization algorithms for endmember extraction, such as the ant colony optimization algorithms [15,42], the bee colony optimization algorithms [43], and the discrete firefly algorithms [44], etc. In addition, refs. [18,22,25–27,29] have turned to multiobjective optimization algorithms to optimize two indicators simultaneously, namely minimizing the RMSE and maximizing the volume.

### 2.2. Brief Introduction of the Surrogate-Assisted Models

The purpose of establishing the surrogate-assisted models is to reduce the expensive evaluation cost of the intelligent optimization algorithms by employing a small amount of expensive real evaluation to construct and update the surrogate-assisted model, which is also known as data-driven optimization. Most of regression or classification techniques, such as PR, RBF, SVM, GP, etc., can be employed as the surrogate-assisted models.

The data-driven surrogate-assisted evolutionary optimization is mainly divided into two major research directions, namely offline and online data-driven optimization. The modeling of the surrogate-assisted model can only rely on offline data in the offline data-driven evolutionary optimization process. While the online data-driven evolutionary optimization is to select the appropriate data in the evolutionary search process to improve the fitting quality of the surrogate-assisted model. Therefore, in the online surrogate management methods, many model management strategies such as population-based, individual-based and generation-based are widely studied. The population-based model management employed multiple populations for evaluation, and each population is evaluated with different fidelity. As its name suggests, the individual-based model management aims to construct and update the surrogate model according to the specific individuals, pre-selection method, clustering based method, uncertainty based method, random strategy and best strategy are often employed in the selection of individuals. The generation-based model management, a relatively simple example, is to use some data to construct the surrogate-assisted model before optimization, and be updated in the number of iterations with the appropriate data to improve the surrogate-assisted model.

In the field of remote sensing, the surrogate-assisted models have also been explored in many research directions in accelerating the convergence and improving the efficiency, such as endmember selection [20], hyperspectral image classification [45] and hyperspectral nonlinear substitution [46], etc.

## 3. Proposed Method

In this section, we first describe the motivation for designing the surrogate-assisted model with the intelligent optimization algorithms to solve the endmember extraction problem. Then initialization mechanism and the objective optimization function will be introduced. Subsequently, the ASAEE framework will be described in detail, including the construction and updating of the surrogate model. Finally, the ASAEE framework with the evolution strategies of GA, PSO and DE are proposed.

### 3.1. Motivation

After determining the optimal endmember set with the intelligent-based optimization algorithms, the extracted endmembers must meet two constraints, i.e., all abundances cannot be negative, and the sum of all abundances is one. In general, the FCLS can be employed to accurately estimate the abundance of inversion. However, mathematically, due to the time spent by FCLS is very expensive, some researches have turned to the NCLS, UCLS and other low-time-cost abundance inversion methods instead. However, it will lead to inaccurate situations where the abundance from the inverse is negative or the sum is not one. Therefore, it is very promising to employ the surrogate model to replace the true FCLS abundance inversion value and reduce the expensive time cost of the intelligent-based endmember extraction algorithms.

### 3.2. Initialization and Objective Optimization Function

Considering that the evolution strategies of different intelligent algorithms have their own advantages and disadvantages, it is very important for the ASAEE framework to design a unified coding and initialization which is suitable for most intelligent algorithms. The encoding of individual or particle is shown in Figure 1. Specifically, the length of the encoded vector is $N$, and the elements in the vector are all binary encoded. Among them, the element of 0 means that the corresponding pixel in the hyperspectral image is not a candidate endmember. On the contrary, if the element is 1, it means that the pixel is the endmember to be extracted. In summary, the number of elements in the vector is equal to the pixels in the hyperspectral image, and the sparsity of the vector (i.e., the number of 1 elements) is equal to the number of endmembers to be extracted.

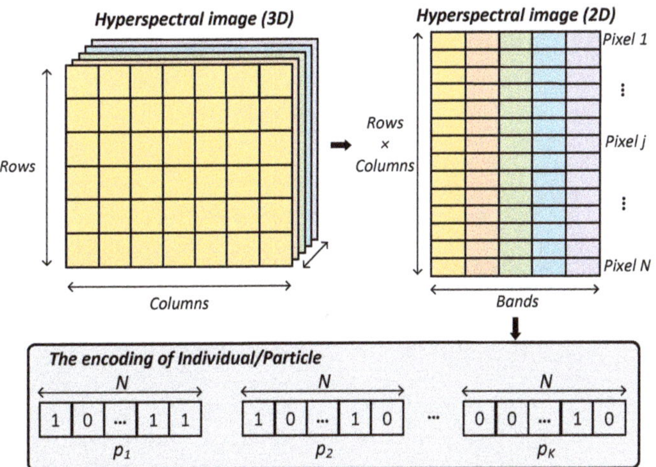


**Figure 1.** The initialization mechanism of the individuals or particles.

In this paper, we mainly focus on the single objective optimization. For the evaluation of individual or particle, as an important measurement index for endmember extraction, RMSE is regarded as the optimization objective, which can be expressed as

$$f_{ASAEE} = RMSE(Y, \hat{Y}) = \frac{1}{N} \sum_{i=1}^{N} \sqrt{\frac{1}{L} \|y_i - \hat{y}_i\|_2^2} \tag{4}$$

where $y_i$ and $\hat{y}_i$ are the pixels in the original image $Y$ and the reconstructed image $\hat{Y}$, respectively. $N$ and $L$ represent the number of pixels and the number of spectral bands, respectively.

### 3.3. ASAEE Framework

The pseudo code of the overall framework is shown in Algorithm 1. The entire ASAEE framework can be divided into three parts, namely the construction of the surrogate-assisted model, evolution strategies and the updating of the surrogate-assisted model. The first step is to generate the initial population and construct the adaptive surrogate-assisted model. Because different surrogate-assisted models have their own characteristics, it is difficult to determine only one surrogate-assisted model to optimize for a specific optimization problem. Therefore, a stable and efficient surrogate-assisted model with adaptive weights is designed to be constructed, which can be expressed as

$$S_{aw} = \sum_{t=1}^{s} w_t S_t \tag{5}$$

where $S_t$ is the $t$-th of all $s$ surrogate-assisted models. Four classical surrogate-assisted models including the PR, SVM, RBF and GP are employed in this paper. $w_t$ is the weight corresponding to the $t$-th surrogate-assisted model, which is defined as

$$w_t = \frac{\sum_{q=1}^{s} r_q - r_t}{2 \sum_{q=1}^{s} r_q} \tag{6}$$

where $r_t$ is defined as: $\sum_{i=1}^{R} \sqrt{\frac{1}{R} \|f_i - \hat{f}_i\|^2}$, $R$ is the number of samples used to construct or update the surrogate model, $f_i$ and $\hat{f}_i$ are the value evaluated by FCLS and the surrogate-assisted model, respectively.

---

**Algorithm 1** The ASAEE Framework

---

**Input:** $Y$: the original hyperspectral image, *Maxgen*: the max generation number, $K$: the population size.

**Output:** $\hat{E}$: the endmember set for reconstructing the remixed image.

1: Generate the initial population $P_0$.
2: %Construct the surrogate-assisted model
3: $D_0 \leftarrow$ Randomly select $K/10$ individuals and Evaluate them with FCLS.
4: Construct the surrogate-assisted model by: $S_{aw} = \sum_{t=1}^{s} w_t S_t$.
5: Evaluate each $p_i$ in $P_0$ with the surrogate model $S_{aw}$.
6: *gen* $\leftarrow 1$.
7: **while** *gen* $<$ *Maxgen* **do**
8:     Evolve the $P_{gen}$ with an intelligent optimization algorithm.
9:     Evaluate each $p_i$ in $P_{gen}$ with the surrogate-assisted model $S_{aw}$.
10:    %Update the surrogate-assisted model
11:    $P'_{gen} \leftarrow$ the reinitialized population with $K/10$ individuals.
12:    $(x^*_{gen}, y^*_{gen}) \leftarrow$ Obtain the individual with the best fitness in $P_{gen}$.
13:    $(x^u_{gen}, y^u_{gen}) \leftarrow$ Obtain the individual with the most uncertain in $P_{gen}$.
14:    $(x^r_{gen}, y^r_{gen}) \leftarrow$ Obtain the individual by randomly selecting in $P'_{gen}$.
15:    Calculate the fitness of $(x^*_{gen}, y^*_{gen})$, $(x^u_{gen}, y^u_{gen})$ and $(x^r_{gen}, y^r_{gen})$ with FCLS.
16:    $D_{gen} \leftarrow D_{gen} \cup (x^*_{gen}, y^*_{gen}) \cup (x^u_{gen}, y^u_{gen}) \cup (x^r_{gen}, y^r_{gen})$.
17:    Update the surrogate-assisted model $S_{aw}$ with $D_{gen}$.
18:    *gen* = *gen* + 1.
19: **end while**

---

The construction of the surrogate-assisted model with adaptive weights is shown in Figure 2. It is difficult to determine which surrogate-assisted model is more suitable for expensive evaluation problems. Therefore, it makes more sense to assign the corresponding weights according to the errors of different surrogate-assisted models.

After constructing the surrogate-assisted model, since the samples selected at the beginning are not enough for the predicted value of the surrogate-assisted model to simulate the whole real abundance inversion value, we design the online data-driven model management strategy to update the surrogate-assisted model. In each generation of evolution, the optimal fitness sample $(x^*_{gen}, y^*_{gen})$, the most uncertain sample $(x^u_{gen}, y^u_{gen})$ and a random sample $(x^r_{gen}, y^r_{gen})$ are selected to update the surrogate-assisted model to ensure accurate and efficient approximation of the expensive evaluation function. It should be noted that the $(x^*_{gen}, y^*_{gen})$ is obtained by selecting the individual with the best fitness value evaluated by the surrogate-assisted model for all the individuals in $P_{gen}$. The $(x^u_{gen}, y^u_{gen})$ is obtained by calculating the maximum neighborhood distance of all individuals evaluation values in $P_{gen}$. The $(x^r_{gen}, y^r_{gen})$ is randomly selected from an initial population $P'_{gen}$ with $K/10$ individuals. With the above designs, the overall computational complexity of the ASAEE framework is almost reduced by $Kgen\mathcal{O}_{FCLS}$ times compared with the traditional intelligent-based algorithms.

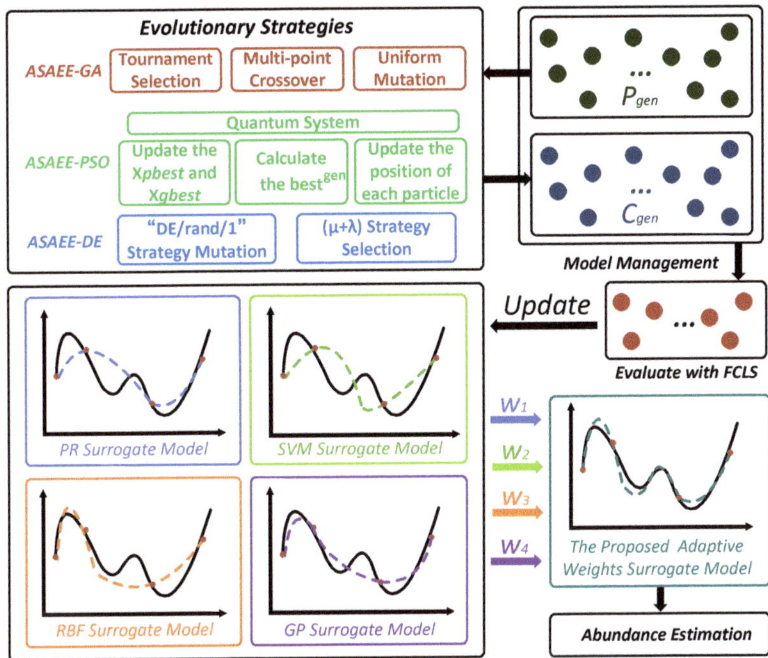


**Figure 2.** Illustration of the proposed ASAEE Framework.

### 3.4. Evolution Strategies

In this section, in order to verify the generality and robustness of our proposed ASAEE framework, three intelligent algorithms, including the GA, PSO and DE, are employed to design with the ASAEE framework for evolution.

#### 3.4.1. ASAEE-GA

For the ASAEE-GA, population $P_{gen}$ will first form the parent population $P'_{gen}$ with the tournament selection, which is used to generate the offspring population $C_{gen}$. Then each two individuals $p_a$ and $p_b$ are selected from $P'_{gen}$ for the crossover and mutation, and their corresponding offspring are generated as: $c_i = Mutation(Crossover(p_a, p_b))$, where the $Crossover()$ and $Mutation()$ are the multi-point crossover operator and the uniform mutation operator, respectively. The crossover operation randomly selects multiple points on the parent chromosome, and then exchanges part of genes in $p_a$ and $p_b$ to obtain two new individuals. In addition, uniform mutation randomly generates a number within the interval $[U_{max}, U_{min}]$ to replace the original variable at the mutated genes. $U_{max}$ and $U_{min}$ are respectively the upper and lower bounds of the decision variable. The offspring population $C_{gen}$ is generated through repeating this step until there is no unselected individual in the population $P'_{gen}$. Then $K$ individuals with the best fitness are selected from $P_{gen} \cup C_{gen}$ to form the next parent population $P_{gen+1}$. As the current generation reaches the maximum number of generations $Maxgen$, the individual with the best fitness is selected as the reconstructed endmember set $\hat{E}$.

#### 3.4.2. ASAEE-PSO

In the ASAEE-PSO, the concept of basic PSO is applied to quantum space, and the wave functions is used to describe the motion states of the particles, which enables the motion of particles in feasible solution space to exhibit global randomness. Instead of the traditional PSO that uses position and velocity to represent the particle state, quantum-behaved PSO (QPSO) evolves only with the particle positions, which also has the advantage of fewer parameters. For the evolution of the ASAEE-PSO, suppose that the $j$-th decision variable of the $i$-th particle at generation *gen* is expressed as $x_{i,j}^{gen}$. In addition, $x_{ibest}^{gen}$ and $x_{gbest}^{gen}$ represent the self-optimum position of the $i$-th particle and the global optimal position of all particles, respectively. The self-optimum position and the global optimal position can be obtained according to the objective function (4), and their update rules are shown as follows

$$x_{ibest}^{gen+1} = \begin{cases} x_i^{gen+1}, & \text{if } f_{ASAEE}(x_i^{gen+1}) < f_{ASAEE}(x_{ibest}^{gen}) \\ x_{ibest}^{gen}, & \text{if } f_{ASAEE}(x_i^{gen+1}) \geq f_{ASAEE}(x_{ibest}^{gen}) \end{cases}$$

$$x_{gbest}^{gen+1} = argmin\{f_{ASAEE}(x_{ibest}^{gen+1}\}, \quad i = 1, 2, ..., K. \tag{7}$$

In a quantum system, particles are attracted by quantum delta potential wells centered on local attractor points. The update of the $x_{i,j}^{gen+1}$ is obtained by solving the probabilistic model of the particle position and then transforming it from the quantum state to the classical state by Monte Carlo Simulation method, which is expressed as follows

$$x_{i,j}^{gen+1} = o_{i,j}^{gen} \pm *\beta * \left| best_j^{gen} - x_{i,j}^{gen} \right| * ln\left[ \frac{1}{u_{i,j}^{gen}} \right]$$

$$u_{i,j}^{gen} = rand(0,1), i = 1, 2, ..., K, \quad j = 1, 2, ..., m \tag{8}$$

$$best_j^{gen} = \frac{1}{K} \sum_{i=1}^{K} x_{ibest,j}^{gen}, j = 1, 2, ..., m,$$

where $\beta$ is the contraction-expansion coefficient, which controls the convergence speed of the algorithm and its value will change linearly with *gen* from 1.0 to 0.5 according to [47]. $best_j$ is the average position of the $j$-th decision variable in the self-optimum particles of all individuals, and $u_{i,j}$ is a random number in the interval [0, 1]. In addition, $o_{i,j}$ is the position of the local attractor, which can be expressed as

$$o_{i,j} = \varphi_{i,j} * x_{ibest,j} + (1 - \varphi_{i,j}) * x_{gbest,j}$$

$$\varphi_{i,j} = rand(0,1), i = 1, 2, ..., K, \quad j = 1, 2, ..., m, \tag{9}$$

where $\varphi_{i,j}$ is also a random number in the interval [0, 1]. After the number of generations reaches the termination condition, the particle of $x_{gbest}^{gen}$ is taken as the optimal endmember set.

### 3.4.3. ASAEE-DE

Different from ASAEE-GA, the mutation in ASAEE-DE adds the difference of the selection vectors to the basis vector to realize the change of the decision variable. The mutation operator adopts the "DE/rand/1" strategy, which can be expressed as follows

$$v_i^{gen} = x_{r_1}^{gen} + F(x_{r_2}^{gen} - x_{r_3}^{gen}), \tag{10}$$

where $v_i$ is the $i$-th vector generated by the individuals $(x_{r_1}, x_{r_2}, x_{r_3})$ through the mutation operator. $F$ is the mutation scaling factor, and $r_1, r_2$ and $r_3$ are three mutex integers randomly selected from the range $[1, K]$. In addition, the binomial crossover is employed after the mutation operation to generate offspring $c$, which is shown as follows

$$c_{i,j}^{gen} = \begin{cases} v_{i,j}^{gen} & \text{if } j = j_{rand} \text{ or } rand_{i,j} \leq Cr \\ x_{i,j}^{gen} & \text{otherwise,} \end{cases} \tag{11}$$

where $j_{rand}$ is an integer randomly selected from the range $[1, m]$ and $rand_{i,j}$ is a random number within the interval [0, 1]. The crossover rate $Cr$ and the mutation scaling factor $F$

will be adaptively updated according to [28]. A ($\mu + \lambda$) strategy [48,49] is applied in the selection stage, which combines $\mu$ parents and $\lambda$ generated offsprings to obtain a population with ($\mu + \lambda$) individuals, and the best $\mu$ individuals are selected to enter the next generation. In ASAEE-DE, the number of parents $\mu$ and the number of offspring $\lambda$ are the same as the size of the population $K$. Finally, the individual with the best fitness in the max generation *Maxgen* is regarded as the final endmember set.

## 4. Experimental Results

In order to verify the effectiveness of the proposed ASAEE framework, a series of experiments are designed and performed on three benchmark data sets, including two simulated data sets and one real data set. In the following, these three data sets are briefly introduced first. Then the ablation experiments are analyzed to prove the rationality of the ASAEE framework design. Then the ablation experiments are analyzed to prove the rationality of the ASAEE framework. The proposed method is compared with other endmember extraction algorithms on different data sets. In addition, three algorithms of the proposed framework are compared with the state-of-the-art peer competitors. Finally, the generality of the proposed framework is reflected in the transfer to some classical intelligent-based endmember extraction algorithms.

### 4.1. Data Sets Description

In these experiments, three widely used endmember extraction benchmark data sets are employed to examine the performance of the proposed ASAEE framework. The first data set (DS1) and the second data set (DS2) are two simulated hyperspectral image from the USGS spectral library [50], which are displayed separately in Figure 3a,b. Five endmembers including the Alunite, Buddingtonite, Calcite, Kaolinite and Muscovite synthesize the DS1 with $80 \times 100$ pixels. On the basis of these five endmembers, five more endmembers (Illite, Jarosite, Nontronite, Halloysite, and Pyrophyllite) constitute a total of ten endmembers to simulate DS2 with $160 \times 160$ pixels. The spectra of DS1 and DS2 are shown in Figure 3c,d, respectively. The third data set (DS3) is a widely used real hyperspectral image [19] (the AVIRIS Cuprite image) with $400 \times 350$ pixels, including 50 spectral bands, which is shown in Figure 4.

In addition, two important indicators are employed to measure the performance of the algorithms, which are the RMSE and the running time. The RMSE is an index to measure the difference between the reconstructed image and the original image, and the running time is an important manifestation of the efficiency of the algorithms. Moreover, for the simulated data sets, experiments will be performed on three different levels of signal-to-noise ratio (SNR), namely, 20, 30 and 40 dB. The endmembers in the simulated data sets are known in advance, so the number of endmembers in DS1 and DS2 is set to 5 and 10, respectively. In the real data set, since the number of endmembers cannot be obatined as the priori knowledge, the number of endmembers is set to 5, 10, 15, 20 respectively as recommended in reference [21,28]. In all the intelligent-based algorithms, the number of individuals or particles is set to 20 and the number of iterations is set to 200. Besides, all the experimental results take the average of 10 independent experiments as the final presentation. All the experiments are implemented on the Matlab 2021 platform using Intel i5-10400 CPU@2.90GHz.

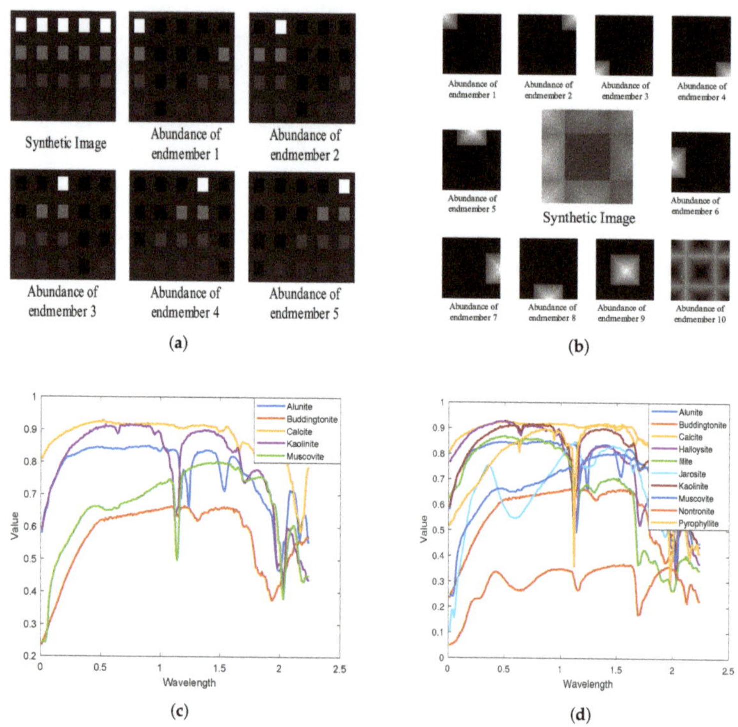

**Figure 3.** Simulated data sets. (**a**) The image and the abundance of 5 endmembers of DS1. (**b**) The image and the abundance of 10 endmembers of DS2. (**c**) Five spectra in DS1. (**d**) Ten spectra in DS2.

**Figure 4.** The AVIRIS Cuprite image.

### 4.2. Experiments on the Proposed ASAEE Framework

In this section, two ablation experiments are performed to prove the rationality and validity of constructing the adaptive surrogate-assisted model. First, we would investigate whether the construction of surrogate-assisted model will affect the endmember extraction performance. Second, we would also explore the performance between the proposed adaptive surrogate-assisted model and other single surrogate-assisted models.

The experimental results on two simulated data sets of with and without surrogate-assisted model are shown in Figure 5. In the results, EE-GA, EE-PSO, and EE-DE represent three classic intelligent-based endmember extraction algorithms, and the FCLS are all employed for their individual evaluations. On the contrary, ASAEE-GA, ASAEE-PSO and ASAEE-PSO are combined with the proposed ASAEE framework. It can be observed from the experimental results that the endmember extraction algorithms based on the ASAEE framework can significantly reduce the expensive time cost compared with ordinary intelligent-based endmember extraction algorithms. In addition, the comparison results on the real data set with and without the ASAEE framework are illustrated in Figure 6. The above experimental results clearly shows that the proposed ASAEE framework improves the expensive cost of previous intelligent-based endmember extracion algorithms on all the benchmarks. To be specific, the ASAEE framework reduces the time in simulated data sets and the real data set by almost thirty times and two thousand times respectively compared with the original algorithms, which coincides with the analysis of the algorithm complexity in Section 3.3.

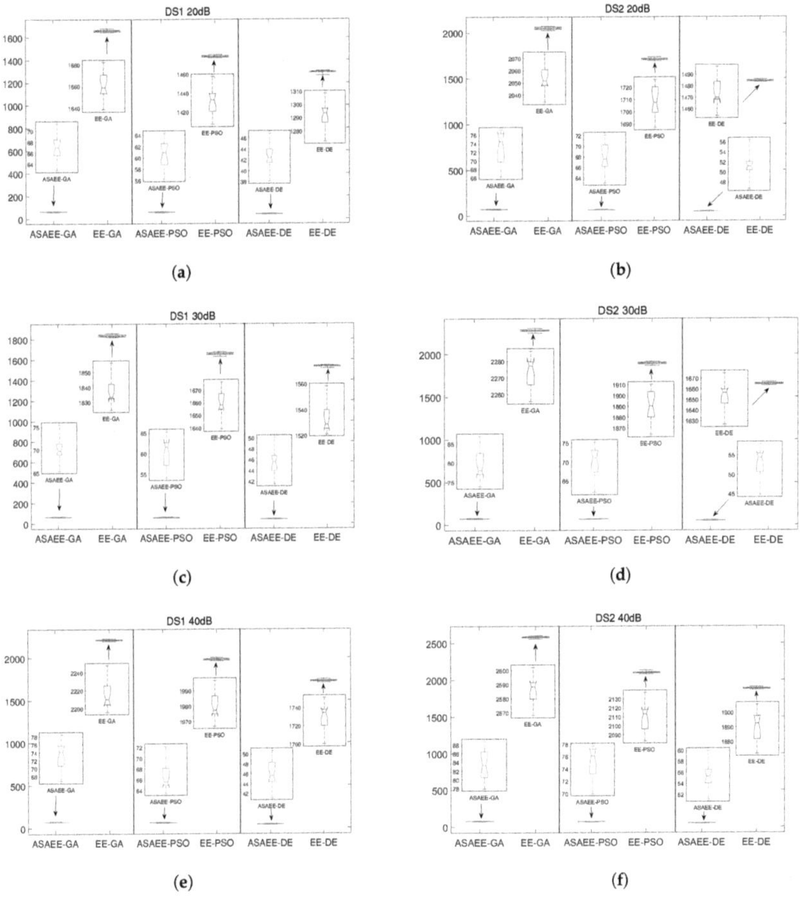

**Figure 5.** Comparison with and without ASAEE framework on DS1 and DS2 under different SNR. (**a**) DS1 20 dB. (**b**) DS2 20 dB. (**c**) DS1 30 dB. (**d**) DS2 30 dB. (**e**) DS1 40 dB. (**f**) DS2 40 dB.

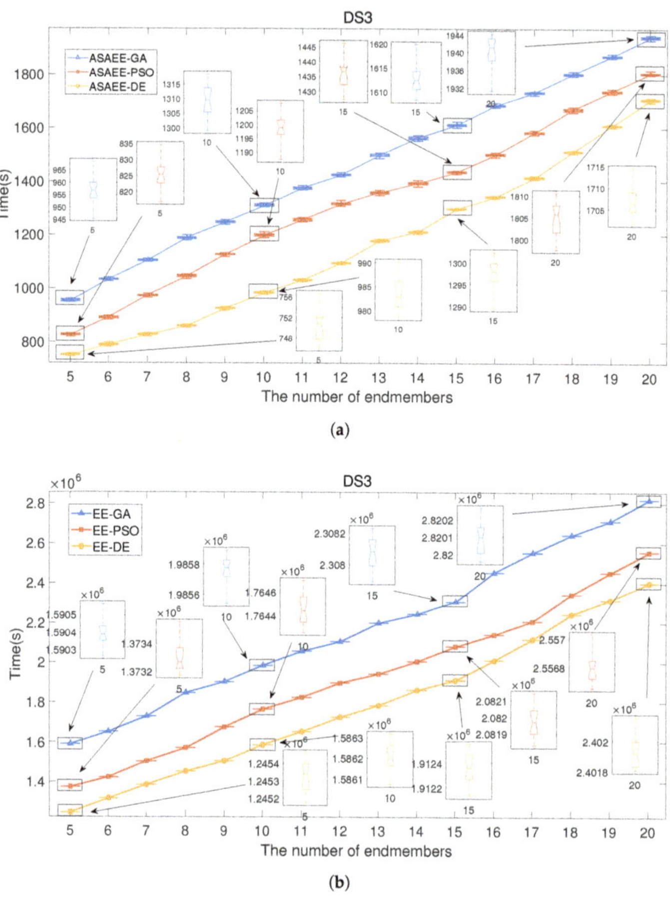

**Figure 6.** Comparison with and without ASAEE framework on DS3. (**a**) Methods with ASAEE framework. (**b**) Methods without ASAEE framework.

Besides, Figure 7 illustrates the experimental results of different surrogate-assisted models and the proposed adaptive surrogate-assisted model. In other word, four representative surrogate models constructed with PR, SVM, RBF and GP are compared with the proposed model. It can be concluded that the performance of different surrogate models has great differences for each hyperspectral data sets, but the proposed adaptive surrogate model strategy can assign different weights to each surrogate model for obtaining a compromise between these surrogate models, which is capable of better approaching the real evaluation results. In summary, the design of adaptive surrogate-assisted model can not only reduce the expensive time cost for the intelligent algorithms, but also allocate the appropriate weights to select the most suitable surrogate models according to their corresponding fitting.

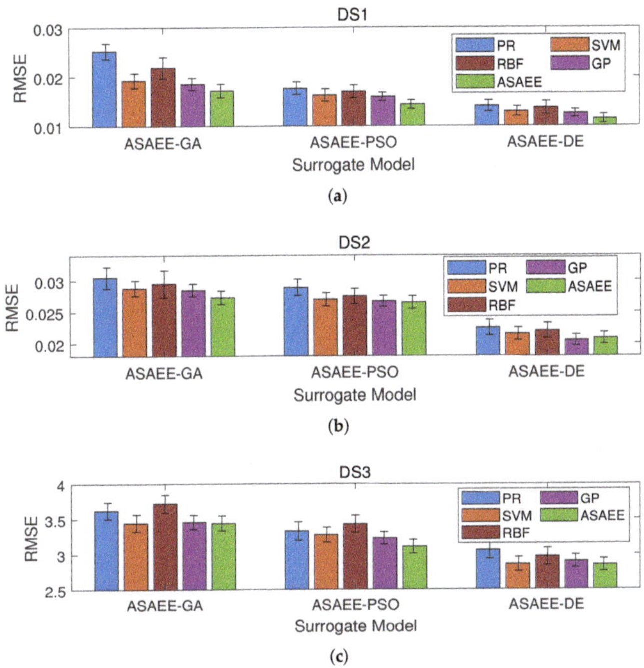

**Figure 7.** Comparison of the proposed adaptive surrogate-assisted model with other surrogate-assisted models in different data sets. (**a**) DS1. (**b**) DS2. (**c**) DS3.

### 4.3. Comparison of the Proposed ASAEE with Other Methods

In this section, three algorithms based on the ASAEE framework are compared with other endmember extraction algorithms. The comparison algorithms employed in this paper include PPI [51], N-FINDR [13], VCA [14], GOP [19], DPSO [21], ADEE [28], QPSO [23] and IQPSO [26]. Specifically, the PPI, N-FINDR and VCA are three classical geometrial-based approaches, and the GOP, DPSO, ADEE, QPSO and IQPSO represent the endmember extraction algorithms based on GA, PSO and DE in the intelligent algorithms. For the sake of fairness, the parameter settings of these intelligent-based algorithms are consistent with their respective papers. In addition, in the experimental results presented in the table, the best data is shown in bold, and the second best data is shown in bold and underlined.

Tables 1–3 present the indicators' values obtained by three ASAEE-based algorithms and the comparison algorithms in identifying the real endmembers on two simulated data sets under different SNR and four endmember situations in the real data set, respectively. In general, in the simulated data sets, the performance of all algorithms are improved as the SNR increases, and the time spent are also increase except for the geometrial-based algorithms. In terms of the indicator of time, although the traditional geometrial-based algorithms have achieved the excellent results, their performance in accurate extraction are not ideal reflected by the RMSE indicator. On the contrary, the intelligent-based algorithms are generally better than the traditional methods in terms of endmember extraction accuracy, but the time of these algorithms is also very expensive. However, the endmember extraction algorithms with the ASAEE framework has achieved a good compromise between these two indicators. Specifically, the ASAEE-DE has the best endmember extraction performance in two simulated data sets, while ASAEE-GA has the excellent performance in the real data sets. In terms of time index, ASAEE-DE takes the shortest time due to the simplicity of its evolutionary steps, followed by ASAEE-PSO and ASAEE-GA takes the longest.

**Table 1.** Comparison of the proposed ASAEE framework with other algorithms on DS1.

| Attributes | SNR | 20 | | 30 | | 40 | |
|---|---|---|---|---|---|---|---|
| | Methods | RMSE | Time (s) | RMSE | Time (s) | RMSE | Time (s) |
| Geometrial-based | PPI | 0.6062 | 3.100 | 0.6069 | 3.365 | 0.6055 | 3.599 |
| | N-FINDR | 0.0823 | 1.436 | 0.0263 | 1.522 | 0.0183 | 1.626 |
| | VCA | 0.0735 | **0.910** | 0.0232 | **0.936** | 0.0173 | **0.980** |
| Intelligent-based | GOP | 0.0784 | 1558.228 | 0.0224 | 1835.942 | 0.0109 | 2212.031 |
| | DPSO | 0.0811 | 1429.519 | 0.0196 | 1701.182 | 0.0115 | 2064.372 |
| | ADEE | 0.0809 | 1291.413 | 0.0171 | 1534.217 | 0.0098 | 1727.190 |
| | QPSO | 0.0739 | 1357.904 | 0.0157 | 1660.213 | 0.0091 | 1882.512 |
| | IQPSO | 0.0717 | 1332.013 | 0.0138 | 1653.510 | 0.0072 | 1861.607 |
| ASAEE-based | ASAEE-GA | 0.0731 | 66.706 | 0.0171 | 70.272 | 0.0095 | 74.264 |
| | ASAEE-PSO | 0.0722 | 59.958 | 0.0143 | 62.391 | 0.0080 | 66.220 |
| | ASAEE-DE | **0.0697** | <u>43.331</u> | 0.0114 | <u>45.447</u> | 0.0061 | <u>47.059</u> |

**Table 2.** Comparison of the proposed ASAEE framework with other algorithms on DS2.

| Attributes | SNR | 20 | | 30 | | 40 | |
|---|---|---|---|---|---|---|---|
| | Methods | RMSE | Time (s) | RMSE | Time (s) | RMSE | Time (s) |
| Geometrial-based | PPI | 0.5132 | 4.522 | 0.5063 | 4.642 | 0.5051 | 4.723 |
| | N-FINDR | 0.0805 | 1.995 | 0.0336 | 2.061 | 0.0218 | 2.102 |
| | VCA | 0.0711 | **1.236** | 0.0306 | **1.309** | 0.0189 | **1.381** |
| Intelligent-based | GOP | 0.0780 | 2050.407 | 0.0295 | 2273.227 | 0.0113 | 2587.485 |
| | DPSO | 0.0802 | 1813.623 | 0.0305 | 2099.171 | 0.0136 | 2392.728 |
| | ADEE | 0.0759 | 1472.874 | 0.0273 | 1651.492 | 0.0101 | 1884.253 |
| | QPSO | 0.0724 | 1668.131 | 0.0262 | 1891.692 | 0.0098 | 2105.269 |
| | IQPSO | 0.0679 | 1613.092 | 0.0240 | 1810.125 | 0.0082 | 2080.572 |
| ASAEE-based | ASAEE-GA | 0.0702 | 74.408 | 0.0273 | 78.559 | 0.0094 | 83.952 |
| | ASAEE-PSO | 0.0684 | 67.945 | 0.0265 | 71.623 | 0.0089 | 75.798 |
| | ASAEE-DE | **0.0658** | <u>54.151</u> | 0.0208 | <u>58.847</u> | 0.0075 | <u>62.801</u> |

**Table 3.** Comparison of the proposed ASAEE framework with other algorithms on DS3.

| Attributes | Endmember | 5 | | 10 | | 15 | | 20 | |
|---|---|---|---|---|---|---|---|---|---|
| | Methods | RMSE | Time (s) | RMSE | Time (s) | RMSE | Time (s) | RMSE | Time (s) |
| Geometrial-based | PPI | 20.7768 | 30.774 | 18.3991 | 42.293 | 16.8536 | 57.495 | 14.3208 | 65.473 |
| | N-FINDR | 5.8611 | 26.633 | 4.0298 | 34.205 | 3.8376 | 48.465 | 3.2275 | 59.217 |
| | VCA | 5.5463 | **25.495** | 3.8370 | **32.197** | 3.5101 | **43.151** | 2.9383 | **57.542** |
| Intelligent-based | GOP | 5.2643 | $1.590 \times 10^6$ | 3.8251 | $1.985 \times 10^6$ | 3.5212 | $2.308 \times 10^6$ | 2.9180 | $2.820 \times 10^6$ |
| | DPSO | 4.5321 | $1.373 \times 10^6$ | 3.3797 | $1.764 \times 10^6$ | 3.0944 | $2.081 \times 10^6$ | 2.7488 | $2.556 \times 10^6$ |
| | ADEE | 4.2970 | $1.24 \times 10^6$ | 3.3102 | $1.586 \times 10^6$ | 3.0206 | $1.912 \times 10^6$ | 2.6831 | $2.401 \times 10^6$ |
| | QPSO | 4.1542 | $1.270 \times 10^6$ | 3.1326 | $1.600 \times 10^6$ | 2.9437 | $2.005 \times 10^6$ | 2.6704 | $2.493 \times 10^6$ |
| | IQPSO | 4.0720 | $1.258 \times 10^6$ | 3.0327 | $1.581 \times 10^6$ | 2.7794 | $1.990 \times 10^6$ | 2.5925 | $2.451 \times 10^6$ |
| ASAEE-based | ASAEE-GA | 4.3364 | 954.296 | 3.4417 | 1309.780 | 3.1561 | 1613.094 | 2.7436 | 1940.092 |
| | ASAEE-PSO | 4.0862 | 826.323 | 3.1058 | 1198.461 | 2.8456 | 1436.977 | 2.6024 | 1805.624 |
| | ASAEE-DE | **3.7321** | <u>751.325</u> | **2.8469** | <u>984.226</u> | **2.5564** | <u>1318.374</u> | 2.2613 | <u>1705.950</u> |

From Tables 1–3, we can find that the indicators' values obtained by ASAEE framework are smaller than those obtained by other comparison algorithms except for one time value compared with geometrial-based algorithms. For the results of real data set, Figure 8 illustrates the comparison of abundance inversion results of some endmembers obtained by the ASAEE framework. Overall, the results are in line with our expectations, the original intention of the ASAEE design is to reduce the expensive time cost while ensuring that the extraction accuracy is not severely affected for the intelligent-based endmember extraction algorithms.

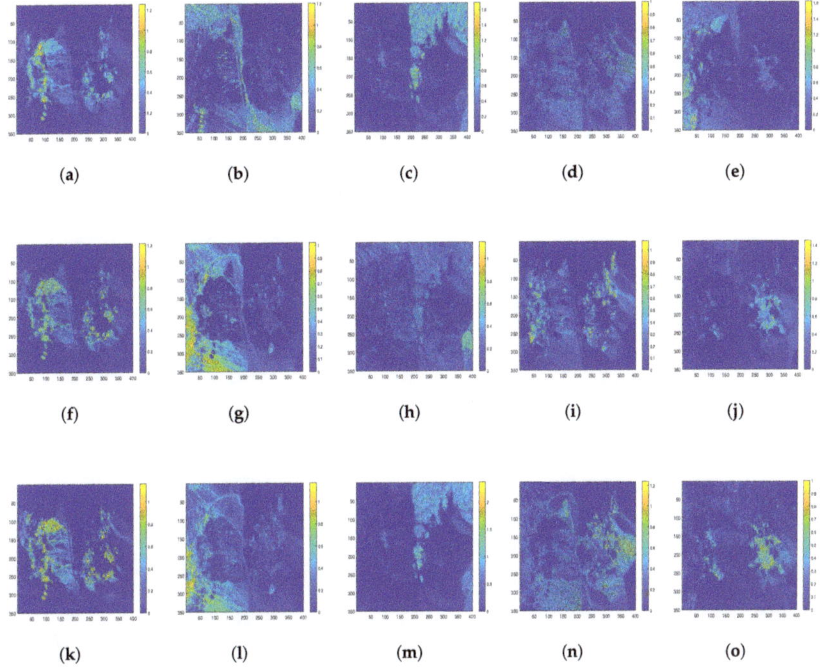

**Figure 8.** The abundance maps of five endmembers obtained from (**a–e**) ASAEE-GA. (**f–j**) ASAEE-PSO. (**k–o**) ASAEE-DE.

### 4.4. Transfer to Other Intelligent-Based Endmember Extraction Algorithms

In this section, we will prove the applicability of the ASAEE framework by transferring it to five intelligent-based comparison methods. As can be seen from Figure 9, the results are the time comparison before and after the transfer of ASAEE framework. It can be concluded that it is very significant with the evaluation from the surrogate-assisted model to the intelligent-based algorithms in endmember extraction. Theoretically, as long as it is an intelligent-based algorithm involving individual or particle evaluation, the ASAEE framework can be transferred and greatly shorten the entire evaluation time.

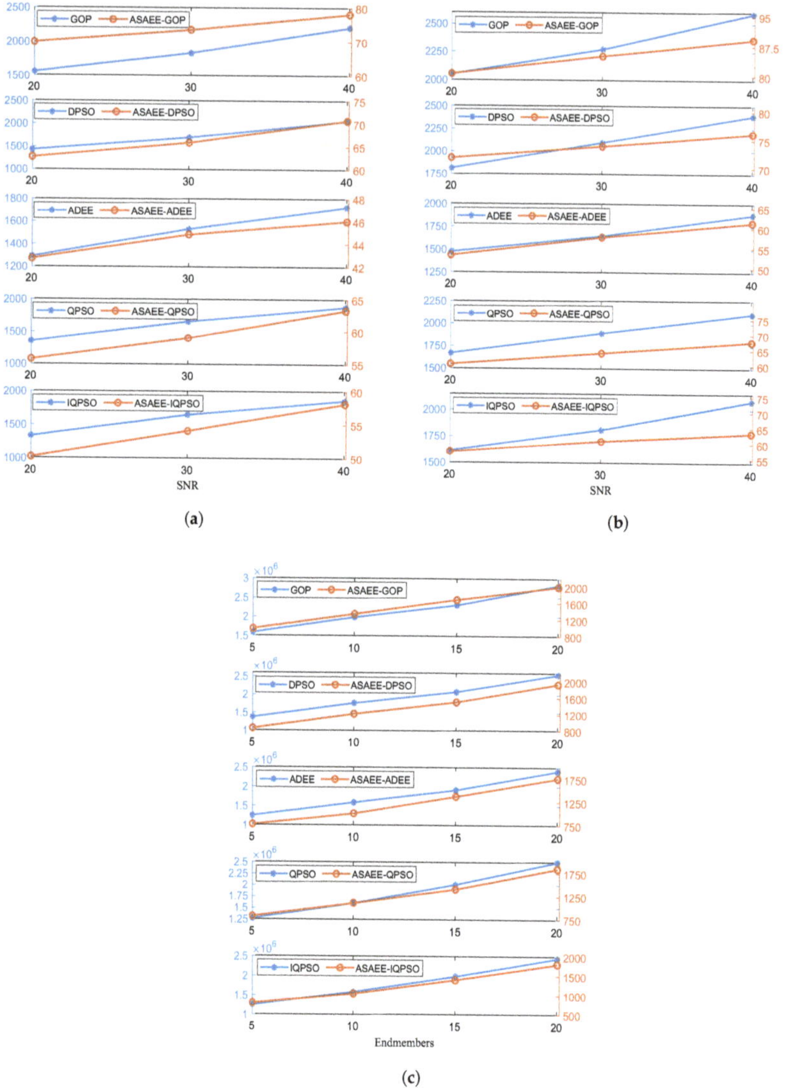

**Figure 9.** Time Comparison of transferring the proposed ASAEE framework to other algorithms on different data sets. (**a**) DS1. (**b**) DS2. (**c**) DS3.

## 5. Conclusions

This paper has proposed an adaptive surrogate-assisted intelligent optimization algorithms framework to deal with the endmember extraction for hyperspectral remote sensing image. Specially, the surrogate-assisted model is established in the abundance inversion stage of the intelligent algorithms, an adaptive weight strategy are designed to automatically assign the weights according to the fitting degree of various surrogate models, so as to reduce the evaluation time and accelerate the convergence of the algorithms under the condition of ensuring certain accuracy. Three intelligent algorithms, ASAEE-GA, ASAEE-PSO, and ASAEE-DE, combined with the design of an adaptive surrogate-assisted model, are proposed to efficiently solve the endmember extraction problem.

In the future work, we will focus on solving more complex endmember extraction scenarios and improve the robustness and practicability of the ASAEE framework. In addition, other intelligent-based algorithms, such as the ant colony algorithms and bee colony algorithms, will also be explored to incorporate into the ASAEE framework. The MOEA-based intelligent endmember extraction algorithms will also be studied.

**Author Contributions:** Conceptualization, Z.W. and J.L.; methodology, Z.W.; validation, Z.W., J.L. and Y.L.; investigation, Y.L.; writing—original draft preparation, Z.W., J.L., F.X.; writing—review and editing, F.X., J.L. and P.L. All authors have read and agreed to the published version of the manuscript.

**Funding:** This work was supported by the Natural Science Foundation of Shaanxi Province (grant No. 2021JQ-210), the National Natural Science Foundation of China (Grant No. 61973249), the Key R & D programs of Shaanxi Province (Grant No. 2021ZDLGY02-06) and Qin Chuangyuan cited the high-level innovative and entrepreneurial talent project (Grant No. 2021QCYRC4-49).

**Institutional Review Board Statement:** Not applicable.

**Informed Consent Statement:** Not applicable.

**Data Availability Statement:** Not applicable.

**Conflicts of Interest:** The authors declare no conflict of interest. The funders had no role in the design of the study; in the collection, analyses, or interpretation of data; in the writing of the manuscript, or in the decision to publish the results.

## References

1. Wu, Y.; Ma, W.; Gong, M.; Su, L.; Jiao, L. A Novel Point-Matching Algorithm Based on Fast Sample Consensus for Image Registration. *IEEE Geosci. Remote Sens. Lett.* **2015**, *12*, 43–47. [CrossRef]
2. Wu, Y.; Li, J.; Yuan, Y.; Qin, A.K.; Miao, Q.G.; Gong, M.G. Commonality Autoencoder: Learning Common Features for Change Detection from Heterogeneous Images. *IEEE Trans. Neural Netw. Learn. Syst.* **2021**, 1–14. [CrossRef]
3. Brown, A.J.; Hook, S.J.; Baldridge, A.M.; Crowley, J.K.; Bridges, N.T.; Thomson, B.J.; Marion, G.M.; de Souza Filho, C.R.; Bishop, J.L. Hydrothermal formation of clay-carbonate alteration assemblages in the Nili Fossae region of Mars. *Earth Planet. Sci. Lett.* **2010**, *297*, 174–182. [CrossRef]
4. Jiao, Q.; Zhang, B.; Liu, J.; Liu, L. A novel two-step method for winter wheat-leaf chlorophyll content estimation using a hyperspectral vegetation index. *Int. J. Remote Sens.* **2014**, *35*, 7363–7375. [CrossRef]
5. Wu, Y.; Xiao, Z.; Liu, S.; Miao, Q.; Ma, W.; Gong, M.; Xie, F.; Zhang, Y. A Two-Step Method for Remote Sensing Images Registration Based on Local and Global Constraints. *IEEE J. Sel. Top. Appl. Earth Obs. Remote Sens.* **2021**, *14*, 5194–5206. [CrossRef]
6. Plaza, A.; Du, Q.; Bioucas-Dias, J.M.; Jia, X.; Kruse, F.A. Foreword to the Special Issue on Spectral Unmixing of Remotely Sensed Data. *IEEE Trans. Geosci. Remote Sens.* **2011**, *49*, 4103–4110. [CrossRef]
7. Plaza, A.; Martinez, P.; Perez, R.; Plaza, J. A quantitative and comparative analysis of endmember extraction algorithms from hyperspectral data. *IEEE Trans. Geosci. Remote Sens.* **2004**, *42*, 650–663. [CrossRef]
8. Chang, X.; Nie, F.; Wang, S.; Yang, Y.; Zhou, X.; Zhang, C. Compound Rank-$k$ Projections for Bilinear Analysis. *IEEE Trans. Neural Netw. Learn. Syst.* **2016**, *27*, 1502–1513. [CrossRef]
9. Bioucas-Dias, J.M.; Plaza, A.; Dobigeon, N.; Parente, M.; Du, Q.; Gader, P.; Chanussot, J. Hyperspectral Unmixing Overview: Geometrical, Statistical, and Sparse Regression-Based Approaches. *IEEE J. Sel. Top. Appl. Earth Observ. Remote Sens.* **2012**, *5*, 354–379. [CrossRef]
10. Marrero, R.; Lopez, S.; Callico, G.M.; Veganzones, M.A.; Plaza, A.; Chanussot, J.; Sarmiento, R. A Novel Negative Abundance-Oriented Hyperspectral Unmixing Algorithm. *IEEE Trans. Geosci. Remote Sens.* **2015**, *53*, 3772–3790. [CrossRef]
11. Ma, W.K.; Bioucas-Dias, J.M.; Chan, T.H.; Gillis, N.; Gader, P.; Plaza, A.J.; Ambikapathi, A.; Chi, C.Y. A Signal Processing Perspective on Hyperspectral Unmixing: Insights from Remote Sensing. *IEEE Signal Process. Mag.* **2014**, *31*, 67–81. [CrossRef]
12. Boardman, J.W.; Kruse, F.A.; Green, R.O. Mapping target signatures via partial unmixing of AVIRIS data. In *Summaries of the Fifth Annual JPL Airborne Earth Science Workshop*; JPL Publication 95–1; NASA Jet Propulsion Laboratory: Pasadena, CA, USA, 1995; pp. 23–26.
13. Winter, M.E. N-FINDR: An algorithm for fast autonomous spectral end-member determination in hyperspectral data. *Proc. SPIE* **1999**, *3753*, 266–275.
14. Nascimento, J.; Dias, J. Vertex component analysis: A fast algorithm to unmix hyperspectral data. *IEEE Trans. Geosci. Remote Sens.* **2005**, *43*, 898–910. [CrossRef]
15. Gao, L.; Gao, J.; Li, J.; Plaza, A.; Zhuang, L.; Sun, X.; Zhang, B. Multiple Algorithm Integration Based on Ant Colony Optimization for Endmember Extraction From Hyperspectral Imagery. *IEEE J. Sel. Top. Appl. Earth Observ. Remote Sens.* **2015**, *8*, 2569–2582. [CrossRef]

16. Yuan, Y.; Feng, Y.; Lu, X. Statistical Hypothesis Detector for Abnormal Event Detection in Crowded Scenes. *IEEE Trans. Cybern.* **2017**, *47*, 3597–3608. [CrossRef]
17. Graña, M.; Veganzones, M.A. Endmember induction by lattice associative memories and multi-objective genetic algorithms. *EURASIP J. Adv. Signal Process.* **2012**, *2012*, 1–12. [CrossRef]
18. Cheng, Q.; Du, B.; Zhang, L.; Liu, R. ANSGA-III: A Multiobjective Endmember Extraction Algorithm for Hyperspectral Images. *IEEE J. Sel. Top. Appl. Earth Observ. Remote Sens.* **2019**, *12*, 700–721. [CrossRef]
19. Rezaei, Y.; Mobasheri, M.R.; Zoej, M.J.V.; Schaepman, M.E. Endmember Extraction Using a Combination of Orthogonal Projection and Genetic Algorithm. *IEEE Geosci. Remote Sens. Lett.* **2012**, *9*, 161–165. [CrossRef]
20. Li, J.; Li, H.; Liu, Y.; Gong, M. Multi-fidelity evolutionary multitasking optimization for hyperspectral endmember extraction. *Appl. Soft Comput.* **2021**, *111*, 107713. [CrossRef]
21. Zhang, B.; Sun, X.; Gao, L.; Yang, L. Endmember Extraction of Hyperspectral Remote Sensing Images Based on the Discrete Particle Swarm Optimization Algorithm. *IEEE Trans. Geosci. Remote Sens.* **2011**, *49*, 4173–4176. [CrossRef]
22. Liu, R.; Du, B.; Zhang, L. Multiobjective Optimized Endmember Extraction for Hyperspectral Image. *Remote Sens.* **2017**, *9*, 558. [CrossRef]
23. Liu, R.; Zhang, L.; Du, B. A Novel Endmember Extraction Method for Hyperspectral Imagery Based on Quantum-Behaved Particle Swarm Optimization. *IEEE J. Sel. Top. Appl. Earth Observ. Remote Sens.* **2017**, *10*, 1610–1631. [CrossRef]
24. Xu, M.; Zhang, L.; Du, B.; Zhang, L.; Fan, Y.; Song, D. A Mutation Operator Accelerated Quantum-Behaved Particle Swarm Optimization Algorithm for Hyperspectral Endmember Extraction. *Remote Sens.* **2017**, *9*, 197. [CrossRef]
25. Tong, L.; Du, B.; Liu, R.; Zhang, L. An Improved Multiobjective Discrete Particle Swarm Optimization for Hyperspectral Endmember Extraction. *IEEE Trans. Geosci. Remote Sens.* **2019**, *57*, 7872–7882. [CrossRef]
26. Du, B.; Wei, Q.; Liu, R. An Improved Quantum-Behaved Particle Swarm Optimization for Endmember Extraction. *IEEE Trans. Geosci. Remote Sens.* **2019**, *57*, 6003–6017. [CrossRef]
27. Liu, R.; Zhu, X. Endmember Bundle Extraction Based on Multiobjective Optimization. *IEEE Trans. Geosci. Remote Sens.* **2021**, *59*, 8630–8645. [CrossRef]
28. Zhong, Y.; Zhao, L.; Zhang, L. An Adaptive Differential Evolution Endmember Extraction Algorithm for Hyperspectral Remote Sensing Imagery. *IEEE Geosci. Remote Sens. Lett.* **2014**, *11*, 1061–1065. [CrossRef]
29. Tong, L.; Du, B.; Liu, R.; Zhang, L.; Tan, K.C. Hyperspectral Endmember Extraction by $(\mu + \lambda)$ Multiobjective Differential Evolution Algorithm Based on Ranking Multiple Mutations. *IEEE Trans. Geosci. Remote Sens.* **2021**, *59*, 2352–2364. [CrossRef]
30. Liu, R.; Du, B.; Zhang, L. Multiobjective endmember extraction for hyperspectral image. In Proceedings of the 2017 IEEE International Geoscience and Remote Sensing Symposium (IGARSS), Fort Worth, TX, USA, 23–28 July 2017; pp. 1161–1164.
31. Yan, C.; Chang, X.; Luo, M.; Zheng, Q.; Zhang, X.; Li, Z.; Nie, F. Self-weighted robust LDA for multiclass classification with edge classes. *ACM Trans. Intell. Syst. Technol.* **2020**, *12*, 1–19. [CrossRef]
32. Lu, X.; Liu, L.; Nie, L.; Chang, X.; Zhang, H. Semantic-Driven Interpretable Deep Multi-Modal Hashing for Large-Scale Multimedia Retrieval. *IEEE Trans. Multimed.* **2021**, *23*, 4541–4554. [CrossRef]
33. Guan, W.; Song, X.; Gan, T.; Lin, J.; Chang, X.; Nie, L. Cooperation Learning From Multiple Social Networks: Consistent and Complementary Perspectives. *IEEE Trans. Cybern.* **2021**, *51*, 4501–4514. [CrossRef]
34. Zhou, Z.; Ong, Y.S.; Nguyen, M.H.; Lim, D. A study on polynomial regression and Gaussian process global surrogate model in hierarchical surrogate-assisted evolutionary algorithm. In Proceedings of the 2005 IEEE Congress on Evolutionary Computation, Edinburgh, UK, 2–5 September 2005; Volume 3, pp. 2832–2839.
35. Loshchilov, I.; Schoenauer, M.; Sebag, M. Self-adaptive surrogate-assisted covariance matrix adaptation evolution strategy. In Proceedings of the 14th Annual Conference on Genetic and Evolutionary Computation, Philadelphia, PA, USA, 7–11 July 2012; pp. 321–328.
36. Herrera, M.; Guglielmetti, A.; Xiao, M.; Coelho, R.F. Metamodel-assisted optimization based on multiple kernel regression for mixed variables. *Struct. Multidiscipl. Optim.* **2014**, *49*, 979–991. [CrossRef]
37. Loshchilov, I.; Schoenauer, M.; Sebag, M. A mono surrogate for multiobjective optimization. In Proceedings of the 12th Annual Conference on Genetic and Evolutionary Computation, Portland, OR, USA, 7–11 July 2010; pp. 471–478.
38. Ong, Y.S.; Nair, P.B.; Keane, A.J. Evolutionary optimization of computationally expensive problems via surrogate modeling. *AIAA J.* **2003**, *41*, 687–696. [CrossRef]
39. Zapotecas Martínez, S.; Coello Coello, C.A. MOEA/D assisted by RBF networks for expensive multi-objective optimization problems. In Proceedings of the 15th Annual Conference on Genetic and Evolutionary Computation, Amsterdam, The Netherlands, 6–10 July 2013; pp. 1405–1412.
40. Buche, D.; Schraudolph, N.; Koumoutsakos, P. Accelerating evolutionary algorithms with Gaussian process fitness function models. *IEEE Trans. Syst. Man Cybern. C Appl. Rev.* **2005**, *35*, 183–194. [CrossRef]
41. Zhang, Q.; Liu, W.; Tsang, E.; Virginas, B. Expensive Multiobjective Optimization by MOEA/D With Gaussian Process Model. *IEEE Trans. Evol. Comput.* **2010**, *14*, 456–474. [CrossRef]
42. Zhang, B.; Sun, X.; Gao, L.; Yang, L. Endmember Extraction of Hyperspectral Remote Sensing Images Based on the Ant Colony Optimization (ACO) Algorithm. *IEEE Trans. Geosci. Remote Sens.* **2011**, *49*, 2635–2646. [CrossRef]
43. Sun, X.; Yang, L.; Zhang, B.; Gao, L.; Gao, J. An Endmember Extraction Method Based on Artificial Bee Colony Algorithms for Hyperspectral Remote Sensing Images. *Remote Sens.* **2015**, *7*, 16363–16383. [CrossRef]

44. Zhang, C.; Qin, Q.; Zhang, T.; Sun, Y.; Chen, C. Endmember extraction from hyperspectral image based on discrete firefly algorithm (EE-DFA). *ISPRS J. Photogramm. Remote Sens.* **2017**, *126*, 108–119. [CrossRef]

45. Damodaran, B.B.; Courty, N.; Lefèvre, S. Sparse Hilbert Schmidt Independence Criterion and Surrogate-Kernel-Based Feature Selection for Hyperspectral Image Classification. *IEEE Trans. Geosci. Remote Sens.* **2017**, *55*, 2385–2398. [CrossRef]

46. Han, T.; Goodenough, D.G. Investigation of Nonlinearity in Hyperspectral Imagery Using Surrogate Data Methods. *IEEE Trans. Geosci. Remote Sens.* **2008**, *46*, 2840–2847. [CrossRef]

47. Sun, J.; Fang, W.; Wu, X.; Palade, V.; Xu, W. Quantum-behaved particle swarm optimization: analysis of individual particle behavior and parameter selection. *Evol. Comput.* **2012**, *20*, 349–393. [CrossRef] [PubMed]

48. Wang, Y.; Cai, Z. Constrained evolutionary optimization by means of $(\mu + \lambda)$-differential evolution and improved adaptive trade-off model. *Evol. Comput.* **2011**, *19*, 249–285. [CrossRef] [PubMed]

49. Jia, G.; Wang, Y.; Cai, Z.; Jin, Y. An improved $(\mu + \lambda)$-constrained differential evolution for constrained optimization. *Inf. Sci.* **2013**, *222*, 302–322. [CrossRef]

50. Xu, M.; Du, B.; Zhang, L. Spatial-Spectral Information Based Abundance-Constrained Endmember Extraction Methods. *IEEE J. Sel. Top. Appl. Earth Observ. Remote Sens.* **2014**, *7*, 2004–2015. [CrossRef]

51. Chang, C.I.; Plaza, A. A fast iterative algorithm for implementation of pixel purity index. *IEEE Geosci. Remote Sens. Lett.* **2006**, *3*, 63–67. [CrossRef]

*Article*

# Borrow from Source Models: Efficient Infrared Object Detection with Limited Examples

Ruimin Chen [1,2,3], Shijian Liu [1,2,*], Jing Mu [1,2,3], Zhuang Miao [1,2,3] and Fanming Li [1,2]

[1]  Key Laboratory of Infrared System Detection and Imaging Technology, Chinese Academy of Sciences, Shanghai 200083, China; chenruimin@mail.sitp.ac.cn (R.C.); mujing@mail.sitp.ac.cn (J.M.); akkomz@mail.sitp.ac.cn (Z.M.); lifanming@mail.sitp.ac.cn (F.L.)
[2]  Shanghai Institute of Technical Physics, Chinese Academy of Sciences, Shanghai 200083, China
[3]  University of Chinese Academy of Sciences, Beijing 100049, China
*  Correspondence: Shj_liu@ustc.edu

**Abstract:** Recent deep models trained on large-scale RGB datasets lead to considerable achievements in visual detection tasks. However, the training examples are often limited for an infrared detection task, which may deteriorate the performance of deep detectors. In this paper, we propose a transfer approach, Source Model Guidance (SMG), where we leverage a high-capacity RGB detection model as the guidance to supervise the training process of an infrared detection network. In SMG, the foreground soft label generated from the RGB model is introduced as source knowledge to provide guidance for cross-domain transfer. Additionally, we design a Background Suppression Module in the infrared network to receive the knowledge and enhance the foreground features. SMG is easily plugged into any modern detection framework, and we show two explicit instantiations of it, SMG-C and SMG-Y, based on CenterNet and YOLOv3, respectively. Extensive experiments on different benchmarks show that both SMG-C and SMG-Y achieve remarkable performance even if the training set is scarce. Compared to advanced detectors on public FLIR, SMG-Y with 77.0% mAP outperforms others in accuracy, and SMG-C achieves real-time detection at a speed of 107 FPS. More importantly, SMG-Y trained on a quarter of the thermal dataset obtains 74.5% mAP, surpassing most state-of-the-art detectors with full FLIR as training data.

**Keywords:** infrared object detection; limited training examples; knowledge transfer

**Citation:** Chen, R.; Liu, S.; Mu, J.; Miao, Z.; Li, F. Borrow from Source Models: Efficient Infrared Object Detection with Limited Examples. *Appl. Sci.* **2022**, *12*, 1896. https://doi.org/10.3390/app12041896

Academic Editor: Yue Wu

Received: 10 January 2022
Accepted: 10 February 2022
Published: 11 February 2022

**Publisher's Note:** MDPI stays neutral with regard to jurisdictional claims in published maps and institutional affiliations.

## 1. Introduction

Recently, thermal infrared cameras have become increasingly popular in security and military surveillance operations [1,2]. Thus, infrared object detection, including both classification and localization of the targets in thermal images, is a critical problem to be invested in. With the advent of Convolution Neural Network (CNN) in many applications [3–7] such as action recognition and target tracking, a number of advanced models [8–10] based on CNN are proposed in object detection. Those detectors lead to considerable achievements in visual RGB detection tasks because they are mainly driven by large training data, which are easily available in the RGB domain. However, the relative lack of large-scale infrared datasets restricts CNN-based methods to obtain the same level of success in the thermal infrared domain [1,11].

One popular solution is finetuning an RGB pre-trained model with limited infrared examples. Many researchers firstly initialize a detection network with parameters trained on public fully-annotated RGB datasets, such as PASCAL-VOC [12] and MS-COCO [13]. Then, the network is finetuned by limited infrared data for specific tasks. To extract infrared object features better, most of the infrared detectors improve existing detection frameworks by introducing some extra enhanced modules such as feature fusion and background suppression. For example, Zhou et al. [14] apply a dual cascade regression mechanism to fuse high-level and low-level features. Miao et al. [15] design an auxiliary foreground prediction loss to reduce background interference. To some extent, the aforementioned modules are effective for infrared object detection. However, it is hard for simple finetuning with inadequate infrared examples to eliminate the difference between thermal and visual images, which hinders the detection of infrared targets.

An alternative solution is to borrow some features from a rich RGB domain. Compared to the finetuning, this method leverages abundant features from the RGB domain to boost accuracy in infrared detection. Konig et al. [16] and Liu et al. [17] combine visual and thermal information by constructing multi-modal networks. They feed paired RGB and infrared examples into the network to detect the objects in thermal images. However, the paired images from two domains are difficult to be obtained, which hampers the development of the multi-modal networks. To tackle this problem, Devaguptapu et al. [1] employ a trainable image-to-image translation framework to generate pseudo-RGB equivalents from thermal images. Although this pseudo multi-modal detector is feasible in the absence of large-scale available datasets, the complicated architecture is difficult to train and thus rarely reaches advanced performance.

In this work, we address this problem from a novel perspective, knowledge transfer. Our proposed approach, named Source Model Guidance (SMG), is the first transfer learning solution for infrared limited-examples detection, to the best of our knowledge. By leveraging existing RGB detection models as source knowledge, we convert recent state-of-the-art RGB detectors to infrared detectors with inadequate thermal data. The basic idea is that if we already have an RGB model with strong ability to distinguish foreground from background, the model can be used as a source model to supervise another network training for infrared detection. Then, the problems becomes how to transfer the source knowledge between different domains and where to add the source supervision.

We first observe modern RGB detection frameworks including anchor-based (Faster RCNN [8], SSD [9], YOLOv3 [18]) and anchor-free (CenterNet [19], CornorNet [20], ExtremeNet [21], FCOS [22]) methods. All of them consist of two main modules, a Feature Extraction Network (FEN) to calculate feature maps and a Detection Head (DH) to generate results. Many researchers have trained those frameworks with large-scale RGB datasets and exposed network weights as common RGB object detection models. Despite the fact that an RGB model is designed for visual images, it still can detect most infrared targets when given a thermal image. However, the precise categories and bounding boxes are hard to be predicted by it due to the difference between two domains. Therefore, we combine all category predictions as a foreground soft label, which is regarded as the source knowledge to be transferred. Then, we look for where to add the source supervision. Different from ground-truth supervision on the final DH, we propose a Background Suppression Module (BSM) to receive the source knowledge. BSM is inserted after FEN to enhance the feature maps and produce a foreground prediction at the same time. By calculating the transfer loss between the foreground prediction and the soft label, we introduce source supervision into the training process of the infrared detector, as shown in Figure 1.

Theoretically, our transfer approach SMG can be implemented in any visual detection networks effortlessly. In this paper, we choose two popular frameworks, CenterNet [19] and YOLOv3 [18], as instantiations, and the frameworks we proposed are named SMG-C and SMG-Y, respectively. To validate the performance of SMG, we conduct extensive experiments on two infrared benchmarks, FLIR [23] and Infrared Aerial Target (IAT) [15]. Experimental results show that SMG is an effective method to boost detection accuracy especially when there are limited training examples. On FLIR, using only a quarter of training data, SMG-Y obtains higher mAP than the original YOLOv3 finetuned on the entire dataset. Furthermore, compared to other infrared detectors, both SMG-C and SMG-Y achieve state-of-the-art accuracy and inference speed.

The main contributions are described as the following three folds:

- First, we propose a cross-domain transfer approach SMG, which easily converts a visual RGB detection framework to an infrared detector.

- Second, SMG decreases the data dependency for an infrared network. The detectors with SMG maintain remarkable performance even if trained on the small-scale datasets.
- Third, two proposed instantiations of SMG, SMG-C and SMG-Y, outperform other advanced approaches in accuracy and speed, showing that SMG is a preferable strategy for infrared detection.

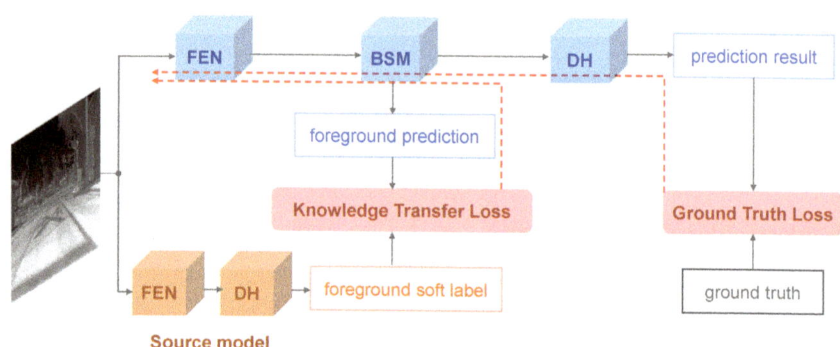


**Figure 1.** The overall framework of SMG, which mainly consists of two parts: a source model to provide source knowledge and a target network to predict infrared detection results. Red arrows indicate the backpropagation pathways.

The structure of this paper is as follows. In Section 2 , we briefly present some aspects related to our work. Section 3 shows the proposed method SMG in detail. Extensive experiments and ablation studies are conducted in Sections 4 and 5, respectively. We explain why SMG works well and analyze the failure cases of our detectors in Section 6. Finally, the summary is drawn in Section 7.

## 2. Related Work

In this section, we briefly introduce recent object detection frameworks including both visual and infrared methods. In addition, we describe the knowledge transfer, which is the inspiration of our method.

### 2.1. Object Detection

Current object detection frameworks can be divided into two groups: anchor-based methods such as Faster RCNN [8], SSD [9], and YOLOv3 [18] and anchor-free methods represented by CenterNet [19], CornorNet [20], ExtremeNet [21], and FCOS [22]. Anchor-based methods firstly define a series of rectangle bounding boxes, called anchors, as proposal candidates. Then, all potential object detections are enumerated exhaustively according to proposed anchors. Finally, additional Non-Maximum Suppression (NMS) [24] is used to remove duplicated locations for the same instance. To avoid the redundant design of anchors and lessen the computation burden, anchor-free methods regard the detection problem as a keypoint estimation without pre-defined anchors. For example, CenterNet [19] predicts the center point of an object and then regresses to other properties such as object size. Although those algorithms achieve remarkable performance, they are mainly driven by extensive public training data and focus on detecting the targets in standard visual RGB images. For infrared detection, the lack of large-scale labeled thermal images hinders the power of detectors based on CNN. Researchers cope with this problem from two aspects: one is finetuning a pre-trained model [14,15], the other is introducing corresponding RGB images as supplements [1,16]. The first strategy hardly makes full use of the information from the RGB domain, and the sophisticated structures in the second method are difficult to be performed. Different from two solutions, our SMG not only leverages existing RGB models as the guidance for infrared detectors but also is easily plugged in any modern detection framework.

### 2.2. Knowledge Transfer

Knowledge transfer is a popular strategy to tackle various problems, such as object classification [25–27], model compression [28,29], and detection [30–32]. It first distills knowledge from a trained model (source) and then transfers the knowledge to another network (target). Hinton et al. [25] introduce the concept of soft label as the guidance in knowledge transfer for classification tasks. In comparison with the hard label such as ground truths, the soft label is a softened version of the final output from the source model. Benefiting from the soft label, the target network can learn how the source model classifies different objects. Many methods [28,29] with soft label obtain achievement in classification and retain accuracy in model compression. However, applying transfer techniques to object detection is challenging because detection is a more complex task that combines regression, region proposals, and classification. To tackle this problem, Chen et al. [31] designed a novel teacher bounded regression loss for knowledge transfer and adaptation layers to better learn from the source model. Although this method is easy to be applied in object detection, the method is driven by large-scale training datasets. Some researchers try to perform transfer learning in few-shot detection and construct a target-domain detector with very few training data. Chen et al. [32] alleviate transfer difficulties in low-shot detection by adding a background-depression regularization and designing a deep architecture, a combination of SSD and Faster RCNN, called LSTD. However, LSTD is suitable for RGB object detection without involving the transfer between different domains. Additionally, it just masks feature maps with the ground-truth bounding boxes in the background-depression regularization, which damages the features extracted from the backbone. Different from LSTD, our SMG introduces an independent block BSM to enhance the foreground features of thermal infrared images by taking advantage of the knowledge from the visual RGB domain.

## 3. Method

In this section, we detail our method Source Model Guidance (SMG). First, we introduce the structure of SMG, including the overall framework and proposed Background Suppression Module (BSM). Then, we describe the training details of SMG, including how to transfer knowledge from the source model to the target network and how to train the whole network. Finally, we show two explicit instantiations of SMG, SMG-C and SMG-Y.

### 3.1. Overall Framework

As illustrated in Figure 1, we train an infrared object detector by using the knowledge of a source model. The source model is a high-capacity RGB detection model, which has been trained with large-scale RGB datasets. The source model is composed of two modules, a Feature Extraction Network (FEN) for feature map calculation and a Detection Head (DH) to generate the prediction. We choose two popular detection models, CenterNet [19] and YOLOv3 [18], as source models to guide different infrared detectors, named SMG-C and SMG-Y, respectively.

Compared to the source model, the infrared detection network not only consists of FEN and DH but also has an extra part, Background Suppression Module (BSM). The structure of FEN is flexible, and it can be the same or different from the source model. The DH in an infrared detection network is similar to the source model except for the predicted category. For BSM, it is a novel part with two functions, predicting the foreground and enhancing the feature map from FEN.

### 3.2. BSM

The BSM in the infrared detection network (target network) is a key module to receive the knowledge transferred from the source model. We describe the principle of BSM, as shown in Figure 2. The idea of BSM is inspired by the concept of attention mechanism [33–37], and thus, its main structure is a transformation mapping from the input $\mathbf{X} \in \mathbb{R}^{H \times W \times C}$ to an enhanced feature map $\mathbf{X}' \in \mathbb{R}^{H \times W \times C}$. In addition, an extra prediction, named foreground prediction $\mathbf{P_{FG}} \in \mathbb{R}^{H \times W \times k}$, is obtained in BSM. The $\mathbf{P_{FG}}$ is defined as the combination of ground-truth targets based on anchors, where $k$ is the number of anchors and $k$ is 1 for anchor-free methods.

To be specific, the input $\mathbf{X}$ first passes two convolutional layers to produce an intermediate feature map. Then, it is fed into to two different branches: one for predicting foreground and the other for feature enhancement. The foreground prediction is achieved by a convolution with sigmoid function to generate a score $\mathbf{P_{FG}}$. The intermediate feature map is also employed to re-weight the input feature map over spatial dimension because it reflects the feature of the foreground. After a $1 \times 1$ convolution for channel transformation, we use an average pooling to squeeze global information into channel-wise weights. Finally, the enhanced feature map branch $\mathbf{X}'$ is obtained by rescaling input $\mathbf{X}$ with the weights.

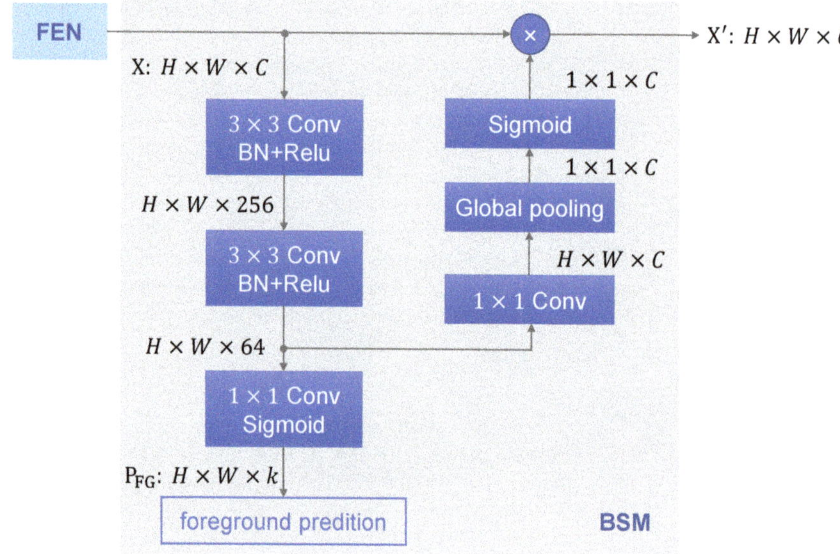


**Figure 2.** The network structure of BSM.

### 3.3. Transfer-Knowledge Regularization

Although the foreground enhancement in BSM can alleviate the disturbance of background, the foreground prediction $\mathbf{P_{FG}}$ from BSM should be supervised in the limited-examples scenario. For this reason, we propose a novel transfer-knowledge regularization by leveraging the source model as a guidance.

In this paper, the foreground prediction $\mathbf{P_{FG}}$ with values within 0 and 1 is supervised by the foreground soft label $\mathbf{S_{FG}}$ generated from the source model. Different from the hard label in ground-truth supervision, we adopt the soft label in knowledge transfer because it contains hidden information about how the source model makes inferences when given samples. In every position of $\mathbf{S_{FG}}$, the value of the soft label is in $[0, 1]$ based on anchor, while the hard label is either 0 or 1.

For different source models, we choose different methods to obtain the foreground soft label $\mathbf{S_{FG}}$. We sum the label prediction (heatmap) for all positions in SMG-C and use the anchor confidence directly in SMG-Y, as shown in Figures 3 and 4. The soft label $\mathbf{S_{FG}}$ is the foreground score based on anchor and has the same size with foreground prediction $\mathbf{P_{FG}}$ from the target network. We take $\mathbf{S_{FG}}$ as source-domain knowledge to regularize the training of target network. Mean Squared Error (MSE) is applied as a transfer-knowledge regularization:

$$\mathcal{L}_{TK} = MSE(\mathbf{S_{FG}}, \mathbf{P_{FG}}). \tag{1}$$

In this case, the trained RGB detection model can be integrated into the training procedure of the infrared detector, which achieves cross-domain transfer in SMG.

### 3.4. Training Algorithm

The whole loss $\mathcal{L}$ of SMG consists of two parts: one is the standard detection loss with ground truth supervision $\mathcal{L}_{GT}$, and the other is the transfer-knowledge loss $\mathcal{L}_{TK}$ mentioned in the above subsection:

$$\mathcal{L} = \mathcal{L}_{GT} + \lambda \mathcal{L}_{TK}. \tag{2}$$

The weight $\lambda$ represents hyper-parameters to control the balance between different losses. We fix it to be 1 in SMG-C. In SMG-Y, $\lambda$ is 0.3 because we introduce 3 BSMs to generate the transfer-knowledge loss in SMG-Y, as explained in the following subsection. During the training, we first initialize the source model with public parameters trained on COCO, which is a large-scale RGB detection dataset. For the target network, the FEN is initialized with ImageNet pretrained parameters, and other modules are randomly initialized. Then, training loss is calculated according to Equation (2). Finally, we update the weights of target network in the back propagation. It is notable that the source model is not updated, and thus, we just employ the target network as an infrared detector in the inference.

### 3.5. Instantiations

SMG can be implemented in standard visual RGB detection networks and convert those networks to infrared detectors. To illustrate this point, we apply SMG in both anchor-free and anchor-based detection frameworks, which is described next.

We first consider CenterNet [19], an anchor-free model, as an instantiation, and the framework we proposed is named SMG-C. As shown in Figure 3, CenterNet predicts center points of targets directly by producing a heatmap $\hat{Y} \in [0,1]^{H \times W \times class}$, where *class* is the number of categories (for RGB models trained on COCO, $class = 80$). Therefore, the sum of the heatmap represents foreground prediction, and we use it as $\mathbf{S_{FG}}$ to transfer knowledge. For the infrared detection network of SMG-C, only a BSM is inserted in between FEN and DH in comparison with CenterNet.

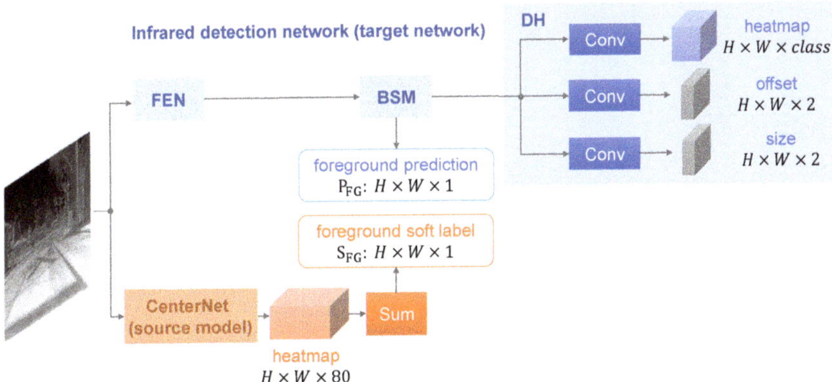


**Figure 3.** The framework of SMG-C.

SMG is also applied in YOLOv3 [18], an anchor-based model, and Figure 4 shows the framework of SMG-Y. YOLOv3 predicts bounding boxes at 3 different scales by extracting features from 3 scales. As a result, we add 3 BSMs in the infrared detection network. Furthermore, YOLOv3 sets $k$ anchors with different sizes, and thus, the prediction in every scale is a $k$-d tensor encoding location, confidence, and class. The confidence reflects whether there is an object in the anchor, and we adapt it as the foreground soft label $\mathbf{S_{FG}}$ directly. In this work, we set $k = 3$ according to the original paper [18].

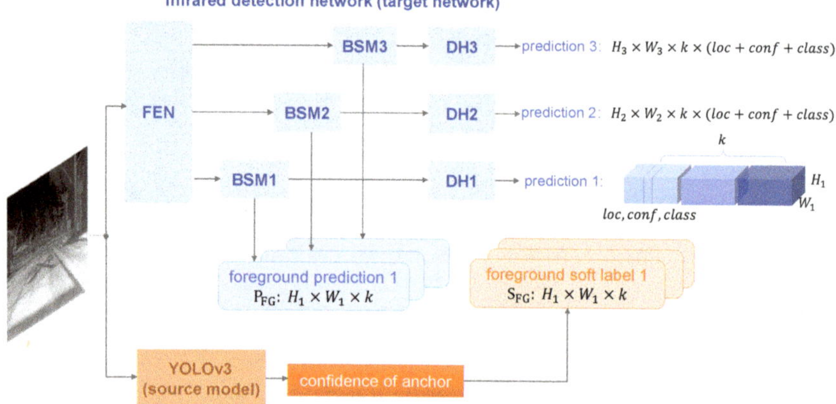


**Figure 4.** The framework of SMG-Y.

## 4. Experiments

In this section, we first introduce experimental details and the training datasets we use in this paper. Then, we conduct extensive experiments to evaluate the detection performance of two frameworks, SMG-C and SMG-Y. Finally, our method is compared with some popular detectors on the public FLIR benchmark.

### 4.1. Dataset and Experimental Setup

We adopt the public FLIR dataset [23] and self-build IAT dataset [15] for our experimental studies.

FLIR [23] collects 9214 infrared images with annotations, where the labeled objects contain a person, car, and bicycle. It is acquired via a thermal camera mounted on a vehicle, and all images are taken on the streets and highways, as illustrated in Figure 5. To evaluate the capability of our method with limited data, we perform experiments with full, half, and one-quarter of training examples in FLIR. The statistics of the training datasets are shown in Table 1. Although the numbers of training images are different in the three datasets, their test sets are the same as those provided in the FLIR benchmark.

**Table 1.** Numbers of instances on FLIR datasets.

| Dataset | Person | Car | Bicycle |
| --- | --- | --- | --- |
| FLIR | 22,372 | 41,260 | 3986 |
| FLIR-1/2 | 10,997 | 20,700 | 1979 |
| FLIR-1/4 | 5574 | 10,286 | 928 |

The IAT [15] consists of 2750 infrared images with aerial targets, including five categories: airline, bird, fighter, helicopter, and trainer. All images are captured by ground-to-air infrared cameras, and some samples on IAT are shown in Figure 6. Different from the images with target occlusions in FLIR, IAT contains small targets in complex aerial backgrounds, and the main challenge of it is background interference. We split IAT with the ratio of 7:3 as the training set and test set, respectively. Similar to FLIR, we use all and half of the training images to implement experiments, as presented in Table 2.

**Figure 5.** Samples on FLIR dataset.

**Table 2.** Numbers of instances on IAT datasets.

| Dataset | Airline | Bird | Fighter | Helicopter | Trainer |
|---|---|---|---|---|---|
| IAT | 121 | 535 | 667 | 310 | 469 |
| IAT-1/2 | 64 | 277 | 321 | 152 | 242 |

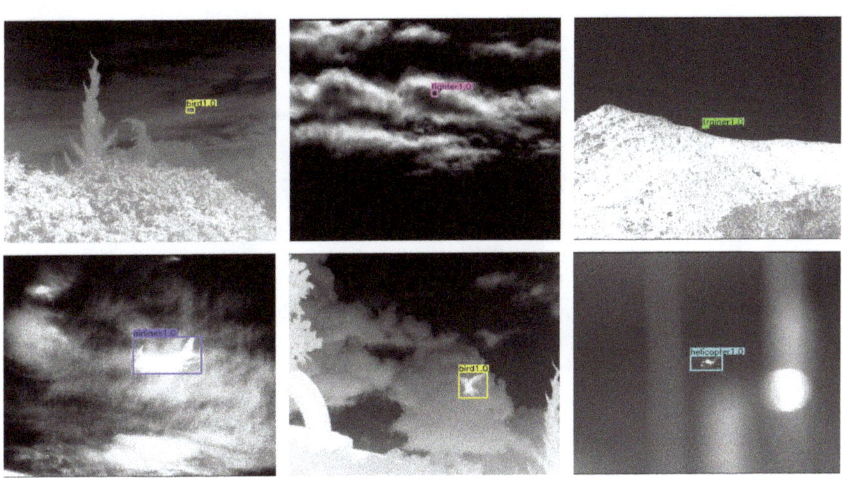

**Figure 6.** Samples on IAT dataset.

All experiments are implemented on a PC with an i7-8700K CPU and a signal GTX1080Ti GPU. For SMG-C, we adopt CenterNet with ResNet-18 [19] as the source model, because it is light-weight and enough to provide the guidance. The FEN of the target network in SMG-C is the fully convolutional upsampling version of Deep Layer Aggregation (DLA-34) [38]. For SMG-Y, YOLOv3 with DarkNet-53 [18] is used as the source model and the backbone of the target network is DarkNet-53. The source models of two frameworks are RGB detection models trained on COCO [13].

The input resolution is set to $512 \times 512$ in SMG-C and $416 \times 416$ in SMG-Y. During the training process of two frameworks, we follow their original papers [18,19] separately for training setting and hyper-parameters, unless specified otherwise. In the inference, we evaluate the performance with the mean Average Precision (mAP) at IoU of 0.5, which is a common metric for object detection tasks.

*4.2. SMG-C Results*

We use SMG-C as the detection framework and implement experiments on both FLIR and IAT benchmarks. The baseline method in this subsection is the original CenterNet [19] without SMG.

Table 3 shows the comparison of AP for each class and mAP of SMG-C against the baseline detection network when trained with different numbers of training examples on the FLIR benchmark. One can see that our SMG-C outperforms the baseline detector across all classes when trained with the same dataset. For example, SMG-C on FLIR obtains 75.6% mAP, which is 4.5% higher than the baseline. This can be attributed to the fact that the source model offers sufficient guidance for the infrared detector in SMG.

More importantly, SMG-C achieves outstanding performance when the training data are insufficient. Taking the bicycle as example, we find that its AP maintains 51.5%, although the training examples are reduced to 1/4 of the original. In contrast, the highest bicycle's AP is 51.2% for the baseline method. Furthermore, the mAP of SMG-C trained on FLIR-1/2 obtains 73.3% mAP, surpassing the original CenterNet trained on the entire FLIR (71.1%).

**Table 3.** Detection results of SMG-C on the FLIR benchmark.

| Dataset | Method | mAP (%) | AP (%) | | |
| --- | --- | --- | --- | --- | --- |
| | | | Person | Car | Bicycle |
| FLIR | Baseline | 71.1 | 76.6 | 85.4 | 51.2 |
| | SMG-C | 75.6 | 79.0 | 85.8 | 62.0 |
| FLIR-1/2 | Baseline | 68.1 | 75.1 | 83.5 | 45.8 |
| | SMG-C | 73.3 | 78.7 | 86.0 | 55.3 |
| FLIR-1/4 | Baseline | 65.8 | 71.5 | 81.8 | 44.1 |
| | SMG-C | 70.9 | 76.7 | 84.5 | 51.5 |

We also report the results on the IAT benchmark in Table 4. All mAPs of SMG-C exceed 95%, while the highest accuracy of CenterNet is only 93%. When we reduce training datasets to half of the original, the accuracy of the baseline drops to 90.6%, while SMG-C maintains 95.2% in mAP. Furthermore, SMG-C trained on IAT-1/2 surpasses the baseline method trained with the entire training dataset. This demonstrates that SMG-C yields an effective infrared detection method even when there are a lack of available training data.

Some results on IAT-1/2 are visualized in Figure 7. When the target is small, some interference from the background may adversely affect the detection especially in the absence of enough training examples. As shown in Figure 7, the baseline CenterNet hardly overcomes this problem so as to generate many wrong detection results. However, SMG-C guided by the high-performance RGB model suppresses the interference from the background and predicts more precisely than the baseline.

**Table 4.** Detection results of SMG-C on the IAT benchmark.

| Dataset | Method | mAP (%) |
| --- | --- | --- |
| IAT | Baseline | 93.0 |
| | SMG-C | 96.8 |
| IAT-1/2 | Baseline | 90.6 |
| | SMG-C | 95.2 |

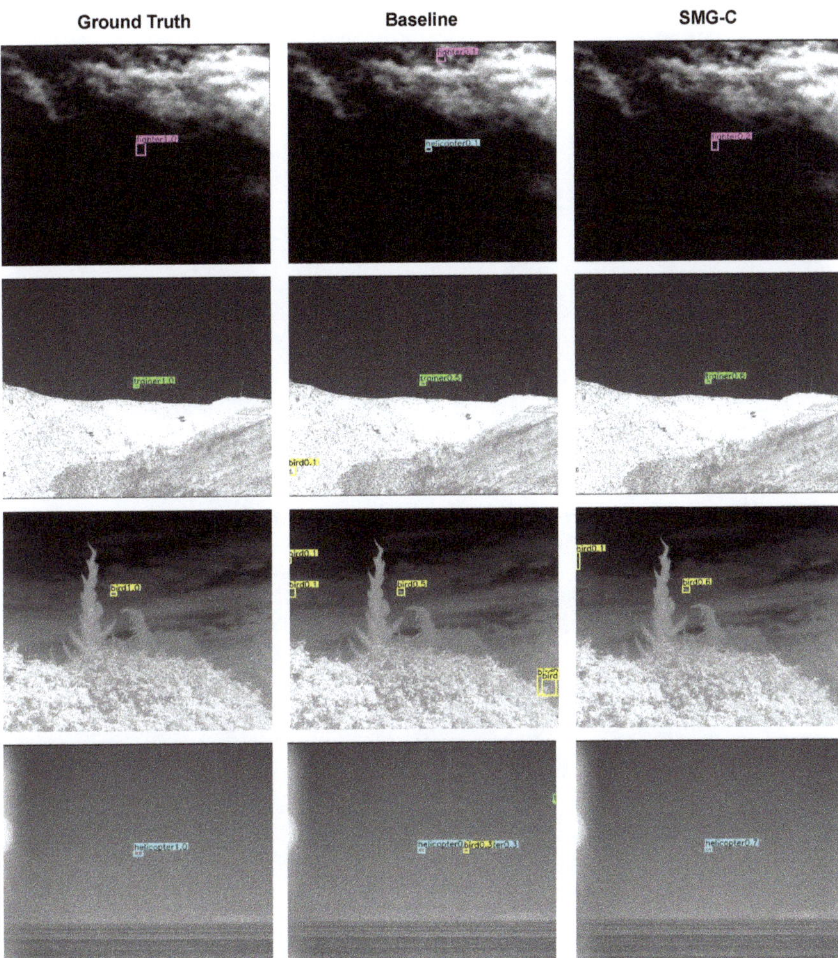

**Figure 7.** Visualization results on IAT-1/2.

### 4.3. SMG-Y Results

Similar to SMG-C, we conduct experiments on both FLIR and IAT datasets to evaluate the performance of SMG-Y. SMG-Y is compared with the baseline detector, YOLOv3 [18].

Table 5 presents the results of SMG-Y on the FLIR benchmark. The mAP of SMG-Y exceeds the baseline method nearly 10% on the same dataset, and the gap of them increases with the decrease of training examples. On FLIR-1/4, SMG-Y achieves 62.5% AP in bicycle detection in comparison with 29.1% for the baseline. We also observe that the accuracy of SMG-Y on FLIR-1/4 (74.5% mAP) outperforms the baseline method trained with full FLIR (69.4% mAP), which demonstrates SMG-Y maintains remarkable accuracy with limited training data. When the dataset is reduced to 1/4 of the original, the mAP of SMG-Y decreases by 2.5% (from 77.0% to 74.5%). However, the mAP of the baseline method drops by 13.2% (from 69.4% to 56.2%). The low reduction of SMG-Y indicates that it can take full advantage of the knowledge from the source model and decrease the data dependency of the network.

**Table 5.** Detection results of SMG-Y on the FLIR benchmark.

| Dataset | Method | mAP (%) | AP (%) | | |
|---|---|---|---|---|---|
| | | | Person | Car | Bicycle |
| FLIR | Baseline | 69.4 | 74.5 | 84.4 | 49.2 |
| | SMG-Y | 77.0 | 78.5 | 86.6 | 65.8 |
| FLIR-1/2 | Baseline | 64.9 | 68.5 | 82.1 | 44.0 |
| | SMG-Y | 75.4 | 76.9 | 86.7 | 62.7 |
| FLIR-1/4 | Baseline | 56.2 | 61.2 | 78.3 | 29.1 |
| | SMG-Y | 74.5 | 76.6 | 84.4 | 62.5 |

We visualize some results of SMG-Y and its baseline YOLOv3 when both of them are trained on FLIR-1/4, as shown in Figure 8. We find that the baseline method hardly predicts the position of the bicycle because it is always obscured by people. Furthermore, due to insufficient training data, YOLOv3 is difficult to recognize objects with special gestures such as the sitting woman in the last row of Figure 8 (note that most people in the training dataset are walking or riding). However, SMG-Y overcomes those problems and detects precisely under the circumstances of severe occlusion and appearance change even if the training examples are limited.

Experiments are also conducted on the IAT benchmark, and the results are shown in Table 6. We witness a sharp fall in the baseline accuracy as the number of training instances decreases. In contrast, SMG-Y trained on IAT-1/2 keeps competitive accuracy with 96.2% mAP, which is slightly lower than that trained on the full IAT dataset.

**Table 6.** Detection results of SMG-Y on the IAT benchmark.

| Dataset | Method | mAP (%) |
|---|---|---|
| IAT | Baseline | 92.5 |
| | SMG-Y | 97.8 |
| IAT-1/2 | Baseline | 88.3 |
| | SMG-Y | 96.2 |

### 4.4. Comparison of SMG-C and SMG-Y

We compare two instantiations and their baseline methods in Figure 9. It is notable that SMG-Y outperforms SMG-C but YOLOv3 is inferior to CenterNet. In other words, the gap between SMG-Y and its baseline is larger in comparison with SMG-C. To be specific, SMG-Y achieves 77.0% mAP, which is 7.6% higher than its baseline when trained on a full FLIR. In contrast, SMG-C obtains 75.6% mAP, exceeding its baseline by 4.5%. We attribute this phenomenon to the fact that three different BSMs are added in SMG-Y to receive knowledge from different scales, and only one BSM is inserted in SMG-C.

Additionally, the data dependency for a detector can be reflected in the performance degradation when we reduce the training examples, which is also the slope of the curves in Figure 9. The decline of CenterNet is less than that of YOLOv3 due to the different principles between two frameworks: one is anchor-free and the other is anchor-based. We observe that the curves of both SMG-Y and SMG-C are smoother than their baselines. For example, a slight reduction in mAP can be witnessed in SMG-Y while its baseline accuracy drops dramatically, which indicates that SMG is an efficient strategy to decrease the data dependency for an infrared detection network.

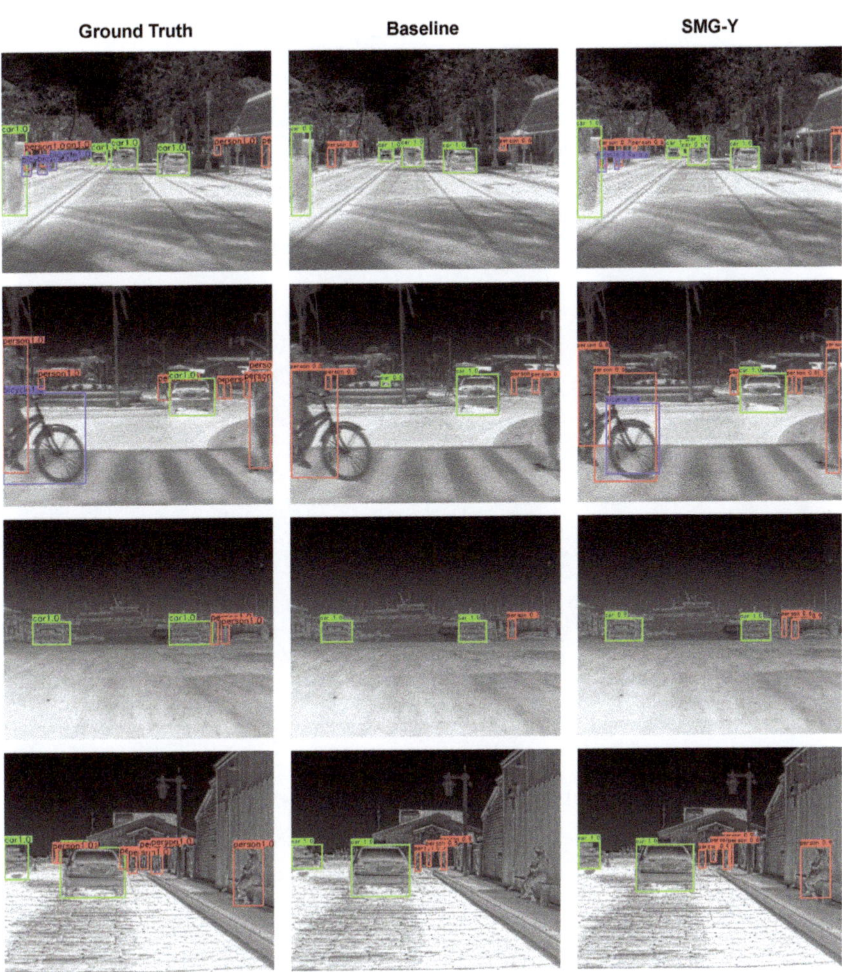

**Figure 8.** Visualization results on FLIR-1/4.

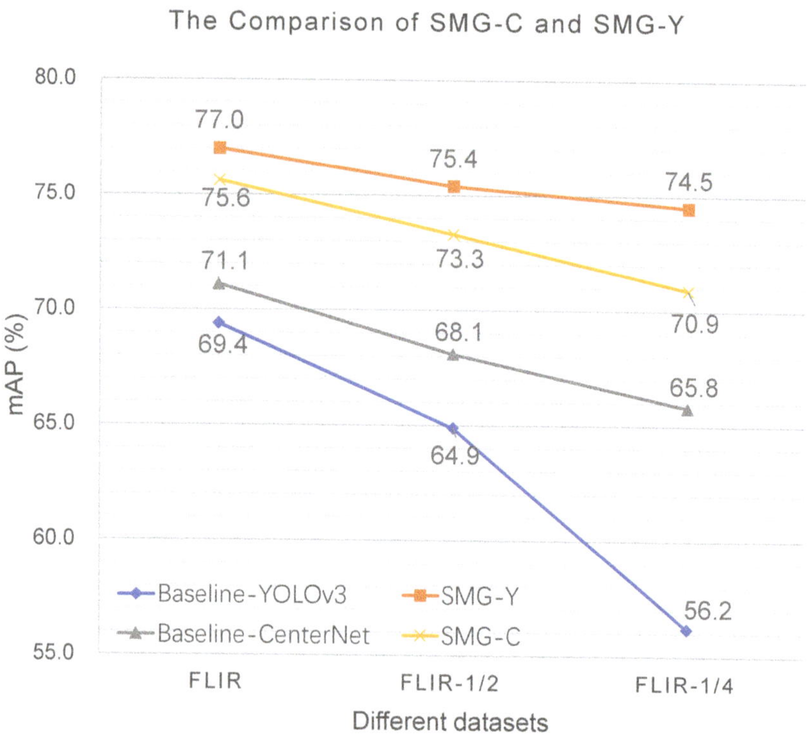

The Comparison of SMG-C and SMG-Y

**Figure 9.** The comparison of SMG-C and SMG-Y in terms of mAP.

*4.5. Comparison with State-of-the-Arts*

Our frameworks (SMG-C and SMG-Y) are compared with some recent state-of-the-art detectors on the FLIR benchmark. The compared trackers are divided into two categories, visual and infrared detectors. The visual detectors such as SSD [9], YOLOv3 [18], Faster-RCNN [8], CenterNet [19], and RefineNet [39] are designed for RGB object detection and finetuned on the training set of FLIR. The infrared detectors, including MMTOD-CG [1], MMTOD-UNIT [1], Effi-YOLOv3 [40] and Pesudo-two-stage [14], are applied for thermal images directly.

We present the qualitative results in Table 7. It is remarkable that two proposed detection frameworks achieve outstanding performance. Specifically speaking, SMG-Y obtains the highest mAP with 77.0% and the AP of person, car, and bicycle are 78.5%, 86.6%, and 65.8%, respectively. It outperforms advanced detectors in mAP, and the speed of it maintains 40 frames per second (FPS), keeping the balance of accuracy and speed. Despite the slightly lower mAP (75.6%) in comparison with SMG-Y, SMG-C runs at the speed of 107 FPS, which is five times faster than other infrared detectors. Compared to the high-speed detector CenterNet [19], SMG-C gains 4.5% improvement in mAP, which shows that SMG-C is an efficient real-time detector.

More importantly, SMG-Y with 1/4 training data also achieves 74.5% mAP, surpassing all visual detectors and most infrared detectors trained on full FLIR. The bicycle accuracy in SMG-Y-1/4 is 62.5% AP, which is on par with that of Pseudo-two-stage [14]. Note that the training dataset of SMG-C-1/4 only contains 928 bicycle instances, while Pseudo-two-stage [14] is trained with 3986 examples for bicycle detection.

**Table 7.** Detection results of different detectors on the FLIR benchmark.

| Category | Model | mAP (%) | AP (%) | | | FPS |
|---|---|---|---|---|---|---|
| | | | Person | Car | Bicycle | |
| Visual detectors | SSD [9] | 62.1 | 63.1 | 75.8 | 47.5 | 24 |
| | YOLOv3 [18] | 69.4 | 74.5 | 84.4 | 49.2 | 42 |
| | Faster-RCNN [8] | 70.9 | 71.3 | 75.8 | 61.8 | 8 |
| | CenterNet [19] | 71.1 | 76.6 | 85.4 | 51.2 | 107 |
| | RefineDet [39] | 74.3 | **79.4** | 85.6 | 58.0 | 22 |
| Infrared detectors | MMTOD-CG [1] | 61.4 | 63.3 | 70.6 | 50.3 | - |
| | MMTOD-UNIT [1] | 61.5 | 64.5 | 70.7 | 49.4 | - |
| | Effi-YOLOv3 [40] | 70.8 | 74.5 | 84.7 | 53.2 | 22 |
| | Pseudo-two-stage [14] | 75.6 | 78.7 | 85.5 | 62.5 | 21 |
| | SMG-C | 75.6 | 79.0 | 85.8 | 62.0 | **107** |
| | SMG-Y | **77.0** | 78.5 | **86.6** | **65.8** | 40 |
| | SMG-Y-1/4 | 74.5 | 76.6 | 84.4 | 62.5 | 40 |

SMG-Y-1/4 is trained on FLIR-1/4, and the other detectors are trained on FLIR.

## 5. Ablation Studies

In this section, we conduct ablation studies with SMG-C to understand the effect of image resolution, guidance, and backbone. All networks are evaluated on the FLIR benchmark, and the source model is CenterNet with ResNet-18 [19].

### 5.1. Effect of Image Resolution

We employ ResNet-18 as the FEN in the target network, and the compared baseline is the original CenterNet without SMG. Table 8 presents the mAP of two methods when the image resolution is changed from $384 \times 384$ to $512 \times 512$. It is obvious that the higher resolution contributes to better accuracy. However, at different resolutions, SMG-C exceeds the baseline more than 5% in mAP. It indicates that the image resolution just affects the performance of the baseline network and has less influence on SMG.

**Table 8.** Detection results on the FLIR benchmark at different image resolutions.

| Input Size | Method | mAP (%) |
|---|---|---|
| $384 \times 384$ | Baseline | 53.9 |
| | SMG-C | 59.0 |
| $512 \times 512$ | baseline | 62.7 |
| | SMG-C | 68.8 |

### 5.2. Guidance with Hard or Soft Label

In SMG, we use the foreground soft label generated from the source model as the guidance. However, the hard label from the ground truth also can be utilized as the guidance. The hard label is the ground-truth foreground score, which is the combination of all ground-truth targets mapped to the heatmap. In every position of heatmap, the value of the hard label is either 0 or 1, which is different from the soft label in $[0, 1]$.

We fix the image resolution at $512 \times 512$ and compare the baseline (no guidance) with three different guidance methods, including hard, soft, and both of them in Table 9. The methods with guidance surpass the baseline more than 5% in mAP, which shows that the guidance is an important factor in performance improvement. Furthermore, the soft guidance obtains higher accuracy than other guidance methods. We attribute it to the fact that the soft label contains hidden information about how the source model distinguishes foreground from background, which is exactly what the target network needs to learn. Therefore, we choose the soft guidance in SMG other than hard guidance.

**Table 9.** Detection results of different guidance methods on the FLIR benchmark.

| Guidance Method | mAP (%) |
| --- | --- |
| No guidance (baseline) | 62.7 |
| Hard | 67.7 |
| Hard and soft | 68.1 |
| Soft | 68.8 |

### 5.3. Effect of Backbone

In this subsection, two different backbones, ResNet-18 [19] and DLA-34 [38], are used as FENs in the target networks. Table 10 shows the comparison of the their mAP with corresponding baselines at the image resolution of $512 \times 512$. The structure of DLA-34 is more complicated than ResNet-18, and thus, higher detection accuracy can be achieved. In spite of different backbones, we observe a significant increase in mAP (over 5%) when SMG is added to the framework. That indicates SMG is an effective strategy no matter which backbone we employ.

**Table 10.** Detection results of SMG-C with different FENs on the FLIR benchmark.

| Backbone | Method | mAP (%) |
| --- | --- | --- |
| ResNet-18 | Baseline | 62.7 |
| | SMG-C | 68.8 |
| DLA-34 | Baseline | 71.1 |
| | SMG-C | 75.6 |

## 6. Discussions

In this section, we give some insights about why our proposed SMG works well when there are limited training examples. Then, we analyze the failure cases of our methods.

### 6.1. Why SMG Works Well

In SMG, we suppress the background disturbances by borrowing the knowledge from the source model so as to reduce the data dependency of the target network (infrared detection network). Taking SMG-C as an example, we visualize the soft label generated from the source model and the heatmap of the target network. Figure 10 shows that the source model can filter out the main background, such as roads, houses, and so on. However, it hardly detects specific targets in heavy occlusion, such as people in the crowd, cyclists, and bicycles. In other words, the soft label from the source model can be viewed as effective knowledge to provide supervision, but it cannot be leveraged directly. We solve this problem by inserting a BSM in the target network to receive the knowledge transferred from the source model and enhance the foreground at the same time. The last column in Figure 10 illustrates that the target network with BSM locates center points of targets more precisely than the source model. As a result, the target network can pay more attention to target objects, which is important for training with limited examples.

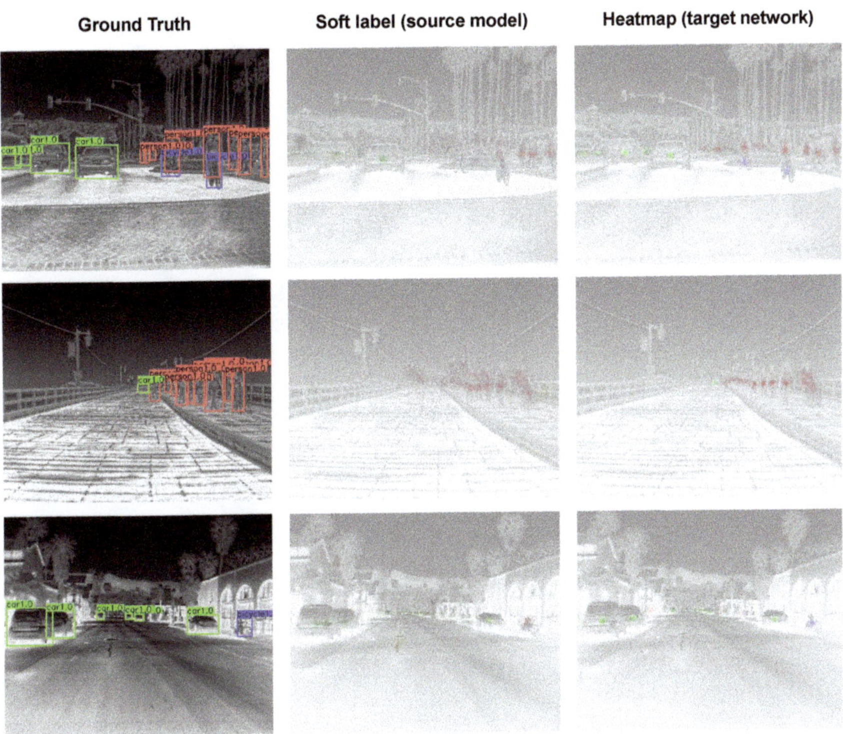



**Figure 10.** The visualization of soft label and heatmap.

## 6.2. Missed Detections

Although SMG promotes accuracy in infrared object detection, the limited-examples detection is still a challenging task. By visualizing the results of SMG-Y trained on FLIR-1/4 and full FLIR in Figure 11, we study the missed detections in absence of training examples. We also represent logarithmic average miss rates of SMG-Y and SMG-Y-1/4 in Table 11. The miss rates of SMG-Y-1/4 are slightly higher than those of SMG-Y. When two objects are close to each other, such as two pedestrians walking together, SMG-Y-1/4 may detect them as a single target, while SMG-Y with sufficient training data easily distinguishes them, as shown in Figure 11. Furthermore, we find that both SMG-Y and SMG-Y-1/4 miss the small objects located far from the camera or obscured by others, such as person and bicycle. We attribute this drawback to the fact that their source model YOLOv3 has poor detection performance for small targets. In the future, we will focus on these challenges and try to cope with them.

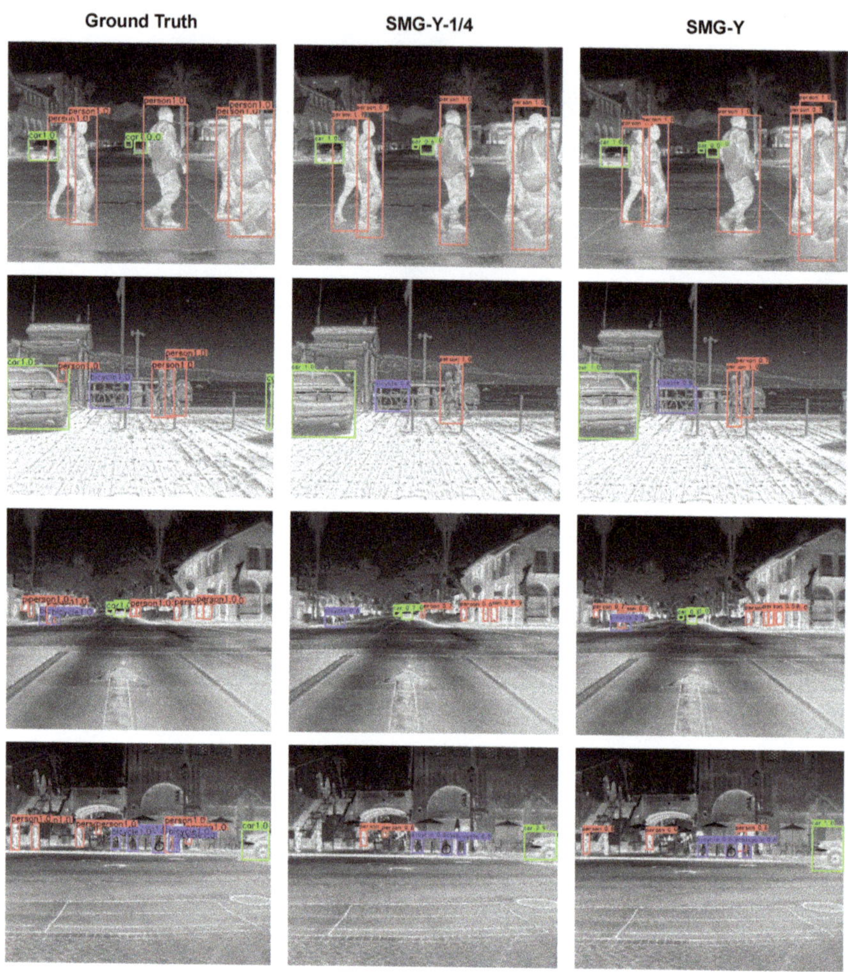

**Figure 11.** Some examples of missed detections. Note that SMG-Y-1/4 represents SMG-Y trained on FLIR-1/4.

**Table 11.** Miss rates of SMG-Y and SMG-Y-1/4 on the FLIR benchmark.

| Method | Person | Car | Bicycle |
| --- | --- | --- | --- |
| SMG-Y | 0.53 | 0.41 | 0.52 |
| SMG-Y-1/4 | 0.55 | 0.43 | 0.55 |

## 7. Conclusions

In summary, we present a novel cross-domain transfer approach SMG to address the problem of infrared detection on small-scale datasets. SMG can convert a visual detection framework into an infrared detector by borrowing the knowledge from the source model, which is a trained RGB detection model. We apply SMG in both anchor-free and anchor-based detection frameworks, named as SMG-C and SMG-Y, respectively. Experiments on FLIR and IAT illustrate that our infrared detectors achieve outstanding performance in lack of available training data. Compared to state-of-the-art detectors, SMG-Y with only 1/4 training data outperforms most of them, demonstrating that SMG is a preferable method for limited-examples infrared detection.

**Author Contributions:** All of the authors contributed to this study. Conceptualization, R.C. and S.L.; methodology, R.C.; software, R.C.; data curation, J.M. and Z.M.; writing—original draft preparation, R.C.; writing—review and editing, R.C., J.M. and Z.M.; funding acquisition, S.L. and F.L. All authors have read and agreed to the published version of the manuscript.

**Funding:** This research was funded by Shanghai Key Laboratory of Criminal Scene Evidence funded Foundation (Grant No. 2017xcwzk08) and the Innovation Fund of Shanghai In-stitute of Technical Physics (Grant No. CX-321).

**Conflicts of Interest:** The authors declare no conflict of interest.

## References


1. Devaguptapu, C.; Akolekar, N.; Sharma, M.M.; Balasubramanian, V.N. Borrow From Anywhere: Pseudo Multi-Modal Object Detection in Thermal Imagery. In Proceedings of the 2019 IEEE/CVF Conference on Computer Vision and Pattern Recognition Workshops (CVPRW), Long Beach, CA, USA, 16–17 June 2019; pp. 1029–1038. [CrossRef]
2. Zhang, L.; Peng, Z. Infrared Small Target Detection Based on Partial Sum of the Tensor Nuclear Norm. *Remote Sens.* **2019**, *11*, 382. [CrossRef]
3. Rashid, M.; Khan, M.A.; Alhaisoni, M.; Wang, S.H.; Naqvi, S.R.; Rehman, A.; Saba, T. A Sustainable Deep Learning Framework for Object Recognition Using Multi-Layers Deep Features Fusion and Selection. *Sustainability* **2020**, *12*, 5037. [CrossRef]
4. Krizhevsky, A.; Sutskever, I.; Hinton, G.E. ImageNet Classification with Deep Convolutional Neural Networks. *Commun. ACM* **2017**, *60*, 84–90. [CrossRef]
5. Masood, H.; Zafar, A.; Ali, M.U.; Hussain, T.; Khan, M.A.; Tariq, U.; Damaševičius, R. Tracking of a Fixed-Shape Moving Object Based on the Gradient Descent Method. *Sensors* **2022**, *22*, 1098. [CrossRef]
6. He, K.; Zhang, X.; Ren, S.; Sun, J. Deep Residual Learning for Image Recognition. In Proceedings of the 2016 IEEE Conference on Computer Vision and Pattern Recognition (CVPR), Las Vegas, NV, USA, 27–30 June 2016; pp. 770–778. [CrossRef]
7. Hussain, N.; Khan, M.; Kadry, S.; Tariq, U.; Mostafa, R.; Choi, J.; Nam, Y. Intelligent Deep Learning and Improved Whale Optimization Algorithm based Framework for Object Recognition. *Hum.-Centric Comput. Inf. Sci.* **2021**, *11*, 1–18.
8. Ren, S.; He, K.; Girshick, R.; Sun, J. Faster R-CNN: Towards Real-Time Object Detection with Region Proposal Networks. *IEEE Trans. Pattern Anal. Mach. Intell.* **2017**, *39*, 1137–1149. [CrossRef]
9. Liu, W.; Anguelov, D.; Erhan, D.; Szegedy, C.; Reed, S.; Fu, C.Y.; Berg, A.C. SSD: Single Shot MultiBox Detector. In *Computer Vision—ECCV 2016*; Leibe, B., Matas, J., Sebe, N., Welling, M., Eds.; Springer International Publishing: Cham, Switzerland, 2016; pp. 21–37.
10. Redmon, J.; Divvala, S.; Girshick, R.; Farhadi, A. You Only Look Once: Unified, Real-Time Object Detection. In Proceedings of the 2016 IEEE Conference on Computer Vision and Pattern Recognition (CVPR), Las Vegas, NV, USA, 27–30 June 2016; pp. 779–788. [CrossRef]
11. Chen, R.; Liu, S.; Miao, Z.; Li, F. Infrared aircraft few-shot classification method based on meta learning. *Infrared Millim. Waves* **2021**, *40*, 554–560. [CrossRef]
12. Everingham, M.; Van Gool, L.; Williams, C.K.; Winn, J.; Zisserman, A. The PASCAL Visual Object Classes (VOC) Challenge. *Int. J. Comput. Vis.* **2010**, *88*, 303–338. [CrossRef]
13. Lin, T.Y.; Maire, M.; Belongie, S.; Hays, J.; Perona, P.; Ramanan, D.; Dollár, P.; Zitnick, C.L. Microsoft COCO: Common Objects in Context. In *Computer Vision—ECCV 2014*; Fleet, D., Pajdla, T., Schiele, B., Tuytelaars, T., Eds.; Springer International Publishing: Cham, Switzerland 2014; pp. 740–755.
14. Zhou, T.; Yu, Z.; Cao, Y.; Bai, H.; Su, Y. Study on an infrared multi-target detection method based on the pseudo-two-stage model. *Infrared Phys. Technol.* **2021**, *118*, 103883. [CrossRef]
15. Miao, Z.; Zhang, Y.; Li, W.H. Real-time infrared target detection based on center points. *Infrared Millim. Waves* **2021**, *40*. [CrossRef]
16. Konig, D.; Adam, M.; Jarvers, C.; Layher, G.; Neumann, H.; Teutsch, M. Fully Convolutional Region Proposal Networks for Multispectral Person Detection. In Proceedings of the IEEE Conference on Computer Vision and Pattern Recognition (CVPR) Workshops, Honolulu, HI, USA, 21–26 July 2017.
17. Liu, J.; Zhang, S.; Wang, S.; Metaxas, D.N. Multispectral deep neural networks for pedestrian detection. *arXiv* **2016**, arXiv:1611.02644.
18. Redmon, J.; Farhadi, A. Yolov3: An incremental improvement. *arXiv* **2018**, arXiv:1804.02767.
19. Zhou, X.; Wang, D.; Krähenbühl, P. Objects as points. *arXiv* **2019**, arXiv:1904.07850.
20. Law, H.; Deng, J. CornerNet: Detecting Objects as Paired Keypoints. In *Computer Vision—ECCV 2018*; Ferrari, V., Hebert, M., Sminchisescu, C., Weiss, Y., Eds.; Springer International Publishing: Cham, Switzerland, 2018; pp. 765–781.
21. Zhou, X.; Zhuo, J.; Krahenbuhl, P. Bottom-Up Object Detection by Grouping Extreme and Center Points. In Proceedings of the IEEE/CVF Conference on Computer Vision and Pattern Recognition (CVPR), Long Beach, CA, USA, 15–20 June 2019.
22. Tian, Z.; Shen, C.; Chen, H.; He, T. FCOS: Fully Convolutional One-Stage Object Detection. In Proceedings of the IEEE/CVF International Conference on Computer Vision (ICCV), Seoul, Korea, 27 October–2 November 2019.
23. Teledyne FLIR. Flir Thermal Dataset for Algorithm Training [DB/OL]. FLIR. 1 September 2018. Available online: https://www.flir.com/oem/adas/adas-dataset-form/ (accessed on 7 January 2022).

24. Hosang, J.; Benenson, R.; Schiele, B. Learning Non-maximum Suppression. In Proceedings of the 2017 IEEE Conference on Computer Vision and Pattern Recognition (CVPR), Honolulu, HI, USA, 21–26 July 2017; pp. 6469–6477. [CrossRef]
25. Hinton, G.; Vinyals, O.; Dean, J. Distilling the knowledge in a neural network. *arXiv* **2015**, arXiv:1503.02531.
26. Zhang, Y.; Xiang, T.; Hospedales, T.M.; Lu, H. Deep Mutual Learning. In Proceedings of the 2018 IEEE/CVF Conference on Computer Vision and Pattern Recognition, Salt Lake City, UT, USA, 18–23 June 2018; pp. 4320–4328. [CrossRef]
27. Huang, Z.; Pan, Z.; Lei, B. Transfer Learning with Deep Convolutional Neural Network for SAR Target Classification with Limited Labeled Data. *Remote Sens.* **2017**, *9*, 907. [CrossRef]
28. Romero, A.; Ballas, N.; Kahou, S.E.; Chassang, A.; Gatta, C.; Bengio, Y. Fitnets: Hints for thin deep nets. *arXiv* **2014**, arXiv:1412.6550.
29. Yim, J.; Joo, D.; Bae, J.; Kim, J. A Gift from Knowledge Distillation: Fast Optimization, Network Minimization and Transfer Learning. In Proceedings of the 2017 IEEE Conference on Computer Vision and Pattern Recognition (CVPR), Honolulu, HI, USA, 21–26 July 2017; pp. 7130–7138. [CrossRef]
30. Li, Q.; Jin, S.; Yan, J. Mimicking Very Efficient Network for Object Detection. In Proceedings of the 2017 IEEE Conference on Computer Vision and Pattern Recognition (CVPR), Honolulu, HI, USA, 21–26 July 2017; pp. 7341–7349. [CrossRef]
31. Chen, G.; Choi, W.; Yu, X.; Han, T.; Chandraker, M. Learning efficient object detection models with knowledge distillation. In Proceedings of the 31st International Conference on Neural Information Processing Systems, Long Beach, CA, USA, 4–9 December 2017; pp. 742–751.
32. Chen, H.; Wang, Y.; Wang, G.; Qiao, Y. LSTD: A Low-Shot Transfer Detector for Object Detection. Proceedings of the AAAI Conference on Artificial Intelligence, New Orleans, LA, USA, 2–7 February 2018; Volume 32.
33. Hu, J.; Shen, L.; Sun, G. Squeeze-and-Excitation Networks. In Proceedings of the 2018 IEEE/CVF Conference on Computer Vision and Pattern Recognition, Salt Lake City, UT, USA, 18–23 June 2018; pp. 7132–7141. [CrossRef]
34. Wang, Q.; Wu, B.; Zhu, P.; Li, P.; Zuo, W.; Hu, Q. ECA-Net: Efficient Channel Attention for Deep Convolutional Neural Networks. In Proceedings of the 2020 IEEE/CVF Conference on Computer Vision and Pattern Recognition (CVPR), Seattle, WA, USA, 13–19 June 2020; pp. 11531–11539. [CrossRef]
35. Woo, S.; Park, J.; Lee, J.Y.; Kweon, I.S. CBAM: Convolutional Block Attention Module. In *Computer Vision—ECCV 2018*; Ferrari, V., Hebert, M., Sminchisescu, C., Weiss, Y., Eds.; Springer International Publishing: Cham, Switzerland, 2018; pp. 3–19. [CrossRef]
36. Zhang, Q.L.; Yang, Y.B. SA-Net: Shuffle Attention for Deep Convolutional Neural Networks. In Proceedings of the ICASSP 2021—2021 IEEE International Conference on Acoustics, Speech and Signal Processing (ICASSP), Toronto, ON, Canada, 6–11 June 2021; pp. 2235–2239. [CrossRef]
37. Wei, D.; Du, Y.; Du, L.; Li, L. Target Detection Network for SAR Images Based on Semi-Supervised Learning and Attention Mechanism. *Remote Sens.* **2021**, *13*, 2686. [CrossRef]
38. Yu, F.; Wang, D.; Shelhamer, E.; Darrell, T. Deep Layer Aggregation. In Proceedings of the 2018 IEEE/CVF Conference on Computer Vision and Pattern Recognition, Salt Lake City, UT, USA, 18–23 June 2018; pp. 2403–2412. [CrossRef]
39. Zhang, S.; Wen, L.; Bian, X.; Lei, Z.; Li, S.Z. Single-Shot Refinement Neural Network for Object Detection. In Proceedings of the 2018 IEEE/CVF Conference on Computer Vision and Pattern Recognition, Salt Lake City, UT, USA, 18–23 June 2018; pp. 4203–4212. [CrossRef]
40. Qin, P.; Tang, C.; Liu, Y.; Zhang, J.; Xu, Z. Infrared target detection method based on improved YOLOv3. *Comput. Eng.* **2021**, 1–12 [CrossRef]

*remote sensing*

*Article*

# An Efficient Algorithm for Ocean-Front Evolution Trend Recognition

**Yuting Yang** [1,2,3], **Kin-Man Lam** [2], **Xin Sun** [3], **Junyu Dong** [3,4,5,*] **and Redouane Lguensat** [6,7]

1    College of Computer Science and Engineering, Shandong University of Science and Technology, Qingdao 266590, China; yangyuting@stu.ouc.edu.cn

2    Department of Electronic and Information Engineering, The Hong Kong Polytechnic University, Hong Kong 999077, China; enkmlam@polyu.edu.hk

3    Department of Information Science and Engineering, Ocean University of China, No. 579, Qianwangang Road, Huangdao District, Qingdao 266590, China; sunxin1984@ieee.org

4    Haide College, Ocean University of China, Qingdao 266100, China

5    Institute of Advanced Ocean Study, Ocean University of China, Qingdao 266100, China

6    Laboratoire des Sciences du Climat et de l'Environnement (LSCE-IPSL), 75020 Paris, France; redouane.lguensat@locean.ipsl.fr

7    LOCEAN-IPSL, Sorbonne Université, 75020 Paris, France

*    Correspondence: dongjunyu@ouc.edu.cn

**Abstract:** Marine hydrological elements are of vital importance in marine surveys. The evolution of these elements can have a profound effect on the relationship between human activities and marine hydrology. Therefore, the detection and explanation of the evolution laws of marine hydrological elements are urgently needed. In this paper, a novel method, named Evolution Trend Recognition (ETR), is proposed to recognize the trend of ocean fronts, being the most important information in the ocean dynamic process. Therefore, in this paper, we focus on the task of ocean-front trend classification. A novel classification algorithm is first proposed for recognizing the ocean-front trend, in terms of the ocean-front scale and strength. Then, the GoogLeNet Inception network is trained to classify the ocean-front trend, i.e., enhancing or attenuating. The ocean-front trend is classified using the deep neural network, as well as a physics-informed classification algorithm. The two classification results are combined to make the final decision on the trend classification. Furthermore, two novel databases were created for this research, and their generation method is described, to foster research in this direction. These two databases are called the Ocean-Front Tracking Dataset (OFTraD) and the Ocean-Front Trend Dataset (OFTreD). Moreover, experiment results show that our proposed method on OFTreD achieves a higher classification accuracy, which is 97.5%, than state-of-the-art networks. This demonstrates that the proposed ETR algorithm is highly promising for trend classification.

**Keywords:** remote sensing; video signal process; sea surface

**Citation:** Yang, Y.; Lam, K.-M.; Sun, X.; Dong, J.; Lguensat, R. An Efficient Algorithm for Ocean-Front Evolution Trend Recognition. *Remote Sens.* **2022**, *14*, 259. https://doi.org/10.3390/rs14020259

Academic Editor: Yue Wu

Received: 1 December 2021

Accepted: 23 December 2021

Published: 6 January 2022

**Publisher's Note:** MDPI stays neutral with regard to jurisdictional claims in published maps and institutional affiliations.

## 1. Introduction

The ocean dynamic process contains essential factors that characterize and reflect the ocean hydrological status and phenomena. Detection, localization, and classification of their formation and interaction processes are essential in various ocean-related fields, such as fisheries and global warming. Several ocean-related variables have been identified for the ocean dynamic process, such as ocean currents, ocean tides, inner waves, ocean fronts, mesoscale vortices [1], etc. The oceanfront is an important branch of the ocean dynamic process [2–4]. Specifically, ocean fronts are located at the boundary between water masses with different properties [5,6], such as density, temperature, salinity, etc. Changes in the strength and scale of ocean fronts are some of the most vital subjects being studied, because they play an important role in the coupling of winds and the ocean processes [7,8]. For example, water masses in the ocean-front system have a great effect on air-sea exchange [9–11], activate the biological activity of the region [12], and absorb atmospheric carbon dioxide [13,14].

For marine fishing and marine environmental protection, it is vitally important to characterize the trend of the ocean front [15–19]. In fact, identifying the trend of an oceanfront is a difficult task, because simply working on short snippets cannot provide sufficient information to recognize it. The key to achieving high trend-recognition accuracy is to extract features from the consecutive frames, i.e., a video clip. The video sequence should include the whole process of an ocean-front trend. Usually, the length of a sequence is no more than 200 frames. In our dataset, we choose videos containing 5 to 200 frames. To a certain extent, action recognition is similar to ocean-front trend recognition. Recognizing the actions in a video, e.g., walking, jumping, etc., requires observing the entire motion process. Similarly, we have to consider a certain number of consecutive frames for recognizing the trend of an oceanfront, which is in either an enhancement or attenuation state.

In our previous work [20–25], both traditional machine-learning methods and deep neural networks were introduced to detect, recognize and predict ocean fronts and eddies. However, to the best of our knowledge, there is little previous work trying to recognize ocean-front trends based on oceanfront video sequences, but there are plenty of works trying to recognize or classify actions based on surveillance video sequences. Action classification [26–28] is an active field of research attracting increasing attention, due to its numerous potential applications in surveillance, video analysis, etc. The long-standing research on this classification task can be roughly divided into two categories. The first category relies on statistical feature extraction, followed by classifiers [29,30], while the second category is based on convolutional neural networks (CNNs). Examples of methods based on statistical features include [31–33]. However, these methods have limited generalization ability compared with CNNs. CNNs, which replace handcrafted features with "learned-from-data" features, have been successfully used for image classification [34,35]. Specifically, deep-learning-based methods [36–40] have achieved remarkable progress in video analysis.

According to our previous work, deep learning models are promising methods for ocean-front recognition and prediction. Thus, in this paper, we propose to use deep learning methods to classify ocean-front evolution trends. However, if deep learning models, such as CNNs, are directly applied to a video sequence, Karpathy et al. [41] found that the recognition performance achieved was inferior, compared with the state-of-the-art statistical features. Besides, inspired by the success of the region-proposal methods for object detection [42,43], some methods have attempted to extract temporal information from short snippets [28,44,45], by sparsely sampling from a long video sequence.

To improve the classification accuracy, a two-stream deep model [46], consisting of a spatial and a temporal CNN, was proposed, which achieved comparable performance with the most representative statistical features. One major limitation of the two-stream CNNs is that the method pays too much attention to the features extracted from a single RGB frame and the short-term motions, rather than the entire temporal information. Those frames, which are not within the selected short snippets of the video, may contain important temporal information, which can help improve the classification accuracy. Therefore, the deep model dismisses some useful temporal information. On the contrary, statistical features have an advantage in extracting the temporal information by using a specifically designed feature extraction algorithm based on prior knowledge. Therefore, in this paper, we propose a new fusion method for recognizing the ocean-front trend. We propose new statistical algorithms, which can extract temporal information from a video sequence, and we also apply a deep learning model to learn the deep feature from the video sequence, we then use weighted fusion to incorporate temporal information to improve classification accuracy. In our experiments, we prove that the proposed method can achieve high classification accuracy, better than using state-of-the-art deep-learning-based methods.

The novelty of this paper is twofold. (1) We introduce an Evolution Trend Recognition (ETR) method, which is based on classifiers with prior physical knowledge. The method not only gets rid of the complex operations required for selecting the frames with ocean fronts from a video sequence but can also aggregate the information extracted from different classification methods. (2) We have created a new database for ocean-front trend recognition, to encourage other researchers to evaluate their methods for ocean-front trend classification and facilitate them in using data-driven methods, especially deep-learning-based methods, to deal with this challenging task.

More specifically, our ETR method uses an effective mechanism to combine results from classification algorithms based on strength and scale, and employs deep-learning-based classification methods, based on the GoogLeNet Inception network [47], to recognize the ocean-front trend. Our experiment results show that the proposed ETR method achieves superior recognition performance over state-of-the-art methods on the Ocean-front Trend Dataset (OFTreD).

The remainder of this paper is organized as follows. The ETR framework and the process of building the OFTreD and Ocean-front Tracking Dataset (OFTraD), used in our experiments, are presented in detail in Section 2. Experimental results are presented in Section 3 and discussed in Section 4, and finally, Section 5 concludes this paper.

## 2. Materials and Methods

### 2.1. The Proposed Method

Extracting representative features from a video sequence is of prime importance for the task of ocean-front trend recognition. In this section, we will describe a novel idea for extracting discriminative features for recognizing the ocean-front trend, based on the analysis of a whole video. The key idea of the proposed method is shown in Figure 1, the proposed trend recognition method relies on the combination of the statistical algorithms and deep learning models. Softmax classifier is then applied for trend recognition of enhancement and attenuation. The proposed method avoids the complex operations required for selecting recommended frames, because the proposed method can extract representative temporal and deep features from the video sequence, and hence, it is efficient and effective.

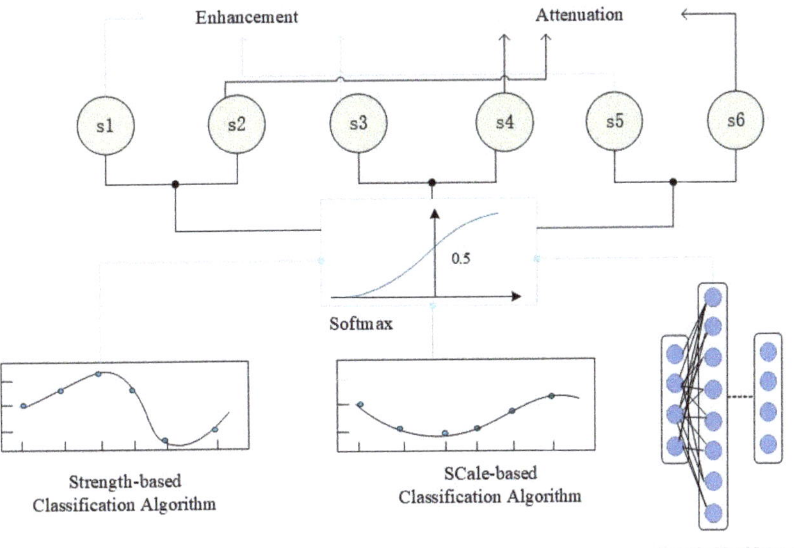


**Figure 1.** The proposed trend recognition method is composed of the strength-based algorithm, the scale-based algorithm, and the GoogLeNet network. Each of these three parts will be processed by the softmax classifier and give 2 scores. The scores then are used to recognize the enhancement and attenuation oceanfront.

In this section, we first described the network structure of the proposed recognition method in Section 2.1.1. Then, we described the ocean-front classification algorithm based on strength and scale feature in Sections 2.1.2 and 2.1.3, respectively. In Section 2.1.4, we explained the feature matrices generation method. Then, we described the ocean-front trend classification algorithm based on the GoogLeNet Inception network in Section 2.1.5. Finally, we described the ocean-front tracking algorithm in Section 2.1.6.

### 2.1.1. Network Structure

The proposed recognition framework, which is composed of three parallel networks, is depicted in Figure 2. The first and second networks are designed for trend classification based on prior physical knowledge, which will be explained in Sections 2.1.2 and 2.1.3. Their inputs are the video sequences from OFTreD. The OFTreD database is proposed for the ocean-front trend recognition task. The third network is also designed for ocean-front trend classification, based on GoogLeNet Inception, whose input is the optical flow images extracted from the video sequences in OFTreD. The first, second and third networks are integrated to classify the ocean-front trends. In this paper, two kinds of ocean-front trends are defined, namely, the enhancement trend and attenuation trend. In Figure 2, Score A and Score B are used to classify the ocean-front trend. The value of Score A denoted as $s_A$, represents the probability that an oceanfront enhancement trend, and that of Score B, denoted as $s_B$, represents the probability that an oceanfront has an attenuation trend. The scores $s_A$ and $s_B$ are computed as follows:

$$s_A = w_1 \times s_1 + w_2 \times s_3 + w_3 \times s_5 \tag{1}$$

$$s_B = w_1 \times s_2 + w_2 \times s_4 + w_3 \times s_6 \tag{2}$$

where $w_i, i = 1, 2, 3$, are the weights, whose values will be discussed in Section 4. $s_j$, $j = 1, ..., 6$, represents the value of Score $j$ in Figure 2. The larger score of $s_A$ and $s_B$ will be used to determine the ocean-front trend category.

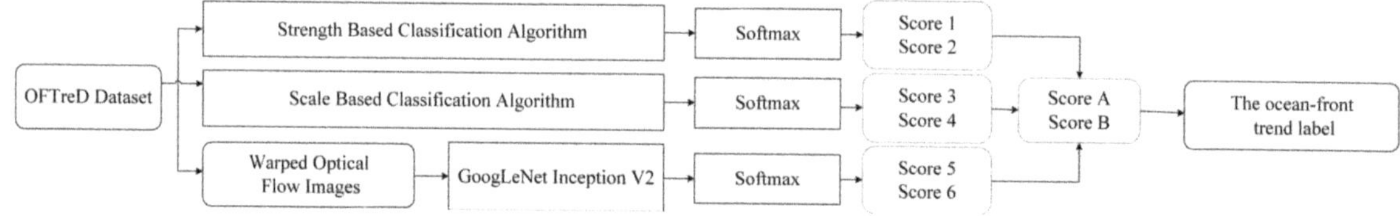


**Figure 2.** The overall network architecture. The input video sequences are fed to three parallel networks. The input frames of the Ocean-Front Trend Database (OFTreD) are fed to the strength and scale-based classification algorithms directly without pre-processing. However, the input frames of the OFTreD dataset are pre-processed to form warped optical flow images, before feeding to a GoogLeNet Inception network. Besides, for the three parallel networks, Scores 1 to 6 are produced. Scores 1, 3, and 5 are combined to obtain the Score A, and scores 2, 4, and 6 are combined to obtain the Score B according to Equations (1) and (2). These two scores are used to determine if the oceanfront is under enhancement or attenuation, Score A is for enhancement and Score B for attenuation.

Each of the three proposed networks ends with a softmax layer, which outputs two scores to represent the probabilities of the input video sequence belonging to the enhancement or the attenuation trend. In total, six classification scores are generated. The six scores, i.e., Score 1 to Score 6, are used to classify whether the oceanfront is enhancing or attenuating. In our experiments, Scores 1, 3, and 5 are used to represent the probabilities of belonging to the enhancement trend, while the Scores 2, 4, and 6 are used to represent the attenuation trend. An ocean front in a video sequence belongs to either the "enhancement" class or the "attenuation" class. Finally, we integrate these six weighted scores to make the final decision on the trend class.

As shown in Figure 3, we also propose an oceanfront tracking algorithm to check whether the current input video sequence contains an oceanfront and where the ocean-front trend is in the video sequence. For this task, we train a GoogLeNet network on the OFTraD dataset.

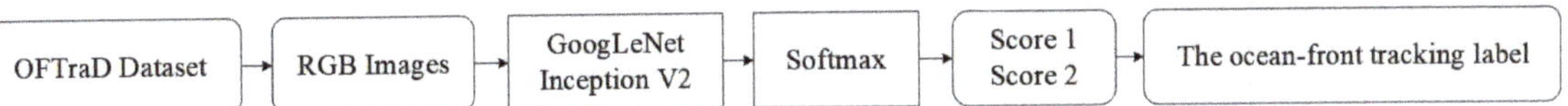


**Figure 3.** The procedure of the proposed ocean-front tracking algorithm. The input RGB images are fed to a GoogLeNet Inception network. Then, the softmax gives the probabilities of the input image belonging to the foreground or the background. The foreground images are used to track the ocean-front location in a video sequence.

The input of this network is the RGB images from OFTraD. The network is used to determine whether the input belongs to the background or the foreground. Those images that contain a tracking target, i.e., an oceanfront, belong to the foreground class, otherwise, they belong to the background class. Based on the location information carried by the input images, the output labeled images can be reconstructed into ocean-front video sequences, and then the ocean-front trend in the video sequences can be tracked.

### 2.1.2. Ocean-Front Classification Algorithm Based on Strength

The ocean-front trend classification algorithms based on strength and scale are trained on OFTreD. As shown in Figure 4, we calculate the mean intensity of the oceanfront to represent the oceanfront strength information of a frame. For the scale, we count the number of pixels of the oceanfront in each frame and use it to represent the scale information for the frame.

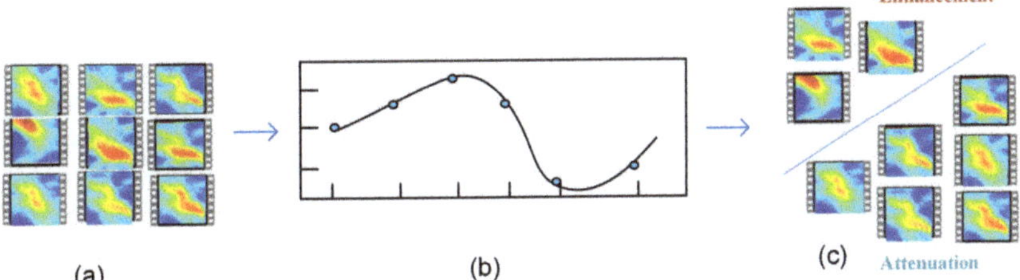


**Figure 4.** The ocean-front classification algorithms are based on strength and scale. These algorithms are very similar. First, the feature values are calculated from the video sequences (**a**). Then, these values are used to fit a curve (**b**). After that, we can extract points from the curve to form a matrix. Then, this matrix is processed and fed to softmax for classification. The scores hence can be acquired and used to label the enhancement and attenuation classes (**c**).

To improve the classification accuracy, we focus on the classification of ocean-front trends in a video sequence, rather than the snippets of a video. We analyze the overall ocean-front trend in a video sequence, based on the ocean-front strength and scale. The ocean-front strength can be represented by the numerical intensity of an oceanfront, while the ocean-front scale can be represented by the area of the existing oceanfront. Since the scale and strength of an oceanfront are highly correlated with the ocean-front trends, they can be used to effectively infer the trend of an oceanfront in a video sequence. Based on this prior knowledge, the scale and strength information of an ocean-front video sequence is used as an important reference for formulating the corresponding feature matrices $B_1$ and $B_2 \in R^{(H-1) \times W}$, where $W$ is the number of representative points extracted from a feature curve, and $H$ represents the number of frames in the video. The details of computing the strength and scale feature matrices for an ocean-front video are shown in Algorithms 1 and 2, respectively. The method of generating a feature curve and extracting representative points from the feature curve will be described later in Section 2.1.4.

---

**Algorithm 1** Classification Algorithm Based on Strength

---

1: **Input:** A video $v_i(x,y), i = 1, ..., H$, where $H$ is the number of frames, and $(x,y)$ are the pixel coordinates

2: **for** $i = 1$ to $H$ **do**

Calculate the mean intensity of the ocean front in the frame $i$, denoted as $m_s(i)$, which is computed as follows:

$$m_s(i) = \frac{1}{n_r \times n_c} \sum_{x=1}^{n_c} \sum_{y=1}^{n_r} v_i(x,y), \tag{3}$$

where $n_r$ and $n_c$ are the number of rows and columns, respectively, in a frame.

3: **end for**

Generate the mean intensity vector $a_1 = [m_s(1)...m_s(H)]^T$ for the video.

Apply a curve fitting technique to $a_1$ to form the feature curve $V_1$, and sample the curve $V_1$ with $(H - 1) \times W$ points, where $W$ is the number of representative points of each frame. In our experiments, $W$ is set at 10. The sampled points then formulate a matrix $B_1 \in R^{(H-1) \times W}$.

Then, use an average pooling filter to process the matrix $B_1$ to generate the resulting vector $c_1$. The resulting elements are denoted as $m_f$, and hence the vector $c_1 = [m_f(1)...m_f(40)]^T$, whose dimension is set at $40 \times 1$ in our implementation. $c_1$ is the feature vector with unified dimension for trend classification.

Use the trained softmax to classify the vector $c_1$

4: **Output** Classification scores $s_1, s_2$

---

---

**Algorithm 2** Classification Algorithm Based on Scale

---

1: **Input:** A video containing $H$ frames

2: **for** $i = 1$ to $H$ **do**

Count the number of ocean-front points in the frame $i$, denoted as $n_s(i)$, which are detected using the oceanfront detection method [22].

3: **end for**

Count the number of ocean-front points for each of the $H$ frames, to form the vector $a_2 = [n_s(1)...n_s(H)]^T$ for the video.

Apply a curve fitting technique to $a_2$ to form the curve $V_2$, and sample the curve $V_2$ with $(H - 1) \times W$ points, where $W$ is the number of representative points of each frame. In our experiments, $W$ is set at 10. The sampled points then formulate a matrix $B_2 \in R^{(H-1) \times W}$.

Then, use an average pooling filter to process the matrix $B_2$, get the resulting vector $c_2$. The resulting elements are denoted as $n_f$, and hence the vector $c_2 = [n_f(1)...n_f(40)]^T$, whose dimension is set at $40 \times 1$ in our implementation. $c_2$ is the feature vector with unified dimension for trend classification.

Use the trained softmax to classify the vector $c_2$

4: **Output** Classification scores $s_3, s_4$

---

### 2.1.3. Ocean-Front Classification Algorithm Based on Scale

With the proposed algorithms, we will illustrate how to extract the strength and scale information about the oceanfront in a video sequence and the databases used for training and testing. Algorithm 1 is designed for recognizing ocean-front trends based on the strength of an oceanfront. To classify the trend, we need to compute the variations of the ocean-front strength. Since the strength of an oceanfront varies from point to point, we propose to use the mean intensity of an oceanfront in a frame to represent its strength. Similarly, Algorithm 2 is designed to classify the ocean-front trend based on its scale. The scale of an oceanfront is calculated based on the number of oceanfront points in a frame. The greater the number of ocean-front points, the larger the ocean-front scale is.

Here, the vectors $a_1$ and $a_2$ represent the strength and scale information, respectively, and the matrices $B_1$ and $B_2$ represent the points extracted from the corresponding curves, and the feature vectors $c_1$ and $c_2$ represent the filtered output from the corresponding matrices $B_1$ and $B_2$. Thus, the feature vectors $c_1$ and $c_2$ represent the processed strength and scale information, respectively. We use the feature vectors $c_1$ and $c_2$ to classify the ocean-front trend. Then, these feature vectors are sent to softmax for classification and generate the output $s_i, i = 1, 2, 3, 4$.

### 2.1.4. Feature Matrices Generation Method

In the ocean-front trend algorithms, the number of frames of different videos may be different, so the dimensions of the strength vector $a_1$ and the scale vector $a_2$ of different videos, as described in Algorithms 1 and 2, respectively, are different. To make the two vectors always have the same length, Algorithms 1 and 2 apply curve fitting to the vectors $a_1$ and $a_2$, then resamples the two curves with a fixed number of points. Specifically, as shown in Figure 5, we use the cubic polynomial interpolation method to fit the curves. With a fixed number of points on the curve, two matrices, $B_1$ and $B_2 \in R^{(H-1) \times W}$, are generated. The matrices generation process is shown in Figure 6, starting from the point representing the strength/scale of the first frame, we sample points on the curve at regular intervals until the point that represents the last frame. We set $W = 10$ in our experiments, because we need to extract more than 40 points from the curve. As analyzed in Section 4, the best vector dimension is $40 \times 1$, too small will not meet the requirement, too large is unnecessary. After that, the matrices $B_1$ and $B_2$ are processed by three pooling filters to obtain fixed-dimensional vectors $c_1$ and $c_2$.

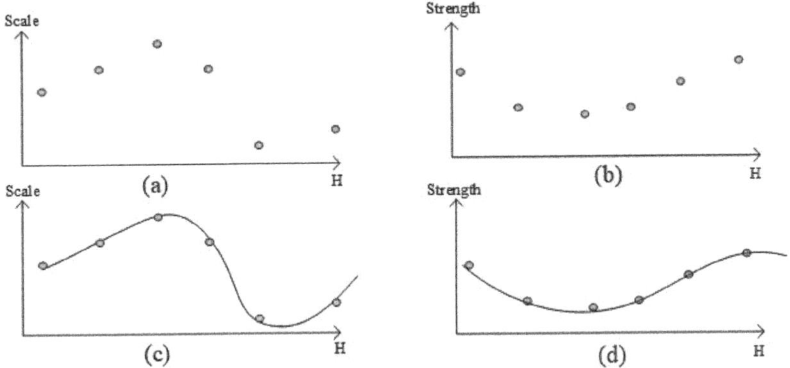

**Figure 5.** The curve fitting technique. The strength and scale features extracted from each frame are represented by a point in (**a,b**). Therefore, the number of the points is equal to the frame number. Then, using the cubic polynomial interpolation method to fit the curves, we get feature curve (**c,d**).

Given matrices $B_1$ and $B_2 \in R^{(H-1) \times W}$, we vectorize the matrices $B_1$ and $B_2$ to acquire the feature vectors $b_1$ and $b_2$. The elements of the matrices $B_1$ and $B_2$ are denoted as $m_p$ and $n_p$, and hence the vector $b_1 = [m_p(1)...m_p((H-1) \times W)]^T$, $b_2 = [n_p(1)...n_p((H-1) \times W)]^T$, whose dimension is $[(H-1) \times W, 1]$. As shown in Algorithm 3, according to the dimension of the matrices $B_1$ and $B_2$, we use different pooling filters. If the dimension of the feature vectors $b_1$ and $b_2$ is greater than $200 \times 1$, average pooling is performed every 5 elements from the first and the last 50 elements in the feature vectors, that is $b_1[1:50,1]$, $b_1[(H-1) \times W - 49 : (H-1) \times W, 1]$, $b_2[1:50,1]$, and $b_2[(H-1) \times W - 49 : (H-1) \times W, 1]$, the filter size is $[5,1]$. Then, We assign $c_1[1:10,1]$, $c_1[31:40,1]$, $c_2[1:10,1]$, and $c_2[31:40,1]$ the value of the processed data. Then, the number of the remaining elements in the feature vectors $b_1$ and $b_2$ is $(H-1) \times W - 100$. The pooling size is set at $((H-1) \times W - 100)/20 \times 1$, the stride is set at $((H-1) \times W - 100)/20$. Average pooling is performed every $((H-1) \times W - 100)/20$ elements from $b_1[51:(H-1) \times W - 50, 1]$

and $b_2[51 : (H-1) \times W - 50, 1]$. And then we assign $c_1[11 : 30, 1]$ and $c_2[11 : 30, 1]$ the value of the processed data.

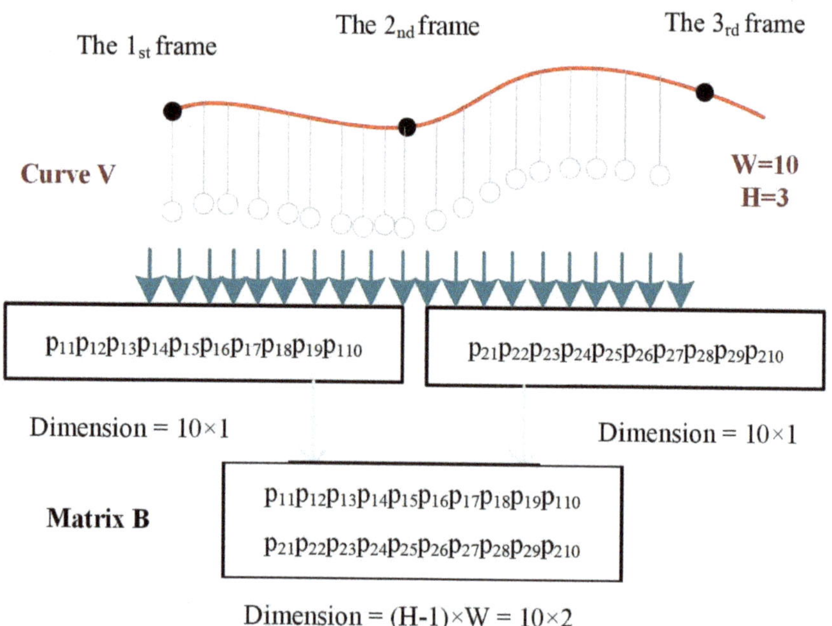


**Figure 6.** The construction of the matrices $B$. 10 points are sampled from every two adjacent frames, there are totally $(H-1) \times W$ points sampled from the curve. The sampled points are sorted into a matrix $B \in R^{(H-1) \times W}$.

---

**Algorithm 3** The matrix processing method

---

1: **Input:** Matrices $B_1$ and $B_2$
    Given matrices $B_1$ and $B_2 \in R^{(H-1) \times W}$, we vectorize them to acquire its feature vectors $b_1$ and $b_2$.
2: **if** the dimension of the feature vectors $b_1$ and $b_2 > [200, 1]$ **do**
    Average pooling is performed every 5 elements from the first 50 elements and the last 50 elements of the matrices, the filter size is $[5, 1]$, the stride is 5. The processed data is assigned to $c_1$ and $c_2$. Average pooling is applied to the remaining elements in the feature vectors $b_1$ and $b_2$, the filter size is set according to the number of the remaining elements.
3: **else if** the dimension of the feature vectors $b_1$ and $b_2 > [100, 1]$ **do**
    Average pooling is performed every 2 elements from the first and the last 30 elements, the filter size is $[2, 1]$, the stride is 2. The processed data is assigned to $c_1$ and $c_2$. Average pooling is applied to the remaining elements in the feature vectors $b_1$ and $b_2$, the filter size is set according to the number of the remaining elements.
4: **else do**
    The first and the last 15 elements of the vectorized matrices $B_1$ and $B_2$ are assigned to $c_1$ and $c_2$. Average pooling is applied to the remaining elements in the feature vectors $b_1$ and $b_2$, the filter size is set according to the number of the remaining elements.
5: **Output** Feature vectors $c_1$ and $c_2$

---

Otherwise, if the the dimension of the feature vectors $b_1$ and $b_2 > [100, 1]$, average pooling is performed every 2 elements from the first and the last 30 elements in the feature vectors, that is $b_1[1 : 30, 1]$, $b_1[(H-1) \times W - 29 : (H-1) \times W, 1]$, $b_2[1 : 30, 1]$, and $b_2[(H-1) \times W - 29 : (H-1) \times W, 1]$, the filter size is $[2, 1]$. We assign $c_1[1 : 15, 1]$,

$c_1[26:40,1]$, $c_2[1:15,1]$, and $c_2[26:40,1]$ the value of the processed data. Then, the number of the remaining elements in the feature vectors $b_1$ and $b_2$ is $(H-1) \times W - 60$. The pooling size is set at $((H-1) \times W - 60)/10 \times 1$, the stride is set at $((H-1) \times W - 60)/10$. Average pooling is performed every $((H-1) \times W - 60)/10$ elements from $b_1[31:(H-1) \times W - 30]$ and $b_2[31:(H-1) \times W - 30]$. We assign $c_1[16:25,1]$ and $c_2[16:25,1]$ the value of the processed data, assign $c_1[11:30,1]$ and $c_2[11:30,1]$ the value of the processed data.

If the dimension of the feature vectors $b_1$ and $b_2 < [100,1]$, we assign $c_1[1:15,1]$, $c_1[26:40,1]$, $c_2[1:15,1]$, and $c_2[26:40,1]$ the value of the first and the last 15 elements in the feature vectors, that is $b_1[1:15,1]$, $b_1[(H-1) \times W - 14:(H-1) \times W,1]$, $b_2[1:15,1]$, and $b_2[(H-1) \times W - 14:(H-1) \times W,1]$. Then, the number of the remaining elements in the feature vectors $b_1$ and $b_2$ is $(H-1) \times W - 30$. The pooling size is set at $((H-1) \times W - 30)/10 \times 1$, the stride is set at $((H-1) \times W - 30)/10$. Average pooling is performed every $((H-1) \times W - 30)/10$ elements of $b_1[16:(H-1) \times W - 15,1]$ and $b_2[16:(H-1) \times W - 15,1]$. Then we assign $c_1[16:25,1]$ and $c_2[16:25,1]$ the value of the processed data. In this way, feature vectors $c_1$ and $c_2$ can be constructed.

### 2.1.5. Ocean-Front Trend Classification Algorithm Based on GoogLeNet

The structure of the GoogLeNet is shown in Figure 7, the Inception block helps to handle the high-dimensional features and balance the width and depth of the network. It also enables the network to perform spatial aggregation in low-dimensional features without worrying about losing too much information. So, we apply this network to recognize the ocean-front trend and track the ocean-front location.

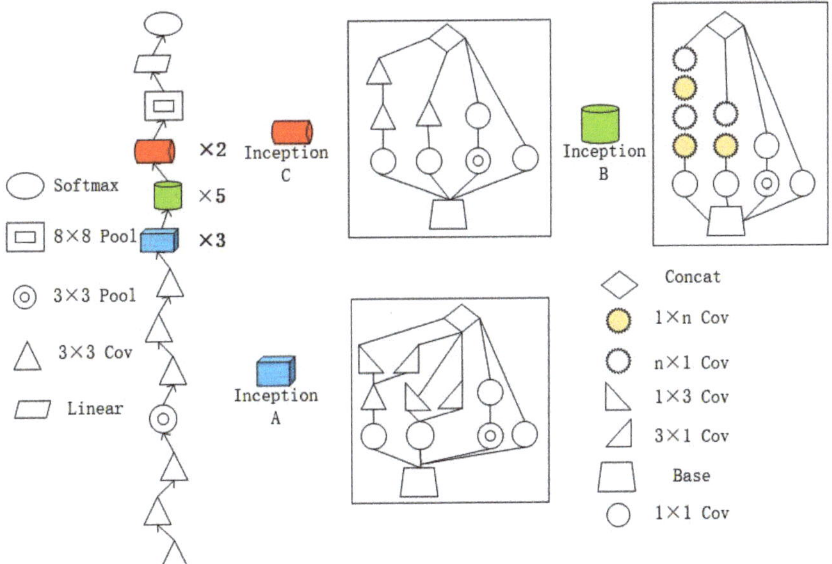


**Figure 7.** The architecture of GoogLeNet Inception V2 network [48]. Its basic convolutional block is named Inception. There are three kinds of Inception blocks in the network, Inception A, Inception B, and Inception C, respectively.

Figure 8 shows the process of the ocean-front trend recognition, GoogLeNet Inception network is employed to classify enhancement and attenuation of an oceanfront. The video input is warped by using the optical flow method. The GoogLeNet Inception network is trained and tested on the OFTreD dataset. The video sequence is first processed into warped optical flow images. Then, these images are sent to the GoogLeNet Inception network for classification. The softmax layer of the network generates the scores $s_i$, $i = 5, 6$, which are used to label the video sequence as an enhancement or attenuation trend.

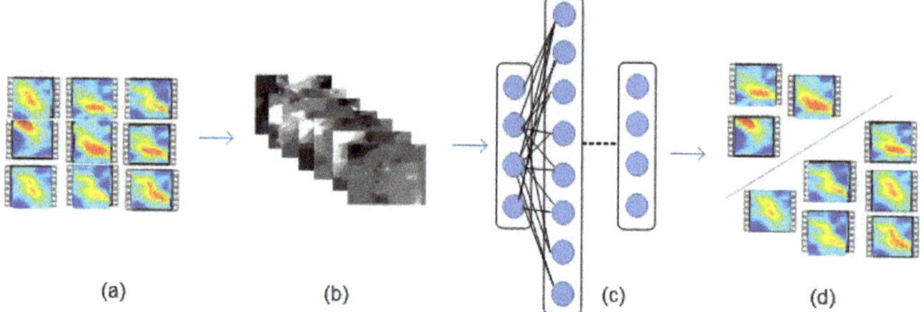

(a)          (b)          (c)          (d)

**Figure 8.** The oceanfront classification network is based on the GoogLeNet Inception network, including the following steps. First, the video sequence (**a**) is processed into warped optical flow images (**b**). Then, these images are sent to the GoogLeNet Inception network (**c**) for classification. The softmax layer (**d**) of the network produces the final scores, which are used to label the video sequence as an enhancement or attenuation trend. The ocean-front tracking algorithm. Firstly, the images are sent to the GoogLeNet Inception network to perform classification. The foreground images are changed to white, and the background image blocks remain unchanged. Finally, the images are used to reconstruct the video sequence.

### 2.1.6. Ocean-Front Tracking Algorithm Based on GoogLeNet

As shown in Figure 9, the ocean-front tracking algorithm is also based on the GoogLeNet Inception network. The network is used to classify image blocks into two classes: the oceanfront and the background. We first colored the oceanfront image blocks in white, and then, we further use the location information and place them back to the same position in the original frame. In this way, we can track the ocean-front location in a video sequence. It is worth noting that this network is trained on OFTraD, with 8000 and 2000 image blocks from the database used for training and testing, respectively.

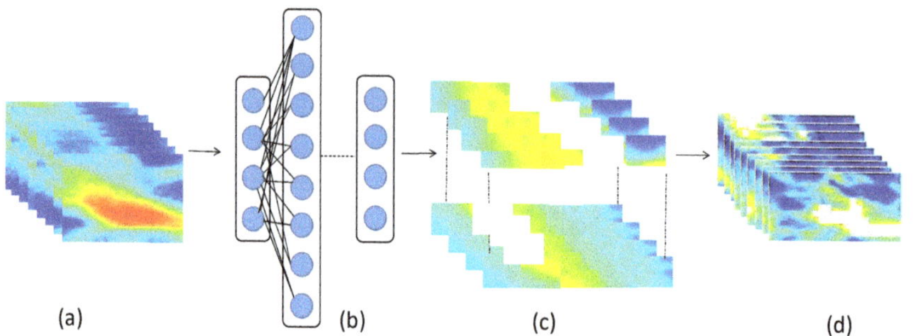

(a)          (b)          (c)          (d)

**Figure 9.** The ocean-front tracking algorithm. Firstly, the images (**a**) are sent to the GoogLeNet Inception network (**b**) to perform classification. The foreground images are changed to white, and the background image blocks remain unchanged (**c**). Finally, the images are used to reconstruct the video sequence (**d**).

The input of our algorithm is the image blocks and their time-position information. Firstly, we extract the RGB image blocks from each video sequence, and feed them into the GoogLeNet Inception network for classification. The color of the image blocks is set to white, if the image block is classified as the oceanfront. In our experiments, the block size is set to $5 \times 5$, because this size can cover mesoscale ocean fronts. If the size of the blocks is too large, it will be hard to find the exact location of the background. If the block is too small, the classification accuracy will be reduced. Then, according to the corresponding time-position information, the blocks are put together to form a video sequence. When dividing an ocean-front frame into image blocks, we label their names with the time-position information, so that when getting their classification labels, we can put them back to their original time-position location. Therefore, the location of the oceanfront in a video sequence can be located.

### 2.2. Construction of the Dataset

To the best of our knowledge, there is no public database available for ocean-front trend classification. This may be one of the reasons why ocean-front trend classification is a difficult task. In this paper, one of our contributions is the creation of the two training databases: OFTreD and OFTraD. The OFTreD contains 1000 video sequences, and the number of image blocks of OFTraD reaches 10,000. 90% of the video sequences are used for training, and 10% are used for testing. 80% of the image blocks are used for training, and 20% are used for testing. We believe that our work will inspire more researchers to research trend classification and will be used as a benchmark for this new research area. The microcanonical multiscale formalism (MMF) will first be described in detail, and then used to detect the ocean front.

#### 2.2.1. Microcanonical Multiscale Formalism

In this paper, we aim to recognize an oceanfront and classify it into either the enhancement or the attenuation type. To recognize an ocean-front trend, we need to detect and locate the oceanfront from remote sensing images. Currently, ocean-front detection methods can be roughly divided into three categories. The methods in the first category are those based on the computation of the vertical and horizontal gradients [49,50]. In the second category, the methods make use of the ocean-water characteristics for ocean-front detection, since ocean fronts are often located at the boundary of two or more ocean waters with different characteristics. These methods include those based on histogram representations [51] and those based on the MMF [22,52]. The third category includes those based on data-driven methods, such as deep neural networks [20]. Each of these categories has its own advantages. In this paper, we use MMF, because it is efficient, accurate, stable, and has been one of the best automatic ocean-front detection approaches.

To extract an oceanfront from a video sequence, we use the mathematical formalism, which is computed based on the strength variations between adjacent pixels. By using MMF [22], physical processes, like ocean fronts and eddies, can be easily recognized, and then a deep neural network [20] can be used to classify them.

The key point of MMF is the accurate computation of the Singular Exponent (SE) value $h(\vec{x})$ at pixel position $x$. In this context, the method proposed in [53] provides numerically stable computation of the SE value at each pixel, as follows:

$$h(\vec{x}) = \frac{\frac{\log(\tau_\psi \mu(\vec{x},r_0))}{<\tau_\psi \mu(.,r_0)>}}{\log r_0} + o\left(\frac{1}{\log r_0}\right) \qquad (4)$$

where $r_0$ is used for image normalization. Given an image with the size of $N \times M$, $r_0 = \frac{1}{N \times M}$. $< \tau_\psi \mu(.,r_0) >$ is the average value of the wavelet coefficients of the whole signal, and $\tau_\psi \mu(x,r_0)$ is the wavelet projection at point $x$. The smallest SE, namely the Most Singular Manifold (MSM), corresponds to the strongest temperature variations in the SST image, i.e., the oceanfront. The MSM is defined as follows:

$$F_\infty = \vec{x} : h(\vec{x}) = h_\infty = \min(h(\vec{x})) \qquad (5)$$

To simplify the detection task, we use the inverse of SE. This is because it is more desirable to recognize and track the more obvious parts of an image.

#### 2.2.2. Ocean-Front Trend Database (OFTreD)

The OFTreD takes the time-space variations of ocean fronts into account. The video sequences were taken from the Advanced Very High-Resolution Radiometer (AVHRR) satellite, which has a high-resolution imaging system and can collect images with a resolution of 5 km. Our databases focus on the videos captured in the Atlantic Ocean and the

Pacific Ocean, from 2010 to 2015. We created a total of 1000 ocean-front video sequences. Then, we divided these video sequences into ocean-front enhancement and attenuation classes, according to the trend of the ocean fronts in the video sequences. In the process of creating the databases, an ocean front is classified to have an enhancement trend, if its tendency is becoming larger and stronger. However, a part of the oceanfront with the enhancement trend may become weaker and smaller in a short snippet of a video sequence. In the same way, an ocean front with the attenuation trend tends to become smaller and weaker. It is also possible that a part of the oceanfront with the attenuation trend becomes larger and stronger in short snippets. The existence of this phenomenon is determined by the variability and irregularity characteristics of the ocean fronts. In OFTreD, the number of frames in the ocean-front video sequences ranges from 5 to 200, and the size of each frame is always larger than $20 \times 40$. These characteristics can ensure the robustness of the database. If the frame number is too short or too long, it is difficult to classify its trend. If the size is too small, it may not be able to cover an ocean front.

This database was created based on the efforts of six graduate students, with expertise in oceanography. Each student labeled about 200 video sequences, and then, checked the correctness of the video sequences labeled by the other five students. On average, it took about 20 min to label one video sequence. In total, the students took two weeks to complete the labeling and checking tasks for this database.

In addition, in order to facilitate calibration, we start by randomly selecting an area of the selected ocean and randomly selecting a frame. Then, we display the ocean-front images of the same area 20 days before and after. Thus, we need to check whether the area contains an oceanfront. If an ocean front exists, we change the time-span and choose the suitable start and end frames of the video sequence. Otherwise, another frame will be chosen randomly. The space-time information of the selected frames is also recorded automatically.

We invited a number of oceanographic experts to check the classification results of the 1200 video sequences created, and eliminated 200 of them, which are hard to classify. The difficult sequences contain many ocean fronts, each ocean-front has its own trend. The variation of the speed of the ocean-front trends is another factor that increases the classification difficulty. However, this is a problem we should solve. Therefore, in this research, we locate the ocean fronts in a video sequence, followed by identifying which parts of the ocean fronts are enhancing and which parts are attenuating.

### 2.2.3. Ocean-Front Tracking Dataset (OFTraD)

The construction procedures of the ocean-front tracking database can be summarized as the following steps. First, we split each frame in an oceanfront video sequence into multiple fixed-size image blocks. The time-space-position information of each image block is also recorded. Then, each image block is sequentially, from left to right and from top to bottom, sent to the GoogLeNet Inception network for classification. The image blocks are rearranged into frames so that we can locate the position of the oceanfront from frame to frame.

### 3. Results

The environment configuration used in our experiments is Ubuntu16.04 + GeForce GTX 1080 GPU card + Caffe deep learning framework [54]. The algorithm proposed in this paper is partly based on the GoogLeNet Inception network. Fine-tuning is performed to the pre-trained GoogLeNet Inception network [48] to reduce the negative impact of using a small dataset and to improve the classification accuracy. Furthermore, we apply the TVL1 method [55] to extract optical-flow images. Every two consecutive frames can generate one warped optical-flow image, and these optical-flow images can be used to capture the tendency of the oceanfront between two consecutive frames. Experiment results show that our algorithm is robust, efficient and effective.

As shown in Figure 10, we use the ocean-front tracking algorithm to obtain the position of the ocean fronts in a video sequence. Sixteen representative frames were selected as examples. Figure 10 displays the frames of an ocean-front sequence with the enhancement trend on the top row, the frames with the attenuation trend on the second row, and their tracking label in the third and the bottom rows, respectively.

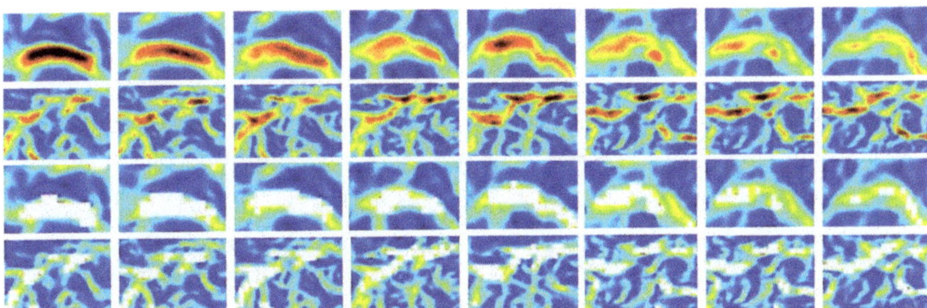

**Figure 10.** The ocean-front trend example. From the top to bottom is an ocean-front attenuation video sequence, an ocean-front enhancement video sequence, an ocean-front attenuation video sequence tracking label, and an ocean-front enhancement video sequence tracking label.

To verify the effectiveness of the ocean-front tracking method, a comparison experiment is carried out. The comparison methods include the traditional method, machine learning method, artificial neural network, and deep learning method. A traditional method, such as BoVW (Bag of Visual Words), learns to classify the foreground and background images by extracting dense sift features from the training data [56]. Different from BoVW, SVM (Support Vector Machine) can simplify the classification task to a minimization problem of loss function [57]. In recent years, CNN (Convolutional Neural Network) has become a classical method in the field of image classification. CNN also relies on extracting features from the training data, but different from BoVW, CNN can extract robust features which are invariant to various degrees of distortions and illumination, the effectiveness of the CNN model has been proved in various recognition and classification tasks. Deep learning is large neural networks. As the development of machine learning, deep learning model, such as GoogLeNet Inception network, has been proposed and gradually become the most widely used machine learning method. It has the advantage of learning from massive amounts of data and has outperformed state-of-the-art machine learning methods, such as SVM and CNN in many domains [58].

As shown in Table 1, we trained the GoogLeNet Inception network on OFTraD. Sufficient training data allows us to train the network to track the position of ocean fronts, with an accuracy of 96%. Compared with BoVW, SVM, and CNN, the GoogLeNet Inception network achieves the highest prediction accuracy. Therefore, we use this network to classify image blocks into the background and foreground classes, and to track the ocean-front location.

**Table 1.** Tracking accuracy using different methods.

| Algorithm | Accuracy | Dataset |
| --- | --- | --- |
| BOVW | 64.5% | OFTraD |
| SVM [58] | 90% | OFTraD |
| CNN | 94.9% | OFTraD |
| GoogLeNet Inception | **96.1%** | OFTraD |

## 4. Discussion

We analyzed the effect of different dimensions of the feature vectors $c_1$ and $c_2$ on classification accuracy. Specifically, we use different pooling operations to produce the feature vectors $c_1$ and $c_2$, whose dimensions are hence different. The experimental results are shown in Table 2. We set the vector dimensions of $c_1$ and $c_2$ to $40 \times 1, 60 \times 1, 80 \times 1$ and $100 \times 1$. As the vector dimension is limited by the number of frames in a video sequence, the largest vector dimension is $100 \times 1$. The experiment results show that the best vector dimension is $40 \times 1$, reaching the highest classification accuracy of 90.96%. This is probably because 40 pixels are enough to represent the strength informantion of a video sequence. Thus, we set the vector dimension at $40 \times 1$.

**Table 2.** Classification results using different feature vector dimensions.

| Dimension | 40 | 60 | 80 | 100 |
|---|---|---|---|---|
| Accuracy | 90.96% | 87.63% | 87.16% | 87.75% |

Then, we compare the classification accuracy and maximum runtime of the classification algorithms based on strength (N1), scale (N2), the GoogLeNet Inception network (N3). As shown in Table 3, the classification algorithm based on strength (N1) achieves the highest accuracy among N1, N2, and N3. Besides, the accuracy of the classification algorithms based on strength (N1) and scale (N2) are both higher than that of the GoogLeNet Inception network (N3). When comparing the runtimes, as shown in Table 3, we found that the training time is only 283 min totally and the testing time of the classification algorithms based on strength and scale is only 0.375 s, twice faster than that of the GoogLeNet Inception network, which is 0.7 s. Therefore, our algorithm is computationally efficient.

**Table 3.** Classification accuracy using different networks of the proposed algorithms.

| Algorithm | Accuracy | Test Time |
|---|---|---|
| N1 | 91.32% | 0.375 s |
| N2 | 87.50% | 0.375 s |
| N3 | 69.90% | 0.7 s |

As shown in Table 4, we tabulate the classification scores of the classification algorithms for strength (N1) and scale (N2), with that of the output of the softmax layer of the GoogLeNet Inception network (N3), which is called the ETR algorithm. Moreover, we conducted comparative experiments to integrate the three classification results, using different weights for the strength, scale, and actual output, i.e., w1, w2, and w3, used to implement the weighted fusion.

**Table 4.** Classification results using different integration weights.

| Algorithm | Integration Weights | | | Accuracy |
|---|---|---|---|---|
| | w1 | w2 | w3 | |
| | 1 | 0 | 0 | 91.3% |
| | 0 | 1 | 0 | 87.5% |
| | 0 | 0 | 1 | 69.9% |
| | 1 | 1 | 0 | 90% |
| ETR | 1 | 1 | 1 | 87.5% |
| | −1 | 1 | 1 | 60% |
| | 1 | −1 | 1 | 65% |
| | 1 | 1 | −1 | 95% |
| | 2 | 1 | −1 | **97.5%** |

The scenarios in this experiment can be divided into the following categories: (1) we use the strength-based classification algorithm only. (2) we use the scale-based classification algorithm only. (3) we use the GoogLeNet network only. (4) we combine the strength-based and scale-based algorithms, and the weight of the two algorithms is 1:1. (5) we use the three algorithms together, and the weight of the strength-based, scale-based algorithms, and GoogLeNet network is 1:1:1. (6) we set the strength-based algorithm weight at −1, and the weight of the strength-based, scale-based algorithms, and GoogLeNet network is $-1{:}1{:}1$. (7) we set the strength-based algorithm weight at $-1$, and the weight of the strength-based, scale-based algorithms, and GoogLeNet network is $1{:}{-}1{:}1$. As the recognition accuracy achieved by the strength-based algorithm is the best, and that of the GoogLeNet network is the worst. We employed two more sets of experiments. (8) we use the three algorithms together, but the weight is $1{:}1{:}{-}1$. (9) we use the three algorithms together, but the weight is $2{:}1{:}{-}1$.

As shown in Table 4, the accuracy of each network in our algorithm can reach, or even exceed, 70%. This indicates that these networks are effective. Furthermore, when we integrate these networks together, we can obtain much better classification accuracy. This proves that the different networks in our algorithm are complementary to each other. Although the classification accuracy of the GoogLeNet Inception network is only about 70%, the final classification accuracy can be improved by integrating with the other two networks.

What's more, the experimental results show that the classification accuracy is the highest, when the weights are in the proportion of $2{:}1{:}{-}1$. From Table 4, we have the following interesting results. (1) The classification accuracy is higher when the weight of the GoogLeNet Inception network is negative, lower when the weight of the GoogLeNet Inception network is 0, proving that the classification results given by the GoogLeNet network are relevant. The reason for this might be that there is a negative correlation between the GoogLeNet Inception network and the classification algorithms based on strength and scale. (2) The classification accuracy is higher when the classification algorithm uses a larger weight for strength. This is probably because the strength information can better represent the ocean-front trend. Therefore, increasing the weight for strength, relative to that for scale, can achieve higher accuracy. (3) When the weights for strength and scale are negative, the classification accuracy is the worst. This indicates that the strength and scale information is closely correlated to ocean-front trends.

As shown in Table 5, we compare the classification accuracy of different learning models on OFTreD. It can be seen that our algorithm can achieve higher classification accuracy than that of SVM, Structured Segment Networks (STN), and GoogLeNet Inception network. This proves that our algorithm is effective, in terms of classification accuracy.

**Table 5.** Classification accuracy compared with other networks.

| Algorithm | Accuracy | Dataset |
|---|---|---|
| SVM | 41% | OFTreD |
| STN [48] | 52% | OFTreD |
| GoogLeNet Inception | 69.90% | OFTreD |
| ETR | **97.50%** | OFTreD |

## 5. Conclusions

In this paper, we proposed a novel and effective algorithm for ocean-front trend recognition, namely Evolution Trend Recognition (ETR), which combines the GoogLeNet Inception network and classification algorithms based on the strength and scale of ocean fronts. For this research, we have also created two novel databases for ocean-front trend recognition and ocean-front tracking. Firstly, we use the Microcanonical Multiscale Formalism (MMF) method to detect the oceanfront in an ocean-front image. Then, we classify the evolution trend in ocean-front video sequences. In our method, we classify the evolution trend of an oceanfront based on its strength, scale, and optical-flow information. The trend classification algorithms are based on strength and scale, and use a curve fitting method to generate feature matrices, which are converted to a specific dimension by using average pooling. Then, based on the feature matrices, the trend category of an oceanfront is determined by the softmax classifier. The trend classification method based on warped optical flow images uses the GoogLeNet Inception network to directly classify the evolution trend of an oceanfront. All of the three trend classification methods have their own advantages.

Finally, a weighted fusion method is used to combine the three trend classification methods to achieve the highest classification accuracy.

Although our proposed method applies to any video classification task, there are still some constraints, which can be reflected in two aspects. First, for complex scenarios creating and labeling a database with a large number of samples is very labor-intensive. Second, feature extraction requires prior knowledge, which may be hard to obtain. These constraints are the shortcomings of our proposed algorithm. Besides, the ocean-front enhancement and attenuation trend recognition is only a simple scenario for the ocean front evolution process, and the proposed fusion method for trend recognition still needs to be improved. In our future research, we will try to analyze more complex scenarios in the oceanfront evolution process, and try to propose a novel end-to-end deep learning network to improve the classification accuracy.

**Author Contributions:** Conceptualisation, X.S. and J.D.; methodology, Y.Y.; software, Y.Y.; validation, R.L.; formal analysis, X.S., J.D. and K.-M.L.; investigation, X.S., J.D. and K.-M.L.; resources, X.S. and J.D.; data creation, Y.Y., X.S. and J.D.; writing—original draft preparation, Y.Y.; writing—review and editing, K.-M.L.; visualization, X.S. and R.L.; supervision, X.S. and J.D.; project administration, X.S. and J.D.; funding acquisition, X.S., J.D. and K.-M.L. All authors have read and agreed to the published version of the manuscript.

**Funding:** This work was jointly supported by the National Natural Science Foundation of China (No U1706218, 61971388).

**Institutional Review Board Statement:** Not applicable.

**Informed Consent Statement:** Not applicable.

**Data Availability Statement:** Publicly available datasets were analyzed in this study. This data can be found here: [https://doi.org/10.21227/902n-yg41], accessed on 28 September 2021.

**Acknowledgments:** The authors are grateful to all the students from the Ocean Group for making the dataset. The numerical calculations in this paper have been done on the server cluster in the Institute of Artificial Intelligence of Ocean University of China.

**Conflicts of Interest:** The authors declare no conflict of interest.

## References

1. Wang, T.; He, H.; Fan, D.; Fu, B.; Dong, S. Global ocean mesoscale vortex recognition based on DeeplabV3plus model. In Proceedings of the IOP Conference Series: Earth and Environmental Science, Chengdu, China, 7–11 September 2021; p. 012001.
2. Priftis, G.; Lang, T.; Garg, P.; Nesbitt, S.; Lindsley, R.; Chronis, T. Evaluating the Detection of Mesoscale Outflow Boundaries Using Scatterometer Winds at Different Spatial Resolutions. *Remote Sens.* **2021**, *13*, 1334. [CrossRef]
3. Azevedo, M.; Rudorff, N.; Aravéquia, J. Evaluation of the ABI/GOES-16 SST Product in the Tropical and Southwestern Atlantic Ocean. *Remote Sens.* **2021**, *13*, 192. [CrossRef]
4. Saldías, G.; Hernández, W.; Lara, C.; Muñoz, R.; Rojas, C.; Vásquez, S.; Pérez-Santos, I.; Soto-Mardones, L. Seasonal Variability of SST Fronts in the Inner Sea of Chiloé and Its Adjacent Coastal Ocean, Northern Patagonia. *Remote Sens.* **2021**, *13*, 181. [CrossRef]
5. Kishcha, P.; Starobinets, B. Spatial Heterogeneity in Dead Sea Surface Temperature Associated with Inhomogeneity in Evaporation. *Remote Sens.* **2021**, *13*, 93. [CrossRef]
6. Wang, Z.; Chen, G.; Han, Y.; Ma, C.; Lv, M. Southwestern Atlantic Ocean Fronts Detected from Satellite-Derived SST and Chlorophyll. *Remote Sens.* **2021**, *13*, 4402. [CrossRef]
7. O'Neill, L.; Chelton, D.; Esbensen, S. Observations of sst-induced perturbations of the wind stress field over the southern ocean on seasonal timescales. *J. Clim.* **2002**, *16*, 2340–2354. [CrossRef]
8. Yu, X.; Naveira, A.; Martin, A.; Evans, D.; Su, Z. Wind-forced symmetric instability at a transient mid-ocean front. *Geophys. Res. Lett.* **2019**, *46*, 11281–11291. [CrossRef]
9. Garabato, A.; Leach, H.; Allen, J.; Pollard, R.; Strass, V. Mesoscale subduction at the antarctic polar front driven by baroclinic. *J. Phys. Oceanogr.* **2001**, *31*, 2087–2107. [CrossRef]
10. D'Asaro, E.; Lee, C.; Rainville, L.; Harcourt, R.; Thomas, L. Enhanced turbulence and energy dissipation at ocean-fronts. *Science* **2011**, *332*, 318. [CrossRef]
11. Ferrari, R. A frontal challenge for climate models. *Science* **2011**, *332*, 316–317. [CrossRef]
12. Ruiz, S.; Claret, M.; Pascual, A.; Olita, A.; Troupin, C.; Capet, A. Effects of oceanic mesoscale and submesoscale frontal processes on the vertical transport of phytoplankton. *J. Geophys. Res.* **2019**, *124*, 5999–6014. [CrossRef]

13. Murphy, P.; Feely, R.; Gammon, R.; Harrison, D.; Kelly, K.; Waterman, L. Assessment of the air-sea exchange of co_2 in the south pacific during austral autumn. *J. Geophys. Res.* **1991**, *96*, 455–465.

14. Currie, K.; Hunter, K. Surface water carbon dioxide in the waters associated with the subtropical convergence, east of new zealand. *Deep-Sea Res. Part I* **1998**, *45*, 1765–1777. [CrossRef]

15. Pan, Y.; Ding, D.; Li, G.; Liu, X.; Liang, J.; Wang, X.; Liu, S.; Shi, J. Potential Temporal and Spatial Trends of Oceanographic Conditions with the Bloom of Ulva Prolifera in the West of the Southern Yellow Sea. *Remote Sens.* **2021**, *13*, 4406. [CrossRef]

16. Liu, S.; Yang, Y.; Tang, D.; Yan, H.; Ning, G. Association between the Biophysical Environment in Coastal South China Sea and Large-Scale Synoptic Circulation Patterns: The Role of the Northwest Pacific Subtropical High and Typhoons. *Remote Sens.* **2021**, *13*, 3250. [CrossRef]

17. Ding, W.; Zhang, C.; Hu, J.; Shang, S. Unusual Fish Assemblages Associated with Environmental Changes in the East China Sea in February and March 2017. *Remote Sens.* **2021**, *13*, 1768. [CrossRef]

18. Belkin, I. Remote Sensing of Ocean Fronts in Marine Ecology and Fisheries. *Remote Sens.* **2021**, *13*, 883. [CrossRef]

19. Hsu, T.; Chang, Y.; Lee, M.; Wu, R.; Hsiao, S. Predicting Skipjack Tuna Fishing Grounds in the Western and Central Pacific Ocean Based on High-Spatial-Temporal-Resolution Satellite Data. *Remote Sens.* **2021**, *13*, 861. [CrossRef]

20. Lima, E.; Sun, X.; Dong, J.; Wang, H.; Yang, Y.; Liu, L. Learning and transferring convolutional neural network knowledge to ocean-front recognition. *IEEE Geosci. Remote Sens. Lett.* **2017**, *14*, 354–358. [CrossRef]

21. Lima, E.; Sun, X.; Yang, Y.; Dong, J. Application of deep convolutional neural networks for ocean-front recognition. *J. Appl. Remote Sens.* **2017**, *11*, 042610. [CrossRef]

22. Yang, Y.; Dong, J.; Sun, X.; Lguensat, R.; Jian, M.; Wang, X. ocean-front detection from instant remote sensing sst images. *IEEE Geosci. Remote Sens. Lett.* **2017**, *13*, 1960–1964. [CrossRef]

23. Yang, Y.; Dong, J.; Sun, X.; Lima, E.; Mu, Q.; Wang, X. A CFCC-LSTM Model for Sea Surface Temperature Prediction. *IEEE Geosci. Remote Sens. Lett.* **2018**, *15*, 207–211. [CrossRef]

24. Sun, X.; Zhang, M.; Dong, J.; Lguensat, R.; Yang, Y.; Lu, X. A Deep Framework for Eddy Detection and Tracking From Satellite Sea Surface Height Data. *IEEE Trans. Geosci. Remote Sens.* **2021**, *59*, 7224–7234. [CrossRef]

25. Sun, X.; Wang, C.; Dong, J.; Lima, E.; Yang, Y. A Multiscale Deep Framework for Ocean Fronts Detection and Fine-Grained Location. *IEEE Geosci. Remote Sens. Lett.* **2018**, *16*, 178–182. [CrossRef]

26. Mettes, P.; Gemert, J.; Cappallo, S.; Mensink, T.; Snoek, C. Bag-of-fragments: Selecting and encoding video fragments for event detection and recounting. In Proceedings of the ACM on International Conference on Multimedia Retrieval, Shanghai, China, 23–26 June 2015; pp. 427–434.

27. Baba, M.; Gui, V.; Cernazanu, C.; Pescaru, D. A sensor network approach for violence detection in smart cities using deep learning. *Sensors* **2019**, *19*, 1676. [CrossRef]

28. Zhi, R.; Zhou, C.; Li, T.; Liu, S.; Jin, Y. Action unit analysis enhanced facial expression recognition by deep neural network evolution. *Neurocomputing* **2021**, *425*, 135–148. [CrossRef]

29. Xie, G.; Zhang, Z.; Liu, L.; Zhu, F.;Zhang, X.; Shao, L.; Li, X. Srsc: Selective, robust, and supervised constrained feature representation for image classification. *IEEE Trans. Neural Netw. Learn. Syst.* **2019**, *31*, 4290–4302. [CrossRef]

30. Xie, G.; Zhang, X.; Yan, S.; Liu, C. Sde: A novel selective, discriminative and equalizing feature representation for visual recognition. *Int. J. Comput. Vis.* **2017**, *124*, 145–168. [CrossRef]

31. Chen, W.; Xiao, G.; Lin, X.; Qiu, K. On a human behaviors classification model based on attribute-bayesian network. *J. Southwest China Norm. Univ.* **2014**, *39*, 7–11.

32. Oneata, D.; Verbeek, J.; Schmid, C. Action and event recognition with fisher vectors on a compact feature set. In Proceedings of the IEEE Conference on Computer Vision, Portland, OR, USA, 23–28 June 2013; pp. 1817–1824.

33. Ruber, H.; Edel, G.; Julián, R.; Nicolás, G. Human action classification using n-grams visual vocabulary. In Proceedings of the Iberoamerican Congress on Pattern Recognition, Puerto Vallarta, Mexico, 2–5 November 2014; pp. 319–326.

34. Lu, X.; Dong, L.; Yuan, Y. Subspace Clustering Constrained Sparse NMF for Hyperspectral Unmixing. *IEEE Trans. Geosci. Remote Sens.* **2020**, *58*, 3007–3019. [CrossRef]

35. Lu, X.; Gong, T.; Zheng, X. Multisource Compensation Network for Remote Sensing Cross-Domain Scene Classification. *IEEE Trans. Geosci. Remote Sens.* **2021**, *58*, 2504–2515. [CrossRef]

36. Qiu, Z.; Sun, J.; Guo, M.; Wang, M.; Zhang, D. Survey on deep learning for human action recognition. In Proceedings of the International Conference of Pioneering Computer Scientists, Engineers and Educators, Guilin, China, 20–23 September 2019; pp. 3–21.

37. Wang, W.; Huang, Z.; Tian, R. Deep Learning Networks Based Action Videos Classification and Search. *Int. J. Pattern Recognit. Artif. Intell.* **2021**, *35*, 2152007. [CrossRef]

38. Le, Q.; Zou, W.; Yeung, S.; Ng, A. Learning hierarchical invariant spatio-temporal features for action recognition with independent subspace analysis. In Proceedings of the Computer Vision and Pattern Recognition, Colorado Springs, CO, USA, 20–25 June 2011; pp. 3361–3368.

39. Li, C.; Chen, H.; Lu, J.; Huang, Y.; Liu, Y. Time and Frequency Network for Human Action Detection in Videos. *arXiv* **2021**, arXiv:2103.04680.

40. Sattar, N.S.; Arifuzzaman, S. Community Detection using Semi-supervised Learning with Graph Convolutional Network on GPUs. In Proceedings of the IEEE International Conference on Big Data (Big Data), Online, 10–13 December 2020; pp. 5237–5246.

41. Karpathy, A.; Toderici, G.; Shetty, S.; Leung, T.; Sukthankar, R.; Li, F. Large-scale video classification with convolutional neural networks. In Proceedings of the IEEE Conference on Computer Vision and Pattern Recognition, Columbus, OH, USA, 24–27 June 2014; pp. 1725–1732.

42. Lee, J.; Lee, S.; Back, S.; Shin, S.; Lee, K. Object Detection for Understanding Assembly Instruction Using Context-aware Data Augmentation and Cascade Mask R-CNN. *arXiv* **2021**, arXiv:2101.02509.

43. Gautam, A.; Singh, S. Deep Learning Based Object Detection Combined with Internet of Things for Remote Surveillance. *Wirel. Pers. Commun.* **2021**, *118*, 2121–2140. [CrossRef]

44. Escorcia, V.; Heilbron, F.C.; Niebles, J.C.; Ghanem, B. Daps: Deep action proposals for action understanding. In Proceedings of the European Conference on Computer Vision, Amsterdam, The Netherlands, 8–16 October 2016; pp. 768–784.

45. Heilbron, F.C.; Niebles, J.C.; Ghanem, B. Fast temporal activity proposals for efficient detection of human actions in untrimmed videos. In Proceedings of the Computer Vision and Pattern Recognition, Las Vegas, NE, USA, 27–30 June 2016; pp. 1914–1923.

46. Simonyan, K.; Zisserman, A. Two-stream convolutional networks for action recognition in videos. In Proceedings of the International Conference on Neural Information Processing Systems, Montreal, QC, Canada, 8–13 December 2014; pp. 568–576.

47. Christian, S.; Vincent, V.; Sergey, I.; Jon, S.; Zbigniew, W. Rethinking the inception architecture for computer vision. In Proceedings of the IEEE Conference on Computer Vision and Pattern Recognition, Las Vegas, NE, USA, 27–30 June 2016; pp. 2818–2826.

48. Zhao, Y.; Xiong, Y.; Wang, L.; Wu, Z.; Tang, X.; Lin, D. Temporal action detection with structured segment networks. In Proceedings of the IEEE International Conference on Computer Vision, Venice, Italy, 24–27 October 2017; pp. 2914–2923.

49. Belkin, I.M.; O'Reilly, J.E. An algorithm for oceanic front detection in chlorophyll and sst satellite imagery. *J. Mar. Syst.* **2009**, *78*, 319–326. [CrossRef]

50. Oram, J.J.; McWilliams, J.C.; Stolzenbach, K.D. Gradient-based edge detection and feature classification of sea-surface images of the southern california bight. *Remote Sens. Environ.* **2008**, *112*, 2397–2415. [CrossRef]

51. Nieto, K.; Demarcq, H.; McClatchie, S. Mesoscale frontal structures in the canary upwelling system: New front and filament detection algorithms applied to spatial and temporal patterns. *Remote Sens. Environ.* **2012**, *123*, 339–346. [CrossRef]

52. Tamim, A.; Yahia, H.; Daoudi, K.; Minaoui, K.; Atillah, A.; Aboutajdine, D.; Smiej, M.F. Detection of moroccan coastal upwelling fronts in sst images using the microcanonical multiscale formalism. *Pattern Recognit. Lett.* **2015**, *55*, 28–33. [CrossRef]

53. Pont, O.; Turiel, A.; Yahia, H. Singularity analysis of digital signals through the evaluation of their unpredictable point manifold. *Int. J. Comput. Math.* **2013**, *90*, 1693–1707. [CrossRef]

54. Jia, Y.; Shelhamer, E.; Donahue, J.; Karayev, S.; Long, J.; Girshick, R.; Guadarrama, S.; Darrell, T. Caffe: Convolutional architecture for fast feature embedding. In Proceedings of the 22nd ACM International Conference on Multimedia, Orlando, FL, USA, 3–7 November 2014; pp. 675–678.

55. Pock, T.; Urschler, M.; Zach, C.; Beichel, R.; Bischof, H. A duality based approach for realtime tv-l 1 optical flow. In Proceedings of the 10th International Conference on Medical Image Computing and Computer-Assisted Intervention, Brisbane, Australia, 29 October–2 November 2007; pp. 214–223.

56. Karim, A.; Sameer, A. Image classification using bag of visual words (bovw). *Al-Nahrain J. Sci.* **2018**, *21*, 76–82. [CrossRef]

57. Kumar, D.; Babaie, M.; Zhu, S.; Kalra, S.; Tizhoosh, R. A comparative study of CNN, BoVW and LBP for classification of histopathological images. In Proceedings of the 2017 IEEE Symposium Series on Computational Intelligence, Honolulu, HI, USA, 27 November–1 December 2017; pp. 1–7.

58. Liu, P.; Choo, K.; Wang, L.; Huang, F. SVM or deep learning? A comparative study on remote sensing image classification. *Soft Comput.* **2017**, *21*, 7053–7065. [CrossRef]

MDPI

St. Alban-Anlage 66

4052 Basel

Switzerland

www.mdpi.com

MDPI Books Editorial Office

E-mail: books@mdpi.com

www.mdpi.com/books